Processes and Design for Manufacturing

Processes and Design for Manufacturing, Fourth Edition, offers a comprehensive and detailed examination of modern manufacturing processes while also delving into the concept of design for manufacturing (DFM) and its application across diverse manufacturing techniques. It examines manufacturing processes from the viewpoint of the product designer, investigating the selection of manufacturing methods in the early phases of design and how this affects the constructional features of a product. The stages from design process to product development are examined, integrating an evaluation of cost factors.

The text emphasizes both a general design orientation and a systems approach and covers topics such as additive manufacturing, concurrent engineering, polymeric and composite materials, cost estimation, design for assembly, and environmental factors. This edition has new and updated chapters, including a detailed chapter focusing on the prominent topic of microchip manufacturing.

This book is essential reading for senior undergraduate students studying manufacturing processes, product design, design for manufacture, and computer-aided manufacturing.

Processes and Design for Manufacturing

Fourth Edition

Sherif D. El Wakil

CRC Press
Taylor & Francis Group
Boca Raton London New York

CRC Press is an imprint of the
Taylor & Francis Group, an **informa** business

Designed cover image: Shutterstock

Fourth edition published 2026
by CRC Press
2385 NW Executive Center Drive, Suite 320, Boca Raton FL 33431

and by CRC Press
4 Park Square, Milton Park, Abingdon, Oxon, OX14 4RN

CRC Press is an imprint of Taylor & Francis Group, LLC

First edition published by Prentice Hall 1989
Second edition published by PWS-Kent 1998
Third edition published by CRC Press 2019

Library of Congress Cataloging-in-Publication Data
Names: El Wakil, Sherif D., 1944- author
Title: Processes and design for manufacturing / Sherif D. El Wakil.
Description: Fourth edition. | Boca Raton, FL : CRC Press, 2025. | Includes bibliographical references and index. |
Identifiers: LCCN 2025005460 (print) | LCCN 2025005461 (ebook) |
ISBN 9781032860022 hbk | ISBN 9781032882994 pbk | ISBN 9781003537106 ebk
Subjects: LCSH: Manufacturing processes | Metal-work | Computer integrated manufacturing systems
Classification: LCC TS183 .E47 2025 (print) | LCC TS183 (ebook) |
DDC 670--dc23/eng/20250327
LC record available at https://lccn.loc.gov/2025005460
LC ebook record available at https://lccn.loc.gov/2025005461

ISBN: 978-1-032-86002-2 (hbk)
ISBN: 978-1-032-88299-4 (pbk)
ISBN: 978-1-003-53710-6 (ebk)

DOI: 10.1201/9781003537106

Typeset in Times
by SPi Technologies India Pvt Ltd (Straive)

Access the Support Material: http://www.routledge.com/9781032860022

Dedication

To my wife, my steadfast partner for half a century, with love and gratitude

Contents

Preface to the Fourth Edition

Engineering is a rapidly evolving profession, necessitating constant updates in perspectives, technologies, and methodologies. With new technologies emerging daily and recent ones quickly becoming obsolete, this dynamic landscape is even more pronounced in the field of Manufacturing Engineering. As a result, there is a continuous demand for the revision and enhancement of publications and textbooks in this area.

Since my retirement in 2018, I have been dedicated to teaching a graduate-level course titled Manufacturing Processes at the University of Massachusetts Lowell. Most of my students are practicing engineers employed by prestigious manufacturing corporations in the Boston area. Through open-ended assignments and engaging discussions, I have gained valuable insights into the cutting-edge technologies utilized by these corporations. Consequently, I have realized the imperative need to update the content of the Third Edition of my book and include coverage of emerging hot topics. The Fourth Edition of my book represents my earnest effort to meet this objective.

Acknowledgments

The author wishes to express his deepest gratitude to his wife, Dr. Fatima El Wakil, for her indispensable support and invaluable assistance in preparing this manuscript. Special thanks are also extended to Mr. Firat Serefoglu for generously providing insightful materials on modern electronics, which greatly contributed to the development of Chapter 16.

The author is profoundly grateful to the reviewers from CRC Press for their constructive feedback and thoughtful criticism of the book proposal, which significantly improved the quality of this work.

A heartfelt thank you goes to the cardiology team at Massachusetts General Hospital—Dr. Tawakul, Dr. Chee Ng, Dr. Rahul Sakhuja, and Dr. Mastoris—for their exceptional care and treatment, which enabled the author to regain his health and complete this book.

Finally, the author extends his sincere appreciation to the staff at CRC Press, particularly Katya Seward and Anahita Nanda, for their patience, encouragement, and unwavering support throughout this journey.

Author

Sherif D. El Wakil currently holds the position of Adjunct Professor of Mechanical Engineering at the University of Massachusetts Lowell. Additionally, he is an Emeritus Professor of Mechanical Engineering at the University of Massachusetts Dartmouth, where he previously held the distinguished rank of Chancellor Professor of Mechanical Engineering. Dr. El Wakil is actively engaged as a consultant specializing in product failure analysis and product design assessment, often providing technical expertise as a witness in legal proceedings.

With over four decades of experience in academia, Dr. El Wakil has taught a wide array of manufacturing engineering courses at both undergraduate and graduate levels. Throughout his esteemed career, he has assumed various administrative roles, including Dean, Chairman of the Academic Council of the College of Engineering, and Chairman of the Mechanical Engineering Department. Dr. El Wakil has spearheaded the development and implementation of numerous academic programs at both undergraduate and graduate levels, significantly contributing to their national recognition and acclaim. Notably, he is the author of the seminal work *Processes and Design for Manufacturing*.

Dr. El Wakil's pioneering research efforts have been recognized with grants from esteemed institutions such as the NSF, NIST, and SME, in addition to being honored with the prestigious USAF Fellowship. He has lent his expertise to a multitude of national and international manufacturing corporations through consulting engagements and has served as an expert witness in legal proceedings. A longstanding member of the Society of Manufacturing Engineers (SME) since 1984, Dr. El Wakil was elected to the North American Manufacturing Research Institute (NAMRI) in the same year. He has played an integral role in shaping manufacturing education through his involvement with the SME Education Foundation, where he served as a member and chair of the Curriculum Committee. Dr. El Wakil's exceptional contributions to SME have been recognized with the President's Award on two occasions, in 1985 and 2006, culminating in his lifelong membership status granted in 2015.

1 Overview

1.1 INTRODUCTION

Before learning about various manufacturing processes and the concept of design for manufacturing, we first must become familiar with some technical terms that are used frequently during planning for and operation of industrial manufacturing plants. We also must understand thoroughly the meaning of each of these terms, as well as their significance to manufacturing engineers. The explanation of the word *manufacturing* and its impact on the lifestyle of the people of industrialized nations should logically come at the beginning. In fact, this chapter will cover all these issues and provide a better understanding of the process of product design, as well as the factors affecting it and the different stages involved in it. Naturally, the concept of *design for manufacturing*, which is the core of this text, will be explained.

1.1.1 Definition of Manufacturing

Manufacturing can be defined as the transformation of raw materials into useful products through the use of the easiest and least expensive method. It is not enough, therefore, to process some raw materials and obtain the desired product. It is, in fact, of major importance to achieve this goal by employing the easiest, fastest, and most efficient methods. If less efficient techniques are used, the production cost of the manufactured part will be high, and the part will not be as competitive as similar parts produced by other manufacturers. Also, the production time should be as short as possible in order to capture a larger market share.

The function of a manufacturing engineer is, therefore, to determine and define the equipment, tools, and processes required to convert the design of the desired product into reality in an efficient manner. In other words, it is the engineer's task to find the most appropriate, optimal combination of machinery, materials, and methods needed to achieve economical and trouble-free production. Thus, a manufacturing engineer must have a strong background in materials and up-to-date machinery, as well as the ability to develop analytical solutions and alternatives for open-ended problems experienced in manufacturing. An engineer must also have a sound knowledge of the theoretical and practical aspects of the various manufacturing methods.

1.1.2 Relationship between Manufacturing and Standard of Living

The standard of living in any nation is reflected in the products and services available to its people. In a nation with a high standard of living, a middle-class family usually owns an automobile, a refrigerator, an electric stove, a dishwasher, a washing machine, a vacuum cleaner, a stereo, and, of course, a television set. Such a family also enjoys health care that involves modern equipment and facilities. All these goods, appliances, and equipment are actually raw materials that have been converted into manufactured products. Therefore, the more active in manufacturing raw materials the people of the nation are, the more plentiful those goods and services become; as a consequence, the standard of living in that nation attains a high level. On the other hand, nations that have raw materials but do not fully exploit their resources by manufacturing those raw materials are usually poor and considered underdeveloped. It is, therefore, the know-how and capability of converting raw materials into useful products, not just the availability of minerals or resources within its territorial land, that basically determines the standard of living of a nation. In fact, many industrial nations, such as Japan and Switzerland, import most of the raw materials that they manufacture and yet still maintain a high standard of living.

DOI: 10.1201/9781003537106-1

1.1.3 Overview of the Manufacturing Processes

The final desired shape of a manufactured component can be achieved through one or more of the following four approaches:

- Changing the shape of the raw stock without adding material to it or taking material away from it. Such change in shape is achieved through plastic deformation, and the manufacturing processes that are based on this approach are referred to as *metal-forming processes*. These processes include bulk-forming processes like rolling, extrusion, forging, and drawing, as well as sheet metal-forming operations like bending, deep drawing, and embossing. Bulk-forming operations are covered in Chapter 5, and the working of sheet metal is covered in Chapter 6.
- Obtaining the required shape by adding metal or joining two metallic parts together, as in welding, brazing, or metal deposition. These operations are covered in Chapter 4.
- Molding molten or particulate metal into a cavity that has the same shape as the final desired product, as in casting and powder metallurgy. These processes are covered in Chapters 3 and 7, respectively.
- Removing portions from the stock material to obtain the final desired shape. A cutting tool that is harder than the stock material and possesses certain geometric characteristics is employed in removing the undesired material in the form of chips. Several chip-making (machining) operations belong to this group. They are exemplified by turning, milling, and drilling operations and are covered in Chapter 11. The physics of the process of chip removal is covered in Chapter 10.

1.1.4 Types of Production

Modern industries can be classified in different ways. These classifications may be by process, by product, or based on production volume and diversity of products. Classification by process is exemplified by casting industries, stamping industries, and the like. Classification by product indicates that industries may belong to the automotive, aerospace, and electronic groups. Classification based on the production volume identifies three distinct types of production: mass, job shop, and moderate. Let us briefly discuss the features and characteristics of each type. We will also discuss the subjects in greater depth in the text.

1.1.4.1 Mass Production

Mass production is characterized by the high production volume of the same (or very similar) parts for a prolonged period of time. An annual production volume of fewer than 50,000 pieces usually cannot be considered mass production. As you may expect, the production volume is based on an established or anticipated sales volume and is not directly affected by the daily or monthly orders. A typical example of mass-produced goods is automobiles. Because that type attained its modern status in Detroit, it is sometimes referred to as the Detroit type.

1.1.4.2 Job Shop Production

Job shop production is based on orders for a variety of small lots. Each lot may consist of up to 200 or more similar parts, depending upon the customer's needs. It is obvious that this type of production is most suitable for subcontractors who produce varying components to supply various industries. The machines employed must be flexible to handle frequent variations in the configuration of the ordered components. Also, the personnel employed must be highly skilled in order to handle a variety of tasks that differ for the different parts that are manufactured.

1.1.4.3 Moderate Production

Moderate production is an intermediate phase between the job shop and the mass production types. The production volume ranges from 10,000 to 20,000 parts, and the machines employed are flexible and multipurpose. This type of production is gaining popularity in industry because of an increasing market demand for customized products.

1.1.5 Fundamentals of Manufacturing Accuracy

Modern manufacturing is based on flow-type "mass" assembly of components into machines, units, or equipment without the need for any fitting operations performed on those components. That was not the case in the early days of the Industrial Revolution, when machines or goods were individually made and assembled and there was always the need for the "fitter" with his or her file to make final adjustments before assembling the components. The concept of mass production and interchangeability came into being in 1798 when the American inventor Eli Whitney (born in Westborough, Massachusetts) contracted with the U.S. government to make 10,000 muskets. Whitney started by designing a new gun and the machine tool to make it. The components of each gun were manufactured separately by different workers. Each worker was assigned the task of manufacturing a large number of the same component. Meanwhile, the dimensions of those components were kept within certain limits so that they could replace each other if necessary and fit their mating counterparts. In other words, each part would fit any of the guns made. The final step was merely to assemble the interchangeable parts. By doing so, Eli Whitney established two very important concepts on which modern mass production is based—namely, interchangeability and fits. Let us now discuss the different concepts associated with manufacturing accuracy required for modern mass-production technologies.

1.1.5.1 Tolerances

A very important fact of manufacturing science is that it is almost impossible to obtain the desired nominal dimension when processing a workpiece. This is caused by the inevitable, though very slight, inaccuracies inherent in the machine tool, as well as by various complicated factors like the elastic deformation and recovery of the workpiece and/or the fixture, temperature effects during processing, and sometimes the skill of the operator. Because it is difficult to analyze and completely eliminate the effects of these factors, it is more feasible to establish a permissible degree of inaccuracy or permissible deviation from the nominal dimension that would not affect the proper functioning of the manufactured part in any way. According to the International Organization for Standardization (ISO) system, the nominal dimension is referred to as the *basic size* of the part.

The deviation from the basic size to each side (in fact, both can be on the same side) determines the high and low limits, respectively, and the difference between these two limits of size is called the *tolerance*. It is an absolute value without a sign and can also be obtained by adding the absolute values of the deviations. As you may expect, the magnitude of the tolerance is dependent upon the basic size and is designated by an alphanumeric symbol called the *grade*. There are 18 standard grades of tolerance in the ISO system, and the tolerances can be obtained from the formulas or tables published by the ISO.

Smaller tolerances, of course, require the use of high-precision machine tools in manufacturing the parts and, therefore, increase production costs. Figure 1.1 indicates the relationship between the tolerance and the production cost. As can be seen, very small tolerances necessitate a very high production cost. Therefore, small tolerances should not be specified when designing a component unless they serve a certain purpose in that design.

1.1.5.2 Fits

Before two components are assembled together, the relationship between the dimensions of the mating surfaces must be specified. In other words, the location of the zero line (i.e., the line indicating the basic size) to which deviations are referred must be established for each of the two mating surfaces. As can be seen in Figure 1.2a,b, this determines the degree of tightness or freedom for relative motion between the mating surfaces. Figure 1.2 also shows that there are basically three types of fits: clearance, transition, and interference.

FIGURE 1.1 The relationship between the tolerance and the production cost.

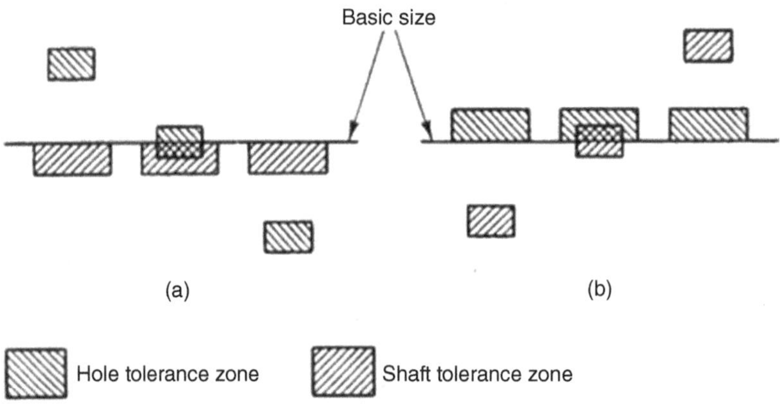

FIGURE 1.2 The two systems of fit according to ISO: (a) shaft-basis system and (b) hole-basis system.

In all cases of *clearance fit*, the upper limit of the shaft is always smaller than the lower limit of the mating hole. This is not the case in *interference fit*, where the lower limit of the shaft is always larger than the upper limit of the hole. The *transition fit*, as the name suggests, is an intermediate fit. According to the ISO, the internal enveloped part is always referred to as the *shaft*, whereas the surrounding surface is referred to as the *hole*. Accordingly, from the fits point of view, a key is the shaft and the keyway is the hole.

It is clear from Figure 1.2a,b that there are two ways for specifying and expressing the various types of fits: the *shaft-basis* and the *hole-basis* systems. The location of the tolerance zone with respect to the zero line is indicated by a letter, which is always capitalized for holes and lowercased for shafts, whereas the tolerance grade is indicated by a number, as previously explained. Therefore, a fit designation can be H7/h6, F6/g5, or any other similar form for a hole basis. Table 1.1 provides the preferred fits and their applications.

1.1.5.3 Interchangeability and Standardization

When the service life of an electric bulb is over, all you do is buy a new one and replace the bulb. This easy operation, which does not need a fitter or a technician, would not be possible without the two main concepts of interchangeability and standardization. *Interchangeability* means that identical parts must be able to replace each other, whether during assembly or subsequent maintenance

TABLE 1.1
Description of the Preferred Fits

ISO Hole Basis	ISO Shaft Basis	American ANSI B4.1	Description
H11/c11	C11/h11	RC 9	Clearance fit. Loose running for wide commercial tolerances on external members
H9/d9	D9/h9	RC 7	Clearance fit. Free running for large temperature variations, high running speeds, or high journal pressure
H8/f7	F8/h7	RC 4	Clearance fit. Close running for accurate location and moderate speeds and journal pressures
H7/g6	G7/h6	RC 2	Clearance fit. Sliding fit for accurate fit and location and free moving and turning, not free running
H7/h6	H7/h6	LC 2	Transition fit. Location clearance for snug fits for parts that can be freely assembled
H7/K6	K7/h6	LT 3	Transition fit. Location transition fit for accurate locations
H7/n6	N7/h6	LT 5	Transition fit. Location transition fit for more accurate locations and greater interference
H7/p6	P7/h6	LN 2	Interference fit. Location interference fit for rigidity and alignment without special bore pressure
H7/s6	S7/h6	FN 2	Interference fit. Medium drive fit for shrink fits on light sections; tightest fit usable for cast iron
H7/u6	U7/h6	FN 4	Interference fit. Force fit for parts that can be highly stressed and for shrink fits

work, without the need for any fitting operations. Interchangeability is achieved by establishing a permissible tolerance, beyond which any further deviation from the nominal dimension of the part is not allowed. *Standardization*, on the other hand, involves limiting the diversity and total number of varieties to a definite range of standard dimensions. An example is the standard gauge system for wires and sheets. Instead of having a very large number of sheet thicknesses in steps of 0.001 inch, the number of thicknesses produced is limited to only 45 (in the U.S. standards). As you can see from this example, standardization has far-reaching economic implications and also promotes interchangeability. Obviously, the engineering standards differ for different countries and reflect the quality of technology and industrial production in each case. Germany established the DIN (Deutsches Institut für Normung), standards that are finding popularity worldwide. The former Soviet Union adopted GOST, standards that were suitable for the period of industrialization of that country.

1.1.6 The Production Turn

In almost all cases, the main goal of a manufacturing project is to make a profit, the exception being projects that have to do with national security or prestige. Let us establish a simplified model that illustrates the cash flow through the different activities associated with manufacturing so that we can see how to maximize profit. As shown in Figure 1.3, the project starts by borrowing money from a bank to purchase machines and raw materials and to pay the salaries of the engineers and other employees. Next, the raw materials are converted into products, which are the output of the manufacturing domain. Obviously, those products must be sold (through the marketing department) in order to get cash. This cash is, in turn, used to cover running costs, as well as required payment to the bank: any surplus money left is the profit.

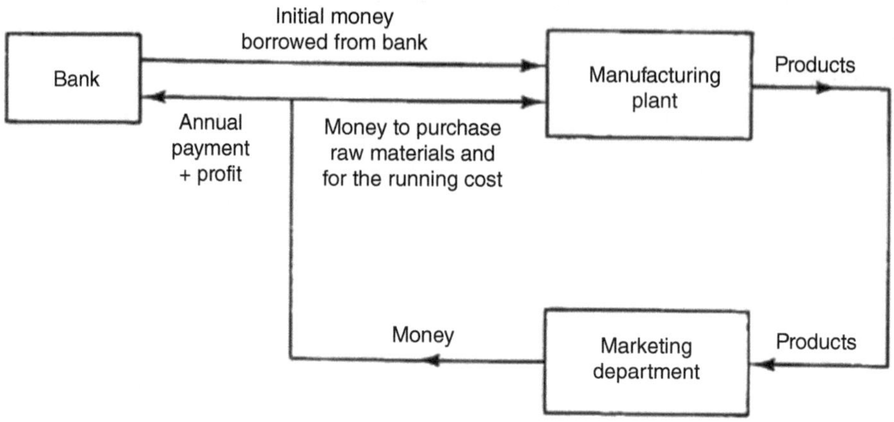

FIGURE 1.3 The production turn.

We can see in this model that the sequence of events forms a continuous cycle (i.e., a closed circuit). This cycle is usually referred to as the *production turn*. We can also realize the importance of marketing, which ensures the continuity of the cycle. If the products are not sold, the cycle is obviously interrupted. Moreover, we can see that maximum profit is obtained through maximizing the profit per run and/or increasing the number of turns per year (i.e., running the cycle faster). Evidently, these two conditions are fulfilled when products are manufactured in the easiest and least expensive way.

1.1.7 Product Life Cycle

It has been observed that all products, from the sales point of view, go through the same product life cycle, no matter how diverse or dissimilar they are. Whether the product is a new-model airplane or a coffeemaker, its sales follow a certain pattern or sequence from the time it is introduced in the market to the time it is no longer sold. The main difference between the cycles of these two products is the span or duration of the cycle, which always depends upon the nature and uses of the particular product. As we will see later in the text, it is very important for the designer and the manufacturing engineer to fully understand that cycle in order to maximize the profits of the production plant.

It is clear from Figure 1.4 that the sales, as well as the rate of increase in sales, are initially low during the *introduction* stage of the product life cycle. The reason is that the consumer is not aware of the performance and unique characteristics of the product.

Through television and newspaper advertisements and word-of-mouth communications, a growing number of consumers learn about the product and its capabilities. Meanwhile, management works on improving performance and eliminating shortcomings through minor design modifications.

It is also the time for some custom tailoring of the product for slightly different customer needs, in order to serve a wider variety of customers. As a result, consumer acceptance is enhanced, and sales accordingly increase at a remarkable rate during this stage, which is known as the *growth* stage. However, this trend does not continue forever, and at a certain point, the sales level out. This is, in fact, the *maturity* stage of the life cycle. During this stage, the product is usually faced with fierce competition, but the sales will continue to be stable if the management succeeds in reducing the cost of the product and/or developing new applications for it. The more successful the management is in achieving this goal, the longer the duration of the maturity stage will be. Finally, the *decline* stage begins, the sales fall at a noticeable rate, and the product is, at some point, completely abandoned. The decrease in sales is usually due to newer and better products that are pumped into the market by competing manufacturers to serve some customer needs. It can also be caused by

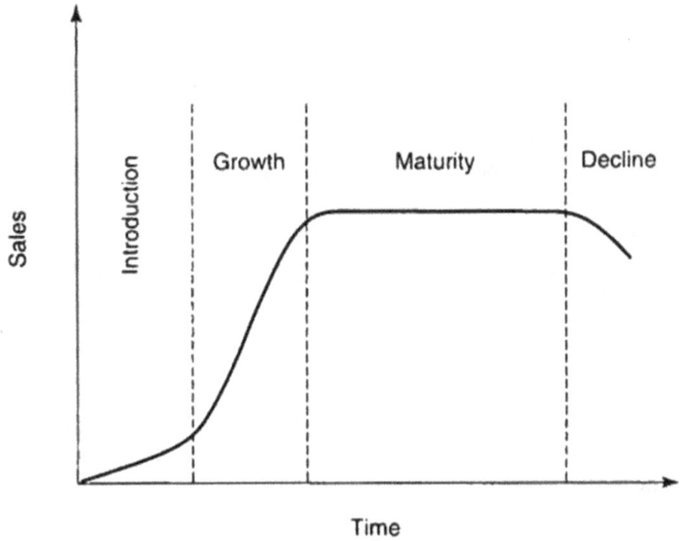

FIGURE 1.4 The product life cycle.

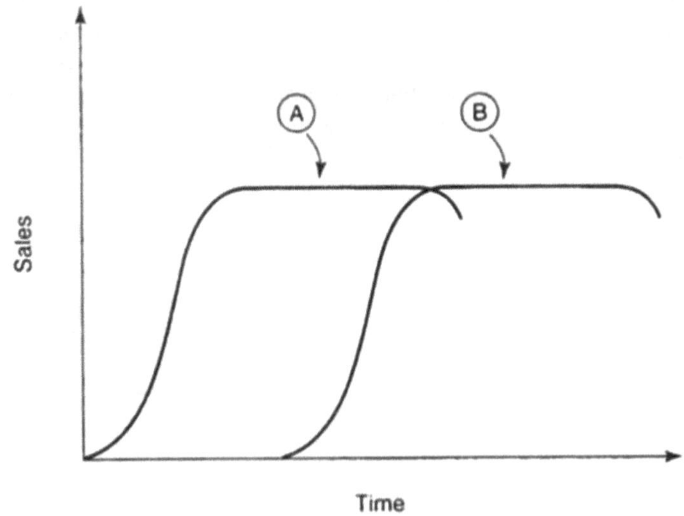

FIGURE 1.5 The proper overlap of products' life cycles.

diminishing the need for the uses and applications of such a product. Clever management would start developing and marketing a new product (B) during the maturity stage of the previous one (A) so as to keep sales continuously high, as shown in Figure 1.5.

1.1.8 Technology Development Cycle

Every now and then, a new technology emerges as a result of active research and development (R&D) and is then employed in the design and manufacture of several different products. It can, therefore, be stated that *technology* is concerned with the industrial and everyday applications of the results of the theoretical and experimental studies that are referred to as *engineering*. Examples of modern technologies include transistors, microchips, and fiber optics.

The relationship between the effectiveness or performance of certain technology and the effort spent to date to achieve such performance is shown graphically in Figure 1.6. This graphical representation is known as the *technology development cycle*. It is also sometimes referred to as the *S curve* because of its shape. As can be seen in Figure 1.6, a lot of effort is required to produce a sensible level of performance at the early stage. Evidently, there is a lack of experimental experience since the techniques used are new. Next, the rate of improvement in performance becomes exponential, a trend that is observed with almost all kinds of human knowledge. At some point, however, the rate of progress becomes linear because most ideas are in place; any further improvement comes as a result of refining the existing ideas rather than adding new ones. Again, as time passes, the technology begins to be "exhausted," and performance levels out. A "ceiling" is reached, above which the performance of the existing current technology cannot go because of social and/or technological considerations.

An enlightened management of a manufacturing facility would allocate resources and devote effort to an active R&D program to come up with a new technology (B) as soon as it realizes that the technology on which the products are based (A) is beginning to mature. The production activities would then be transferred to another S curve, with a higher ceiling for performance and greater possibilities, as shown in Figure 1.7. Any delay in investing in R&D for developing new technology may result in creating a gap between the two curves (instead of continuity with the overlap shown in Figure 1.7), with the final outcome being to lose the market to competing companies that possess newer technology. In fact, the United States dominated the market of commercial airlines because companies like Boeing and McDonnell Douglas knew exactly when to switch from propeller-driven airplanes to jet propulsion commercial airliners. This is contrary to what some major computer companies did when they continued to develop and produce mainframe computers and did not recognize when to make the switch to personal computers. Current examples of technological discontinuity include the change from conventional telecommunications cables to fiber optics for communication and information transfer.

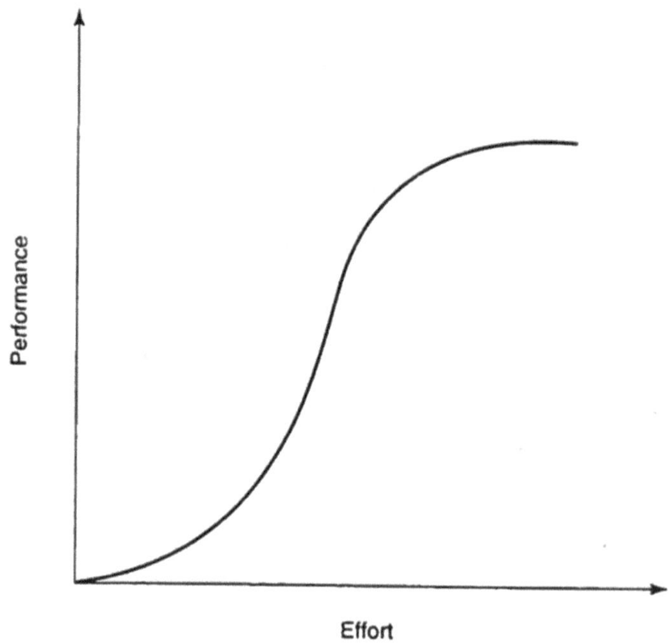

FIGURE 1.6 The technology development cycle (or S curve).

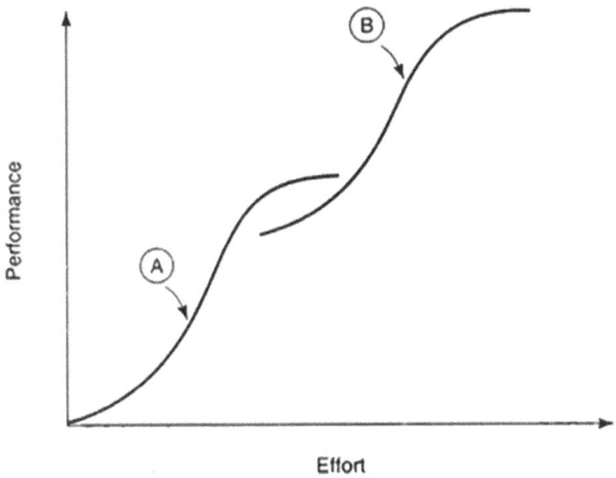

FIGURE 1.7 Transfer from one S curve to another.

1.1.9 The Design Process

An engineer is a problem solver who employs his or her scientific and empirical knowledge together with inventiveness and expert judgment to obtain solutions for problems arising from societal needs. These needs are usually satisfied by some physical device, structure, or process. The creative process by which one or more of the fruits of the engineer's effort are obtained is referred to as *design*. It is, indeed, the core of engineering that provides the professional engineer with the chance to create original designs and watch them become realities. The satisfaction that the engineer feels following the implementation of his or her design is the most rewarding experience in the engineering profession. Because design is created to satisfy a societal need, there can be more than one way to achieve that goal. In other words, several designs can address the same problem. Which one is the best and most efficient design? Only time will tell because it is usually the one that would be favored by the consumers and/or the society as a whole.

Although there is no single standard sequence of steps to create a workable design, E. V. Krick has outlined the procedure involved in the design process, and his work has gained widespread acceptance. Following is a discussion of the stages of the design process according to Krick (see the references at the end of the book for more detailed information).

1.1.9.1 Problem Formulation

As illustrated in Figure 1.8, problem formulation is the first stage of the design process. This phase comes as a result of recognizing a problem and involves defining that problem from a broad perspective without getting deep into the details. It is also at this stage that the engineer decides whether or not the problem at hand is worth solving. In other words, this stage basically constitutes a feasibility study of the problem arising from a recognized need. The designer should, therefore, realize the importance of this stage. Neglecting it may result in wasting money in an effort to solve a problem that is not worth solving or wasting time on details that make it extremely difficult to get a broad view of the problem so as to select the appropriate path for solving it. The formulation of a problem can take any form that is convenient to the designer, although diagrammatic sketching (in particular, the black-box method) has proven to be a valuable tool.

1.1.9.2 Problem Analysis

The second stage involves much information gathering and processing in order to come up with a detailed definition of the problem. Such information may come from handbooks, manufacturers'

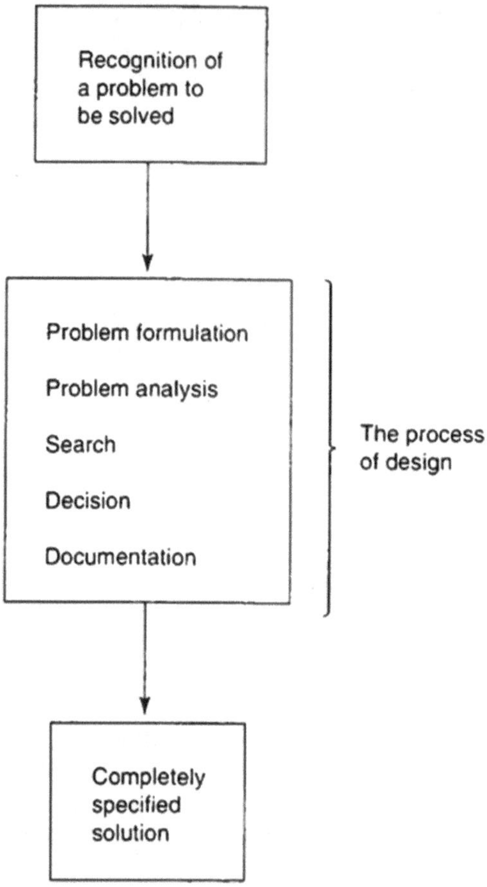

FIGURE 1.8 The design process. (Adapted from Krick, *An Introduction to Engineering and Engineering Design*, 2nd ed., New York: John Wiley, 1969.)

catalogs, leaflets, brochures, the Internet, and personal contacts. You are strongly advised to seek information wherever you can find it; workers at all levels of a company may have some key information that you can use. The end product should be a detailed analysis of the qualitative and quantitative characteristics of the input and output variables and constraints, as well as the criteria that will be used in selecting the best design.

1.1.9.3 Search for Alternative Solutions

In the third stage, the designer actively seeks alternative solutions. A good practice is to make a neat sketch for each preliminary design with some notes about its pros and cons. All sketches should be kept even after a different final design is selected so that if that final design is abandoned for some reason, the designer does not have to start from the beginning again. It is also important to remember not to end the search for alternative solutions prematurely before it is necessary or desirable to do so. Sometimes, a designer gets so involved with details of what he or she thinks is a good idea or solution that he or she will become preoccupied with these details, spending time on them instead of searching for other good solutions. Therefore, you are strongly advised to postpone working out the details until you have an appropriate number of viable solutions.

It is, indeed, highly recommended to employ collaborative methods to enable the mind to penetrate into domains that might otherwise remain unexplored. A typical example is the technique of brainstorming, where a few or several people assemble to produce a solution for a problem by creating an atmosphere that encourages everyone to contribute with whatever comes to mind. After the problem is explained, each member comes up with an idea that is, in turn, recorded on a blackboard, thus making all ideas evident to all team members.

1.1.9.4 Decision Making

The fourth stage involves the thorough weighing and judging of the different solutions with the aim of being able to choose the appropriate one. That is, *trade-offs* have to be made during this stage. They can be achieved by establishing a decision matrix, as shown in Figure 1.9.

As can be seen in Figure 1.9, each of the major design objectives is in a column, and each solution is allocated a row. Each solution is evaluated with regard to how it fulfills each of the design objectives and is, therefore, given a grade (on a scale of 1 to 10) in each column. Because the design objectives do not have the same weight, each grade must be multiplied by a factor representing the weight of the design function for which it was given. The total of all the products of multiplication is the score of that particular solution and can be considered a true indication of how that solution fulfills the design objectives. As you can see, this technique provides a mechanism for rating the various solutions, thus eliminating most and giving further consideration to only a few.

The chosen design is next subjected to a thorough analysis in order to optimize and refine it. Detailed calculations, whether manual or computational, are involved at this point. Both analytical and experimental modeling are also extensively employed as tools in refining the design. It is important, therefore, to now discuss modeling and simulation. A *model* can be defined as a simplified representation of a real-life situation that aids in the analysis of an associated problem. There are many ways to classify and identify models. For example, models can be descriptive, illustrating a real-world counterpart, or prescriptive, helping to predict the performance of the actual system. They can also be deterministic or probabilistic (used when making decisions under uncertainty). A simple example of a model is the free-body diagram used to determine the internal tensile force acting on a wire with a weight attached to its end. Many computer tools (software) are employed by designers to create models easily and quickly. Examples include geometric modeling and finite element analysis software packages. On the other hand, *simulation* can be defined as the process of experimenting with a model by subjecting it to various values of input parameters and observing the output, which can be taken as an indication of the behavior of the real-world system under the tested conditions. As you can see, simulation can save a lot of time and effort that could be spent on experimental models and prototypes. This saving is particularly evident when computer simulation is employed. Still, simulation would not eliminate design iterations but rather would minimize their number. You are, therefore, urged to make use of these tools whenever possible.

1.1.9.5 Documentation

In the fifth and last stage of the design process, the designer organizes the material obtained in the previous stage and puts it in shape for presentation to his or her superiors. The output of this stage should include the attributes and performance characteristics of the refined design, given in sufficient detail. Accordingly, the designer must communicate all information in the form of clear and

Design Objectives	Production Cost			Reliability			Durability			
Alternatives	Level of satisfaction	Weight factor	Weight	Level of satisfaction	Weight factor	Weight	Level of satisfaction	Weight factor	Weight	Overall Score
Design 1	9.0	0.5	4.5	7.0	0.3	2.1	8.0	0.2	1.6	8.2
Design 2	8.0	0.5	4.0	8.0	0.3	2.4	8.0	0.2	1.6	8.0
Design 3	9.0	0.5	4.5	9.0	0.3	2.7	9.0	0.2	1.8	9.0

FIGURE 1.9 A simple decision matrix.

easy-to-understand documents. Documentation consists of carefully prepared, detailed, and dimensioned engineering drawings (in the case of a product, assembly drawings and workshop drawings or blueprints), a written report, and possibly an iconic model. With the recent development in rapid prototyping techniques, a prototype can certainly be a good substitute for an iconic model. This approach has the advantage of revealing problems that may be encountered during manufacturing.

1.2 PRODUCT DESIGN

An engineer is a problem solver to meet societal needs. The duties of a mechanical engineer (design/manufacturing) are to design and manufacture products to make the lives of people easier by solving problems occurring during their daily lives. A new product can come as a result of realizing a problem that can either be a new one or might have been there for a while but gained importance lately. New products should also come when the existing technology is exhausted and the need to adopt an emerging technology becomes a must in order to be able to compete in the market. The manufacturing corporation should be able to provide the market with a new product that has reached its maturity phase, in order to supersede an existing one that is about to reach the end of its life cycle.

Several factors that affect the product design process should seriously be considered by the product design team, which consists of engineers specialized in different engineering disciplines. While we will provide a bird's-eye view of all such factors, we will leave the deep detailed coverage of some topics that fall beyond the field of mechanical/manufacturing engineering to specialized books. The following are the main factors that affect product design.

1.2.1 Listening to the User (Quality Function Deployment)

Since the consumers are the ones who are going to use the product, it is, therefore, a must to have the product design process focused on meeting the needs, expectations, and satisfaction of the potential consumers. If the product to be designed is an upgrade of (or replacing) an existing one, compiled data of the feedback about the performance of the product as well as any suggestions by the consumers should be considered seriously. Examples include limitations observed, areas in which the product failed to meet expectations, and suggestions for design modifications in order to enhance the performance of the product. Consumers may also have additional needs and expectations for additional features as a result of their satisfaction with the existing product.

On the other hand, if a totally new product is to be designed, the design process should be started only after compiling and analyzing data gathered through questionnaires that would have been addressed to all potential consumers. An example of a case where that proved to be very beneficial came when an aircraft manufacturer was designing a new model and decided to consult the potential clients (i.e., airline representatives in that case). A very vital suggestion came from a maintenance person who suggested lowering the level of the wings above ground in order to be reachable to a person standing on the ground, thus facilitating the inspection and maintenance of the components embedded in the lower surface of the wing. Such a simple suggestion resulted in appreciable savings in the time, effort, and cost of frequent inspection and maintenance of the wing equipment.

In fact, all of the abovementioned activities fall under an important area of industrial engineering, entitled quality function deployment or QFD, which is beyond the scope of this book. That method was developed in the shipyards of Kyoto, Japan, by Yoji Akao to help transform the voice of the customer into engineering characteristics for a product. Yoji Akao described QFD as a "method to transform qualitative user demands into quantitative parameters, to deploy the functions forming quality, and to deploy methods for achieving the design quality into subsystems and component parts, and ultimately to specific elements of the manufacturing process." The house of quality, a part of QFD, identifies customer desires, the importance of those desires, and engineering characteristics that may be relevant to those desires. It takes the form of a matrix with consumer desires on one dimension and correlated nonfunctional requirements on the other dimension, while the cells of that matrix are assigned to the weights of the stakeholders. Each column is summed at the bottom of the matrix, thus allowing the systems' characteristics to be weighed according to the stakeholders' desires. Again, such effort is going to be the responsibility of the product design team member specialized in industrial engineering.

1.2.2 Conditions to Which the Product Would Be Subjected during Its Service Life

Those would include the loading condition—namely, the magnitude of the load applied and how it is applied (i.e., whether static, cyclic, pulsating, or random). They would also include the nature of the environment surrounding the product during its service life. Examples of environments that would have a major impact on the design of the product include a corrosive environment having high humidity and accumulation of salts, corrosive fumes, and acidic or alkaline solutions. Furthermore, high temperatures and excessively low temperatures are conditions that significantly affect the design of a product. All these factors would affect the selection of the material of the product, as well as the geometry and dimensions of the body. Such topics are covered in traditional mechanical design courses.

1.2.3 Cost

Cost is a major factor that should be considered during the design process and must be included in the decision matrix as previously explained. It can be estimated with reasonable accuracy (as will be explained later in a dedicated chapter) for the different alternative designs and then plugged into the decision matrix to enable making a rational decision. It has, however, to be emphasized that the weight of cost as a factor may vary for various products. For military products, for example, quality (or performance) is the most important factor while cost has less weight, unlike domestic products where cost weighs the same as or even more than quality. Moreover, the weight of cost may vary depending upon the niche, which the manufacturing corporation selected for its product. Accordingly, we should expect that the weight of cost when designing a luxury motorcar (where the niche is performance and quality) is less than that when designing a popular motorcar (where affordability is the major factor).

1.2.4 The Concept of Design for Manufacturing

The conventional procedure for product design, as illustrated in Figure 1.10, used to start with an analysis of the desired function, which usually dictated the form as well as the materials of the product to be made. The design (blueprint) was then sent to the manufacturing department, where the kind and sequence of production operations were determined mainly by the form and materials of the product. In fact, that old design procedure had several disadvantages and shortcomings:

- In some cases, nice-looking designs were impossible to make; in many other cases, the designs had to be modified so that they could be manufactured.
- Preparing the design without considering the manufacturing process to be carried out and/or the machine tools available would sometimes result in a need for special-purpose, expensive machine tools. The final outcome was an increase in the production cost.
- When the required production volume was large, parts had to be specially designed to facilitate operations involved in mass production (such as assembly).
- A group of different products produced by the same manufacturing process has common geometric characteristics and features that are dictated by the manufacturing process employed (forgings, for example, have certain characteristic design features that are different from those of castings, extrusions, or stampings). Ignoring the method of manufacturing during the design phase would undermine these characteristic design features and thus result in impractical or faulty design.

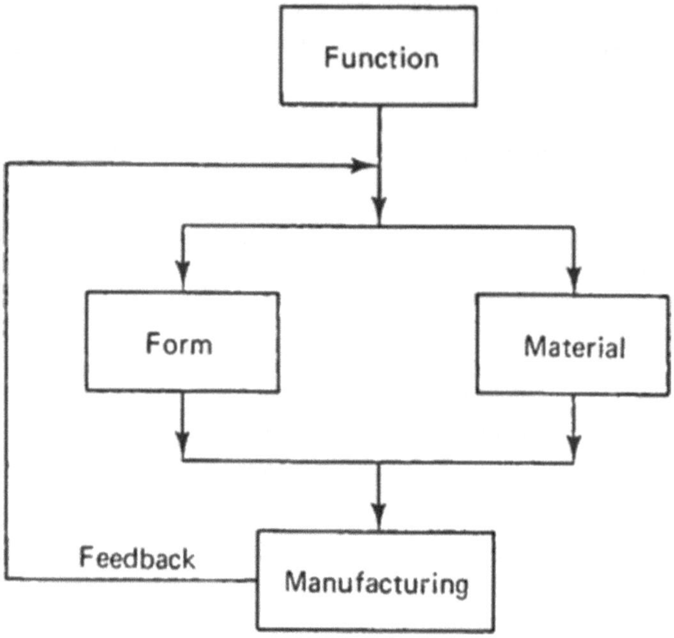

FIGURE 1.10 The old procedure for product design.

Because of these reasons and also because of the trend toward integrating the activities in a manufacturing corporation, modern design procedure takes into consideration the method of manufacturing during the design phase. As can be seen in Figure 1.11, design, material, and manufacturing are three interactive, interrelated elements that form the manufacturing system, whose prime inputs are conceptual products (and/or functions) and whose outputs are manufactured products. In fact, the barriers and borders between the design and the manufacturing departments are fading out and will eventually disappear. The tasks of the designer and those of the manufacturing engineer are going to be combined and done by the same person. It is, therefore, the mission of this text to emphasize concepts like *design for manufacturing* and to promote the systems for product design.

1.2.5 Ergonomics and Industrial Design

Ergonomics, also known as comfort design or functional design, is the practice of designing products to take proper account of the interaction between them and the people who use them. On the other hand, industrial design bridges the gap between what is and what's not possible. It is a transdisciplinary profession that harnesses creativity to resolve problems and co-create solutions with the intent of making a product. It links innovation, technology, research, business, and customers to provide new value and places the human in the center of the process. Think of the way you hold a teacup or a cell phone. In fact, I once purchased an artistic-looking teacup but had to abandon it because I could not hold it without burning my fingers. It is, therefore, a must to have an industrial design/ergonomic specialist as a member of the product design team. Neglecting industrial design would make a product less appealing to the users, who would realize that their comfort was not taken into consideration. This, in turn, would result in a reduction in sales.

1.2.6 Periodic Cleaning and Maintenance

Any product needs to be periodically cleaned and undergo maintenance according to a schedule during its service life. Accordingly, the design of a product should have inherent features to

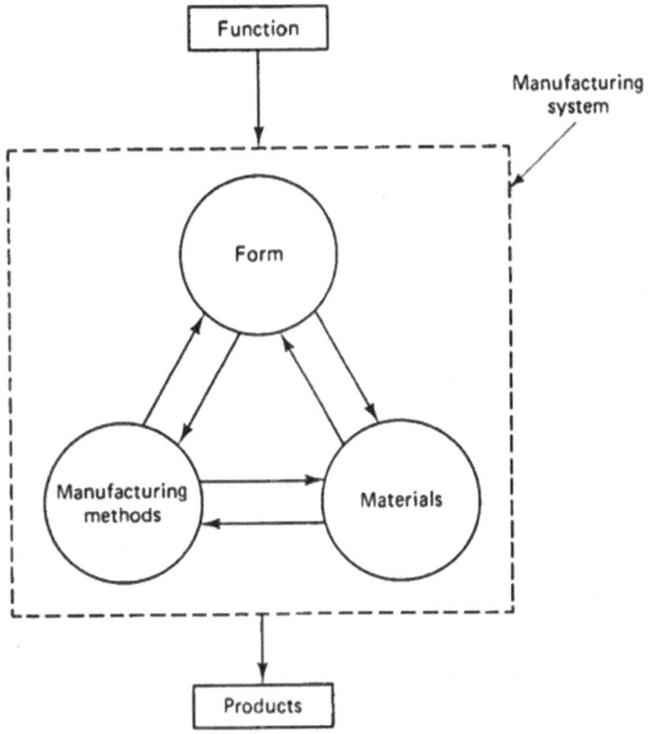

FIGURE 1.11 The new concept of a manufacturing system for achieving rational product designs.

facilitate these two operations. All areas that require periodic cleaning must be easily reachable. In order to realize the importance of this issue, consider the food-processing products that we encounter in our daily lives having areas that are difficult to reach and clean. The outcome would be an accumulation of remnants of food that eventually rot and endanger the health of the user. The design of a product should ensure easy removal of any part (or subassembly) that malfunctions or its service life is over and replacing it with a spare part that is easily fitted in its place. For lubrication, the orifices into which oil is to be injected must be clearly identified and reachable. Some products may have batteries that need to be changed after a period of time. The location should be clearly identified, and easy removal and placement of the battery must be ensured.

1.2.7 Sustainability and Environmental Issues

The increasing problems of landfill usage, the rising cost of energy and raw material, the greenhouse effect, and the decay of the ozone layer are among the major environmental concerns that prompted addressing waste reduction at the source. The key to the solution of these problems lies in the policy of adopting environmentally friendly products and production operations—what environmentalists refer to as the concept of the *eco-factory*. This goal can be achieved through the design of products that promote recycling as well as through the design of manufacturing processes that minimize waste, by-products, and emissions and, therefore, utilize resources more efficiently. In other words, when preparing the design of a product, the impact of that product on the environment must be taken into consideration, whether during its manufacture, its service life, or when it is scrapped. Following are some guidelines for environmentally conscious product design.

1.2.8 Design for Disassembly

It became evident from the industrial experience that some slight changes in the design of a product can easily promote and facilitate recyclability. Dismantling a product for recycling has cost implications (e.g., the cost of labor required to take apart the different components). Accordingly, components and products have to be designed so that they can be disassembled with ease, thus reducing the cost of disassembly and making recycling economically attractive. This leads us to the concept of *design for disassembly* (DFD), which involves designing the product to be amenable to extremely rapid disassembly. Fortunately, most of the guidelines adopted in *design for assembly* (DFA), which will be covered later in the text, hold true for DFD. Nevertheless, some rules for DFD are not compatible with those of DFA. For example, joining two components with an adhesive may be an easy way to assemble them but would create problems when trying to disassemble them, whether for maintenance or at the end of the service life of the product. The analysis and selection of joining methods that promote recycling will be covered in more detail later.

1.2.9 Material Selection

The ultimate goal of a designer, though unrealistic in most cases, is to come up with a one-material product. The practical alternative is to minimize the number of different materials used and make them clearly identifiable. In order to realize the extent of this problem, you have to bear in mind that there are about a hundred different kinds of plastics in an automobile and that most of them look similar to one another. A successful approach for cutting down the variety of plastics used in a product is to design complex components out of only one kind of plastic that has different properties depending on the molecular weight or the degree of polymerization. For example, a plastic part with a label usually requires three different materials: the part, the label, and an adhesive. All these items, however, can be made from the same material but having a different degree of polymerization in each case so as to promote recycling. In fact, an appropriate candidate here would be polycarbonate.

Another good practice when designing plastic components is to try to employ thermoplastic materials rather than thermosets whenever the functional requirements permit it because thermoplastics are easier to recycle. It is also possible in many cases to use a thermoplastic elastomer as an alternative to difficult-to-recycle thermosets. It is also advisable to try to avoid secondary finishing operations like painting and coating for plastic components.

1.2.10 Fastening and Joining Considerations

There are basically two methods of disassembling a product for recycling; the selection of one of them depends upon the method of combining the components together to form a product. The first method is reversed assembly, which involves following the same steps included in the assembly process in reverse order. For example, if two components are snap-fitted together during assembly, they would be similarly separated during dismantling for recycling. This, obviously, will not work for dissimilar plastic parts that are welded or adhesively bonded together. The second approach for disassembly involves dismantling using brute force. The designer has to decide upon the method of disassembly during the early design phase and promote that method when preparing the design. It is also important to remember to strive to make fastening points accessible. Here are the pros and cons of the different fastening and joining methods from the viewpoint of product design for the environment:

- *Welded parts* may or may not be easily recycled. Metals are recycled effectively after use, but this is not the case with plastics, where two dissimilar resins are joined together by stacking, ultrasonic welding, and so on. Brute force is usually required to separate the components but may create problems unless considered during the design.

- *Screws* are undesirable for both assembly and disassembly. You are advised to replace them with snap fits whenever possible. If these are not feasible, standardization of screw types, sizes, and head shapes is strongly recommended to facilitate disassembly.
- *Adhesives*, although they facilitate assembly, are undesirable for disassembly because they create many problems. Although brute force is an efficient way of dismantling glued parts, adhesives are considered contaminants when used to join plastic parts. Solvents constitute another method for dismantling adhesive-bonded parts, but here again, the disposal of these solvents may create environmental problems.
- *Snap-fit latches* are ideal for both DFA and DFD. They certainly facilitate recycling as additional parts or dissimilar materials are not required. The design should, however, ensure that snaps are easy to "unsnap," can withstand the anticipated service conditions, and will not inadvertently unsnap during use. Note that snaps generally increase the complexity and cost of the mold, which must be considered in the economic analysis for recycling feasibility.

In the abovementioned discussion, we adopted the commonly used term *recycling*, whereas a more precise expression should be *solid-waste management*. The latter encompasses recycling as well as other methods for making use of solid waste. Following are the definitions of all these methods as originally given by M. Grayson (see the references at the end of the book):

- *Reuse* refers to further or repeated use of waste in its original form. An example is the refilling of cigarette igniters with fuel.
- *Recycling* is the use of waste, or waste-driven material, as a raw material for products or fuel that may or may not be similar to the original. A typical example is the shredding of soda bottles and then recycling the material into outdoor furniture and the like. Although, in this case, the final product after recycling is obviously different from the original one, this may not be the case when, for example, recycling paper.
- *Recovery* is the processing of waste to prepare a usable material or fuel in a form in which and to the specification by which it can be recycled. An example is the processing of scrap iron into pig-iron ingots that can be used as raw material and further processed into steel.

1.2.11 Design for Assembly for Multicomponent Products

While a product can be a simple single component such as a spanner or a plastic cup, we are witnessing nowadays a large-scale use of sophisticated products; each involves a large number of individual components that are assembled together. The designer of a multicomponent product should, therefore, be concerned with the ease and cost of assembly, in order to minimize the cost of assembly and thus the production cost. The concept of DFA will be discussed later in Chapter 12.

1.2.12 Legal and Ethical Issues

As previously mentioned, mechanical/manufacturing engineers are concerned with the design and manufacture of products to solve problems in order to meet societal needs, thus improving the lot of society. They would, therefore, encounter situations that require a basic understanding of the law as well as making ethical judgments.

The law can be defined as a system of rules that regulate and ensure that individuals or a community adhere to the will and the values of the society. On the other hand, ethics involves the rules of behavior based on ideas about what is good and bad behavior. Ethical values and legal principles are usually closely related, but ethical obligations typically exceed legal duties. Here are some considerations you must always keep in mind and follow:

- Do not infringe on the intellectual properties of others when preparing a product design.
- Do not release any information about a product you are designing or you designed in the past when working for a company, even if you are not employed by that company anymore, because that would be a case of *a breach of contract*.
- Avoid taking a job with a competitor company in the area you are working or were recently working.
- Never, ever think of committing "fraud" by falsely certifying that a product complies with governmental regulations and engineering codes. Unfortunately, a few cases have surfaced in the past couple of years where some automobile manufacturers falsified the results of the emission tests.

Be aware of the product liability law under which the designer and not just the company would be held responsible for a defect in the product design. Remember that if you fail to carefully follow the well-established standards and practices of the profession, you will not be released from negligence, even if you are well-intentioned.

REVIEW QUESTIONS

1. What is the definition of *manufacturing*?
2. Is there any relationship between the status of manufacturing in a nation and the standard of living of the people in that nation? Explain why.
3. Explain the different approaches for obtaining a desired shape and give examples of some manufacturing processes that belong to each group.
4. List the different types of production and explain the main characteristics of each. Also, mention some suitable applications for each type.
5. Explain the meaning of the term *tolerance*.
6. How do we scientifically describe the tightness or looseness of two mating parts?
7. What concepts did Eli Whitney establish to ensure the trouble-free running of mass production of multicomponent products?
8. What is meant by the *production turn*? What role does marketing play in this cycle?
9. Using the concept of production turn, how can we maximize the profits of a company by two different methods?
10. Explain the stages involved in the life cycle of a product.
11. What is the significance of the product life cycle during the phase of planning for the production of new products?
12. What is the *S curve*? Explain an American success story in employing it.
13. Give some examples of transfer from one technology development curve to another.
14. What are the stages involved in the design process? Explain each briefly.
15. What is meant by *trade-offs*? How can these be achieved during the decision-making stage?
16. Explain the old approach to product design. What are its disadvantages?
17. Explain the concept of design for manufacturing. Why is it needed in modern industries?
18. List at least five factors that should be considered when designing a new product.
19. What is QFD? Explain how it is vital to include it in the process of designing a new product.
20. Name a product of your choice and list some of the conditions to which it will be subjected during its service life.
21. Why would the weight of cost as a factor affecting the design vary for different types of products? Explain.
22. What is industrial design?
23. How does sustainability play a role when designing a new product?
24. What consideration should be taken when selecting material for a new product?
25. Why should the engineer understand the basics of law and be able to make ethical judgments, even though his or her work is purely technical?

2 Product Cost Estimation

2.1 INTRODUCTION

As mentioned in Chapter 1, the production turn cannot continue unless the manufactured products are successfully marketed, the fixed and working capital is recovered, and a profit is made. Cost plays a vital role in the marketing process because it provides the information required to set up the selling price of a product. An overpriced product cannot penetrate the market and will eventually lose out to similar but more competitively priced products. Underestimation of the production cost may result in products sold at a loss and, consequently, financial problems for the manufacturing corporation.

Because our main concern is design for manufacturing and because design is an open-ended process that yields more than one workable solution, a logical criterion for evaluating these "solutions" or "designs" would certainly be the cost required to bring each design into being and manufacture the product. Therefore, it is fair to state that cost estimation is initiated by, is linked to, and follows the product design in order to ensure the profitability of new products. Cost is also used to determine the most economical operation or sequence of operations for manufacturing a product, and it can be used as a means for establishing a cost-reduction program aimed at manufacturing the product so that it can be priced more competitively.

2.2 COSTS: CLASSIFICATION AND TERMINOLOGY

Costs can be classified in different ways based on their relationship to the production volume and the nature of the manufacturing operations. The first, and most logical, way to classify costs is to split them into two groups: capital costs and operating costs. As the name suggests, capital costs are incurred because of buildings, production machinery, and land. It is important to remember, when carrying out cost estimation, that buildings and machinery are depreciable (i.e., they tend to lose most of their value with time), whereas land is not. Operating costs are "running" costs that recur as long as the plant is in operation.

Another way to classify costs is to view them as belonging to one of two categories: fixed costs, which are independent of the production volume, and variable costs, which are dependent on it. Here are some examples of cost elements in the fixed-cost category:

- Depreciation on buildings, machinery, and equipment
- Insurance premiums (fire, theft, flooding, and occupational hazards)
- Property taxes (sometimes states and communities give tax breaks to industrial corporations to attract them to a region)
- Interest on investment (money borrowed from a bank, as explained in Chapter 1)
- Factory indirect-labor cost (wages of security, personnel, secretarial, clerical, janitorial, and financial staff)
- Engineering cost (high-level engineering jobs and R&D expenses)
- Cost of rentals, if any (sometimes the building itself is rented, or some equipment may be rented for a short term)
- Cost of general supplies (supplies used by the factory indirect-labor force)
- Management and administrative expenses (salaries paid to corporate staff, plus legal expenses or salaries paid to legal staff)
- Marketing and sales expenses (salaries and wages paid to marketing and sales staff, transportation and delivery expenses, and rentals of warehouses, if any)

DOI: 10.1201/9781003537106-2

The following cost elements fall in the variable-cost category:

- Cost of materials
- Cost of labor (including production supervision)
- Cost of power (electricity, gas, or fuel oil) and utilities (water, sewer, etc.)
- Cost of maintenance of production equipment

The logical way to determine the total cost of a product is to add up all the cost elements, as indicated by the bar diagram in Figure 2.1.

In product cost estimation, the use of spreadsheets is extremely important, and the student is, therefore, encouraged to learn about and practice using them. The value of each cost element can be inserted in a spreadsheet cell, and as an alternative, the formula for computing a cost element can be employed in the cell if the value of that element is not known. A further advantage of using spreadsheets is the ease with which alternative designs can be compared and evaluated from the point of view of cost. Table 2.1 shows a spreadsheet where the specific cost elements and the total cost of each design are shown in columns that make comparisons and conclusions easy. From a quick look at the figure, for example, it is not difficult to realize that design 4 is the optimal choice based on cost.

As easy as it may look, however, it is impossible to carry out the process of determining the total cost of a product unless rational procedures and analyses are employed to overcome two main problems. First, some costs cannot be directly assigned or traced to any particular product but rather are spread over the entire factory; they are, therefore, labeled as "indirect" costs. In other words, the problem is how to calculate the cost share of a product from the salary of a receptionist or secretary. Second, we do not actually know the time taken to produce a design because that design has not been manufactured. It is the objective of this chapter to provide adequate answers to these two problems and to show the student how to independently carry on an engineering cost analysis for any desired design.

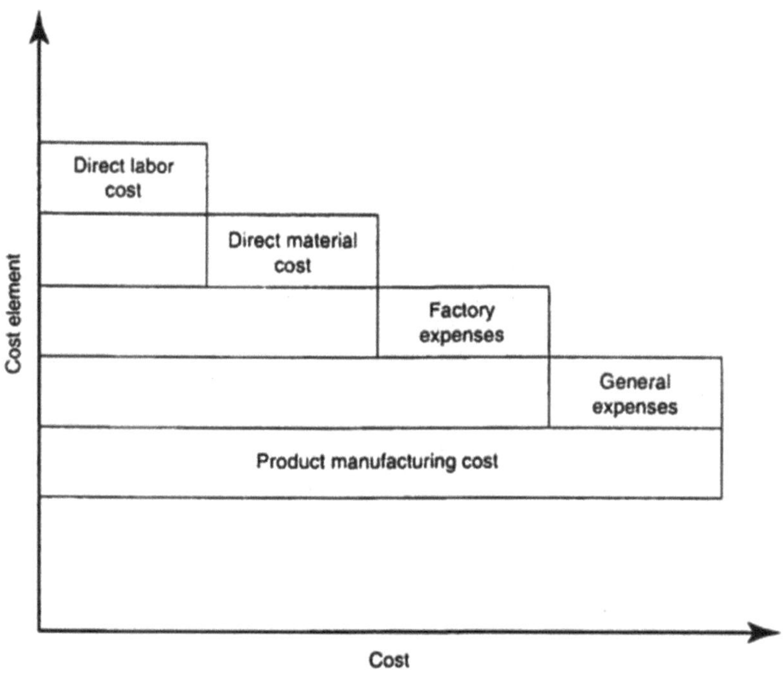

FIGURE 2.1 Elements contributing to the total cost of a product.

TABLE 2.1
A Spreadsheet That Compares the Cost of Four Alternative Designs

Design Alternative			Elements of Product Cost		
Direct Labor Cost ($)	Direct Material Cost ($)	Factory Expenses ($)	General Expenses ($)	Total Manufacturing Cost ($)	
Design 1	2.4	37	8	7.0	54.4
Design 2	2.6	36	8.5	7.1	54.2
Design 3	2.5	36.5	9	7.3	55.3
Design 4	2.3	35.5	7.5	7.2	52.5

Now, with our stated goal as product cost estimation that is based on and begins after a detailed product design is available, we must develop highly accurate cost estimates that are suitable for submission on a bid or purchase order. This type of estimate is referred to as a detailed estimate and must have a level of accuracy of ±5 percent. The American Association of Cost Engineers came up with a list of five types of cost estimates, each having a certain level of accuracy, a different approach, and recommended applications. For example, the first type, a rough estimate, has an accuracy of ±40 percent and is based on indexing and modifying the cost of existing similar designs. It is, therefore, recommended for initial feasibility studies that are used to decide whether or not a probable profit justifies pursuing a project any further. Other types of cost estimates fall between the two extremes of rough to detailed and are, consequently, recommended for applications that depend upon their level of accuracy.

Before we attempt to gain a deeper insight into each of the elements that contribute to the total cost of a product, we must consider some factors that, if overlooked, may adversely affect the accuracy and validity of the estimate. For instance, the cost estimate cannot be held valid for more than a few months if the inflation rate in the country of production is noticeable. Further complications arise when the time taken to construct the plant and manufacture the products is so long that initial costs are affected by inflation (meaning that the money loses its purchasing power). Also, there are sometimes uncertain and unforeseen expenses or contingency factors, a typical example being the escalation of R&D costs when developing new technology for manufacturing the products.

2.3 LABOR COST ANALYSIS

Labor can be either direct or indirect: direct labor is explicitly related to the process of building the design, whereas indirect labor involves the work of foremen, stockroom keepers, and so on. We will be concerned here with the cost analysis of direct labor because the indirect labor cost is generally covered by factory overhead costs in the form of a percentage of the cost of direct labor hours. At this point, our goal is to estimate the labor time for building a design and then multiply that time by the combined value of wages and fringe benefits, which is usually called a gross hourly cost. Note, however, that those wages are sometimes not based just on attendance but also on performance (i.e., incentives are given when the hourly output exceeds a certain established goal).

2.3.1 Methods for Measurement of Time

Although there are quite a few approaches for the measurement and estimation of labor time, two methods are well accepted in the industry and will, therefore, be covered here. The first method is based on a time and motion study, a modern subject that was established by the eminent American engineer Frederick W. Taylor of Pennsylvania in the early twentieth century. This method, which is favored by industrial engineers, involves breaking down the manual work of an operation into individual simple motions. A typical motion is, for example, "reach and grab" (i.e., the worker stretches his or her hand to reach a tool and grab it). The operation is then converted into a tabular form that includes the entire sequence of basic motions that comprise the desired manual operation. Because these basic motions were thoroughly studied by industrial engineers and because time measurements were taken and standardized for each basic motion, our job is fairly easy. It is just to read, from published data that are readily available, the standard time unit for each motion included in the manual operation and insert it into a time and motion study table. By summing up all the time values, the total time required by an average worker to carry out the operation can be obtained. This time is modified by dividing it by the efficiency or the "rating" of the actual worker to account for interruptions and fatigue. This approach has the clear advantage of including a mechanism for rationalization of the operation by eliminating unnecessary and wasteful motions. The procedure described here is used to estimate the time for a single operation only and must be carried out for all operations required to produce a design. Consequently, our first step (after the design is available) is actually to prepare detailed process routing sheets indicating all operations included in the production of the part. It is clear that the time and motion study method requires a considerable amount of work, but it has usually been found that the effort and time spent are well worthwhile. The second method is based on the historical value of time. Time cards for a similar design that has already been built are obtained and studied in order to determine the number of "man-hours" required to do the job (a man-hour is a unit indicating the output of one person working for one hour). Data analysis using spreadsheets is then employed to make the necessary adjustments, taking into consideration such factors as the skill level of the workers, the workplace environment, and cost escalation, if any.

2.3.2 The Learning Curve

It is a well-known fact that doing a job for the first time requires more time from a worker than when doing it for the fifth time, for example. This is evidently due to the phenomenon of self-teaching while performing the work, which, in turn, leads to a gain in work experience and thus a shorter time for doing that job. This is what is usually referred to as the learning curve or product improvement curve. As can be seen in Figure 2.2, the learning curve indicates the relationship between the production time (say, per unit) and the total cumulative production. Learning curves are usually labeled by percentages. In Figure 2.2, for example, we have a "70% learning curve." This simply means that whenever the total cumulative production doubles, the production time per unit decreases by 30 percent. This description suggests that the learning curve is exponential and can, therefore, be expressed by the following equation:

$$t = t_0 \times p^n \qquad (2.1)$$

where

t is the production time per unit after producing a number of units equal to p
t_0 is the time taken to produce the first unit
p is the total cumulative production (i.e., total number of units produced)
n is a constant that depends on the constant percentage reduction characterizing a learning curve (e.g., −0.5146 for 70 percent curve, −0.322 for 80 percent curve, −0.152 for 90 percent curve)

Taking the concept of the learning curve into consideration results in reducing the total production time (and cost) as estimated by the previously mentioned conventional methods, and adjustments have to be made. By simple mathematics, the total production time T is given by

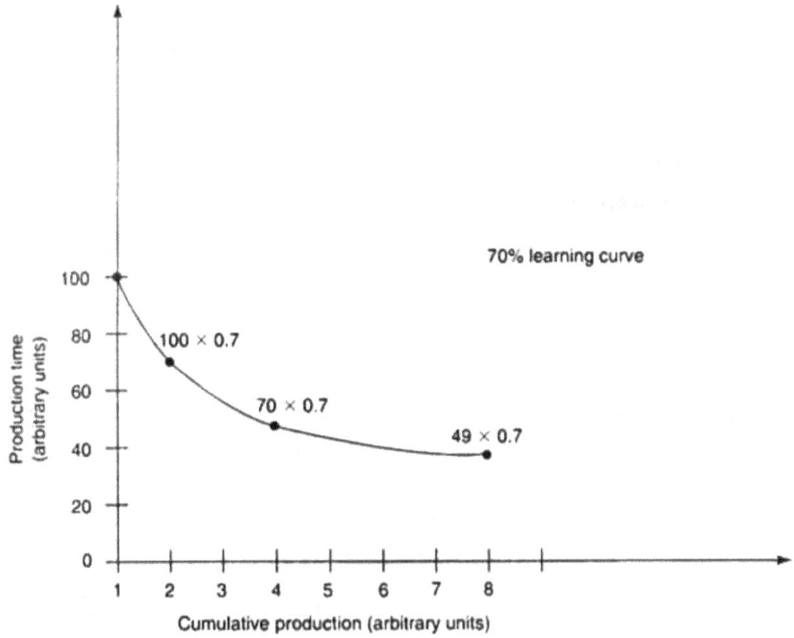

FIGURE 2.2 The learning curve.

$$T = t \times p \tag{2.2}$$

Substituting the value of t from Equation (2.1), we obtain

$$T = t_0 \times p^n \times p = t_0 \times p^{n+1} \tag{2.3}$$

2.3.3 Labor Laws

Some legal aspects must be considered when estimating the cost of labor. Federal and state laws regulate wages, for instance, the minimum wage is \$7.25 per hour. Also, the number of regular working hours per week is 40, and per day is limited to 8. If either (or both) of these is exceeded, a production worker must be paid at a rate equal to 150 percent of his or her regular hourly wage for the number of working hours that exceed the limits. Labor hourly wage rates in the United States, as well as other important relevant information, are compiled and published by the Bureau of Labor Statistics for various industrial sectors and can be obtained from the Department of Labor.

It is also important to remember that the labor cost is not limited to the money spent on wages but must include the fringe benefits paid to workers. Fringe benefits differ for different companies and may include any of the following:

- Health, dental, and life insurance premiums
- Expenses of insurance against job injury and hazards
- Holidays, paid vacations, and sick leave when actually taken
- The company's share in pension plans
- Payments to union stewards (if the company is unionized)
- Profit-sharing bonus money (that part of the company's profits paid to workers)

Fringe benefits can amount to as much as 30 percent or more of wages. It is, therefore, fairly common to combine wages with fringe benefits into the so-called gross hourly cost in order to avoid repetitive calculations.

2.4 MATERIAL COST ANALYSIS

2.4.1 Amount of Material Used

In order to carry out material cost analysis for a product, the amount of material used to manufacture that product first must be determined and then multiplied by the price of material (or materials) in the form of dollars per unit weight or volume. Consequently, some documents must be available before this cost-estimating operation is initiated. These include, for example, the bill of materials, the engineering design documents, and printouts of inventory data. The engineer is then in a position to determine the bill of material required to build the design, a process that is sometimes referred to as the quantity survey. It is not difficult to see that the raw material required is more than the amount of material indicated in the design blueprints. This difference includes but is not limited to the waste during manufacturing, which, in turn, depends upon the specific production processes employed. For example, when the part is to be produced by machining, the amount of material removed from the stock in the form of chips must be added to the amount calculated from the design drawing. When the part is to be produced by casting, the material in the risers, sprues, and gating system must be added, as well as the material that is removed by machining (the "skin," drilled holes, slots, recesses, etc.). The same rule of adding the waste applies to the various manufacturing processes of forging, press working, and extrusion. Note that the waste is sometimes sold to junk dealers for recycling, and the money paid for it must accordingly be subtracted from the cost of the material. In some cases, however, the waste is valueless and is disposed of at a cost that has to be added to the cost of material. In addition to waste, losses due to scrap (i.e., defective parts that are functionally obsolete) must be added. Again, scrap can be sold at a cost for recycling or may require disposal and additional expenditure. Losses may also include "shrinkage," which is a loss caused by environmental conditions (e.g., oxidation of steel or decrease in the volume of lumber when it dries).

The amount of raw material required to build the design and calculated according to the approach just described is part of a cost analysis category known as direct material. This category can also include standard-purchased items like nuts, bolts, springs, and washers. These items have no labor cost in our cost estimate, and the purchase price is considered as material cost (the company that makes these items has to consider the labor cost when estimating the selling price). Direct material also includes subcontracted items, which are assemblies or subassemblies manufactured outside the company and are supplied by an external subcontractor. Again, this does not include any labor cost in our estimate and must be categorized as material cost.

Determination of the amount of direct material required to build a design can sometimes be complicated. Consider, for example, the case when plastic is injection molded into a mold that has several dissimilar cavities for producing different parts. The sprues and runners form the waste in the injection molding operation. The question is, how do we determine the share of each part from that waste? Let us consider dividing the amount of material of the runners equally between the parts in order to get the amount of waste for each product. Unfortunately, the results may be totally misleading, especially when some of these products are very small while others are large. In fact, this case is referred to as one of joint material costs and arises whenever there is a multiple-product manufacturing process where the tracing of the raw material share of each individual product is difficult. A well-accepted approach in this case is to agree upon a primary product and attribute most of the untraceable common expenditures to it. In addition to direct material, material is consumed during the process of transforming the raw stock into useful products. It is usually both necessary and untraceable to a particular design and is, therefore, referred to as indirect material. Typical examples include lubricating oils, soaps, and coolants. As it is clear that direct mathematics will not work here, a simpler method, which has gained acceptance in industrial cost-estimating practice, is to add this item to factory overhead costs.

2.4.2 Purchasing Price of Material

As previously mentioned, the purchasing price of material has to be known, in the form of dollars per unit weight or volume, in order to be able to estimate the total material cost to build the design. When the material used is contractual (i.e., purchased specifically for manufacturing a certain product), the actual purchase price can be directly employed in estimating the cost of the material. However, when the material is taken from inventory, it is difficult to find its value and use it in the cost analysis because inventories usually contain various lots of the same material that have been purchased at different times at different prices. So, the question is, which price do we use in cost estimation? In fact, nobody can provide a precise answer to this question and, therefore, a number of approaches have been adopted by different schools of thought in industry. Following is a brief summary of some of the commonly used methods.

2.4.2.1 First-In-First-Out Method

The first-in-first-out method is based on following the rule of issuing the material that was purchased first (i.e., having the longest time in stock) to the factory for processing and using its purchase price in the cost analysis. A clear drawback of this method arises when the time between purchasing and processing of the material becomes long. The original price may not be a true representation of the current value of the material, thus resulting in an inaccurate cost estimate.

2.4.2.2 Last-In-First-Out Method

In the last-in-first-out method, the material that was added last to the inventory is issued first to the factory. The material used can be from more than one purchased lot, with different costs for each of those lots, a fact that complicates the process of estimating the cost of materials.

2.4.2.3 Current-Cost Method

The approach taken by the current-cost method is to use the cost of materials corresponding to the time when the estimate is prepared. Once again, material issued from the inventory that would have been purchased earlier or later may have a cost different from the current cost.

2.4.2.4 Actual-Price Method

The actual-price method is based on calculating the amount of money originally spent to purchase the material used. If the material issued from the inventory belongs to the same lot, then the calculations are easy and simply take the original purchase price of the material. However, when the material is taken from two (or more) lots having different purchase prices, an average or equivalent cost has to be used in the cost-estimating process. Here is the applicable equation:

$$C_{\text{equivalent}} = \frac{\sum_{i=1}^{n} c_i a_i}{\sum_{i=1}^{n} a_i} \tag{2.4}$$

where

c_i is the cost of material of lot i
a_i is the amount of material taken from lot i
i to n are use-lot serial numbers

2.5 EQUIPMENT COST ANALYSIS

The cost of equipment belongs to the fixed-cost category, and the depreciation of machinery as well as the interest on investment must be taken into account. It is often difficult, however, to get a quote for the cost of equipment, especially during the early phase of a feasibility study. Published data about the cost or equipment are unfortunately not directly applicable due to inflation and devaluation of the purchasing power of the dollar, as well as the difference in capacities or ratings given in the published data and those of the required equipment. Adjustments must, therefore, be made to account for factors that affect the validity of the published cost-of-equipment data. Following are some of the methods used.

2.5.1 Cost Indexing

A cost index is an indication of the buying power of money at a particular time for a certain category of equipment and machinery. Accordingly, if the cost of equipment and the corresponding index are known at some initial time, the current cost can be determined if the current index is obtained:

$$C_c = C_i\left(\frac{I_c}{I_i}\right) \tag{2.5}$$

where

C_c is the current cost
C_i is the cost at some initial time
I_c is the current index
I_i is the index at the same initial time as C_i

There are many published indexes, and each pertains to a certain area of application (e.g., chemical plants, consumer price). The Marshall and Swift cost index is the one to use for industrial equipment. It is readily available and has different values for the various kinds of industry. Care must, therefore, be taken not to use the index for the paper industry to calculate the cost of a steam turbine, for example. More important, indexing does not hold true when there is a radical change in the technology. This is evidenced by the fact that the prices of many electronic products have actually decreased as a result of technological change. Also, further adjustments are sometimes needed to account for regional conditions because published indexes are indications of national averages.

2.5.2 Size Effect

Sometimes, it is possible to only get hold of the cost of a machine that is similar to the required one but having a different size or rating. Corrections must, therefore, be made to that cost in order to obtain the cost of the desired machine. Consequently, a mathematical relationship between the cost and the capacity (size or rating) of capital equipment must first be established. As you may have guessed, the relationship is not linear due to the effect of the economy of scale on engineering design and production. The relationship can be given by the following empirical formula, which is usually referred to as the six-tenths rule:

$$C_2 = C_1\left(\frac{S_2}{S_1}\right)^{0.6} \tag{2.6}$$

where

C_2 is the cost of capital equipment 2
C_1 is the cost of capital equipment 1
S_2 is the capacity (size or rating) of capital equipment 2
S_1 is the capacity (size or rating) of capital equipment 1

2.5.3 Regression Analysis

Statistical techniques are used for collecting factual data that are, in turn, subjected to mathematical analysis and curve fitting in order to establish a relationship between cost and the various parameters that affect it. Here is the general equation:

$$C = C_0 \sum_{i=1}^{i=n} P_i^{m_i} \tag{2.7}$$

where

C is the cost
C_0 is a constant
P_i is the parameter i affecting the cost
m_i is a constant exponent for parameter i
C_0, P_i, and m_i are determined by mathematical and statistical methods. Although the formula is limited in scope to the specific equipment or system for which it is developed, it has proven to be very useful in cost models because it elaborates the elements and parameters that contribute most to the cost. It also leads to the ability to minimize and optimize costs using simple mathematical manipulations such as those of differential calculus.

2.6 ENGINEERING COST

Engineering cost includes salaries for high-level engineering jobs as well as expenditures (whether salaries or general expenses) for R&D. Usually, the engineering cost for a product is considered part of the factory overheads (or even part of the corporate overheads in some cases). Nevertheless, it is sometimes estimated separately based on previous records of existing similar products. In some cases (especially when the product is supplied to the federal government), a firm is hired to do the engineering on a contractual basis. The contract may specify a lump sum for the engineering cost or may involve the true engineering cost plus a negotiated fee or profit of, say, 15 percent of the cost. In this latter case, the engineering cost must be accurately determined.

2.7 OVERHEAD COST

Overhead costs are usually viewed by cost engineers as a burden because such costs cannot be directly or specifically related to the manufacturing of any particular product or even to a particular category of the company's production. Overhead costs can be divided into two main groups: factory overheads and corporate overheads.

2.7.1 Factory Overheads

Factory overheads include the previously mentioned engineering costs as well as other factory expenses that are not related to direct labor or material. An example would be the wages paid to personnel for security, safety, shipping and receiving, storage, and maintenance. The challenge here is how to calculate the "share" of each different product from these expenses. There are many approaches for charging these expenses to the cost of the various products. The following are three bases upon which factory overhead costs can be allocated:

- The ratio between the direct labor hours required to manufacture the product and the total number of direct labor hours spent on the factory floor (this ratio, when multiplied by the total overhead expenses, yields the share of that product from the overhead cost)
- The ratio between the material cost of the product and the total cost of material consumed on the factory floor (again, the share of a product from overhead cost is the product of the multiplication of this ratio by the total overhead expenses)
- The ratio between the space occupied by the production equipment (e.g., furnace or machine tool) and the total area of the factory floor

The direct-labor-hours method is by far the most commonly used approach for allocating factory overhead costs. As is clear, the production volume has a tangible effect on the factory overhead cost per product. If the production is reduced to half its normal level, for example, without reducing the total overhead expenses, the overhead share of a product will automatically double. Consequently, it is always a good idea to carry out cost analysis for any potential product at different production volumes (i.e., percentages of full capacity of production lines). Note that increasing productivity results in a reduced number of direct labor hours. This is sometimes misinterpreted by management, and decisions may be made to reduce the budget for maintenance and other factory overhead items. It is the duty of the manufacturing engineer to eliminate any misinterpretations on the part of management. An alternative, in this case, is to use a different basis for allocating the overhead costs and requesting budgetary funds.

2.7.2 Corporate Overheads

Corporate overheads basically involve the cost of daily operations of the company beyond the factory floor throughout the year. These expenses include, for example, the salaries and fringe benefits of corporate executives as well as those of the business, administrative, and legal staff. Again, the commonly adopted approach is to obtain an overhead rate that is the product of dividing the total corporate overhead expenses by the total cost of direct labor. Knowing the direct labor time and cost for manufacturing a product, you can easily calculate the corporate overhead cost using this overhead rate. It is worth mentioning that corporations may operate more than one plant from the corporate headquarters, a fact that has to be taken into consideration when calculating both the corporate and the direct labor costs in order to obtain the overhead rate.

2.8 DESIGN TO COST

The preceding discussions reflect the usual or conventional sequence of preparing the design and then costing the product based on the information provided in that design. With increasing global competition, however, cost is becoming more and more the driving force. Consequently, a need arises for costing a potential product before its design is completed or even made. This unusual approach is aimed at continuously improving the design in order to manufacture the desired product at a designated price that is equal to or less than the market price of the competitor's product. This "reverse" procedure is known as design to cost and is gaining popularity in the industry, especially with newly emerging methodologies such as reengineering.

The process starts with benchmarking a given product, taking market price and quality as the judging criteria. By removing the retail profit, the manufacturing cost is obtained. Next, the various overhead rates that are well established in the company are employed to remove the different overhead cost items, yielding the prime cost. Then comes the difficult task of meticulously breaking down the prime cost among components, assemblies, and subassemblies. Favoring one component at the expense of another is a big mistake, as setting a target cost below reasonable limits will make the design of the component virtually impossible. Once the target cost for a component is allocated, the design begins using that target cost as an incentive for continuously improving the design. If the direct labor cost, for example, is found to be less than the target, this will give some relief in the process of selecting materials. The same rule is applicable to subassemblies and assemblies (i.e., if the cost of a component is less than the target, this will give more flexibility when designing other components). When the design is finalized, it must be subjected to the conventional and accurate cost-estimating process.

REVIEW QUESTIONS

1. Why is cost estimation of vital importance for a design engineer?
2. What role does cost play in process planning?
3. List two methods for classifying costs.
4. List some important elements of fixed cost.
5. List some important elements of variable cost.
6. What are the two main problems that complicate cost estimation?
7. Do all types of cost-estimating methods have the same accuracy? Explain why.
8. Can you rely upon a cost estimate that was done last year? Why not?
9. Assuming that the construction of the plant takes a long time, what effects would this have on the cost-estimating process?
10. What is meant by direct labor?
11. Is there any indirect labor? Explain.
12. What is the gross hourly cost?
13. Explain the difference between the two cases of wages based on attendance and wages based on performance.
14. How can we measure the direct labor time before the product is actually manufactured?
15. What are the pros and cons of the industrial engineering approach for measuring the direct labor time?
16. Explain the timecard method for estimating the direct labor time.
17. What is the learning curve? What effects does it have on cost-estimating results?
18. List some important labor laws that must be considered when estimating the cost of a product.
19. List some common fringe benefits.
20. What is the quantity survey?
21. What are the sources of difference between the materials in a product as indicated by the design drawing and the material actually consumed in the manufacture of that product?
22. Explain the term *indirect material*.
23. What is the joint material cost? Give examples.
24. Why is it difficult to get the cost of unit material when the material is issued from inventory?
25. Explain briefly the different methods used to obtain the cost of unit material when the material is issued from inventory. Give the pros and cons of each.
26. What is a cost index? Why is it important in cost estimation?
27. How can you calculate the cost of a machine with a known capacity if you know the cost and the capacity of another machine?
28. Show how regression analysis and statistics can be employed in cost estimation.
29. What is meant by engineering cost? How is it estimated?

30. What are the different types of overhead costs?
31. On what bases are factory overhead costs allocated?
32. In some cases, increasing productivity might have an adverse effect on budget allocations. Explain how.
33. Explain the concept of design to cost.
34. What is the driving force for design to cost?
35. What is the main problem encountered in the procedure of design to cost?

EXAMPLE PROBLEM

A number of stock bars, each 3.25 inches (81 mm) in diameter and 12 feet (3.6 m) in length, are to be used to produce 2000 bars, each 2.75 inches (69 mm) in diameter and 12 inches (300 mm) in length. The material cost is \$0.14 per pound (\$0.3/kg), and the density is 0.282 pounds per cubic inch (789 kg/m^3). The total overhead and other expenses are \$95,000. The total direct labor expense for the plant is \$60,000. Estimate the production cost for a piece.

SOLUTION

First, we have to calculate the production time per piece. Consequently, technical production data have to be either obtained or assumed, in order to be able to calculate the machining time.

We will take the machining time to be 1254 seconds (the calculation procedure is detailed later in Chapter 11).

$$\text{Number of pieces produced from a single bar} = \frac{12\times 12}{12.25}$$

$$= \frac{\text{Total length}}{\text{Length / piece}} = 11\,\text{pieces}$$

$$\text{Number of stockbars} = \frac{2000}{11} = 181.8 = 182$$

$$\text{Total loading time} = \frac{2\times 182\times 60}{2000}$$

$$= 10.9 \approx 11\,\text{seconds}\left(\text{assuming loading time / bar} = 2\,\text{minutes}\right)$$

$$\text{Total average production time / piece} = 1254 + 11$$

$$= 1265\text{ seconds}\left(\text{direct labor}\right)$$

$$\text{Cost of labor / piece} = \frac{1265}{3600}\times \$30\,/\,\text{hr}$$

$$= \$10.15\left(\text{assuming Computerized Numerical Control}\left(\text{CNC}\right)\text{machine is used}\right)$$

Cost of Material/Piece

We have to consider the waste. Assume no scrap, as the operation is simple:

$$\text{Cost of material / piece} = 182 \times \frac{\pi}{4} \frac{(3.25)^2 \times 12 \times 12 \times 0.282 \times 0.14}{2000} = \$\, 4.29$$

Cost of Overhead/Piece

In this, all other costs are included:

$$\text{Overhead rate} = \frac{95{,}000 \times 100}{60{,}000} = 158.33\,\text{percent}$$

$$\text{Overhead cost / piece} = 10.15 \times \frac{158.33}{100} = \$\, 16.07$$

$$\begin{aligned}\text{Total cost / piece} &= \text{direct labor cost} + \text{material cost} + \text{overhead cost}\\ &= 10.15 + 4.29 + 16.07 = \$30.51\end{aligned}$$

DESIGN PROJECT

Select a few of the design projects that supplement Chapters 3–7, preferably a project for each manufacturing process, and then carry out cost estimation for the product. You are strongly advised to obtain real values for the different cost elements (e.g., material) by contacting industrial companies and obtaining quotations.

3 Casting and Foundry Work

At the dawn of the metal age, human knowledge was not advanced enough to generate the high temperatures necessary for smelting metals. Therefore, because casting was not possible, metals were used as found or heated to a soft state and worked into shapes. The products of that era are exemplified by the copper pendant from Shanidar Cave (northeast of Iraq), which dates back to 9500 BC and was shaped by hammering a piece of native metal and finishing with abrasives. Later, copper-smelting techniques were developed, and copper castings were produced in Mesopotamia as early as 3000 BC. The art of casting was then refined by the ancient Egyptians, who innovated the "lost-wax" molding process. During the Bronze Age, foundry work flourished in China, where high-quality castings with intricate shapes could be produced. The Chinese developed certain bronze alloys and mastered the lost-wax process during the Shang dynasty. Later, that art found its way to Japan with the introduction of Buddhism in the sixth century. There were also some significant achievements in the West, where the Colossus of Rhodes, a statue of the sun god Helios weighing 360 tons, was considered to be one of the Seven Wonders of the World. That bronze statue was cast in sections, which were assembled later, and stood 105 feet high at the entrance of the harbor of Rhodes.

Although iron was known in Egypt as early as 4000 BC, the development of cast iron was impossible because the high melting temperature needed was not achievable then and pottery vessels capable of containing molten iron were not available. The age of cast iron finally arrived in AD 1340 when a flow oven (a crude version of the blast furnace) was erected at Marche-les-Dames in Belgium. It was capable of continuous volume production of molten iron. Ferrous foundry practice developed further with the invention of the cupola furnace by John Wilkinson in England. This was followed by the production of black-heart malleable iron in 1826 by Seth Boyden and the development of metallography by Henry Sorby of England. The relationship between the properties and the microstructure of alloys became understood, and complete control of the casting process became feasible based on this knowledge. Nevertheless, forming processes developed more rapidly than foundry practice because wrought alloys could better meet a wider range of applications. Nodular cast iron, which possesses both the castability of cast iron and the impact strength of steel, was introduced in 1948, thus paving the way for castings to compete more favorably with wrought alloys.

3.1 METALLURGICAL ASPECTS OF METAL CASTING

3.1.1 Solidification of Liquid Metal in a Mold

As previously mentioned in Chapter 1, the casting process basically involves molding of molten metal into a cavity and allowing it then to solidify or "freeze." It is therefore important to have a basic understanding of the thermodynamics as well as the metallurgical aspects of the transformation of liquid metal into a solid. We know that liquid metal alloys occur as a single homogeneous liquid phase, but in order to simplify the discussion, let us take the special case of a pure metallic element that would yield a single solid phase after freezing. The solid is a crystalline phase in which the atoms are aligned in space in a definite pattern to form a lattice. On the contrary, the liquid phase does not possess the long-range order of the solid, and atoms are unrestricted. Accordingly, when the liquid phase freezes, it has to lose some energy, called *the latent heat of fusion.* On the other hand, the movement of the interface separating a liquid from a solid crystalline phase, under a temperature gradient normal to the interface, can be viewed as the resultant of two atom movements. The first involves atoms leaving the liquid and joining the solid, while the atoms travel in the opposite

DOI: 10.1201/9781003537106-3

direction in the second. The direction of movement of the interface is determined by which of the two will prevail. When the temperature in the liquid in advance of the interface falls, the latter becomes unstable, and nuclei would grow out from the general interface into the liquid. The resulting structure would have secondary and tertiary branches forming on the primary one. The resulting crystal would look like a miniature pine tree, and it is referred to as a *dendrite*, from the Greek word *dendrites*, meaning "of a tree."

Solidification of liquid metals occurs by nucleation and growth. When a molten metal is poured into a mold or an ingot, the liquid metal in a narrow band that follows the contour of the mold comes in contact with the surface of the mold, which is at room temperature. As a consequence, the temperature of the liquid metal drops rapidly below the equilibrium freezing temperature. A considerable magnitude of supercooling takes place, which in turn results in a high rate of nucleation in that narrow layer. The grains in that zone are, therefore, characterized as being small, uniform in shape and size, equiaxed, and randomly oriented. This zone is referred to as the *chill zone*.

In the next zone, which is referred to as the *columnar zone* (see design considerations later in the text), crystal growth predominates and very little nucleation is observed. As soon as nucleation starts in the chill zone, the temperature in this region begins to rise again toward the equilibrium freezing temperature, as a result of the release of the latent heat of fusion. The columnar zone is thus composed of grains that start at the chill zone and become larger in size and elongated in shape, with the direction of the dendritic growth parallel to the heat flow direction (normal to the contour of the mold). These crystals may continue to grow in this manner until the solid–liquid interfaces that start from opposite sides of a mold approach each other and their zones of supercooling overlap.

3.1.2 Castability (Fluidity)

The ability of the molten metal to flow easily without premature solidification is a major factor in determining the proper filling of the mold cavity. This important property is referred to as *castability* or, more commonly, *fluidity*. The higher the fluidity of a molten metal, the easier it is for that molten metal to fill thin grooves in the mold and exactly reproduce the shape of the mold cavity, thereby successfully producing castings with thinner sections. Poor fluidity leads to casting defects such as incomplete filling or misruns, especially in the thinner sections of a casting. Because fluidity is dependent mainly upon the viscosity of the molten metal, it is clear that higher temperatures improve the fluidity of molten metal and alloys, whereas the presence of impurities and nonmetallic inclusions adversely affects it.

Several attempts have been made to quantify and measure the fluidity of metals. A commonly used standard test involves pouring the molten metal into a basin so that it flows along a spiral channel of a particular cross section, as shown in Figure 3.1. Both the basin and the channel are molded in sand, and the fluidity value is indicated by the distance traveled by the molten metal before it solidifies in the spiral channel.

3.2 CLASSIFICATION OF CASTING PROCESSES

The word *casting* is used both for the process and for the product. The process of casting is the manufacture of metallic objects (castings) by melting the metal, pouring it into a mold cavity, and allowing the molten metal to solidify as a casting whose shape is a reproduction of the mold cavity. This process is carried out in a *foundry*, where either ferrous (i.e., iron-base) or nonferrous metals are cast.

Casting processes have found widespread application, and the foundry industry is considered the sixth largest in the United States because it produces hundreds of intricately shaped parts of various sizes like plumbing fixtures, furnace parts, cylinder blocks of automobile and airplane engines, pistons, piston rings, machine tool beds and frames, wheels, and crankshafts. In fact, the foundry industry includes a variety of casting processes that can be classified in one of the following three ways:

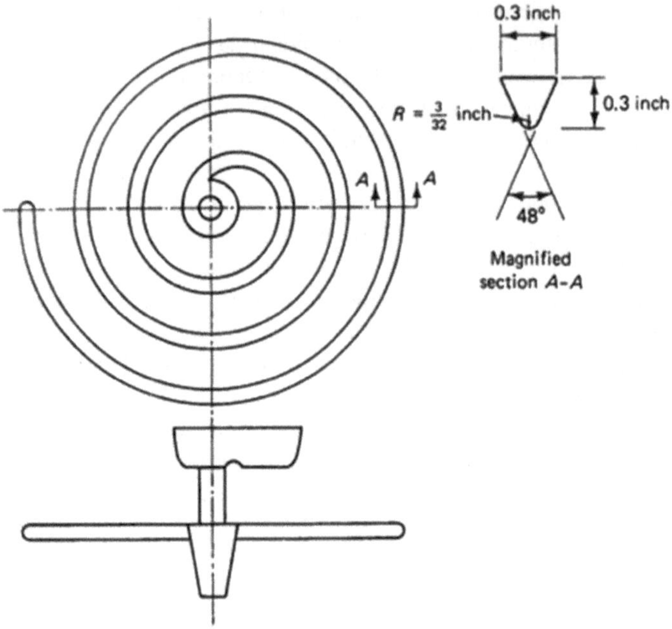

FIGURE 3.1 Details of the test for measuring fluidity.

- By the mold material and/or procedure of mold production
- By the method of filling the mold
- By the metal of the casting itself.

3.2.1 Classifications of Casting by Mold Material

Molds can be either *permanent* or *nonpermanent*. Permanent molds are made of steel, cast iron, and even graphite. They allow large numbers of castings to be produced successively without changing the mold. A nonpermanent mold is used for one pouring only. It is usually made of a silica sand mixture but sometimes of other refractory materials like chromite and magnetite.

3.2.1.1 Green Sand Molds

3.2.1.1.1 Molding Materials

Natural deposits taken from water or riverbeds are used as molding materials for low-melting-point alloys. Thus, the material is called *green* sand, meaning unbaked or used as found. These deposits have the advantages of availability and low cost, and they provide smooth as-cast surfaces, especially for light, thin jobs. However, they contain 15–25 percent clay, which, in turn, includes some organic impurities that markedly reduce the fusion temperatures of the natural sand mixture, lower the initial binding strength, and require a high moisture content (6–8 percent). Therefore, synthetic molding sand has been developed by mixing a cleaned pure silica sand base, in which grain structure and grain-size distribution are controlled, with up to 18 percent combined fireclay and bentonite and only about 3 percent moisture. Because the amount of clay used as a binding material is minimal, synthetic molding sand has higher refractoriness, higher green (unbaked) strength, better permeability, and lower moisture content. The latter advantage results in the evolution of less steam during the casting process. Thus, control of the properties of the sand mixture is an important condition for obtaining good castings. For this reason, a sand laboratory is usually attached to the foundry to determine the properties of molding sands prior to casting. The following are some important properties of a green sand mixture.

- *Permeability*. Permeability is the most important property of the molding sand and can be defined as the ability of the molding sand to allow gases to pass through. This property depends not only on the shape and size of the particles of the sand base but also on the amount of the clay binding material present in the mixture and on the moisture content. The permeability of molds is usually low when casting gray cast iron and high when casting steel.
- *Green Compression Strength of a Sand Mold*. Green strength is mainly due to the clay (or bentonite) and the moisture content, which both bind the sand particles together. Molds must be strong enough not to collapse during handling and transfer and must also be capable of withstanding pressure and erosion forces during the pouring of the molten metal.
- *Moisture Content*. Moisture content is expressed as a percentage and is important because it affects other properties, such as permeability and green strength. Excessive moisture content can result in entrapped steam bubbles in the casting.
- *Flowability*. Flowability is the ability of sand to flow easily and fill the recesses and the fine details in the pattern.
- *Refractoriness*. Refractoriness is the resistance of the molding sand to elevated temperatures; that is, the sand particles must not melt, soften, or sinter when they come in contact with the molten metal during the casting process. Molding sands with poor refractoriness may burn when the molten metal is poured into the mold. Usually, sand molds should be able to withstand up to 3000°F (1650°C).

3.2.1.1.2 Sand Molding Tools

Sand molds are made in *flasks*, which are bottomless containers. The function of a flask is to hold and reinforce the sand mold to allow handling and manipulation. A flask can be made of wood, sheet steel, or aluminum and consists of two parts: an upper half called the *cope* and a lower half called the *drag*. The standard flask is rectangular, although special shapes are also in use. For proper alignment of the two halves of the mold cavity when putting the cope onto the drag prior to casting, flasks are usually fitted with guide pins. When the required casting is high, a middle part, called the *cheek*, is added between the drag and the cope. Also, when a large product is to be cast, a pit in the ground is substituted for the drag; the process is then referred to as *pit molding*. Other sand molding tools can be divided into two main groups:

- Tools (such as molders, sand shovels, bench rammers, and the like) used for filling the flask and ramming the sand
- Tools (such as draw screws, draw spikes, trowels, slicks, spoons, and lifters) used for releasing and withdrawing the pattern from the mold and for making required repairs on or putting finishing touches to the mold surfaces.

3.2.1.1.3 Patterns for Sand Molding

The mold cavity is the impression of a *pattern*, which is an approximate replica of the exterior of the desired casting. Permanent patterns (which are usually used with sand molding) can be made of softwood like pine, hardwood like mahogany, plastics, or metals like aluminum, cast iron, or steel. They are made in special shops called *pattern shops*. Wood patterns must be made of dried or seasoned wood containing less than 10 percent moisture to avoid warping and distortion of the pattern if the wood dries out. They should not absorb any moisture from the green molding sand. Thus, the surfaces of these patterns are painted and coated with a waterproof varnish. A single-piece wood pattern can be used for making 20–30 molds, a plastic pattern can be used for 20,000 molds, and a metal pattern can be used for up to 100,000 molds, depending upon the metal of the pattern. In fact, several types of permanent patterns are used in foundries. They include the following:

- *Single or Loose Pattern*. This pattern is actually a single copy of the desired casting. Loose patterns are usually used when only a few castings are required or when prototype castings are produced.
- *Gated Patterns*. These are patterns with gates in a runner system. They are used to eliminate the hand-cutting of gates.
- *Match-Plate Patterns*. Such patterns are used for large-quantity production of smaller castings, where machine molding is usually employed. The two halves of the pattern, with the line of separation conforming to the parting line, are permanently mounted on opposite sides of a wood or metal plate. This type of pattern always incorporates the gating system as a part of the pattern.
- *Cope-and-Drag Pattern Plates*. The function of this type of pattern is similar to that of the match-plate patterns. Such a pattern consists of the cope and drag parts of the pattern mounted on separate plates. It is particularly advantageous for preparing molds for large and medium castings, where the cope and drag parts of the mold are prepared on different molding machines. Therefore, accurate alignment of the two halves of the mold is necessary and is achieved through the use of a guide and locating pins and bushings in the flasks. In order for a pattern to be successfully employed in producing a casting having the desired dimensions, it must not be an exact replica of the part to be cast. A number of allowances must be made on the dimensions of the pattern.
- *Pattern Drafts*. This is a taper of about 1 percent that is added to all surfaces perpendicular to the parting line in order to facilitate the removal of the pattern from the mold without ruining the surfaces of the cavity. Higher values of pattern draft are employed in the case of pockets or deep cavities.
- *Shrinkage Allowance*. Because molten metals shrink during solidification and contract with further cooling to room temperature, linear dimensions of patterns must be made larger to compensate for that shrinkage and contraction. The value of the shrinkage allowance depends upon the metal to be cast and, to some extent, on the nature of the casting. The shrinkage allowance is usually taken as 1 percent for cast iron, 2 percent for steel, 1.5 percent for aluminum, 1.5 percent for magnesium, 1.6 percent for brass, and 2 percent for bronze. In order to eliminate the need for recalculating all the dimensions of a casting, pattern makers use a shrink rule. It is longer than the standard 1-foot rule; its length differs for the different metals of the casting.
- *Machine Finish Allowance*. The dimensions on a casting are oversized to compensate for the layer of metal that is removed through subsequent machining to obtain a better surface finish.
- *Distortion Allowance*. Sometimes, intricately shaped or slender castings distort during solidification, even though reproduced from a defect-free pattern. In such cases, it is necessary to distort the pattern intentionally to obtain a casting with the desired shape and dimensions.

3.2.1.1.4 Cores and Core Making

Cores are the parts of the molds that form desired internal cavities, recesses, or projections in castings. A core is usually made of the best quality of sand to have the shape of the desired cavity and is placed into position in the mold cavity. Figure 3.2 shows the pattern, mold, and core used for producing a short pipe with two flanges. As you can see, projections, called *core prints*, are added to both sides of the pattern to create impressions that allow the core to be supported and held at both ends. When the molten metal is poured, it flows around the core to fill the rest of the mold cavity. Cores are subjected to extremely severe conditions, and they must, therefore, possess very high resistance to erosion, exceptionally high strength, good permeability, good refractoriness, and adequate collapsibility (i.e., the rapid loss of strength after the core comes in contact with the molten metal). Because a core is surrounded by molten metal from all sides (except the far ends) during

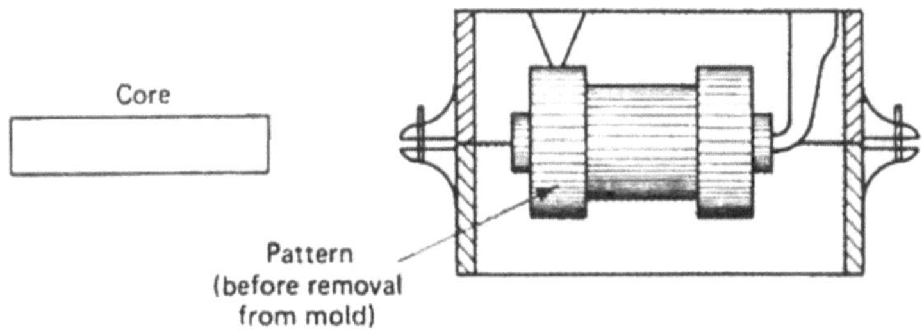

FIGURE 3.2 The pattern, mold, and core used for producing a short pipe.

casting, gases have only a small area through which to escape. Therefore, good permeability is sometimes assisted by providing special vent holes to allow gases to escape easily. Another required characteristic of a core is the ability to shrink in volume under pressure without cracking or failure. The importance of this characteristic is obvious when you consider a casting that shrinks onto the core during solidification. If the core is made hard enough to resist the shrinkage of the casting, the latter would crack as a result of being hindered from shrinking.

Core sand is very pure, fine-grained silica sand that is mixed with different binders depending on the casting metal with which it is going to be used. The binder used with various castings includes materials such as fireclay, bentonite, and sodium silicate (inorganic binders), as well as oils (cottonseed or linseed oil), molasses, dextrin, and polymeric resins (organic binders).

Cores are usually made separately in *core boxes*, which involve cutting or machining cavities into blocks of wood, metal, or plastic. The surfaces of each cavity must be very smooth, with ample taper or draft, to allow easy release of the green (unbaked) core. Sometimes, a release agent is applied to the surfaces of the cavity. Core sand is rammed into the cavity, and the excess is then struck off evenly with the top of the core box. Next, the green core is carefully rolled onto a metal plate and is baked in an oven. Intricate cores are made of separate pieces that are pasted together after baking. Sometimes, cores are reinforced with annealed low-carbon steel wires or even cast-iron grids (in the case of large cores) to ensure coherence and stability. Figure 3.3 illustrates a simple core and its corresponding core box.

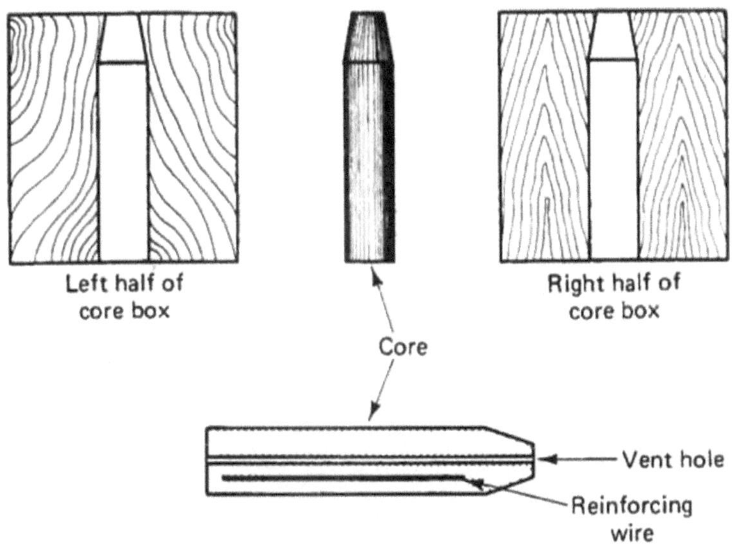

FIGURE 3.3 A simple core and its corresponding core box.

Large, round cores can be made by means of sweeps or templates, and drawing sweeps are employed to produce large cores that are not bodies of revolution. Various machines may also be employed in the core-making process. These include die extruders, jolt-squeeze machines, sand-slingers, and pneumatic core blowers. Large cores are handled in the foundry and placed into the mold by means of a crane.

3.2.1.1.5 Gating Systems

Molds are filled with molten metal by means of channels, called *gates*, cut in the sand of the mold. Figure 3.4 illustrates a typical *gating system*, which includes a pouring basin, a down sprue, a sprue base (well), a runner, and in-gates. The design of the gating system is sometimes critical and should, therefore, be based on the theories of fluid mechanics, as well as the recommended industrial practice. In fact, a gating system must be designed so that the following are ensured:

- A continuous, uniform flow of molten metal into the mold cavity must be provided without any turbulence.
- A reservoir of molten metal that feeds the casting to compensate for the shrinkage during solidification must be maintained.
- The molten metal stream must be prevented from separating from the wall of the sprue.

Let us now break down the gating system into its components and discuss the design of each of them. The *pouring basin* is designed to reduce turbulence. The molten metal from the ladle must be poured into the basin at the side that does not have the tapered sprue hole. The hole should have a projection with a generous radius around it, as shown in Figure 3.4, in order to eliminate turbulence as the molten metal enters the sprue. Next, the *down sprue* should be made tapered (its cross-sectional area should decrease when going downward) to prevent the stream of molten metal from separating from its walls, which may occur because the stream gains velocity as it travels downward and, therefore, contracts (remember the continuity equation in fluid mechanics, $A1 \cdot V1 = A2 \cdot V2$). The important and critical element of the gating system is the *in-gate*, whose dimensions affect those of all other elements. Sometimes, the cross-sectional area of the in-gate is reduced in the zone adjacent

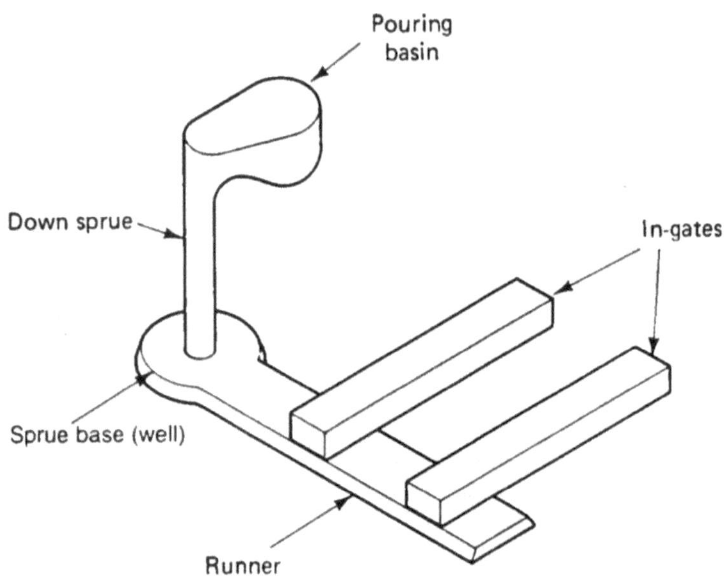

FIGURE 3.4 A typical gating system.

to the *sprue base* to create a "choke area" that is used mainly to control the flow of molten metal and, consequently, the pouring time. In other words, it serves to ensure that the rate of molten metal flow into the mold cavity is not higher than that delivered by the ladle and, therefore, keeps the gating system full of metal throughout the casting operation. On the other hand, gas contamination, slag inclusions, and the like should be eliminated by maintaining laminar flow. Accordingly, the Reynolds number (Rn) should be checked throughout the gating system (remember that the flow is laminar when $Rn < 2000$). Use must also be made of Bernoulli's equation to calculate the velocity of flow at any cross section of the gating system.

In some cases, when casting heavy sections or high-shrinkage alloys, extra reservoirs of molten metal are needed to compensate continually for the shrinkage of the casting during solidification. These molten-metal reservoirs are called *risers* and are attached to the casting at appropriate locations to control the solidification process. The locations of the feeding system and the risers should be determined based on the phenomenon that sections most distant from those molten-metal reservoirs solidify first. Risers are molded into the cope half of the mold to ensure gravity feeding of the molten metal and are usually open to the top surface of the mold. In that case, they are referred to as *open risers*. When they are not open to the top of the mold, they are then called *blind risers*. Risers can also be classified as *top risers* and *side risers*, depending upon their location with respect to the casting.

Another way to achieve directional solidification is the use of *chills*; these involve inserts of steel, cast iron, or copper that act as a "heat sink" to increase the solidification rate of the metal at appropriate regions of the casting. Depending upon the shape of the casting, chills can be external or internal.

3.2.1.1.6 Molding Processes

Green sand can be molded by employing a variety of processes, including some that are carried out both by hand and with molding machines. Following is a brief survey of the different green sand molding methods:

Flask Molding. Flask molding is the most widely used process in both hand- and machine-molding practices. Figure 3.5 illustrates the procedure for simple hand-molding using a single (loose) pattern. First, the lower half of the pattern is placed on a molding board and surrounded by the drag. The drag is then filled with sand (using a shovel) and rammed very firmly. Ventilation holes are made using a steel wire, but these should not reach the pattern. The drag is turned upside down to bring the parting plane up so that it can be dusted. Next, the other half of the pattern is placed in position to match the lower half, and the cope is located around it, with the eyes of the cope fitted to the pins of the drag. Sand is shoveled into the cope and rammed firmly, after using a sprue pin to provide for the feeding passage. Ventilation holes are made in the cope part of the mold in the same way they were made in the other half. The pouring basin is cut around the head of the sprue pin using a trowel, and the sprue pin is pulled out of the cope. The cope is then carefully lifted off the drag and turned so that the parting plane is upward. The two halves of the pattern are removed from both the cope and the drag. The runner and/or gate are cut from the mold cavity to the sprue in the drag part of the mold. Then, any damages are repaired by slightly wetting the location and using a slick. The cope is then carefully placed on the drag to assemble the two halves of the mold. Finally, the cope and the drag are fastened together by means of shackles or bolts to prevent the pressure created by the molten metal (after pouring) from separating them. Enough weight can be placed on the cope as an alternative to using shackles or bolts. In fact, the pressure of the molten metal after casting can be given by the following equation:

$$p = w \times h$$

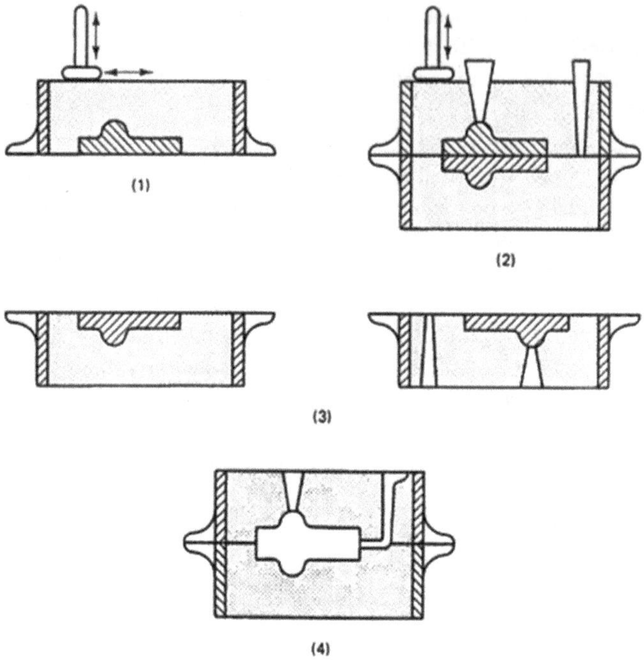

FIGURE 3.5 The procedure for simple hand-molding using a single (loose) pattern.

where

p is the pressure
w is the specific weight of the molten metal
h is the height of the cope

The force that is trying to separate the two halves of the mold can, therefore, be given by the following equation:

$$F = p \times A$$

where

F is the force
A is the cross-sectional area of the casting (including the runner, gates, etc.) at the parting line

Stack Molding. Stack molding is best suited for producing a large number of small, light castings while using a limited amount of floor space in the foundry. As can be seen in Figure 3.6a,b, there are two types of stack molding: *upright* and *stepped*. In upright stack molding, 10–12 flask sections are stacked up. They all have a common sprue that is employed in feeding all cavities. The drag cavity is always molded in the upper surface of the flask section, whereas the cope cavity is molded in the lower surface. In stepped stack molding, each section has its own sprue and is, therefore, offset from the one under it to provide for the pouring basin. In this case, each mold is cast separately.

Sweep Molding. Sweep molding is used to form the surfaces of the mold cavity when a large-size casting must be produced without the time and expenses involved in making a pattern.

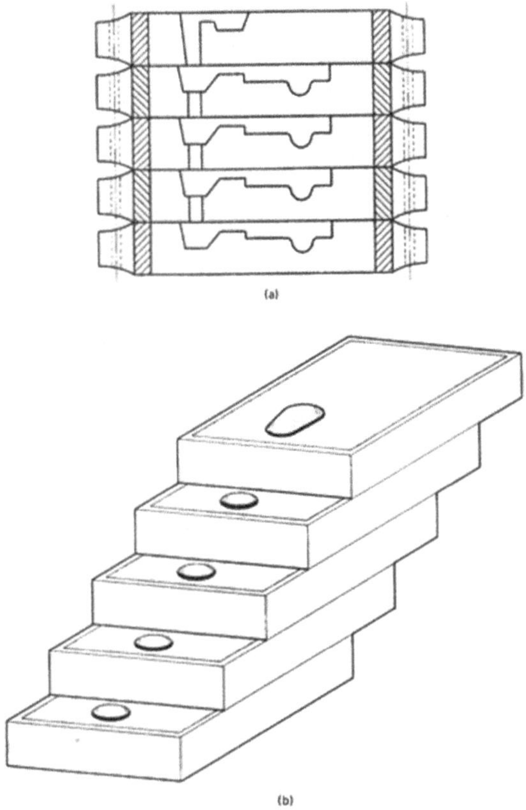

FIGURE 3.6 The two types of stack molding: (a) upright and (b) stepped.

A sweep that can be rotated around an axis is used for producing a surface of revolution, contrary to a drawing sweep, which is pushed axially while being guided by a frame to produce a surface having a constant section along its length (see discussion of the extrusion process in Chapter 5).

Pit Molding. Pit molding is usually employed for producing a single piece of a large casting when it would be difficult to handle patterns of that size in flasks. Molding is done in specially prepared pits in the ground of the foundry. The bottom of the pit is often covered with a layer of coke that is 2–3 inches (50–75 mm) thick. Then, a layer of sand is rammed onto the coke to act as a "bed" for the mold. Vent pipes connect the coke layer to the ground surface. Molding is carried out as usual, and molds are almost always dried before pouring the molten metal. This drying is achieved by means of a portable mold drier. A cope that is also dried is then placed on the pit, and a suitable weight or a group of weights is located on the cope to prevent it from floating when the molten metal is poured.

Molding Machines. The employment of molding machines results in an increase in the production rate, a marked increase in productivity, and a higher and more consistent quality of molds. The function of these machines is to pack the sand onto the pattern and draw the pattern out from the mold. There are several types of molding machines, each with a different way of packing the sand to form the mold. The main types include squeezers, jolt machines, and sandslingers. There are also some machines, such as jolt-squeeze machines, that employ a combination of the working principles of two of the main types. Following is a brief discussion of the three main types of molding machines (see Figure 3.7).

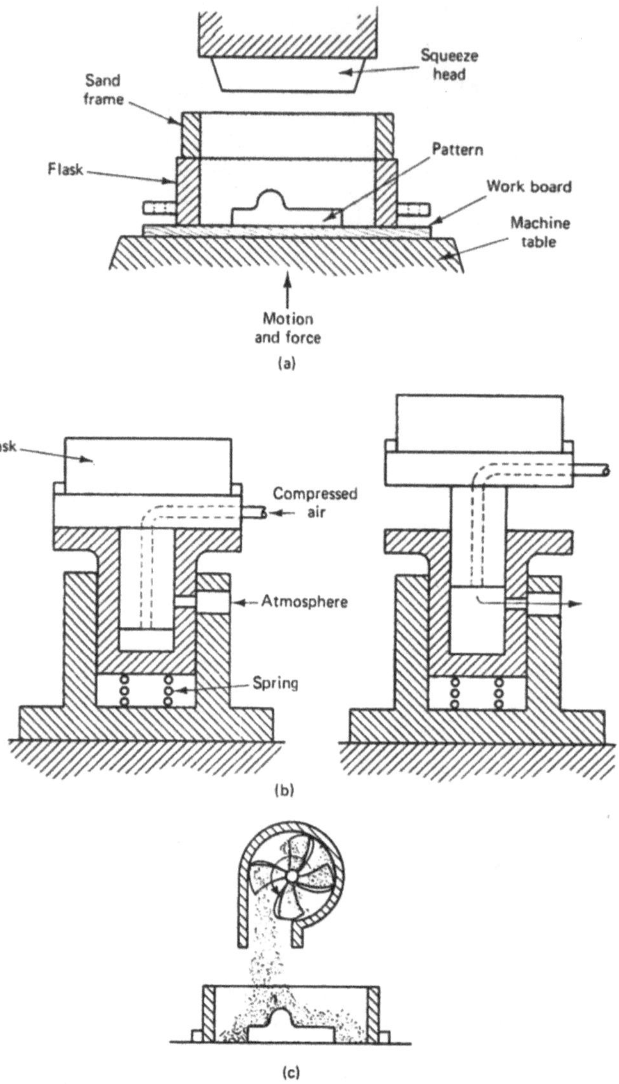

FIGURE 3.7 Molding machines: (a) squeezer, (b) jolt machine, and (c) sandslinger.

Squeezers. Figure 3.7a illustrates the working principle of the squeezer type of molding machine. The pattern plate is clamped on the machine table, and a flask is put into position. A sand frame is placed on the flask, and both are then filled with sand from a hopper. Next, the machine table travels upward to squeeze the sand between the pattern plate and a stationary head. The squeeze head enters into the sand frame and compacts the sand so that it is level with the edge of the flask.

Jolt Machines. Figure 3.7b illustrates the working principle of the jolt type of molding machine. As can be seen, compressed air is admitted through the hose to a pressure cylinder to lift the plunger (and the flask, which is full of sand) up to a certain height, where the side hole is uncovered to exhaust the compressed air. The plunger then falls down and strikes the stationary guiding cylinder. The shock wave resulting from each of the successive impacts contributes to packing the molding sand in the flask.

Sandslingers. Figure 3.7c shows a sandslinger. This type of machine is employed in molding sand in flasks of any size, whether for individual or mass production of molds. Sandslingers are characterized by their high output, which amounts to 2500 cubic feet (more than 60 cubic meters) per hour. As can be seen, molding sand is fed into a housing containing an impeller that rotates rapidly around a horizontal axis. Sand particles are picked up by the rotating blades and thrown at a high speed through an opening onto the pattern, which is located in the flask. No matter what type of molding machine is used, special machines are employed to draw the pattern out of the mold. Basically, these machines achieve that goal by turning the flask (together with the pattern) upside down and then lifting the pattern out of the mold. Examples of these machines include roll-over molding machines and rock-over pattern-draw machines.

3.2.1.1.7 Sand Conditioning

The molding sand, whether new or used, must be conditioned before being used. When used sand is to be recycled, lumps should be crushed and then metal granules or small parts removed (a magnetic field is employed in a ferrous foundry). Next, sand (new or recycled) and all other molding constituents must be screened in shakers, rotary screens, or vibrating screens. Molding materials are then thoroughly mixed in order to obtain a completely homogeneous green sand mixture. The more uniform the distribution, the better the molding properties (like permeability and green strength) of the sand mixture will be.

Mixing is carried out in either continuous-screw mixers or vertical-wheel mullers. The mixers mix the molding materials by means of two large screws or worm gears; the mullers are usually used for batch-type mixing. A typical muller is illustrated in Figure 3.8. It consists primarily of a pan in which two wheels rotate about their own horizontal axis as well as about a stationary vertical shaft. Centrifugal mullers are also in use, especially for high production rates.

3.2.1.2 Dry Sand Molds

As previously mentioned, green sand molds contain up to 8 percent water, depending upon the kind and percentage of the binding material. Therefore, this type of mold can be used only for small castings with thin walls; large castings with thick walls would heat the mold, resulting in the vaporization of water, which would, in turn, lead to bubbles in the castings. For this reason, molds for large

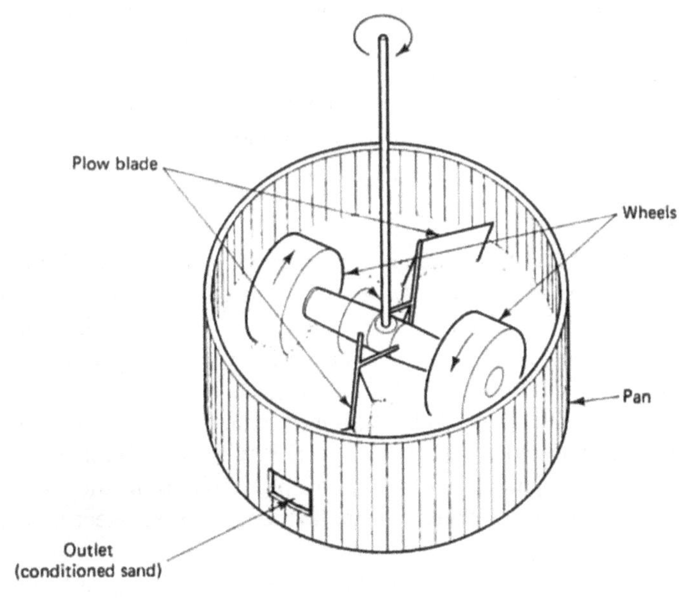

FIGURE 3.8 A muller for sand conditioning.

castings should be dried after they are made in the same way as green sand molds. The drying operation is carried out in ovens at temperatures ranging from 300°F to 650°F (150°C to 350°C) for 8–48 hours, depending upon the kind and amount of binder used.

3.2.1.3 Core-Sand Molds

When the mold is too big to fit in an oven, molds are made by assembling several pieces of sand cores. Consequently, patterns are not required, and core boxes are employed instead to make the different sand cores necessary for constructing the mold. Because core-sand mixtures (which have superior molding properties) are used, very good quality and dimensional accuracy of the castings are obtained.

3.2.1.4 Cement-Bonded Sand Molds

A mixture of silica sand containing 8–12 percent cement and 4–6 percent water is used. When making the mold, the cement-bonded sand mixture must be allowed to harden first before the pattern is withdrawn. The obtained mold is then allowed to cure for about 3–5 days. Large castings with intricate shapes, accurate dimensions, and smooth surfaces are usually produced in this way, the only shortcoming being the long time required for the molding process.

3.2.1.5 Carbon Dioxide Process for Molding

Silica sand is mixed with a binder involving a solution of sodium silicate (water glass) amounting to 6 percent. After the mold is rammed, carbon dioxide is blown through the sand mixture. As a result, the gel of silica binds the sand grains together, and no drying is needed. Because the molds are allowed to harden while the pattern is in position, high dimensional accuracy of molds is obtained.

3.2.1.6 Plaster Molds

A plaster mold is appropriate for casting silver, gold, magnesium, copper, and aluminum alloys. The molding material is a mixture of fine silica sand, asbestos, and plaster of Paris as a binder. Water is added to the mixture until a creamy slurry is obtained, which is then employed in molding. The drying process should be very slow to avoid the cracking of the mold.

3.2.1.7 Loam Molds

The loam mold is used for very large jobs. The basic shape of the desired mold is constructed with bricks and mortar (just like a brick house). A loam mixture is then used as a molding material to obtain the desired fine details of the mold. Templates, sweeps, and the like are employed in the molding process. The loam mixture used in molding consists of 50 percent or more of loam, with the rest being mainly silica sand. Loam molds must be thoroughly dried before pouring the molten metal.

3.2.1.8 Shell Molds

In shell molding, a thin mold is made around a heated-metal pattern plate. The molding material is a mixture of dry, fine silica sand (with a very low clay content) and 3–8 percent of a thermosetting resin like phenol formaldehyde or urea formaldehyde. Conventional dry-mixing techniques are used to obtain the molding mixture. Specially prepared resin-coated sands are also used.

When the molding mixture drops onto the pattern plate, which is heated to a temperature of 350°F to 700°F (180°C to 375°C), a shell about 1/4 inch (6 mm) thick is formed. In order to cure the shell completely, it must be heated at 450°F to 650°F (230°C to 350°C) for about 1–3 minutes. The shell is then released from the pattern plate by ejector pins. To prevent sticking of the baked shell, sometimes called the biscuit, to the pattern plate, a silicone release agent is applied to the plate before the molding mixture drops onto it.

Shell molding is suitable for mass production of thin-walled, gray cast-iron (and aluminum-alloy) castings having a maximum weight between 35 and 45 pounds (15 and 20 kg). However, castings weighing up to 1000 pounds (450 kg) can be made by employing shell molding on an individual basis. The advantages of shell molding include a good surface finish, few restrictions on casting design, and the fact that this process renders itself suitable for automation.

3.2.1.9 Ceramic Molds

In the ceramic molding process, the molding material is actually a slurry consisting of refractory grains, ceramic binder, water, alcohol, and an agent to adjust the pH value (see discussion of slurry casting in Chapter 7). The slurry is poured around the permanent (reusable) pattern and is allowed to harden when the pattern is withdrawn. Next, the mold is left to dry for some time and then is fired to gain strength. In fact, ceramic molds are usually preheated before pouring the molten metal. For this reason, they are suitable for casting high-pouring-temperature alloys. Excellent surface finish and very close tolerances of the castings are among the advantages of this molding process and lead to the elimination of the machining operations that are usually performed on castings. Therefore, ceramic molds are certainly advantageous when casting precious or difficult-to-machine metals as well as for making castings with great shape intricacy.

3.2.1.10 Precision Molds (Investment Casting)

Precision molding is used when castings with intricate shapes, good dimensional accuracy, and very smooth surfaces are required. The process is especially advantageous for high-melting-point alloys as well as for difficult-to-machine metals. It is also most suitable for producing small castings having intricate shapes, such as the group of investment castings shown in Figure 3.9. A nonpermanent pattern that is usually made of wax must be prepared for each casting. Therefore, the process is sometimes referred to as the *lost-wax process*. Generally, the precision molding process involves the following steps (see Figure 3.10).

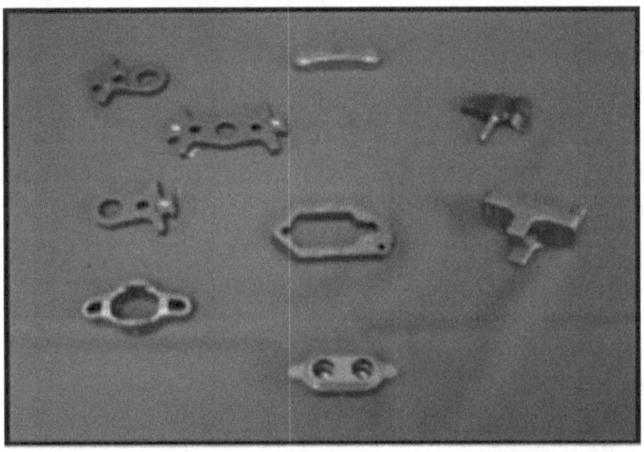

FIGURE 3.9 A group of investment castings. (Courtesy of Intercast, McAllen, TX.)

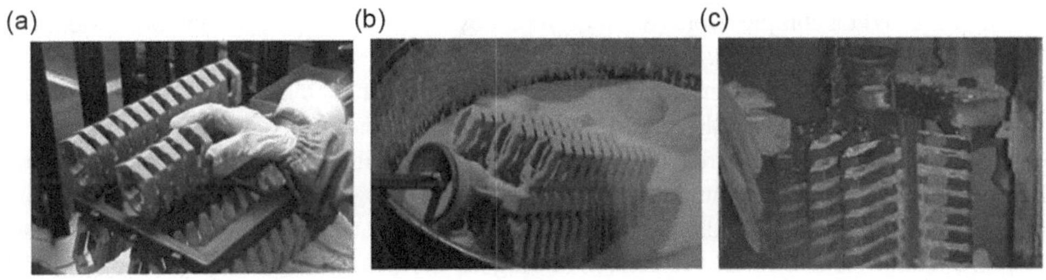

FIGURE 3.10 Steps involved in investment casting: (a) a cluster of wax patterns, (b) a cluster of ceramic shells, and (c) a cluster of castings. (Courtesy of Intercast, McAllen, TX.)

- A heat-disposable pattern, together with its gating system, is prepared by injecting wax or plastic into a die cavity.
- A pattern assembly that is composed of a number of identical patterns is made. Patterns are attached to a runner bar made of wax or plastic in much the same manner as leaves are attached to branches. A ceramic pouring cup is also attached to the top of the pattern assembly, which is sometimes referred to as the *tree* or *cluster* (see Figure 3.10a).
- The tree is then invested by separately dipping it into a ceramic slurry that is composed of silica flour suspended in a solution of ethyl silicate and sprinkling it with very fine silica sand. A self-supporting ceramic shell mold about 1/4 inch (6 mm) thick is formed all around the wax assembly (see Figure 3.10b). Alternatively, a thin ceramic precoating is obtained, and then the cluster is placed in a flask and a thick slurry is poured around it as a backup material.
- The pattern assembly is then baked in an oven or a steam autoclave to melt out the wax (or plastic). Therefore, the dimensions of the mold cavity precisely match those of the desired product.
- The resulting shell mold is fired at a temperature ranging from 1600°F to 1800°F (900°C to 1000°C) to eliminate all traces of wax and to gain reasonable strength.
- The molten metal is poured into the mold while the mold is still hot, and a cluster of castings is obtained (see Figure 3.10c).

Today, the lost-wax process is used in manufacturing large objects like cylinder heads and camshafts. The modern process, which is known as the *lost-foam method*, involves employing a Styrofoam replica of the finished product, which is then coated with a refractory material and located in a box, where sand is molded around it by vibratory compaction. When the molten metal is finally poured into the mold, the Styrofoam vaporizes, allowing the molten metal to replace it.

3.2.1.11 Graphite Molds

Graphite is used in making molds to receive an alloy (such as titanium) that can be poured only into inert molds. The casting process must be performed in a vacuum to eliminate any possibility of contaminating the metal. Graphite molds can be made either by machining a block of graphite to create the desired mold cavity or by compacting a graphite-base aggregate around the pattern and then sintering the obtained mold at a temperature of 1800°F to 2000°F (1000°C to 1120°C) in a reducing atmosphere (see Chapter 7). In fact, graphite mold liners have found widespread industrial applications in the centrifugal casting of brass and bronze.

3.2.1.12 Permanent Molds

A permanent mold can be used repeatedly for producing castings of the same form and dimensions. Permanent molds are usually made of steel or gray cast iron. Each mold is generally made of two or more pieces that are assembled together by fitting and clamping. Although the different parts of the mold can be cast to their rough contours, subsequent machining and finishing operations are necessary to eliminate the possibility of the casting's sticking to the mold. Simple cores made of metal are frequently used. When complex cores are required, they are usually made of sand or plaster, and the mold is said to be *semipermanent*.

Different metals and alloys can successfully be cast in permanent molds. They include aluminum alloys, magnesium alloys, zinc alloys, lead, copper alloys, and cast irons. It is obvious that the mold should be preheated to an appropriate temperature prior to casting. In fact, the operating temperature of the mold, which depends upon the metal to be cast, is a very important factor in successful permanent-mold casting. Based on the preceding discussion, we can expect the mold life to be dependent upon a number of interrelated factors, including the mold material, the metal to be cast, and the operating temperature of the mold. Nevertheless, it can be stated that the life of a permanent mold is about 100,000 pourings or more when casting zinc, magnesium, or aluminum alloys and not more than 20,000 pourings for copper alloys and cast irons. However, spraying the surface of the mold cavity with colloidal refractories suspended in liquids can extend mold life.

The advantages of permanent-mold casting include substantial increases in productivity (a mold does not have to be made for each casting), close tolerances, superior surface finish, and improved mechanical properties of the castings. A further advantage is the noticeable reduction in the percentage of rejects when compared with the conventional sand-casting processes. Nevertheless, the process is economically feasible for mass production only. There is also a limitation on the size of parts produced by permanent-mold casting. A further limitation is that not all alloys are suited to this process.

3.2.2 Classifications of Castings by Method of Filling the Mold

For all types of molds that we have discussed, the molten metal is almost always fed into the mold only by the action of gravity. Therefore, the casting process is referred to as *gravity casting*. There are, however, other special ways of pouring or feeding the molten metal into the desired cavities. These casting methods are generally aimed at forcing the molten metal to flow and fill the fine details of the mold cavity while eliminating the internal defects experienced in conventional gravity casting processes. Following is a survey of the commonly used special casting processes.

3.2.2.1 Die Casting

Die casting involves forcing the molten metal into the cavity of a steel mold, called a *die*, under very high pressure (1000–30,000 pounds per square inch, or about 70–2000 times the atmospheric pressure). In fact, this characteristic is the major difference between die casting and permanent-mold casting, where the molten metal is fed into the mold either by gravity or at low pressures. Die casting may be classified according to the type of machine used. The two principal types are *hot-chamber* machines and *cold-chamber* machines.

3.2.2.1.1 Hot-Chamber Machines

The main components of the hot-chamber die casting machine include a steel pot filled with the molten metal to be cast and a pumping system that consists of a pressure cylinder, a plunger, a gooseneck passage, and a nozzle. With the plunger in the up position, as shown in Figure 3.11a, the

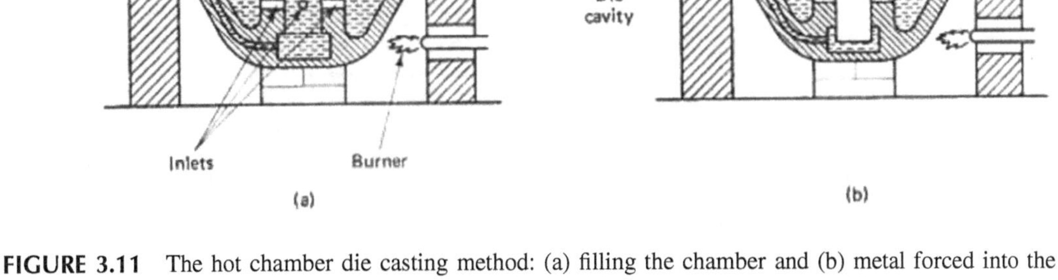

FIGURE 3.11 The hot chamber die casting method: (a) filling the chamber and (b) metal forced into the die cavity.

molten metal flows by gravity through the intake ports into the submerged hot chamber. When the plunger is pushed downward by the power cylinder (not shown in the figure), it shuts off the intake port. Then, with further downward movement, the molten metal is forced through the gooseneck passage and the nozzle into the die cavity, as shown in Figure 3.11b. Pressures ranging from 700 to 2000 pounds per square inch (50–150 atmospheres) are quite common to guarantee complete filling of the die cavity. After the cavity is full of molten metal, the pressure is maintained for a preset dwell time to allow the casting to solidify completely. Next, the two halves of the die are pushed apart, and the casting is knocked out by means of ejector pins. The die cavity is then cleaned and lubricated before the cycle is repeated. The advantages of hot-chamber die casting are numerous. They include high production rates (especially when multicavity dies are used), improved productivity, superior surface finish, very close tolerances, and the ability to produce intricate shapes with thin walls. Nevertheless, the process has some limitations. For instance, only low-melting-point alloys (such as zinc, tin, lead, and the like) can be cast because the components of the pumping system are in direct contact with the molten metal throughout the process. Also, die casting is usually only suitable for producing small castings that weigh less than 10 pounds (4.5 kg).

3.2.2.1.2 Cold-Chamber Machines

In the cold-chamber die casting machine, the molten-metal reservoir is separate from the casting machine, and just enough for one shot of molten metal is ladled every stroke. Consequently, the relatively short exposure of the chamber and the plunger to the molten metal allows die casting of aluminum, magnesium, brass, and other alloys having relatively high melting points. In the sequence of operations in cold-chamber die casting, the molten metal is first ladled through the pouring hole of the chamber while the two halves of the die are closed and locked together, as shown in Figure 3.12. Next, the plunger moves forward to close off the pouring hole and then forces the molten metal into the die cavity. Pressures in the chamber may go over 30,000 pounds per square inch (2000 atmospheres). After the casting has solidified, the two halves of the die are opened, and the casting, together with the gate and the slug of excess metal, is ejected from the die. It is not difficult to see that large parts weighing 50 pounds (23 kg) can be produced by cold-chamber die casting. The process is very successful when casting aluminum alloys, copper alloys, and high-temperature aluminum–zinc alloys. However, this process has a longer cycle time when compared with hot-chamber die casting. A further disadvantage is the need for an auxiliary system for pouring the molten metal. It is mainly for this reason that vertical cold-chamber machines were developed. As can be seen in Figure 3.13, such a machine has a transfer tube that is submerged in molten metal. It is fed into the chamber by connecting the die cavity to a vacuum tank by means of a special valve. The molten metal is forced into the die cavity when the plunger moves upward.

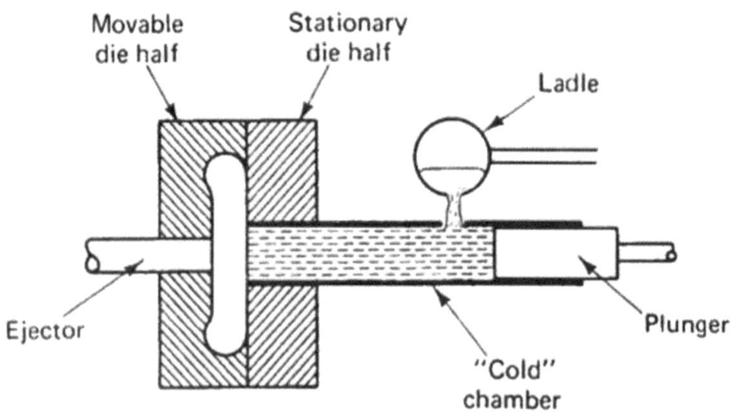

FIGURE 3.12 The cold-chamber die casting method.

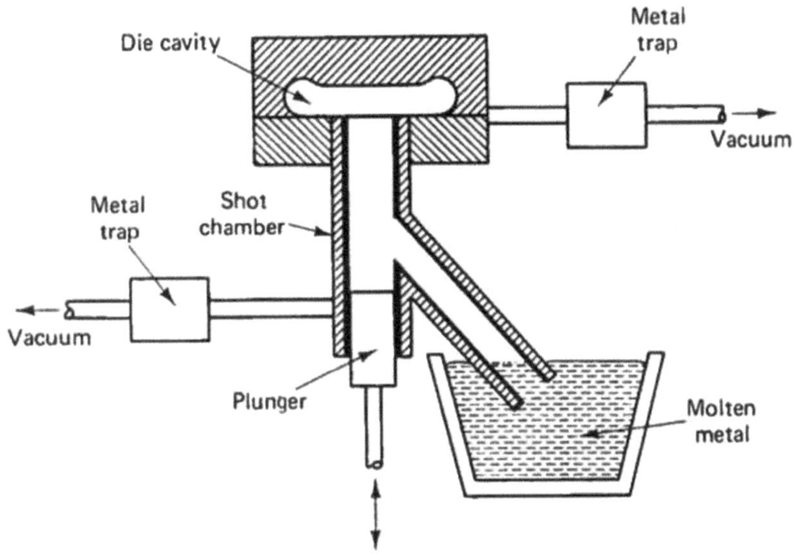

FIGURE 3.13 A vertical cold-chamber die casting machine.

3.2.2.2 Centrifugal Casting

Centrifugal casting refers to a group of processes in which the forces used to distribute the molten metal in the mold cavity (or cavities) are caused by centrifugal acceleration. Centrifugal casting processes can be classified as *true centrifugal* casting, *semicentrifugal* casting, and the *centrifuging* method. Each of these processes is briefly discussed next.

3.2.2.2.1 True Centrifugal Casting

True centrifugal casting involves rotating a cylindrical mold around its own axis, with the revolutions per minute high enough to create an effective centrifugal force, and then pouring molten metal into the mold cavity. The molten metal is pushed to the walls of the mold by centrifugal acceleration (usually 70–80 times that of gravity), where it solidifies in the form of a hollow cylinder. The outer shape of the casting is given by the mold contour, while the diameter of the inner cylindrical surface is controlled by the amount of molten metal poured into the mold cavity. The machines used to spin the mold may have either horizontal or vertical axes of rotation. Short tubes are usually cast in vertical-axis machines, whereas longer pipes, like water supply and sewer pipes, are cast using horizontal-axis machines. The basic features of a true centrifugal casting machine with a horizontal axis are shown in Figure 3.14.

Centrifugal castings are characterized by their high density, refined fine-grained structure, and superior mechanical properties, accompanied by a low percentage of rejects and, therefore, a high production output. A further advantage of the centrifugal casting process is the high efficiency of metal utilization due to the elimination of sprues and risers and the small machining allowance used.

3.2.2.2.2 Semicentrifugal Casting

Semicentrifugal casting is quite similar to the preceding type, the difference being that the mold cavity is completely filled with the molten metal. However, because centrifugal acceleration is dependent upon the radius, the central core of the casting is subjected to low pressure and is, therefore, the region where entrapped air and inclusions are present. For this reason, the semicentrifugal casting process is recommended for producing castings that are to be subjected to subsequent machining to remove their central cores. Examples include cast track wheels for tanks, tractors, and

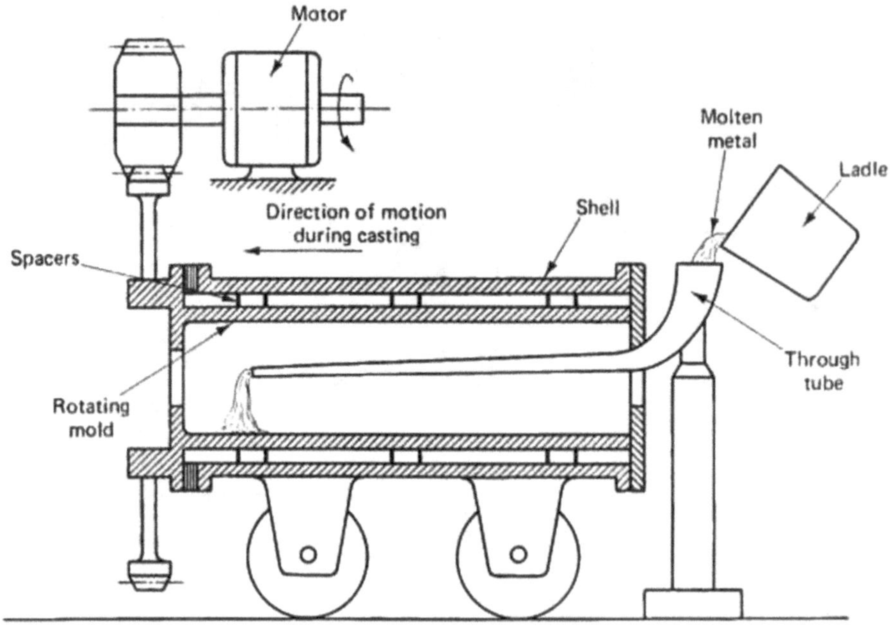

FIGURE 3.14 The basic features of a true centrifugal casting machine.

the like. A sand core is sometimes used to form the central cavity of the casting in order to eliminate the need for subsequent machining operations.

3.2.2.2.3 Centrifuging

In the centrifuging method, a number of mold cavities are arranged on the circumference of a circle and are connected to a central down sprue through radial gates. Next, molten metal is poured, and the mold is rotated around the central axis of the sprue. In other words, each casting is rotated around an axis off (shifted from) its own center axis. Therefore, mold cavities are filled under high pressure, so the process is usually used for producing castings with intricate shapes; the increased pressure on the casting during solidification allows the fine details of the mold to be obtained.

3.2.2.3 Continuous Casting

The continuous casting process is gaining widespread industrial use, especially for high-quality alloy steel. In fact, the process itself passed through a few evolutionary stages. Although it was originally developed for producing cast iron sheets, an up-to-date version is now being used for casting semi-finished products that are to be processed subsequently by piercing, forging, extrusion, and the like.

The continuous casting process basically involves controlling the flow of a stream of molten metal that comes out from a water-cooled orifice in order to solidify and form a continuous strip (or rod). The new version of this process is usually referred to as *rotary continuous casting* because the water-cooled mold (orifice) is always oscillating and rotating at about 120 revolutions per minute during casting. Figure 3.15 illustrates the principles of rotary continuous casting. The steel is melted, refined, and degassed and its chemical composition is controlled before it is transferred and poured into the caster (tundish). The molten metal then enters the rotating mold tangent to the edge through the bent tube. The centrifugal force then forces the steel against the mold wall, while lighter inclusions and impurities remain in the center of the vortex, where the operator removes them. Solidification of the metal flowing out of the mold continues at a precalculated rate. The resulting bar is then cut by a circular saw that is traveling downward at the same speed as the bar. The bar is tilted and loaded onto a conveyor to transfer it to the cooling bed and the rolling mill.

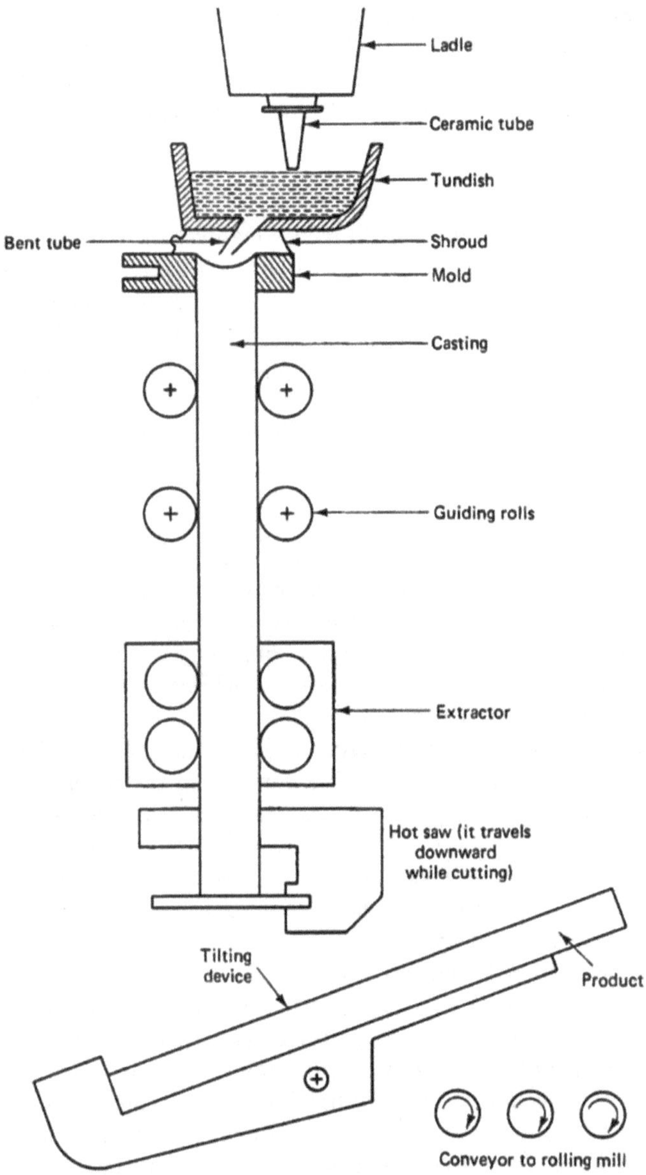

FIGURE 3.15 The principles of rotary continuous casting.

The continuous casting process has the advantages of a very high metal yield (about 98 percent, compared with 87 percent in conventional ingot-mold practice), excellent quality of cast, controlled grain size, and the possibility of casting special cross-sectional shapes.

3.2.2.4 The V-Process

The vacuum casting process (V-process for short) involves covering the two halves of the pattern with two plastic films that are 0.005 inch (0.125 mm) thick by employing vacuum forming (see Chapter 8). The pattern is then removed, and the two-formed plastic sheets are tightened together to form a mold cavity that is surrounded by a flask filled with sand (there is no need for a binder). This mold cavity is kept in a vacuum as the molten metal is poured to assist and ensure easy flow.

The V-process, developed in Japan in the early 1970s, offers many advantages, such as the elimination of the need for special molding sands with binders and the elimination of the problems associated with green sand molding (like gas bubbles caused by excess humidity). Also, the size of risers, vents, and sprues can be reduced markedly, thus resulting in an increase in the efficiency of material utilization.

3.2.2.5 Mega Casting

The construction of a motor car chassis traditionally involves the sequential assembly of panels and sheets at various stations along an assembly line, facilitated by robotic systems positioned on both sides. These robots not only accurately position the panels but also perform spot welding to join them together. This method, however, presents challenges due to its time-intensive nature and the necessity for precise panel positioning within tight tolerances. To mitigate these challenges, principles of "Design for Assembly" can be applied, aiming to consolidate panels, that do not have relative motion between them, wherever possible to streamline the assembly process.

A revolutionary breakthrough in chassis manufacturing was introduced by Alan Musk, proposing the fabrication of Tesla motor car chassis as a single unified structure, termed "one piece" or unibody construction. Unlike traditional framed vehicles, unibody cars lack separate frames, instead featuring a single metal body engineered to support the vehicle's drivetrain and battery, and ensure occupant safety in the event of a collision. Musk's vision required a departure from conventional manufacturing techniques, particularly in the casting process, he decided that full-size cars should be made by casting in the same manner as little cars (hot wheels) are made.

Conventional cold chamber die casting methods proved inadequate for producing such large-scale chassis due to the prolonged duration required to fill the chamber. Thus, a novel approach utilizing high-pressure die casting was devised. The Idra Group in Italy pioneered the development of specialized high-pressure die casting machines, each weighing approximately 400 tons, with clamping forces initially set at 6000 tons, later increased to 8000 tons. Molten aluminum alloy, weighing around 180 lbs (80 kg), is injected into the casting mold at velocities of approximately 33 ft/s (10 m/s). With a cycle time of 80–90 seconds, these machines achieve production rates of 40:54 completed castings per hour or roughly 1000 chassis per day.

Following casting, each chassis undergoes trimming and machining processes to attain precise dimensions. The casting cycle involves melting the aluminum alloy, transferring it via heated pipes to a temporary warm storage container with a nitrogen atmosphere to prevent oxidation, and then injecting the precise amount of molten alloy into the mold using a high-speed plunger. The mold is subsequently opened, and the newly formed casting is ejected and quenched in water.

Careful consideration was given to the selection of chassis material, aiming to ensure optimal performance and ease of operation. The chosen alloy, Al Si8 (8% silicon), is heat-treatable, allowing for control over its hardness and strength characteristics. With a composition close to the eutectic alloy of the Al-Si system (12.2% Si and a temperature of 577°C), this alloy completely melts at a temperature of 610 °C thus making it easy to store molten in the warm container and it exhibits a narrow freezing range of 30°C, facilitating rapid solidification within the mold and consequently reducing cycle times. Post-casting, the chassis undergoes heat treatment to achieve the desired mechanical properties.

3.2.3 Classifications of Castings by Metal to Be Cast

When classified by metal, castings can be either *ferrous* or *nonferrous*. The ferrous castings include cast steels and the family of cast irons, whereas the nonferrous castings include all other metals, such as aluminum, copper, magnesium, titanium, and their alloys. Each of these metals and alloys is melted in a particular type of foundry furnace that may not be appropriate for melting other metals and alloys. Also, molding methods and materials, as well as fluxes, degassers, and additives, depend upon the metal to be cast. Therefore, this classification method is popular in foundry work. Following is a brief discussion of each of these cast alloys.

3.2.3.1 Ferrous Metals

3.2.3.1.1 Cast Steels

Steels are smelted in open-hearth furnaces, convertors, electric-arc furnaces, and electric-induction furnaces. Cast steels can be either plain-carbon, low-alloy, or high-alloy steel. However, plain-carbon cast steel is the most commonly produced type. When compared with cast iron, steel certainly has poorer casting properties, namely, higher melting points, higher shrinkage, and poorer fluidity. Steels are also more susceptible to hot and cold cracks after the casting process. Therefore, cast steels are almost always subjected to heat treatment to relieve the internal stresses and improve the mechanical properties. In order to control the oxygen content of molten steels, aluminum, silicon, or manganese is used as a deoxidizer. Aluminum is the most commonly used of these elements because of its availability, low cost, and effectiveness. There is an important difference between cast-steel and wrought products. This involves the presence of a "skin" or thin layer, just below the surface of a casting, where scales, oxides, and impurities are concentrated. Also, this layer may be chemically or structurally different from the base metal. Therefore, it has to be removed by machining in a single deep cut, which is achieved by reducing the cutting speed to half of the conventionally recommended value.

3.2.3.1.2 Gray Cast Iron

Gray cast iron is characterized by the presence of free graphite flakes when its microstructure is examined under the microscope. This kind of microstructure is, in fact, responsible for the superior properties possessed by gray cast iron. For instance, this dispersion of graphite flakes acts as a lubricant during the machining of gray cast iron, thus eliminating the need for machining lubricants and coolants. When compared with any other ferrous cast alloy, gray cast iron certainly possesses superior machinability. The presence of those graphite flakes is also the reason for its ability to absorb vibrations. The compressive strength of this iron is normally four times its tensile strength. Thus, gray cast iron has found widespread application in machine tool beds (bases) and the like. On the other hand, gray cast iron has some disadvantages and limitations, such as its low tensile strength, brittleness, and poor weldability. Nevertheless, gray cast iron has the lowest casting temperature, least shrinkage, and the best castability of all cast ferrous alloys.

The cupola is the most widely used foundry furnace for producing and melting gray cast iron. The chemical composition, microstructure, and, therefore, the properties of the obtained castings are determined by the constituents of the charge of the cupola furnace. Thus, the composition and properties of gray cast iron are controlled by changing the percentages of the charge constituents and also by adding inoculants and alloying elements. Commonly used inoculants include calcium silicide, ferrosilicon, and ferromanganese. An inoculant is added to the molten metal (either in the cupola spout or ladle) and usually amounts to between 0.1 and 0.5 percent of the molten iron by weight. It acts as a deoxidizer and also hinders the growth of precipitated graphite flakes. It is important for a product designer to remember that the properties of a gray cast-iron product are also dependent upon the dimensions (the thicknesses of the walls) of that product because the cooling rate is adversely affected by the cross section of the casting. Actually, the cooling rate is high for small castings with thin walls, sometimes yielding white cast iron. For this reason, gray cast iron must be specified by the strength of critical cross sections.

3.2.3.1.3 White Cast Iron

When the molten cast iron alloy is rapidly chilled after being poured into the mold cavity, dissolved carbon does not have enough time to precipitate in the form of flakes; instead, it remains chemically combined with iron in the form of cementite. This material is primarily responsible for the whitish crystalline appearance of a fractured surface of white cast iron. Cementite is also responsible for the high hardness, extreme brittleness, and excellent wear resistance of this kind of cast iron. Industrial applications of white cast iron involve components subjected to abrasion. Sometimes, gray cast iron can be chilled to produce a surface layer of white cast iron in order to combine the advantageous properties of the two types of cast iron. In this case, the product metal is usually referred to as *chilled* cast iron.

3.2.3.1.4 Ductile Cast Iron

Ductile cast iron is also called *nodular* cast iron and *spheroidal graphite* cast iron. It is obtained by adding trace amounts of magnesium to a very pure molten alloy of gray cast iron that has been subjected to desulfurization. Sometimes, a small quantity of cerium is also added to prevent the harmful effects of impurities like aluminum, titanium, and lead. The presence of magnesium and cerium causes the graphite to precipitate during the solidification of the molten alloy in the form of small spheroids, rather than flakes as in the case of gray cast iron. This microstructural change results in a marked increase in ductility, strength, toughness, and stiffness of ductile iron, as compared with gray cast iron because the stress concentration effect of a flake is far higher than that of a spheroid (remember what you learned in fracture mechanics). The disadvantages of ductile iron, as compared with gray cast iron, include lower damping capacity and thermal conductivity. Ductile iron is used for making machine parts like axles, brackets, levers, crankshafts, housings, die pads, and die shoes.

3.2.3.1.5 Compacted-Graphite Cast Iron

Compacted-graphite (CG) cast iron falls between gray and ductile cast irons, in both its microstructure and mechanical properties. The free graphite in this type of iron takes the form of short, blunt, and interconnected flakes. The mechanical properties of CG cast iron are superior to those of gray cast iron but are inferior to those of ductile cast iron. The thermal conductivity and damping capacity of CG cast iron approach those of gray cast iron. Compacted-graphite cast iron has some application in the manufacture of diesel engines.

3.2.3.1.6 Malleable Cast Iron

Malleable cast iron is obtained by two-stage heat treatment of white cast iron having an appropriate chemical composition. The hard white cast iron becomes malleable after the heat treatment due to microstructural changes. The combined carbon separates as free graphite, which takes the form of nodules. Because the raw material for producing malleable iron is actually white cast iron, there are always limitations on casting design. Large cross sections and thick walls are not permitted because it is difficult to produce a white cast iron part with these geometric characteristics.

The two basic types of malleable cast iron are the *pearlitic* and the *ferritic* (black heart). Although the starting alloy for both types is the same (white cast iron), the heat treatment cycle and the atmosphere of the heat-treating furnace are different in each case. Furnaces with oxidizing atmospheres are employed for producing pearlitic malleable cast iron, whereas furnaces with neutral atmospheres are used for producing ferritic malleable cast iron. When comparing the properties of these two types, the ferritic grades normally have higher ductility and better machinability but lower strength and hardness. Pearlitic grades can, however, be subjected to further surface hardening when the depth of the hardened layer is controlled.

Figure 3.16 shows the heat treatment sequence for producing malleable cast iron. Referred to as the *malleabilizing cycle*, it includes two stages, as shown in Figure 3.16. In the first stage, the casting is slowly heated to a temperature of about 1700°F (950°C) and is kept at that temperature for about 24 hours. In the second stage, the temperature is decreased very slowly at a rate of 5°F to 9°F (3°C to 5°C) per hour from a temperature of 1400°F (800°C) to a temperature of 1200°F (650°C), where the process ends and the casting is taken out of the furnace. The whole malleabilizing cycle normally takes about 100 hours.

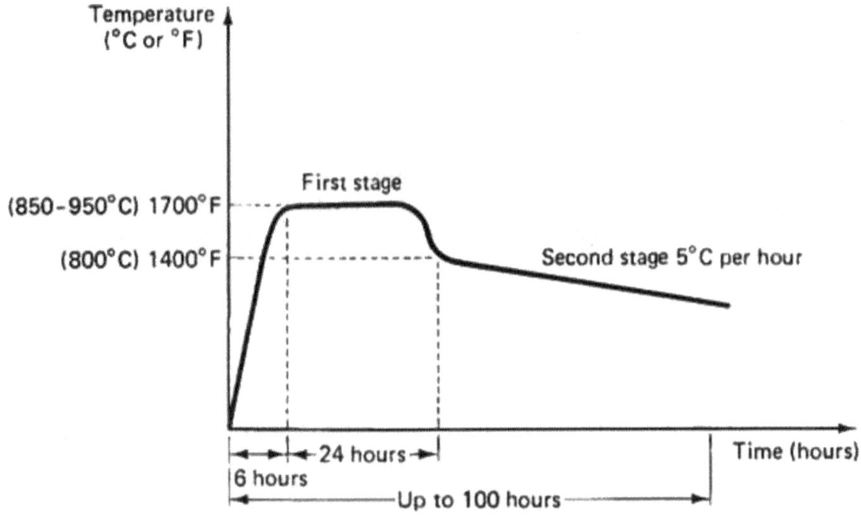

FIGURE 3.16 The heat treatment sequence for producing malleable cast iron.

Malleable cast iron is usually selected when the engineering application requires good machinability and ductility. Excellent castability and high toughness are other properties that make malleable cast iron attractive as an engineering material. Typical applications of malleable cast iron include flanges, pipe fittings, and valve parts for pressure service at elevated temperatures; steering-gear housings; mounting brackets; and compressor crankshafts and hubs.

3.2.3.1.7 Alloyed Cast Irons

Alloying elements like chromium, nickel, and molybdenum are added to cast irons to manipulate the microstructure of the alloy. The goal is to improve the mechanical properties of the casting and also to impart some special properties to it, like resistance to wear, corrosion, and heat. A typical example of alloyed irons is the white cast iron containing nickel and chromium that is used for corrosion-resistant (and abrasion-resistant) applications like water pump housings and grinding balls (in a ball mill).

3.2.3.2 Nonferrous Metals

3.2.3.2.1 Cast Aluminum and Its Alloys

Aluminum continues to gain wide industrial applications, especially in the automotive and electronics industries, because of its distinguished strength-to-weight ratio and its high electrical conductivity. Alloying elements can be added to aluminum to improve its mechanical properties and metallurgical characteristics. Silicon, magnesium, zinc, tin, and copper are the elements most commonly alloyed with aluminum. In fact, most metallic elements can be alloyed with aluminum, but commercial and industrial applications are limited to those just mentioned.

A real advantage of aluminum is that it can be cast by almost all casting processes. Nevertheless, the common methods for casting aluminum include die casting, gravity casting in sand and permanent molds, and investment casting (the lost-foam process). The presence of hydrogen when melting aluminum always results in unsound castings. Typical sources of hydrogen are the furnace atmosphere and the charge metal. When the furnace has a reducing atmosphere because of incomplete combustion of the fuel, carbon monoxide and hydrogen are generated and absorbed by the molten metal. The presence of contaminants like moisture, oil, or grease, which are not chemically stable at elevated temperatures, can also liberate hydrogen. Unfortunately, hydrogen is highly soluble in molten aluminum but has limited solubility in solidified aluminum. Therefore, any hydrogen that is absorbed by the molten metal is liberated or expelled during solidification, causing porosity. Hydrogen may also react with (and reduce) metallic oxides to form water vapor, which again causes porosity. Thus, hydrogen must be completely removed from molten aluminum before casting. This is achieved by using appropriate *degassers*. Chlorine and nitrogen are considered the traditional degassers for aluminum. Either of these is blown through the molten aluminum to eliminate any hydrogen. However, because chlorine is toxic and nitrogen is not that efficient, organic chloride fluxing compounds (chlorinated hydrocarbons) are added to generate chlorine within the melt. They are commercially available in different forms, such as blocks, powders, and tablets; the most commonly used fluxing degasser is perhaps hexachlorethane. Another source of problems when casting aluminum is iron, which dissolves readily in molten aluminum. Therefore, care must be taken to spray (or cover) iron ladles and all iron surfaces that come into direct contact with the molten aluminum with a ceramic coating. This extends the service life of the iron tools used and also results in sound castings.

The most important cast aluminum alloys are those containing silicon, which serves to improve the castability, reduce the thermal expansion, and increase the wear resistance of aluminum. Small additions of magnesium make these alloys heat-treatable, thus allowing the final properties of the castings to be controlled. Aluminum–silicon alloys (with 5–13 percent silicon) are used in making automobile parts (e.g., pistons) and aerospace components.

Aluminum–copper alloys are characterized by their very high tensile strength-to-weight ratio. They are, therefore, mainly used for the manufacture of premium-quality aerospace parts. Nevertheless, these alloys have poorer castability than the aluminum–silicon alloys. Also, amounts of the copper constituent in excess of 12 percent make the alloy brittle. Copper additions of up to 5 percent are usually used and result in improved high-temperature properties and machinability.

Additions of magnesium to aluminum result in improved corrosion resistance and machinability, higher strength, and attractive appearance of the casting when anodized. However, aluminum–magnesium alloys are generally difficult to cast. Zinc is also used as an alloying element, and the aluminum–zinc alloys have good machinability and moderately high strength. However, these alloys are generally prone to hot cracking and have poorer castability and high shrinkage. Therefore, zinc is usually alloyed with aluminum in combination with other alloying elements and is employed in such cases for promoting very high strength. Aluminum–tin alloys are also in use. They possess high load-carrying capacity and fatigue strength and are, therefore, used for making bearings and bushings.

3.2.3.2.2 Cast-Copper Alloys

The melting temperatures of cast-copper alloys are far higher than those of aluminum, zinc, or magnesium alloys. Cast-copper alloys can be grouped according to their composition as follows:

- Pure copper and high-copper alloys
- Brasses (alloys including zinc as the principal alloying element)
- Bronzes (alloys including tin as the principal alloying element)
- Nickel silvers, including copper–nickel alloys and copper–nickel–zinc alloys

Cast-copper alloys are melted in crucible furnaces, open-flame furnaces, induction furnaces, or indirect-arc furnaces. The selection of a furnace depends upon the type of alloy to be melted, as well as the purity and quantity required. In melting pure copper, high-copper alloys, bronzes, or nickel silver, precautions must be taken to prevent contamination of the molten metal with hydrogen. It is recommended that the atmosphere of the furnace be slightly oxidizing and also that a covering flux be used. Prior to casting, however, the molten metal should be *deoxidized* by adding phosphorus in the form of a phosphorous copper flux. On the other hand, brass is usually not susceptible to hydrogen porosity. The problem associated with melting brass is the vaporization and oxidation of the zinc. As a remedy, the atmosphere of the furnace should be slightly reducing. Also, a covering flux should be used to prevent vaporization of the zinc; a deoxidizing flux (like phosphorous copper) is then added immediately prior to pouring. The applications of cast-copper alloys include pipe fitting, ornaments, propeller hubs and blades, steam valves, and bearings.

3.2.3.2.3 Zinc Alloys

The family of zinc alloys is characterized by low melting temperatures. Zinc alloys also possess good fluidity. Therefore, they can be produced in thin sections by submerged hot-chamber die casting. Alloying elements employed include aluminum, copper, and magnesium.

3.2.3.2.4 Magnesium Alloys

The main characteristic of magnesium is its low density, which is lower than that of any other commercial metal. The potential uses of magnesium are many because it is readily available as a component of seawater and most of its disadvantages and limitations can be eliminated by alloying. Magnesium alloys usually are cast in permanent molds or are produced by hot-chamber die casting.

3.3 FOUNDRY FURNACES

Various furnaces are employed for smelting different ferrous and nonferrous metals in foundry work. The type of foundry furnace to be used is determined by the kind of metal to be melted, the hourly output of molten metal required, and the purity desired. Following is a brief review of each of the commonly used foundry furnaces.

3.3.1 Cupola Furnaces

3.3.1.1 Structure

The cupola is the most widely used furnace for producing molten-gray cast iron. A sketch of a cupola furnace is given in Figure 3.17. As can be seen, the cupola is a shaft-type furnace whose height is three to five times its diameter. It is constructed of a steel plate that is about 3/8 inch (10 mm) thick and that is internally lined with refractory fireclay bricks. The whole structure is erected on legs or columns. Toward the top of the furnace is an opening through which the charge is fed. Air, which is needed for the combustion, is blown through the tuyeres located about 36 inches (900 mm) above the bottom of the furnace. Slightly above the bottom and in the front is a tap hole and spout to allow molten cast iron to be collected. There is also a slag hole located at the rear and above the level of the tap hole (because slag floats on the surface of molten iron). The bottom of the cupola is closed with drop doors to dump residual coke or metal and also to allow for maintenance and repair of the furnace lining.

3.3.1.2 Operation

A bed of molding sand is first rammed on the bottom to a thickness of about 6 inches (150 mm) or more. A bed of coke about 40 inches (1.0 m) thick is next placed on the sand. The coke is then ignited, and air is blown at a lower-than-normal rate. Next, the charge is fed into the cupola through the charging door. Many factors, such as the charge composition, affect the final structure of the gray cast iron obtained. Nevertheless, it can generally be stated that the charge is composed of 25 percent pig iron, 50 percent gray cast-iron scrap, 10 percent steel scrap, 12 percent coke as fuel, and 3 percent limestone as flux. These constituents form alternate layers of coke, limestone, and metal. Sometimes, ferromanganese briquettes and inoculants are added to the charge to control and improve the structure of the cast iron produced.

3.3.2 Direct Fuel-Fired (Reverberatory) Furnaces

The direct fuel-fired furnace, or reverberatory furnace, is used for the batch-type melting of bronze, brass, or malleable iron. The burners of the furnace are fired with pulverized coal or another liquid petroleum product. Figure 3.18 shows that the roof of the reverberatory furnace reflects the flame onto the metal placed on the hearth, thus heating the metal and melting it. The gaseous products of

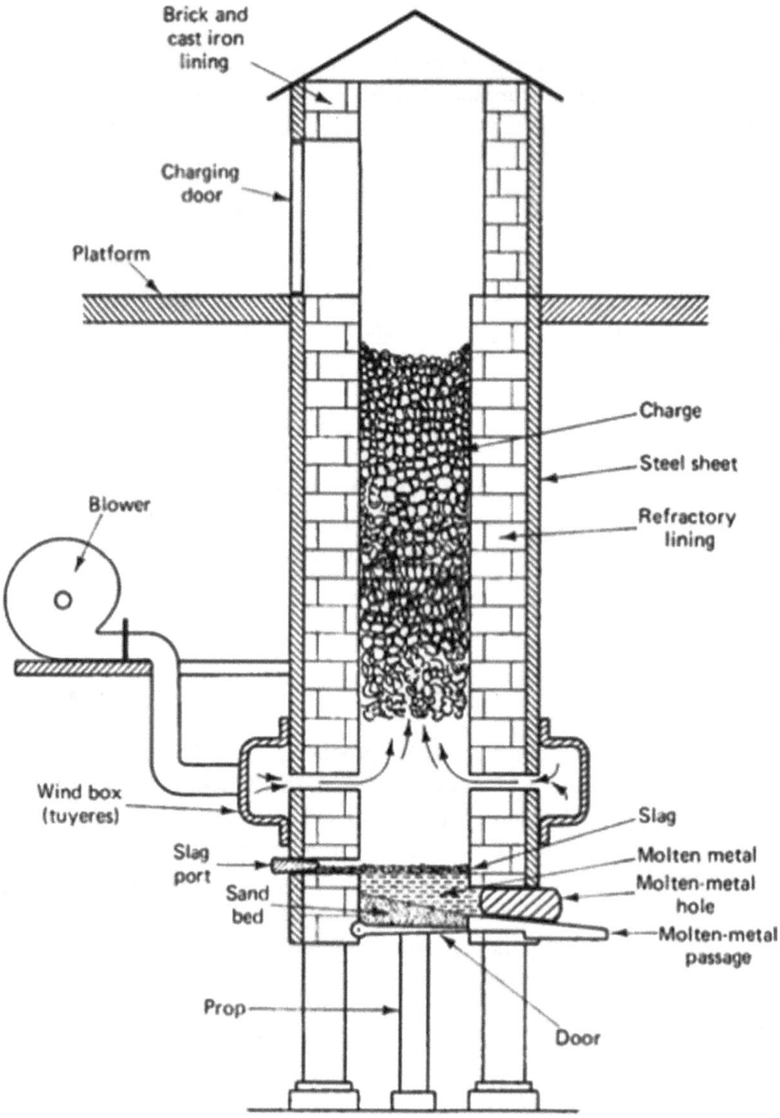

FIGURE 3.17 A sketch of a cupola furnace.

combustion leave the furnace through the flue duct. The internal surface of the furnace is lined with firebricks, and there are charging and tap holes. When iron is melted, the fuel-air ratio is adjusted to produce a completely white iron without free graphite flakes because they lower the properties of the resulting malleable iron.

3.3.3 Crucible (Pot) Furnaces

Nonferrous metals like bronzes, brasses, aluminum, and zinc alloys are usually melted in a crucible, or pot furnace. Crucible furnaces are fired by liquid, gaseous, or pulverized solid fuel. Figure 3.19 shows that the products of combustion in a crucible furnace do not come in direct contact with the molten metal, thus enabling the production of quality castings. Crucible furnaces can be stationary or tilting. When the stationary type is employed, crucibles are lifted out by tongs and are then carried in shanks. On the other hand, crucibles with long pouring lips are always used with the tilting type.

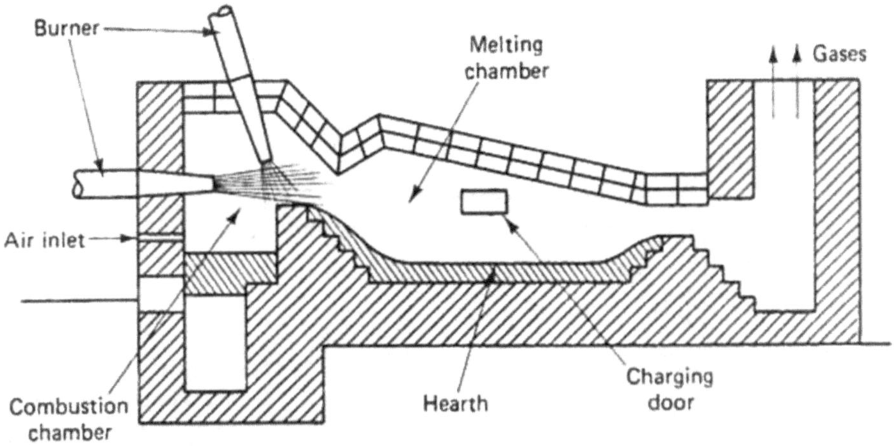

FIGURE 3.18 A reverberatory furnace.

FIGURE 3.19 A crucible furnace.

Crucibles are made of either refractory material or alloy steels (containing 25 percent chromium). Refractory crucibles can be of the clay-graphite ceramic-bonded type or the silicon-carbide carbon-bonded type. The first type is cheaper, while the second one is more popular in industry. Ceramic crucibles are used when melting aluminum, bronze, or gray cast iron, whereas brasses are melted in alloy steel crucibles. Different alloys must not be melted in the same crucible to avoid contamination of the molten metal.

3.3.4 Electric Furnaces

An electric furnace is usually used when there is a need to prevent the loss of any constituent element from the alloy and when high purity and consistency of casting quality are required. An electric furnace is also employed when melting high-temperature alloys. In all types of electric furnaces, whether they are electric-arc, resistance, or induction furnaces, the electric energy is converted into heat.

3.3.4.1 Electric-Arc Furnace

The electric-arc furnace is the most commonly used type of electric furnace. Figure 3.20 is a sketch of an electric-arc furnace. The heat generated by an electric arc is transferred by direct radiation or by reflected radiation off the internal lining of the furnace. The electric arc is generated about

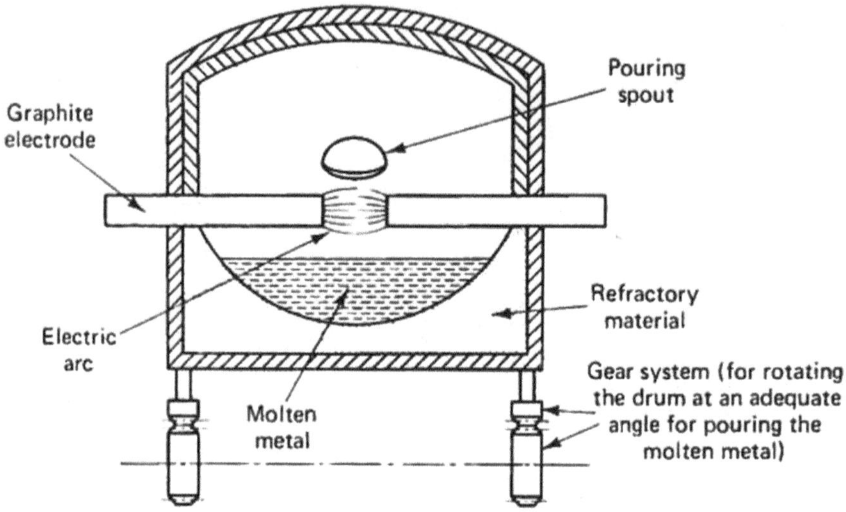

FIGURE 3.20 A sketch of an electric-arc furnace.

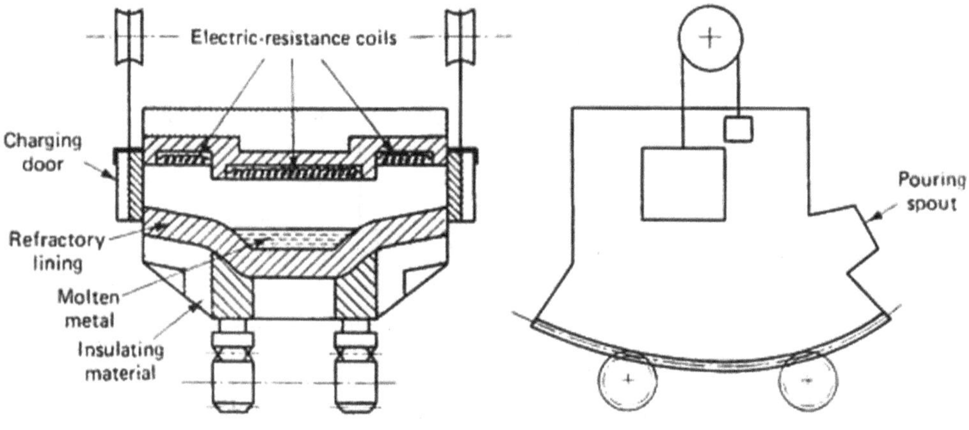

FIGURE 3.21 The basic features of an electric-resistance furnace.

midway between two graphite electrodes. In order to control the gap between the two electrodes and, accordingly, control the intensity of heat, one electrode is made stationary and the other one movable. Electric-arc furnaces are used mainly for melting steels and, to a lesser extent, gray cast iron and some nonferrous metals.

3.3.4.2 Resistance Furnace

The resistance furnace is employed mainly for melting aluminum and its alloys. Figure 3.21 indicates the basic features of a typical resistance furnace. The solid metal is placed on each of the two inclined hearths and is subjected to heat radiation from the electric-resistance coils located above it. When the metal melts, it flows down into a reservoir. The molten metal can be poured out through the spout by tilting the whole furnace.

3.3.4.3 Induction Furnace

The induction furnace has many advantages, including evenly distributed temperatures within the molten metal, flexibility, and the possibility of controlling the atmosphere of the furnace. In

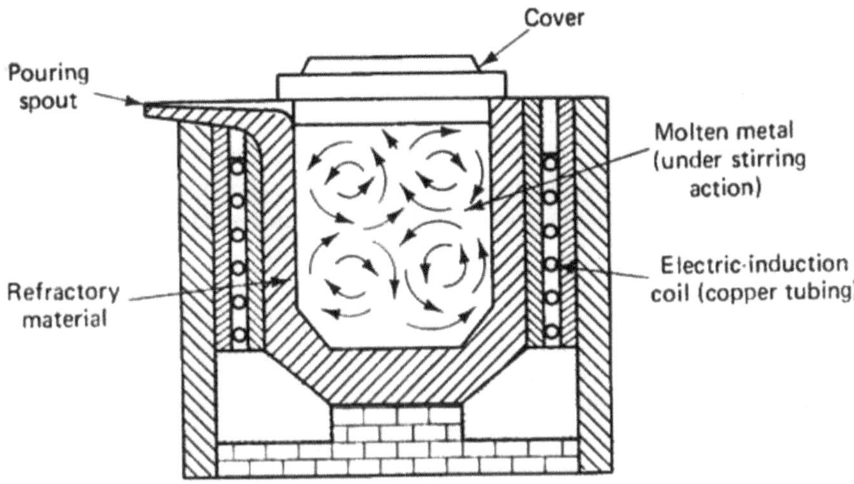

FIGURE 3.22 An electric-induction furnace.

addition, the motor effect of the electromagnetic forces helps to stir the molten metal, thus producing a more homogeneous composition. Induction furnaces are used to melt steel and aluminum alloys. Figure 3.22 shows the construction of a typical induction furnace. It basically involves an electric-induction coil that is built into the walls of the furnace. An alternating current in the coil induces a current in any metallic object that obstructs the electromagnetic field.

3.4 CASTING DEFECTS AND DESIGN CONSIDERATIONS

3.4.1 Common Defects in Castings

In order to obtain a sound casting, it is necessary to control adequately the various factors affecting the casting process. Casting and pattern designs, molding procedure, and melting and pouring of molten metal are among the factors affecting the soundness of a casting. Following is a survey of the commonly experienced defects in castings.

3.4.1.1 Hot Tears

Hot tears can appear on the surface or through cracks that initiate during the cooling of the casting. They usually are in locations where the metal is restrained from shrinking freely, such as a thin wall connecting two heavy sections.

3.4.1.2 Cold Shut

A cold shut is actually a surface of separation within the casting. It is believed to be caused by two "relatively cold" streams of molten metal meeting each other at that surface.

3.4.1.3 Sand Wash

A sand wash can be described as rough, irregular surfaces (hills and valleys) of the casting that result from erosion of the sand mold. This erosion is, in turn, caused by the metal flow.

3.4.1.4 Sand Blow

A sand blow is actually a surface cavity that takes the form of a very smooth depression. It can be caused by insufficient venting, lack of permeability, or a high percentage of humidity in the molding sand.

3.4.1.5 Scab

A scab is a rough "swollen" location in the casting that has some sand embedded in it. Such a defect is usually encountered when the molding sand is too fine or too heavily rammed.

3.4.1.6 Shrinkage Porosity (or Cavity)

A shrinkage porosity is a microscopic or macroscopic hole formed by the shrinkage of spots of molten metal that are encapsulated by solidified metal. It is usually caused by poor design of the casting.

3.4.1.7 Hard Spots

Hard spots are hard, difficult-to-machine areas that can occur at different locations.

3.4.1.8 Deviation of the Chemical Composition from the Desired One

Deviation may be due to the loss of a constituent element (or elements) during the melting operation. It may also be caused by contamination of the molten metal.

3.4.2 Design Considerations

A product designer who selects casting as the primary manufacturing process should make a design not only to serve the function (by being capable of withstanding the loads and the environmental conditions to which it is going to be subjected during its service life) but also to facilitate or favor the casting process. Following are some design considerations and guidelines.

3.4.2.1 Promote Directional Solidification

When designing the mold, be sure that the risers are properly dimensioned and located to promote directional solidification of the casting toward the risers. In other words, the presence of large sections or heat masses in locations distant from the risers should be avoided, and good rising practice as previously discussed should be followed. Use can also be made of chills to promote directional solidification. Failure to do so may result in shrinkage cavities (porosity) or cracks in those large sections distant from the risers. It is also very important to remember that a riser will not feed a heavy section through a lighter section.

3.4.2.2 Ensure Easy Pattern Drawing

Make sure that the pattern can easily be withdrawn from the nonpermanent mold (this does not apply to investment casting). This can be achieved through rational selection of the parting line as well as by providing an appropriate pattern draft wherever needed. In addition, undercuts or protruding bosses (especially if their axes do not fall within the parting plane) and the like should be avoided. Nevertheless, remember that undercuts can be obtained, if necessary, by using cores.

3.4.2.3 Avoid the Shortcomings of Columnar Solidification

Dendrites often start to form on the cold surface of a mold and then grow to form a columnar casting structure. This almost always results in planes of weakness at sharp comers, as illustrated in Figure 3.23a. Therefore, rounding the edges is a must for eliminating the development of planes of weakness, as shown in Figure 3.23b. Rounded edges are also essential for the smooth laminar flow of the molten metal.

3.4.2.4 Avoid Hot Spots

Certain shapes, because of their effect on the rate of heat dissipation during solidification, tend to promote the formation of shrinkage cavities. This is always the case at any particular location where the rate of solidification is slower than that at the surrounding regions of the casting. The rate of solidification (and the rate of heat dissipation to start with) is slower at locations having a low ratio

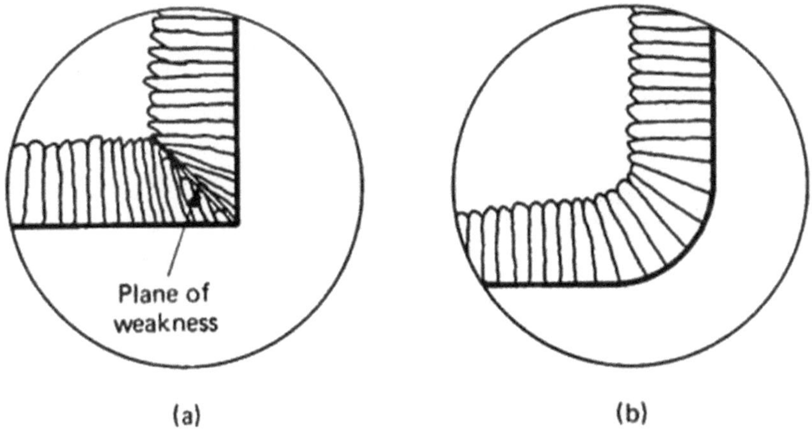

FIGURE 3.23 Columnar solidification and planes of weakness: (a) poor design (sharp corners) and (b) rounded edges to eliminate planes of weakness.

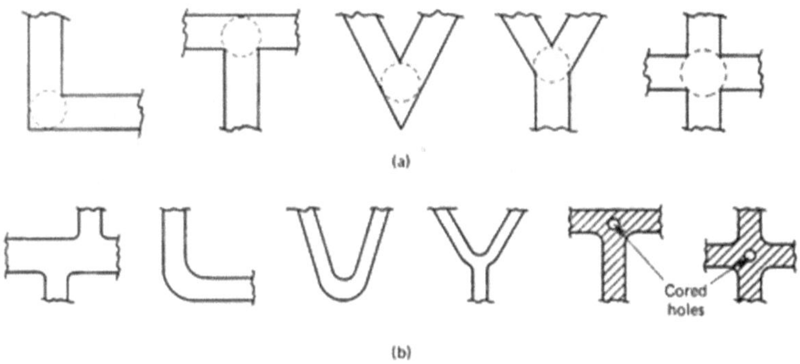

FIGURE 3.24 Hot spots: (a) poor design, yielding hot spots, and (b) better design, eliminating hot spots.

of surface area to volume. Such locations are usually referred to as *hot spots* in foundry work. Unless precautions are taken during the design phase, hot spots and, consequently, shrinkage cavities are likely to occur at the L, T, V, Y, and + junctions, as illustrated in Figure 3.24a. Shrinkage cavities can be avoided by modifying the design, as shown in Figure 3.24b. Also, it is always advisable to avoid abrupt changes in sections and to use taper (i.e., make the change gradual), together with generous radii, to join thin to heavy sections, as shown in Figure 3.25.

3.4.2.5 Avoid the Causes of Hot Tears

Hot tears are casting defects caused by tensile stresses as a result of restraining a part of the casting. Figure 3.26a,b shows locations where hot tears can occur and a recommended design that would eliminate their formation.

3.4.2.6 Distribute the Masses of a Section to Save Material

Cast metals are generally weaker in tension in comparison with their compressive strengths. Nonetheless, the casting process offers the designer the flexibility of distributing the masses of a section with a freedom not readily available when other manufacturing processes are employed. Therefore, when preparing a design of a casting, try to distribute masses in such a manner as to lower the magnitude of tensile stresses in highly loaded areas of the cross section and to reduce material in lightly loaded areas. As can be seen in Figure 3.27, a T section or an I-beam is more advantageous than just a round or square one when designing a beam that is to be subjected to bending.

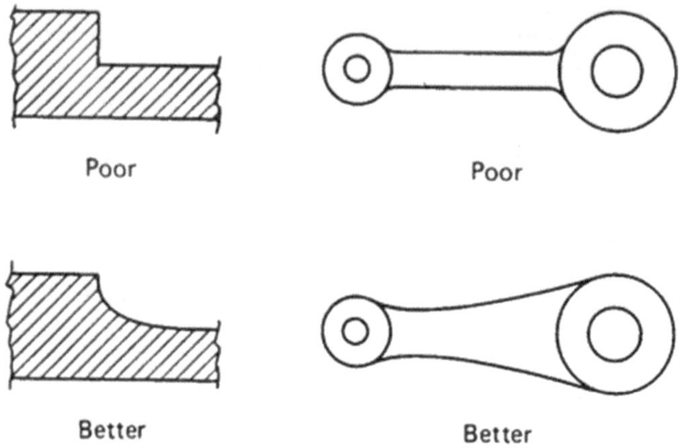

FIGURE 3.25 Avoid abrupt changes in sections.

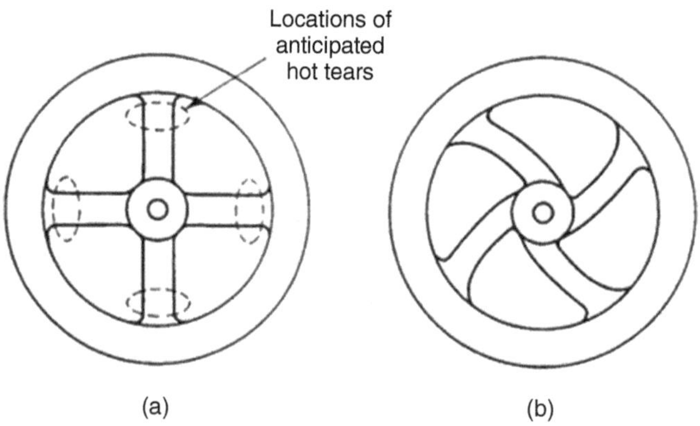

FIGURE 3.26 Hot tears: (a) a casting design that promotes hot tears and (b) a recommended design to eliminate hot tears.

3.4.2.7 Avoid Thicknesses Lower than the Recommended Minimum Section Thickness

The minimum thickness to which a section of a casting can be designed depends upon such factors as the material, the size, and the shape of the casting as well as the specific casting process employed (i.e., sand casting, die casting, etc.). In other words, strength and rigidity calculations may prove a thin section to be sufficient, but casting considerations may require adopting a higher value for the thickness so that the cast sections will fill out completely. This is a consequence of the fact that a molten metal cools very rapidly as it enters the mold and may become too cold to fill a thin section far from the gate. A minimum thickness of 0.25 inch (6 mm) is suggested for design use when conventional steel casting techniques are employed, but wall thicknesses of 0.060 inch (1.5 mm) are quite common for investment castings. Figure 3.28 indicates the relationship between the minimum thickness of a section and its largest dimension. It should be pointed out that for a given thickness, steel flows best in a narrow rather than in a wide web. For cast iron and nonferrous castings, the recommended values for minimum thicknesses are much lower than those for steel castings having the same shape and dimensions.

3.4.2.8 Strive to Make Small Projections in a Large Casting Separate

As can be seen in Figure 3.29a, a small projection may be subjected to more accidental knocks than a large casting, and if it gets broken, the whole casting will be scrapped. It is, therefore, highly

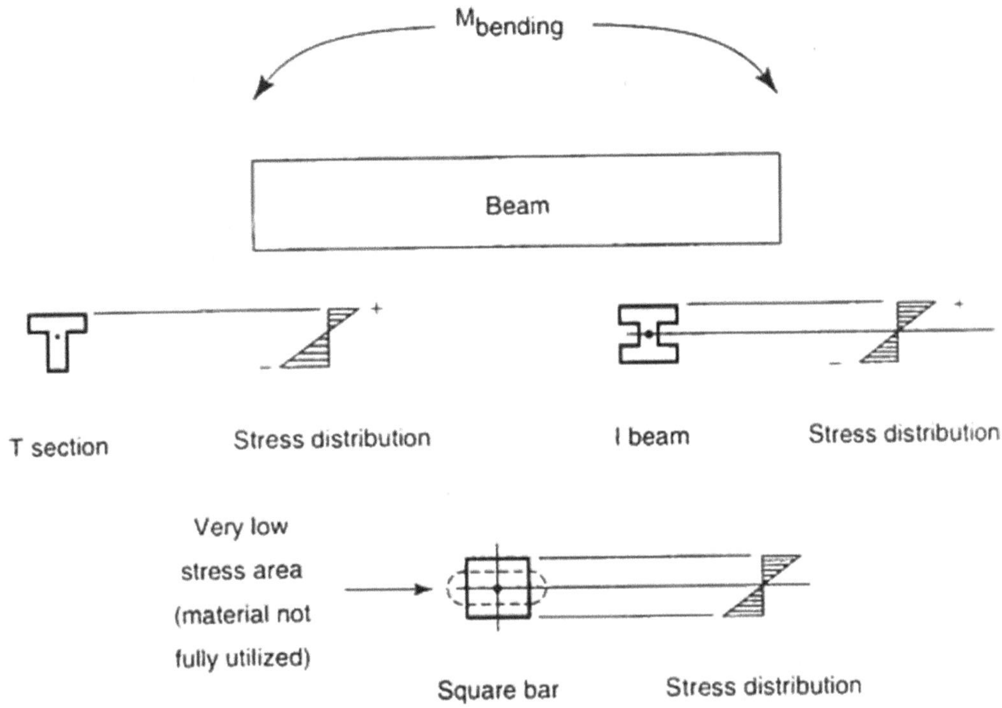

FIGURE 3.27 Distribution of masses to reduce weight.

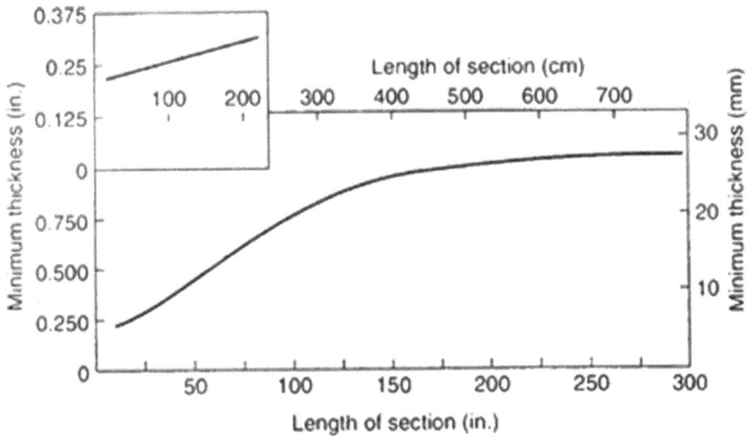

FIGURE 3.28 Minimum thickness of cast steel sections as a function of their largest dimension. (Adapted from *Steel Castings Handbook*, 5th ed. Rockery River, OH: Steel Founders Society of America, 1980.)

recommended to make the small projection separate and attach it to the large casting by an appropriate mechanical joining method, as shown in Figure 3.29b.

3.4.2.9 Strive to Restrict Machined Surfaces

Whereas some castings are used in their entirely as-cast condition, some others may require one or more machining operations. It is the task of the designer to ensure that machining is performed only on areas where it is absolutely necessary. An example of cases where the machining needed involves bearing surfaces is shown in Figure 3.30.

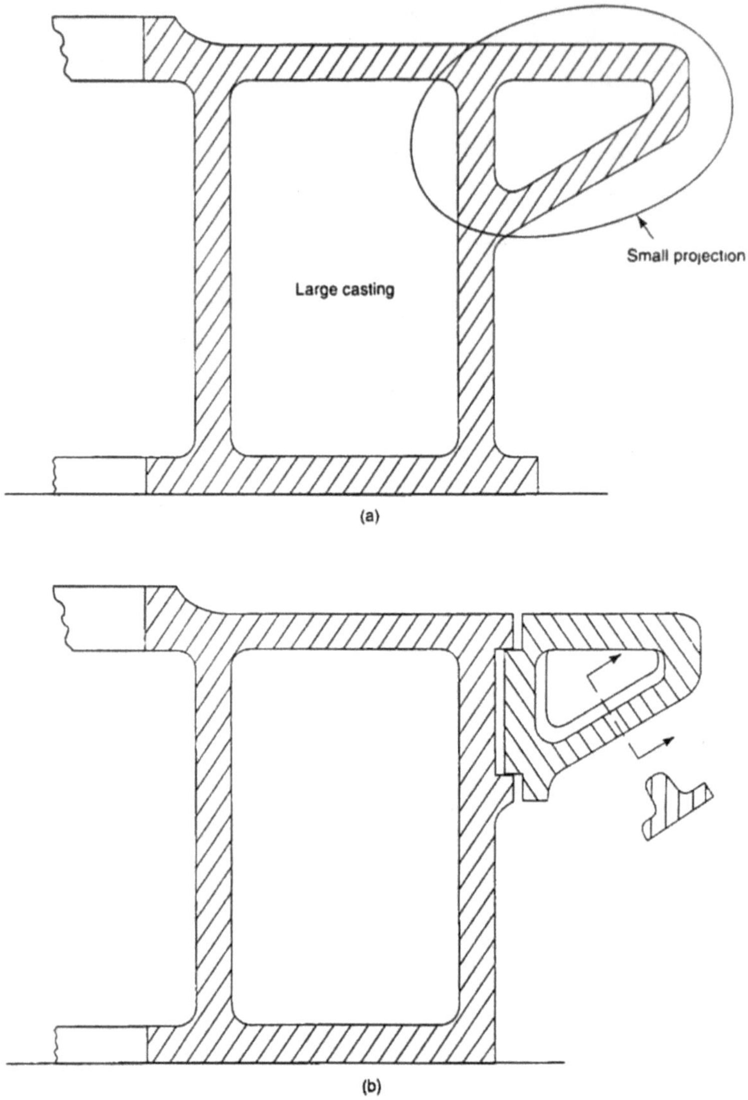

FIGURE 3.29 Large casting with a small projection: (a) as an integral part and (b) two separate parts.

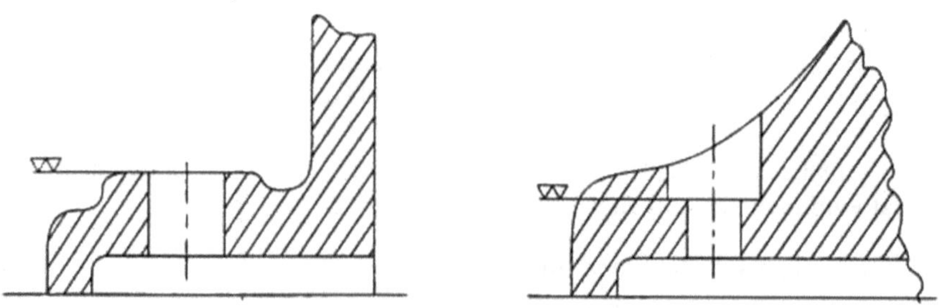

FIGURE 3.30 Restriction of surfaces to be machined.

3.4.2.10 Use Reinforcement Ribs to Improve the Rigidity of Thin, Large Webs

A common use of brackets or reinforcement ribs is to provide rigidity to thin, large webs (or the like) as an alternative to increasing the thickness of the webs. The ribs should be as thin as possible (i.e., minimum permissible thickness) and should also be staggered, as shown in Figure 3.31. Always remember that parabolic ribs are better than straight ribs in terms of economy and uniformity of stress.

3.4.2.11 Consider the Use of Cast-Weld Construction to Eliminate Costly Cored Design

The design of some products necessitates the use of complicated steel-wire-reinforced cores that are difficult to reach and remove after casting, thus leaving the surfaces unclean. An example, a steam ring, is shown in Figure 3.32a. The alternative design would be to employ a simple cut plate that is welded into the casting to produce the cast-weld construction shown in Figure 3.32b.

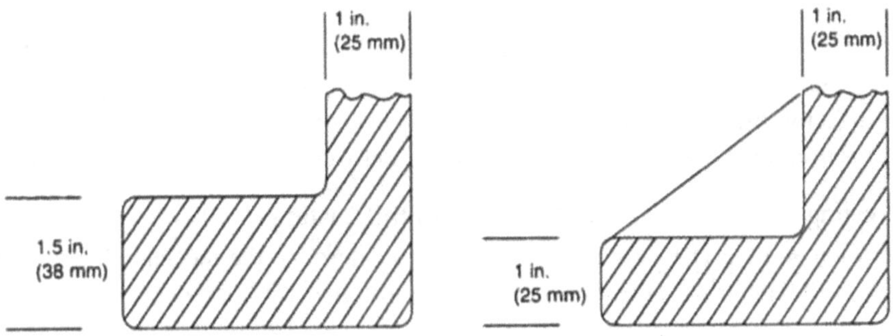

FIGURE 3.31 Use of reinforcement ribs.

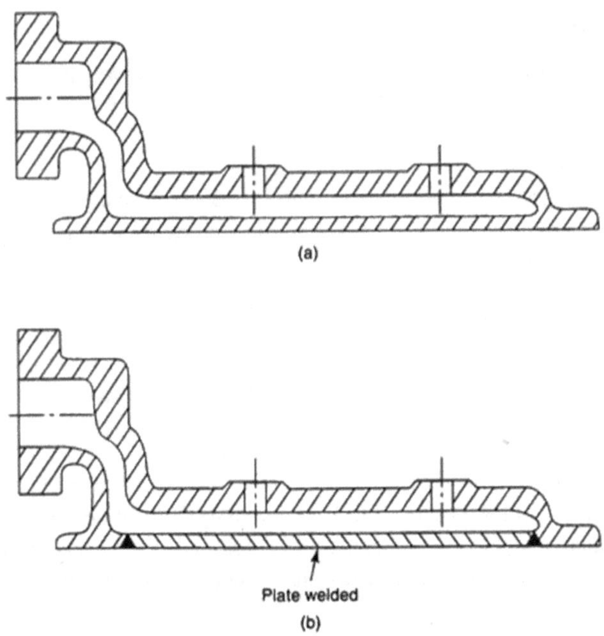

FIGURE 3.32 The design of a steam ring: (a) cast construction and (b) cast-weld construction.

3.5 CLEANING, TESTING, AND INSPECTION OF CASTINGS

3.5.1 Cleaning

The cleaning process involves the removal of the molding sand adhering to a casting. It also includes the elimination of gates, runners, and risers. Generally, surface cleaning can be carried out in rotary separators or by employing sandblasting and/or metallic shot-blasting machines. The latter two machines use sand particles or shots traveling at high velocities onto the surface of the casting to loosen and remove the adhering sand. As you may expect, these machines are particularly suitable when cleaning medium and heavy castings. On the other hand, rotary separators are advantageous for cleaning light castings. A separator is actually a long, large-diameter drum that rotates around its horizontal axis into which the castings are loaded together with jack stars made of white cast iron. A further advantage of rotary separators is that they automatically break off gate and runner systems and, often, risers.

3.5.2 Testing and Inspection

Like any other manufactured parts, castings must be subjected to thorough quality control in order to separate defective products and to reduce the percentage of rejects by identifying the defects and tracing their sources. Following are some of the commonly used tests and inspection methods.

3.5.2.1 Testing of the Mechanical Properties of the Casting

Standard tension and hardness tests are carried out to determine the mechanical properties of the metal of the casting in order to make sure that they conform to the specifications.

3.5.2.2 Inspection of the Dimensions

Dimensions must fall within the specified limits. Therefore, measuring tools and different kinds of gauges (e.g., snap, progressive, plug, template) are used to check that the dimensions conform to the blueprint.

3.5.2.3 Visual Examination

Visual inspection is used to reveal only very clear defects. However, it is still commonly used in foundries.

3.5.2.4 Hydraulic Leak Testing

The hydraulic leak test is used to detect microscopic shrinkage porosity. Various penetrants and testing methods are now available. Details are given in the American Society for Testing and Materials (ASTM) standards, designation E165.

3.5.2.5 Nondestructive Testing

Several nondestructive testing methods detect microscopic and hair cracks. They involve ultrasonic testing, magnetic particle inspection, eddy current testing, and radiography.

3.5.2.6 Testing for Metal Composition

Several methods are employed to determine the chemical composition accurately and to ensure product quality. The classical method used to be "wet analysis" (i.e., employing acids and reagents in accurate chemical analysis). However, because this method is time-consuming, it is being replaced by methods like emission spectroscopy, X-ray fluorescence, and atomic absorption spectroscopy (Table 3.1.).

TABLE 3.1
Comparison between Different Casting Processes

Process	Mold Material and Method of Filling	Metal Being Cast	Dimensional Accuracy and Surface Finish	Examples of Products and Production Volume
Green Sand Molding	Natural deposits from river beds. Gravity filling	Ferrous & nonferrous metals	Castings need machining to remove the hard skin.	Small articles, machine parts. Relatively large batches
Dry Sand Molding	Sand mixture. Gravity filling	Ferrous & nonferrous metals, alloyed cast iron	Better than green sand molding. May need some machining.	Large castings with thick walls, machine parts, flanges, pipe fittings, valve parts, pump housing, steering gear housing
Cement-Bonded Sand Mold	Mixture of silica sand & cement. Gravity filling	Ferrous & nonferrous metals	Smooth surfaces, accurate dimensions	Large castings with intricate shapes
Carbon Dioxide Process	Mixture of silica sand and sodium silicate. Gravity filling	Ferrous & nonferrous metals	High dimensional accuracy, very good surface finish	Articles where only minimum machining of casting is permitted
Plaster Molds	Mixture of silica sand, asbestos, and plaster of Paris. Gravity filling	Silver, gold, magnesium, copper, & aluminum alloys	Extremely smooth surfaces, needs no machining	Small articles of precious metals & nonferrous metals
Loam Molds	Basic shape built with bricks & mortar. Fine details by a mixture of loam & silica sand. Gravity filling	Usually ferrous metals	Castings require machining	Very large castings like water turbines
Shell Molding	Mixture of dry silica sand & phenol formaldehyde. Shell made on a heated metal pattern. Gravity filling	Grey cast iron, ductile cast iron, aluminum alloys	Good surface finish, few restrictions on casting design	Suitable for automation, mass production of thin-walled ductile cast iron and aluminum alloys weighing 25 to 35 lbs (10 to 15 kg), axles, brackets, levers, housing, crankshafts
Investment Casting (Precision molds or Lost wax)	Self-supporting ceramic shell 0.25 inch thick. Gravity filling	High-melting-point alloys, difficult-to-machine metals	Excellent, needs very little or no machining	Small objects with intricate shapes
Die Casting (Cold Chamber)	Tool steel. Pressure applied by a plunger.	Aluminum and its alloys, magnesium, copper and its alloys	Good. Little or no machining	Parts weighing up to 50 lbs (23 kg)
Die Casting (Hot Chamber)	Tool steel. Action of a pump immersed in a pot of molten metal	Low-melting-point alloys (e.g., zinc, tin, lead)	Superior surface finish, very close tolerances	High production rates, ability to produce intricate shapes with thin sections, objects weighing less than 10 lbs (4.5 kg)

(*Continued*)

TABLE 3.1 (Continued)

Process	Mold Material and Method of Filling	Metal Being Cast	Dimensional Accuracy and Surface Finish	Examples of Products and Production Volume
Mega Casting	Tool steel. High-pressure large plunger	Aluminum alloys (AlSi8)	Very good	The chassis of the Tesla motor car weighing 500 lbs (200 kg). Production rate: 1,000 chassis per day
Centrifugal Casting	Dry silica sand mixture, effective centrifugal forces	Cast iron, brass	Good, superior mechanical properties	High production rate, sewer & water pipes, bushings
Continuous Casting	Special alloy steel or ceramic. Gravity filling	High-quality alloy steel	Very good	Round stock

3.6 COMPARISON BETWEEN DIFFERENT CASTING PROCESSES

REVIEW QUESTIONS

1. What is meant by the word *casting?*
2. What are the constituents of green molding sand?
3. List some of the important properties that green sand must possess.
4. What is a flask? What is its function? List the parts that form a flask.
5. Explain the meaning of the word *pattern*.
6. List some of the materials used in making patterns.
7. List the different types of permanent patterns used in foundries. What are the different pattern allowances? Discuss the function of each.
8. What are cores? How are they made?
9. What is meant by a gating system? What functions does it serve?
10. What are the components of a gating system?
11. What are risers? What function do they serve?
12. List the various green sand properties and discuss each briefly.
13. Why should weights be located on the cope in pit molding?
14. List the various molding machines and discuss the operation of each briefly.
15. Explain sand conditioning and how it is done.
16. What advantages does dry sand molding have over green sand molding?
17. When are cement-bonded sand molds recommended?
18. What is the main advantage of the carbon dioxide process for molding?
19. What metals can be cast in plaster molds?
20. When are loam molds used?
21. Describe shell molding. What are its advantages?
22. When are ceramic molds recommended?
23. Explain investment casting and why it is sometimes called the lost-wax process.
24. Name a metal that should be cast in a graphite mold.
25. What are the advantages of employing permanent molds? Why?
26. Can molten metals be cast directly into cavities of cold permanent molds? Why?
27. What is the main difference between the hot-chamber and the cold-chamber methods of die casting?
28. List some metals that you think can be cast by the hot-chamber method. Justify your answer.
29. List some metals that you think can be cast by the cold-chamber method. Justify your answer.

30. What are the types of centrifugal casting?
31. Differentiate between the different types of centrifugal casting and discuss the advantages and shortcomings of each type.
32. What are the products that can be manufactured by continuous casting?
33. What does the continuous casting process involve?
34. Discuss some advantages of the continuous casting process.
35. What does the V-process involve?
36. List some of the merits and advantages of the V-process.
37. Discuss some of the problems encountered in casting steels.
38. What precautions should be taken to eliminate the problems in casting steels?
39. What is gray cast iron?
40. Discuss some of the properties that make gray cast iron attractive for some engineering applications.
41. Why are inoculants added to gray cast iron?
42. Differentiate between gray cast iron and white cast iron.
43. What is meant by compacted-graphite cast iron?
44. What is ductile cast iron? How can it be obtained?
45. What is malleable cast iron? How can it be obtained? What are the limitations of producing it?
46. List some alloying elements that are added to cast iron. List some applications for alloyed cast iron.
47. What are the problems caused by hydrogen when melting and casting aluminum and how can these problems be eliminated?
48. What are the sources of hydrogen when melting aluminum?
49. List some cast aluminum alloys and discuss their applications.
50. How are cast-copper alloys classified?
51. What is meant by a deoxidizer? Give an example.
52. List some of the characteristics and applications of cast zinc alloys.
53. List some of the characteristics and applications of cast magnesium alloys.
54. For what purpose is the cupola furnace used?
55. Describe briefly the operation and charge of the cupola furnace.
56. For what purpose is the reverberatory furnace used?
57. List some of the metals that can be melted in crucible furnaces.
58. What are the main differences in construction between the stationary and the tilting crucible furnaces?
59. List the different types of electric furnaces and mention the principles of operation in each case.
60. List the main advantages and applications of electric furnaces.
61. List some of the common defects of castings and discuss the possible causes of each defect.
62. List and discuss the main design considerations for castings.
63. List and discuss the various testing and inspection methods used for the quality control of castings.

DESIGN PROBLEM

Your company has received an order to manufacture wrenches for loosening and tightening nuts and bolts of large machines. The plant of the company involves a foundry and a machining workshop with a few basic machine tools. Here are the details of the order.

Lot size: 500 wrenches
Nut size: 2 inches (50 mm)
Required torque: About 20 lb·ft (27.12 N·m)

You are required to provide a design and a production plan (see the explanation of the word *design* in the design projects section that appears later).

Solution

Before we start solving this design problem, we should make some assumptions. For instance, consider the force that can be generated by the ordinary human hand. It will allow us to determine the length of the wrench using the following equation:

$$T = F \times l$$

where

T is the torque
F is the force
l is the length

As can be seen from the equation, a low value of F would make the length large and thus make the handling of the wrench impractical because of the weight. On the other hand, a high value of F is not practical and may not be generated by an ordinary person.

Let us take $F = 15$ pounds. Therefore, and apparently, the force acts at the middle of the fist, and we have to add a couple of inches for proper holding.

$$\text{Length of wrench} = 18 \text{ in.}$$

Let us now design the section where the maximum bending moment occurs. You can assume some dimensions and determine the stress, which will serve as a guide in selecting material. Take the section as shown in Figure 3.33a. The moment of inertia of the section is

$$\begin{aligned} I &= \frac{1(0.25)(0.75)^3}{12} + 2\left[\frac{1(0.375)(0.25)^3}{12} + 0.375 \times 0.25 \times (0.5)^2\right] \\ &= 0.008789 + 0.00098 + 0.046875 \\ &= 0.056655 \text{ in.}^4 \end{aligned}$$

Note that the minimum thickness for steel casting was adhered to. Now, determine the stress:

$$\text{Max .stress} = \frac{20 \times 12 \times 1.25}{2 \times 0.056655} = 2648 \text{ lb / in.}^2$$

That value is very low, and we should try to reduce the section and save material. It is always a good idea to make use of spreadsheets to change the dimensions and get the stresses acting in each case. Now, take the section as shown in Figure 3.33b:

$$\begin{aligned} l &= \frac{1}{12}(0.25)(^{0.5})^3 + 2\left[\frac{1}{12}(0.375)(0.25)^3 + 0.375 \times 0.25 \times (0.375)^2\right] \\ &= 0.002604 + 0.00098 + 0.026367 \\ &= 0.038771188 \text{ in.}^4 \end{aligned}$$

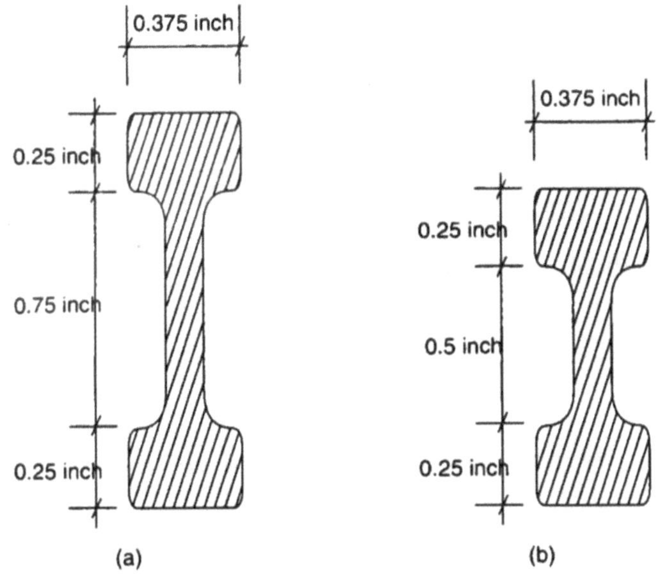

FIGURE 3.33 Cross section of the wrench: (a) first attempt and (b) second attempt.

$$\text{Max .stress} = \frac{20 \times 12 \times 1.25}{0.038771188 \times 2} = 3868 \text{ lb / in.}^2$$

As can be seen, we took the minimum thickness to be 0.25 inch, which is the recommended value for conventional castings of steels. The material should be low-carbon steel having 0.25 percent carbon in order to possess enough ductility. Also, the steel should be thoroughly killed. A recommended material is ASTM A27-77, grade U60-30, which has a yield strength of 30 ksi. When taking a factor of safety of 4, the allowable stress would be 7500 lb/in.^{2}, which is higher than the obtained value of the working stress.

Now, in order to calculate the thickness of the wrench, let us calculate the bearing stress on the nut. A reasonable estimate of the force on the surface of the nut is

$$\frac{20 \times 12}{0.75} = 320 \text{ lb}$$

This is based on the assumption that the torque is replaced by two opposite forces having a displacement of 0.75 inch between the lines of action. Thus,

$$\text{Bearing stress} = \frac{320}{0.75 \times t} = 7500$$

$$t = 0.056 \text{ in.}$$

Take it as 0.5 inch to facilitate casting the part.

Because all dimensions are known, a detailed design can be prepared, as shown in Figure 3.34. Notice the surface finish marks indicating the surfaces to be machined.

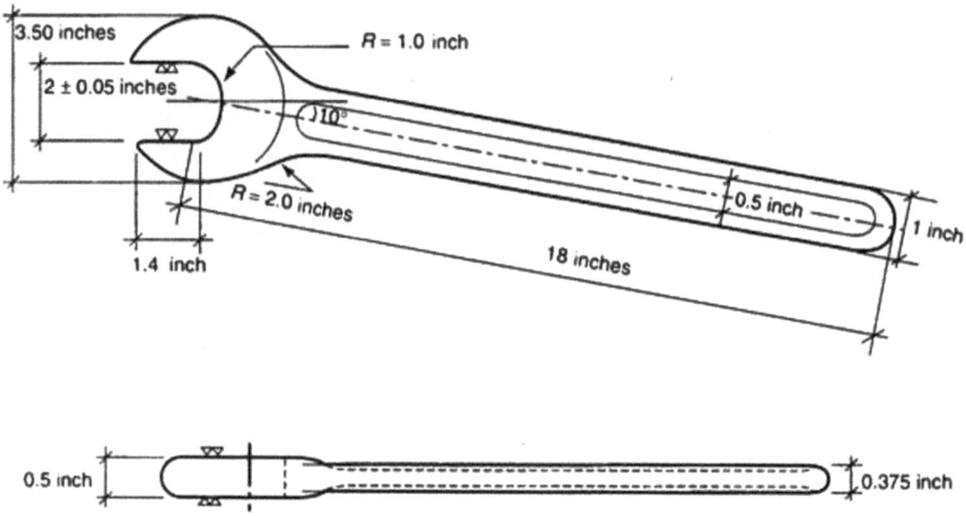

FIGURE 3.34 A wrench manufactured by casting.

DESIGN PROJECTS

Whenever the word *design* is mentioned hereafter, you should provide, *at least*, the following:

- Two neatly dimensioned graphical projections of the product (i.e., a blueprint ready to be released to the workshop for actual production), including fits (if applicable), tolerances, surface finish marks, and so on
- Material selection with rational justification
- Selection of the specific manufacturing processes required, as well as their sequence in detail
- Simple but necessary calculations to check the stresses at the critical sections

1. Design a bracket for a screw C-clamp that has the following characteristics.
   Maximum clamping force: 22 lb (100 N)
   Clamping gap: 3 inches (7.5 cm)
   Distance between the centerline of the screw and the inner surface of the bracket: 2 inches (5 cm)
   Assume that manufacturing is by casting and that the production volume is 4000 pieces.
2. Design a flat pulley. Its outer diameter is 36 inches (90 cm), and it is to be mounted on a shaft that is 2½ inches (6.25 cm) in diameter. Its width is 10 inches (25 cm), and it has to transmit a torque of 3000 lb·ft (4000 N·m). Assume that 500 pieces are required. Will the design change if only 3 pieces are required?
3. A connecting lever has two short bosses, each at one of its ends and each with a vertical hole that is 3/4 inch (19 mm) in diameter. The lever is straight, and the horizontal distance between the centers of the holes is 8 inches (200 mm). The lever during functioning is subjected to a bending moment of 50 lb·ft (67.8 N·m) that acts in the plane formed by the two vertical axes. Provide a detailed design for this lever if it is to be produced by casting and
   i. When only 100 pieces are required.
   ii. When 10,000 pieces are required.

4. Design a micrometer frame for each of the following cases:
   i. The gap of the micrometer is 1.0 inch (25 mm), and the distance from the axis of the barrel to the inner side of the frame is 1.5 inches (37.5 mm). The maximum load on the anvil is 22 lb (100 N).
   ii. The gap of the micrometer is 6 inches (150 mm), and the distance from the axis of the barrel to the inner side of the frame is 4.0 inches (100 mm). The maximum load on the anvil is 22 lb (100 N).
   iii. Assume that the production volume is 4000 pieces and that one of the various casting processes is used.
   iv. **Tip**: Base your design on rigidity. The maximum deflection must not exceed 0.1 of the smallest reading of the micrometer.
5. A pulley transmits a torque of 600 lb·ft (813.6 N·m) to a shaft that is 1¼ inches (31 mm) in diameter. The outer diameter of the pulley is 10 inches (250 mm), and it is to be driven by a flat belt that is 2 inches (50 mm) in width. Design this pulley if it is to be manufactured by casting and 500 pieces are required.
6. Design a hydraulic jack capable of lifting 1 ton and having a stroke of 6 inches (150 mm). The jack is operated by a manual displacement (plunger) pump that pumps oil from a reservoir into the high-pressure cylinder through two spring-actuated nonreturn valves to push the ram upward. The reservoir and the high-pressure cylinder are also connected by a conduit, but the flow of oil is obstructed by a screw that, when unscrewed, relieves the pressure of the cylinder by allowing high-pressure oil to flow back into the reservoir and for the ram to be then pushed downward. Provide a workshop drawing for each component, as well as an assembly drawing for the jack. Steel balls and springs are to be purchased. Assume the production volume is 5000 pieces.
7. Design a table for the machine shop. That table should be 4 feet (1.2 m) in height, with a surface area of 3 × 3 feet (0.9 × 0.9 m), and should be able to carry a load of half a ton. Assume the production volume is 2000 pieces.
8. Design a little wrench for loosening and tightening the nuts and bolts of a bicycle. The nut size is 5/8 inch (15 mm), and the required torque is about 1.0 lb·ft (1.356 N·m). Assume the production volume is 10,000 pieces.
9. A straight-toothed spur-gear wheel transmits 1200 lb·ft (1627 N·m) of torque to a steel shaft that is 2 inches (50 mm) in diameter. The pitch diameter of the gear is 8 inches (200 mm), its width is 3 inches (75 mm), and the base diameter is 7.5 inches (187.5 mm). Design this gear's blank. Assume the production volume is 4000 pieces.
10. Design a frame for an open-arch (C-type) screw press that can deliver a load of up to 2 tons. The open gap is 2 feet (600 mm), and the bed on which workpieces are placed is 12 × 12 inches (300 × 300 mm). Assume that the base diameter of the screw thread is 1½ inches (37.5 mm).
11. Design a hydraulic cylinder for earth-moving equipment. It can generate a maximum force of 2 tons and has a stroke of 4 feet (1200 mm). Although the maximum force is generated only when the plunger rod is moving out, the cylinder is double-acting and generates a force of 1 ton during its return stroke. The expected production volume is 2000 pieces, and the pistons, oil rings, and so on are going to be purchased from vendors.
12. Design a safety valve to be mounted on a high-pressure steam boiler. The pipe on which it will be mounted has a bore diameter of 2 inches (50 mm). The pressure inside the boiler should not exceed 50 folds of the atmospheric pressure. The expected production volume is 5000 pieces, and the stems, springs, bolts, and gaskets are going to be purchased from vendors.

4 Joining of Metals

When two parts of metal are to be attached together, the resulting joint can be made dismountable (using screws and the like), or it can be made permanent by employing riveting, welding, or brazing processes. The design of dismountable joints falls beyond the scope of this text and is covered in machine design. It is, therefore, the aim of this chapter to discuss the design and production of permanent joints when various technologies and methods are applied. Because the same equipment used in welding is also sometimes employed in the cutting of plates, thermal cutting processes will also be discussed in this chapter.

4.1 RIVETING

The process of *riveting* involves inserting a ductile metal pin through holes in two or more sheet metals and then forming over (heading) the ends of the metal pin so as to secure the sheet metals firmly together. This process can be performed either cold or hot, and each rivet is usually provided with one preformed head. Figure 4.1a,b indicates the sequence of operations in riveting, while Figure 4.2 illustrates different shapes of preformed rivet heads.

4.2 WELDING

Welding is the joining of two or more pieces of metal by creating atom-to-atom bonds between the adjacent surfaces through the application of heat, pressure, or both. In order for a welding technique to be industrially applicable, it must be reasonable in cost, yield reproducible or consistent weld quality, and, more importantly, produce joints with properties comparable to those of the base material. Various welding techniques have been developed that are aimed at achieving these three goals. However, no matter what welding method is used, the interface between the original two parts must disappear if a strong joint is to be obtained. Before we discuss the different methods employed to make those surfaces disappear, let us discuss joint types and preparation.

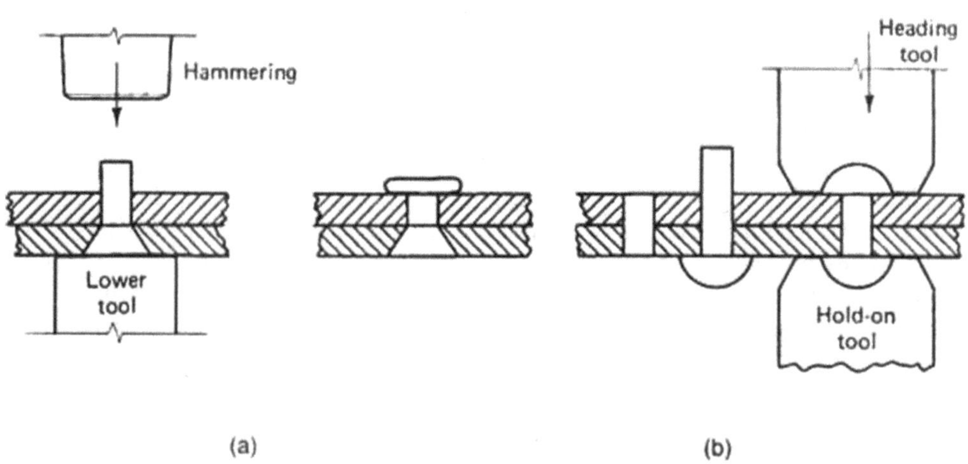

FIGURE 4.1 The sequence of operations in riveting: (a) flat-head rivet and (b) regular rivet.

DOI: 10.1201/9781003537106-4

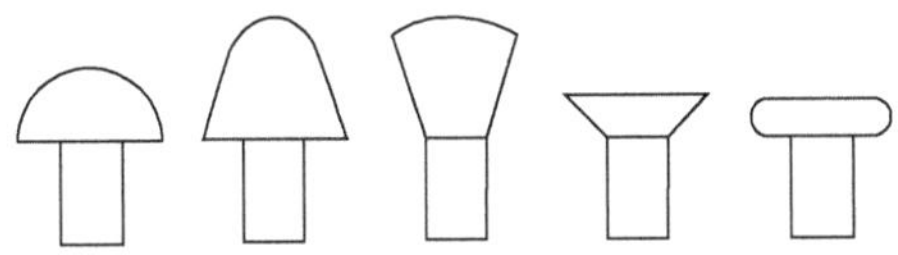

FIGURE 4.2 Different shapes of preformed rivet heads.

A weld joint must be designed to withstand the forces to which it is going to be subjected during its service life. Therefore, the joint design is determined by the type and magnitude of the loading that is expected to act on the weldment. In other words, the selection of the type of joint has to be made primarily on the basis of load requirement. As Figure 4.3a–e shows, there are five types of weld joints: butt, lap, corner, T, and edge. Following is a discussion of each of these different types of joints.

- ***Butt Joint.*** The butt joint involves welding the edges or end faces of the two original parts, as shown in Figure 4.3a. Therefore, the two parts must be aligned in the same plane. Usually, when the thickness of the parts falls between 1/8 and 3/8 inch (about 3 and 9 mm), the two parts are welded without any edge preparation. This type of weld is referred to as a *square* weld and can be either single or double, depending upon the thickness of the metal, as shown in Figure 4.4a. As can be seen in Figure 4.4b–e, the edges of thicker parts should be prepared with single or double bevels or V-, J-, or U-grooves to allow adequate access to the root of the joint. Usually, it is recommended to adopt the single or double U-groove when the thickness of the parts is more than 0.8 inch (20 mm).
- ***Lap Joint.*** We can see in Figure 4.3b that the lap joint is produced by fillet welding overlapping members: the amount of overlap is normally taken to be about three to five times the thickness of the member. The fillet weld can be continuous and may also be of the plug or slot type, as shown in Figure 4.5.
- ***Corner Joint.*** Figure 4.3c illustrates the corner joint, which can be welded with or without edge preparation (see Figure 4.4 for the various possible edge preparations).
- ***T-Joint.*** The T-joint is shown in Figure 4.3d. T-joints that will be subjected to light static loads may not require edge preparation. On the other hand, edge preparations (again see Figure 4.4) are often necessary for greater metal thicknesses or when the joint is to be subjected to relatively high, alternating, or impulsive loading.
- ***Edge Joint.*** The edge joint is usually used when welding thin sheets of metal with a thickness of up to 1/8 inch (3 mm). Notice in Figure 4.3e that the edges of the members must be bent before the welding process is carried out.

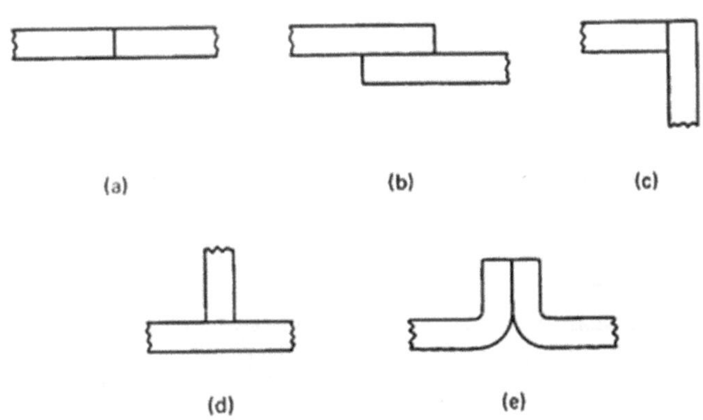

FIGURE 4.3 Types of weld joints: (a) butt, (b) lap, (c) corner, (d) T, and (e) edge.

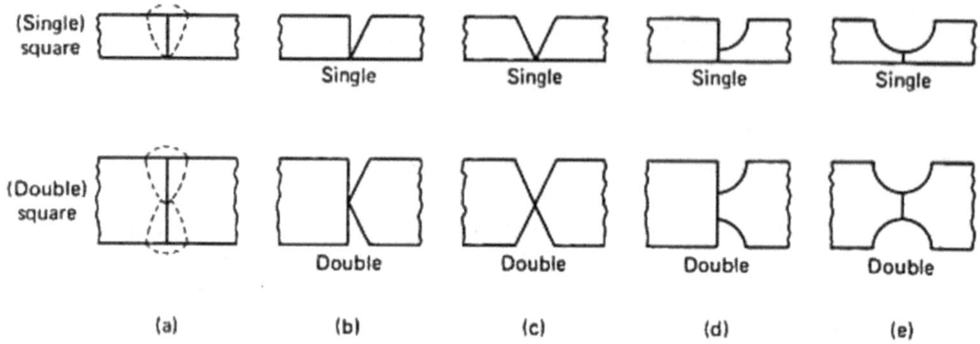

FIGURE 4.4 Different edge preparation for butt welding: (a) square, (b) bevel, (c) V-groove, (d) J-groove, and (e) U-groove.

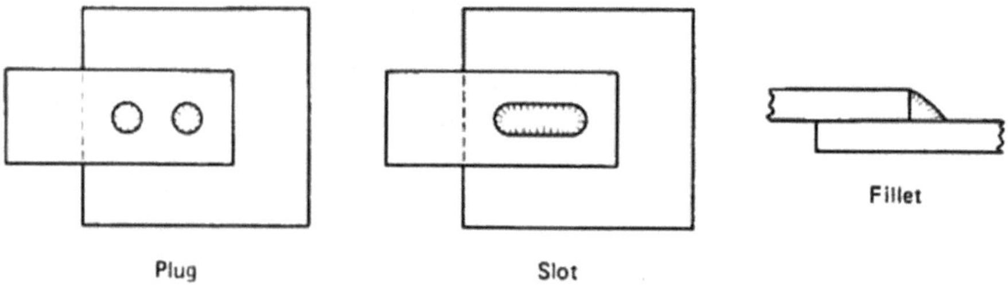

FIGURE 4.5 Basic types of fusion lap welds.

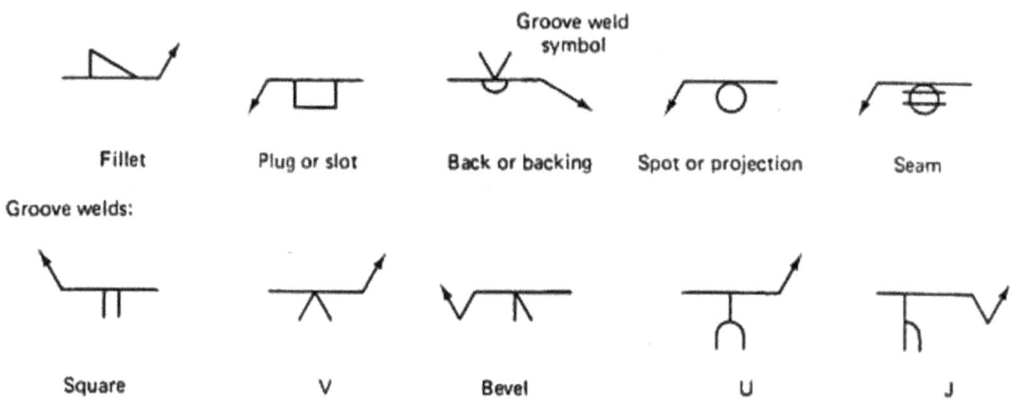

FIGURE 4.6 The different weld symbols.

Figure 4.6 shows the different weld symbols, whereas Figure 4.7 shows the standard identification of welds employed in design drawings.

In order to provide a comprehensive coverage of the various welding processes, we must divide them into groups of similar processes, with each group having something in common. For instance, welding processes can be classified according to the source of energy required to accomplish welding. In such a case, it is obvious that there are four main groups: mechanical, electrical, chemical, and optical. Welding processes can also be classified by the degree of automation adopted, which yields three groups: manual, semiautomatic, and automatic. The most commonly used method of classification is according to the state of the metal at the locations being welded, thus splitting the

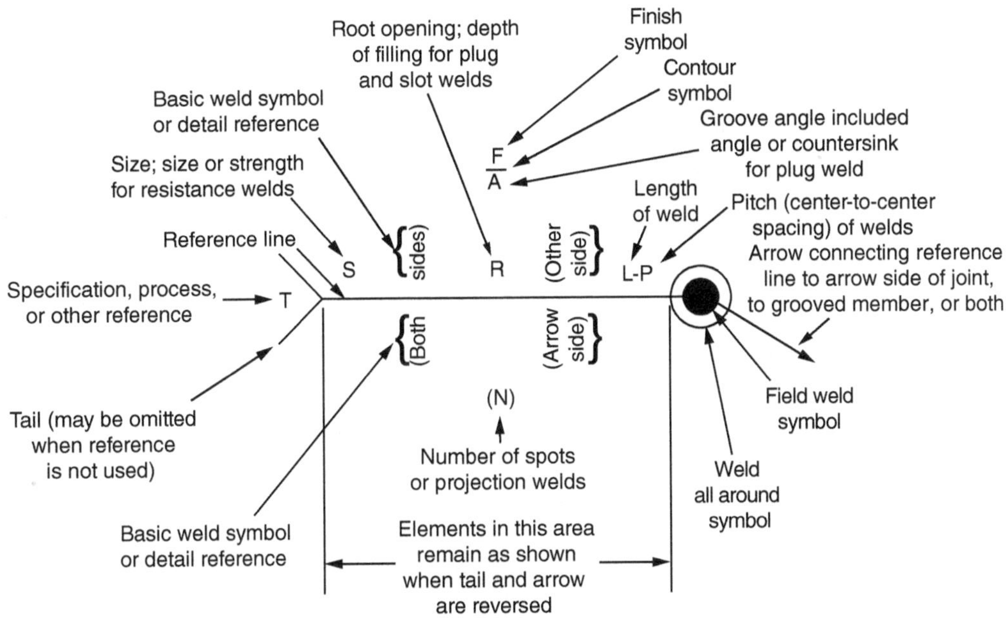

FIGURE 4.7 The standard identification of welds employed in design drawings.

welding processes into two main categories: pressure welding and fusion welding. We now discuss each of these two categories in detail.

4.2.1 Pressure Welding Processes

Pressure welding involves processes in which the application of external pressure is indispensable to the production of weld joints formed either at temperatures below the melting point (solid-state welding) or at temperatures above the melting point (fusion welding). In both cases, it is important to have very close contact between the atoms of the parts that are to be joined. The atoms must be moved together to a distance that is equal to or less than the equilibrium interatomic-separation distance. Unfortunately, two obstacles must be overcome so that successful pressure welding can be carried out and a sound weldment can be obtained. First, surfaces are not flat when viewed on a microscopic scale. Consequently, intimate contact can be achieved only where peaks meet peaks, as can be seen in Figure 4.8, and the number of bonds would not be enough to produce a strong welded joint. Second, the surfaces of metals are usually covered with oxide films that inhibit direct contact between the two metal parts to be welded. Therefore, those oxide and nonmetallic films must be removed (cleaned with a wire brush) before welding in order to ensure a strong welded joint. Pressure welding processes are applied primarily to metals possessing high ductility or those whose ductility increases with increasing temperatures; thus, the peaks that keep the surfaces of the two metallic members apart are leveled out under the action of mechanical stresses or the combined effect of high temperatures and mechanical stresses. In fact, a wide variety of pressure welding processes are used in industry. The commonly used ones are listed in Figure 4.9.

4.2.1.1 Cold-Pressure Welding

Cold-pressure welding is a kind of solid-state welding used for joining sheets, wires, and small electric components. As previously discussed, the surfaces to be welded must be cleaned with a wire brush to remove the oxide film and must be carefully degreased before welding. As Figure 4.10 shows, a special tool is used to produce localized plastic deformation, which results in coalescence

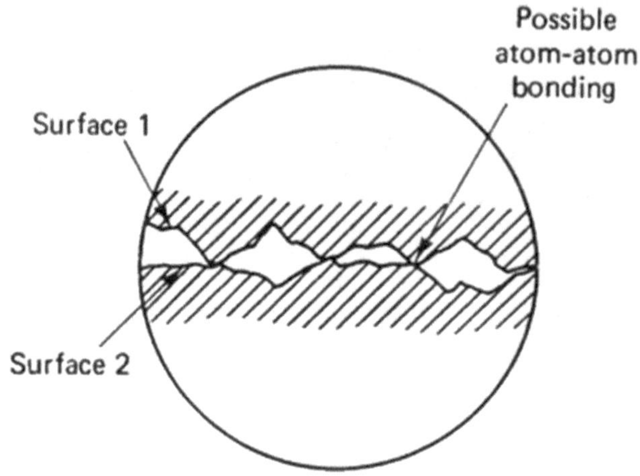

FIGURE 4.8 A microscopic view of two mating surfaces.

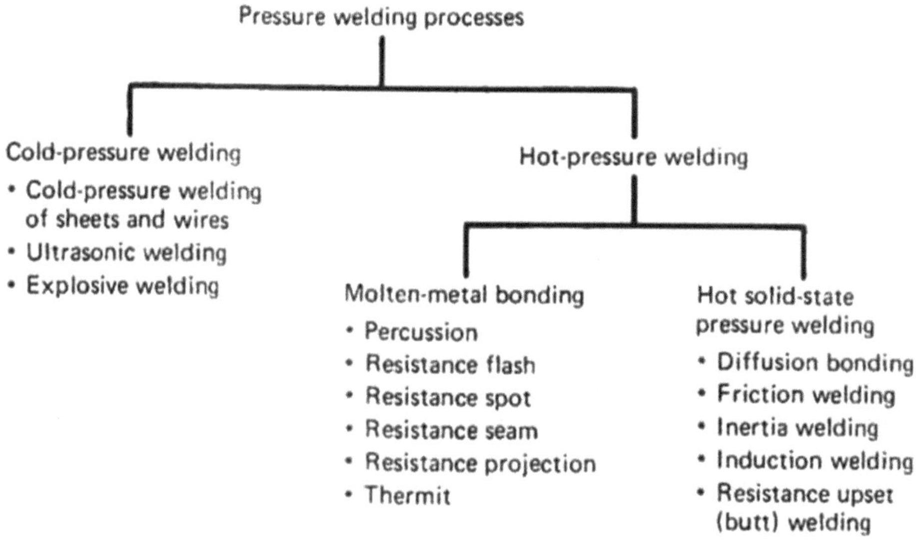

FIGURE 4.9 Classification of the commonly used pressure welding processes.

between the two parts. This process, which can replace riveting, is usually followed by annealing of the welded joint. Figure 4.10 also shows that recrystallization takes place during the annealing operation. This is added to diffusion, which finally results in the complete disappearance of the interface between the two parts.

Cold-pressure welding of wires is performed by means of a special-purpose machine. Figure 4.11 illustrates the steps involved in this process. As can be seen, the wires' ends are clamped and pressed repeatedly against each other in order to ensure adequate plastic deformation. The excess upset metal is then trimmed by the sharp edges of the gripping jaws. This technique is used when welding wires of nonferrous metals such as aluminum, copper, or aluminum-copper alloys.

4.2.1.2 Explosive Welding

Explosive welding is another technique that produces solid-state joints and, therefore, eliminates the problems associated with fusion welding methods, like the *heat-affected zone* (HAZ) and the microstructural changes. The process is based on using high explosives to generate extremely high

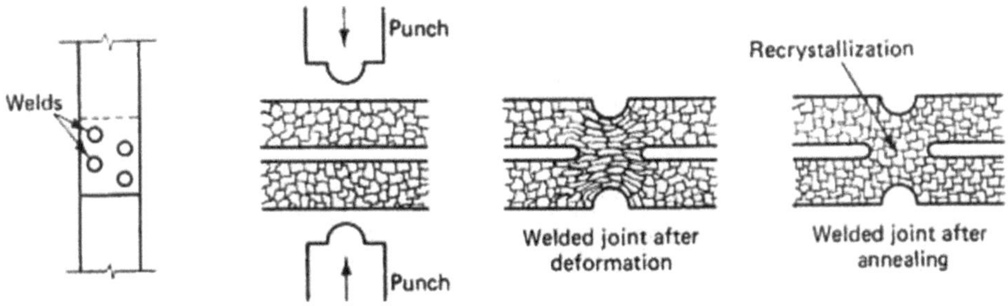

FIGURE 4.10 Cold-pressure welding of sheets.

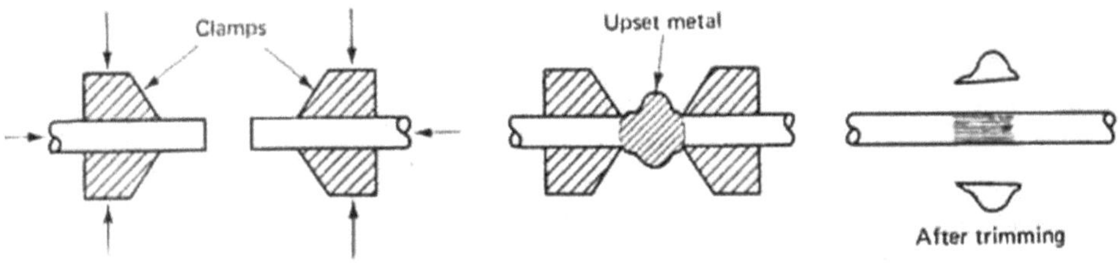

FIGURE 4.11 Cold-pressure welding of wires.

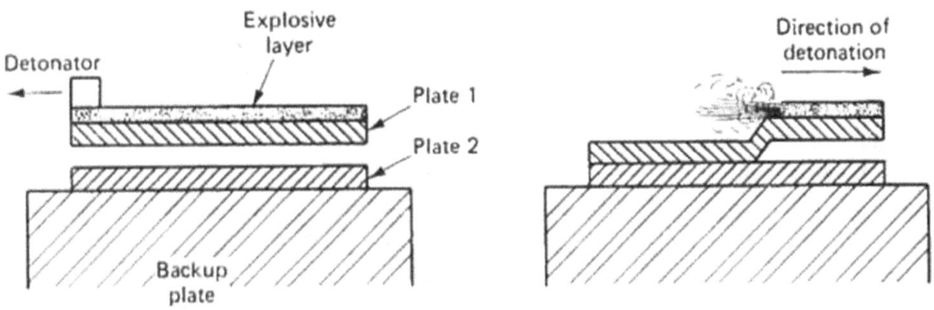

FIGURE 4.12 An arrangement for explosive welding of two flat plates.

pressures that are, in turn, used to combine flat plates or cylindrical shapes metallurgically. Joints of dissimilar metals and/or those that are extremely difficult to combine using conventional methods can easily be produced by explosive welding.

During explosive welding, a jet of soft (or fluid-like) metal is formed (on a microscopic scale) and breaks the oxide film barrier to bring the two metal parts into intimate contact. That metal jet is also responsible for the typical wavy interface between the two metal parts, thus creating mechanical interlocking between them and, finally, resulting in a strong bond. Figure 4.12 illustrates an arrangement for explosive welding of two flat plates, and Figure 4.13 is a magnified sketch of the wavy interface between explosively welded parts.

Explosive welding and explosive cladding are popular in the manufacture of heat exchangers and chemical-processing equipment. Armored and reinforced composites with a metal matrix are also produced by explosive welding. Nevertheless, a clear limitation is that the process cannot be used successfully for welding hard, brittle metals. Research is being carried out in this area, and new applications are continuously introduced.

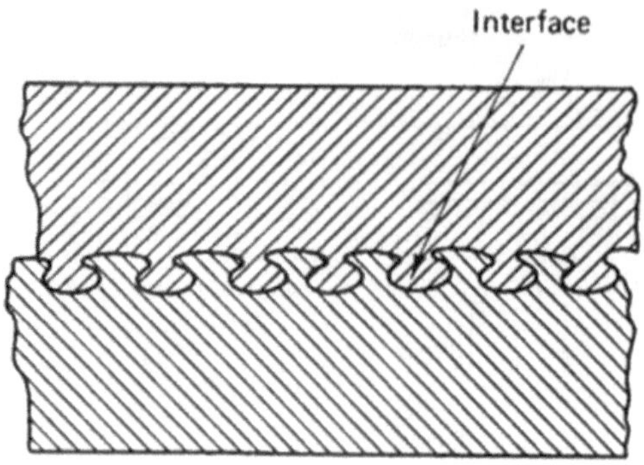

FIGURE 4.13 A magnified sketch of the wavy interface between explosively welded parts.

4.2.1.3 Ultrasonic Welding

The ultrasonic welding method of solid-state welding is commonly used for joining thin sheets or wires of similar or dissimilar metals in order to obtain lap-type joints. Mechanical vibratory energy with ultrasonic frequencies is applied along the interfacial plane of the joint, while nominal static stress is applied, normal to that interface, to clamp the two components together. Oscillating shear stresses are, therefore, initiated at the interface and disperse surface oxide films, allowing intimate contact between the two metals and, consequently, producing a strong joint. Ultrasonic welding does not involve the application of high pressures or temperatures and is accomplished within a short time. Therefore, this process is especially suitable for automation and has found widespread application in the electrical and microelectronics industries in the welding of thin metal foils for packaging and splicing and in the joining of dissimilar materials in the fabrication of nuclear reactor components. It must be noted, however, that the process is restricted to joining thin sheets or fine wires. Nevertheless, this restriction applies only to thinner pieces, and the process is often used in welding thin foils to thicker sheets.

Different types of ultrasonic welding machines are available, each constructed to produce a certain type of weld, such as spot, line, continuous seam, or ring. A sketch of a spot-type welding machine that is commonly used in welding microcircuit elements is illustrated in Figure 4.14. As we can see, the machine consists basically of a frequency converter that transforms the standard 60-Hz (or 50-Hz in Europe) electric current into a high-frequency current (with a fixed frequency in the range of 15–75 kHz), a transducer that converts the electrical power into elastic mechanical vibrations, and a horn that magnifies the amplitude of these vibrations and delivers them to the weld zone. Other associated elements include the anvil, a force-application and clamping device, a sonotrode (as compared with the electrode in resistance welding), and appropriate controls to set up optimal values for the process variables, such as vibratory power and weld time.

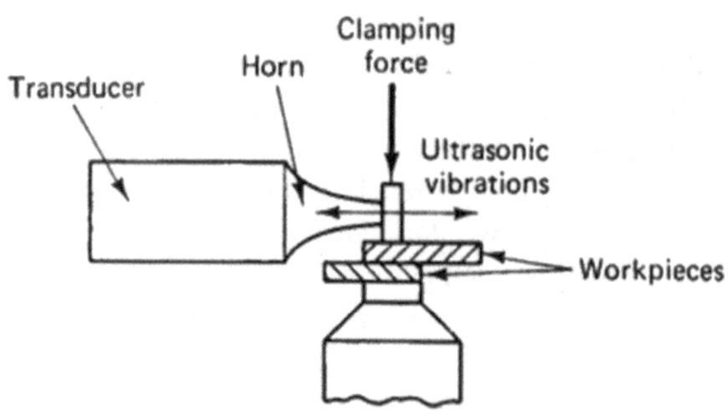

FIGURE 4.14 A sketch of an ultrasonic spot-type welding machine.

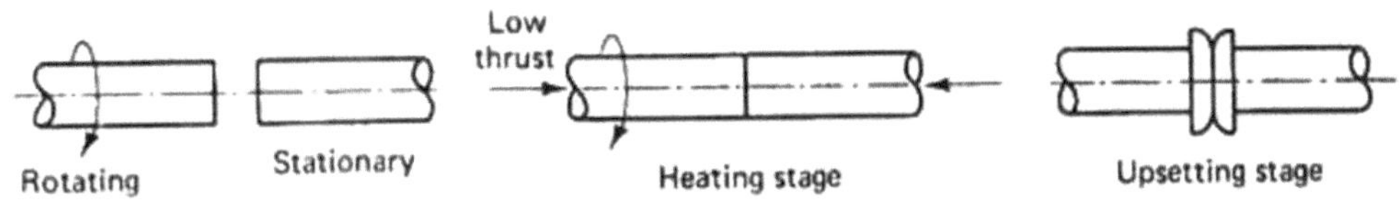

FIGURE 4.15 Stages involved in friction welding.

4.2.1.4 Friction Welding

In friction welding, a type of hot solid-state welding, the parts to be welded are tightly clamped, one in a stationary chuck and the other in a rotatable chuck that is mounted on a spindle. External power is employed to drive the spindle at a constant speed, with the two parts in contact under slight pressure. Kinetic energy is converted to frictional heat at the interface. When the mating edges of the workpiece attain a suitable temperature (in the forging range) that permits easy plastic flow, the spindle rotation is halted, and high axial pressure is applied to plastically deform the metal, obtain intimate contact, and produce a strong, solid weld. This is clearly shown in Figure 4.15, which indicates the stages involved in friction welding.

Several advantages have been claimed for the friction welding process. These include simplicity, high efficiency of energy utilization, and the ability to join similar as well as dissimilar metal combinations that cannot be joined by conventional welding methods (e.g., aluminum to steel or aluminum to copper). Also, since contaminants and oxide films are carried away from the weld area where grain refinement takes place, a sound bond is obtained and usually has the same strength as the base metal. Nevertheless, a major limitation of the process is that at least one of the two parts to be joined must be a body of revolution around the axis of rotation (like a round bar or tube). A further limitation is that only forgeable metals that do not suffer from hot shortness can successfully be friction welded. Also, care must be taken during welding to ensure the squareness of the edges of workpieces as well as the concentricity of round bars or tubes.

4.2.1.5 Inertia Welding

Inertia welding is a version of friction welding that is recommended for larger workpieces or where high-strength alloys (i.e., superalloys) are to be joined together. Inertia welding, as the name suggests, efficiently utilizes the kinetic energy stored in a rotating flywheel as a source for heating and for much of the forging of the weld. As is the case with friction welding, the two workpieces to be inertia welded are clamped tightly in stationary and rotatable chucks, the difference being that the rotatable chuck is rigidly coupled to a flywheel in the case of inertia welding. The process involves rotating the flywheel at a predetermined angular velocity and then converting the kinetic energy of the freely rotating flywheel to frictional heat at the weld interface by applying an axial load to join the abutting ends under controlled pressure. The process requires a shorter welding time than that taken in conventional friction welding, especially for larger workpieces. Examples of inertia-welded components include hydraulic piston rods for agricultural machinery, carbon steel shafts welded to superalloy turbocharger wheels, and bar stock welded to small forgings.

4.2.1.6 Induction Welding

As the name suggests, induction welding is based on the phenomenon of induction. We know from physics (electricity and magnetism) that when an electric current flows in an inductor coil, another electric current is induced in any conductor that intersects with the magnetic flux. In induction welding, the source of heat is the resistance, at the abutting workpieces' interface, to the flow of current induced in the workpieces through an external induction coil. Figure 4.16 illustrates the principles of induction welding. For efficient conversion of electrical energy into heat energy, high-frequency current is employed, and the process is usually referred to as *high-frequency induction welding* (HFIW). Frequencies in the range of 300–450 kHz are commonly used in industry, although

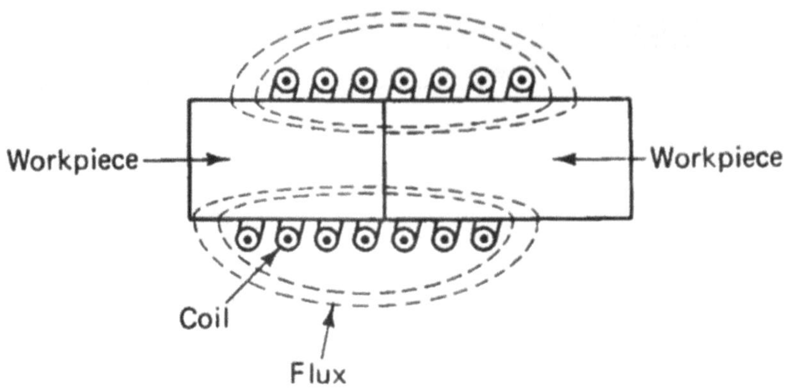

FIGURE 4.16 The principles of induction welding.

frequencies as low as 10 kHz are also in use. It is always important to remember the "skin effect" when designing an induction-welded joint. This effect refers to the fact that the electric current flows superficially (i.e., near the surface). In fact, the depth of the layer through which the current flows is dependent mainly upon the frequency and the electromagnetic properties of the workpiece metal. Industrial applications of induction welding include butt welding of pipes and continuous-seam welding for the manufacture of seamed pipes.

4.2.1.7 Thermit Welding

Thermit welding makes use of an exothermic chemical reaction to supply heat energy. That reaction involves the burning of thermit, which is a mixture of fine aluminum powder and iron oxide in the form of a rolling-mill scale, mixed at a ratio of about 1:3 by weight. Although a temperature of 5400°F (3000°C) may be attained as a result of the reaction, localized heating of the thermit mixture up to at least 2400°F (1300°C) is essential in order to start the reaction, which can be given by the following chemical formula.

$$8cAl + 3Fe_3O_4 \rightarrow 9Fe + 4Al_2O_3 + heat \tag{4.1}$$

As we can see from the formula, the outcome is very pure molten iron and slag. In fact, other oxides are also used to produce pure molten metals; these include chromium, manganese, or vanadium, depending upon the parent metals to be welded.

Usually, the thermit welding process requires the application of pressure in order to achieve proper coalescence between the parts to be joined. However, fusion thermit welding is also used; it does not require the application of force. In this case, the resulting molten metal is a metallurgical joining agent and not just a means for heating the weld area.

Thermit welding is used in joining railroad rails, pipes, and thick steel sections, as well as in repairing heavy castings. The procedure involves fitting a split-type refractory mold around the abutting surfaces to be welded, igniting the thermit mixture using a primer (ignition powder) in a special crucible, and, finally, pouring the molten metal (obtained as a result of the reaction) into the mold. Because the temperature of the molten metal is about twice the melting point of steel, the heat input is enough to fuse the abutting surfaces, which are usually pressed together to give a sound weld.

4.2.1.8 Diffusion Bonding

Diffusion bonding is a solid-state welding method in which the surfaces to be welded are cleaned and then maintained at elevated temperatures under appropriate pressure for a long period of time. No fusion occurs, deformation is limited, and bonding takes place principally due to diffusion.

As we know from metallurgy, the process parameters are pressure, temperature, and time, and they should be adjusted to achieve the desired results.

4.2.1.9 Butt Welding

Butt welding belongs to the resistance welding group, which also consists of the spot, seam, projection, percussion, and flash welding processes. All of these operate on the same principle, which involves heating the workpieces as a result of being a part of a high-amperage electric circuit and then applying external pressure to accomplish strong bonding. Consequently, all the resistance welding processes belong to the larger, more general group of pressure welding; without the application of external pressure, the weld joint cannot be produced.

In butt welding, sometimes called *upset-butt* welding or just *upset* welding, the parts are clamped and brought in solid contact, and low-voltage (1–3 V) alternating current is switched on through the contact area, as illustrated in Figure 4.17. As a result of the heat generated, the metal in the weld zone assumes a plastic state (above the solidus) and is gradually squeezed and expelled from the contact area. When enough upset metal becomes evident, the current is switched off and the welded parts are released. Figure 4.18 indicates a typical upset welding cycle. Note that upset welding would not be successful for larger sections because these cannot be uniformly heated and require extremely high-amperage current. Therefore, the process is limited to welding wires and rods up to 3/8 inch (about 10 mm) in diameter. Also, a sound joint can be ensured only when the two surfaces being welded together have the same cross-sectional area as well as negligible or no eccentricity.

4.2.1.10 Flash Welding

Flash welding is somewhat similar to upset welding. The equipment for flash welding includes a low-voltage transformer (5–10 V), a current timing device, and a mechanism to compress the two workpieces against each other. Figure 4.19 illustrates the different stages involved in a flash welding cycle. We can see that the pressure applied at the beginning is low. Therefore, a limited number of contact points act as localized bridges for the electric current. Consequently, metal is heated at those points when the current is switched on, and the temperature increases with the increasing current until it exceeds the melting point of the metal. At this stage, the molten metal is expelled from the

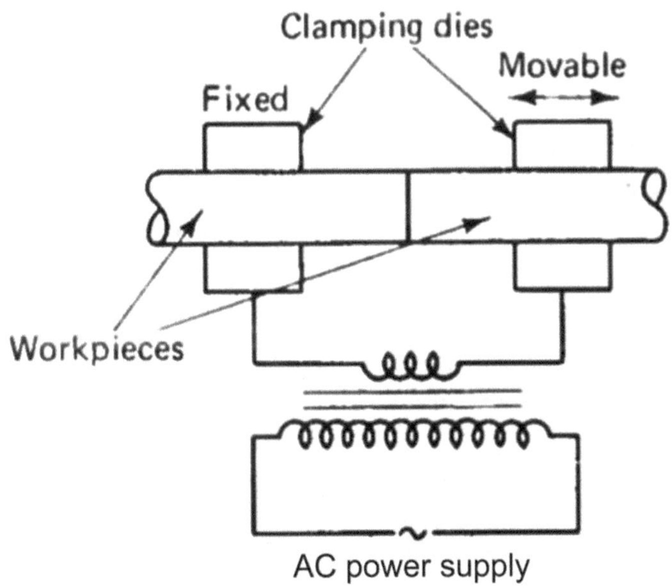

FIGURE 4.17 Upset-butt welding.

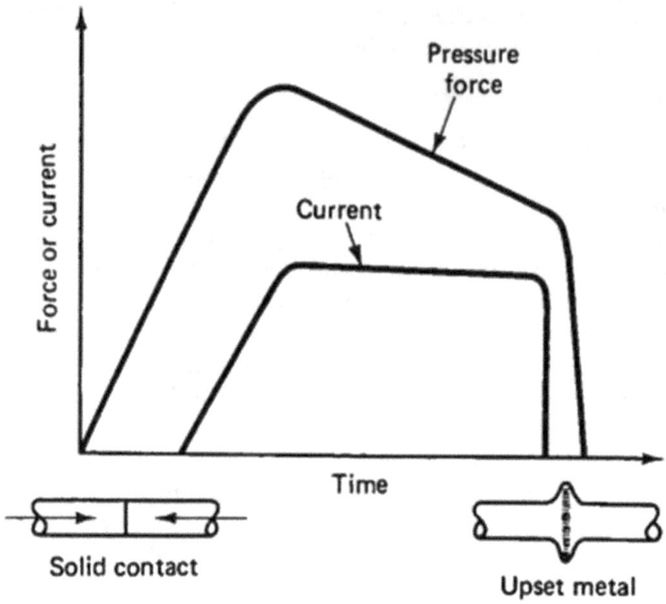

FIGURE 4.18 A typical upset welding cycle.

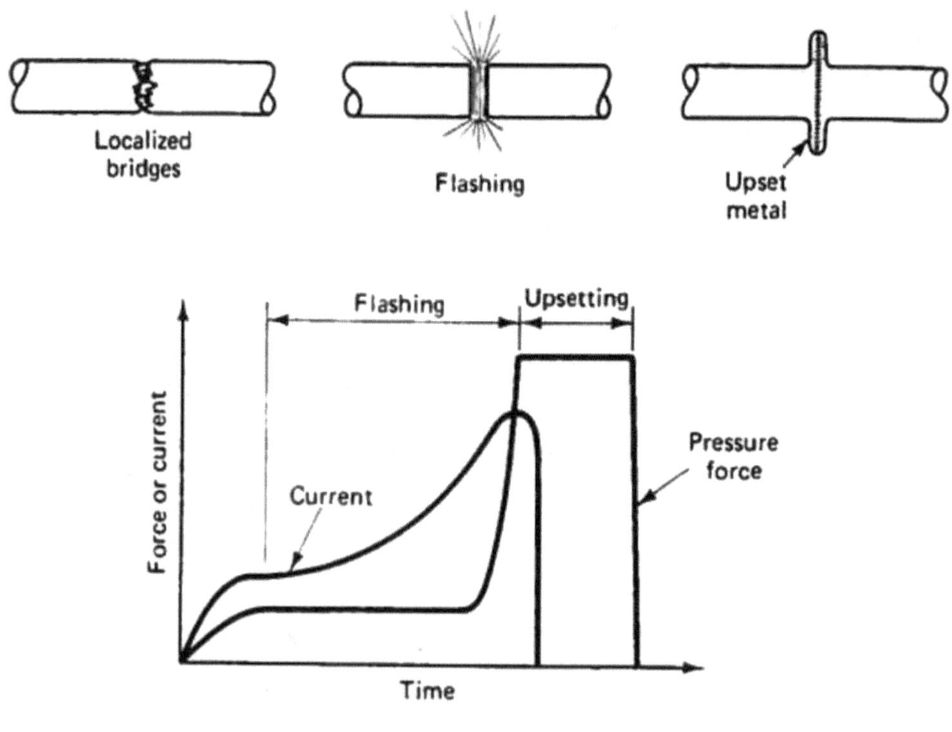

FIGURE 4.19 The different stages involved in a flash welding cycle.

weld zone, causing "flashing." New bridges are formed and move quickly across the whole interface, resulting in uniform heating all over. When the whole contact area is heated above the liquidus line, the electric current is switched off, and the pressure is suddenly increased to squeeze out the molten metal, upset the abutted parts, and weld them together.

Flash welding is used for joining large sections, rails, chain links, tools, thin-walled tubes, and the like. It can also be employed for welding dissimilar metals. The advantages claimed for the process include its higher productivity and its ability to produce high-quality welds. The only disadvantage is the loss of some metal in flashing.

4.2.1.11 Percussion Welding

In percussion welding, a method of resistance welding, a high-intensity electric current is discharged between the parts before they are brought into solid contact. This results in an electric arc in the gap between the two surfaces. That electric arc lasts only for about 0.001 seconds and is enough to melt the surfaces to a depth of a few thousandths of an inch. The two parts are then impacted against each other at a high speed to obtain a sound joint. The major limitation of this process is the cross-sectional area of the welded joint. It should not exceed 0.5 square inches in order to keep the intensity of current required at a practical level. In industry, percussion welding is limited to joining dissimilar metals that cannot be welded otherwise.

4.2.1.12 Spot Welding

Figure 4.20a illustrates the principles of operation of spot welding, a resistance welding process. Electric current is switched on between the welding electrodes to flow through the lapped sheets (workpieces) that are held together under pressure. As can be seen in Figure 4.20b, the metal fuses in the central area of the interface between the two sheets and then solidifies in the form of a nugget, thus providing the weld joint. Heat is also generated at the contact areas between the electrodes and

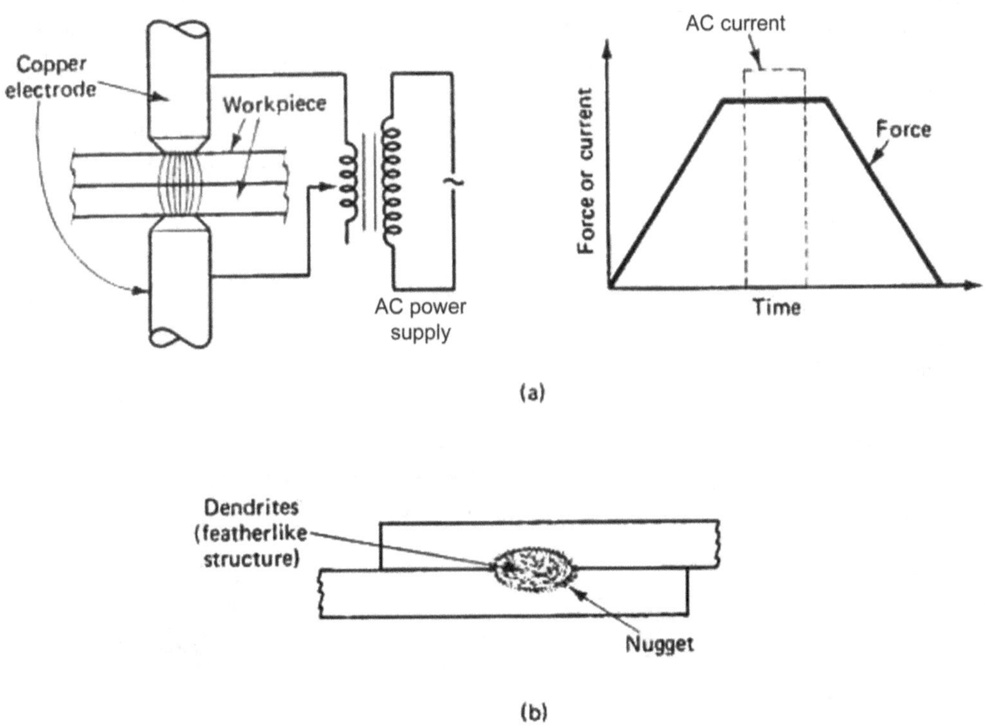

FIGURE 4.20 Resistance spot welding: (a) principles of operation and (b) a cross section through a spot weld.

the workpieces. Therefore, some precautions must be taken to prevent excessive temperatures and fusing of the metal at those spots. The electrodes used must possess good electrical and thermal conductivities. They are usually hollow and are water-cooled. In addition, areas of workpieces in contact with the electrodes must be cleaned carefully.

Spot welding is the most widely used resistance welding process in the industry. Carbon steel sheets having a thickness of up to 0.125 inches (4 mm) can be successfully spot-welded. Spot welding machines have ratings up to more than 600 kVA and use a voltage of 1–12 V obtained from a step-down transformer. Multispot machines are used, and the process can be fully automated. Therefore, spot welding has found widespread application in the automobile, aircraft, and electronics industries, as well as in sheet metal work.

4.2.1.13 Seam Welding

Seam welding and projection welding are modifications of spot welding. In seam welding, the lapped sheets are passed between rotating circular electrodes through which the high-amperage current flows, as shown in Figure 4.21. Electrodes vary in diameter from less than 2 up to 14 inches (40–350 mm), depending upon the curvature of the workpieces being welded. Welding current as high as 5000 A may be employed, and the pressing force acting upon the electrodes can go up to 6 kN (more than half a ton). A welding speed of about 12 feet per minute (4 m/min) is quite common. Seam welding is employed in the production of pressure-tight joints used in containers, tubes, mufflers, and the like. Advantages of this process include low cost, high production rates, and suitability for automation. Nevertheless, the thickness of the sheets to be seam welded is limited to 0.125 inch (4 mm) in the case of carbon steels and much less for more conductive alloys due to the extremely high amperage required (0.125-inch-thick steel sheets require 19,000 A, whereas aluminum sheets having the same thickness would require 76,000 A).

4.2.1.14 Projection Welding

In projection welding, one of the workpieces is purposely provided with small projections so that current flow and heating are localized at those spots. The projections are usually produced by die pressing, and the process calls for the use of a special upper electrode. Figure 4.22 illustrates an arrangement of two parts to be projection welded, as well as the resulting weld nugget. As you may expect, the projections collapse under the externally applied force after sufficient heating, thus yielding a well-defined, fused weld nugget. When the current is switched off, the weld cools down

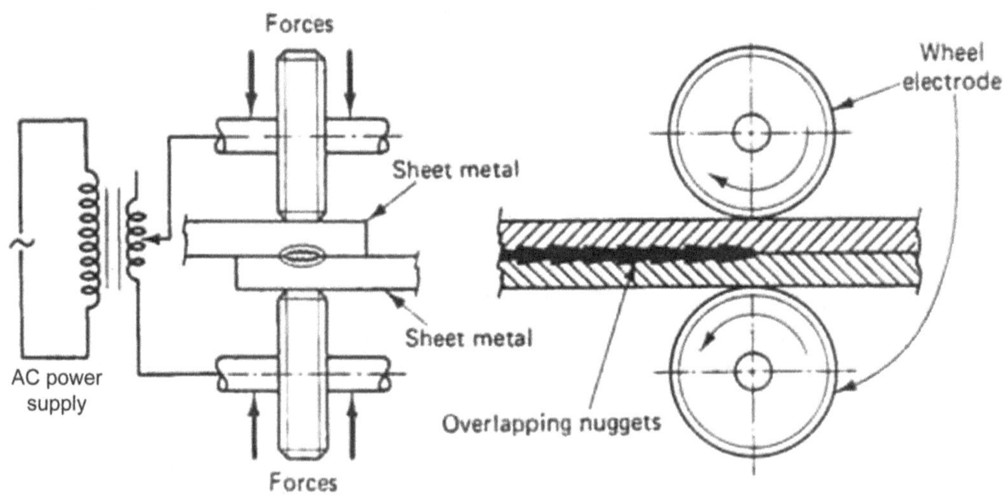

FIGURE 4.21 Principles of seam welding.

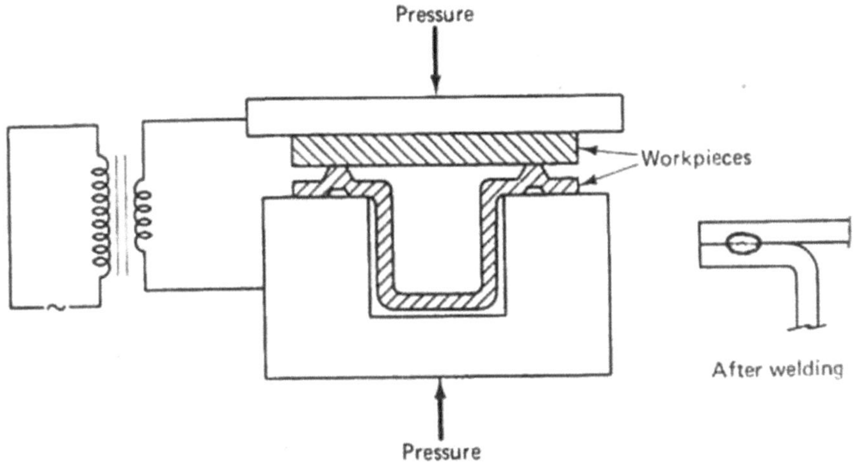

FIGURE 4.22 An arrangement for projection welding two parts.

and solidification takes place under the applied force. The electrode force is then released, and the welded workpiece is removed. As is the case with spot welding, the entire projection welding process takes only a fraction of a second.

Projection welding has some advantages over conventional spot welding. For instance, sheets that are too thick to be joined by spot welding can be welded using this process. Also, the presence of grease, dirt, or oxide films on the surface of the workpieces has less effect on the weld quality than in the case of spot welding. A further advantage of projection welding is the accuracy of locating welds inherent in that process.

4.2.2 Fusion Welding Processes

Fusion welding includes a group of processes that all produce welded joints as a result of localized heating of the edges of the base metal above its melting temperature, wherein coalescence is produced. A filler metal may or may not be added, and no external pressure is required. The welded joint is obtained after the solidification of the fused weld pool.

In order to join two different metals together by fusion welding, they must possess some degree of mutual solubility in the solid state. In fact, metals that are completely soluble in the solid state exhibit the highest degree of weldability. Metals with limited solid solubility have lower weldability, and metals that are mutually insoluble in the solid state are completely unweldable by any of the fusion welding methods. In that case, an appropriate pressure welding technique should be employed. An alternative solution is to employ an intermediate metal that is soluble in both base metals.

Before surveying the different fusion welding processes, let us discuss the metallurgy of fusion welding. Important microstructural changes take place in and around the weld zone during and after the welding operation. Such changes in the microstructure determine the mechanical properties of the welded joint. Therefore, a study of the metallurgy of fusion welding is essential for the good design of welded joints, as well as for the optimization of the process parameters.

During fusion welding, three zones can be identified, as shown in Figure 4.23, which indicates a single V-weld in steel after solidification and the corresponding temperature distribution during welding. In the first zone, called *the fusion zone*, the base metal and deposited metal (if a filler rod is used) are brought to the molten state during welding. Therefore, when this zone solidifies after welding, it generally has a columnar dendritic structure with haphazardly oriented grains. In other words, the microstructure of this zone is quite similar to that of the cast metal. Nevertheless, if the molten metal is overheated during welding, this results in an acicular structure that is brittle, has low

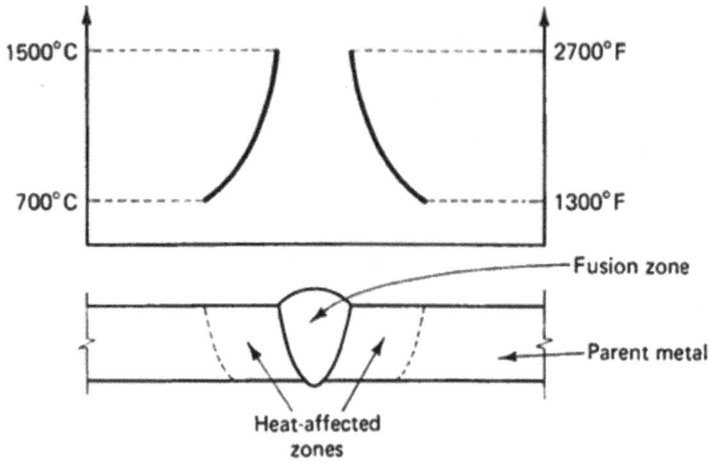

FIGURE 4.23 The three zones in a fusion-welded joint and the temperature distribution during welding.

strength, and is referred to as the *Widmanstätten structure*. Also, the chemical composition of the fusion zone may change, depending on the kind and amount of filler metal added.

The second zone, which is referred to as the *heat-affected zone*, is that portion of the base metal that has not been melted. Therefore, its chemical composition before and after welding remains unchanged. Nevertheless, its microstructure is always altered because of the rapid heating during welding and subsequent cooling. In fact, the HAZ is subjected to a normalizing operation during welding and may consequently undergo phase transformations and precipitation reactions, depending upon the nature (chemical composition and microstructure) of the base metal. The size of the HAZ is dependent upon the welding method employed and the nature of the base metal. This can be exemplified by the fact that the HAZ is 0.1 inch (2.5 mm) when automatic submerged arc welding is used, ranges from 0.2–0.4 inches (5–10 mm) for shielded-metal arc welding, and may reach 1.0 inch (25 mm) in conventional gas welding. This evidently affects the microstructure of the weld, which is generally fine-grained. The effect of these structural changes on the mechanical properties of the weld differs for different base metals. For instance, the structural changes have a negligible effect on the mechanical properties of low-carbon steel, regardless of the welding method used. On the contrary, when welding high-carbon alloy steel, hardened structures like martensite are formed in the HAZ of the weld, resulting in a sharp reduction in the ductility of the welded joint and/or crack formation (remember the effect of alloying elements on the critical cooling rate in the TTT diagram that you studied in metallurgy).

The third zone involves the unaffected *parent metal* adjacent to the HAZ that is subjected to a temperature below AC_3 (a critical temperature) during welding. In this zone, no structural changes take place unless the base metal has been subjected to plastic deformation prior to welding, in which case recrystallization and grain growth would become evident.

4.2.2.1 Arc Welding

Arc welding is based on the thermal effect of an electric arc that acts as a powerful heat source to produce localized melting of the base metal. The electric arc is, in fact, a sustained electrical discharge (of electrons and ions in opposite directions) through an ionized, gaseous path between two electrodes (i.e., the anode and the cathode). In order to ionize the air in the gap between the electrodes so that the electric arc can consequently be started, a certain voltage is required. (The voltage required depends upon the distance between the electrodes.) The ionization process results in the generation of electrons and positively charged ions. Next, the electrons impact on the anode, and the positively charged ions impact on the cathode. The collisions of these particles, which are

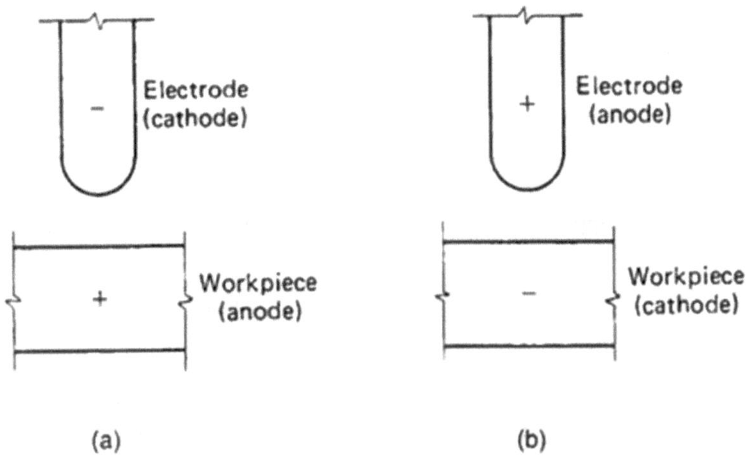

FIGURE 4.24 Straight and reversed polarity in direct current (DC) arc welding: (a) DC straight polarity (DCSP) and (b) DC reversed polarity (DCRP).

accelerated by the arc voltage, transform the kinetic energy of the particles into thermal and luminous energy, and the temperature at the center of the arc can reach as high as 11,000°F (6000°C). Actually, only a comparatively low potential difference between the electrodes is required to start the arc. For instance, about 45 V is usually sufficient for direct current (DC) welding equipment and up to 60 V for an alternating current (AC) welder. Also, the voltage drops after the arc is started, and a stable arc can then be maintained with a voltage in the range of 15–30 V. Generally, arc welding involves using a metal electrode rod and attaching the other electrode to the workpiece. The electrode rod either melts during the process (consumable electrode) and provides the necessary filler metal for the weld, or the electrode does not melt and the filler metal is provided separately.

As just mentioned, either alternating current or direct current can be used in arc welding, although each has its distinct advantages. While arc stability is much better with alternating current than with direct current, AC welding equipment is far less expensive, more compact in size, and simpler to operate. A further advantage of AC arc welding is the high efficiency of the transformer used, which goes up to 85 percent, whereas the efficiency of DC welding systems usually varies between only 30 and 60 percent.

In DC arc welding, the degree to which the work is heated can be regulated by using either straight or reversed polarity. As can be seen in Figure 4.24a,b, the cathode is the electrode rod and the anode is the workpiece in direct current straight polarity (DCSP), whereas it is the other way around in direct current reversed polarity (DCRP). When using DCSP, more heat is concentrated at the cathode (the electrode rod) than at the anode (the workpiece). Therefore, melting and deposition rates (of consumable electrodes) are high, but penetration in the workpiece is shallow and narrow. Consequently, DCSP is recommended when welding sheet metal, especially at higher welding speeds. With DCRP, heat is concentrated at the cathode (the workpiece) and results in deeper penetration for a given welding condition. It is, therefore, preferred for groove welds and similar applications.

During the welding operation, heat is generated in the transformer as well as in other elements of the welding circuit, resulting in a temperature rise that may cause damage to those elements. There is, therefore, a time limitation when using the welding equipment at a given amperage. That time limitation is usually referred to as the *rated duty cycle*. Consider the following numerical example. A power supply for arc welding rated at a 150-A 40 percent duty cycle means that it can be used only 40 percent of the time when welding at 150 A. The idle or unused time is required to allow the equipment to cool down. The percentage of duty cycles at currents other than the rated current can be calculated using the following equation:

$$\% \text{ duty cycle} = \left(\frac{\text{rated current}}{\text{load current}}\right)^2 \times \text{rated duty cycle} \tag{4.2}$$

Therefore, for this power supply, the percentage of the duty cycle at 100 A is as follows:

$$\text{Duty cycle at } 100\text{A} = \left(\frac{150}{100}\right)^2 \times 40 = 90$$

In addition, there is another welding process, electroslag welding (EW), which is not based on the phenomenon of the electric arc but, nevertheless, employs equipment similar to that used in gas–metal arc, flux-cored arc, or submerged arc welding.

4.2.2.1.1 Shielded-Metal Arc Welding

Shielded-metal arc welding is a manual arc welding process that is sometimes referred to as *stick welding*. The source of heat for welding is an electric arc maintained between a flux-covered, consumable metal electrode and the workpiece. As can be seen in Figure 4.25, which indicates the operating principles of this process, shielding of the electrode tip, weld puddle, and weld area in the base metal are ensured through the decomposition of the flux covering. A blanket of molten slag also provides shielding for the molten-metal pool. The filler metal is provided mainly by the metal core of the electrode rod. Shielded-metal arc welding can be used for joining thin and thick sheets of plain-carbon steels, low-alloy steels, and even some alloy steels and cast iron, provided that the electrode is properly selected and also that preheating and postheating treatments are performed. It is actually the most commonly used welding process and has found widespread application in steel construction and shipbuilding. Nevertheless, it is uneconomical and/or impossible to employ shielded-metal arc welding to join some alloys, such as aluminum alloys, copper, nickel, copper–nickel alloys, and low-melting-point alloys such as zinc, tin, and magnesium alloys. Another clear shortcoming of the process is that welding must be stopped each time an electrode stick is consumed to allow mounting a new one. This results in idle time and, consequently, a drop in productivity. The core wires of electrodes used for shielded-metal arc welding have many different compositions. The selection of a particular electrode material depends upon the application for which it is going to be used and the kind of base metal to be welded. Consumable electrodes are usually coated with flux but can also be uncoated. The metal wire can have a diameter of up to 15/32 inch (12 mm) and a length of about 18 inches (450 mm). Although various metals are used as wire materials, by far the

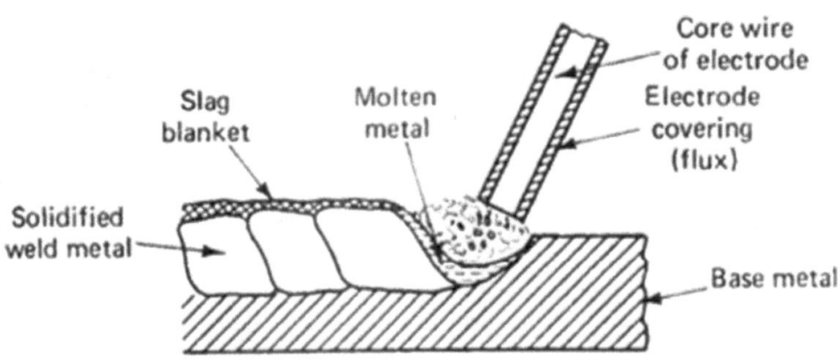

FIGURE 4.25 Operating principles of shielded-metal arc welding.

TABLE 4.1
The Constituents of Typical Electrode Coatings and Their Functions

Main Function	Constituent	Percentage
Gas generating	Starch, cellulose calcium carbonate	25–40
Slag forming	Kaolin, titanium dioxide, feldspar, asbestos	20–40
Binding	Sodium silicate, potassium silicate	20–30
Deoxidizing	Ferrosilicon, aluminum	5–10
Arc stabilizing	Potassium titanate, titanium oxide	5–10
Increasing deposition rate	Iron powder	0–40
Improving weld strength	Different alloying elements	5–10

most commonly used electrode materials involve low-carbon steel (for welding carbon steels) and low-alloy steel (for welding alloy steels). Electrodes can be bare, lightly coated, or heavily coated. The electrode covering, or coating, results in better-quality welds as it improves arc stability, produces gas shielding to prevent oxidation and nitrogen contamination and also provides slag, which, in turn, retards the cooling rate of the weld's fusion zone. Therefore, electrode coatings are composed of substances that serve these purposes. Table 4.1 indicates the composition of typical electrode coatings, together with the function of each constituent.

4.2.2.1.2 Carbon Arc Welding

In carbon arc welding (CAW), nonconsumable electrodes made of carbon or graphite are used. Only a DC power supply can be employed, and the electric arc is established either between a single carbon electrode and the workpiece (Bernardos method) or between two carbon electrodes (independent arc method). In both cases, no shielding is provided. A filler metal may be used, especially when welding sheets with thicknesses of more than 1/8 inch (3 mm). The carbon electrodes have diameters ranging from 3/8 to 1 inch (10–25 mm) and are used with currents that range from 200 to 600 A.

Carbon arc welding is not commonly used in industry. Its application is limited to the joining of thin sheets of nonferrous metals and to brazing.

4.2.2.1.3 Flux-Cored Arc Welding

Flux-cored arc welding is an arc welding process in which the consumable electrode takes the form of a tubular, flux-filled wire that is continuously fed from a spool. Shielding is usually provided by the gases evolving during the combustion and decomposition of the flux contained within the tubular wire. The process is, therefore, sometimes called *inner-shielded*, or *self-shielded*, arc welding. Additional shielding may be acquired through the use of an auxiliary shielding gas, such as carbon dioxide, argon, or both. In the latter case, the process is a combination of the conventional flux-cored arc welding and the gas–metal arc welding methods and is referred to as *dual-shielded* arc welding.

Flux-cored arc welding is generally applied on a semiautomatic basis, but it can also be fully automated. In that case, the process is normally used to weld medium-to-thick steel plates and stainless steel sheets. Figure 4.26 illustrates the operating principles of flux-cored arc welding.

4.2.2.1.4 Stud Arc Welding

Stud arc welding is a special-purpose arc welding process by which studs are welded to flat surfaces. This facilitates fastening and handling of the components to which studs are joined and meanwhile eliminates the drilling and tapping operations that would have been required to achieve the same goal. Only DC power supplies are employed, and the process also calls for the use of a special

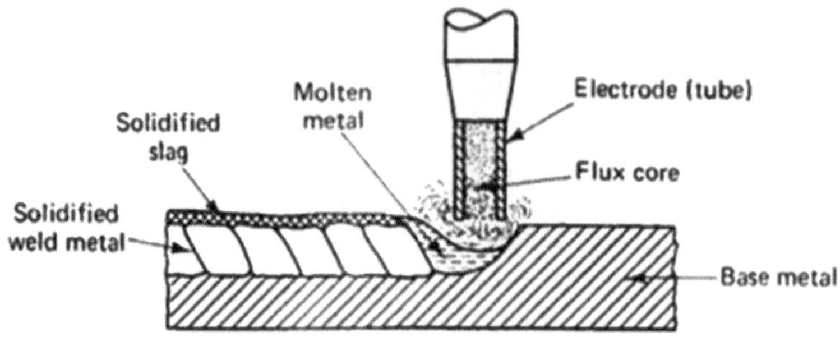

FIGURE 4.26 The operating principles of flux-cored arc welding.

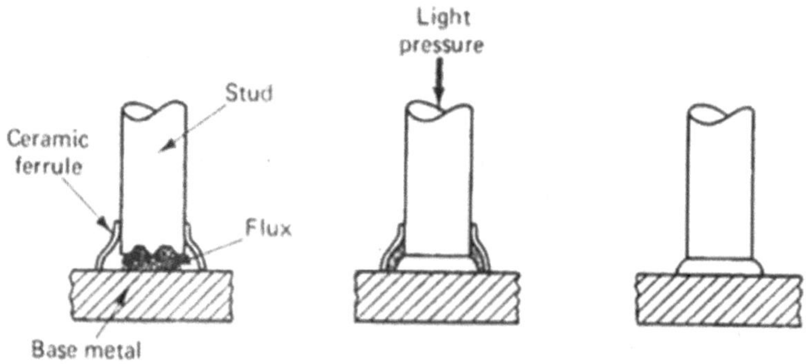

FIGURE 4.27 Stages involved in stud arc welding.

welding gun that holds the stud during welding. Figure 4.27 shows the stages involved in stud arc welding, a process that is entirely controlled by the timer of the gun. As can be seen in the figure, shielding is accomplished through the use of a ceramic ferrule that surrounds the end of the stud during the process. Stud arc welding requires a low degree of welding skill, and the whole welding cycle usually takes less than a second.

4.2.2.1.5 Submerged Arc Welding

Submerged arc welding is a fairly new automatic arc welding method in which the arc and the weld area are shielded by a blanket of fusible granular flux. A bare electrode is used and is continuously fed by a special mechanism during welding. Figure 4.28 shows the operating principles of submerged arc welding. As can be seen from the figure, the process is used to join flat plates in the horizontal position only. This limitation is imposed by the nature of the flux and the way it is fed.

As is the case with previously discussed arc welding processes, gases evolve as a result of combustion and decomposition of the flux, due to the high temperature of the arc, and form a pocket, or gas bubble, around the arc. As Figure 4.29 shows, this gas bubble is sealed from the arc by a layer of molten flux. This isolates the arc from the surrounding atmosphere and, therefore, ensures proper shielding. The melting temperature of the flux must be lower than that of the base metal. As a result, the flux always solidifies after the metal, thus forming an insulating layer over the solidifying molten-metal pool. This retards the solidification of the fused metal and, therefore, allows the slag and nonmetallic inclusions to float off the molten pool. The final outcome is always a weld that is free of nonmetallic inclusions and entrapped gases and has a homogeneous chemical composition. The flux should also be selected to ensure proper deoxidizing of the fused metal and should contain additives that make up for the elements burned and lost during the welding process. Electric currents commonly used with submerged arc welding range between 3000 and 4000 A. Consequently, the arc

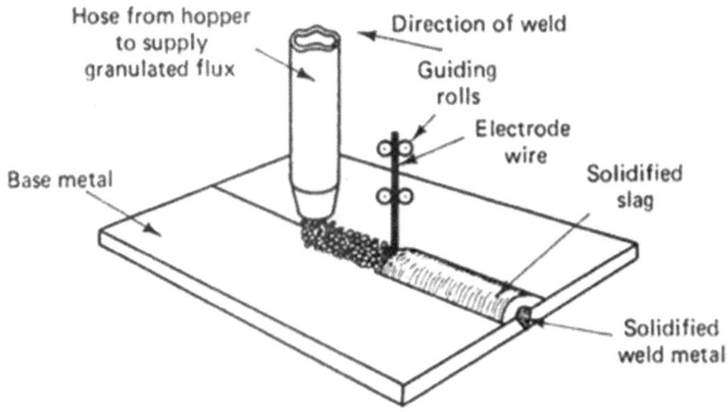

FIGURE 4.28 The operating principles of submerged arc welding.

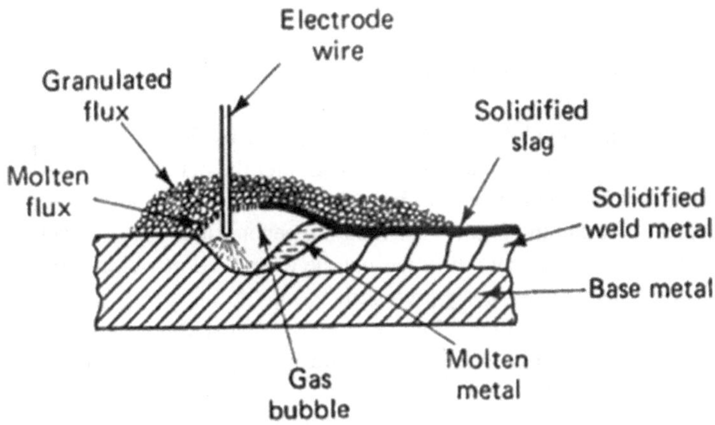

FIGURE 4.29 The mechanics of shielding in submerged arc welding.

obtained is extremely powerful and is capable of producing a large molten-metal pool as well as achieving deeper penetration. Other advantages of this process include its high welding rate, which is five to ten times that produced by shielded-metal arc welding, and the high quality of the welds obtained.

4.2.2.1.6 Gas–Metal Arc Welding

The gas–metal arc welding process is commonly called *metal–inert-gas* (MIG) welding. It employs an electric arc between a solid, continuous, consumable electrode and the workpiece. As can be seen in Figure 4.30, shielding is obtained by pumping a stream of chemically inert gas, such as argon or helium, around the arc to prevent the surrounding atmosphere from contaminating the molten metal. (The electrode is bare, and no flux is added.) Dry carbon dioxide can sometimes be employed as a shielding gas, yielding fairly good results. Gas–metal arc welding is generally a semiautomatic process. However, it can also be applied automatically by machine. In fact, welding robots and numerically controlled MIG welding machines have gained widespread industrial application. The gas–metal arc welding process can be used to weld thin sheets as well as relatively thick plates in all positions, and the process is particularly popular when welding nonferrous metals such as aluminum, magnesium, and titanium alloys. The process is also used for welding stainless steel and critical steel parts.

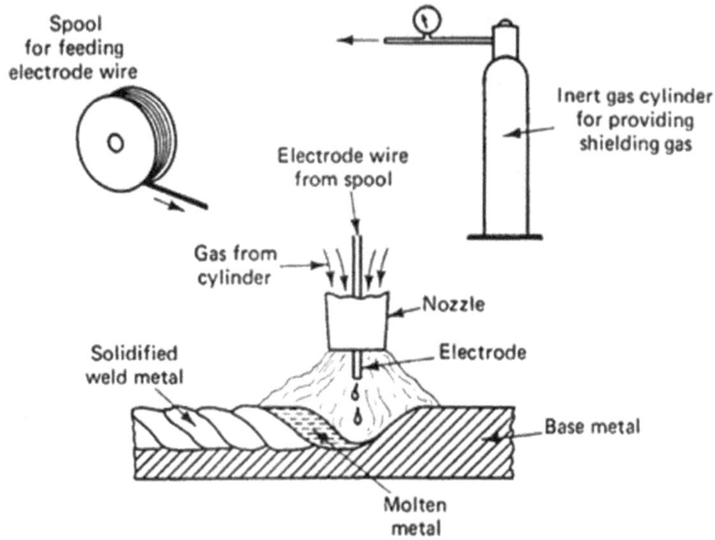

FIGURE 4.30 Operating principles of gas–metal arc welding.

The penetration for gas–metal arc welding is controlled by adopting DCRP and adjusting the current density. The higher the current density is, the greater the penetration is. The kind of shielding gas used also has some effect on the penetration. For instance, helium gives the maximum penetration; carbon dioxide, the least; and argon, intermediate penetration. Thus, it is clear that higher current densities and the appropriate shielding gas can be employed in welding thick plates, provided that the edges of these plates are properly prepared.

The electrode wires used for MIG welding must possess close dimensional tolerances and a consistent chemical composition appropriate for the desired application. The wire diameter varies between 0.02 and 0.125 inches (0.5 and 3 mm). Usually, MIG wire electrodes are coated with a very thin layer of copper to protect them during storage. The electrode wire is available in the form of a spool weighing from 2.5 to 750 lb (1–300 kg). As you may expect, the selection of the composition of the electrode wire for a given material depends upon other factors, such as the kind of shielding gas used, the conditions of the metal being welded (i.e., whether there is an oxide film, grease, or contaminants), and, finally, the required properties of the weldment.

4.2.2.1.7 Gas–Tungsten Arc Welding

Gas–tungsten arc welding, which is usually called *Tungsten–inert-gas* (TIG) welding, is an arc welding process that employs the heat generated by an electric arc between a nonconsumable tungsten electrode and the workpiece. Figure 4.31 illustrates the operating principles of this process. As can be seen, a filler rod may (or may not) be fed to the arc zone. The electrode, arc, weld puddle, and adjacent areas of the base metal are shielded by a stream of either argon or helium to prevent any contamination from the atmosphere. TIG welding is normally applied manually and requires a relatively high degree of welder skill. It can also be fully automated, in which case the equipment used drives the welding torch at a preprogrammed path and speed, adjusts the arc voltage, and starts and stops it.

Gas–tungsten arc welding is capable of welding nonferrous and exotic metals in all positions. The list of metals that can be readily welded by this process is long and includes alloy steels, stainless steels, heat-resisting alloys, refractory metals, aluminum alloys, magnesium alloys, titanium alloys, copper and nickel alloys, and steel coated with low-melting-point alloys. The process is recommended for welding very thin sheets, as thin as 0.005 inch (about 0.125 mm), for the root and

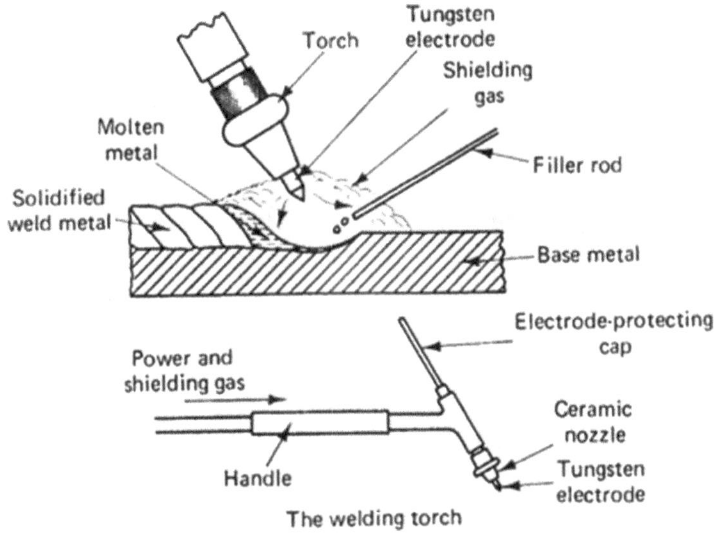

FIGURE 4.31 Operating principles of gas tungsten arc welding.

hot pass on tubing and pipes, and wherever smooth, clean welds are required (e.g., in food-processing equipment). Ultra-high-quality welds can be obtained in the nuclear, rocketry, and submarine industries by employing a modified version of TIG welding that involves placing carefully selected and prepared inserts in the gap between the sections to be joined and then completely fusing the inserts together with the edges of base metal using a TIG torch.

All three types of current supplies (i.e., AC, DCSP, and DCRP) can be used with gas–tungsten arc welding, depending upon the metal to be welded. Thin sheets of aluminum or magnesium alloys are best welded by using DCRP, which prevents burns-through, as previously explained. Nevertheless, it is recommended that an AC power supply be used when welding normal sheets of aluminum and magnesium. DCSP is best suited for welding high-melting-point alloys such as alloy steels, stainless steels, heat-resisting alloys, copper alloys, nickel alloys, and titanium. In addition to these considerations, DCRP is also helpful in removing surface oxide films due to its cleaning action (the impacting of ions onto the surface like a grit blasting).

4.2.2.1.8 Plasma Arc Welding

Figure 4.32 is a sketch of the torch employed in plasma arc welding (PAW). The electric arc can take either of two forms: a transferred arc that is a constricted arc between a tungsten electrode and the workpiece or a nontransferred arc between the electrode and the constricting nozzle. The gas flowing around the arc heats up to extremely high temperatures like 60,000°F (33,000°C) and becomes, therefore, ionized and electrically conductive; it is then referred to as *plasma*. The main shielding is obtained from the hot ionized gas emerging from the nozzle. Additional inert gas shielding can be used when high-quality welds are required. In fact, plasma arc welding can be employed to join almost all metals in all positions, although it is usually applied to thinner metals. Generally, the process is applied manually and requires some degree of welder skill; however, the process is sometimes automated in order to increase productivity.

4.2.2.1.9 Electroslag Welding

Electroslag welding, which was developed by the Russians, is not an arc welding process but requires the use of equipment similar to that used in arc welding. Although an electric arc is used to start the process, heat is continuously generated as a result of the current flow between the electrode

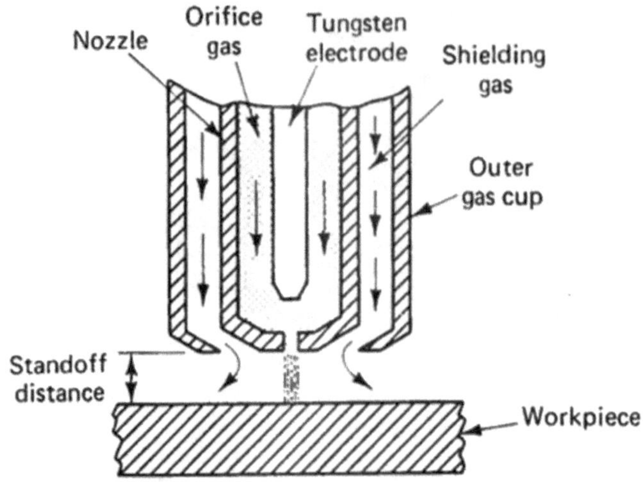

FIGURE 4.32 A sketch of the torch employed in plasma arc welding.

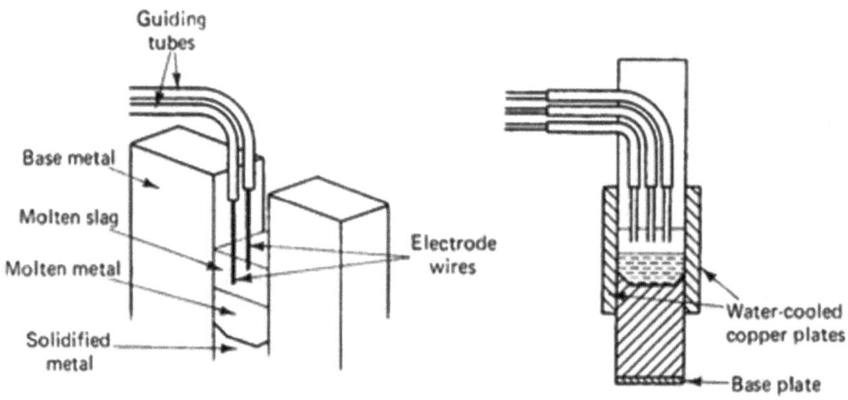

FIGURE 4.33 The electroslag welding process.

(or electrodes) and the base metal through a pool of molten slag (flux). As we will see later, the molten-slag pool also serves as a protective cover for the fused-metal pool.

The electroslag welding process is shown in Figure 4.33. As can be seen in the figure, the parts to be joined are set in the vertical position, with a gap of 1/2–1.5 inches (12–37 mm) between their edges. (The gap is dependent upon the thickness of the parts.) The welding electrode (or electrodes) and the flux are fed automatically into the gap, and an arc is established between the electrodes and the steel backing plate to provide the initial molten-metal and slag pools. Next, the electrical resistivity of the molten slag continuously produces the heat necessary to fuse the flux, the filler, and the base metals. Water-cooled copper plates travel upward along the joint, thus serving as dams and cooling the fused metal in the cavity to form the weld.

Electroslag welding is very advantageous in joining very thick parts together in a single pass without any need for beveling the edges of those parts. Therefore, the process is widely used in industries that fabricate beds and frames for heavy machinery, drums, boilers, and the like.

4.2.2.2 Gas Welding

Gas welding refers to a group of oxyfuel gas processes in which the edges of the parts to be welded are fused together by heating them with a flame obtained from the combustion of a gas (such as acetylene) in a stream of oxygen. A filler metal is often introduced into the flame to melt and, together with the base metal, form the weld puddle. Gas welding is usually applied manually and requires good welding skills. Common industrial applications involve welding thin-to-medium sheets and sections of steel and nonferrous metals in all positions. Gas welding is also widely used in repair work and in restoring cracked or broken components.

The fuel gases used for producing the flame during the different gas welding processes include acetylene, hydrogen, natural gas (94 percent methane), petroleum gas, and vaporized gasoline and kerosene. However, acetylene is the most commonly used gas for gas welding because it can provide a flame temperature of about 5700°F (3 150°C). Unfortunately, acetylene is ignited at a temperature as low as 790°F (420°C) and becomes explosive in nature at pressures exceeding 1.75 atmospheres. Therefore, it is stored in metal cylinders, in which it is dissolved in acetone under a pressure of about 19 atmospheres. For more safety, acetylene cylinders are also filled with a porous filler (such as charcoal) in order to form a system of capillary vessels that are then saturated with the solution of acetylene in acetone.

The oxygen required for the gas welding process is stored in steel cylinders in a liquid state under a pressure of about 150 atmospheres. It is usually prepared in special plants by liquefying air and then separating the oxygen from the nitrogen.

The equipment required in gas welding, as shown in Figure 4.34, includes oxygen and acetylene cylinders, regulators, and the welding torch. The regulators serve to reduce the pressure of the gas in the cylinder to the desired working value and keep it that way throughout the welding process. Thus, the proportion of the two gases is controlled, which determines the characteristics of the flame. Next, the welding torch serves to mix the oxygen and the acetylene together and discharges the mixture out at the tip, where combustion takes place.

Depending upon the ratio of oxygen to acetylene, three types of flames can be obtained: neutral, reducing, and oxidizing. Figure 4.35 is a sketch of a typical oxyacetylene welding flame. As can be seen, the welding flame consists of three zones: the inner luminous cone at the tip of the torch, the reducing zone, and the oxidizing zone. The first zone, the *luminous cone*, consists of partially decomposed acetylene as a result of the following reaction:

$$C_2H_2 \rightarrow 2C + H_2$$

The carbon particles obtained are incandescent and are responsible for the white luminescence of the brightest part of the flame. Those carbon particles are partly oxidized in the second zone, the *reducing zone*, yielding carbon monoxide and a large amount of heat that brings the temperature up to about 5400°F (3000°C). Gases like hydrogen and carbon monoxide are capable of reducing oxides. Next, complete combustion of those gases yields carbon dioxide and water vapor that, together with the excess oxygen (if any), result in the third zone, the *oxidizing zone*. Those gases,

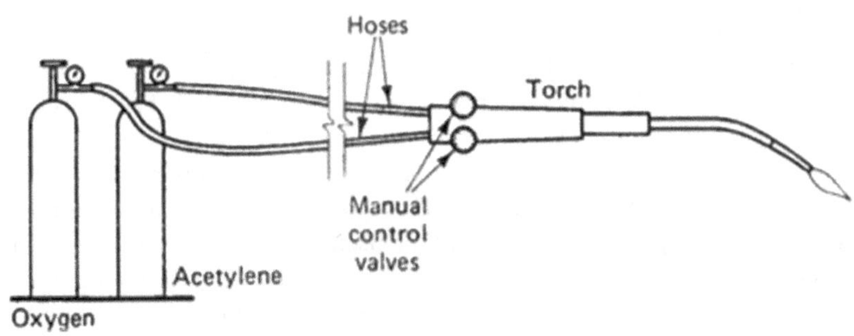

FIGURE 4.34 The equipment required in gas welding.

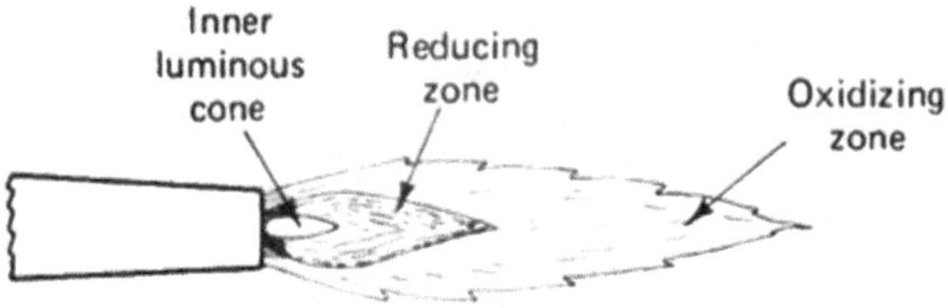

FIGURE 4.35 A sketch of a typical oxyacetylene welding flame.

however, form a shield that prevents the atmosphere from coming in contact with the molten-metal pool.

As can be expected, the extent (as well as the appearance) of each zone depends upon the type of flame (i.e., the oxygen-to-acetylene ratio). When the ratio is about 1, the flame is *neutral* and distinctively has the three zones just outlined. If the oxygen-to-acetylene ratio is less than 1, a reducing, or carbonizing, flame is obtained. In this case, the luminous cone is longer than that obtained with the neutral flame, and the outline of the flame is not sharp. This type of flame is employed in welding cast iron and in hard surfacing with HSS and cemented carbides. The third type of flame, the *oxidizing* flame, is obtained when the oxygen-to-acetylene ratio is higher than 1. In this case, the luminous cone is shorter than that obtained with the neutral flame, and the flame becomes light blue in color. The oxidizing flame is employed in welding brass, bronze, and other metals that have great affinity to hydrogen.

Another method that utilizes the heat generated as a result of the combustion of a fuel gas is known as pressure-gas welding. As the name suggests, this is actually a pressure welding process in which the abutting edges to be welded are heated with an oxyacetylene flame to attain a plastic state; then, coalescence is achieved by applying the appropriately high pressure. In order to ensure uniform heating of the sections, a multiple-flame torch that surrounds the sections is used. The shape of that torch is dependent upon the outer contour of the sections to be welded, and the torch is usually made to oscillate along its axis. Upsetting is accomplished by a special pressure mechanism. This method is sometimes used for joining pipeline mains, rails, and the like.

4.2.2.3 Electron-Beam Welding

Electron-beam welding was developed by Dr. Jacques Stohr (CEA-France, the Atomic Energy Commission) in 1957 to solve a problem in the manufacturing of fuel elements for atomic power generators. The process is based upon the conversion of the kinetic energy of a high-velocity, intense beam of electrons into thermal energy as the accelerated electrons impact the joint to be welded. The generated heat then fuses the interfacing surfaces and produces the desired coalescence.

Figure 4.36 shows the basic elements and working principles of an electron-beam welding system. The system consists of an electron-beam gun (simply an electron emitter such as a hot filament) that is electrically placed at a negative potential with respect to an anode and that, together with the workpiece, is earth-grounded. A focus coil (i.e., an electromagnetic lens) is located slightly below the anode in order to bring the electron beam into focus upon the work. This is achieved by adjusting the current of the focus coil. Additional electromagnetic coils are provided to deflect the beam from its neutral axis as required. Because the electrons impacting the work travel at an ultra-high velocity, the process should be carried out in a vacuum in order to eliminate any resistance to the traveling electrons. Pressures on the order of 10^{-4} torr (1 atmosphere = 760 torr) are commonly employed, although pressures up to almost atmospheric can be used. Nevertheless, it must be noted that the higher the pressure is, the wider and more dispersed the electron beam becomes, and the lower the energy density is. (Energy density is the number of kilowatts per unit area of the spot being welded.)

Electron-beam welding machines can be divided into two groups: low-voltage and high-voltage machines. Low-voltage machines are those operating at accelerating voltages up to 60 kV, whereas

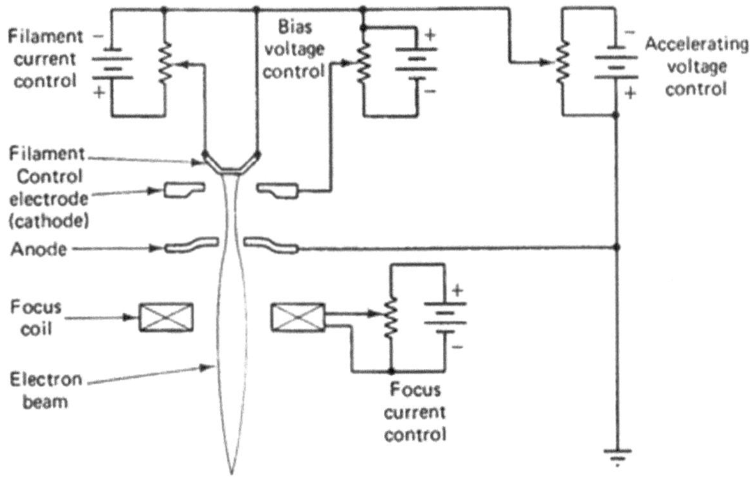

FIGURE 4.36 The basic elements and working principles of an electron-beam welding system.

high-voltage machines operate at voltages up to 200 kV. Although each of these two types has its own merits, the main consideration should be the beam-power density, which is, in turn, dependent upon the beam power and the (focused) spot size. In the early days of electron-beam welding, machines were usually built to have a rating of 7.5 kW or less. Today, a continuous-duty rating of 60 kW is quite common, and the trend is toward still higher ratings.

There are several advantages to the electron-beam welding process. They include the following five:

- Because of the high intensity of the electron beam used, the welds obtained are much narrower, and the penetration in a single pass is much greater than that obtained by conventional fusion welding processes.
- The high intensity of the electron beam can also develop and maintain a borehole in the workpiece, thus yielding a parallel-sided weld with a very narrow HAZ. As a consequence, the welds produced by this method have almost no distortion, have minimum shrinkage, and are stronger than welds produced by conventional fusion welding processes.
- Because this process obtains parallel-sided welds, there is no need for edge preparation of the workpieces (such as V- or J-grooves). Square butt-type joints are produced by electron-beam welding.
- High welding speeds can be obtained with this process. Speeds up to 200 inches per minute (0.09 m/s) are common, resulting in higher productivity.
- Because the process is usually performed in a vacuum chamber at pressures on the order of 10^{-4} torr, the resulting weld is excellent, is metallurgically clean, and has an extremely low level of atmospheric contamination. Therefore, electron-beam welding is especially attractive for joining refractory metals whose properties are detrimentally affected by even low levels of contamination.

Because of the ultra-high quality of the joints produced by electron-beam welding, the process has found widespread use in the atomic power, jet engine, aircraft, and aerospace industries. Nevertheless, the time required to vacuum the chamber before each welding operation results in reduced productivity, and, therefore, the high cost of the electron-beam welding equipment is not easily justified. This apparently kept the process from being applied in other industries until it was automated. Today, electron-beam welding is becoming popular for joining automotive parts such as gear clusters, valves, clutch plates, and transmission components.

4.2.2.4 Laser-Beam Welding

The term *laser* stands for light amplified by stimulated emission of radiation. It is, therefore, easy to see that a laser beam is actually a controlled, intense, highly collimated, and coherent beam of light. In fact, a laser beam proved to be a unique source of high-intensity energy that can be used in fusing metals to produce welded joints having very high strength.

Figure 4.37 shows the working principles of laser-beam welding. In this laser system, energy is pumped into a laser medium to cause it to fluoresce. This fluorescence, which has a single wavelength or color, is trapped in the laser medium (laser tube) between two mirrors. Consequently, it is reflected back and forth in an optical resonator path, resulting in more fluorescence, which amplifies the intensity of the light beam. The amplified light (i.e., the laser beam) finds its way out through the partly transparent mirror, which is called the *output mirror*. The laser medium can be a solid, such as a crystal made of yttrium aluminum garnet (YAG). It can also be a gas, such as carbon dioxide, helium, or neon. In the latter case, the pumping energy input is usually introduced directly by electric current flow.

Let us now consider the mechanics of laser-beam welding. The energy intensity of a laser beam is not high enough to fuse a metal such as steel or copper. Therefore, it must be focused by a highly transparent lens at a very tiny spot, 0.01 inch (0.25 mm) in diameter, in order to increase the intensity of energy up to a level of 10 million W converted into heat as it strikes the surface of a metal, causing instantaneous fusion of the metal at the focal point. Next, a cylindrical cavity, known as a *keyhole*, that is full of vaporized, ionized metallic gas is formed and is surrounded by a narrow region of molten metal. As the beam moves relative to the workpiece, the molten metal fills behind the keyhole and subsequently cools and solidifies to form the weld. It is worth mentioning that a stream of a cooling (and shielding) gas should surround the laser beam to protect the focusing lens from vaporized metal. Usually, argon is used for this purpose because of its low cost, although helium is actually the best cooling gas.

Despite the high initial capital cost required, laser-beam welding has gained widespread industrial application because of several advantages that the process possesses. Among these advantages are the following six:

- Based on the preceding discussion of the mechanics of laser-beam welding, we would always expect to have a very narrow HAZ with this welding method. Consequently, the chemical, physical, and mechanical properties of the base metal are not altered, thus eliminating the need for any postwelding heat treatment.

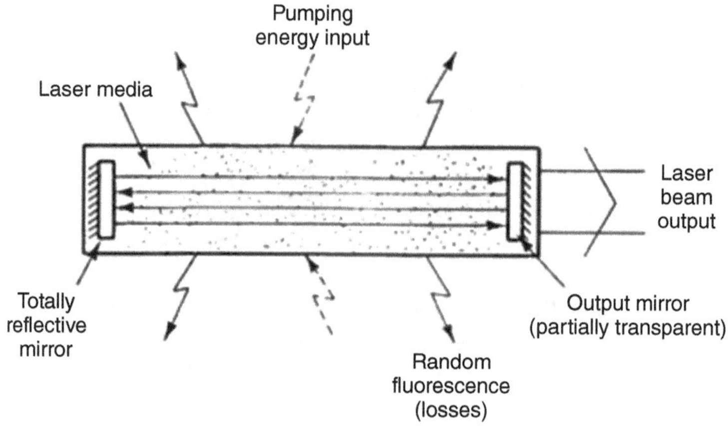

FIGURE 4.37 The working principles of laser-beam welding.

- The ultra-high intensity of energy of the laser beam at the focal point allows metals having high melting points (refractory metals) to be welded.
- The process can be successfully used to weld both nonconductive as well as magnetic materials that are almost impossible to join even with electron-beam welding.
- The laser beam can be focused into a chamber through highly transparent windows, thus rendering laser-beam welding suitable for joining radioactive materials and for welding under sterilized conditions.
- The process can be used for welding some materials that have always been considered unweldable.
- The process can be easily automated. Numerically controlled laser-beam welding systems are quite common and are capable of welding along a complex contour.

Since the Apollo project, laser-beam welding has become popular in the aerospace industry. Today, the process is mainly employed for joining exotic metals such as titanium, tantalum, zirconium, columbium, and tungsten. The process is especially advantageous for making miniature joints as in tiny pacemaker cans, integrated-circuit packs, camera parts, and batteries for digital watches. Nevertheless, laser-beam welding is not recommended for joining brass, zinc, silver, gold, or galvanized steel.

4.2.2.5 Welding Defects

In fusion welding processes, considerable thermal stresses develop during the heating and subsequent cooling of the workpiece, especially with those processes that result in large HAZs. Also, metallurgical changes and structural transformations take place in the weld puddle as well as in the HAZ, and these may be accompanied by changes in the volume. Therefore, if no precautions are taken, defects that are damaging to the function of the weldment may be generated. It is the combined duty of the manufacturing engineer, the welder, and the inspector to make sure that all weldments are free from all kinds of defects. Following is a brief survey of the common kinds of welding defects.

4.2.2.5.1 Distortion

Distortion, warping, and buckling of the welded parts are welding defects involving deformation (which can be plastic) of the structures as a result of residual stresses. They come as a result of restraining the free movement of some parts or members of the welded structure. They can also result from nonuniform expansion and shrinkage of the metal in the weld area as a consequence of uneven heating and cooling. Although it is possible to predict the magnitude of the residual stresses in some simple cases (e.g., butt welding of two plates), an analysis to predict the magnitude of these stresses and to eliminate distortion in the common case of a welded three-dimensional structure is extremely complicated. Nevertheless, here are some recommendations and guidelines to follow to eliminate distortion:

- Preheat the workpieces to a temperature dependent on the properties of the base metal in order to reduce the temperature gradient.
- Clamp the various elements (to be welded) in a specially designed rigid welding fixture. Although no distortion occurs with this method, there are always inherent internal stresses. The internal stresses can be eliminated by subsequent stress-relieving heat treatment.
- Sometimes, it is adequate just to tack-weld the elements securely in the right position (relative to each other) before an actual-strength welding arc is applied. It is also advisable to start by welding the section least subject to distortion first in order to form a rigid skeleton that contributes to the balance of assembly.
- Create a rational design of weldments (e.g., apply braces to sections most likely to distort).

4.2.2.5.2 Porosity

Porosity can take the form of elongated blowholes in the weld puddle, which is known as *wormhole porosity*, or of scattered tiny spherical holes. In both cases, porosity is due mainly to either the evolution of gases during welding or the release of gases during solidification as a result of their decreasing solubility in the solidifying metal. Excess sulfur or sulfide inclusions in steels are major contributors to porosity because they generate gases that are often entrapped in the molten metal. Other causes of porosity include the presence of hydrogen (remember the problem caused by hydrogen in casting), contamination of the joint, and contaminants in the flux. Porosity can be eliminated by maintaining clean workpiece surfaces, by properly conditioning the electrodes, by reducing welding speed, by eliminating any moisture on workpieces, and, most importantly, by avoiding the use of a base metal containing sulfur or electrodes with traces of hydrogen.

4.2.2.5.3 Cracks

Welding cracks can be divided into two main groups: fusion zone cracks and HAZ cracks. The first group includes longitudinal and transverse cracks as well as cracks appearing at the root of the weld bead. This type of cracking is sometimes called *hot cracking* because it occurs at elevated temperatures just after the molten metal starts to solidify. It is especially prevalent in ferrous alloys with high percentages of sulfur and phosphorus and in alloys having large solidification ranges.

The second type of cracking, HAZ cracks, is also called *cold cracking*. This defect is actually due to aggravation by excessive brittleness of the HAZ that can be caused by hydrogen embrittlement or by martensite formation as a result of rapid cooling, especially in high-carbon and alloy-steel welded joints. (Remember the effect of alloying elements on the TTT curve; they shift it to the right, thus decreasing the critical cooling rate.) Cold cracks can be eliminated by using a minimum potential source of hydrogen and by controlling the cooling rate of the welded joint to keep it at a minimum (e.g., keep joints in a furnace after welding or embed them in sand).

The use of multiple passes in welding can sometimes eliminate the need for prewelding or postwelding heat treatment. Each pass would provide a sort of preheating for the pass to follow. This technique is often effective in the prevention of weld cracks.

4.2.2.5.4 Slag Inclusions

Slag entrapment in the weld zone can occur in single-pass as well as in multipass welds. In single-pass arc welding, slag inclusions are caused by improper manipulation of the electrode and/or factors such as too high a viscosity of the molten metal or too rapid solidification. Some slag pushed ahead of the arc is drawn down by turbulence into the molten-metal pool, where it becomes entrapped in the solidifying weld metal. In multipass welds, slag inclusions are caused by improper removal of the slag blanket after each pass.

4.2.2.5.5 Lack of Fusion

Lack of fusion, shown in Figure 4.38, can result from a number of causes. These include inadequate energy input, which leads to insufficient temperature rise; improper electrode manipulation; and failure to remove oxide films and clean the weld area prior to welding.

4.2.2.5.6 Lack of Penetration

Lack of penetration, shown in Figure 4.39, is due to a low energy input, the wrong polarity, or a high welding speed.

4.2.2.5.7 Undercutting

Undercutting, shown in Figure 4.40, is a result of a high-energy input (excessive current in arc welding), which, in turn, causes the formation of a recess. As we know, such sharp changes in the weld contour act as stress raisers and often cause premature failure.

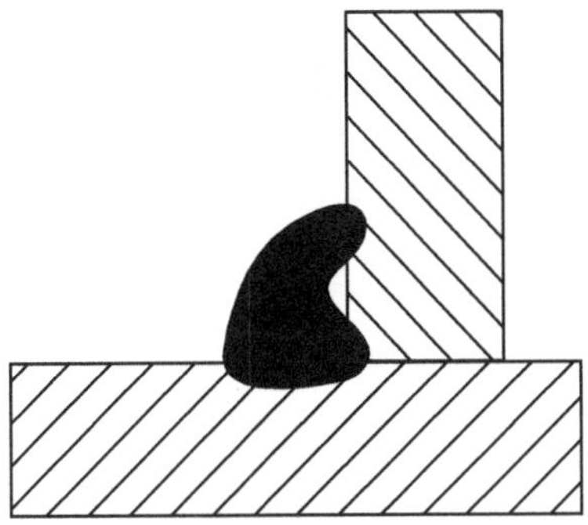

FIGURE 4.38 Lack of fusion.

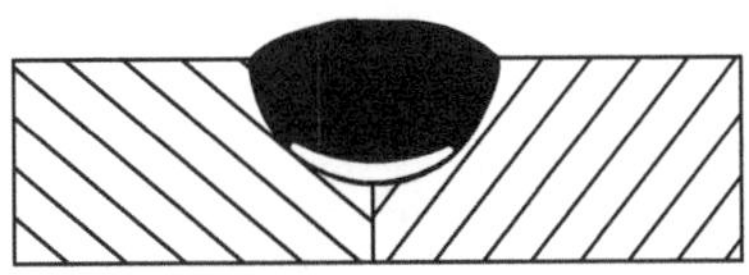

FIGURE 4.39 Lack of penetration.

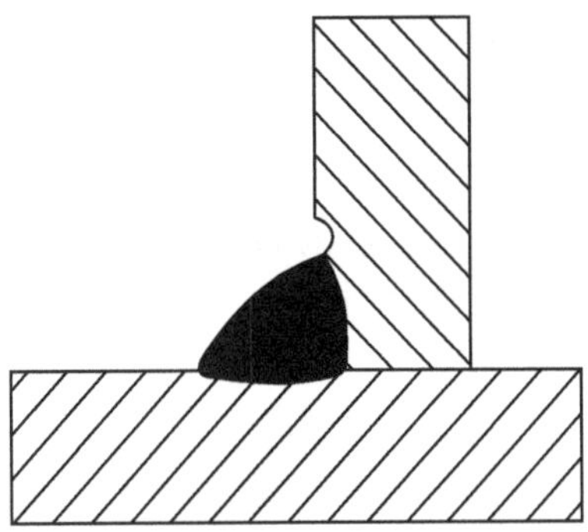

FIGURE 4.40 Undercutting.

4.2.2.5.8 Underfilling

Underfilling, shown in Figure 4.41, involves a depression in the weld face below the surface of the adjoining base metal. More filler metal has to be added in order to prevent this defect.

4.2.2.6 Testing and Inspection of Welds

Welds must be evaluated by being subjected to testing according to codes and specifications that are different for different countries. The various types of tests can be divided into two groups: destructive and nondestructive. Destructive testing always results in destroying the specimen (the welded joint) and rendering it unsuitable for its design function. Destructive tests can be mechanical, metallurgical, or chemical. We next review various destructive and nondestructive testing methods.

4.2.2.6.1 Visual Inspection

Visual inspection involves an examination of the weld by the naked eye and checking its dimensions by employing special gauges. Defects such as cracks, porosity, undercuts, underfills, or overlaps can be revealed by this technique.

FIGURE 4.41 Underfilling.

4.2.2.6.2 Mechanical Tests

Mechanical tests are generally similar to conventional mechanical tests, the difference being the shape and size of the test specimen. Tensile, bending, impact, and hardness tests are carried out. Such tests are conducted either on the whole welded joint or on the deposited metal only.

4.2.2.6.3 Metallurgical Tests

Metallurgical tests involve metallurgical microstructure and macrostructure examination of specimens. Macrostructure examination reveals the depth of penetration, the extent of the HAZ, the weld bead shape, hidden cracks, porosity, and slag inclusions. Microstructure examination can show the presence of nitrides, martensite, or other structures that cause metallurgically oriented welding problems.

4.2.2.6.4 Chemical Tests

Chemical tests are carried out to ensure that the composition of the filler metal is identical to that specified by the manufacturing engineer. Some are crude tests, such as spark analysis or reagent analysis; however, if accurate data are required, chemical analysis or spectrographic testing must be carried out.

4.2.2.6.5 Radiographic Inspection

Radiographic inspection is usually performed by employing industrial X-rays. This technique can reveal hidden porosity, cracks, and slag inclusions. It is a nondestructive test that does not destroy the welded joint. High-penetration X-rays are sometimes also employed for inspecting weldments having thicknesses up to 1½ inches (37 mm).

4.2.2.6.6 Pressure Test

Hydraulic (or air) pressure is applied to welded conduits that are going to be subjected to pressure during their service lives to check their tightness and durability.

4.2.2.6.7 Ultrasonic Testing

Ultrasonic waves with frequencies over 20 kHz are employed to detect various kinds of flaws in the weld, such as the presence of nonmetallic inclusions, porosity, and voids. This method is reliable even for testing very thick parts.

4.2.2.6.8 Magnetic Testing

As we know from physics, the lines of magnetic flux are distorted in such a way as to be concentrated at the sides of a flaw or a discontinuity, as seen in Figure 4.42a,b. This test, therefore, involves magnetizing the part and then using fine iron-powder particles that were uniformly dispersed on the surface of the part to reveal the concentration of the flux lines at the location of the flaw. This method is successful in detecting superficial hair cracks and pores in ferrous metal.

4.2.2.6.9 Ammonia Penetrant Test

The ammonia penetrant test is used to detect any leakage from welded vessels. It involves filling the vessel with a mixture of compressed air and ammonia and then wrapping it with paper that has been

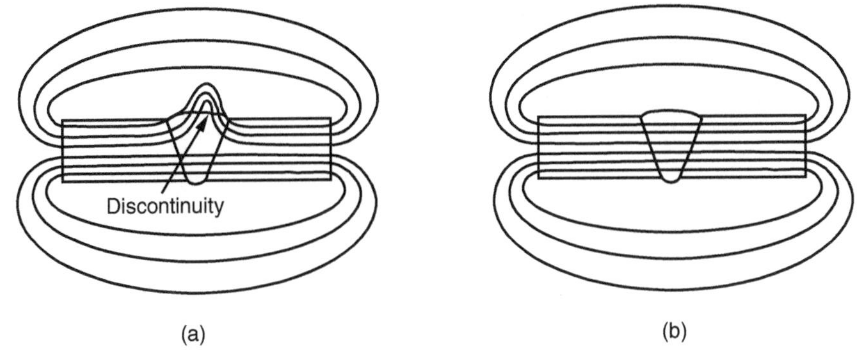

FIGURE 4.42 Magnetic testing of welds: (a) defective weld and (b) sound weld.

impregnated in a 5 percent solution of mercuric nitrate. Any formation of black spots is an indication of leakage.

4.2.2.6.10 Fluorescent Penetrant Test

The part is immersed for about half an hour in oil (or an oil mixture) and then dipped in magnesia powder. The powder adheres at any crack location.

4.2.2.7 Design Considerations

As soon as the decision is made to fabricate a product by welding, the next step is to decide which welding process to use. This decision should be followed by the selection of the types of joints, by determination of the locations and distribution of the welds, and, finally, by making the design of each joint. Following is a brief discussion of the factors to be considered in each design stage.

4.2.2.7.1 Selection of the Joint Type

We have previously discussed the various joint designs and realized that the type of joint depends upon the thickness of the parts to be welded. In fact, other factors should also affect the process of selecting a particular type of joint. For instance, the magnitude of the load to which the joint is going to be subjected during its service life is one other important factor. The manner in which the load is applied (i.e., impact, steady, or fluctuating) is another factor. Whereas the square butt, simple-V, double-V, and simple-U butt joints are suitable only for usual loading conditions, the double-U butt joint is recommended for all loading conditions. On the other hand, the square-T joint is appropriate for carrying longitudinal shear under steady-state conditions. When severe longitudinal or transverse loads are anticipated, other types of joints (e.g., the single-bevel-T, the double-bevel-T, and the double-J) have to be considered. In all cases, it is obvious that cost is the decisive factor whenever there is a choice between two types of joints that would function equally well.

4.2.2.7.2 Location and Distribution of Welds

It has been found that the direction of the linear dimension of the weld with respect to the direction of the applied load has an effect on the strength of the weld. In fact, it has been theoretically and experimentally proven that a lap weld, whose linear direction is normal to the direction of the applied load, as is shown in Figure 4.43a, is 30 percent stronger than a lap weld whose linear direction is parallel to the direction of the applied load, as shown in Figure 4.43b. In the first case, the maximum force F that the joint can carry without any signs of failure can be approximated by the following equation:

$$F = 0.707 \times l \times W \times \sigma_{\text{allowable}} \tag{4.3}$$

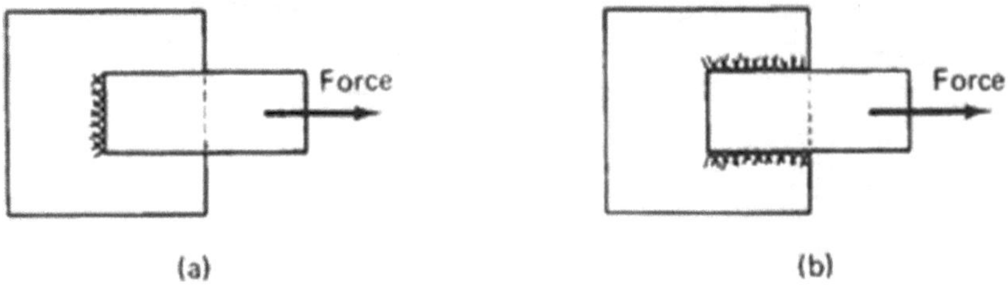

FIGURE 4.43 Location and distribution of welds: (a) weld linear direction is normal to the applied load, and (b) weld linear direction is parallel to the applied load.

where

l is the weld leg
W is the length of the weld
$\sigma_{allowable}$ is the allowable tensile strength of the filler material (e.g., electrode)

In the second case (Figure 4.43b), the strength of the joint is based on the fact that the throat plane of the weld is subjected to pure shear stress and is given by the following equation:

$$F = 0.707 \times l \times W \times \tau_{\text{allowable}} \tag{4.4}$$

where

l is the weld leg
W is the length of the weld
$\tau_{allowable}$ is the allowable shear stress of the electrode

From the theory of plasticity, assuming you adopt the same safety factor in both cases, it is easy to prove that

$$\tau_{\text{allowable}} = \frac{1}{\sqrt{3}}\sigma_{\text{allowable}} = 0.565\sigma_{\text{allowable}} \tag{4.5}$$

On the other hand, the strength of a butt-welded joint can be given by the following equation:

$$F = l \times W \times \sigma_{\text{allowable}} \tag{4.6}$$

where l, W, and $\sigma_{allowable}$ are as previously mentioned. A product designer should, therefore, make use of this characteristic when planning the location and distribution of welds.

Another important point to consider is the prevention of any tendency of the welded elements to rotate when subjected to mechanical loads. A complete force analysis must be carried out in order to determine the proper length of each weld. Let us now consider a practical example to see the cause and the remedy for this tendency to rotate. Figure 4.44 shows an L angle welded to a plate. Any load applied through the angle will pass through its center of gravity.

Therefore, the resisting forces that act through the welds will not be equal; the force closer to the center of gravity of the angle will always be larger. Consequently, if any tendency to rotate is to be prevented, the weld that is closer to the center of gravity must be longer than the other one. Using simple statics, it can easily be seen that

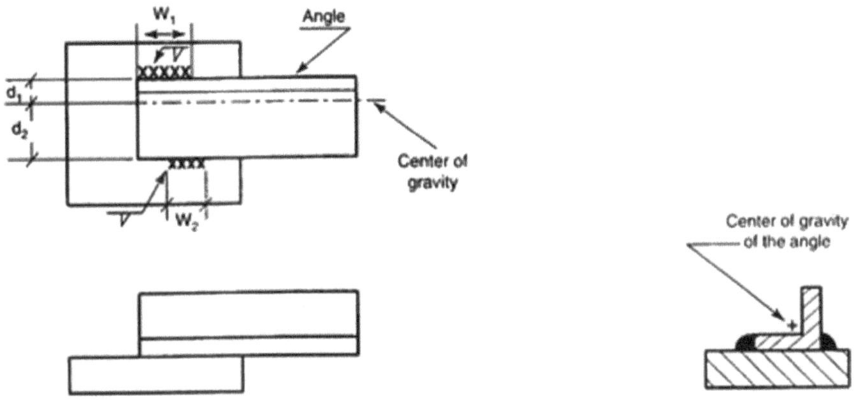

FIGURE 4.44 Preventing the tendency of the weld element to rotate by appropriate distribution of welds.

$$\frac{W_1}{W_2} = \frac{d_2}{d_1} \tag{4.7}$$

It is also recommended that very long welds be avoided. It has been found that two small welds, for example, are much more effective than a single long weld.

4.2.2.7.3 Joint Design

In addition to the procedures and rules adopted in common design practice, there are some guidelines that apply to joint design:

- Try to ensure accessibility to the locations where welds are to be applied.
- Try to avoid overhead welding.
- Consider the heating effect on the base metal during the welding operation. Balance the welds to minimize distortion. Use short, intermittent welds. Figure 4.45a shows distortion caused by an unbalanced weld, whereas Figure 4.45 b,c shows methods for reducing that distortion.
- Avoid crevices around welds in tanks as well as grooves (and the like) that would allow dirt to accumulate. Failure to do so may result in corrosion in the welded joint.
- Do not employ welding to join steels with high hardenability.
- Do not weld low-carbon steels to alloy steels by the conventional fusion welding methods because they have different critical cooling rates and hence cannot be successfully welded.
- When employing automatic welding (e.g., submerged arc), the conventional joint design of manual welding should be changed. Wider Vs (for butt joints) are used, and single-pass welds replace multipass welds.

4.3 SURFACING AND HARDFACING

Surfacing involves the application of a thin deposit on the surface of a metallic workpiece by employing a welding method such as oxyacetylene-gas welding, shielded-metal arc welding, or automatic arc welding. The process is carried out to increase the strength, the hardness, and the resistance to corrosion, abrasion, or wear. For the last reason, the process is commonly known as *hardfacing*.

Good hardfacing practice should be aimed at achieving a strong bond between the deposit and the base metal and also at preventing the formation of cracks and other defects in the deposited layers.

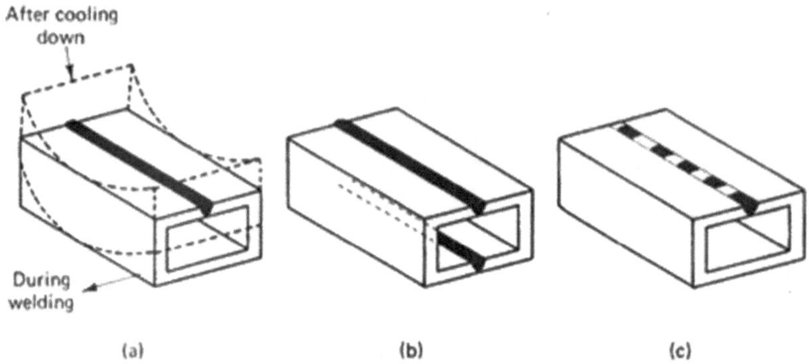

FIGURE 4.45 Designs that promote or eliminate distortion in welding: (a) distortion caused by unbalanced weld; (b) and (c) methods for reducing distortion.

Therefore, the deposited layer should not generally exceed 3/32 inch (2 mm) and will rarely exceed 1/4 inch (6 mm). Also, the base metal should be heated to a temperature of 500°F to 950°F (350°C to 500°C) to ensure a good metallurgical bond and to allow the deposited layer to cool down slowly.

Hardfacing permits the use of very hard wear- and corrosion-resisting compounds. The materials used in this process are complex. They involve hard compounds, like carbides and borides, that serve as the wear-resisting elements, and a tough matrix composed of air-hardening steel or iron-base alloys. Such deposited materials increase the service life of a part three- or fourfold. The process is also employed in restoring worn parts.

The process of hardfacing has found widespread application in the heavy construction equipment industry, in mining, in agricultural machinery, and in the petroleum industry. The list of parts that are usually hardfaced is long and includes, for example, the vulnerable surfaces of chemical-process vessels, pump liners, valve seats, drive sprockets, ripper teeth, shovel teeth, chutes, and the edges of coal recovery augers.

4.4 THERMAL CUTTING OF METALS

In this section, we discuss the *thermal cutting* of metals, specifically oxyfuel flame cutting, and the different arc-cutting processes. Although all these processes do not, by any means, fall under the topic of *joining* (the action involved is opposite to that of joining), they employ the same equipment as the corresponding welding process in each case. The thermal cutting processes are not alternatives to sawing but rather are used for cutting thick plates, 1–10 inches (25–250 mm) thick, as well as for difficult-to-machine materials. Thermal cutting may be manual, using a hand-operated cutting torch (or electrode), or the cutting element can be machine-driven by a numerically controlled system or by special machines called *radiographs*.

4.4.1 Oxyfuel Cutting

Oxyfuel cutting is similar to oxyfuel welding except that an oxidizing flame must always be used. The process is extensively used with ferrous metal having thicknesses up to 10 inches (250 mm). During the process, red-hot iron, directly subjected to the flame, is oxidized by the extra oxygen in the flame; it then burns up, leaving just ashes or slag. Also, the stream of burning gases washes away any molten metal in the region being cut. Generally, there is a relationship between the speed of travel of the torch or electrode and the smoothness of the cut edge: the higher the speed of travel, the coarser the cut edge.

Although acetylene is commonly used as a fuel in this process, other gases are also employed, including butane, methane, propane, natural gas, and a newly developed gas with the commercial name Mapp. Hydrogen is sometimes used as a fuel, especially underwater, to provide a powerful preheating flame. In this case, compressed air is used to keep water away from the flame.

The oxyfuel-cutting process can be successfully employed only when the ignition temperature of the metal being cut is lower than its melting point. Another condition involves ensuring that the melting points of the formed oxides are lower than that of the base metal itself. Therefore, oxyfuel cutting is not recommended for cast iron because its ignition temperature is higher than its melting point. The process is also not appropriate for cutting stainless steel, high-alloy chromium, chrome–nickel alloys, and nonferrous alloys because the melting points of the oxides of these metals are higher than the melting points of the metals themselves.

4.4.2 Arc Cutting

There are several processes based upon the utilization of the heat generated by an electric arc. These arc-cutting processes are generally employed for cutting nonferrous metals, medium-carbon steel, and stainless steel.

4.4.2.1 Conventional Arc Cutting

Conventional arc cutting is similar to shielded-metal arc welding. It should always be remembered, however, that the electrode enters the gap of the cut, so the coating must serve as an insulator to keep the electric arc from shorting out. Consequently, electrodes with coatings containing iron powder are not recommended for use with this process.

4.4.2.2 Air Arc Cutting

Air arc cutting involves preheating the metal to be cut by an electric arc and blowing out the resulting molten metal by a stream of compressed air. The arc-air torch is actually a steel tube through which compressed air is blown.

4.4.2.3 Oxygen Arc Cutting

Oxygen arc cutting is similar to air arc cutting except that oxygen is blown instead of air. The process is capable of cutting cast irons and stainless steels with thicknesses up to 2 inches (50 mm).

4.4.2.4 Carbon Arc Cutting

In carbon arc cutting, a carbon or graphite electrode is used. The process has the disadvantage of consuming that electrode quickly, especially if continuous cutting is carried out.

4.4.2.5 Tungsten Arc Cutting

The electrode used in tungsten arc cutting is made of tungsten and has, therefore, a service life that is far longer than that of the carbon or graphite electrodes. Tungsten arc cutting is commonly employed for stainless steel, copper, magnesium, and aluminum.

4.4.2.6 Air-Carbon Arc Cutting

Air-carbon arc cutting is quite similar to carbon arc cutting, the difference being the use of a stream of compressed air to blow the molten metal (that has been fused by the arc) out of the kerf (groove). The process cuts almost all metals because its mechanics involve the oxidation of the metal. Its applications involve the removal of welds, the removal of defective welds, and the dismantling of steel structures.

4.4.3 Plasma Cutting

A plasma arc is employed to cut metals in plasma arc cutting. The temperature of the plasma jet is extremely high (ten times higher than that obtained with oxyfuel), thus enabling high-speed cutting rates to be achieved. Also, as a consequence, the HAZ formed along the edge of the kerf is usually less than 0.05 inch (1.3 mm). Plasma arc cutting can be used for cutting stainless steel as well as hard-to-cut alloys. A modification of the process involves using a special nozzle to generate a whirlpool of water on the workpiece, thus increasing the limit on the thickness of the workpiece up to 3 inches (75 mm) and meanwhile improving the quality of the cut. The only limitation of plasma arc cutting is that the workpiece must be electrically conductive.

4.4.4 Laser-Beam Cutting

The basic principles of laser-beam cutting are similar to those of laser-beam welding. Nevertheless, laser cutting is achieved by the pressure from a jet of gas that is coaxial with the laser beam, as shown in Figure 4.46. The function of the gas jet is to blow away the molten metal that has been fused by the laser beam. Laser beams can be employed in cutting almost any material, including nonconductive polymers and ceramics. Also, the process is usually automated by using computerized numerical control systems to control the movements of the machine table under the laser beam so that workpieces can be cut to any desired contour. Other advantages of the laser-beam cutting process include the straight-edged kerfs obtained, the very narrow HAZ that results, and the elimination of the part distortion experienced with other conventional thermal cutting processes.

4.5 BRAZING AND SOLDERING

Brazing and *soldering* are processes employed for joining solid metal components by heating them to the proper temperature and then introducing between them a molten filler alloy (brazing metal or solder). The filler alloy must always have a melting point lower than that of the base metal of the components. The filler alloy must also possess high fluidity and wettability (i.e., be able to spread and adhere to the surface of the base metal). As you may expect, the mechanics of brazing or soldering are different from those of welding. A strong brazed joint is obtained only if the brazing metal can diffuse into the base metal and form a solid solution with it. Figure 4.47a,b is a sketch of the microstructure of two brazed joints and is aimed at clarifying the mechanics of brazing and soldering.

Brazing and soldering can be employed to join carbon and alloy steels, nonferrous alloys, and dissimilar metals. The parts to be joined together must be carefully cleaned, degreased, and clamped. Appropriate flux is applied to remove any remaining oxide and to prevent any further oxidation of

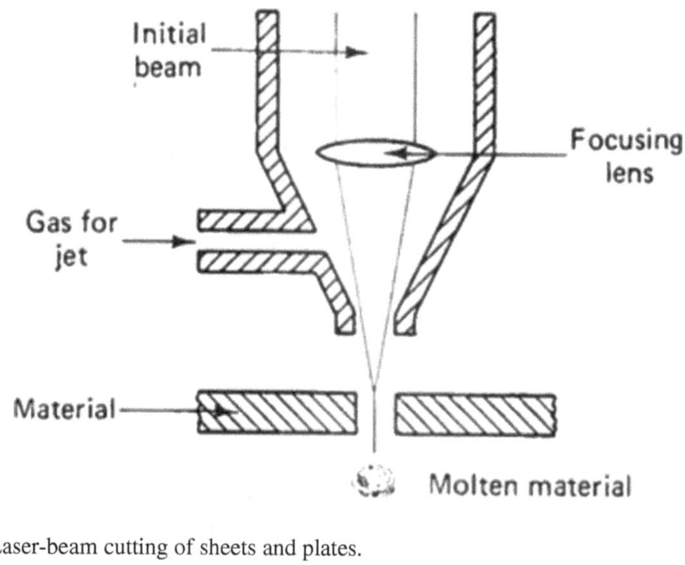

FIGURE 4.46 Laser-beam cutting of sheets and plates.

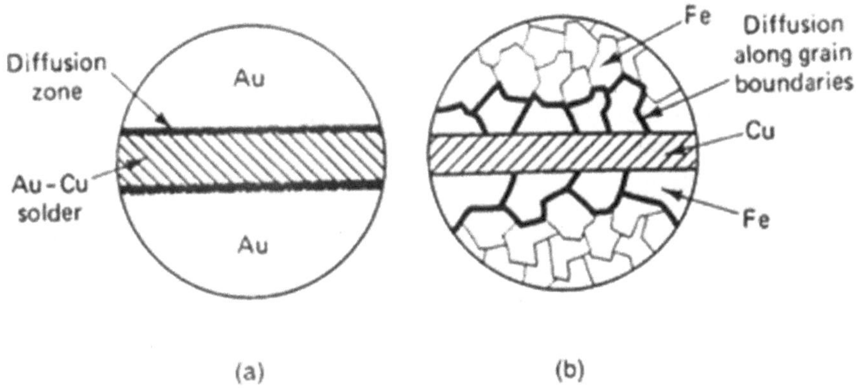

FIGURE 4.47 A sketch of the microstructure of two brazed joints: (a) gold base metal and (b) low carbon steel base metal.

TABLE 4.2
Recommended Filler Alloys for Different Base Metals

Base Metal		Filler Alloys			
Tin–Lead	**Silver Solder**	**40% Zinc Brass**	**Nickel Silver**	**Copper**	
Steel	*	*	X	*	X
Stainless steel	*	X	*	X	
Nickel alloys	*	*	X	X	
Copper (pure)	*	X	X		
Brass or bronze	*	X			
Silver (pure)	*	X			
Aluminum	Sn-Zn 10 or				
Al-Mg 3, Al-Si 5	Al-Si 12				

* May be used.
X = Best use.

the metals. It is only under such conditions that the filler metal can form a strong metallic bond with the base metal.

The main difference between soldering and brazing is the melting point of the filler metal in each case. Soft solders used in soldering have melting points below 930°F (500°C) and produce joints with relatively low mechanical strength, whereas hard solders (brazing metals) have higher melting points, up to 1650°F (900°C), and produce joints with high mechanical strength.

Soft solders are low-melting-point eutectic alloys. They are basically tin, lead, cadmium, bismuth, or zinc alloys. On the other hand, brazing filler metals are alloys consisting mainly of copper, silver, aluminum, magnesium, or nickel. Table 4.2 gives the recommended filler alloys for different base metals. The chemical composition and the field of application for the commonly used soft and hard solders are given in Table 4.3.

4.5.1 Fluxes

Fluxes are employed in soft soldering as well as in brazing in order to protect the cleaned surfaces of the base metal against oxidation during those processes. In addition, fluxes enable proper wetting of the surfaces of the base metals by the molten filler solders.

TABLE 4.3
Most Commonly Used Soft and Hard Brazing Filler Metals

Type	Approximate Chemical Composition	Brazing or Soldering Temperature	Application
Tin solder	Sn-Pb 70	360–490°F (183–255°C)	General-purpose solder
	Sn-Pb 50	360–420°F (183–216°C)	Electrical application
	Sn-Zn 10	509–750°F (265–400°C)	Soft soldering of aluminum
Silver solder	Ag 25–50, Cu 20–40, Sn 0–35 Cd 0–20, Zn 0–20	1175–1550°F (635–845°C)	Brazing of copper alloys and silver in electronics
Brass solder	Cu-Zn 40	1620–1750°F (910–955°C)	General-purpose hard solder
Nickel silver	Cu balance, Zn 20–30 Ni 10–20	1720–1800°F (938–982°C)	Nickel alloys and steel brazing
Copper	99.9% Copper	2000–2100°F (1093–1150°C)	Brazing of steel
Silumin	Al-Si 12	1080–1120°F (582–605°C)	Brazing of all aluminum allays except silumin

There are two kinds of fluxes for soft soldering operations: organic and inorganic. The inorganic fluxes are mostly aqueous solutions of zinc and/or ammonium chlorides. They must, however, be completely removed after the soldering operation because of their corrosive effect. It is, therefore, completely forbidden to use inorganic fluxes in soldering electronic components. On the other hand, organic fluxes do not have such corrosive effects and are, therefore, widely used for fine soldering in electronic circuits. The commonly used organic fluxes involve colophony, a kind of resin with a melting point between 350°F and 390°F (180°C and 200°C), as well as some fats.

The fluxes employed in brazing include combinations of borax, boric acid, borates, fluorides, and fluoroborates together with a wetting agent. The flux can be in the form of a liquid, slurry, powder, or paste, depending upon the brazing method used.

4.5.2 Soldering Techniques

The manual soldering method involves using a hand-type soldering iron that is made of copper and has to be tinned each time before use. The iron is first heated to a temperature of about 570°F (300°C), and its tip is then dipped into the flux and tinned with the solder. Next, the iron is used for heating the prepared surfaces of the base metal and for melting and distributing the soft solder. When the solder solidifies, it forms the required solder seam.

Several other methods are also used for soldering. These include dip soldering and induction soldering as well as the use of guns (blowtorches). Nevertheless, electric soldering irons are still quite common.

4.5.3 Brazing Techniques

The selection of a preferred brazing method has to be based on the size and shape of the joined components, the base metal of the joint, the brazing filler metal to be used, and the production rate. When two brazing techniques are found to be equally suitable, cost is the deciding factor. The following brazing methods are commonly used in industry.

4.5.3.1 Torch Brazing

Torch brazing is still the most commonly used method. It is very similar to oxyfuel flame welding in that the source of heat is a flame obtained from the combustion of a mixture of a fuel gas (e.g., acetylene) and oxygen. The process is very popular for repair work on cast iron and is usually applied manually, although it can be used on a semiautomatic basis. In this process, however, a reducing flame should be used to heat the joint area to the appropriate brazing temperature. A flux is then applied, and as soon as it melts, the filler metal (brazing alloy) is hand-fed to the joint area. When the filler metal melts, it flows into the clearance between the base components by capillary attraction. The filler metal should always be melted by the heat gained by the joint and not by directly applying the flame.

4.5.3.2 Furnace Brazing

Furnace brazing is performed in either a batch or a continuous conveyor-type furnace and is, therefore, best suited for mass production. The atmosphere of the furnace is controlled to prevent oxidation and to suit the metals involved in the process. That atmosphere can be dry hydrogen, dissociated ammonia, nitrogen, argon, or any other inert gas. Vacuum furnaces are also employed, especially with brazing materials containing titanium or aluminum. Nevertheless, a suitable flux is often employed. The filler metal must be placed in the joint before the parts go inside the furnace. The filler metal can, in this case, take the form of a ring, washer, wire, powder, or paste.

4.5.3.3 Induction Brazing

In induction brazing, the components to be brazed are heated by placing them in an alternating magnetic field that, in turn, induces an alternating current in the components that rapidly reverses its direction. Special coils made of copper, referred to as *inductors*, are employed for generating the magnetic field. The filler metal is often placed in the joint area before brazing but can also be hand-fed by the operator. This technique has a clear advantage, which is the possibility of obtaining a very closely controlled heating area.

4.5.3.4 Dip Brazing

Dip brazing involves dipping the joint to be brazed in a molten filler metal. The latter is maintained in a special externally heated crucible and is covered with a layer of flux to protect it from oxidation. Because the filler metal coats the entire workpiece, this process is used only for small parts.

4.5.3.5 Salt-Bath Brazing

The source of heating in salt-bath brazing is a molten bath of fluoride and chloride salts. The filler metal is placed in the joint area before brazing and is also sometimes cladded. Next, the whole assembly is preheated to an appropriate temperature and then dipped for 1–6 minutes in the salt bath. Finally, the hot brazed joint is rinsed thoroughly in hot and cold water to remove any remaining flux or salt. Generally, this process is employed for brazing aluminum and its alloys. There is, however, a problem associated with the process, and that is the pollution caused by the effluent resulting from the rinsing operation.

4.5.3.6 Resistance Brazing

Low-voltage, high-amperage current is used as the source of energy in resistance brazing, as is the case with spot welding. In fact, a spot welder can be employed to carry out this process, provided that the pressure is carefully adjusted so as to be just enough to secure the position of the contact where heat develops. The workpiece is held between the two electrodes, with the filler metal preloaded at the joint area. This process is normally used for the brazing of electrical contacts and in the manufacture of copper transformer leads.

4.5.4 Design of Brazed Joints

For the proper design of brazed joints, two main factors have to be taken into consideration. The first factor involves the mechanics of the process in that the brazing filler metal flows through the joint

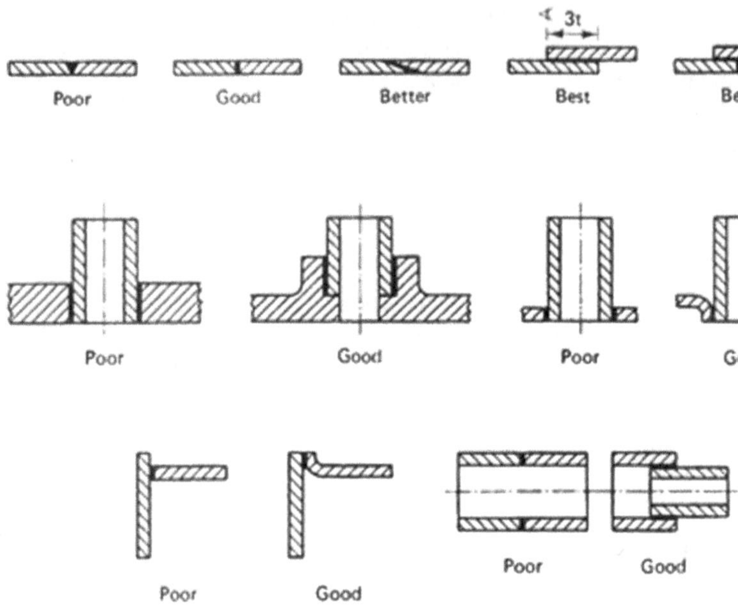

FIGURE 4.48 Good and poor practices in the design of brazed joints.

by capillary attraction. The second factor is that the strength of the filler metal is poorer than that of the base metals. The product designer should aim for the following:

- Ensuring that the filler metal is placed on one side of the joint and allocating space for locating the filler metal before (or during) the process
- Adjusting the joint clearance in order to ensure optimum conditions during brazing. That clearance is dependent upon the filler metal used and normally takes a value less than 0.005 inches (0.125 mm), except for silumin, in which case, it can go up to 0.025 inches (0.625 mm)
- Ensuring that the distance to be traveled by the filler metal is shorter than the limit distant, as dictated by the physics of capillarity
- Providing enough filler metal
- Increasing the area of the joint because the filler metal is weaker than the base metal

There are three types of joint-area geometrics: butt, scarf, and lap. The butt joint is the weakest, and the lap is the strongest. Nevertheless, when designing lap joints, make sure that the joint overlap is more than 3 *t*, where *t* is the thickness of the thinner parent metal. Examples of some good and poor practices in the design of brazed joints are shown in Figure 4.48 as guidelines for beginners in product design. Also, always remember that brazed joints are designed to carry shear stress and not tension.

4.6 ADHESIVE BONDING OF METALS

Sticking, or *adhesive bonding*, of metals is becoming very popular in the automotive, aircraft, and packaging industries because of the advantages that this technique can offer. Thanks to the recent development in the chemistry of polymers, adhesives are now cheap, can be applied easily and quickly, and can produce reasonably strong joints. Adhesive bonding can also be employed in producing joints of dissimilar metals or combinations of metals and nonmetals like ceramics or polymers. This certainly provides greater flexibility when designing products and eliminates the need for complicated, expensive joining processes.

As we know, it is possible to stick entirely smooth metal surfaces together. It is obvious, therefore, that the sticking action is caused by adhesive forces between the sticking agent and the workpiece and not by the flowing and solidification of the sticking agent into the pores of the workpiece as occurs, for instance, with wood. In other words, adhesion represents attractive intermolecular forces under whose influence the particles of a surface adhere to those of another one. There are also many opinions supporting the theory that mechanical interlocking plays a role in bonding.

4.6.1 Adhesives

Structural adhesives are normally systems including one or more polymeric materials. In their unhardened state (i.e., before they are applied and cured), these adhesives can take the form of viscous liquids or solids with softening temperatures of about 212°F (100°C). The unhardened adhesive agents are often soluble in ketones, esters, and higher alcohols, as well as in aromatic and chlorine hydrocarbons. The hardened adhesives, however, resist nearly all solvents. Adhesives that find industrial application in bonding two nonmetallic workpieces include cyanoacrylates, acrylics, and polyurethanes. Following is a brief description of the adhesives that are commonly used in industry.

4.6.1.1 Epoxies

Epoxies are thermosetting polymers (see Chapter 8) that require the addition of a hardener or the application of heat so that they can be cured. Epoxies are considered the best sticking agents because of their versatility, their resistance to solvents, and their ability to develop strong and reliable joints.

4.6.1.2 Phenolic

Phenolics are characterized by their low cost and heat resistance of up to about 930°F (500°C). They can be cured by a hardener or by heat or can be used in solvents that evaporate and thus allow setting to occur. Like epoxies, phenolics are thermosetting polymers with good strength, but they generally suffer from brittleness.

4.6.1.3 Polyamide

The polyamide group of polymers is characterized by its oil and water resistance. Polyamides are usually applied in the form of hot melts but can also be used by evaporation of solvents in which they have been dissolved. Polyamides are normally used as can-seam sealants and the like. They are also used as hot-melt for shoes.

4.6.1.4 Silicones

Silicones can perform well at elevated temperatures; however, cost and strength are the major limitations. Therefore, silicones are usually used as high-temperature sealants.

4.6.2 Joint Preparation

The surfaces to be bonded must be clean and degreased because most adhesives do not amalgamate with fats, oils, or wax. Joint preparation involves grinding with sandpaper, grinding and filling, sandblasting, and pickling and degreasing with trichloroethylene. Oxide films, electroplating coats, and varnish films need not be removed (as long as they are fixed to the surface). Roughening of the surface is advantageous, provided that it is not overdone.

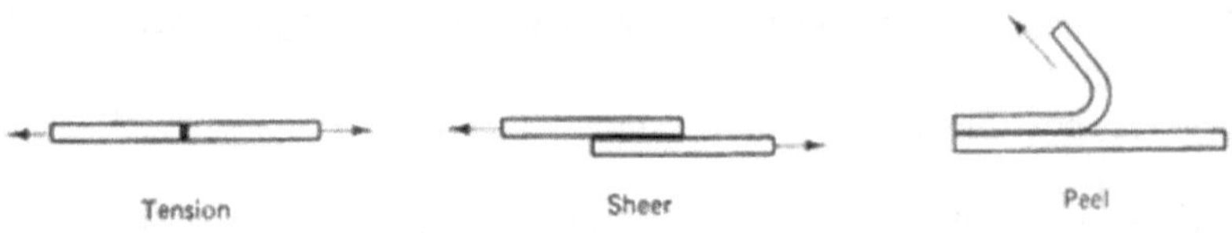

FIGURE 4.49 The three types of adhesive-bonded joints.

4.6.3 Joint Design

There are basically three types of adhesive-bonded joints. They are shown in Figure 4.49 and include tension, shear, and peel joints. Most of the adhesives are weaker in peel and tension than in shear. Therefore, when selecting an adhesive, you should always keep in mind the types of stresses to which the joint is going to be subjected. It is also recommended that you avoid using tension and peel joints and change the design to replace these with shear joints whenever possible.

REVIEW QUESTIONS

1. What does the riveting process involve?
2. What are rivets usually made of?
3. List some applications of riveting.
4. Spot welding could not completely replace riveting. List some applications of riveting that cannot be done by spot welding.
5. How would you define *welding?*
6. List five types of welded joint designs and discuss suitable applications for each type.
7. What are the types of different methods for classifying the welding processes?
8. How would you break all the manufacturing methods into groups according to each of these classifying methods?
9. Explain briefly the mechanics of solid-state welding.
10. What are the two main obstacles that must be overcome so that successful pressure welding can be achieved?
11. What is cold-pressure welding? Give two examples using sketches.
12. Discuss briefly the mechanics of explosive welding and draw a sketch to show the interface between the welded parts.
13. List some industrial applications for explosive welding.
14. Discuss briefly the mechanics of ultrasonic welding.
15. What are the typical applications of ultrasonic welding?
16. What are the different types of ultrasonic welding machines? List the main components common to all these machines.
17. What is the basic idea on which friction welding is based?
18. Explain briefly the stages involved in a friction welding operation.
19. List the various advantages claimed for friction welding.
20. List the limitations of friction welding.
21. What is the difference between friction welding and inertia welding?
22. What advantages does inertia welding have over friction welding?
23. Give examples of some parts that are fabricated by inertia welding.
24. Explain briefly the mechanics of induction welding.
25. List some of the common industrial applications of the induction welding process.
26. What is the source of energy in thermit welding? Explain.
27. How is thermit welding performed?

28. What are the common applications of thermit welding?
29. How does bonding take place in diffusion bonding?
30. List the different processes that belong to resistance welding.
31. Explain briefly the stages involved in a resistance-butt welding process.
32. Using a sketch, explain the pressure-time and current-time relationships in resistance-butt welding.
33. List some of the applications of resistance-butt welding.
34. Clarify the difference between flash welding and butt welding.
35. Draw a graph illustrating current versus time and pressure versus time in flash welding.
36. When is flash welding recommended over butt welding?
37. What is the major disadvantage of flash welding?
38. Explain the basic idea of percussion welding.
39. Explain briefly the mechanics of spot welding.
40. Draw a sketch of a section through a spot-welded joint.
41. Draw a sketch to show a typical cycle for a spot welding machine.
42. How do you compare seam welding with spot welding?
43. List some of the advantages of seam welding.
44. What are the industrial applications of seam welding?
45. What is the basic idea of projection welding?
46. What are the advantages of projection welding?
47. What is the condition for two metals to be joined together by fusion welding?
48. How many zones can be identified in a joint produced by a conventional fusion welding process? Discuss briefly the microstructure in each of these zones.
49. For what do the letters HAZ stand?
50. Explain briefly the phenomenon of the electric arc and how it can be employed in welding.
51. What are the advantages of alternating current over direct current in arc welding?
52. What is the difference between DCSP and DCRP? When would you recommend using each of them?
53. What is meant by the *rated duty cycle?*
54. What shields the molten metal during shielded-metal arc welding?
55. What is the main shortcoming of shielded-metal arc welding?
56. List some of the functions of electrode coatings.
57. Explain briefly the Bernardos welding method.
58. What is the main feature of the electrodes in flux-cored arc welding?
59. What provides the shielding in flux-cored arc welding?
60. What is stud arc welding? How is it performed? List the main applications of this process.
61. How is shielding achieved in submerged arc welding?
62. Why must the plates to be joined by submerged arc welding be horizontal only?
63. Why does submerged arc welding always yield very high-quality welds?
64. List some of the advantages of submerged arc welding.
65. What provides shielding in MIG welding?
66. Why does the MIG welding process render itself suitable for automation?
67. How can the penetration for gas–metal arc welding be controlled?
68. What is the main difference between the MIG and the TIG welding processes?
69. List some of the applications of TIG welding.
70. In TIG welding, when would you use an AC power supply and when would you use DCSP and DCRP?
71. Explain briefly the mechanics and the basic idea of plasma arc welding.
72. When is plasma arc welding most recommended?

73. Do you consider elcctroslag welding to be a true arc welding process?
74. How does welding take place in the electroslag welding process?
75. When is electroslag welding usually recommended?
76. What is the source of energy in oxyacetylene flame welding?
77. How is acetylene stored for use in welding operations?
78. What does the equipment required in gas welding include? Explain the function of each component.
79. What are the types of flames that can be obtained in gas welding? How is each one obtained?
80. What are the zones of a neutral flame? Discuss the effect of the oxygen-to-acetylene ratio on the nature of the flame obtained.
81. Explain briefly the operating principles of electron-beam welding.
82. What are the major limitations of the electron-beam welding process?
83. List some of the advantages of electron-beam welding.
84. What are the major applications of electron-beam welding?
85. For what do the letters in the word *laser* stand?
86. Using a sketch, explain how a laser beam capable of carrying out welding can be generated.
87. Explain briefly the mechanics of laser-beam welding.
88. What are the main advantages of laser-beam welding?
89. List some of the applications of laser-beam welding.
90. Using sketches, illustrate the commonly experienced welding defects. How can each be avoided?
91. What are the main tests for the inspection of welds? Discuss each briefly.
92. What are the factors affecting the selection of the joint type?
93. On what basis are the location and distribution of welds planned? What rules would you consider when designing a welded joint?
94. What is meant by *hardfacing?*
95. What are the main applications of hardfacing?
96. List the main types of thermal cutting processes. Discuss briefly the advantages and limitations of each.
97. How do the mechanics of brazing differ from those of welding?
98. What is the main difference between brazing and soft soldering?
99. List some of the alloys used as brazing fillers and mention the base metals that can be brazed with each one.
100. List some of the commonly used soft solders.
101. What is the main function of brazing fluxes?
102. List some of the fluxes used in brazing.
103. List some of the fluxes used in soft soldering. Discuss the limitations and applications of each.
104. In soft soldering, how should the solder be fused?
105. List the different brazing techniques used in industry. Discuss the advantages and limitations of each.
106. As a product designer, what factors should you take into consideration when designing a brazed joint?
107. In what case can the sticking of metals not be replaced by other welding and brazing techniques?
108. List some of the commonly used adhesives. Discuss the characteristics and common applications of each.
109. What are the types of adhesive-bonded joints? Which one is usually the strongest?

PROBLEMS

1. Two steel slabs, each 1/4 inch (6.35 mm) thick, are to be joined by two fillet welds (i.e., at both edges). If the width of each slab is 2.5 inches (62.5 mm) and the joint is to withstand a load of 35,000 lb (156,000 N), determine the allowable tensile strength of the electrode type to be used in welding.
2. Two steel plates, each 1/4 inch (6.35 mm) thick, are to be joined by two fillet welds. If the joint is to withstand a load of 50,000 lb (222,500 N) and an E7014 electrode (allowable tensile strength = 21,000 lb/in.^{2}; i.e., 145,000 KN/m^2) is used, determine the length of weld at each edge. If the plate width is 10 inches (254 mm), how would you distribute the weld? Draw a sketch.
3. Two steel plates, each 5/16 inch thick (7.9 mm), are to be fillet-welded to a third one that is sandwiched between them. The width of each of the first two plates is 4 inches (100 mm), whereas the width of the third one is 6 inches (150 mm). The two plates overlap the third one by 6 inches (150 mm), and an E7014 electrode (allowable tensile strength = 21,000 lb/in.^{2}; i.e., 145,000 kN/m^2) is to be used. If the joint is to withstand a load of 190,000 pounds (846 kN), use a sketch to illustrate a design for this joint and provide all calculations.
4. Two mild steel pipes, each having a 3/4-inch (19-mm) outer diameter, are to be joined together by brazing. Assuming that the joint is to withstand an axial load of 6 tons, give a detailed design of this joint. (Take the allowable shear stress of copper to be 6000 lb/in.^{2}; i.e., 41,430 kN/m^2.)
5. Two mild steel sheets, each 3/32 inch (2.4 mm) thick, are to be brazed together using copper as a filler material. Calculate the strength of the joint when it is manufactured according to each of the designs given in Figure 4.48. Compare the results and recommend the design that gives maximum strength.
6. A power supply for arc welding is rated at a 150-A 30 percent duty cycle. What will be the percentage of actual time utilized in welding to the total time the power supply is on if the current employed in welding is only 125 A?

DESIGN PROBLEM

You are required to design a flat-belt pulley so that it can be fabricated by welding. The pulley is to be mounted on a shaft that is 1¼ inches (31 mm) in diameter, and the outside diameter of the rim is 10 inches (250 mm). The rim of the pulley is to provide a surface to transmit a torque of 600 lb·ft (816 N·m) from a 2-inch-wide (50-mm) flat belt to the shaft. The number of pulleys required is only 5.

SOLUTION

It is advisable to start by gathering information about guidelines for the constructional features of flat-belt pulleys (e.g., width of rim for a certain belt width and thickness of rim). Information about the safe speeds of various sizes of pulleys should also be collected.

Key. The best strategy is to design the key so that it will be the weak link in the pulley-key-shaft assembly because it is easy to replace. A suitable key material is AISI 1020 CD steel, which is commercially available as a key stock material. It has the following mechanical properties:

Ultimate tensile strength (UTS) = 78,000 lb/in.^{2}
Yield stress = 66,000 lb/in.^{2}
Yield stress in shear = 38,000 lb/in.^{2}

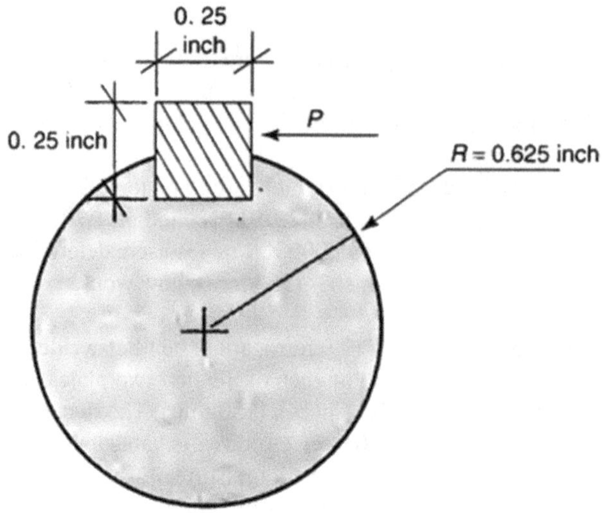

FIGURE 4.50 Forces acting on the key.

Consider Figure 4.50. The force acting on the key is given by

$$P = \frac{T}{r} = \frac{600 \text{ lb ft} \times 12 \text{in/ft}}{0.625} = 11{,}520 \text{ lb}$$

Take the key cross section to be 1/4 × 1/4 inch (6 × 6 mm) and its length l in inches:

$$\text{Shear stress in the key} = \frac{11{,}520 \text{ lb}}{0.25\ l} \le \tau_{\text{allowable}}$$

Take a safety factor of 2 for the key:

$$\tau_{\text{allowable}} = \frac{38{,}000}{2} = 19{,}000 \text{ lb/in.}^2$$

Therefore,

$$\frac{11{,}520}{0.25l} = 19{,}000$$

and

$$l = 2.4\text{in.}(60 \text{ mm})$$

But,

$$\text{Bearing stress} = \frac{11{,}520}{0.25 \times 0.5 \times l} \le \text{allowable compressive stress}$$

$$\le \frac{66{,}000}{2} = 33{,}000 \text{ lb/in.}^2$$

Therefore,

$$\frac{6400}{0.25 \times 0.707 \times 0.57 \times 21{,}000} = 3.03 \text{ in.} (75\text{mm})$$

We take this value to ensure safety against both shearing and compressing loads. We should, however, round it, so the length of the key is to be 3 inches (75 mm).

Hub. Use a round seamless tube having a 2.25-inch outer diameter and 9/16-inch wall. A suitable material is AISI 1020 CD steel. Again, the length of the hub must not be less than 2.8 inches to keep the bearing stress below the allowable value. Take it as 2.875 inches.

Rim. Use a round seamless tube having a 10-inch outer diameter and 1/4-inch wall. Again, a suitable material is AISI 1020 CD steel because of its availability and ability to withstand the rubbing effect of the moving belt.

Spokes. The positioning and welding of four or five spokes would create a serious problem and necessitate the use of a complicated welding fixture. Therefore, the spokes are to be replaced in the design by a web. Use a 5/16-inch flat plate, machined to have an outer diameter of 9.43 inches and an inner diameter of 2.31 inches. An appropriate material is AISI 1020 HR steel. Because weight can be a factor, it is good practice to provide six equally spaced holes in the web by machining. These can also serve as an aid in the handling and positioning of the web during welding.

Welding. Use conventional arc welding; an E7014 electrode (allowable tensile stress = 21,000 lb/in.^{2}) can be used. A fillet weld with a leg of 0.25 inches is adopted. The force is given by

$$F = \frac{\text{Torque}}{\text{Radius}} = \frac{7200}{1.125 \text{ in.}} = 6400 \text{ lb}$$

The required length of the weld is

$$\frac{6400}{0.25 \times 0.707 \times 0.57 \times 21{,}000} = 3.03 \text{ in.} (75\text{mm})$$

Use 2.00 inches (50 mm) of weld on each side of the hub.

The circumference of the hub equals π times 2.25, or 7.06 inches. Space four welds, each 0.5 inches (12.5 mm) in length, equally, 90° apart around the circumference of the hub. Welds on both sides of the web should be staggered. Adopt the same welds at the rim. They should be safe because the shearing force is much lower (the radius is larger than that of the hub).

Once all the dimensions and details are known, we are in a position to construct the pulley as shown in the workshop drawing in Figure 4.51.

DESIGN PROJECTS

1. Design a table for the machine shop. The table should be 4 feet (1200 mm) in height, with a surface area of 3 × 3 feet (900 × 900 mm), and should be able to carry a load of half a ton. Because only two tables are required, the design should involve the use of steel angles and a plate that are to be joined together by welding.
2. Design a tank for compressed air. It has a capacity of 100 cubic feet (2.837 m^3) and can withstand an internal pressure of 40 atmospheres. The number of tanks required is 50, and the tanks are going to be placed in a humid environment.
3. Design a compressed-air reservoir (tank) that is to be subjected to an extremely corrosive environment. The capacity of the tank is 30 cubic feet (0.85 m^3), and the maximum gauge pressure is 70 atmospheres, but the pressure is pulsating from zero to the maximum value about once every 5 minutes. The number of tanks required is 100.

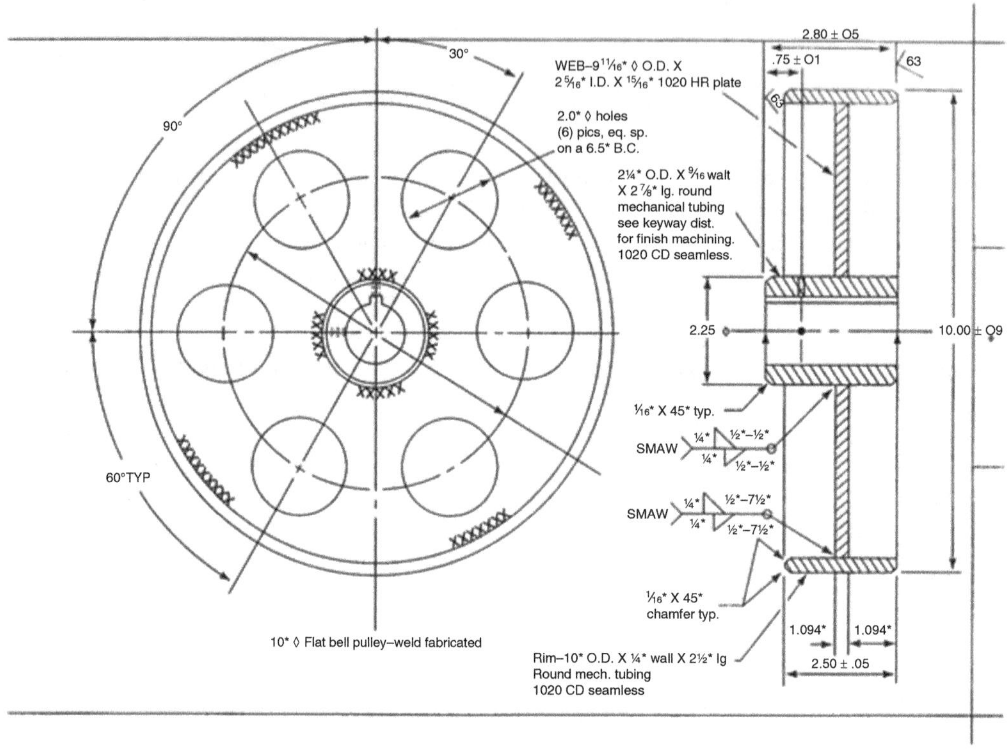

FIGURE 4.51 Detailed workshop drawing of the pulley.

4. A straight-toothed spur-gear wheel transmits a torque of 1200 lb·ft (1632 N·m) to a 2-inch-diameter (50 mm) steel shaft (AISI 1045 CD steel). The pitch diameter of the gear is 8 inches (200 mm), its width is 3 inches (75 mm), and the base diameter is 7.5 inches (187.5 mm). Make a detailed design for the gear's blank (i.e., before the teeth are cut).
5. A mobile winch (little crane) can be moved on casters. It has a capacity of lifting 1 ton for 3 feet (0.9 m) above ground. The lifting arm can be extended, and the winch can then lift 1/2 ton for up to 6 feet (1.8 m). Knowing that the production volume is 4000 units and that casters and hydraulic pressure cylinders are to be purchased from vendors, provide a detailed design and include full specifications of the parts to be purchased.
6. The lifting arm for a crane is 60 feet (about 20 m), and its lifting capacity is 1 ton. It is to be used in construction work and to be subjected to humidity, dirt, and so on. Provide a detailed design for this arm using steel angles that are to be welded together.
7. Design a frame for a hydraulic press for fabrication by welding. The height of the cross arm is 12 feet (about 4 m). The cross arm is mounted (by welding) on two vertical columns that are, in turn, welded to the base. The press can produce a maximum load of 200 tons by means of a hydraulic cylinder attached to the cross arm (below it), and the stroke is 12 inches (300 mm).

   Tip: The energy absorbed when the frame deforms should not exceed 2 percent of the total energy output of the press.

5 Metal Forming

Metal-forming processes have gained significant attention since World War II as a result of the rapid increase in the cost of raw materials. Whereas machining processes involve the removal of portions of the stock material (in the form of chips) in order to achieve the required final shape, metal-forming processes are based upon the plastic deformation and flow of the billet material in its solid state so as to take the desired shape. Consequently, metal-forming processes render themselves more efficient with respect to raw material utilization than machining processes, which always result in an appreciable material waste.

In fact, although metal-forming techniques were employed in manufacturing only semifinished products (like sheets, slabs, and rods) in the past, finished products that require no further machining can be produced today by these techniques. This was brought about by the recent developments in working methods, as well as by the construction features of the forming machines employed. Among the advantages of these up-to-date forming techniques are high productivity and very low material waste. Therefore, more designers tend to modify the construction of the products manufactured by other processes to use forming. Also, bearing in mind that metal-forming methods are still being used for producing semifinished products, it is evident that the vast majority of all metal products are subjected to forming, at least at one stage during their production. This latter fact clearly manifests the importance of the metal-forming methods.

Generally, metal forming involves both billet and sheet metal forming. However, it has been a well-accepted convention to divide those processes into two main groups: bulk (or massive) forming and sheet metal working. In this chapter, only bulk-forming processes (e.g., forging, cold forming, and rolling) are covered; Chapter 6 deals with the working of sheet metal.

5.1 FUNDAMENTAL ANALYSIS AND METALLURGY OF METAL FORMING

5.1.1 Factors Affecting Plastic Deformation

During any forming process, the material plastically flows while the total volume of the workpiece remains substantially constant. However, some marked changes take place on a microscopic scale within the grains and the crystal lattice of the metal, resulting in a corresponding change in the properties of the material. This latter change can be explained in view of the dislocation theory, which states that the plastic deformation and flow of metal are caused by movement and transfer of dislocations (defects in the crystal lattice) through the material with the final outcome of either piling up or annihilating them. Following are some factors that affect plastic deformation by influencing the course of dislocations.

Impurities and Alloying Additives. It is well known that pure metals possess higher plasticity than their alloys. The reason is that the presence of structural components and chemical compounds impedes the transfer and migration of dislocations, resulting in lower plasticity.

Temperature at Which Deformation Takes Place. As a rule, the plasticity of a metal increases with temperature, whereas its resistance to deformation decreases. The higher the temperature, the higher the plasticity and the lower the yield point. Moreover, no work hardening occurs at temperatures above the recrystallization temperature. This should be expected because recrystallization denotes the formation and growth of new grains of metal from the fragments of the deformed grains, together with restoring any distortion in the crystal lattice.

DOI: 10.1201/9781003537106-5

Consequently, strength values drop to the level of a nonwork-hardened state, whereas plasticity approaches that of the metal before deformation. In fact, a forming process is termed hot if the temperature at which deformation takes place is higher than the recrystallization temperature. Lead that is formed at room temperature in summer actually undergoes *hot forming* because the recrystallization temperature for lead is 39.2°F (4°C). When deformation occurs at a temperature below the recrystallization temperature of the metal, the process is termed *cold forming*. Cold-forming processes are always accompanied by work hardening due to the piling up of dislocations. As a result, strength and hardness increase while both ductility and notch toughness decrease. These changes can be removed by heat treatment (annealing). On the other hand, when hot forming a metal, the initial dendritic structure (the primary structure after casting) disintegrates and deforms, and its crystals elongate in the direction of the metal flow. The insoluble impurities like nonmetallic inclusion (around the original grain boundaries) are drawn and squeezed between the elongated grains. This texture of flow lines is usually referred to as the *fibrous macrostructure*. This fibrous macrostructure is permanent and cannot be removed by heat treatment or further working. As a result, there is always anisotropy of mechanical properties; strength and toughness are better in the longitudinal direction of fibers. Also, during hot forming, any voids or cracks around grain boundaries are closed, and the metal welds together, which, in turn, results in improvements in the mechanical properties of the metal.

Rate of Deformation. It can generally be stated that the *rate of deformation* (strain rate) in metalworking adversely affects the plasticity of the metal (i.e., an increase in the deformation rate is accompanied by a decrease in plasticity). Because it takes the process of recrystallization some time to be completed, that process will not have enough time for completion when deformation occurs at high strain rates. Therefore, greater resistance to metal deformation should be expected. This does not mean that the metal becomes brittle.

State of Stress. A *state of stress* at a point can be simply described by the magnitudes and directions of the principal stresses (stress is a force per unit area) acting on planes that include the point in question. The state of stress is, in fact, a precise and scientific expression for the magnitudes and the directions of the external forces acting on the metal. All possible states of stress can be reduced to only nine main systems, as shown in Figure 5.1. These nine cases can, in turn, be divided into three groups. The first group includes two systems that are characterized by the absence of stress (forces) along two directions, and the stress system is therefore called *uniaxial*. This is the case when stretching sheet metal has a length that considerably exceeds its width. In each of the three systems included in the second group, it is clear that a stress along only one of the directions is absent. Because the other two directions (stresses) form a plane, each of these systems is referred to as a *plane-stress state*. It may approximately be represented by the stretching of a thin sheet in two or more directions. The remaining group indicates the state of stress of a body, where there are stresses acting along all three directions in space, yielding the term *triaxial*. In fact, most of the bulk-forming operations (forging, rolling, and wiredrawing) cause states of stress that belong to this latter group.

5.1.2 Estimating the Force Required for Metal Forming

The force required for deforming a given metal (at any unchanged desired temperature and at usual strain-rate levels) is dependent upon the *degree of deformation*, which is the absolute value of the natural logarithm of the ratio of the final length of the billet to its original length. On the other hand, the energy consumed throughout the forming process is equivalent to the area under the load-deformation curve for that forming process. Therefore, that energy can be calculated if the relationship between the load and the deformation is known. Figure 5.2 shows the degree of deformation and the energy consumed in an upsetting operation. It must be noted that in both hot and cold

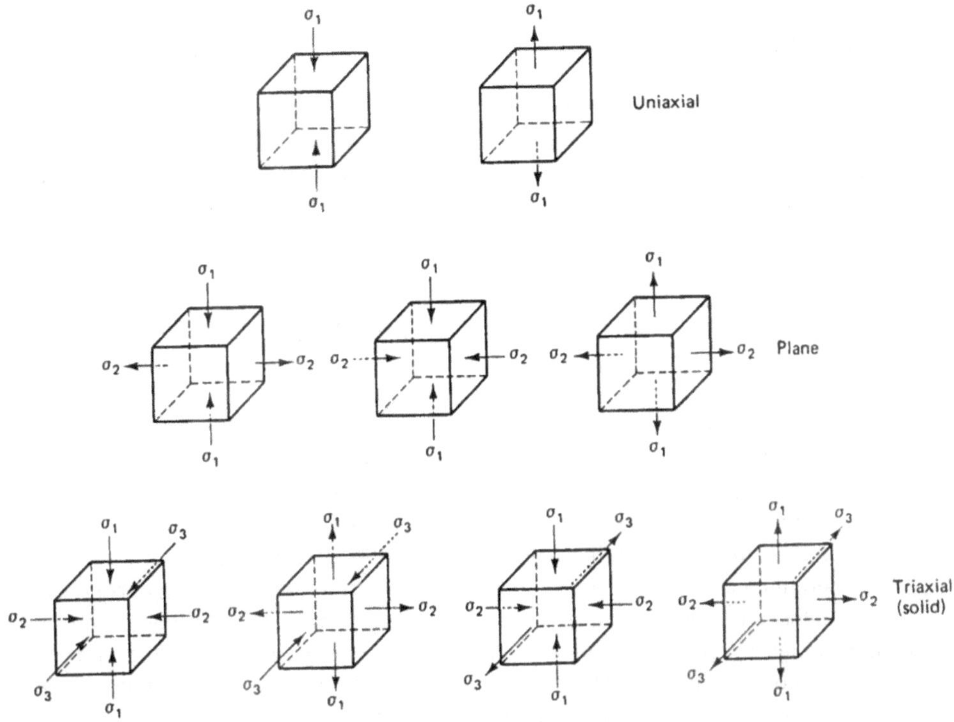

FIGURE 5.1 The nine main systems of states of stress.

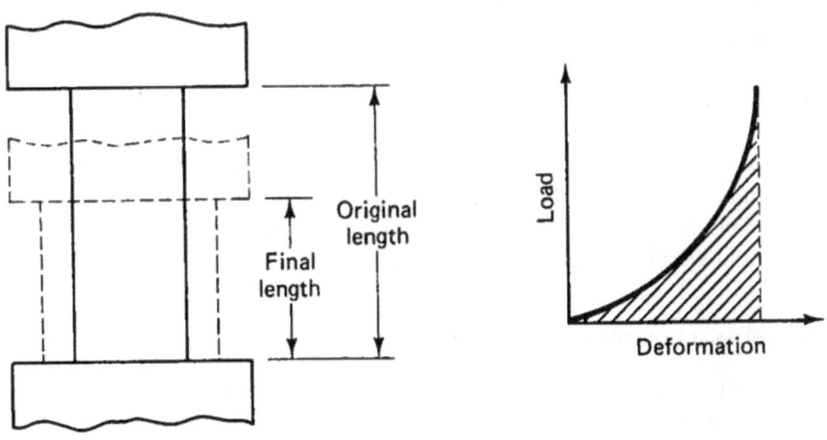

$$\phi = \left| \ln \frac{l_F}{l_0} \right|$$

FIGURE 5.2 The degree of deformation and the energy consumed in an upsetting.

working, there is an upper limit for the degree of deformation (especially in cold working) above which cracks and discontinuities in the workpiece initiate.

In the past, there were four methods of estimating the force required in any desired metal-forming process. These include the slab, the lower bound, the upper bound, and slip-line field methods. The outcomes were, however, "guestimates" rather than estimates in most cases. But fortunately, the finite element method (FEM) that was suitable for the analysis of elastically deforming bodies has been expanded to deal with the plastic deformation of a material, provided that the constitutive equation for that material is known. In fact, the stress distribution, the force required for deforming the metal, and the mode of deformation can be determined with excellent accuracy. There are commercially available FEM software packages for that purpose. By constitutive equation, we mean the relationship between the true stress and true strain for a billet undergoing plastic deformation under uniaxial compression.

5.1.3 Heating the Metal for Hot Forming

Before being subjected to hot-forming processes, ingots (or billets) should be uniformly heated throughout their cross sections, without overheating or burning the metal at the surface. This is particularly important when forming steels. Attention must also be given to the problems of decarburization and formation of scale in order to bring them to a minimum. The thermal gradient is another important factor that affects the soundness of the deformed part. If the temperature gradient is high, thermal stresses may initiate and can cause internal cracks. This usually happens when a portion of the metal is above the critical temperature of metal (AC_1 or AC_3) while the rest of the billet is not. The larger the cross section of the billet and the lower its coefficient of thermal conductivity, the steeper the temperature gradient will be, and the more liable to internal cracking during heating the billet becomes. In the latter case, the rate of heating should be kept fairly low (about 2 hours per inch of section of the billet) in order not to allow a great difference to occur between the temperatures at the surface and the core of the billet. The metal must then be "soaked" at the maximum temperature for a period of time long enough to ensure uniformity of temperature.

The maximum temperature to which the billet is heated before forming differs for different metals. There is usually an optimum range of temperatures within which satisfactory forming is obtained because of increased plasticity and decreased resistance to deformation. Nevertheless, any further increase in temperature above that range may, on the contrary, result in a defective product. Burned metal and coarse grain structure are some of the defects encountered when a metal is excessively heated.

The ingots may be heated in soaking pits, forge hearths, chamber furnaces, or car-bottom furnaces, which are all heated with gas. Rotary hearth furnaces represent another type of heating furnace. In mass production or automated lines, small objects (billets) are heated using electric current and the phenomenon of induction. This induction-heating method is quick and keeps the surfaces of the billets clean, and temperatures can be accurately controlled. Moreover, physical equipment requires limited floor space and can be fully automated.

5.1.4 Friction and Lubrication in Working of Metals

Friction plays an important role in all metal-forming processes and is generally considered undesirable because it has various harmful effects on the forming processes, on the properties of products, and on the tool life. During the deformation of a metal, friction occurs at the contact surface between the flowing metal and the tool profile. Consequently, the flow of the metal is not homogeneous, which leads to the initiation of residual stresses, with the final outcome being an unsound product with inferior surface quality. Also, friction increases the pressure acting on the forming tool (as well as the power and energy consumed) and thus results in greater wear of the tools.

Friction in metal forming is drastically different from the conventional Coulomb's friction because extremely high pressure between the mating bodies (tool and workpiece) is involved. Recent theories on friction in metal forming indicate that it is actually the resistance to shear of a layer, where intensive shear stress is generated as a result of relative displacement between two bodies. When these bodies have direct metal-to-metal contact, slipping and shear flow occur in a layer adjacent to the contact interface. But, if a surface of contact is coated with a material having low shear

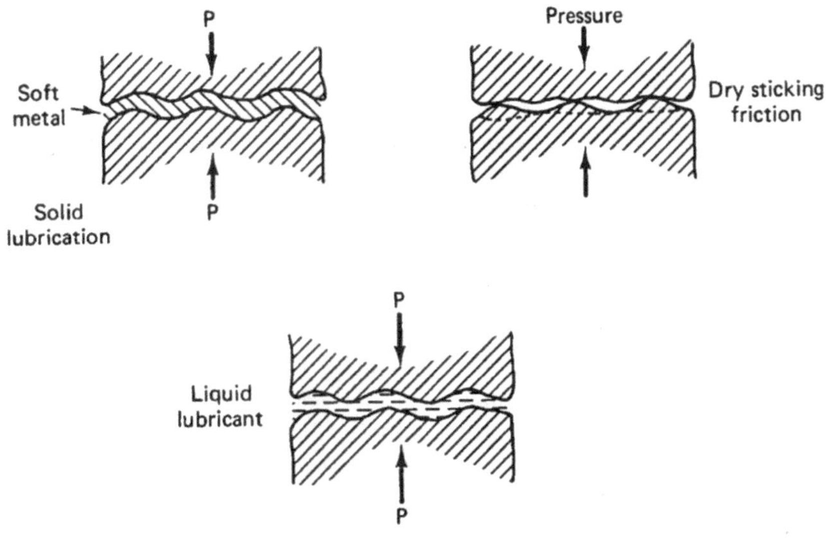

FIGURE 5.3 The shear layer in three different cases.

resistance (a lubricant that can be solid or liquid), slipping takes place through the lubricant and, therefore, has low resistance. This discussion indicates clearly that the magnitude of the friction force is determined by the mechanical properties (yield point in shear) of the layer where actual slipping occurs. Hence, it is evident that a metal having a low yield point in shear, such as lead, can be used as a lubricant when forming metals having relatively high yield strength in compression. Figure 5.3 shows the shear layer in three different cases: solid lubrication, dry sticking friction, and hydrodynamic (liquid) lubrication.

In order to reduce friction and thus eliminate its harmful effects, lubricants are applied to the tool–workpiece interface in metal-forming processes. The gains include lower load and energy requirement, prevention of *galling* or sticking of the workpiece metal onto the tool, better surface finish of products, and longer tool life. An important consideration when selecting a lubricant is its activity (i.e., its ability to adhere strongly to the surface of the metal). The activity of a lubricant can be enhanced, however, by adding material with a high capability of adsorption, such as fat acids. Among other factors to be considered are thermal stability, absence of poisonous fumes, and complete burning during heat treatment of the products.

In cold-forming processes, vegetable and mineral oils, as well as aqueous emulsions, are employed as lubricants. These have the advantage of acting as coolants, eliminating excessive heat and thus reducing the temperature of the tool. Solid polymers, waxes, and solid soaps (sodium or calcium stearates) are also widely used in cold metalworking.

For relatively high-temperature applications, chlorinated organic compounds and sulfur compounds are used. Solid lubricants like molybdenum disulfide and graphite possess low-friction properties up to elevated temperatures and are, therefore, used as solid lubricants in hot forming. Graphite is sometimes dispersed in grease, especially in hot forging ferrous materials. Lately, use has been made of molten glass as a lubricant when alloy steels and special alloys are hot formed. The glass is added in the form of powder between the die and a hot billet. The advantages of molten glass include low friction, excellent surface finish, and improved tool life.

5.1.5 Cold Forming versus Hot Forming

Cold forming has its own set of advantages and disadvantages, as does hot forming, and, therefore, each renders itself appropriate for a certain field of applications. For instance, cold forming will enhance the strength of the workpiece metal, improve the quality of the surface, and provide good dimensional accuracy, but the plastic properties of the metal (elongation percentage and reduction-in-area percentage) and the impact strength drop. Therefore, the final properties of cold-formed products are obtained as required by adjusting the degree of deformation and the parameters of the postheating treatment process. Because the loads involved in cold forming are high, this technique is generally employed in the manufacture of small parts of soft, ductile metals, such as low-carbon steel. Also, large quantities must be produced to justify the high cost of tooling involved. Nevertheless, if the products are to be further processed by machining, the increased hardness caused by cold working is a real advantage because it results in better machinability. Therefore, cold-rolled plates and cold-drawn bars are more suitable for machining purposes than hot-formed ones.

On the other hand, the yield strength of a metal drops significantly at elevated temperatures, and no work hardening occurs. Consequently, hot-forming processes are used when high degrees of deformation are required and/or when forming large ingots or billets because the loads and energies needed are far lower than those required in cold forming. Moreover, hot forming refines the grain structure, thus producing softer and more ductile parts suitable for further processing by cold-forming processes. However, high temperatures affect the surface quality of products, giving oxidation and scales.

Decarburization may also occur in steels, especially when hot-forming high-carbon steel. The scales, oxides, and decarburized layers must be removed by one or more machining processes. This slows down the production, adds machining costs, and yields waste material, resulting in lower efficiency of material utilization. A further limitation of hot forming is reduced tool life due to the softening of tool surfaces at elevated temperatures and the rubbing action of the hot metal while flowing. This actually subjects the tools to thermal fatigue, which shortens their life.

5.2 ROLLING

5.2.1 Fundamentals

Hot rolling is the most widely used metal-forming process because it is employed to convert metal ingots to simple stock members called *blooms* and *slabs*. This process refines the structure of the cast ingot, improves its mechanical properties, and eliminates hidden internal defects. The process is termed *primary rolling* and is followed by further hot rolling into plates, sheets, rods, and structural shapes. Some of these may be subjected to cold rolling to enhance their strength, obtain a good surface finish, and ensure closer dimensional tolerances. Figure 5.4 illustrates the sequence of operations involved in manufacturing rolled products.

The process of *rolling* consists of passing the metal through a gap between rolls rotating in opposite directions. That gap is smaller than the thickness of the part being worked. Therefore, the rolls compress the metal while simultaneously shifting it forward because of the friction at the roll–metal interfaces. When the workpiece completely passes through the gap between the rolls, it is considered fully worked. As a result, the thickness of the work decreases while its length and width increase. However, the increase in width is insignificant and is usually neglected. As can be seen in Figure 5.5, which shows the rolling of a plate, the decrease in thickness is called *draft*, whereas the increase in length and the increase in width are termed *absolute elongation* and *absolute spread*, respectively. Two other terms are the *relative draft* and the *coefficient of elongation*, which can be given as follows:

$$\text{Relative draft } \in = \frac{\Delta h \times 100}{h_o} = \frac{h_o - h_f}{h_o} \times 100 \tag{5.1}$$

$$\text{Coefficient of elongation } \eta = \frac{l_f}{l_o} \tag{5.2}$$

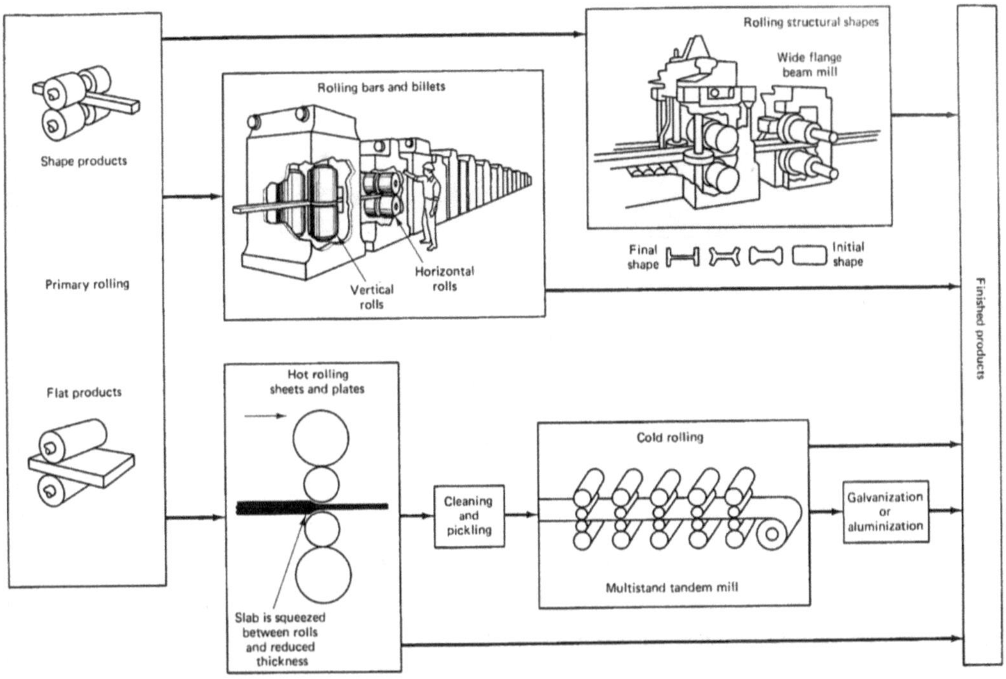

FIGURE 5.4 The sequence of operations involved in manufacturing rolled products.

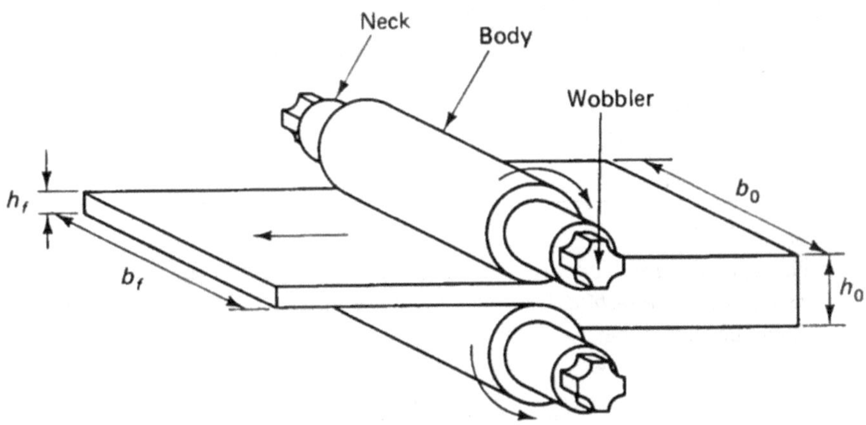

FIGURE 5.5 Simple rolling of a plate.

But because the volume of the work is constant, it follows that

$$\eta = \frac{h_{o \times b_o}}{h_{f \times b_f}} \tag{5.3}$$

Equation (5.3) indicates that the coefficient of elongation is adversely proportional to the ratio of the final to the original cross-sectional areas of the work.

As can be seen in Figure 5.6, the metal is deformed in the shaded area, or *deformation zone*. The metal remains unstrained before this area and does not undergo any further deformation after it. It can also be seen that the metal undergoing deformation is in contact with each of the rolls along the

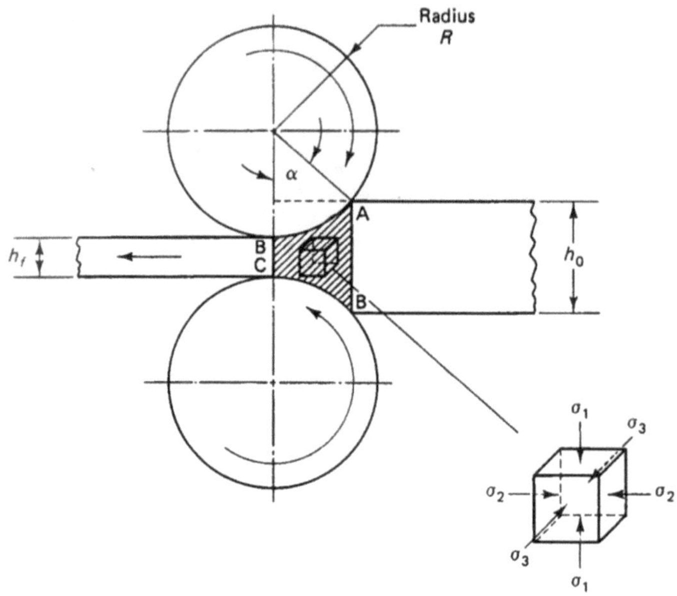

FIGURE 5.6 The deformation zone, state of stress, and angle of contact in rolling.

TABLE 5.1
Maximum Allowable Angle of Contact for Rolling

Rolling Process	Maximum Allowable Angle of Contact
Rolling of blooms and heavy sections	24° to 30°
Hot rolling of sheets and strips	15° to 20°
Cold rolling of lubricated sheets	2° to 10°

arc AB, which is called the *arc of contact*. It corresponds to a central angle, ∝, that is, in turn, called the *angle of contact*, or *angle of bite*. From the geometry of the drawing and by employing simple trigonometry, it can be shown that

$$\cos \propto = 1 - \frac{h_o - h_f}{2R} = 1 - \frac{\Delta h}{2R} \tag{5.4}$$

Equation (5.4) gives the relationship between the geometrical parameters of the rolling process, the angle of contact, the draft, and the radius of the rolls. Note that in order to ensure that the metal will be shifted by friction, the angle of contact must be less than β, the angle of friction, where tan β = μ (the coefficient of friction between roll surface and metal). In fact, the maximum permissible value for the angle of contact depends upon other factors, such as the material of the rolls, the work being rolled, and the rolling temperature and speed. Table 5.1 indicates the recommended maximum angle of contact for different rolling processes.

5.2.2 Load and Power Requirement

As can also be seen in Figure 5.6, the main stress system in the deformation zone in a rolling process is triaxial compression, with the maximum (principal) stress acting normal to the direction of rolling. The deformed metal exerts an equal counterforce on each of the rolls to satisfy the equilibrium conditions. Therefore, this force normal to the direction of rolling is important when doing the design calculations for the rolls as well as the mill body. It is also important in determining the power consumption in a rolling process. Unfortunately, the exact determination of that rolling load and power consumption is complicated and requires knowledge of the theory of plasticity as well as calculus. Nevertheless, a first approximation of the roll load can be given by the following simple equation:

$$F = Y^{-} \times b \times \sqrt{R \times \Delta h} \tag{5.5}$$

where Y^{-} is the average (plane-strain) yield stress assuming no spread and is equal to 1.15 Y, where Y is the mean yield stress of the metal.

Therefore, Equation (5.5) should take the following form:

$$F = 1.15 \times Y \times b \times \sqrt{R \times \Delta h} \tag{5.6}$$

Equation (5.6) neglects the effect of friction at the roll-work interface and, therefore, gives lower estimates of the load. Based on experiments carried out on a wide range of rolling mills, this equation can be modified to account for friction by multiplying by a factor of 1.2. The modified equation is

$$F = 1.2 \times 1.15 \times Y \times b \times \sqrt{R \times \Delta h} \tag{5.7}$$

The power consumed in the process cannot be obtained easily; however, a rough estimate in low-friction conditions is given by

$$h_p = \frac{Y^{-} \times b \times R \times \Delta h \times \omega}{550} \tag{5.8}$$

where

ω is the angular velocity of rolls in radians per second
Y^{-}, b, R, and Δh are all in English units

5.2.3 Rolling Mills

A rolling mill includes one or more roll stands and a main drive motor, reducing gear, stand pinion, flywheel, and coupling gear between the units. The roll stand is the main part of the mill, where the rolling process is actually performed. It basically consists of housings in which antifriction bearings that are used for carrying (mounting) the rolls are fitted. Moreover, there is a screw-down mechanism to control the gap between the rolls and thus the required thickness of the product.

Depending upon the profile of the rolled product, the body of the roll may be either smooth for rolling sheets (plates or strips) or grooved for manufacturing shapes such as structural members. A roll consists of a body, two necks (one on each side), and two wobblers (see Figure 5.5). The body is the part that contacts and deforms the metal of the workpiece. The necks rotate in bearings that act as supports, while the wobblers serve to couple the roll to the drive. Rolls are usually made from high-quality steel and sometimes from high-grade cast iron to withstand the very severe service conditions to which the rolls are subjected during the rolling process, such as combined bending and torque, friction and wear, and thermal effects. Gray cast-iron rolls are employed in roughing passes when hot rolling steel. Cast- or forged-steel rolls are used in blooming, slabbing, and section mills as well as in cold-rolling mills. Forged rolls are stronger and tougher than cast rolls. Alloy-steel rolls made of chrome–nickel or chrome–molybdenum steels are used in sheet mills.

5.2.4 Classification of Rolling Mills

Rolling mills are classified according to the number and arrangement of the rolls in a stand. Following are the five main types of rolling mills, as shown in Figure 5.7a–e.

1. *Two-High Rolling Mills*. Two-high rolling mills, the simplest design, have a two-high stand with two horizontal rolls. This type of mill can be nonreversing (unidirectional), where the rolls have a constant direction of rotation, or reversing, where the rotation and direction of metal passage can be reversed.
2. *Three-High Rolling Mills*. Three-high rolling mills have a three-high stand with three rolls arranged in a single vertical plane. This type of mill has a constant direction of rotation, and it is not required to reverse that direction.
3. *Four-High Rolling Mills*. In sheet rolling, the rolls should be designed as small as possible in order to reduce the rolling force F of the metal on the rolls and the power requirement. If such small-diameter rolls are used alone, they will bend and result in nonuniform thickness distribution along the width of the sheet, as shown in Figure 5.8. For this reason, another two backup rolls are used to minimize bending and increase the rigidity of the system. The four rolls are arranged above one another in a vertical plane. Also, the backup rolls always have larger diameters than those of the working rolls.
4. *Multihigh Rolling Mills (Sendzimir Mills)*. Multihigh rolling mills are used particularly in the manufacture of very thin sheets, those with a thickness down to 0.0005 inches (0.01 mm) and a width up to 80 inches (2000 mm), into coils. In this case, the working rolls must have very small diameters (to reduce load and power consumption, as explained before), usually in the range of 3/8 inch (10 mm) to 1.25 inches (30 mm). Such small-diameter working rolls make a drive practically impossible. They are, therefore, driven by friction through an

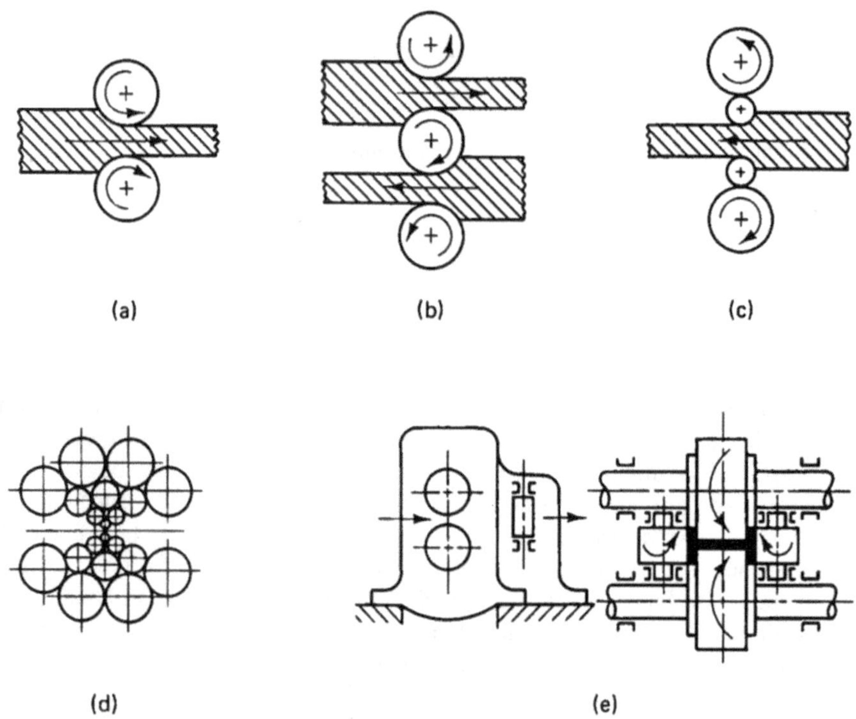

FIGURE 5.7 The five main types of rolling mills: (a) two-high rolling mill, (b) three-high rolling mill, (c) four-high rolling mill, (d) multihigh rolling mill, and (e) universal rolling mill.

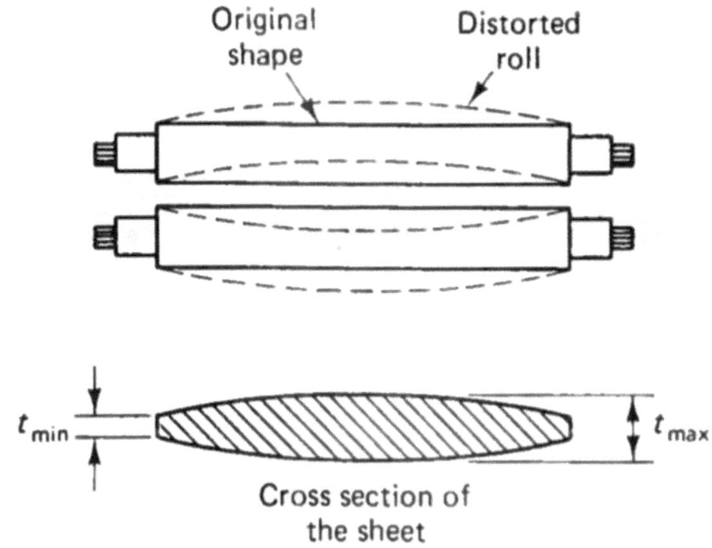

FIGURE 5.8 Rolling thin sheets with small-diameter rolls.

intermediate row of driving rolls that are, in turn, supported by a row of backup rolls. This arrangement involves a cluster of 12 or 20 rolls, resulting in exceptional rigidity of the whole roll system and an almost complete absence of working-roll deflections. An equivalent system that is sometimes used is the planetary rolling mill, in which a group of small-diameter working rolls rotate around a large, idle supporting roll on each side of the work.

5. *Universal Rolling Mills*. Universal rolling mills are used for producing blooms from ingots and for rolling wide-flange H beams (Gray's beams). In this type of mill, there are vertical rolls in addition to the horizontal ones. The vertical rolls of universal mills (for producing structural shapes) are idle and are arranged between the bearing chocks of the horizontal rolls in the vertical plane.

5.2.5 The Range of Rolled Products

The range of rolled products is standardized in each country in the sense that the shape, dimensions, tolerances, properties, and the like are given in a standard specifications handbook that differs from country to country. The whole range of rolled products can generally be divided into the following four groups.

1. *Structural Shapes or Sections*. The first group includes general-purpose sections like round and square bars, angles, and channels, H- and I-beams, as well as special sections (with intricate shapes) like rails and special shapes used in construction work and industry. Figure 5.9 shows a variety of sections that belong to this group. These products are rolled in either rail mills or section mills, where the body of each roll has grooves called *passes* that are made in the bodies of the upper and lower rolls in such a manner as to lie in the same vertical plane. They are used to impart the required shape to the work. This process is carried out gradually (i.e., the stock is partly deformed at each stand, or pass, in succession). The skill of a rolling engineer is to plan and construct the details of a system of successive passes that ensures the adequate rolling of blanks into the desired shape.

   This operation is called *roll pass design*. Figure 5.10a illustrates the roll passes for producing rails; Figure 5.10b, those for producing an I-beam.

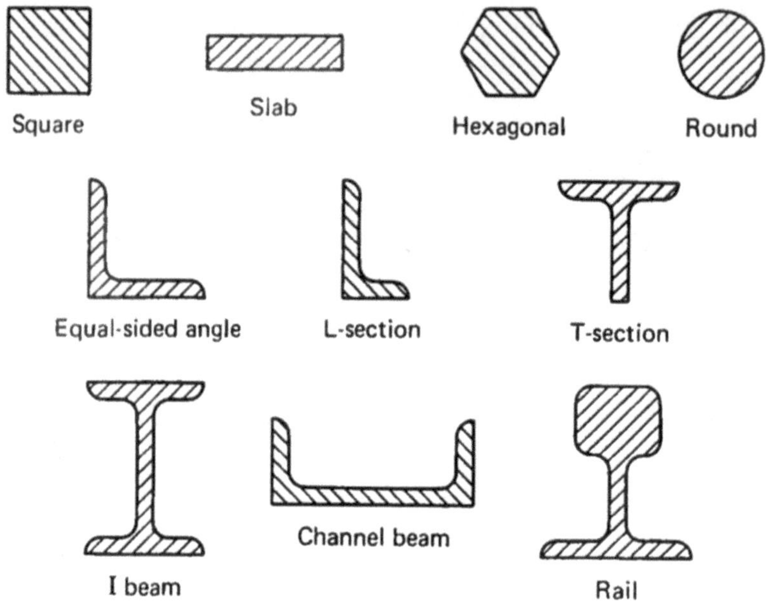

FIGURE 5.9 Some structural shapes or sections produced by rolling.

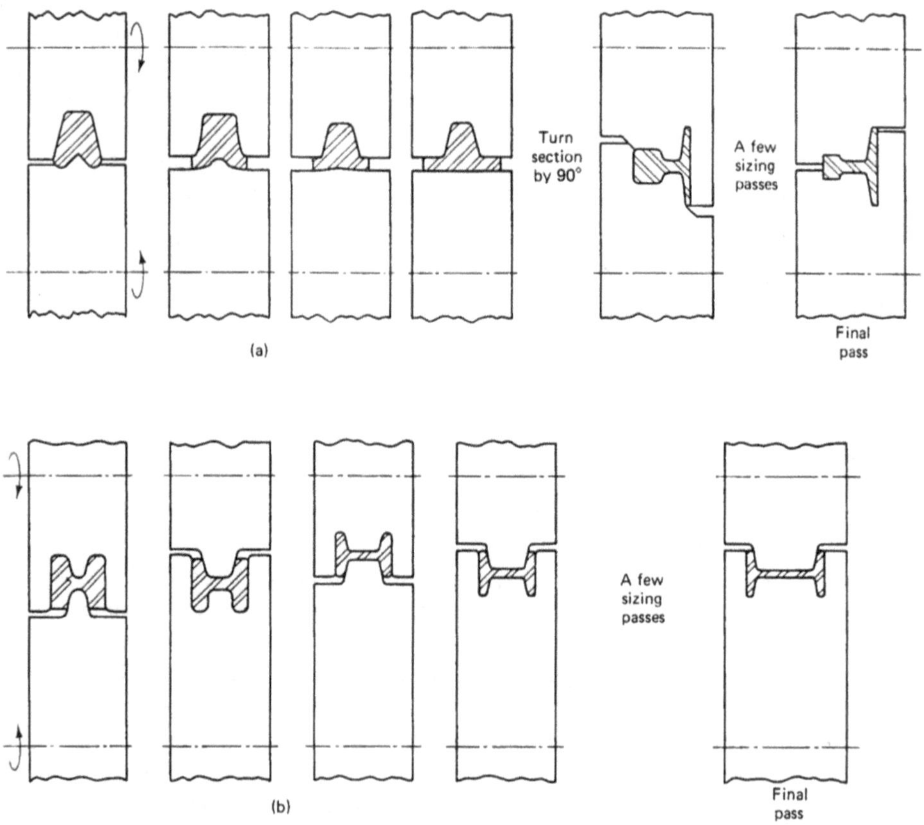

FIGURE 5.10 Roll passes: (a) for producing rails and (b) for producing an I-beam.

2. *Plates and Sheets*. Plates and sheets are produced in plate and sheet mills for the hot rolling of metal and in cold reduction mills for the production of cold-rolled coils, where multihigh rolling mills are employed, as previously mentioned. This group of products is classified according to thickness. A flat product with a width ranging from 5/32 inch (4 mm) to 4 inches (100 mm) is called a *plate*, whereas wider and thinner flat stocks are called *sheets*.
3. *Special-Purpose Rolled Shapes*. This group includes special shapes, one-piece rolled wheels, rings, balls, ribbed tubes, and die-rolled sections in which the cross section of the bar varies periodically along its length. These kinds of bars are used in the machine-building industry and in the construction industry for reinforcing concrete beams and columns. Figure 5.11a shows the sequence of operations in manufacturing a rolled wheel for railway cars; Figure 5.11b, the wheel during the final stage in the rolling mill.
4. *Seamless Tubes*. The process of manufacturing seamless tubes involves two steps:
   a. Piercing an ingot or a roughened-down round blank to form a thick-walled shell
   b. Rolling the obtained shell into a hollow thin-walled tube having the desired diameter and wall thickness

In the first step, the solid blank is center-drilled at one end, heated to the appropriate temperature, and then placed in the piercing mill and forced into contact with the working rolls. There are several types of piercing mills, but the commonly used one has barrel-shaped rolls. As Figure 5.12 shows, the axes of the two rolls are skew lines, each deviating with a small angle from the direction of the blank axis. Also, the two rolls rotate in the same direction, forcing the blank to rotate and proceed against a mandrel. A hole is formed and becomes larger; finally, a rough tube is obtained. The milling stand is provided with side rollers for guiding the blank and the formed rough tube during this operation. In the second step, the hollow shell (rough tube) is usually forced over another mandrel, and the combination is longitudinally rolled at their state between grooved rolls. Mills of different types are used, including continuous, automatic, and pilger mills. Finally, a sizing operation may be performed, between sizing rolls and without the use of a mandrel, at room temperature in order to improve the properties and finish of the tubes.

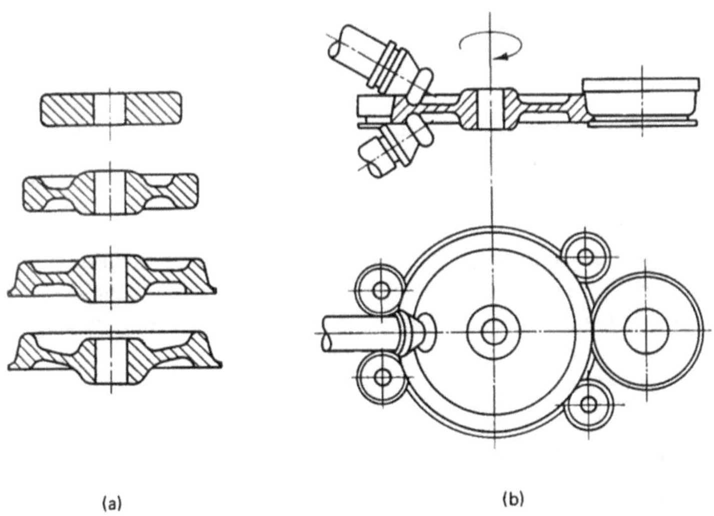

FIGURE 5.11 The production of a rolled railway car wheel: (a) sequence of stages and (b) wheel in the final stage.

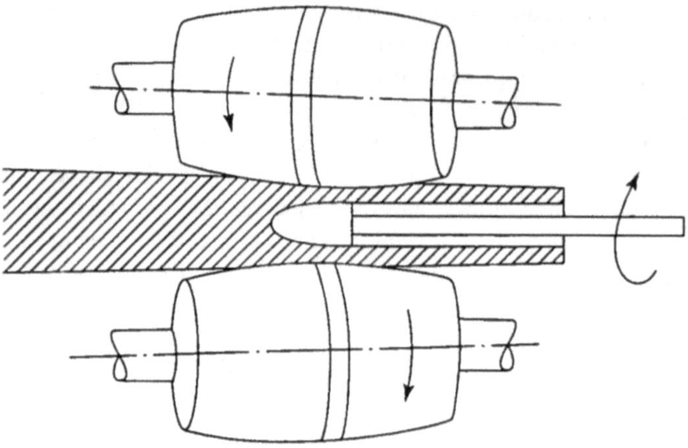

FIGURE 5.12 The production of seamless tube by rolling.

5.2.6 Lubrication in Rolling Processes

Friction plays a very important role in a rolling process and has some beneficial effects, provided that it is not excessive. In fact, it is responsible for shifting the work between the rolls and should not, therefore, be eliminated or reduced below an appropriate level. This is an important point to be taken into account when choosing a lubricant for a rolling process.

In the cold rolling of steel, fluid lubricants of low viscosity are employed, but paraffin is suitable for nonferrous materials like aluminum or copper alloys to avoid staining during subsequent heat treatment. On the other hand, hot rolling is often carried out without lubricants but with a flood of water to generate steam and break up the scales that are formed. Sometimes, an emulsion of graphite or graphited grease is used.

5.2.7 Defects in Rolled Products

A variety of defects in the products arise during rolling processes. A particular defect is usually associated with a particular process and does not arise in other processes. Following are some of the common defects in rolled products.

1. *Edge Cracking*. Edge cracking occurs in rolled ingots, slabs, or plates and is believed to be caused by either limited ductility of the work metal or uneven deformation, especially at the edges.
2. *Alligatoring*. Figure 5.13 shows the defect of *alligatoring*, which is less common than it used to be. It usually occurs in the rolling of slabs (particularly aluminum alloys), where the workpiece splits along a horizontal plane on exit, with the top and bottom parts following the rotation of their respective rolls. This defect always occurs when the ratio of slab thickness to the length of contact falls within the range of 1.4–1.7.
3. *Folds*. Folds are defects occurring during plate rolling when the reduction per pass is too small.
4. *Laminations*. Laminations associated with cracking may develop when the reduction in thickness is excessive.

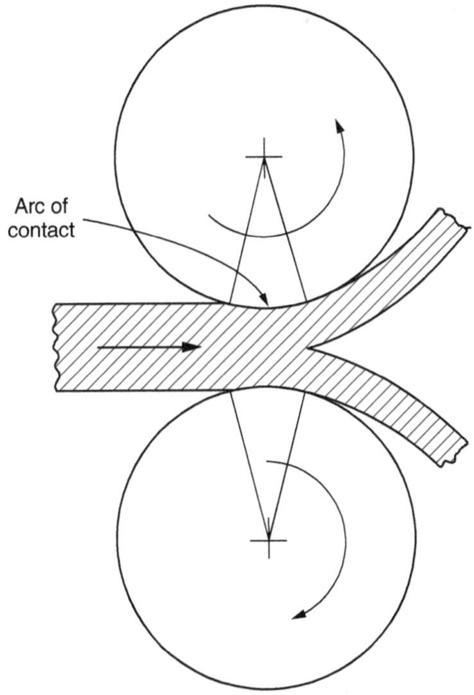

FIGURE 5.13 Alligatoring when rolling aluminum slabs.

5.3 METAL DRAWING

Drawing is basically a forming process that involves pulling a slender semifinished product (like wire, bar stock, or tube) through a hole of a drawing die. The dimensions of that hole are smaller than the dimensions of the original material. Metals are usually drawn in their cold state, and the required shape may be achieved in a single drawing operation or through several successive drawing operations, in which case the diameters of the holes are successively decreasing. Sometimes, annealing is carried out between the drawing operations to relieve the metal from work hardening. Accurate dimensions, good surface quality, increased strength and hardness, and the possibility of producing very small sections are some advantages of the drawing process. The drawing process has, therefore, wide industrial applications and is used for manufacturing thin wires, thin-walled tubes, and components with sections that cannot be made except by machining. It is also used for sizing hot-rolled sections.

5.3.1 Preparing the Metal for Drawing

Before being subjected to the drawing process, metal blanks (wires, rods, or tubes) are heat treated and then cleaned of scales that result from that operation. Descaling is usually done by pickling the heat-treated metal in acid solutions. Steels are pickled in either sulfuric or hydrochloric acid or a mixture of both; copper and brass blanks are treated in sulfuric acid, whereas nickel and its alloys are cleaned in a mixture of sulfuric acid and potassium bichromate. After pickling, the metal is washed to remove any traces of acid or slag from its surface. The final operation before drawing is drying the washed blanks at a temperature above 212°F (100°C). This eliminates the moisture and a great deal of the hydrogen dissolved in the metal, thus helping to avoid pickling brittleness.

If steel is to be subjected to several successive drawing passes, its surface should then be conditioned for receiving and retaining the drawing lubricant. Conditioning is performed directly after pickling and can take the form of sulling, coppering, phosphating, or liming. In sulling, the steel rod is given a thin coat of iron hydroxide, which combines with lime and serves as a carrier for the lubricant. Phosphating involves applying a film of iron, manganese, or zinc phosphates to which lubricants stick very well. Liming neutralizes the remaining acid and forms a vehicle for the lubricant. Coppering is used for severe conditions and is achieved by immersing the steel rods (or wires) in a solution of vitriol. All conditioning operations are followed by drying at a temperature of about 650°F (300°C) in special chambers.

5.3.2 Wiredrawing

1. *Drawing Dies*. A *die* is a common term for two parts: the die body and the die holder. Die bodies are made of cemented carbides or hardened tool steel, whereas die holders are made of good-quality tool steel that possesses high toughness. The constructional details of a die are shown in Figure 5.14. It can be seen from the figure that the die opening involves four zones: entry, working zone, die bearing, and exit. The *entry zone* allows the lubricant to reach the working zone easily and also protects the wire (or rod) against scoring by sharp edges. The *working zone* is conical in shape and has an apex angle that ranges between 6° and 24°, depending upon the type of work and the metal being drawn. The die bearing, sometimes called the land, is a short cylindrical zone in which a sizing operation is performed to ensure the accuracy of the shape and dimensions of the end product. The exit zone provides back relief to avoid scoring of the drawn wire (or rod). In a wiredrawing operation, the end of the wire is pointed by swaging and then fed freely into the die hole so that it appears behind the die. This pointed end is gripped by the jaws of a carriage that pull the wire through the die opening, where it undergoes a reduction in cross-sectional area and elongation in length.
2. *Draw Benches*. A wiredrawing operation usually involves the use of multidie draw benches, where the wire passes through a series of draw plates. First, the wire leaves the coil and passes through the first drawing die. Then, it is wound two or three turns around a capstan (drum) before it enters the next drawing die. A typical draw bench of this type with six draw plates is shown in Figure 5.15. In practice, a bench may include from 2 to 22 draw

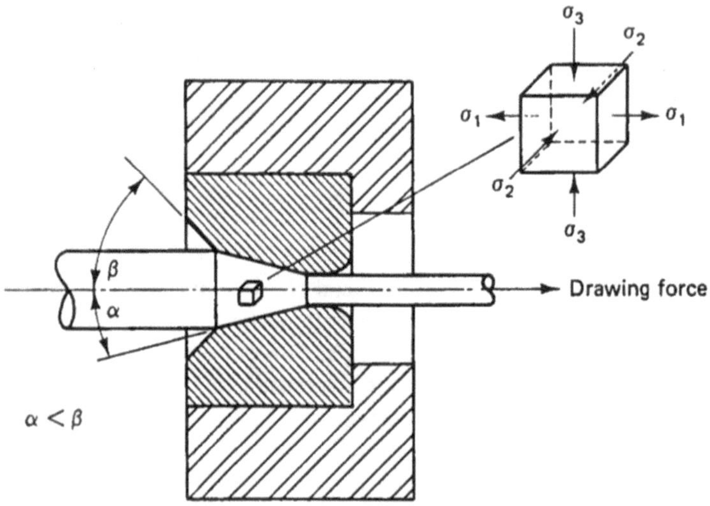

FIGURE 5.14 The constructional details of a drawing die.

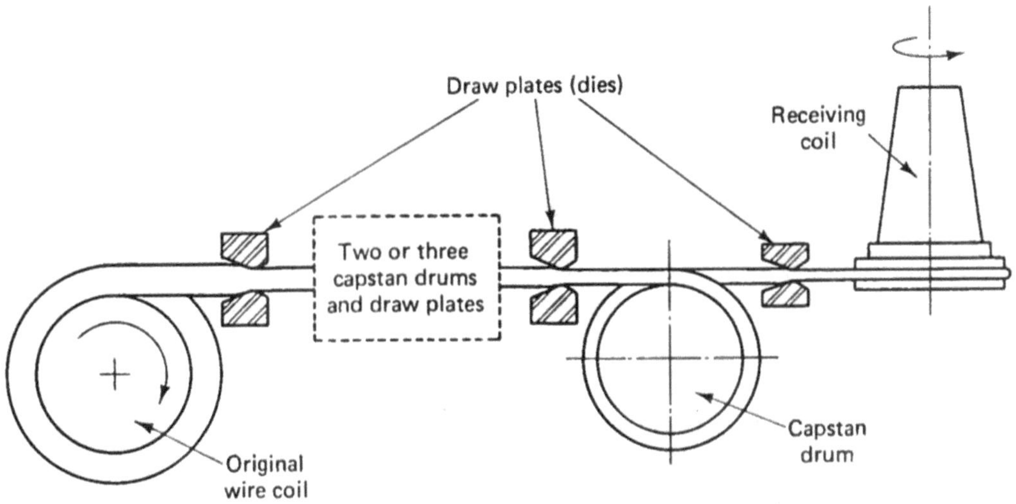

FIGURE 5.15 A typical multidie draw bench.

plates, and the wire leaving the last die may attain a velocity of 9800 feet per minute (50 mis). The capstan drives are designed to provide not only a forward pull after each pass but also a backward pull to the wire before it enters the next drawing die.

3. *Lubrication*. Lubrication reduces the required drawing force and the energy consumed during the process, increases the service life of the die, and allows a smoother wire surface to be obtained. Various kinds of soap are used as lubricants in wiredrawing processes. Examples are sodium soap or calcium stearate, which is picked up by the wire from a soap box adjacent to the die. Although they are difficult to apply and remove, polymers are also used as solid lubricants, especially in severe conditions, as in the case of drawing hard alloys or titanium. Various kinds of mineral and vegetable oils containing fatty or chlorinated additives are also used as drawing lubricants.
4. *Mechanics of Wiredrawing*. The state of stress during the wiredrawing process (see Figure 5.14) involves compressive forces along two of the directions and tension along the third one. An approximate but simple estimate of the drawing force can be given by the following equation:

$$F = a_f \times Y \times \ln\left(\frac{a_o}{a_f}\right) \tag{5.9}$$

where

a_o is the original area
a_f is the final area
Y is the mean yield stress of the metal

$$\text{Reduction } r = \frac{a_o - a_f}{a_o} \times 100 \tag{5.10}$$

The theoretically obtained maximum value for the reduction is 64 percent; however, it usually does not exceed about 40 percent in industry.

5. *Defects in Wiredrawing*. Structural damage in the form of voids or cracks occurs in different forms in wiredrawing processes under certain conditions. Following are some of the defects encountered:
   a. Internal bursts in wire, taking the form of repeating internal cup and cone fractures (cuppy wire), usually occur when drawing heavily cold-worked copper under conditions of light draft and very large die angles.
   b. Similar centerline arrowhead fractures occur if the blank is a sheet and when the die angle and reduction produce severe tension on the centerline.
   c. Transverse surface cracking may occur as a result of longitudinal tension stresses in the surface layers.

5.3.3 Tube Drawing

Drawing can reduce the diameter and thickness of pipes. Figure 5.16 illustrates the simplest type of tube drawing. The final tube thickness is affected by two contradicting factors. The longitudinal stress tends to make the wall thinner, whereas the circumferential stress thickens it. If a large die angle is used, the thinning effect will dominate.

The technique shown in Figure 5.17 of using a fixed plug reduces the tube diameter and controls its thickness. However, a disadvantage of this type of tube drawing is the limitation imposed on the

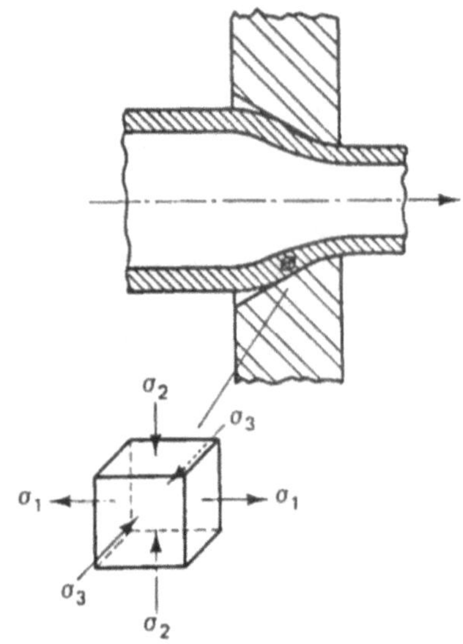

FIGURE 5.16 The simplest type of tube drawing.

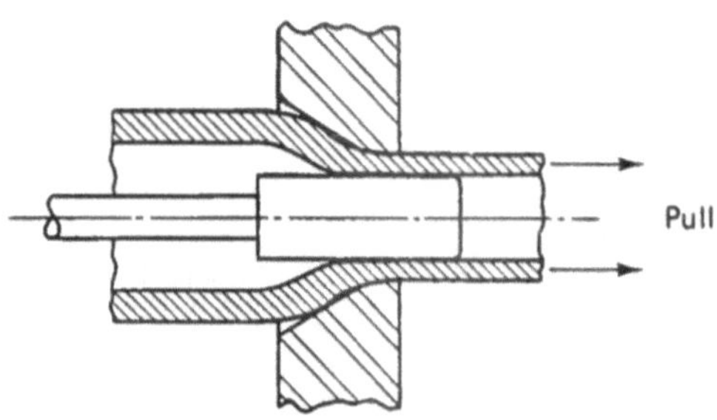

FIGURE 5.17 Tube drawing using a fixed plug.

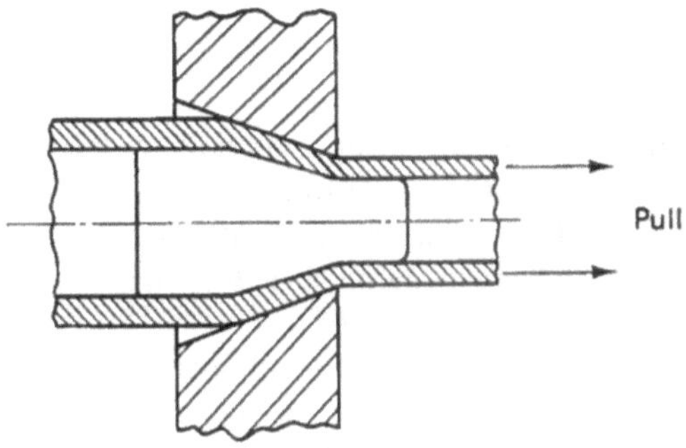

FIGURE 5.18 Tube drawing using a floating mandrel.

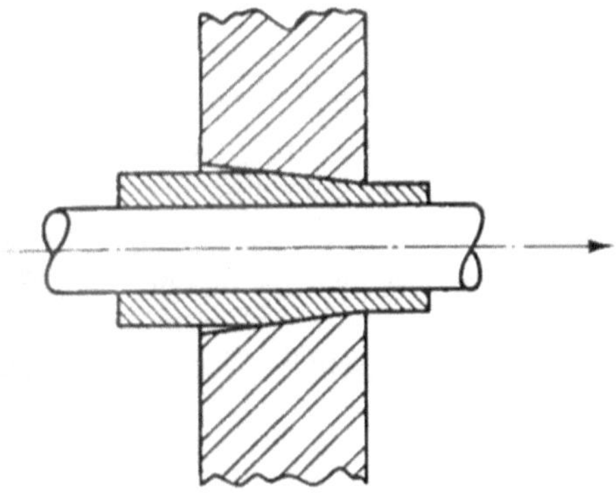

FIGURE 5.19 Tube drawing using a removable mandrel.

length of the tube by the length of the mandrel. When tubes having longer lengths are to be drawn, a floating mandrel like that shown in Figure 5.18 is then employed. Another method that has gained widespread application is using a removable mandrel like that shown in Figure 5.19.

5.4 EXTRUSION

Extrusion involves forcing a billet that is enclosed in a container through an opening whose cross-sectional area and dimensions are smaller than those of the original billet. The cross section of the extruded metal will conform to that of the die opening.

Historically, extrusion was first used toward the end of the eighteenth century for producing lead pipes. It later gained widespread industrial application for processing nonferrous metals and alloys like copper, brass, aluminum, zinc, and magnesium. Recently, with the modern developments in extrusion techniques, lubricants, and tooling, other metals, such as steels, titanium, refractory metals, uranium, and thorium, can also be extruded successfully. The stock used for extrusion is mainly a cast ingot or a rolled billet. Any surface defects in the original billets must be removed by sawing, shearing, turning, or any other appropriate machining operation before the extrusion process is performed. Extrusion carried out when the billets are at their cold state is known as *cold extrusion*; when they are at elevated temperatures, it is known as *hot extrusion*. In this latter case, the container, the die, and the pressing plunger must be heated to a temperature of about 650°F (350°C) prior to each extrusion cycle.

5.4.1 Types of Extrusion

5.4.1.1 Direct Extrusion

Direct extrusion is used in the manufacture of solid and hollow slender products and for structural shapes that cannot be obtained by any other metal-forming process. Figure 5.20 illustrates the working principles of this method, and Figure 5.21 shows the details of an extrusion die arrangement for producing channel sections. As can be seen, during an extrusion process, a billet is pushed out of the die by a plunger and then slides along the walls of the container as the operation proceeds. At the end of the stroke, a small piece of metal (stub-end scrap) remains unextruded in the container.

The extruded product is separated by shearing, and the stub-end is then ejected out of the container after the plunger is withdrawn. Also, the leading end of the extruded product does not undergo enough deformation. It is, therefore, poorly shaped and must be removed as well. Obviously, the efficiency of material utilization in this case is low, and the waste can amount to 10 percent or even 15 percent, as opposed to rolling, where the waste is only 1–3 percent. This makes the productivity of direct extrusion quite inferior to that of rolling Figure 5.22 illustrates the technique used for producing hollow sections and tubes. As can be seen, a mandrel or a needle passes freely through a hole in the blank and the die opening. If the die opening is circular, an annular clearance between the die opening and the mandrel results. When the metal is extruded through the annular clearance, it forms a tube. A hole has to be pierced or drilled into the original blank before it is extruded.

Based on this discussion, it is clear that the conventional extrusion process has the advantages of high dimensional accuracy and the possibility of producing complex sections from materials having poor plasticity. On the other hand, its disadvantages include low productivity, short tool life, and expensive tooling. Therefore, the process is usually employed for the manufacture of complex shapes with high dimensional accuracy, especially when the material of the product has a low ductility.

5.4.1.2 Indirect Extrusion

In indirect extrusion, the extrusion die is mounted on a hollow ram that is pushed into the container. Consequently, the die applies pressure to the billet, which undergoes plastic deformation. As shown in Figure 5.23, the metal flows out of the die opening in a direction opposite to the ram motion.

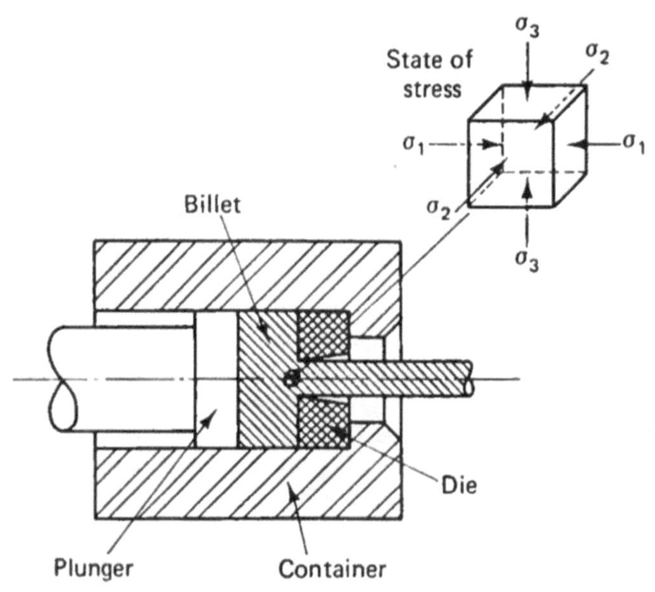

FIGURE 5.20 Principles of direct extrusion for producing solid objects.

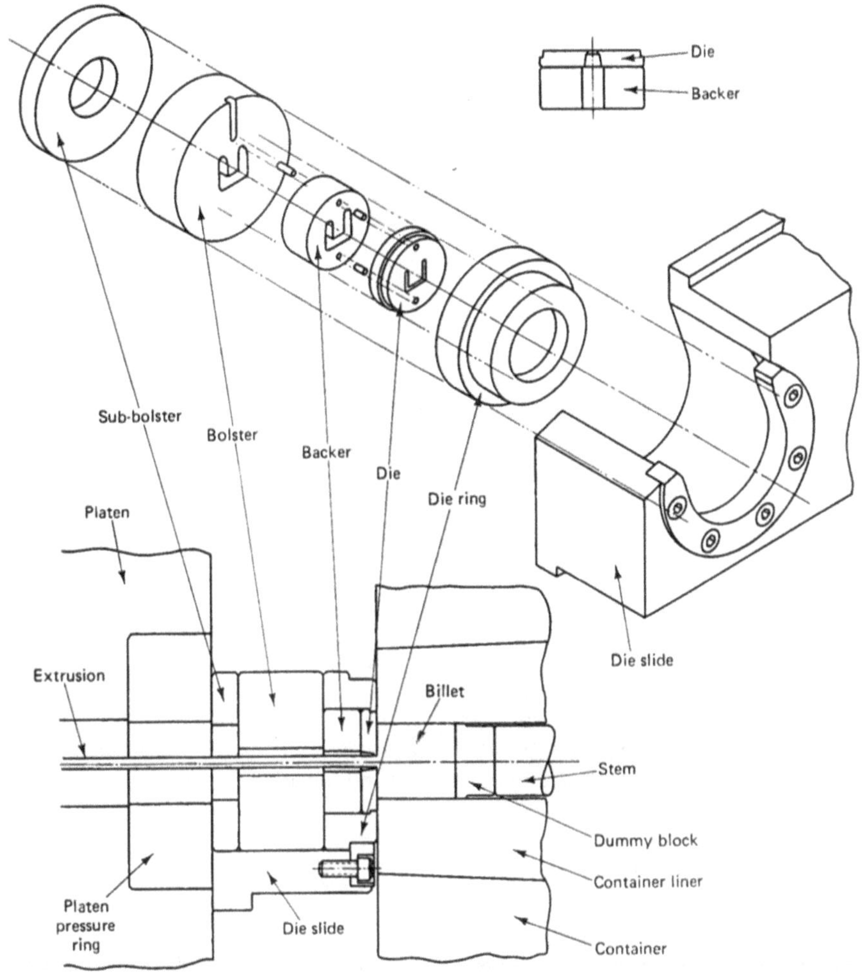

FIGURE 5.21 Typical extrusion die arrangement for producing channel sections.

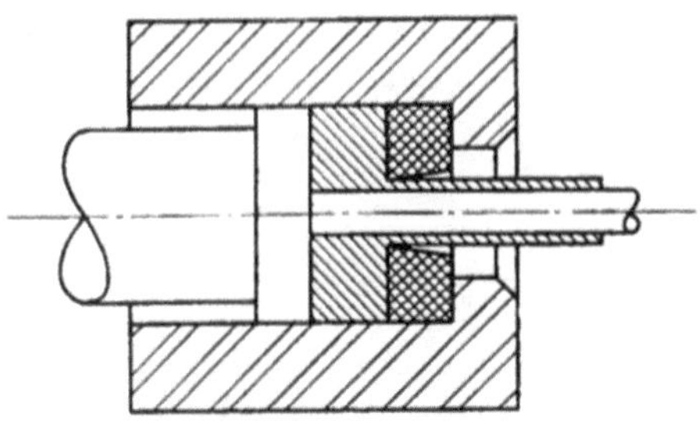

FIGURE 5.22 Direct extrusion for producing hollow objects.

There is almost no sliding motion between the billet and the container walls. This eliminates friction, and the extrusion load will be lower than that required in forward direct extrusion by about 30 percent. Also, the amount of waste scrap is reduced to only 5 percent. Nevertheless, indirect extrusion finds only limited application due to the complexity and the cost of tooling and press arrangement required.

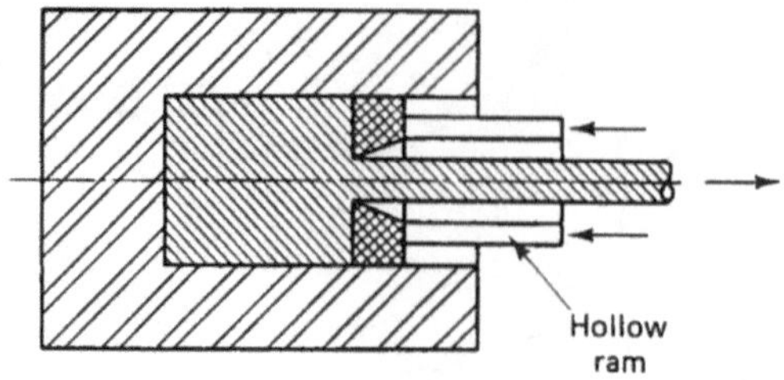

FIGURE 5.23 Indirect extrusion for producing solid objects.

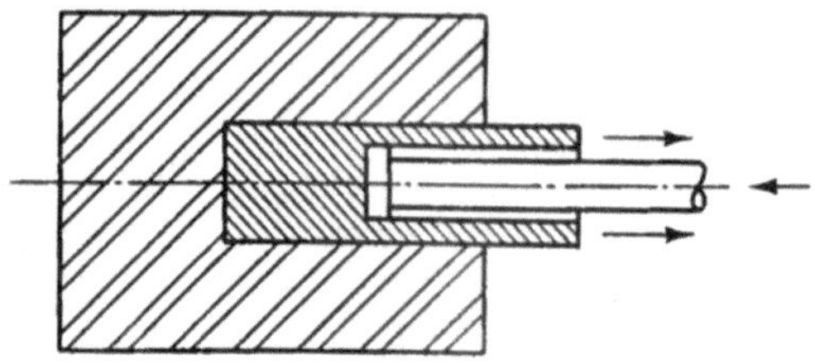

FIGURE 5.24 Indirect (backward) extrusion for producing hollow objects.

Another indirect extrusion method, usually called backward or reverse extrusion, used in manufacturing hollow sections is shown in Figure 5.24. In this case, the metal is extruded through the gap between the ram and the container. As in indirect extrusion for solid objects, the ram and the product travel in opposite directions.

5.4.1.3 Hydrostatic Extrusion

A radical development that eliminates the disadvantages of cold extrusion (like higher loads) involves hydrostatic extrusion. Figure 5.25 illustrates the basic principles of this process, where the billet is shaped to fit the die and surrounded by a high-pressure hydraulic fluid in a container. When the plunger is pressed, it increases the pressure inside the container, and the resulting high pressure forces the billet to flow through the die. Friction between the billet and the container is thus eliminated, whereas friction between the billet and the die is markedly reduced. Also, the buckling effect of longer billets is eliminated because virtually the entire length of the billet is subjected to hydrostatic pressure. This makes it possible to extrude very long billets.

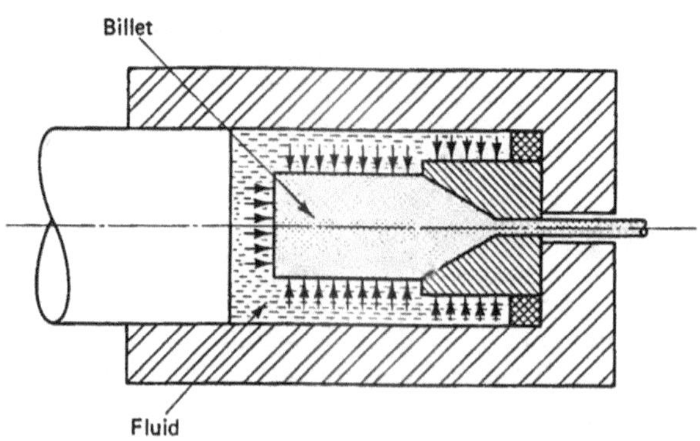

FIGURE 5.25 Principles of hydrostatic extrusion.

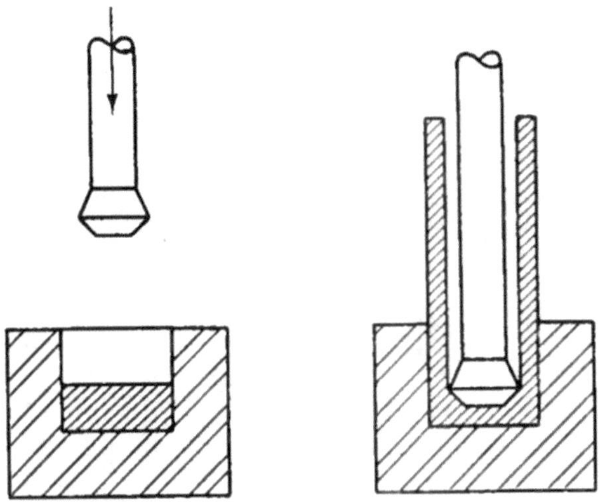

FIGURE 5.26 Principles of reverse impact extrusion.

5.4.1.4 Impact Extrusion

Impact extrusion involves striking a cold *slug* of soft metal (like aluminum) that is held in a shallow die cavity with a rapidly moving punch, thus causing the metal to flow plastically around the punch or through the die opening. The slug itself is a closely controlled volume of metal that is lubricated and located in the die cavity. The press is then activated, and the high-speed punch strikes the slug. A finished impacted product is extruded with each stroke of the press. These products are not necessarily cylindrical with a circular cross section. In fact, the range of shapes possible is very broad, including even irregular symmetrical shapes. There are three types of impact extrusion processes: forward, reverse, and combination (the names referring to the direction of motion of the deforming metal relative to that of the punch).

Figure 5.26 illustrates the basic principles of *reverse impact* extrusion. It is used for manufacturing hollow parts with forged bases and extruded sidewalls. The flowing metal is guided only initially; thereafter, it goes by its own inertia. This results in the elimination of friction and, therefore, an appreciable reduction in the load and energy required. A further advantage is the possibility of producing thinner walls.

The principles of *forward* impact extrusion are illustrated in Figure 5.27. It is mainly employed in producing hollow or semihollow products with heavy flanges and multiple diameters formed on the inside and outside. Closer wall tolerances, larger slenderness ratios, better concentricities, and sound thinner sections are among the advantages of this process.

Complex shapes can be produced by a *combination* of the two preceding processes, which are performed simultaneously in the same single stroke, as shown in Figure 5.28. Like the other impact extrusion methods, this process has the advantage of cleaner product surfaces, elimination of trimming or further machining operations, and higher strength of the parts obtained.

5.4.2 Load Requirement

For the sake of simplicity, it is sometimes assumed that the processes involve ideal deformation without any friction. The extrusion pressure can then be given by the following equation:

$$P_{\text{extrusion}} = Y \times \ln\frac{a_o}{a_f} = Y \times \ln R \tag{5.11}$$

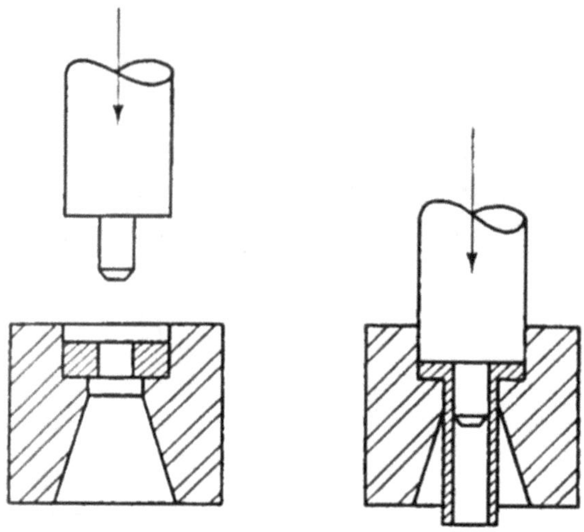

FIGURE 5.27 The principles of forward impact extrusion.

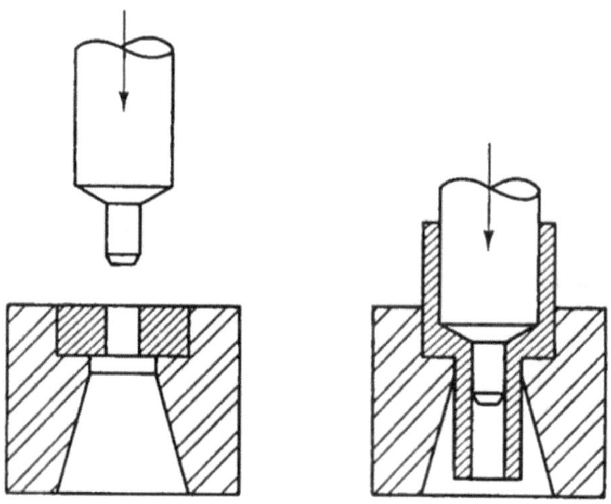

FIGURE 5.28 Combination impacting.

where

a_o is the original cross-sectional area
a_f is the final cross-sectional area after extrusion
R is the extrusion ratio
Y is the mean yield stress of the metal

The extrusion load is, therefore,

$$F = P \times a_o \tag{5.12}$$

These equations are used to give only rough estimates because actual extrusion processes involve friction and the lack of homogeneous deformation of the metal, as will be seen later. Therefore, research workers developed several empirical formulas to give the extrusion pressure as a function of the extrusion ratio and the mechanical properties of the metal. A convenient formula was proposed by W. Johnson (the eminent British researcher in the area of metal forming) as follows:

$$P_{\text{extrusion}} = Y^{-}\left(0.8 + 1.5\ln\frac{1}{1-r}\right) \tag{5.13}$$

$$r = \frac{a_o - a_f}{a_o} \tag{5.14}$$

5.4.3 Metal Flow and Deformation

To study metal flow, let us consider extruding a billet involving two identical halves, with a rectangular grid engraved on the meridional plane of each half. The separation surface is covered with lanolin or similar appropriate material to prevent welding or sticking of the two halves during the process. After extruding the split billet, the two halves are separated, and the distortion of the grid can be investigated. Figure 5.29 shows the grid after extrusion. We can see that the units of the grid, which were originally square in shape, became parallelograms, trapezoids, and other shapes. The following can also be observed:

- The velocity of the core is greater than that of the outer layers.
- The outer layers are deformed to a larger degree than the core.
- The leading end of the extruded part is almost unreformed.
- The metal adjacent to the die does not flow easily, leading to the initiation of zones where little deformation occurs. These zones are called dead-metal zones.

In fact, the preceding method for studying the metal flow is usually used with models made of wax, plasticine, and lead for predicting any defect that may occur during the actual process so that appropriate precautions can be taken in advance.

5.4.4 Lubrication in Extrusion

Friction at the billet–die and billet–container interfaces increases the load and the power requirement and reduces the service life of the tooling. For these reasons, lubricants are applied to the die and container walls.

As in wiredrawing, soaps and various oils containing chlorinated additives or graphite are used as lubricants in the cold extrusion of most metals, whereas lanolin is usually used for the softer ones. For hot extrusion of mild steel, graphite is an adequate lubricant. It is not, however, recommended for high-temperature extrusions, such as extruding molybdenum at 3250°F (1800°C); in this case, glass is the most successful lubricant.

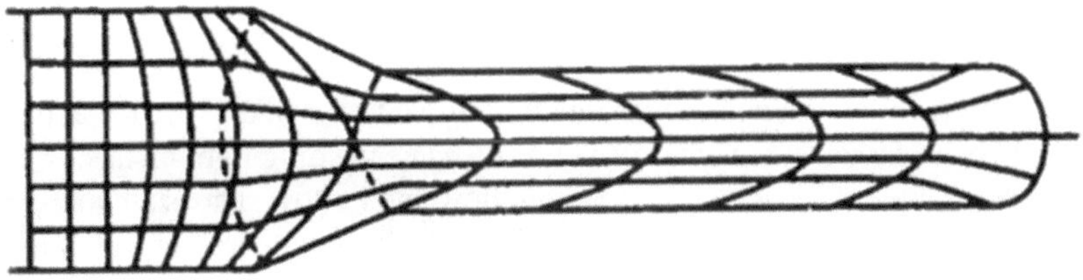

FIGURE 5.29 Distorted grid indicating metal flow in extrusion.

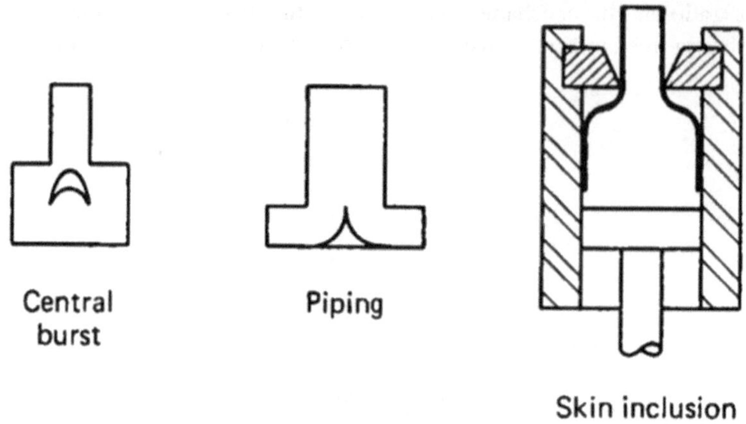

FIGURE 5.30 Three different defects occurring in an extrusion process.

5.4.5 Defects in Extruded Products

Defects in extruded parts usually fall into one of three main categories: surface or internal cracking, sinking (piping), and skin-inclusion defects.

Cracking is caused by secondary tensile stresses acting within a material having low plasticity. Cracking can occur on the surface of a relatively brittle material during the extrusion process, and it may also occur in the form of fire-tree or central bursts when extruding materials like bismuth, magnesium, 60/40 brass, steel, and brittle aluminum alloys.

Piping involves the sinking of the material at the rear of the stub-end. This defect is usually encountered toward the end of the extrusion stroke, especially when the original billets are relatively short.

Skin-inclusion defects may take different forms, depending upon the degree of lubrication and the hardness of the surface layer of the original stock. When extruding lubricated billets of high-copper alloys, the surface skin will slide over the container wall and then penetrate the billet, as illustrated in Figure 5.30, where the three different extrusion defects are sketched.

5.4.6 Design Considerations

5.4.6.1 Conventional Extrusions

When making parts that have constant cross sections, the extrusion process is usually more economical and faster than machining, casting, or fabricating the shapes by welding (or riveting). Also, the designer of the extruded section is relatively free to put the metal where he or she wants. Nevertheless, some design guidelines must be taken into consideration when designing an extruded section:

- The *circle size* (i.e., the diameter of the smallest circle that will enclose the extrusion cross section) can be as large as 31 inches (775 mm) when extruding light metals.
- Solid shapes are the easiest to extrude. Semihollow and hollow shapes are more difficult to extrude, especially if they have thin walls or include abrupt changes in wall thickness.
- Wall thicknesses must be kept uniform. If not, all transitions must be streamlined by generous radii at the thick-thin junctions.

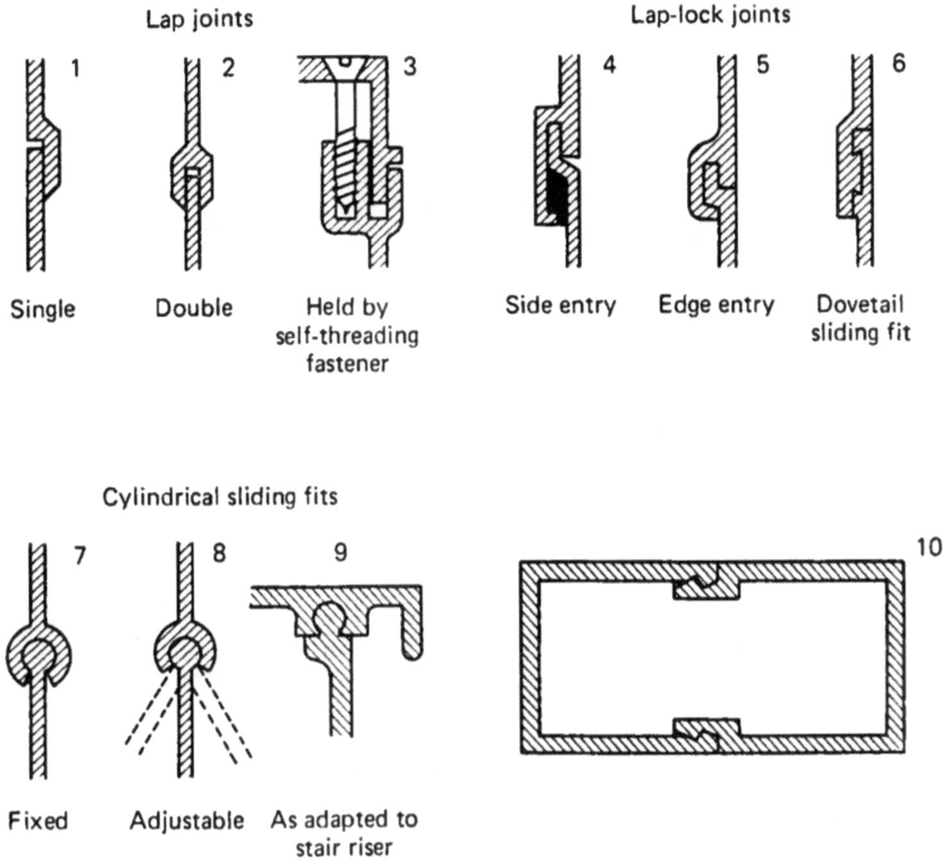

FIGURE 5.31 Some recommended designs for assembling extruded aluminum sections. (Courtesy of Aluminum Association, Arlington, VA.)

- Sharp corners at the root of a die tongue should be avoided when extruding semihollow sections.
- A complicated section should be broken into a group of simpler sections that are assembled after the separate extrusion processes. In such a case, the sections should be designed to simplify assembly; for example, they should fit, hook, or snap together. Screw slots or slots to receive other tightening material, such as plastic, may also be provided.

Figure 5.31 illustrates some recommended designs for assembling extruded aluminum sections. Figure 5.32 illustrates and summarizes some recommended designs as well as those to be avoided as general guidelines for beginning designers.

5.4.6.2 Aluminum Impact Extrusions

In order to accomplish good designs of aluminum impact extrusions, all factors associated with and affecting the process must be taken into account. Examples are alloy selection, tool design, lubrication, and, of course, the general consideration of mechanical design. Following are some basic guidelines and design examples:

- Use alloys that apply in the desired case and have the lowest strength.
- An impact extrusion should be symmetrical around the punch.

Poor	Good	Reason
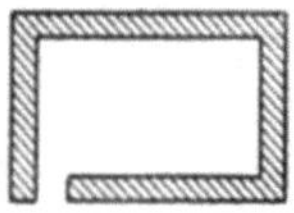	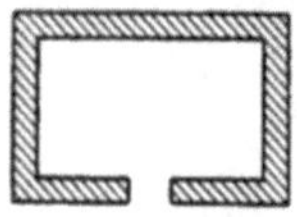	SYMMETRY PREFERRED IN SEMI-HOLLOW AREAS When designing, visualize the die and tongue that will be necessary to produce a semi-hollow shape. By keeping the void symmetrical, you lessen the chances that the die tongue may break.
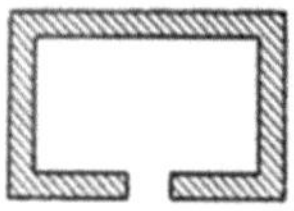	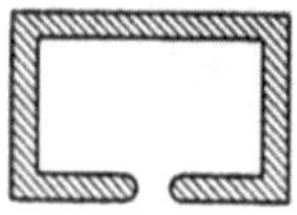	ROUNDED CORNER STRENGTHENS TONGUE The preceding cross section has been further improved. The die tongue is now less likely to snap off.
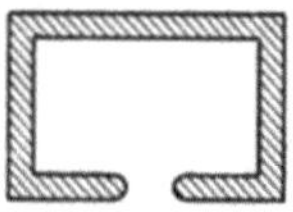	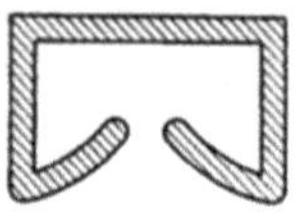	REDUCE AREA OF VOID–1 Further improvement in results if outline can be changed to reduce area enclosed. Reduced area means less pressure on the tongue; easier extrusion.
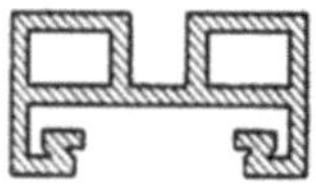	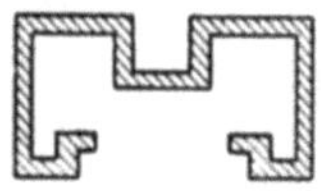	AVOID HOLLOW SHAPE Hollow and multi-hollow extruded shapes are usually much more costly than the simple solid shape. Also less metal has been used.
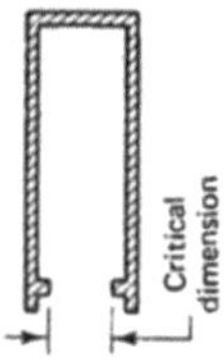	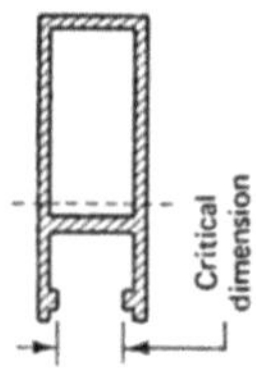	WEB GIVES BETTER DIMENSIONAL CONTROL Metal dimensions are more easily held than gap or angle dimensions. Web also allows thinner wall sections in this example. The hollow condition of the "redesigned" part can be avoided by making the component in two pieces as shown by the dotted line.
	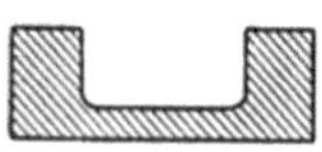	SMOOTH ALL TRANSITIONS Transitions should be streamlined by a generous radius at any thick–thin junction.
		KEEP WALL THICKNESS UNIFORM The preceding shape can be further improved by maintaining uniform wall thickness. In addition to using more metal, thick–thin junctions giv rise to distortion, die breakage or surface defects on the extrusion.
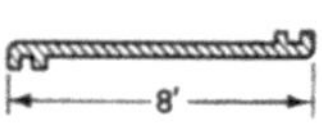	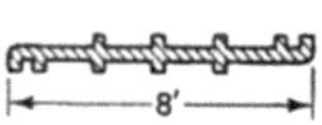	RIBS HELP STRAIGHTENING OPERATION Wide, thin sections can be hard to straighten after extrusion. Ribs help prevent twisting.

FIGURE 5.32 Some design considerations for conventional extrusions. (Courtesy of Aluminum Association, Arlington, VA.)

- Threads, cuts, projections, and the like are made by subjecting the impact extrusions to further processing.
- For reverse extrusions, the ratio of maximum length to internal diameter must not exceed 8 to avoid failure of long punches.
- A small outer-corner radius must be provided for a reverse extrusion, but the inner-corner radius must be kept as small as possible (see Figure 5.33a).
- The thickness of the bottom near the wall must be 15 percent greater than the thickness of the wall itself to prevent shear failure (see Figure 5.33b).
- The inside bottom should not be completely flat. To avoid the possibility of the punch skidding on the billet, only 80 percent of it at most can be flat (see Figure 5.33c).

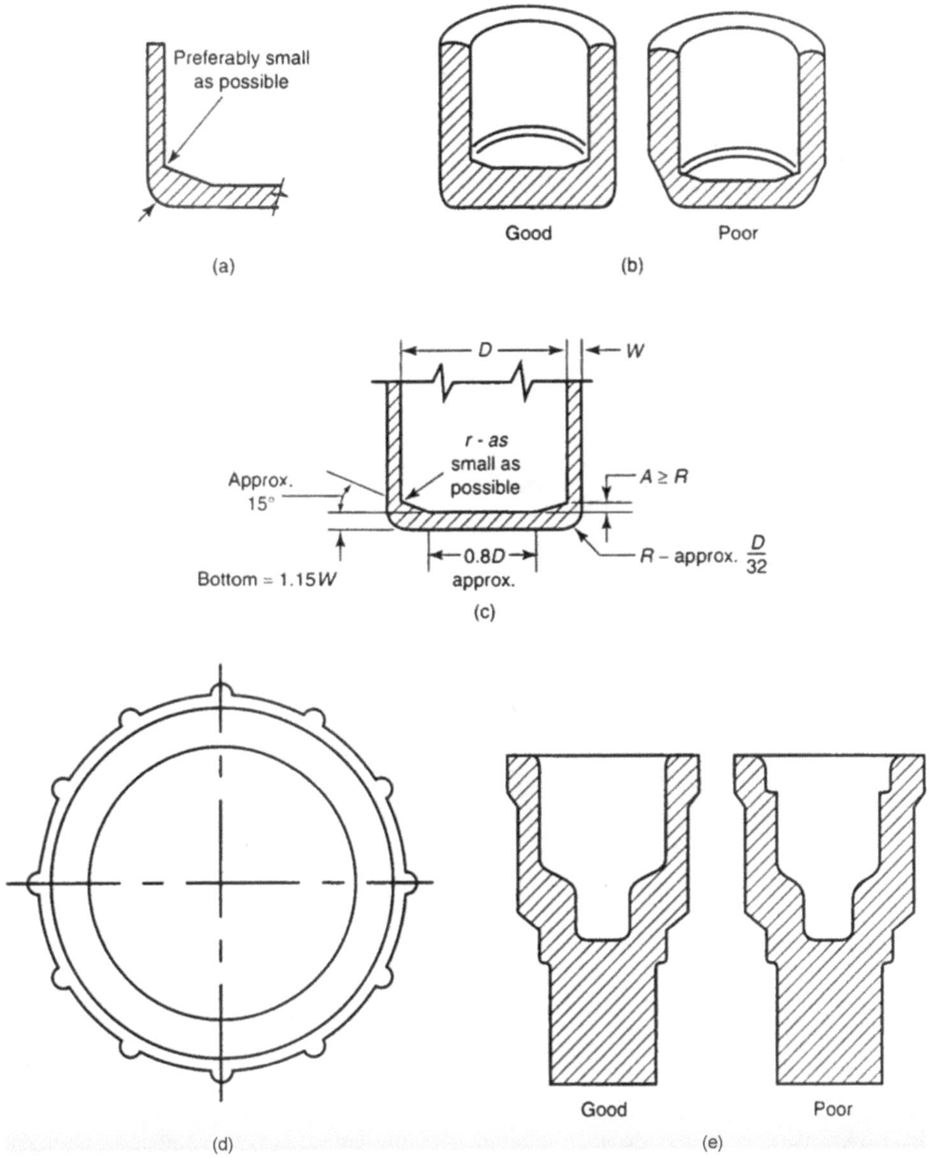

FIGURE 5.33 Some design considerations for impact extrusion: (a) corner radii for reverse extrusion, (b) thickness of the bottom near the wall, (c) inside bottom, (d) ribs, and (e) multiple-diameter parts. (Courtesy of Aluminum Association, Arlington, VA.)

- External and internal bosses are permitted, provided that they are coaxial with the part. However, the diameter of the internal boss should not be more than 1/4 of the internal diameter of the shell.
- Longitudinal ribs, whether external or internal, on the full length of the impact extrusion are permitted. They should preferably be located in a symmetrical distribution. However, the height of each rib must not exceed double the thickness of the wall of the shell (see Figure 5.33d). The main function of ribs is to provide stiffness to the walls of shells. They are also sometimes used for other reasons, such as to provide locations for drilling and tapping (to assemble the part) to enhance cooling by radiation, and to provide an appropriate gripping surface.

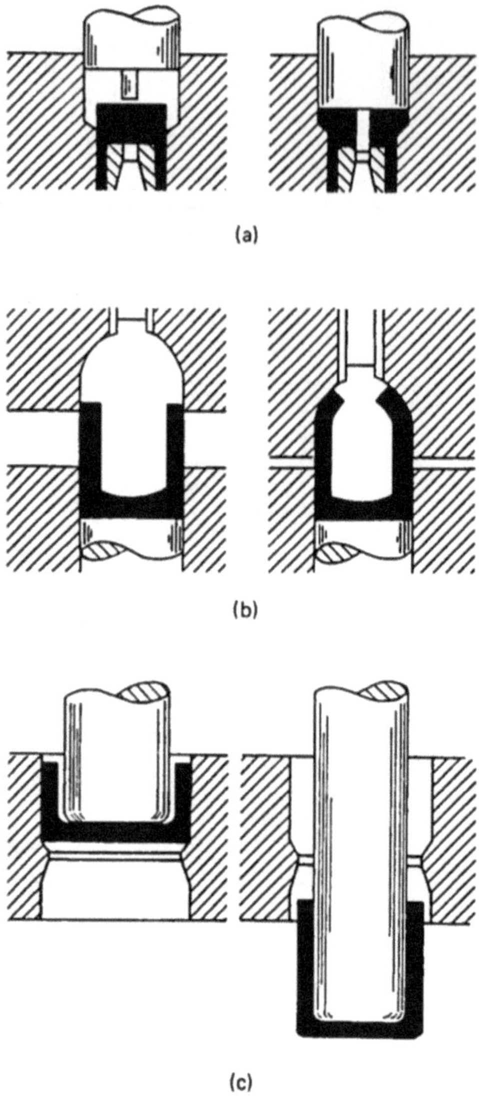

FIGURE 5.34 Some secondary operations after impact extrusion: (a) flange upsetting, (b) nosing, and (c) ironing. (Courtesy of Aluminum Association, Arlington, VA.)

- An impact extrusion can have a wall with varying thickness along its length (i.e., it can be a multiple-diameter part). However, internal steps near the top of the product should be avoided because they cause excessive loading and wear of the punch (see Figure 5.33e).
- Remember that it is sometimes impossible to obtain the desired shape directly by impacting. However, an impact extrusion can be considered an intermediate product that can be subjected to further working or secondary operations like machining, flange upsetting, nosing and necking, or ironing (see Figure 5.34a–c).
- Again, in addition to the preceding guidelines, general rules of mechanical design, as well as common engineering sense, are necessary for obtaining a successful design for the desired product. It would also be beneficial for the beginner to look at various designs of similar parts and to consult with experienced people before starting the design process. Given in Figure 5.35 are sketches reflecting good design practice for some impact-extruded tubular parts and shells.

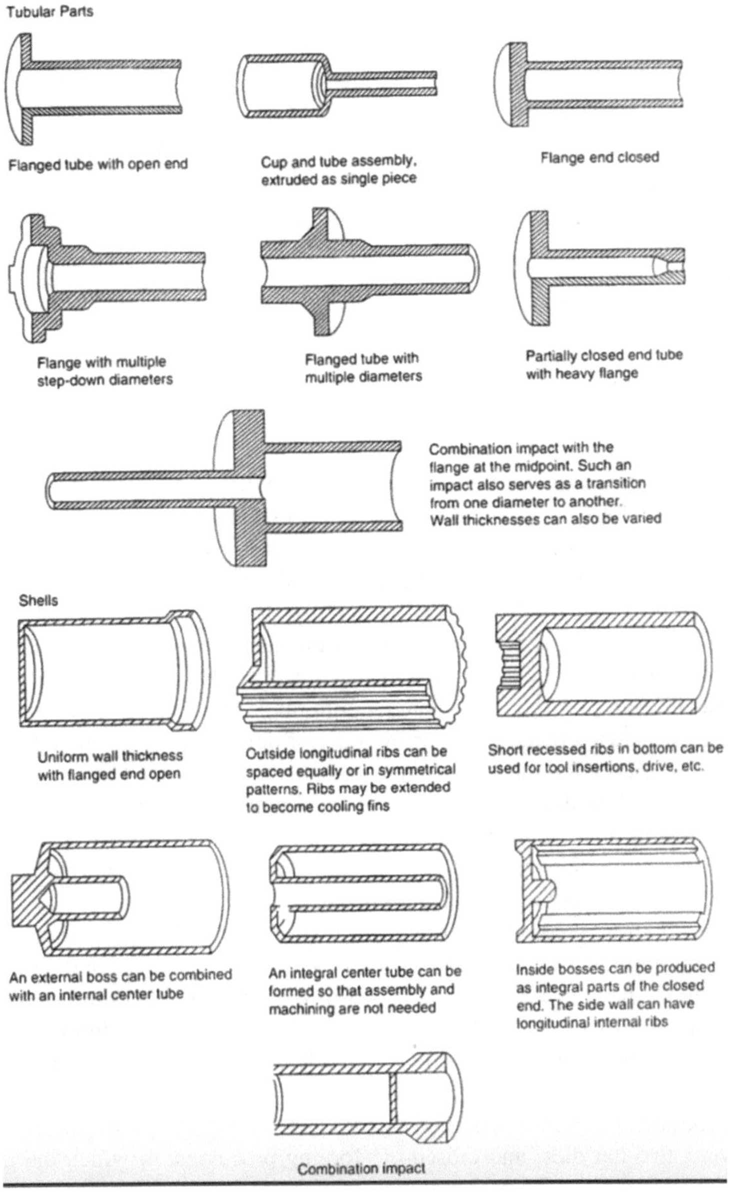

FIGURE 5.35 Sketches reflecting good design practice for some impact-extruded tubular parts and shells. (Courtesy of Aluminum Association, Arlington, VA.)

5.5 FORGING

5.5.1 General

The term *forging* is used to define the plastic deformation of metals at elevated temperatures into predetermined shapes using compressive forces that are exerted through dies by means of a hammer, a press, or an upsetting machine. Like other metal-forming processes, forging refines the microstructure of the metal, eliminates hidden defects such as hair cracks and voids, and rearranges the fibrous macrostructure to conform with the metal flow. It is mainly the latter factor that gives forging its merits and advantages over casting and machining. By successful design of the dies, the metal flow during the process can be employed to promote the alignment of the fibers with the anticipated

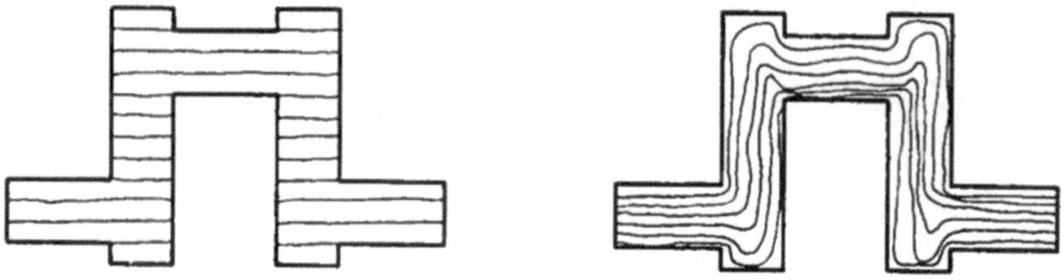

Produced by machining from a bar stock

Produced by forging

FIGURE 5.36 The fibrous macrostructure in two different crankshafts produced by machining and by forging.

direction of maximum stress. A typical example is shown in Figure 5.36, which illustrates the fibrous macrostructure in two different crankshafts produced by machining from a bar stock and by forging. As can be seen, the direction of the fibers in the second case is more favorable because the stresses in the webs when the crankshaft is in service coincide with the direction of fibers where the strength is maximum.

A large variety of materials can be worked by forging. These include low-carbon steels, aluminum, magnesium, and copper alloys, as well as many of the alloy steels and stainless steels. Each metal or alloy has its own plastic forging temperature range. Some alloys can be forged in a wide temperature range, whereas others have narrow ranges, depending upon the constituents and the chemical composition. Usually, the forging temperatures recommended for nonferrous alloys and metals are much lower than those required for ferrous materials. Table 5.2 indicates the range of forging temperatures for the commonly used alloys.

Forged parts vary widely in size, ranging from a few pounds (less than a kilogram) to 300 tons (3 MN), and can be classified into small, medium, and heavy forgings.

Small forgings are illustrated by small tools such as chisels and tools used in cutting and carving wood; medium forgings include railway-car axles, connecting rods, small crankshafts, levers, and hooks. Among the heavier forgings are shafts of powerplant generators, turbines, and ships, as well as columns of presses and rolls for rolling mills. Small and medium forgings are forged from rolled sections (bar stocks and slabs) and blooms, whereas heavier parts are worked from ingots.

All forging processes fall under two main types: open-die forging processes, in which the metal is worked between two flat dies, and closed-die forging processes, in which the metal is formed while being confined in a closed impression of a die set.

TABLE 5.2
Range of Forging Temperatures for Commonly Used Alloys

Metal	Forging Temperature
Low-carbon steel	1450–2550°F (800–1400°C)
Aluminum	645–900°F (340–480°C)
Magnesium	645–800°F (340–430°C)
Copper	800–1900°F (430–1040°C)
Brass	1100–1700°F (590–930°C)

5.5.2 Open-Die Forging

Open-die forging is sometimes referred to as *smith forging* and is actually a development or a modern version of a very old type of forging, blacksmithing, that was practiced by armor makers and craftspeople. Blacksmithing required hand tools and was carried out by striking the heated part repeatedly with a hammer on an anvil until the desired shape was finally obtained. Nowadays, blacksmith forging is used only when low production of light forgings is required, which is mainly in repair shops. Complicated shapes having close tolerances cannot be produced economically by this process.

The modern version of blacksmithing, open-die forging involves the substitution of a power-actuated hammer or press for the arm, hand hammer, and anvil of the smith. This process is used for producing heavy forgings weighing up to more than 300 tons, as well as for producing small batches of medium forgings with irregular shapes that cannot be produced by modern closed-die forging. The skill of the operator plays an important role in achieving the desired shape of the part by manipulating the heated metal during the period between successive working strokes. Accordingly, the shape obtained is just an approximation of the required one, and subsequent machining is always used in order to produce the part that accurately conforms to the blueprint provided by the designer.

5.5.2.1 Open-Die Forging Operations

A smith-forging process usually consists of a group of different operations. Among the operations employed in smith forging are upsetting, drawing out, fullering, cutting off, and piercing. The force and energy required differ considerably from one operation to another, depending upon the degree of "confinement" of the metal being worked. Following is a brief description of some of these operations.

1. *Upsetting*. Upsetting involves squeezing the billet between two flat surfaces, thus reducing its height due to the increase in the cross-sectional area. As can be seen in Figure 5.37a, the state of stress is uniaxial compression. In practice, however, the billets' surfaces in contact with the die are subjected to substantial friction forces that impede the flow of the neighboring layers of metal. This finally results in a heterogeneous deformation and in the barreling of the deformed billet. To obtain uniform deformation, the billet–die interfaces must be adequately lubricated.
2. *Drawing Out*. In drawing out, the workpiece is successively forged along its length between two dies having limited width. This results in reducing the cross-sectional area of the workpiece while increasing its length, as shown in Figure 5.37b. This operation can be performed by starting either at the middle or at the end of the workpiece. A large reduction in the cross-sectional area can be achieved by reducing the feed of the workpiece. The bite (i.e., the length of feed before the working stroke) ranges between 40 and 75 percent of the width of the forging die.
3. *Piercing Operation*. A piercing operation is performed in order to obtain blind or through holes in the billet. A through hole can be pierced directly in a short billet in a single stroke by employing a punch and a supporting ring, as shown in Figure 5.37c. On the other hand, billets with large height-to-diameter ratios are pierced while located directly on the die with the help of a piercer and possibly an extension piece as well, as shown in Figure 5.37d. In this latter case, the diameter of the piercer must not exceed 50 percent of that of the billet. For larger holes, hollow punches are employed. Also, holes can be enlarged by tapered punches.
4. *Cutting Off*. Cutting off involves cutting the workpiece into separate parts using a forge cutter or a suitable chisel. This is usually done in two stages, as can be seen in Figure 5.37e.
5. *Bending*. In bending, thinning of the metal occurs on the convex side at the point of localized bending (where bending actually takes place). It is, therefore, recommended to upset the metal at this location before bending is performed, as shown in Figure 5.37f, in order to obtain a quality bend.

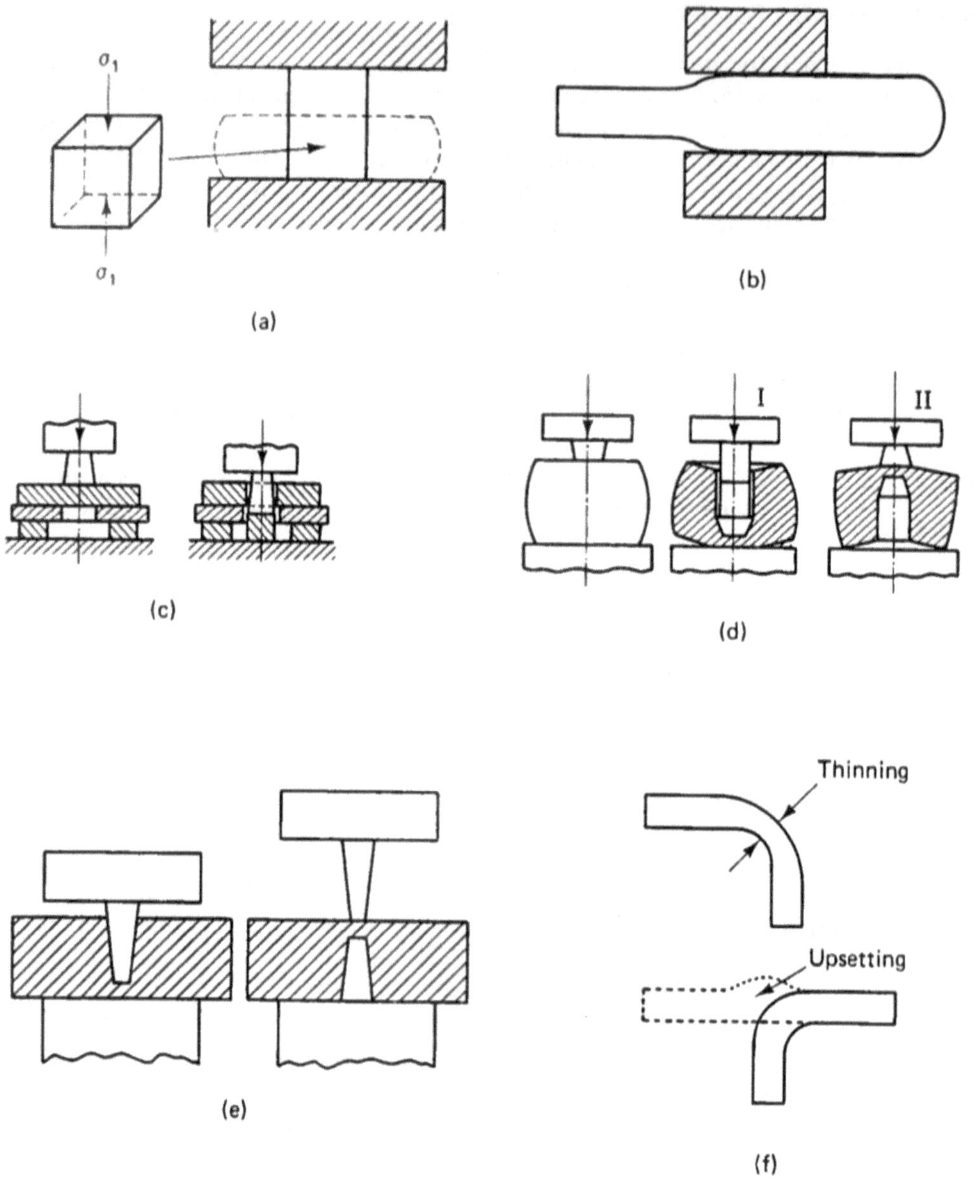

FIGURE 5.37 Various smith-forging operations: (a) upsetting, (b) drawing out, (c) piercing a short billet, (d) piercing a long billet, (e) cutting off, and (f) bending.

5.5.2.2 Examples of Open-Die Forged Parts

As mentioned before, a part may require a series of operations so that it can be given the desired shape by smith forging. Following are some examples of smith-forged industrial components, together with the steps involved in the manufacture of each part:

1. *Large Motor Shaft*. First, 24-inch-square (60 cm) steel ingots are rolled into square blooms, each having a 12-inch (30-cm) side. The blooms are then heated and hammered successively across the corners until the workpiece is finally rounded to a diameter of 10 inches (25 cm). These steps are illustrated in Figure 5.38.

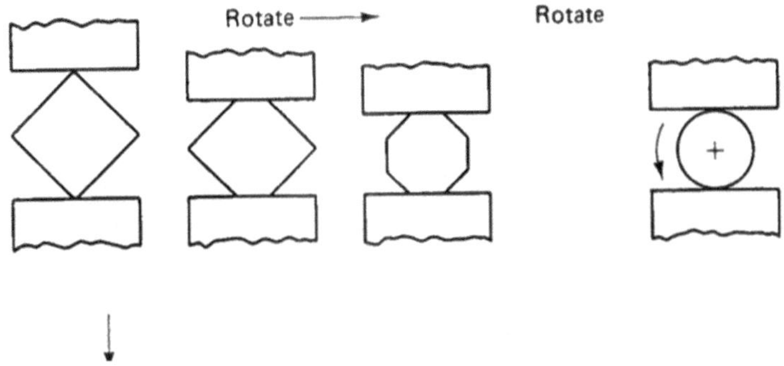

FIGURE 5.38 The production of large motor shaft by smith forging.

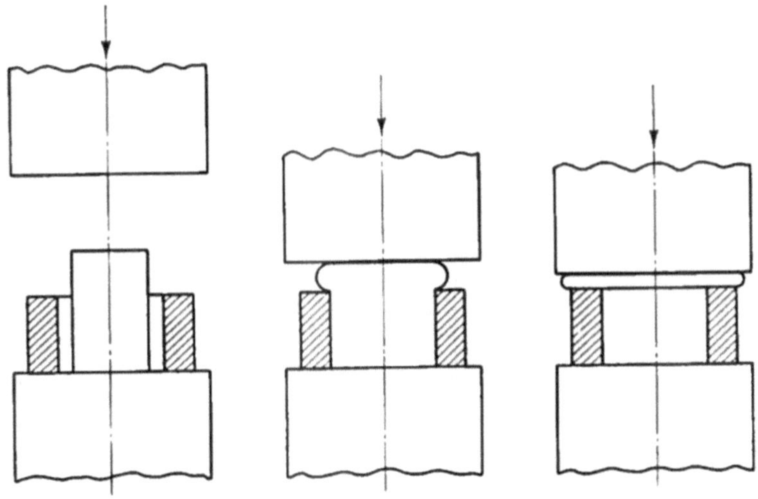

FIGURE 5.39 The production of flange coupling by smith forging.

2. *Flange Coupling*. The sequence of operations is illustrated in Figure 5.39. There are two operations or stages involved, upsetting and heading. In heading, the flow of metal of most of the billet is restricted by using a ring-shaped tool. This process allows excellent grain flow to be obtained, which is particularly advantageous in carrying tangential loads.
3. *Rings*. A billet is first upset and is then subjected to a piercing operation. This is followed by an expanding operation using a mandrel to reduce the thickness of the ring and increase its diameter as required. Larger rings are usually expanded on a saddle. The steps involved in the process of ring forging are illustrated in Figure 5.40.

5.5.2.3 Equipment for Smith Forging

Smaller billets are usually smith-forged using pneumatic-power hammers. Larger components are worked in steam-power hammers (or large pneumatic hammers), whereas very large and heavy parts are produced by employing hydraulic presses. Following is a brief description of smith-forging equipment.

1. *Steam-Power Hammers*. A steam-power hammer consists mainly of the moving parts (including the ram, the rod, and the piston); a lifting and propelling device, which is a double-acting high-pressure steam cylinder; the housing or frame, which can be either an

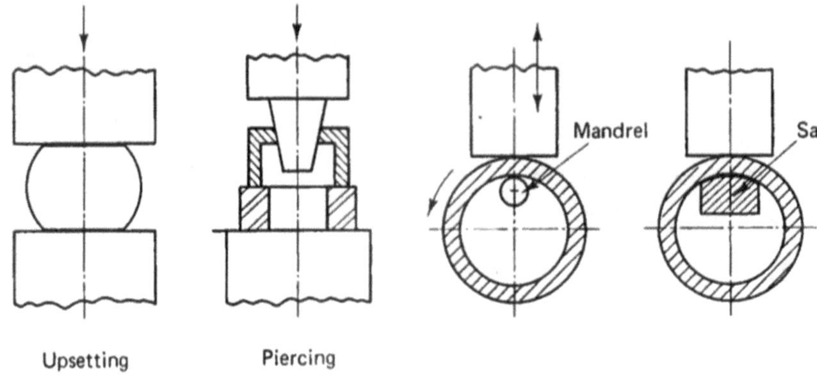

FIGURE 5.40 The production of large rings by smith forging.

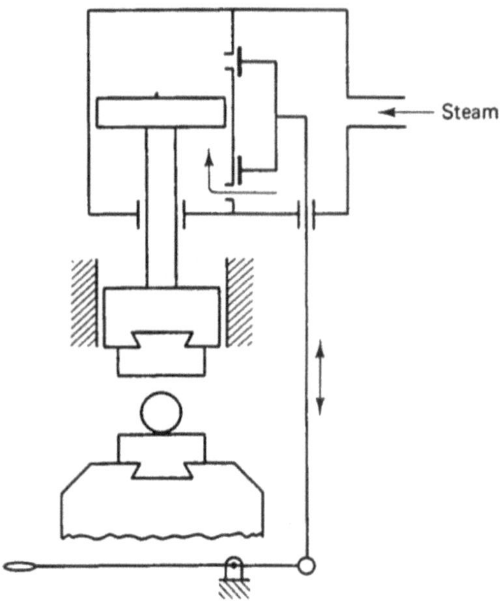

FIGURE 5.41 The working principles of a steam-power hammer.

arch or an open type; and the anvil. Figure 5.41 illustrates the working principles. First, the piston and the other moving parts are raised by admitting steam into the lower side of the cylinder (under the piston) through the sliding valve. When a blow is required, the lever is actuated; the sliding valve is accordingly shifted to admit steam to the upper side of the cylinder (above the piston) and exhaust the steam that was on the lower side, thus pushing the moving parts downward at a high speed. In steam-power hammers, the velocity of impact can be as high as 25 feet per second (3 mis), whereas the mass of the moving parts can be up to 11,000 slugs (5000 kg). The amount of energy delivered per blow is, therefore, extremely large and can be expressed by the following equation:

$$E = \frac{1}{2}\mathrm{m}V^2 \tag{5.15}$$

where
- E is the energy
- m is the mass of the moving parts
- V is the impact velocity

$$\eta = \frac{M}{M+m}\left(1-K^2\right) \quad (5.16)$$

The harder and more elastic the billet is, the higher that factor will be, and the lower the efficiency becomes. In addition, the hammer efficiency depends upon the ratio $M/(M+m)$, or actually the ratio between the masses of the anvil and the moving parts, which is taken in practice between 15 and 20. On the other hand, the value of K ranges between 0.05 and 0.25.

2. *Pneumatic-Power Hammers*. There are two kinds of pneumatic-power hammers. The first kind includes small hammers in which the air compressor is built in: they usually have open frames because their capacity is limited. The second kind of pneumatic hammer is generally similar to a steam-power hammer in construction and operation, the only difference being that steam is replaced by compressed air (seven to eight times the atmospheric pressure). As is the case with steam, this necessitates separate installation for providing compressed air. Pneumatic hammers do not have some of the disadvantages of steam hammers, such as the dripping of water resulting from the condensation of leakage steam onto the hot billet. This may result in cracking of the part, especially when forging steel.
3. *Hydraulic Presses*. Heavy forgings are worked in hydraulic presses. The press installation is composed of the press itself and the hydraulic drive. Presses capable of providing a force of 75,000 tons (750 MN) are quite common. Still, hydraulic presses that are commonly used in the forging industry have capacities ranging between 1000 tons (10 MN) and 10,000 tons (100 MN). These presses can successfully fully handle forgings weighing between 8 and 250 tons. The large-capacity presses require extremely high oil pressure in the hydraulic cylinders (200–300 times the atmospheric pressure). Because no pump can deliver an adequate oil discharge at that pressure level, this process is usually overcome by employing accumulators and intensifiers that magnify the oil pressure delivered by the pump by a factor of 40 or even 60.

5.5.2.4 Planning the Production of a Smith-Forged Part

Before actually smith-forging a part, all the details of the process must be thoroughly planned. This involves preparation of the design details, calculation of the dimensions and the weight of the stock and of the product, choosing the forging operations as well as their sequence, choosing tools and devices that will be used, and thinking about the details of the heating and cooling cycles.

The first step in the design process is to draw the finished part and then obtain the drawing of the forging by adding a machining as well as a forging allowance all around. The machining allowance is the increase in any dimension to provide excess metal that is removed by machining. This subsequent machining is required to remove scales and the chilled, defected surface layers. The forging allowance is added mainly to simplify the shape of the as-forged part. It is always recommended to make the shape of a forging symmetrical and confined by plane and cylindrical surfaces. At this stage, a suitable tolerance is assigned to each dimension to bring the design process to an end.

The next step is to choose the appropriate equipment. Two factors affect the decision: the size of the forging and the rate of deformation (strain rate). Usually, forgings weighing 2 tons or more are forged in hydraulic presses. Also, small forging made of high-alloy steels and some nonferrous alloys must be forged on a press because they are sensitive to high strain rates that arise when using power-hammer forging. At this point, the manufacturing engineer is in a position to decide upon operations, tools, devices, and the like needed to accomplish the desired task.

5.5.3 Closed-Die Forging

Closed-die forging involves shaping the hot forging stock in counterpart cavities or impressions that have been machined into two mating halves of a die set. Under impact (or squeezing), the hot metal plastically flows to fill the die cavity. Because the flow of metal is restricted by the shape of the impressions, the forged part accurately conforms to the shape of the cavity, provided that complete filling of the cavity is achieved. Among the various advantages of closed-die forging are the greater consistency of product attributes than in casting, the close tolerances and good surface finish with minimum surplus material to be removed by machining, and the greater strength at lower unit weight compared with castings or fabricated parts. In fact, the cost of parts produced by machining (only) is usually two to three times the cost of closed-die forgings. Nevertheless, the high cost of forging dies (compared with patterns, for example) is the main shortcoming of this process, especially if intricate shapes are to be produced. Therefore, the process is recommended for mass or large-lot production of steel and nonferrous components weighing up to about 900 lb (350 kg).

Generally, there are two types of closed-die forging: conventional (or flash) die forging and flashless die forging. In *conventional flash* die forging, the volume of the slug has to be slightly larger than that of the die cavity. The surplus metal forms a flash (fin) around the parting line. In *flashless* forging, no fin is formed, so the process consequently calls for accurate control of the volume of the slug. If the slug is smaller than the required final product, proper filling of the die cavity is not achieved. On the other hand, when the size of the slug is bigger than that of the desired forging, excessive load buildup will eventually result in the breaking of the tooling and/or equipment. Accordingly, flashless-forging dies are fitted with load-limiting devices to keep the generated load below a certain safe value in order to avoid breakage of the tooling.

In addition to shaping the metal in die cavities, the manufacturing cycle for a die-forged part includes some other related operations, such as cutting or cropping the rolled stock into slugs or billets, adequately heating the slugs, forging the slugs, trimming the flash (in conventional forging), heat treating the forgings, descaling, and, finally, inspecting or quality controlling. The forging specifications differ from one country to another; however, in order to ensure product quality, one or more of the following acceptance tests must be passed:

- Chemical composition midway between the surface and the center
- Mechanical properties
- Corrosion tests
- Nondestructive tests like magnetic detection of surface or subsurface hair cracks
- Visual tests such as macroetch, macroexamination, and sulfur painting for steel

Closed-die forging processes can be carried out using drop forging hammers, mechanical crank presses, and forging machines. Factors such as product shape and tolerances, quantities required, and forged alloys play an important role in determining the best and most economical equipment to be employed in forging a desired product as each of the processes has its own advantages and limitations. Following is a brief description of the different techniques used in closed-die forging.

5.5.3.1 Drop Forging

In drop forging, a type of closed-die forging, the force generated by the hammer is caused by gravitational attraction resulting from the freefall of the ram. The ram may be lifted by a single-acting steam (or air) cylinder or by friction rollers that engage a board tightly fastened to the ram. In this latter type, called a *board hammer*, once the ram reaches a predetermined desired height, a lever is actuated, the rollers retract, and the board and ram fall freely to strike the workpiece. Figure 5.42 illustrates the working principles. Whether a board hammer or single-acting steam hammer is used, accurate matching of the two halves of the die (i.e., the impressions) must be ensured. Therefore, the hammers employed in drop forging are usually of the double-housing (or arch) type and are

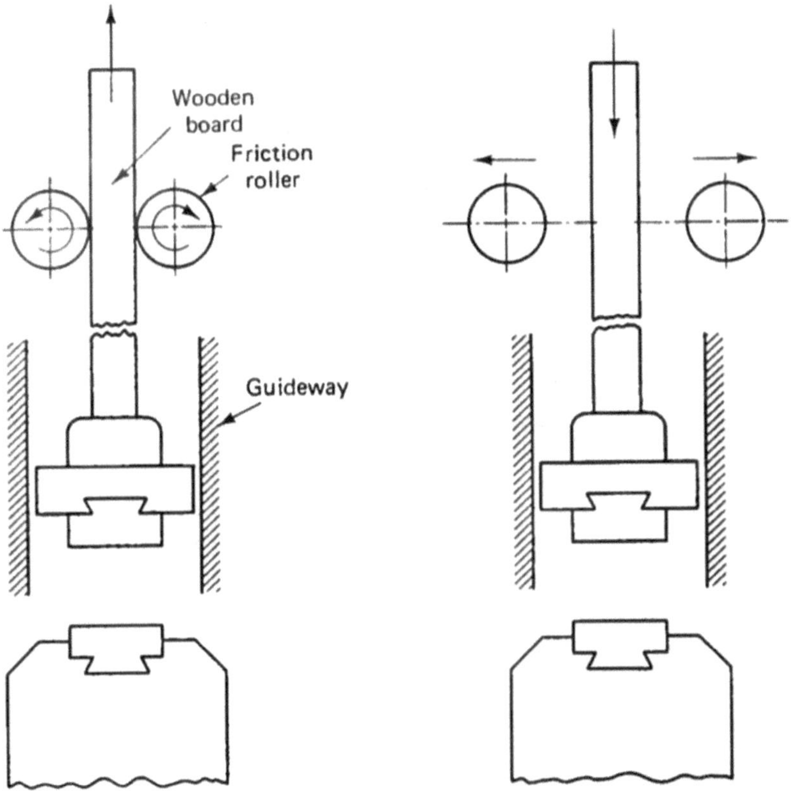

FIGURE 5.42 The working principles of a board hammer.

provided with adequate ram guidance. The desired alignment of the two halves of the die is then achieved by wedging the upper half of the die onto the ram and securing the lower half onto a bolster plate that is, in turn, tightly mounted on the anvil. Also, the ratio of the weights of the anvil and the moving parts can go as high as 30:1 to ensure maximum efficiency and trouble-free impact.

Drop-forging dies can have one, two, or several impressions, depending upon the complexity of the required product. Simple shapes like gears, small flywheels, and straight levers are usually forged in dies with one or two impressions, whereas products with intricate shapes are successively worked in multiple-impression dies, thus making it possible to preshape a forging before it is forged into its final form. Operations like edging, drawing out, fullering, and bending are performed, each in its assigned impression. Finally, the desired shape is imparted to the metal in a finishing impression that has exactly the same shape as the desired product; its dimensions are slightly larger because shrinkage due to cooling down must be taken into account. As can be seen in Figure 5.43, a gutter for flash is provided around the finishing impressions. When properly designed, the gutter provides resistance to the flow of metal into it, thus preventing further flow from the impression and forcing the metal to fill all the details, such as corners (which are the most difficult portions to fill). The drop-forging process may involve several blows so that the desired final shape of the forged part can be obtained. Lubricants are applied to ensure easy flow of the metal within the cavity and to reduce friction and die wear. As many as four blows may be needed while the part is in the finishing impressions, and the part should be lifted slightly between successive blows to prevent overheating of the die. Finally, the gas pressure forces the part out of the die. The number of blows delivered when the part is in the different preshaping impressions is 1½ to 2 times the number of blows while the part is in the finishing impression. This sequence of drop-forging operations is shown when forging a connecting rod. As can be seen in Figure 5.44, the heated stock is first placed in the fullering impression

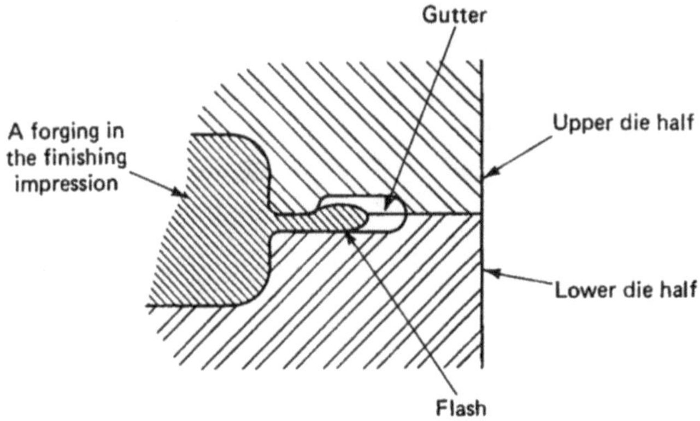

FIGURE 5.43 A gutter providing space for excess metal.

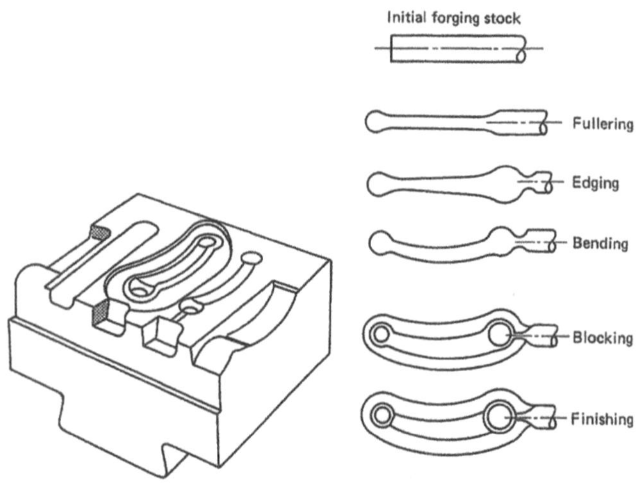

FIGURE 5.44 A multiple-impression die and the forging sequence for a connecting rod.

and then hammered once or twice to obtain local spreading of the metal on the expanse of its cross section. The stock is then transferred to the edging impression, where the metal is redistributed along its length in order to properly fill the finishing die cavities (i.e., metal is "gathered" at certain predetermined points and reduced at some other ones). This is usually achieved through a series of blows, together with turnovers of the metal, as required. The next operation in this sequence is bending, which may or may not be needed, depending upon the design of the product. The stock is then worked in the semifinishing, or blocking, impression before it is finally forged into the desired shape in the finishing impression. We can see that the blocking operation contributes to reducing the tool wear in the finishing impression by giving the part its general shape.

5.5.3.2 Press Forging

Press forging, which is usually referred to as *hot pressing*, is carried out using mechanical (crank-type) or hydraulic presses. These exert force at relatively slow ram travel, resulting in steadily applied pressure instead of impacting pressure.

The nature of metal deformation during hot pressing is, therefore, substantially different from that of drop forging. Under impact loading, the energy is transmitted into only the surface layers of the workpiece, whereas, under squeezing (steadily applied pressure), deformation penetrates deeper

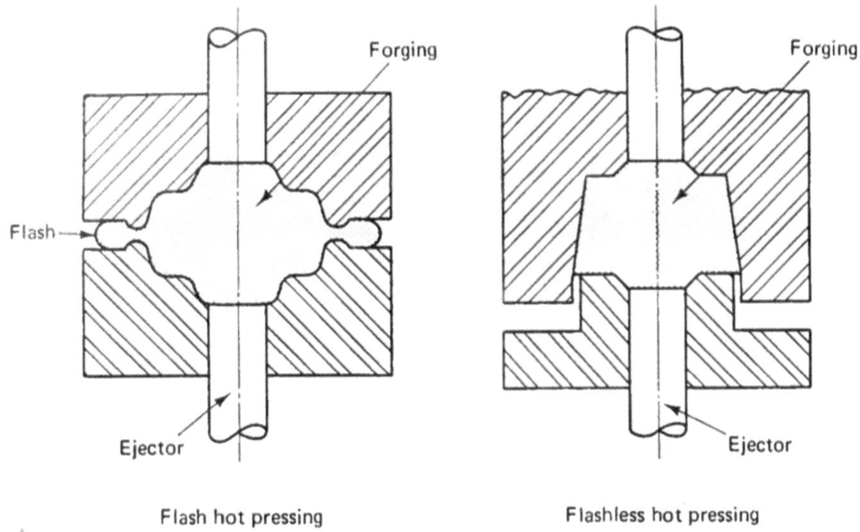

FIGURE 5.45 Flash and flashless hot pressing.

so that the entire volume of the workpiece simultaneously undergoes plastic deformation. Although multiple-impression dies are used, it is always the goal of a good designer to minimize the number of impressions in a die. It is also considered good industrial practice to use shaped blanks or preforms, thus enabling the part to be forged in only a single stroke.

Hot pressing involves both flash as well as flashless forging. In both cases, the forged part is pushed out of the die cavity by means of an ejector, as illustrated in Figure 5.45. Examples of some hot-pressed parts are shown in Figure 5.46, which also shows the sequence of operations, the production rate, and the estimated die life.

A characterizing feature of hot pressing is the accurate matching of the two halves of a die due to the efficient guidance of the ram. Also, the number of working strokes per minute can be as high as 40 or even 50. There is also the possibility of automating the process through the mechanization of blank feeding and of forging removal. It can, therefore, clearly be seen that hot pressing has higher productivity than drop forging and yields parts with greater accuracy in terms of tolerances within 0.010–0.020 inch (0.2–0.5 mm), less draft, and fewer design limitations. Nevertheless, the initial capital cost is higher compared with drop forging because the cost of a crank press is always higher than that of an equivalent hammer and because the process is economical only when the equipment is efficiently utilized. The difficulty of descaling the blanks is another shortcoming of this process. However, this disadvantage can be eliminated by using hydraulic descaling (using a high-pressure water jet) or can be originally avoided by using heating furnaces with inert atmosphere.

5.5.3.3 Die Forging in a Horizontal Forging Machine

Although originally developed for heading operations, the purpose of this machine has been broadened to produce a variety of shapes. For instance, all axisymmetric parts such as rods with flanges (with through and blind holes) and/or side projections are commonly produced on horizontal forging machines. A rolled stock is cut to length, heated in a heating unit, and automatically fed to the machine. As can be seen in Figure 5.47, the hot part is then held by stationary grips (actually a split die) and upset by an upsetting ram or header. The process involves mainly upsetting and gathering where the blank is first upset: then metal flows to fill the die cavity, as opposed to drop forging, where it is spread or flattened. In the return stroke, the upsetting ram retracts, and the part is removed or transferred to the next impression of the horizontal forging machine. It is obvious that a part can be forged in one or several cavities, depending upon the complexity of its shape.

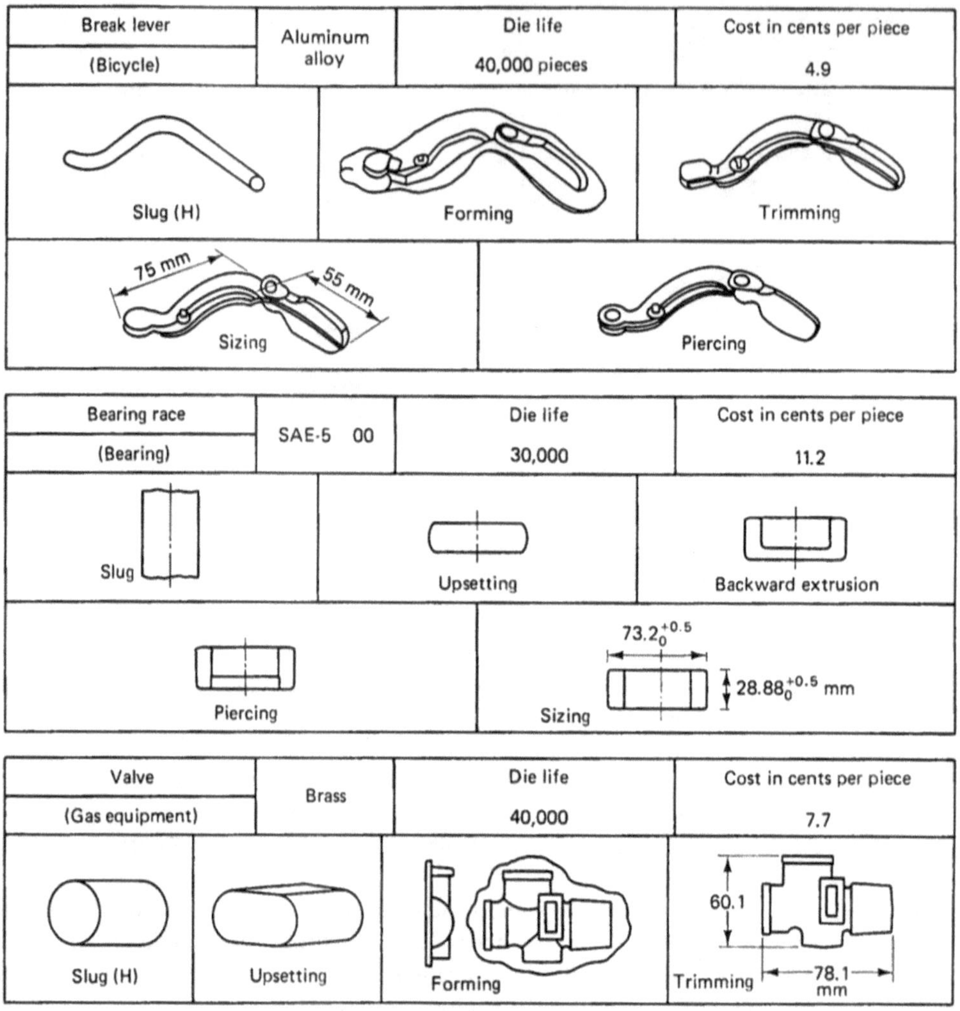

FIGURE 5.46 Examples of hot-pressed parts.

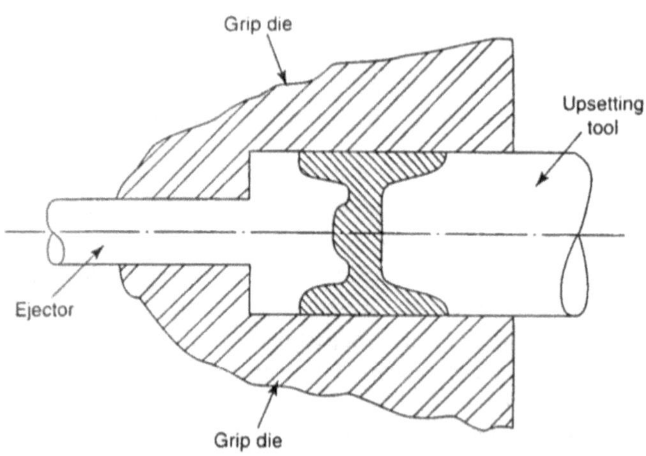

FIGURE 5.47 Die forging in a horizontal forging machine.

The main advantage of this process is the high production rate (up to 5000 parts per hour) due to the fact that it can be fully automated. Further advantages include the elimination of the flash and the forging draft and the high efficiency of material utilization because the process involves little or no waste.

5.5.3.4 Recent Developments in Forging

Warm forging, high-energy-rate forging (HERF), and forming of metals in their mushy state are among the important developments in forging technology. These newly developed processes are usually carried out to obtain intricate shapes or unique structures that cannot be obtained by conventional forging processes. Following is a brief description of each of these processes, together with their advantages and disadvantages.

- *Warm Forging*
  Warm forging involves the forging of the metal at a temperature somewhat below the recrystallization temperature. This process combines some advantages of both the hot- and the cold-forming processes while eliminating their shortcomings. On the one hand, increased plasticity and lower load requirements are caused by the relatively high forging temperature. On the other hand, improved mechanical properties, less scaling, and longer die life are due to the lower temperatures used as compared with those used with hot forging.
- *High-Energy-Rate Forging*
  The conventional forging process takes some time, during which the hot metal cools down and its resistance to deformation increases. As this does not occur with HERF, where the whole process is performed within a few thousandths of a second, the hot metal does not have enough time to cool down and heat is not dissipated into the surroundings. Therefore, HERF is very successful when forging intricate shapes with thin sections. A special HERF machine must be used. In fact, the Petro-Forge machine was developed at the Mechanical Engineering Department of Birmingham University in England for this reason, and a bulky machine with the name Dynapak was developed in the United States. In the first case, the machine consists mainly of an internal-combustion (IC) cylinder integrated into the structure of a high-speed press. The IC cylinder is provided with a sudden-release valve that allows the platen attached to the piston to be fired instantaneously when the combustion pressure reaches a preset level. In the case of the Dynapak, high-pressure nitrogen in a power cylinder is used to push the platen downward. Installations to produce and keep high-pressure gas are, therefore, required in this case.
- *Forging of Alloys in Their Mushy State*
  Forging alloys in their *mushy state* involves plastically forming alloys in the temperature range above the solidus line. Because an alloy at that temperature consists partly of a liquid phase, a remarkable decrease in the required forging load is experienced. The process also has some other merits, such as the high processing rate and the high quality of products compared with castings. Moreover, the friction at the billet–container interface has been found to be almost negligible. Nevertheless, the process is still considered in its experimental stage because of the instability of alloys having low solid fractions. Recently, it was reported that progress has been made toward solving this problem at the Institute of Industrial Science, Tokyo University, where the instability was overcome by dispersing a very fine alumina powder. This also yielded improved mechanical properties of forgings.

5.5.4 Forgeability

5.5.4.1 Tests

For the proper planning of a forging process, it is important to know the deformation behavior of the metal to be forged with regard to the resistance to deformation and any anticipated adverse effects, such as cracking. For this reason, the term *forgeability* was introduced and can be defined as the tolerance of a metal for deformation without failure. Although there is no commonly accepted standard test, a quantitative assessment of the forgeability of a metal (or an alloy) can be obtained through one of the following tests.

Upsetting Test. The upsetting test involves upsetting a series of cylindrical billets having the same dimensions to different degrees of deformation (reductions in height). The maximum limit of upsettability without failure or cracking (usually peripheral cracks) is taken as a measure of forgeability.

Notched-Bar Upsetting Test. The notched-bar upsetting test is basically similar to the first test, except that longitudinal notches or serrations are made prior to upsetting. It is believed that this test provides a more reliable index of forgeability.

Hot-Impact Tensile Test. A conventional impact-testing machine fitted with a tension-test attachment is employed. A hot bar of the metal to be studied is tested, and the impact tensile strength is taken as a measure of forgeability. This test is recommended when studying the forgeability of alloys that are sensitive to high strain rates.

Hot Twist Test. The hot twist test involves twisting a round, hot bar and counting the number of twists until failure. The greater the number of twists, the better the forgeability is considered to be. Using the same bar material, this test can be performed at different temperatures in order to obtain the forging temperature range in which the forgeability of a metal is maximum.

5.5.4.2 Forgeability of Some Alloys

It is obvious that the results of any of the preceding tests are affected by factors like the composition of an alloy, the presence of impurities, the grain size, and the number of phases present. These are added to the effect of temperature, which generally improves forgeability up to a certain limit, where other phases start to appear or where grain growth becomes excessive. At this point, any further increase in temperature is accompanied by a decrease in forgeability. Following is a list indicating the relative forgeability of some alloys in descending order (i.e., alloys with better forgeability are mentioned first):

- Aluminum alloys
- Magnesium alloys
- Copper alloys
- Plain-carbon steels
- Low-alloy steels
- Martensitic stainless steel
- Austenitic stainless steel
- Nickel alloys
- Titanium alloys
- Iron-base superalloys
- Cobalt-base superalloys
- Molybdenum alloys
- Nickel-base superalloys
- Tungsten alloys
- Beryllium

5.5.5 Lubrication in Forging

In hot forging, the role of lubricants is not just limited to eliminating friction and ensuring easy flow of metal. A lubricant actually prevents the hot metal from sticking to the die and meanwhile prevents the surface layers of the hot metal from being chilled by the relatively cold die. Therefore, water spray, sawdust, or liners of relatively soft metals are sometimes employed to prevent adhesion. Mineral oil alone or mixed with graphite is also used, especially for aluminum and magnesium alloys. Graphite and/or molybdenum disulfide are widely used for plain-carbon steels, low-alloy steels, and copper alloys, whereas melting glass is used for difficult-to-forge alloys like alloy steel, nickel alloys, and titanium.

5.5.6 Defects in Forged Products

Various surface and body defects may be observed in forgings. The kind of defect depends upon many factors, such as the forging process, the forged metal, the tool design, and the temperature at which the process is carried out. Cracking, folds, and improper sections are generally the defects observed in forged products. Following is a brief description of each defect and its causes.

Cracking. Cracking is due to the initiation of tensile stresses during the forging process. Examples are *hot tears*, which are peripheral longitudinal cracks experienced in upsetting processes at high degrees of deformation, and *corner cavities*, which occur in the primary forging of low-ductility steels. Thermal cracks may also initiate in cases when nonuniform temperature distribution prevails.

Folds. In upsetting and heading processes, folding is a common defect that is obviously caused by buckling. Folds may also be observed at the edges of parts produced by smith forging if the reduction per pass is too small.

Improper Sections. Improper sections include dead-metal zones, piping, and turbulent (i.e., irregular or violent) metal flow. They are basically related to and caused by poor tool design.

5.5.7 Forging Die Materials

During their service life, forging dies are subjected to severe conditions such as high temperatures, excessive pressures, and abrasion. A die material must, therefore, possess adequate hardness at high temperatures as well as high toughness to be able to withstand severe conditions. Special tool steels (hot-work steels including one or more of the following alloying additives: chromium, nickel, molybdenum, and vanadium) are employed as die materials. Die blocks are annealed, machined to make the shanks, hardened, and tempered; then, impression cavities are sunk by toolmakers.

5.5.8 Fundamentals of Closed-Die Forging Design

The range of forged products with respect to size, shape, and properties is very wide indeed. For this reason, it is both advisable and advantageous for the product designer to consider forging in the early stages of planning the processes for manufacturing new products. The forging design is influenced not only by its function and the properties of the material being processed but also by the kind, capabilities, and shortcomings of the production equipment available in the manufacturing facilities. Therefore, it is impossible to discuss in detail all considerations arising from the infinite combinations of the various factors. Nevertheless, some general guidelines apply in all cases and should be strictly adhered to if a sound forging is to be obtained. Following are some recommended forging design principles.

Parting Line. The plane of separation between the upper and lower halves of a closed-die set is called the *parting line*. The parting line can be straight, whether horizontal or inclined, or irregular, including more than one plane. The parting line must be designated on all forging drawings as it affects the initial cost and wear of the forging die; the grain flow that, in turn, affects the mechanical properties of the forging; and, finally, the trimming procedure and/or subsequent machining operations on the finished part. Following are some considerations for determining the shape and position of the parting line:

- The parting line should usually pass through the maximum periphery of the forging mainly because it is always easier to spread the metal laterally than to force it to fill deep, narrow die impressions (see Figure 5.48).
- It is always advantageous, whenever possible, to try to simplify the die construction if the design is to end up with flat-sided forgings (see Figure 5.49). This will markedly reduce the die cost because machining is limited to the lower die half. Also, the possibility of a mismatch between die halves is eliminated.
- If an inclined parting line must exist, it is generally recommended to limit the inclination so that it does not exceed 75°. The reason is that inclined flashes may create problems in trimming and subsequent machining.
- A parting line should be located so that it promotes alignment of the fibrous macrostructure to fulfill the strength requirement of a forging. Because excess metal flows out of the die cavity into the gutter as the process proceeds, mislocating the parting line will probably result in irregularities, as can be seen in Figure 5.50, which indicates the fibrous macrostructures resulting from different locations of the parting line.
- When the forging comprises a web enclosed by ribs, as illustrated in Figure 5.51, the parting line should preferably pass through the centerline of the web. It is also desirable, with respect to the alignment of fibers, to have the parting line either at the top or at the bottom surfaces. However, that desirable location usually creates manufacturing problems and is not used unless the direction of the fibrous macrostructure is critical.

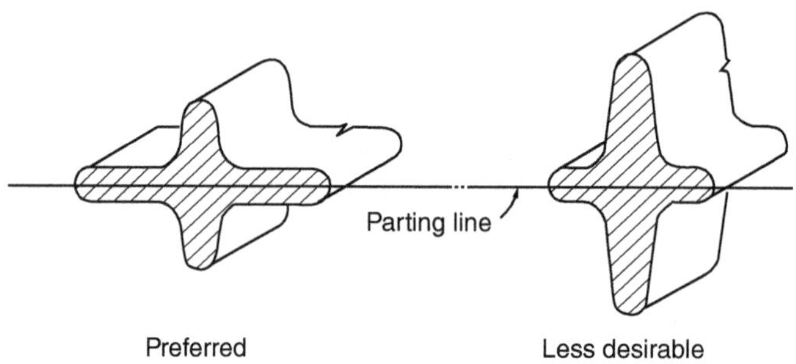

FIGURE 5.48 Recommended locations for the parting line. (Courtesy of Aluminum Association, Arlington, VA.)

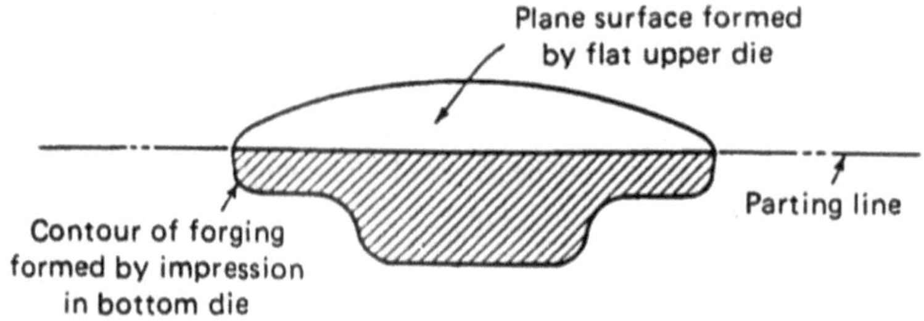

FIGURE 5.49 Flat-sided forging for simplifying the die construction. (Courtesy of Aluminum Association, Arlington, VA.)

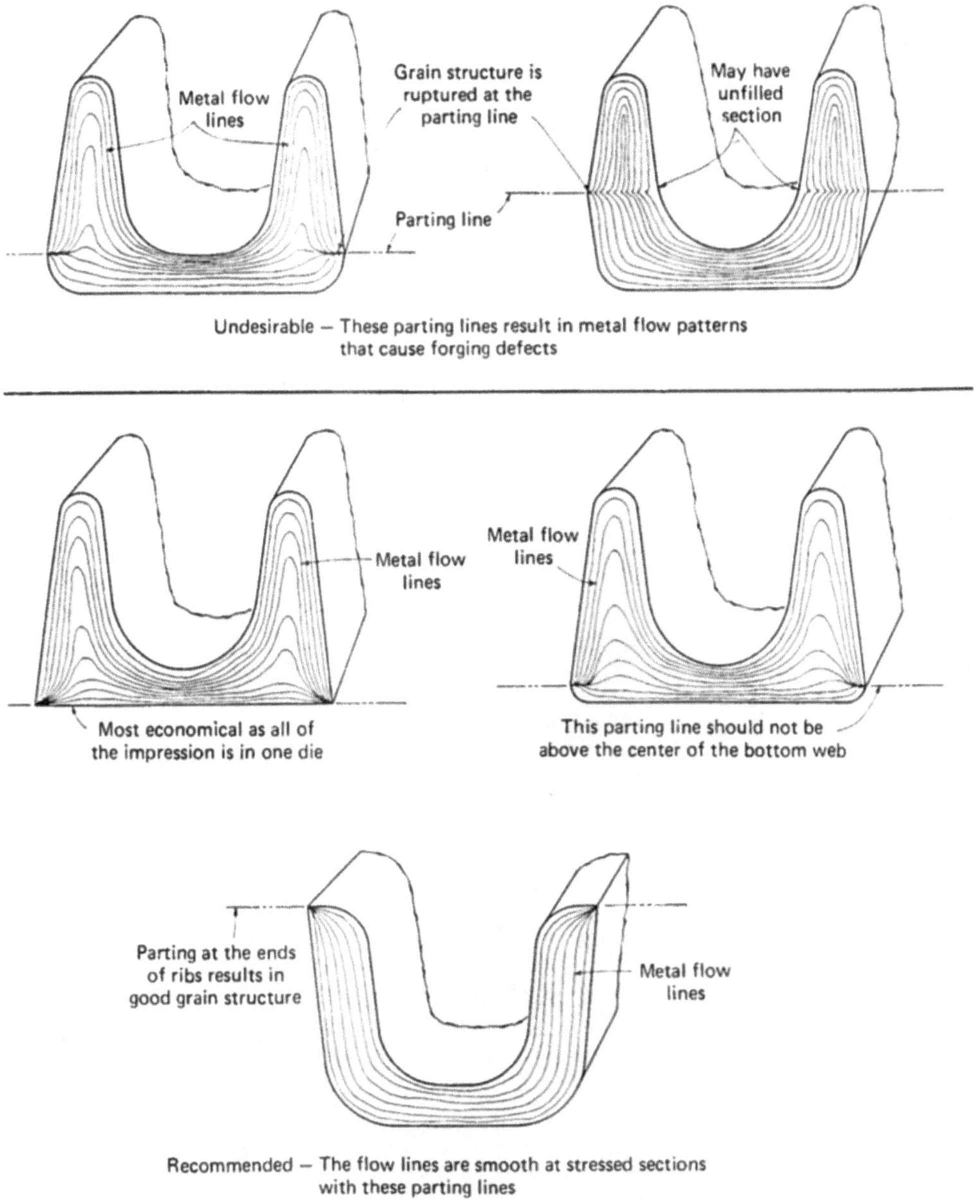

FIGURE 5.50 Using the parting line to promote the alignment of the fibrous macrostructures. (Courtesy of Aluminum Association, Arlington, VA.)

- If an irregular parting line exists, avoid the side thrust of the die, which will cause the die halves to shift away from each other sideways, resulting in matching errors. Figure 5.52 illustrates the problem of side thrust accompanying irregular parting lines, together with two suggested solutions.

Draft. Draft refers to the taper given to the internal and external sides of a closed-die forging and is expressed as an angle from the direction of the forging stroke. A draft is required on the vast majority of forgings to avoid production difficulties, to aid in achieving desired metal flow, and to allow easy removal of the forging from the die cavity. It is obvious that

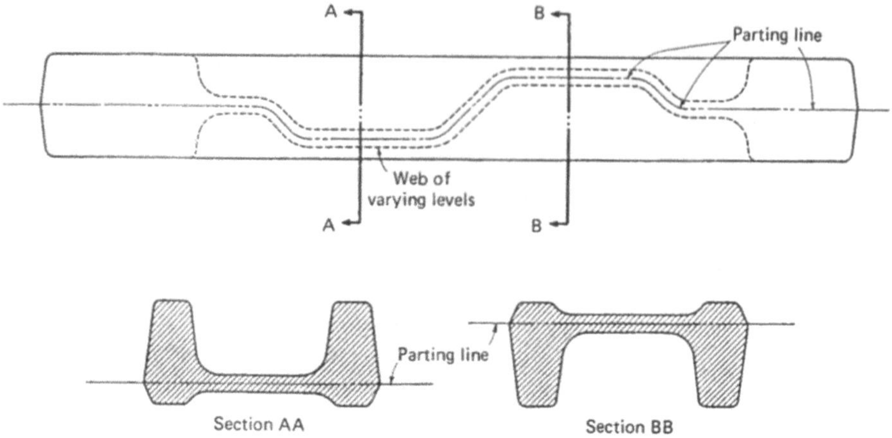

FIGURE 5.51 Location of the parting line with respect to a web. (Courtesy of Aluminum Association, Arlington, VA.)

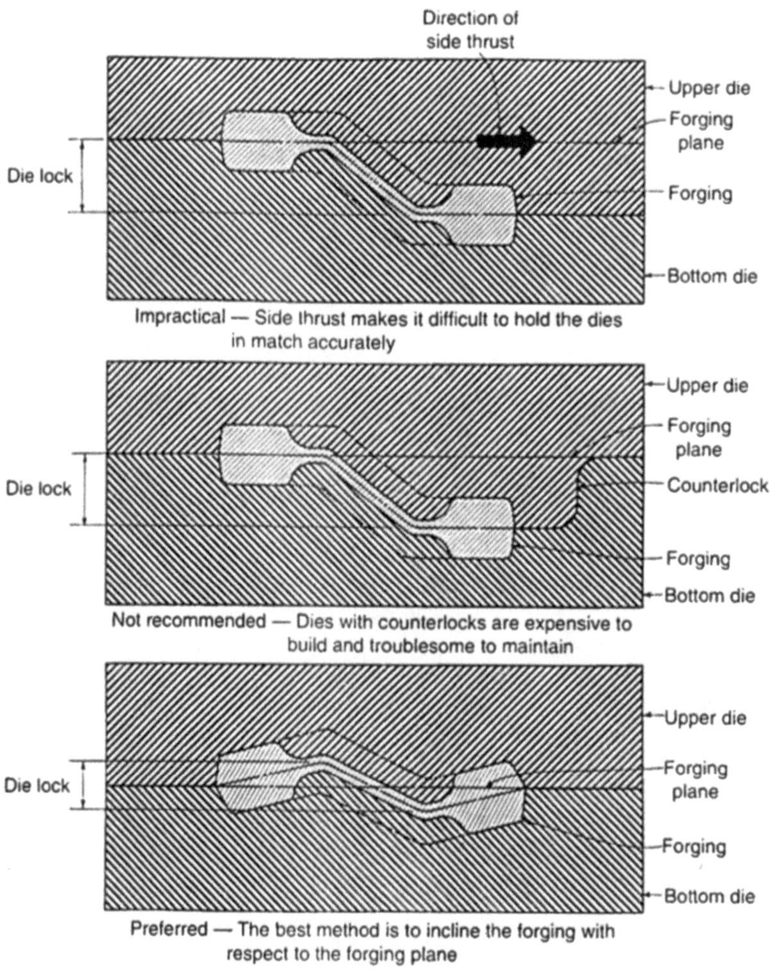

FIGURE 5.52 The problem of side thrust accompanying irregular parting lines and two suggested solutions. (Courtesy of Aluminum Association, Arlington, VA.)

the smaller the draft angle, the more difficult it is to remove the forging out of the die. For this reason, draft angles of less than 5° are not permitted if the part is to be produced by drop forging (remember that there is no ejector to push the part out). Standard draft angles are 7°, 5°, 3°, 1°, and 0°. A draft angle of 3° is usually used for metal having good forgeability, such as aluminum and magnesium, whereas 5° and 7° angles are used for steels, titanium, and the like. It is a recommended practice to use a constant draft all over the periphery of the forging. It is also common to apply a smaller draft angle on the outside periphery than on the inside one. This is justified in that the outer surface will shrink away from the surface of the die cavity as a result of the part's cooling down, thus facilitating the removal of the forging. Following are some useful examples and guidelines:

- When designing the product, try to make use of the natural draft inherent in some shapes, such as curved and conical surfaces (see Figure 5.53).
- In some cases, changing the orientation of the die cavity may result in a natural draft, thus eliminating the need for any draft on the surfaces (see Figure 5.54).
- Sometimes, the cavity in one of the die halves (for instance, the upper) is shallower than that in the other half. This may create problems in matching the contours of the two die halves at the parting line. It is, therefore, recommended that one of the three methods illustrated in Figure 5.55a,b,c be used. The first method involves keeping the draft the same as in the lower cavity but increasing the dimension of the upper surface of the cavity. This results in an increase in weight and this solution is limited to smaller cavities. The second method is based on keeping the draft constant in both halves by introducing a "pad" whose height varies between 0.06 inches (1.5 mm) and 0.5 inches (12.5 mm), depending upon the size of the forging. The third method, which is more common, is to provide a greater draft on the shallower die cavity; this is usually referred to as a matching draft.

Ribs. A rib is a thin part of the forging that is normal to (or slightly inclined to) the forging plane. It is obvious that optimized lighter weight of the forging calls for reducing the thickness of long ribs. However, note that the narrower and longer the rib is, the higher the forging pressure is and the more difficult it is to obtain a sound rib. It is actually a common practice to keep the height-to-thickness ratio of a rib below 6, preferably at 4. The choice of a value for this ratio depends upon many factors, such as the kind of metal being processed and the forging geometry (i.e., the location of the rib, the location of the parting line, and

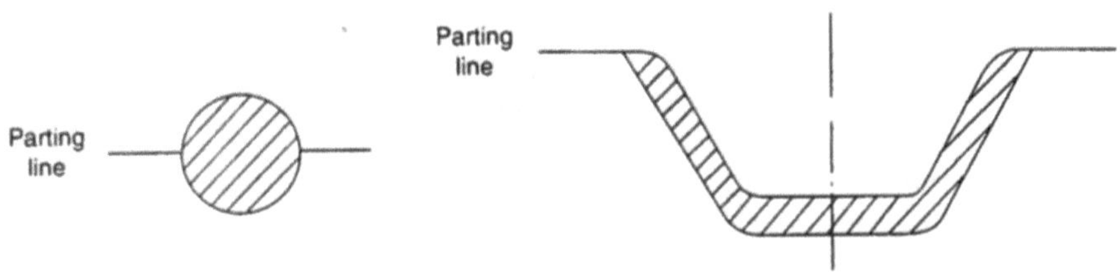

FIGURE 5.53 Example of the natural draft inherent in some shapes.

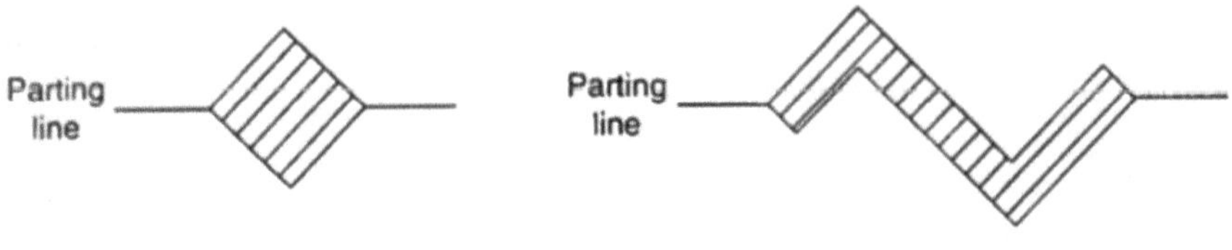

FIGURE 5.54 Examples of changing the orientation of the impression to provide natural draft.

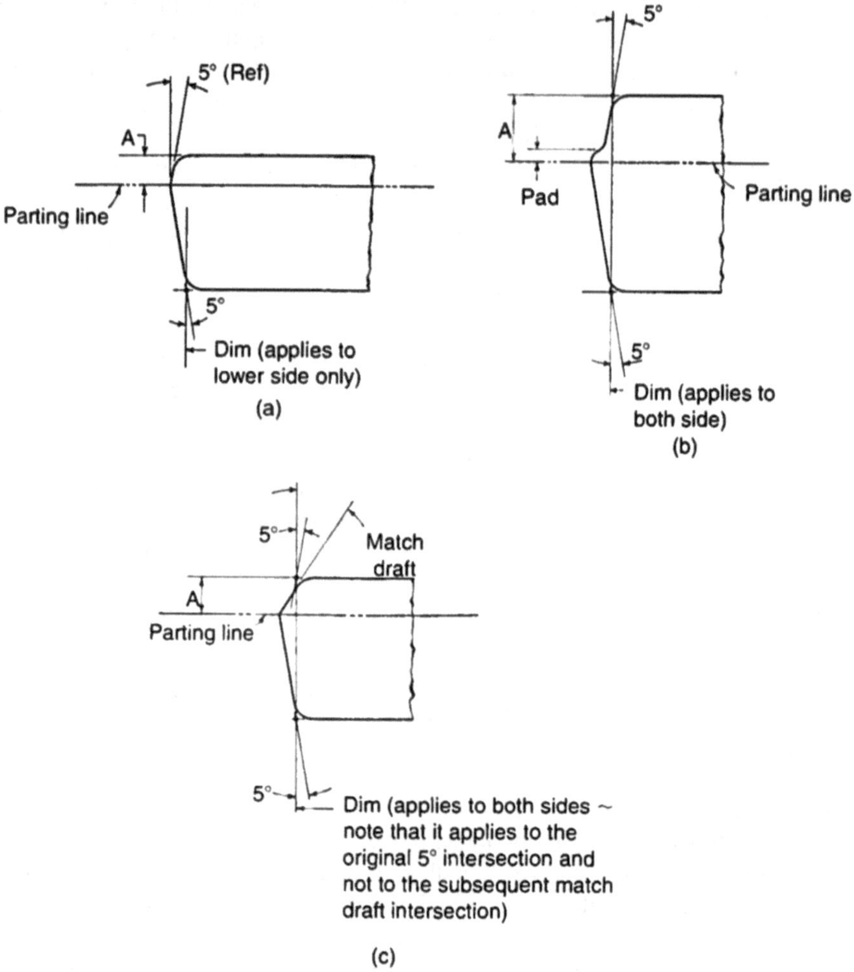

FIGURE 5.55 Methods for matching the contours of the two die impressions having different depths: (a) increasing the dimensions of the upper surface, (b) using a pad, and (c) employing a matching draft. (Courtesy of Aluminum Association, Arlington, VA.)

the fillet radii). Figure 5.56 indicates the desirable rib design as well as limitations imposed on possible alternatives.

Webs. A web is a thin part of the forging that passes through or parallel to the forging plane (see Figure 5.57). Although it is always desirable to keep the thickness of a web at a minimum, there are practical limits to this. The minimum thickness of webs depends on the kind of material being worked (actually on its forging temperature range), the size of forging (expressed as the net area of metal at the parting line), and the average width. Table 5.3 indicates recommended web thickness values applicable to precision and conventional aluminum forgings. For blocking cavities, the values given in Table 5.3 must be increased by 50 percent. Also, for steels and other metals having poorer forgeability than aluminum, it is advisable to increase the values for web thickness.

Thin webs may cause unfilled sections, may warp in heat treatment, and may require additional straightening operations; they even cool faster than the rest of the forging after the forging process, resulting in shrinkage, possible tears, and distortion.

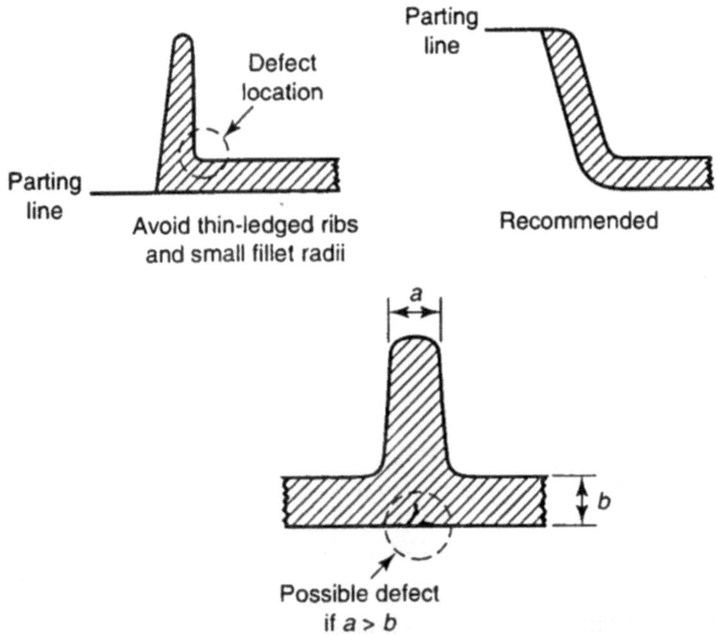

FIGURE 5.56 Recommended rib design. (Courtesy of Aluminum Association, Arlington, VA.)

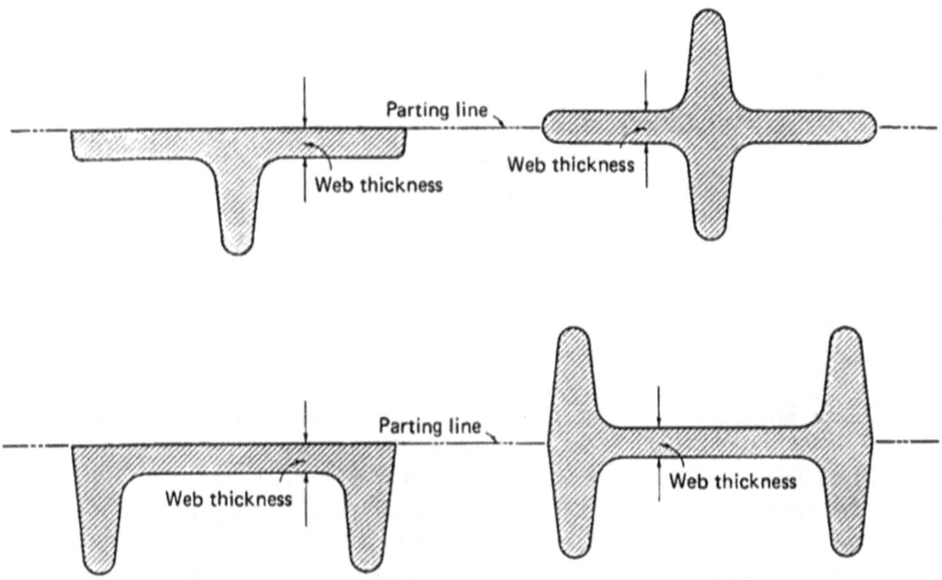

FIGURE 5.57 The shape of a web in forging. (Courtesy of Aluminum Association, Arlington, VA.)

(Continued)

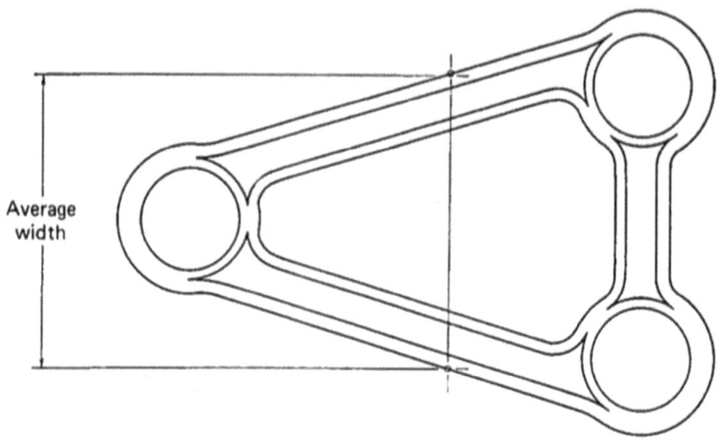

FIGURE 5.57 (Continued)

TABLE 5.3
Recommended Size of Minimum Web Thickness

Up to Average Width, in. (m)	Up to Cross-Sectional Area, in.^{2} (m^2)	Web Thickness, in. (mm)
3 (0.075)	10 (0.00625)	0.09 (2.25)
4 (0.1)	30 (0.01875)	0.12 (3)
6 (0.15)	60 (0.0375)	0.16 (4)
8 (0.2)	100 (0.0625)	0.19 (4.75)
11 (0.275)	200 (0.125)	0.25 (6.25)
14 (0.35)	350 (0.21875)	0.31 (7.75)
18 (0.45)	550 (0.34375)	0.37 (9.25)
22 (0.55)	850 (0.53125)	0.44 (11)
26 (0.65)	1200 (0.75)	0.50 (12.5)
34 (0.85)	2000 (1.25)	0.62 (15.5)
41 (1.025)	3000 (1.875)	0.75 (18.75)
47 (1.1175)	4000 (2.5)	1.25 (31.25)
52 (1.3)	5000 (3.125)	2.00 (50)

Corner Radii. Two main factors must be taken into consideration when selecting a small value for a corner radius. First, a small corner radius requires a sharp fillet in the die steel, which acts as a stress raiser; second, the smaller the corner radius, the higher the forging pressure required to fill the die cavity. In addition, some other factors affect the choice of the corner radius, such as the distance from the corner to the parting line and the forgeability of the metal being worked. The larger the distance from the parting line, the larger the corner radius should be. Also, whereas a corner radius of 0.0625 inches (1.5 mm) is generally considered adequate for aluminum forging, a corner radius of at least 0.125 inches (3 mm) is used for titanium forgings of similar shape and size. In addition, the product designer should try to keep the corner radii as consistent as possible and avoid blending different values for a given shape in order to reduce the die cost (because there will be no need for many tool changes during die sinking). Corner radii at the end of high, thin ribs are critical. A rule of thumb states that it is always desirable to have the rib thickness equal to twice the value of the corner radius. A thicker rib may have a flat edge with two corner

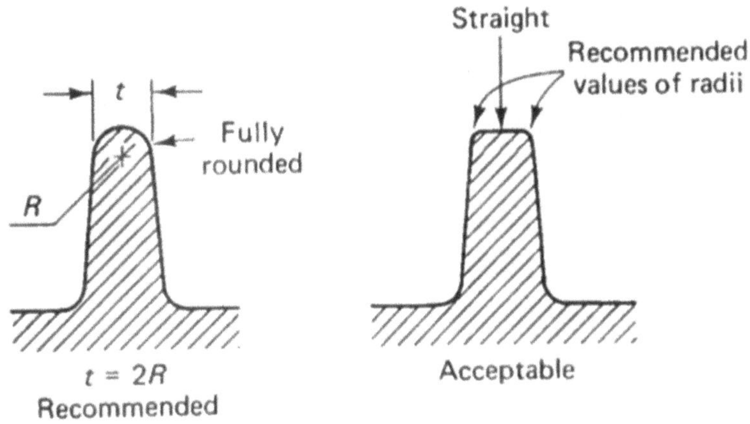

FIGURE 5.58 Recommendations regarding corner radii for ribs. (Courtesy of Aluminum Association, Arlington, VA.)

radii, each equal to the recommended value. Figure 5.58 illustrates these recommendations regarding corner radii for ribs.

Fillet Radii. It is of supreme importance that the product designer allows generous radii for the fillets because the abrupt diversion of the direction of metal flow can result in numerous defects in the product. Figure 5.59 indicates the step-by-step initiation of forging defects and shows that small fillets result in the separation of the metal from the die and the initiation of voids. Although these can be filled at a later stage, laps and cold shuts will replace these voids. When the shape of the part to be forged is intricate (i.e., involving thin ribs and long, thin webs), the metal may preferentially flow into the gutter rather than into the die cavity. This results in a shear in the fibrous macrostructure and is referred to as flow-through. This latter defect can be avoided by using larger-than-normal fillets.

Punchout Holes. Punchout holes are through holes in a thin web that are produced during, but not after, the forging process. Punchouts reduce the net projected area of the forging, thus reducing the forging load required. If properly located and designed, they can be of great assistance in producing forgings with thin webs. In addition to the manufacturing advantages of punchouts, they serve functional design purposes, such as reducing the mass of a forging and/or providing clearance. Following are some guidelines regarding the design of punchouts:

- Try to locate a punchout around the central area of a thin web, where the frictional force that impedes the metal flow is maximum.
- Whenever possible, use a gutter around the interior periphery of a punchout. This provides a successful means for the surplus metal to escape.
- A single large punchout is generally more advantageous than many smaller ones that have the same area. Accordingly, try to reduce the number of punchouts unless more are dictated by functional requirements.
- Although punchouts generally aid in eliminating the problems associated with the heat treatment of forgings, it may prove beneficial to take the limitations imposed by heat treatment processes into account when designing the contour of a punchout (i.e., try to avoid irregular contours with sharp corners).

Pockets and Recesses. Pockets and recesses are used to save material, promote the desirable alignment of the fibrous macrostructure, and improve the mechanical properties by reducing the thickness, thus achieving a higher degree of deformation. Following are some guidelines:

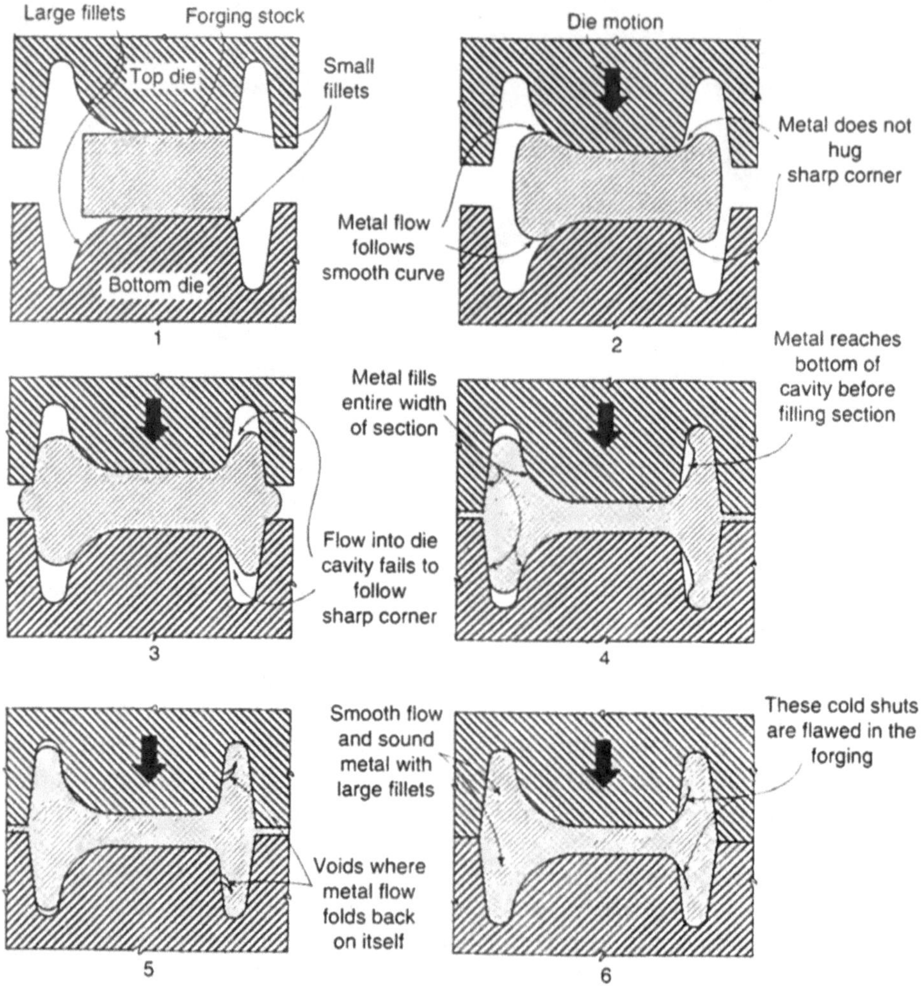

FIGURE 5.59 Defects caused by using smaller fillet radii. (Courtesy of Aluminum Association, Arlington, VA.

- Recesses should never be perpendicular to the direction of metal flow.
- Recesses are formed by punches or plugs in the dies. Therefore, the recess depth is restricted to the value of its diameter (or to the value of minimum transverse dimension for noncircular recesses).
- Simple contours for the recesses, together with generous fillets, should be tried.

5.6 COLD-FORMING PROCESSES

Cold-forming processes are employed mainly to obtain improved mechanical properties, better surface finish, and closer tolerances. Several cold-forming techniques have found wide industrial applications. Among these are sizing, swaging, coining, and cold heading. Following is a brief description of each of them.

5.6.1 Sizing

Sizing (see Figure 5.60a) is a process in which the metal is squeezed in the forming direction but flows unrestricted in all transverse directions. This process is used primarily for straightening forged

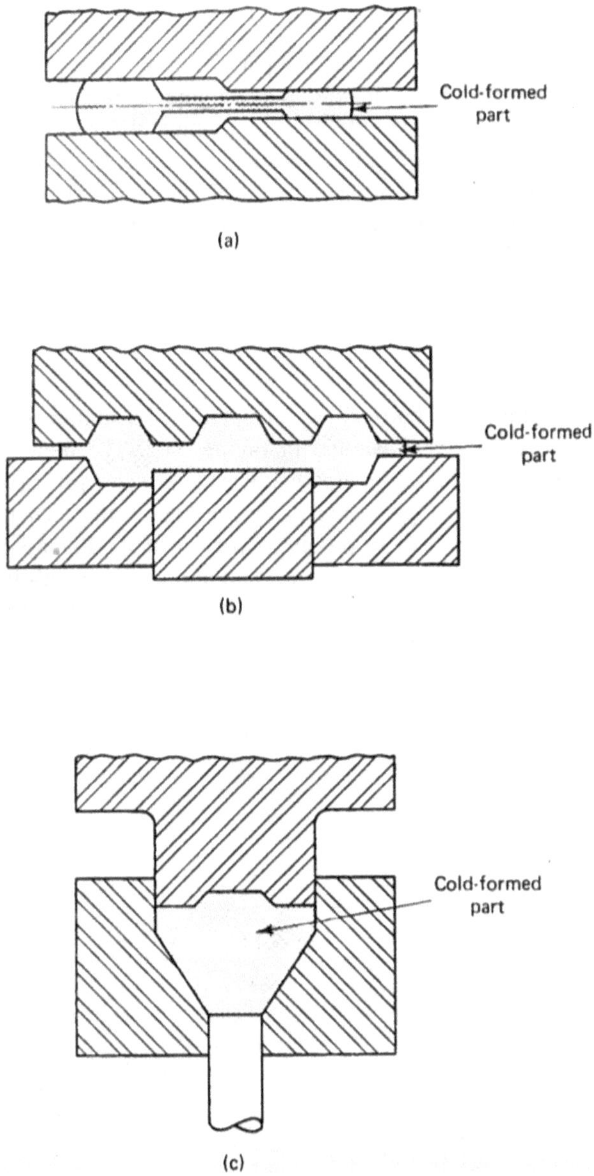

FIGURE 5.60 Cold-forming processes: (a) sizing, (b) swaging, and (c) coining.

parts, improving the surface quality, and obtaining accurate dimensions. A sizing operation can ensure accuracy of dimensions within 0.004–0.010 inches (0.1–0.25 mm). Meanwhile, the pressure generated on the tools can go up to 180,000 lb per square inch (1300 MN/m^2).

5.6.2 Swaging

Swaging (see Figure 5.60b) involves imparting the required shape and accurate dimensions to the entire forging (or most of it). Usually, swaging is carried out in a die where a flash is formed and subsequently removed by abrasive wheels or a trimming operation. Note that the flow of metal in the swaging process is more restricted than in sizing. Accordingly, higher forming pressures are experienced and can go up to 250,000 lb per square inch (1800 MN/m^2).

5.6.3 Coining

Coining (see Figure 5.60c) is a process in which the part subjected to coining is completely confined within the die cavity (by the die and the punch). The volume of the original forging must be very close to that of the finished part. Any tangible increase in that volume may result in excessive pressures and the breakage of tools. Still, common pressures (even when no problems are encountered) are on the order of 320,000 lb per square inch (2200 MN/m^2). For this reason, coining processes (also including sizing and swaging) are carried out on special presses called *knuckle presses*. The main mechanism of a knuckle press is shown in Figure 5.61. It is characterized by the ability to deliver a large force with a small stroke of the ram.

5.6.4 Cold Heading

Cold heading is used to manufacture bolts, rivets, nuts, nails, and similar parts with heads and collars. A group of typical products is illustrated in Figure 5.62. The main production equipment involves a multistage automatic cold header that operates on the same principle as a horizontal forging machine. Full automation and high productivity are among the advantages of this process. Products having accurate dimensions can be produced at a rate of 30–300 pieces per minute. Starting from coiled wires or rods made of plain-carbon steel and nonferrous metals with diameters ranging from 0.025 to 1.6 inches (0.6–40 mm), blanks are processed at different stations. Feeding, transfer, and ejection of the products are also automated. Figure 5.63 illustrates the different stages involved in a simple cold heading operation.

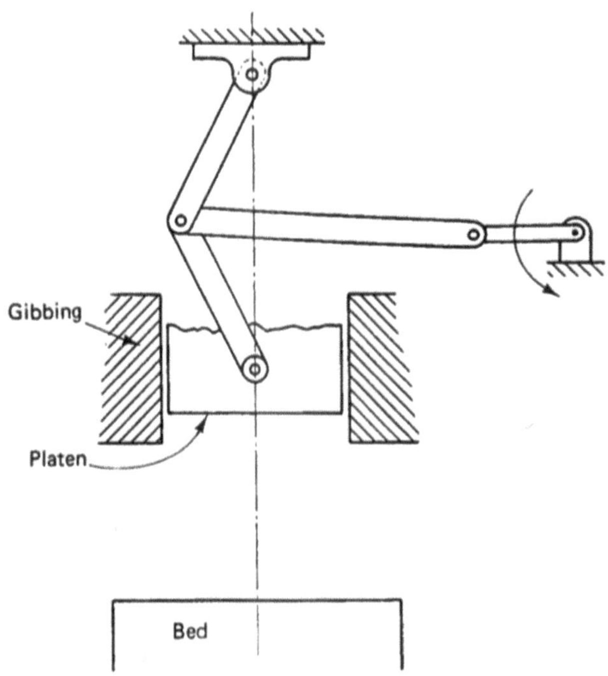

FIGURE 5.61 The working principles of a knuckle press for cold-forming processes.

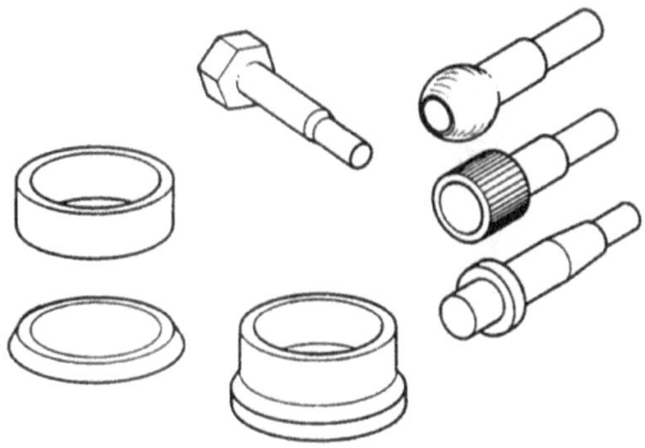

FIGURE 5.62 Some products manufactured using an automatic cold header.

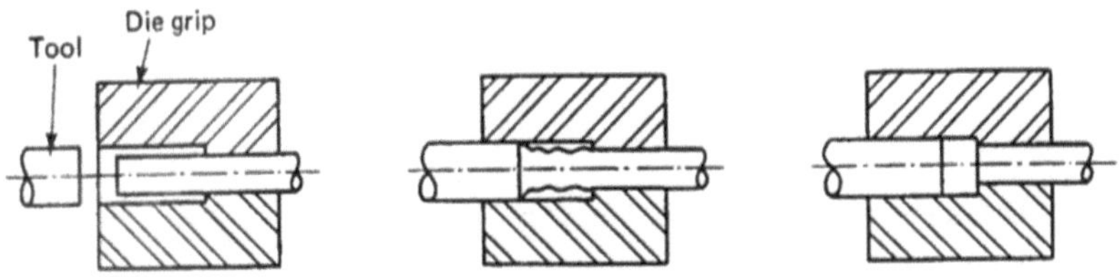

FIGURE 5.63 Different stages of a simple cold heading operation.

5.6.5 Lubrication in Cold Forming

Lubricants employed in cold forming are similar to those used in heavy wiredrawing processes. Phosphating followed by soap dipping is successful with steels, whereas only soap is considered adequate for nonferrous metals.

5.7 MODELING OF METAL-FORMING PROCESSES

As previously discussed in Chapter 1, a model is a simplified representation of a real-world problem designed to facilitate analysis and resolution. This methodology is particularly useful in metal forming for estimating the forces required to complete the process. To develop a model, certain assumptions are necessary.

1. *Material Properties*. The metal of the billet is considered rigid-plastic. This implies that any elastic deformation is disregarded, and the metal does not undergo work hardening post-yield.
2. *Yield Criteria*. Upon yielding and starting to deform, the metal adheres to the yield criteria established by Tresca and Von Mises.
3. *Deformation Homogeneity*. The deformation throughout the metal is uniform.
4. *Friction*. Friction is limited to the interfaces between the metal and the die.
5. *Plane Strain Condition*. For simplicity, the analysis assumes that the process occurs under plane strain conditions, meaning strains are only present within the plane of the paper, with no strain occurring perpendicular to this plane.

We will now examine two practical applications of the simple "slab analysis" (also known as force balance) approach to modeling: in the processes of sheet drawing and forging.

EXAMPLE 5.1: SHEET DRAWING

Figure 5.64 illustrates a diagrammatic sketch of the sheet drawing process featuring a differential slab (hatched) in the area where deformation occurs. For simplicity, we assume the width of the slab is unity.

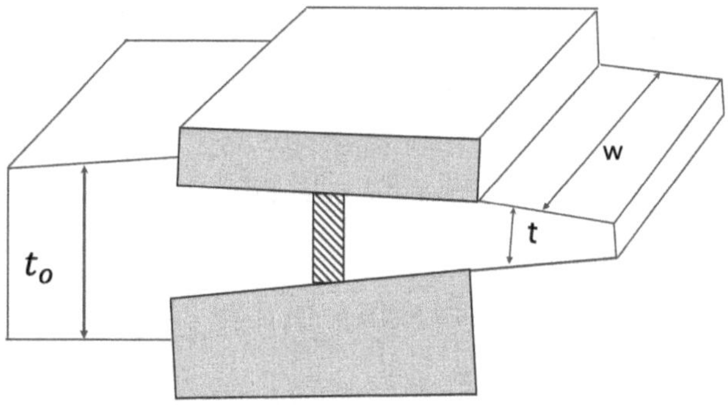

FIGURE 5.64 Diagrammatic sketch of the sheet drawing process.

Figure 5.65 shows a magnified view of the slab with all acting stresses.

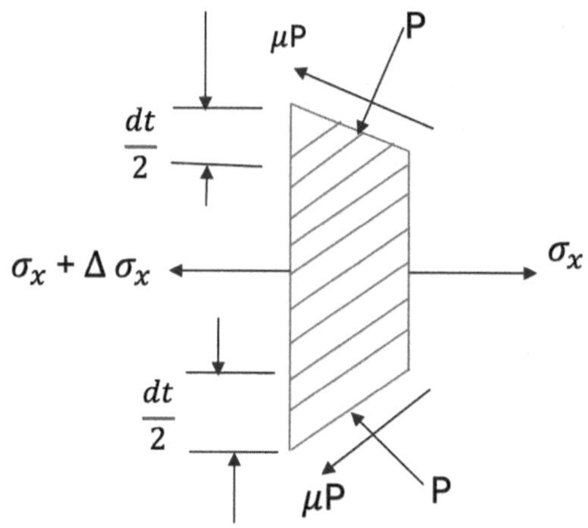

FIGURE 5.65 Magnified view of the slab with all acting forces.

***Equations of Equilibrium*:**

X-Axis Equilibrium: The equilibrium equation along the X-axis can be written as

$$(\sigma_x + d\sigma_x)(t + dt).1 + 2\mu p\cos\alpha.ds.1 + 2p\sin\alpha.1.\,ds = 0$$

Through mathematical manipulation and using the relationships between dx and dt, and dx and ds, we derive

$$d\sigma_x.t + \sigma_x.dt + \mu p \cot\alpha.dt + p.dt = 0$$

To simplify further, let B = μ cot α. We then obtain

$$d\sigma_x.t + dt\left[\sigma_x + p\left(B+1\right)\right] = 0 \tag{5.17}$$

Z-Axis Equilibrium: Considering the equilibrium along the Z direction,

$$\sigma_z.dx + p\cos\alpha.ds = \mu p \sin\alpha.ds$$

or,

$$\sigma_z + p = \mu p \tan\alpha \approx 0 \text{ or } \sigma_z = -p \tag{5.18}$$

Applying the Tresca Yield Criterion:

$$\sigma_1 - \sigma_3 = 2k \text{ or } \sigma_x - (-p) = 2k \tag{5.19}$$

Using equations (5.17) and (5.19), we find

$$d\sigma_x.t + dt\left[\sigma_x + \left(2k - \sigma_x\right)\left(B+1\right)\right] = 0$$

or,

$$\frac{dx}{2k\left(B+1\right) - \sigma_x B} = \frac{dt}{t}$$

Integrating between initial and final conditions:

$$\int_0^{\sigma_d} \frac{d\sigma_x}{2k\left(B+1\right) - \sigma_x B} = \int_{t_0}^{t_f} \frac{dt}{t}$$

We finally obtain

$$\frac{\sigma_d}{2k} = \frac{\sigma_d}{Y} = \frac{1+B}{B}\left[1 - \left(\frac{t_f}{t_0}\right)^B\right] \tag{5.20}$$

Equation (5.20) elucidates the variation in σ_d as a function of the reduction in thickness and can also be utilized to estimate the drawing force:

$$F_d = \sigma_d.t_f.1 \tag{5.21}$$

EXAMPLE 5.2: OPEN-DIE FORGING OF A BILLET

Figure 5.66 illustrates the forging process where a billet is squeezed between two platens, necessitating an estimation of the forging load. The mechanics of deformation involve a reduction in height "h" and a lateral spread of the metal, while the centerline remains fixed.

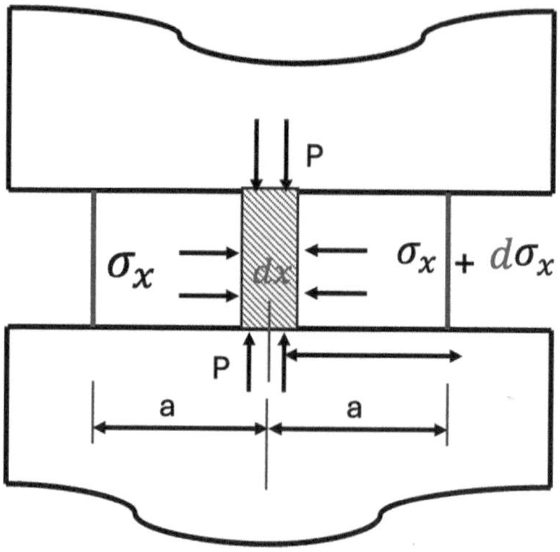

FIGURE 5.66 The process of open-die forging of a Billet.

To address this problem, we'll formulate equations for the balance of forces along the X direction, accompanied by a yield criterion equation (either Tresca or Von Mises). The equations are as follows:

$$\sigma_x.\ \mathrm{h} - (\sigma_x + d\sigma_x).\mathrm{h} - 2\tau_{xy}\, dx = 0$$

This simplifies to

$$\frac{d\sigma_x}{dx} = -\frac{2\tau_{xy}}{h} \qquad (5.22)$$

Assuming $\mathrm{p} = \sigma_x$ as the positive principal compression stress, and using the Von Mises yield criterion: $\sigma_1 - \sigma_3 = \frac{2}{\sqrt{3}}\mathrm{Y}$ where Y is the constant yield stress in tension, resulting in

$$\frac{dY}{dx} = 0 \text{ and } \frac{dp}{dx} - \frac{d\sigma_x}{dx} = 0 \qquad (5.23)$$

Considering shear stress related to normal stress via Coulomb's sliding friction: $\tau_{xy} = \mu p$, and equation (5.22) becomes $\frac{dp}{dx} = \frac{2\mu p}{h}$ or $\frac{dp}{p} = -\frac{2\mu}{h}dx$

Integrating both sides, we derive:

$$\ln p = \frac{2\mu x}{h} + \mathrm{h} \ln \mathrm{C}$$

where C is determined by the boundary conditions.

At x = a and x = −a, where $\sigma_x = 0$ and $p = \frac{2}{\sqrt{3}}\mathrm{Y}$

we have $\frac{C}{\frac{2}{\sqrt{3}}\mathrm{Y}} = e^{\frac{2\mu a}{h}}$.

Expanding e and omitting higher order terms, we find:

$$p = \frac{2}{\sqrt{3}}\mathrm{Y}\left[1 + \frac{2\mu}{h}(\mathrm{a} - \mathrm{x})\right] \tag{5.24}$$

This value is the localized pressure at any point on the billet's surface, with pressure peaking at the centerline and minimizing at x = a and x = −a

The mean pressure can be calculated as

$$p_{mean} = \int_0^A \frac{p.dA}{A} = \frac{2}{\sqrt{3}}Y\frac{e^{2\mu\frac{a}{h}} - 1}{2\frac{\mu a}{h}} \tag{5.25}$$

Finally, the forging force is determined by multiplying the mean pressure by the area.

5.8 COMPARISON BETWEEN THE DIFFERENT METAL-FORMING PROCESSES

The following table organizes the information more systematically, categorizing it by process, materials used, temperature conditions, and product features (Table 5.4).

TABLE 5.4
Comparison between the Different Metal-Forming Processes

Process	Material Being Formed	Temperature of the Billet during Forming	Geometric Features of the Product/ Examples
Rolling (General)	Steel, low-alloy steel, nonferrous metals	Hot, Cold	Flat or special shapes, general-purpose sections, angles, H, I, channel bars, structural shapes, cold rolled coils
Rolling (Special Purpose)	Steel, alloy steel	Hot	One-piece rolled wheels, rings, balls, ribbed tubes
Drawing	Steel, aluminum, copper alloys, titanium	Hot	Wires, tubes
	Low-carbon steel, aluminum, copper	Cold	Wires, tubes
Extrusion (Forward)	Low-ductility metals	Hot, Cold	Hollow, slender shapes with constant cross section, complex structural shapes
Extrusion (Backward)	Same as forward extrusion	Hot, Cold	Hollow sections, cup-like shapes

(*Continued*)

TABLE 5.4 (Continued)

Process	Material Being Formed	Temperature of the Billet during Forming	Geometric Features of the Product/ Examples
Impact Extrusion	Soft metals like aluminum	Cold	Cylindrical or irregular symmetrical shapes, soda containers, aerosol cans
Cold Forming	Plain (low) carbon steel, nonferrous metals	Cold	Parts with heads and collars, bolts, nuts, nails
Forging (Open Die)	Low-carbon steel, aluminum, magnesium, copper alloys, low-alloy steel, stainless steel	Hot	Heavy forgings up to 300 tons, medium forgings with irregular shapes, large motor shafts, flange couplings, rings, crankshafts
Forging (Drop Forging/ Closed Die)	Steel and nonferrous metals	Hot	Parts weighing up to 900 lbs (approx. 350 kg)

REVIEW QUESTIONS

1. Why have metal-forming processes gained widespread industrial application since World War II?
2. What are the two main groups of metal-forming processes?
3. List the different factors affecting the deformation process. Tell how each influences deformation.
4. Why are cold-forming processes always accompanied by work hardening, whereas hot-forming processes are not?
5. What is meant by the fibrous macrostructure?
6. Are the mechanical properties of a rolled sheet isotropic? Why?
7. What is meant by the state of stress? List the three general types.
8. List some advantages of hot forming. What are some disadvantages?
9. List some advantages of cold forming. What are some disadvantages?
10. Are the mechanical properties of a rolled sheet isotropic? Why?
11. What is meant by the state of stress? List the three general types.
12. List some advantages of hot forming. What are some disadvantages?
13. List some advantages of cold forming. What are some disadvantages
14. What may happen when a large section of steel is heated at a rapid rate? Why?
15. What should be avoided when heating large steel sections prior to hot forming?
16. Where does friction occur in metal forming?
17. What are the harmful effects of friction on the forming process?
18. Is friction always harmful in all metal-forming processes?
19. Can lead be used as a lubricant when forming copper? Why?
20. When forming lead at room temperature, do you consider it cold forming? Why?
21. Why are lubricants used in metal-forming processes? List some useful effects.
22. List some lubricants used in cold-forming processes.
23. List some lubricants used in hot-forming processes.
24. Which do you recommend for further processing by machining, a cold-worked part or a hot-worked part?
25. Is hot rolling the most widely used metal-forming process? Why?
26. List some of the useful effects of hot rolling.
27. Define *rolling*.
28. What is the angle of contact?
29. For heavier sections, would you recommend larger angles of contact in rolling? Why?

30. What is the state of stress in rolling?
31. List the different types of rolling mills.
32. What are the different parts of a roll? What is the function of each?
33. Explain why Sendzimir mills are used.
34. What are universal mills used for?
35. List three groups included in the range of rolled products.
36. Explain, using sketches, how seamless tubes are manufactured.
37. What is alligatoring? What causes it?
38. Define *wiredrawing*.
39. Which mechanical property should the metal possess if it is to be used in a drawing process? Why?
40. What is the state of stress in drawing?
41. List some advantages of the drawing process.
42. How is a metal prepared for a drawing process?
43. What are the different zones in a drawing die?
44. Mention the range of the apex angles (of conical shapes) used in drawing dies.
45. What material do you recommend to be used in making drawing dies?
46. Describe a draw bench.
47. What kinds of lubricants are used in drawing processes?
48. What is the drawing ratio?
49. Give an expression indicating the reduction achieved in a wiredrawing process.
50. Why do internal bursts occur in wiredrawing processes?
51. What are arrowhead fractures and why do they occur?
52. What is the state of stress in tube drawing?
53. Using sketches, illustrate the different techniques used in tube drawing.
54. Define *extrusion*.
55. Why can extrusion be used with metals having relatively poor plasticity?
56. List some advantages of the extrusion process.
56. What are the shortcomings and limitations of the extrusion process?
57. Using sketches, differentiate between the direct and indirect extrusion techniques.
58. Although indirect extrusion almost eliminates friction, it is not commonly used in industry. Why?
59. List the advantages of hydrostatic extrusion.
60. Compare extrusion with rolling with respect to the efficiency of material utilization.
61. When is conventional direct extrusion recommended as a production process?
62. Describe impact extrusion.
63. Why is the leading end of an extruded section always sheared off?
64. What are dead-metal zones?
65. If hardness measurements are taken across the section (say, circular) of an extruded part, what locations will have higher hardness values?
66. What lubricants can be used in cold extrusion?
67. What material do you recommend as a lubricant when hot extruding stainless steel?
68. What defect may occur when extruding magnesium at low extrusion ratios?
69. What is piping and why does it occur?
70. In extrusion dies, what is meant by the circle size?
71. List some considerations that must be taken into account when designing a section for extrusion.
72. Why should a designer try to avoid sharp corners at the root of a die tongue? Explain using neat sketches.

73. As a product designer, you are given a very intricate section for production by extrusion. Is there any way around this problem without being forced to use a die with a very intricate construction? How?
74. List some considerations for the design of impact extrusions.
75. How can you avoid shear failure at the bottom of the wall of an impact extrusion?
76. Does forging involve just imparting a certain shape to a billet?
77. Is it just a matter of economy to produce a crankshaft by forging rather than by machining from a solid stock? Why?
78. Can a metal such as aluminum be forged at any temperature? Why?
79. List the main types of forging processes.
80. Which process is suited for the production of small batches of large parts?
81. Give examples of parts produced by each type of forging process. Support your answer with evidence. What is the modern version of blacksmithing? What are the different operations involved in that process?
82. When do you recommend using a power-actuated hammer as a forging machine? Mention the type of forging process.
83. For which type of forging is a drop hammer employed?
84. For which type of forging is a crank press employed?
85. Using sketches, illustrate the different stages in manufacturing a ring by forging.
86. List the advantages that forging has over casting when producing large numbers of small parts having relatively complex shapes.
87. In comparison with question 83, what are the shortcomings of forging? Why don't they affect your decision in that particular case?
88. List some of the specified acceptance tests to be performed on forgings.
89. What is a board hammer used for?
90. Is it true that a closed type of forging die can have only one impression? Explain why.
91. What does hot pressing mean?
92. What is the advantage of HERF?
93. What are the advantages of warm forging?
94. What is meant by a mushy state?
95. Define *forgeability*. How can it be quantitatively assessed?
96. What is the most forgeable metal?
97. What is the main role of lubricants in hot forging?
98. As a product designer, how can you manipulate the alignment of the fibrous macrostructure?
99. List some guidelines regarding the location of the parting line between the upper and lower halves of a die set.
100. What is meant by the term *draft in forging?*
101. A die was designed to forge an aluminum part. Can the same design be used to forge a similar part made of titanium? Why?
102. Explain the meaning of *matching draft* using sketches.
103. Differentiate between a web and a rib in a forging.
104. What is the difference between a corner radius and a fillet radius? Use sketches.
105. What are punchout holes in a forging?
106. List some advantages of including punchout holes in a forging design.
107. Why are recesses sometimes included in a forging design?
108. List the different cold-forming processes and use sketches to illustrate how they differ.

PROBLEMS

1. In hot rolling, determine the load on each roll of a two-high rolling mill, given the following:
   a. Diameter of the roll: 20 inches (500 mm)

b. Stock width: 48 inches (1020 mm)
c. Initial thickness: 0.08 inch (2 mm)
d. Final thickness: 0.04 inch (1 mm)
e. Flow stress of rolled material: 14,200 lb/in.^{2} (100 MN/m^2)

2. In hot rolling a low-carbon steel plate 48 inches (1200 mm) in width, given the roll diameter as 20 inches (500 mm), initial thickness as 1.5 inches (37.5 mm), final thickness as 0.4 inch (10 mm), and the flow stress of steel as 28,400 lb/in.^{2} (200 MN/m^2), calculate the number of rolling passes if the maximum load on the roll in each pass is not to exceed 225,000 lb force (1.0 MN).
3. Write a computer program to solve problem 2, assuming that all the data are variables to be given for each design.
4. Calculate the maximum achievable reduction in a single drawing of a lead wire.
5. Estimate the largest possible extrusion ratio of a 2.0-inch (50-mm) aluminum bar having a mean flow stress of 21,900 lb/in.^{2} (150 MN/m^2) if the press available has a capacity of only 45,000 lb force (200 kN).
6. Plot a curve indicating the efficiency of a drop hammer versus the ratio between the weights of the anvil and the moving parts if the value of K that represents the elasticity of the billet is taken as 0.1. What ratio do you suggest? Why should it not be justified to take large ratios?

DESIGN PROBLEM

Design a simple wrench that measures 1/2 inch (12.5 mm) across bolt-head flats and is used for loosening nuts and bolts. The torque required to loosen (or tighten) a bolt (or a nut) is 1 lb·ft (1.35 N·m). The production volume is 25,000 pieces per year. Forging is recommended as a manufacturing process.

Solution

Because the wrench is going to be short, it cannot be held by the full hand but probably by only three fingers. The force that can be exerted is to be taken, therefore, as 4 lb. The arm of the lever is equal to (1 × 12)/4, or 3 inches (75 mm). Add on allowance for the holding fingers. The shape of the wrench will be as shown in Figure 5.67.

Now, let us select the materials. A suitable material would be AISI 1045 CD steel to facilitate machining (sawing) of the stock material. Closed-die forging of the billets is recommended, as well as employing drop-forging hammers. To facilitate the withdrawal of the part, the cross section of the handle should be elliptical (see Figure 5.67). The parting line should coincide with the major axis of the ellipse.

Let us check the stress due to bending:

$$I = \frac{1}{4}\pi a^3 b = \frac{1}{4}(\pi)(0.375)^3(0.1875) = 7.7 \times 10^{-3}\,\text{in.}^4$$

where

a is half the major axis
b is half the minor axis

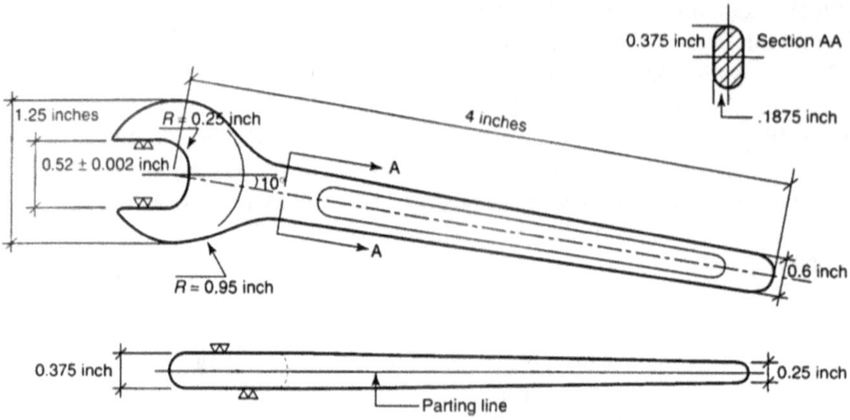

FIGURE 5.67 A wrench manufactured by forging.

$$\text{Stress} = \frac{My}{I} = \frac{Ma}{I} = \frac{5 \times 12 \times 0.375}{7.7 \times 10^{-3}} = 2922 \text{ lb / in.}^2$$

It is less than the allowable stress for 1045 CD steel, which is

$$\frac{60,000}{2} = 30,000 \text{ lb / in.}^2$$

In order to check the bearing stress, let us assume a shift of 0.25 inch between the forces acting on the faces of the nut to form a couple (this assumption can be verified if we draw the nut and the wrench to scale):

$$\text{Each force} = \frac{60}{0.25} = 240 \text{ lb}$$

Further, assume that the bearing area is 0.375 × 0.25 inch. The bearing stress is, therefore,

$$\frac{240}{0.375 \times 0.25} = 2560 \text{ lb / in.}^2$$

It is less than the allowable stress of the 1045 CD steel.

The forged wrench finally has to be trimmed and then machined on the surfaces indicated in Figure 5.67 An allowance of 1/64 inch should be provided between the wrench open-head and the nut. Now, our design is complete and ready to be released to the workshop.

DESIGN PROJECTS

1. A clock frame 3 × 5 inches (75 × 125 mm) is manufactured by machining an aluminum-alloy stock. Make a design and a preliminary feasibility study so that it can be produced by extrusion. Assume the production volume is 20,000 pieces per year.
2. A motor frame that has a 6-inch (150-mm) internal diameter and that is 10 inches (250 mm) long is currently produced by casting. That process yields a high percentage of rejects, and the production cost is relatively high. Knowing that the production volume is 20,000 pieces per year, redesign the part so that it will be lighter and can be easily produced by an appropriate metal-forming operation that has a high efficiency of material utilization.

3. A pulley transmits a torque of 600 lb·ft (816 N·m) to a shaft that is 1¼ inches (31 mm) in diameter. It is to be driven by a flat belt that is 2 inches (50 mm) in width. Provide a detailed design for the pulley if the production volume is 10,000 pieces per year and the pulley is manufactured by forging.
4. A connecting lever is to be manufactured by forging. The estimated production volume is 50,000 pieces per year. The lever has two short bosses, each at one of its ends, and each has a vertical hole 3/4 inch (19 mm) in diameter. The horizontal distance between the centers of the two holes is 12 inches (300 mm), and the vertical difference in levels is 3 inches (75 mm). The lever during its functioning is subjected to a bending moment of 200 lb·ft (272 N·m). Make a detailed design for this lever.
5. If the lever in problem 4 is to be used in a space vehicle, would you use the same material? What are the necessary design changes? Make a design appropriate for this new situation.
6. Design a gear blank that transmits a torque of 200 lb·ft (272 N·m) to a shaft that is 3/4 inch (19 mm) in diameter. The pitch diameter of the gear is 8 inches (200 mm), and 40 teeth are to be cut in that blank by machining. Assume the production volume is 10,000 pieces per year.
7. A straight-toothed spur-gear wheel transmits a torque of 1200 lb·ft (1632 N·m) to a steel shaft (AISI 1045 CD steel) that is 2 inches (50 mm) in diameter. The pitch diameter of the gear is 16 inches (400 mm), its width is 4 inches (100 mm), and the base diameter is 15 inches (375 mm). Make a complete design for this gear's blank (i.e., before teeth are cut) when it is to be manufactured by forging. Assume the production volume is 10,000 pieces per year.
8. A shaft has a minimum diameter of 1 inch (25 mm) at both of its ends, where it is to be mounted in two ball bearings. The total length of the shaft is 12 inches (300 mm). The shaft is to have a gear at its middle, with 40 teeth and a pitch circle diameter of 1.9 inches (47.5 mm). The width of the gear is 2 inches (50 mm). Make a design for this assembly if the production volume is 50,000 per year.

6 Sheet Metal Working

The processes of *sheet metal working* have recently gained widespread industrial application. Their main advantages are their high productivity, close tolerances, and excellent surface finish of the products (which usually require no further machining). The range of products manufactured by these processes is vast, but, in general, all of these products have thin walls (relative to their surface area) and relatively intricate shapes. Sheets made from a variety of metals (e.g., low-carbon steel, high-ductility alloy steel, copper and some of its alloys, and aluminum and some of its alloys) can successfully be worked into useful products. Therefore, these processes are continually becoming more attractive to the automotive, aerospace, electrical, and consumer goods industries. Products that had in the past always been manufactured by processes like casting and forging have been redesigned so that they can be produced by sheet metal working. Components like pulleys, connecting rods for sewing machines, and even large gears are now within the range of sheet metal products. Sheet metals are usually worked while in their cold state. However, when processing thick sheets, which are at least 0.25 inch (6 mm) and are referred to as *plates*, thermal cutting is employed to obtain the required blank shape, and the blank is then hot-worked in a hydraulic or friction screw press. Thus, the fabrication of boilers, tanks, ship hulls, and the like would certainly require hot working of thick plates.

By far, the most commonly used operations in sheet metal working are those performed in a press. For this reason, they are usually referred to as *press working*, or simply *stamping*, operations. Other techniques involve *high-energy-rate forming* (HERF), like using explosives or impulsive discharges of electrical energy to form the blank, and *spinning* of the sheet metal on a form mandrel. This chapter will describe each of the various operations employed in sheet metal working.

6.1 PRESS WORKING OPERATIONS

All press working operations of sheet metals can be divided into two main groups: *cutting* operations and *shape-forming* operations. Cutting operations involve separating a part of the blank, whereas forming operations involve nondestructive plastic deformation, which causes relative motion of parts of the blank with respect to each other. Cutting operations include shearing, cutoff, parting, blanking, punching, and notching. Shape-forming operations include various bending operations, deep drawing, embossing, and stretch forming.

6.1.1 Cutting Operations

The mechanics of separating the metal are the same in all sheet metal cutting operations. Therefore, the operations are identified according to the shape of the curve along which cutting takes place. When the sheet metal is cut along a straight line, the operation is called *shearing* and is usually performed using inclined blades or guillotine shears in order to reduce the force required (see Figure 6.1). Cutting takes place gradually, not all at once, over the width of the sheet metal because the upper blade is inclined. The angle of inclination of the upper blade usually falls between 4° and 8° and must not exceed 15° so that the sheet metal is not pushed out by the horizontal component of the reaction.

When cutting takes place along an open curve (or on an open corrugated line), the operation is referred to as *cutoff*, provided that the blanks match each other or can be fully nested, as shown in Figure 6.2. The cutoff operation results in almost no waste of stock and is, therefore, considered very

DOI: 10.1201/9781003537106-6

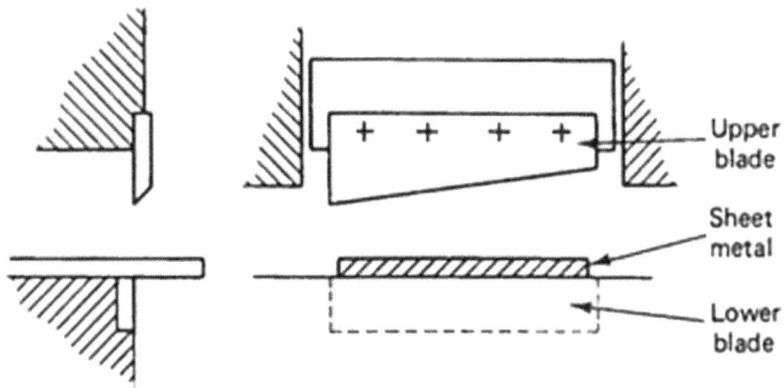

FIGURE 6.1 Shearing operation with inclined blades.

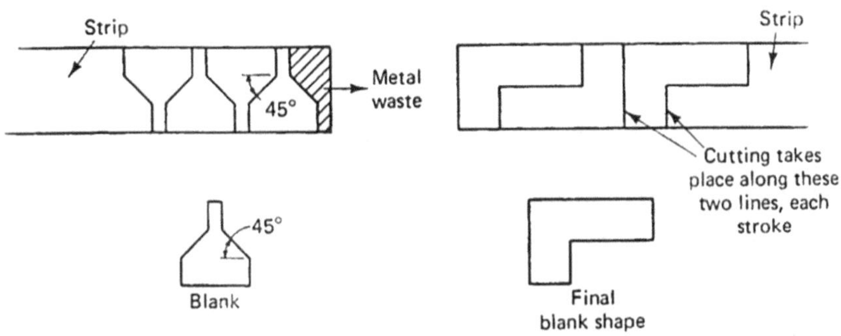

FIGURE 6.2 Examples of cutoff operations.

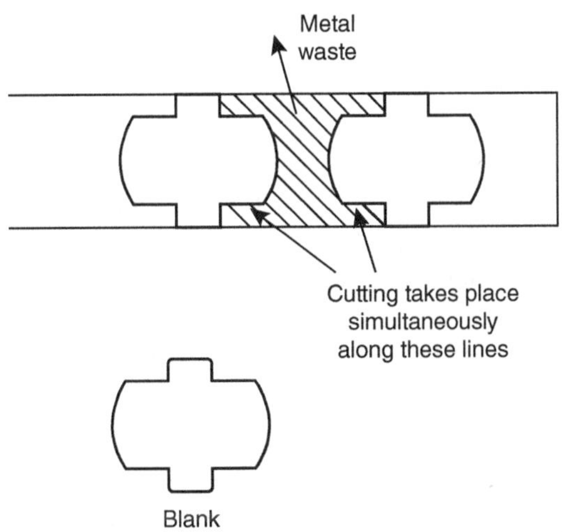

FIGURE 6.3 An example of a parting operation.

efficient with respect to material utilization. This operation is usually performed in a die that is mounted on a crank press. If the blanks do not match each other, it is necessary for cutting to take place along two open curves (or lines), as shown in Figure 6.3. In this case, the operation is called *parting*. It is clear from the figure that a parting operation results in some waste of stock and is, therefore, less efficient than shearing and cutoff operations. In *blanking* operations, cutting occurs along a closed contour and results in a relatively high percentage of waste in stock metal, a fact that

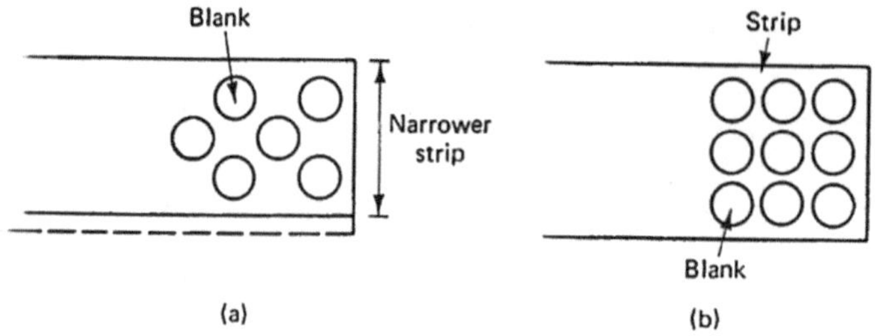

FIGURE 6.4 Two methods for laying out circular blanks for blanking operations: (a) staggered layout and (b) in-line arrangement.

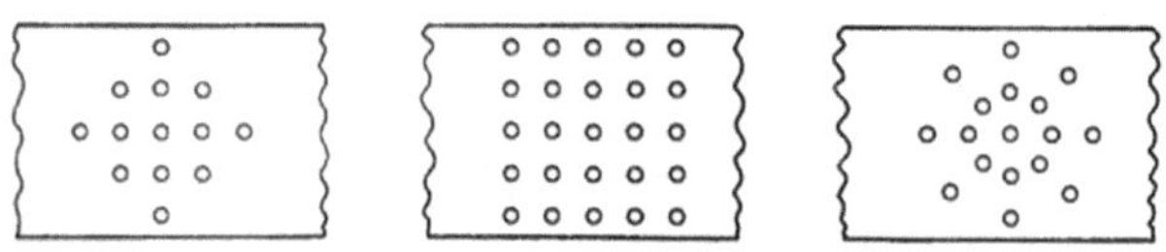

FIGURE 6.5 Different patterns of holes produced by perforating operations.

makes blanking operations less efficient than other cutting operations. Nevertheless, this process is used for the mass production of blanks that cannot be manufactured by any of the preceding operations. An efficient layout of blanks on the strip of sheet metal can result in an appreciable saving of material. An example of a good layout is shown in Figure 6.4a, where circular blanks are staggered. The in-line arrangement shown in Figure 6.4b is less efficient in terms of material utilization. Because a blanking operation is performed in a die, there is a limit to the minimum distance between two adjacent blanks. It is always advantageous to keep this minimum distance larger than 70 percent of the thickness of the sheet metal. In blanking, the part separated from the sheet metal is the product, and it is usually further processed. But if the remaining part of the sheet is required as a product, the operation is then termed *punching*. Sometimes, it is required to simultaneously punch a pattern of small holes as an ornament, for light distribution, or for ventilation; the operation is then referred to as *perforating*. Figure 6.5 illustrates some patterns of perforated holes.

A *notching* operation is actually a special case of punching, where the removed part is adjacent to the edge of the strip. It is clear that any required shape can be obtained by carrying out several notching operations. For this reason, notching is usually employed in progressive dies. A similar operation, called *seminotching*, in which the separated part is not attached to the side of the strip, is also used in the progressive working of sheet metals. In Figure 6.6, we can see both of these operations and how they can be employed progressively to produce a blank with an intricate shape.

6.1.1.1 Mechanics of Sheet Metal Cutting

Let us now look further at the process of cutting sheet metal. For simplicity, consider the simple case where a circular punch, together with a matching die, is employed to punch a hole. Figure 6.7 shows the punch, die, and sheet metal during a punching operation. When a load is applied through the punch, the upper surface of the metal is elastically bent over the edge of the punch, while the lower surface is bent over the edge of the die. With a further increase in the punch load, the elastic curvature becomes permanent or plastic and is referred to as the rollover. Next, the punch sinks into the upper surface of the sheet, while the lower surface sinks into the die hole. This stage involves mainly plastic flow of metal by shearing as there are two forces equal in magnitude and opposite in direction, subjecting the cylindrical surface within the metal to intense shear stress. The result will be a cylindrical smooth surface in contact with the cylindrical surface of the punch as it sinks into the sheet metal. Also, a similar surface forms the border of the part of the metal sinking into the die hole. Each of these smooth surfaces is called a burnish. The extent of a burnish depends upon the metal of

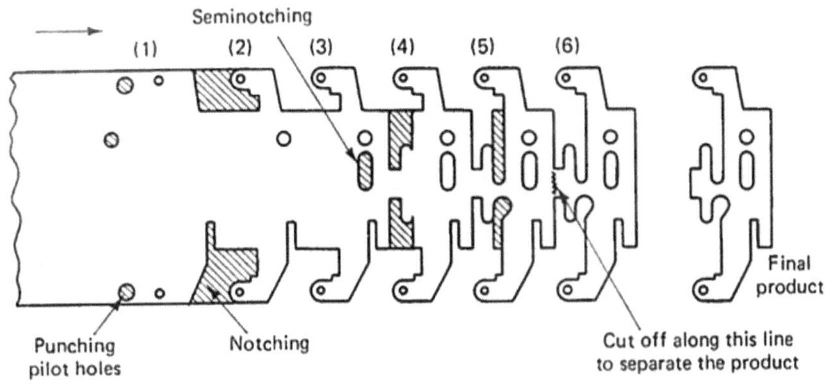

FIGURE 6.6 Progressive working operations.

the sheet as well as on the design features of the die. The burnish ranges approximately between 40 and 60 percent of the stock thickness, the higher values being for soft ductile materials like lead and aluminum. At this stage, two cracks initiate simultaneously in the sheet metal, one at the edge of the punch and the other at the edge of the die. These two cracks propagate and finally meet each other to allow the separation of the blank from the sheet metal. This zone has a rough surface and is called the fracture surface (break area). Finally, when the newly formed blank is about to be completely separated from the stock, a burr is formed all around its upper edge. Thus, the profile of the edge of a blank involves four zones: a rollover, a burnish, a fracture surface, and a burr. In fact, the profile of the edge of the generated hole consists of the same four zones, but in reverse order.

We are now in a position to discuss the effects of some process parameters, such as the punch-die clearance. Figure 6.8a illustrates the case where the punch-die clearance is excessive and is almost equal to the thickness of the sheet. Initially, the metal is bent onto the round edges of the punch and the die, and it then forms a short circular wall connecting the flat bottom and the bulk of the sheet. With a further increase in the applied load, the wall elongates under the tensile stress, and tearing eventually occurs. As can be seen in Figure 6.8a, the blank resulting in this case has a bent, torn edge all around and, therefore, has no value. On the other hand, if the punch-die clearance is too tight, as shown in Figure 6.8b, the two cracks that initiate toward the end of the operation do not meet, and another shearing must take place so that the blank can be separated. This operation is referred to as the *secondary shearing*. As can be seen, the obtained blank has an extremely rough side. In addition, the elastically recovering sheet stock tends to grip the punch, as shown in Figure 6.9, thus increasing the force required to withdraw the punch from the hole, which is usually called the *stripping force*. This results in excessive punch wear and shorter tool life. On the other hand, the blank undergoes elastic recovery, and it is, therefore, necessary to provide relief by enlarging the lower part of the die hole, as shown in Figure 6.10. Between these two extremes for the punch-die clearance, there exists an optimum value that reduces or minimizes the stripping force and the tool wear and also gives a blank with a larger burnish and smaller fracture surface. This recommended value for the punch-die clearance is usually taken as about 10–15 percent of the thickness of the sheet metal, depending upon the kind of metal being punched.

6.1.1.2 Forces Required

Based on the preceding discussion, the force required for cutting sheet metal is equal to the area subjected to shear stress (the product of the perimeter of the blank multiplied by the thickness of the sheet metal) multiplied by the ultimate shear strength of the metal being cut. The blanking force can be expressed by the following equation:

$$F = K \times Q \times t \times \tau_{\text{ultimate}} \tag{6.1}$$

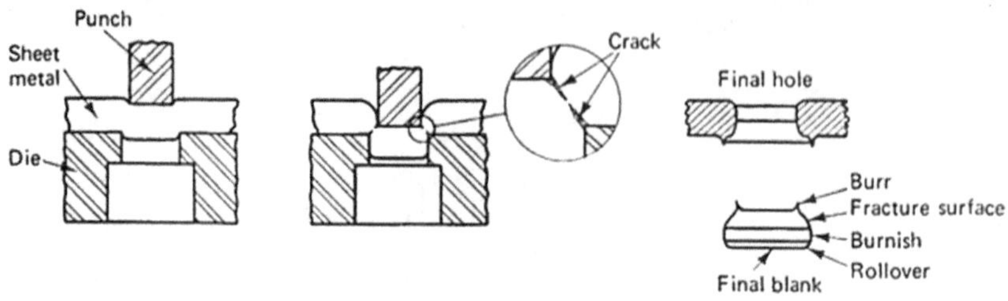

FIGURE 6.7 Stages of a blanking operation.

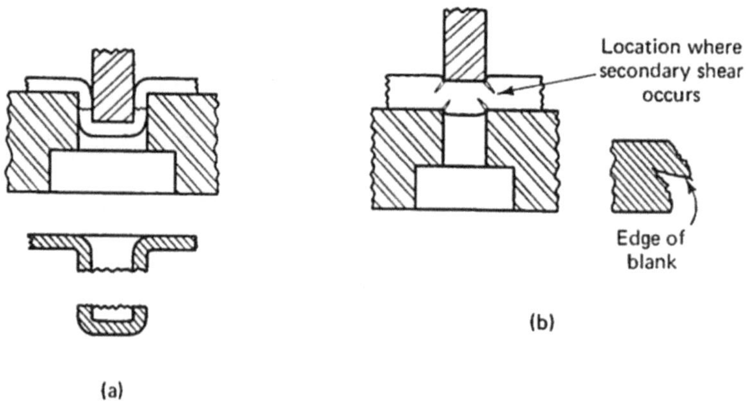

FIGURE 6.8 Blanking operations where the punch-die clearance is (a) excessive and (b) too tight.

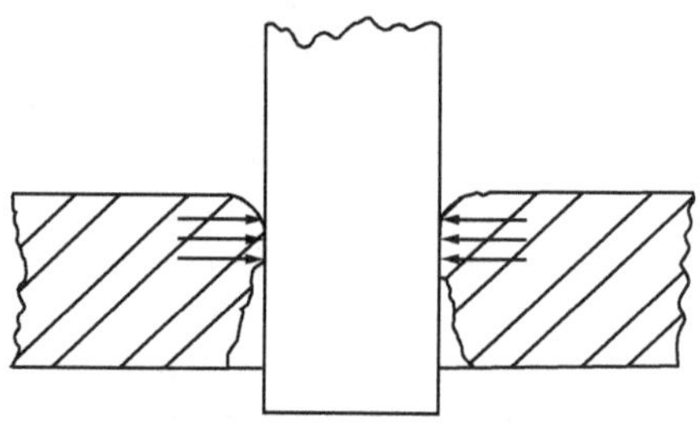

FIGURE 6.9 Elastic recovery of the metal around the hole gripping the punch.

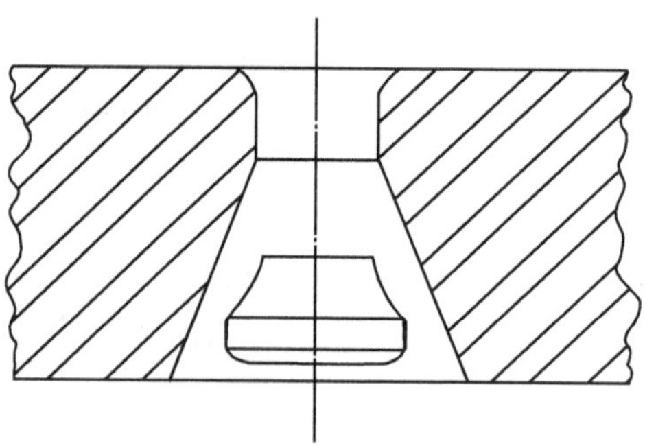

FIGURE 6.10 Elastic recovery of the blank necessitating die relief.

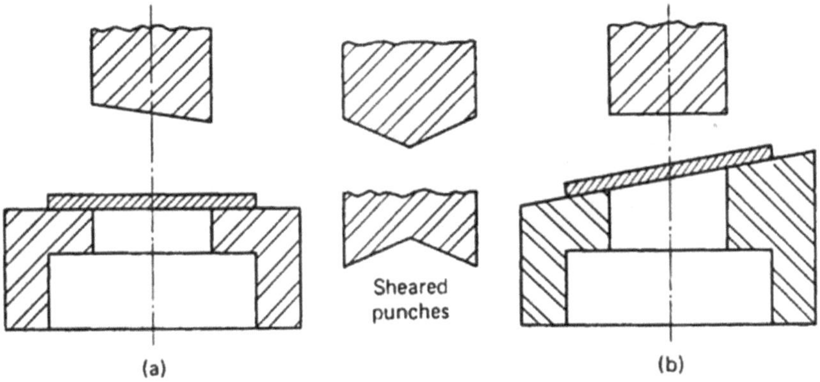

FIGURE 6.11 Shearing of the punch and die: (a) sheared punch resulting in distorted blanks and (b) sheared die resulting in distorted holes.

where

Q is the perimeter
t is the thickness
τ_{ultimate} is the ultimate shear stress

Note that K is an experimentally determined factor to account for the deviation of the stress state from that of pure shear and is taken as about 1.3. The ultimate shear stress can either be obtained from handbooks or be taken as approximately 0.8 of the ultimate tensile strength (UTS) of the same metal.

We can now see that one of the tasks of a manufacturing engineer is to calculate the required force for blanking (or punching) and to make sure that it is below the capacity of the available press. This is particularly important in industries that involve blanking relatively thick plates. There is, however, a solution to the problem when the required force is higher than the capacity of the available press. It is usually achieved by beveling (or shearing) the punch face in punching operations and the upper surface of the die steel in blanking operations. Shearing the punch results in a perfect hole but a distorted blank, whereas shearing the die yields a perfect blank but a distorted hole. Nevertheless, in both cases, cutting takes place gradually, not all at once, along the contour of the hole (or the blank), with the final outcome being a reduction in the required blanking force. The shear angle is usually taken proportional to the thickness of the sheet metal and ranges between 2° and 8°. Double-sheared punches are quite common and are employed to avoid the possibility of horizontal displacement of sheet metals during punching. Figure 6.11 illustrates the basic concept of punch and die shearing. It also provides a sketch of a double-sheared punch.

Another important aspect of the punching (or blanking) operation is the stripping force (i.e., the force required to pull the punch out of the hole). It is usually taken as 10 percent of the cutting force, although it depends upon some process parameters, such as the elasticity and plasticity of the sheet metals, the punch-die clearance, and the kind of lubricant used.

6.1.1.3 Bar Cropping

Bar cropping is similar to sheet metal cutting. Although bars, not sheets, are cut, the mechanics of the process are similar to those of sheet metal cutting, and the separation of the cropped part is due to plastic flow caused by intense shear stress. The process is used for mass production of billets for hot forging and cold-forming processes. Nevertheless, the distortion and work hardening at the sheared cross section limit the application of bar cropping when the billets are to be cold formed.

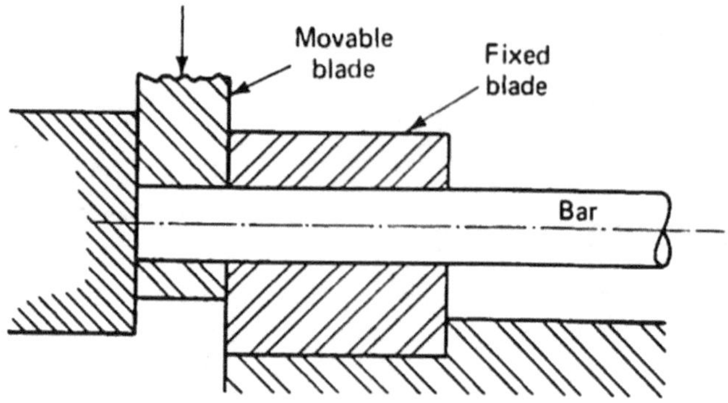

FIGURE 6.12 Bar cropping with workpiece totally confined.

Therefore, a modified version of the cropping operation has to be used. It involves completely confining the cropped billet and applying an axial stress of approximately 20 percent of the tensile strength of the bar material. This bar-cropping technique, which is shown in Figure 6.12, yields a very smooth-cropped surface and distortion-free billets.

6.1.1.4 Fine Blanking

As we saw previously, the profile of the edge of a blank is not smooth but consists of four zones: the rollover, the burnish, the fracture surface (break area), and the burr. Sometimes, however, the blank must have a straight, smooth side for some functional reasons. In this case, an operation called fine blanking is employed, as Figure 6.13 shows. This operation necessitates the use of a triple-action press and a special die with a very small punch-die clearance. As can be seen in the figure, the metal is squeezed and restrained from moving in the lateral directions in order to control the shear flow along a straight vertical direction. A variety of shapes can be produced by this method. They can have any irregular outer contour and a number of holes as well. The fine-blanking operation has found widespread application in precision industries.

6.1.1.5 Miscellaneous Cutting Operations

The primary operation that is used for preparing strips for blanking is needed because the available sheets vary in width between 32 and 80 inches (800–2000 mm), a range that is usually not suitable because of the dimensions of the die and the press. Therefore, coils having a suitable width have to be obtained first. The operation performed is called slitting, and it employs two circular cutters for

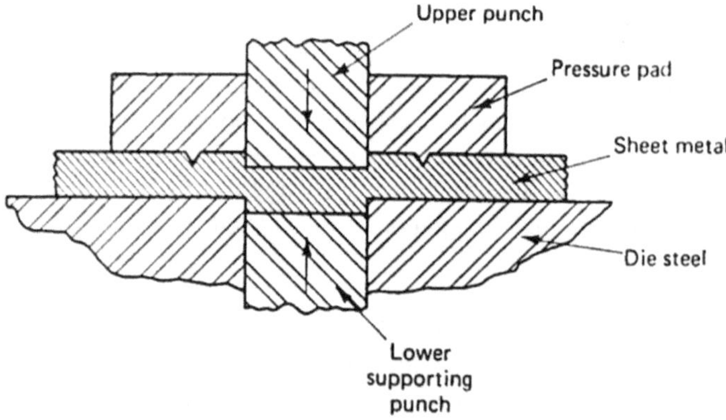

FIGURE 6.13 Fine-blanking operation.

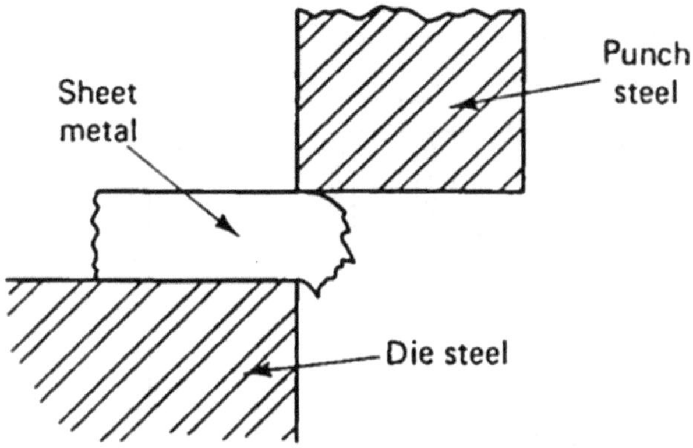

FIGURE 6.14 The shaving operation.

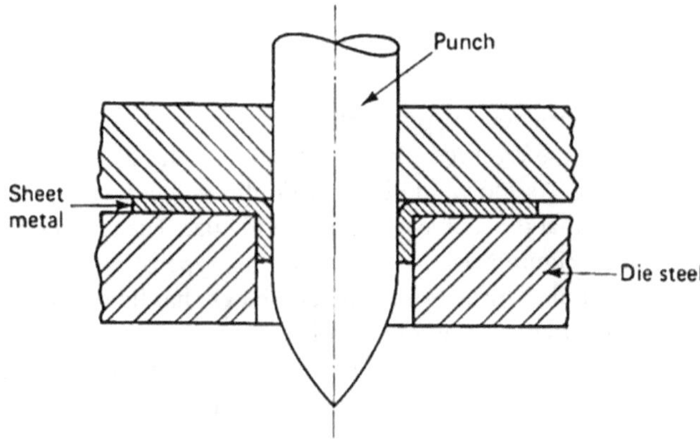

FIGURE 6.15 The piercing operation.

each straight cut. Sometimes, slitting is carried out in a rolling plant, and coils are then shipped ready for blanking.

A secondary operation that is sometimes carried out on blanks (or holes) to eliminate rough sides and/or to adjust dimensions is the *shaving* operation. The excess metal in this case is removed in the form of chips. As can be seen in Figure 6.14, the punch-die clearance is very small. For this reason, the die must be rigid, and the matching of its two halves must be carefully checked.

Sometimes, punching operations are mistakenly called *piercing*. In fact, the mechanics of sheet metal cutting in the two operations are completely different. We can see in Figure 6.15 that piercing involves a tearing action. We can also see the pointed shape of the punch. Neither blanks nor metal waste result from the piercing operation. Instead, a short sleeve is generated around the hole, which sometimes has functional applications in toy construction and the like.

6.1.1.6 Cutting-Die Construction

The construction of cutting dies may take various forms. The simplest one is the drop-through die, which is shown in Figure 6.16. In addition to the punch and die steels, the die includes the upper and lower shoes, the guideposts, and some other auxiliary components for guiding and holding the metal strip. The stripper plate touches the strip first and holds it firmly during the blanking operation; it then continues to press it until the punch is totally withdrawn from the hole made in the strip. The generated blanks fall through the die hole, which has a relief for this reason and are collected in a container located below the bed of the press.

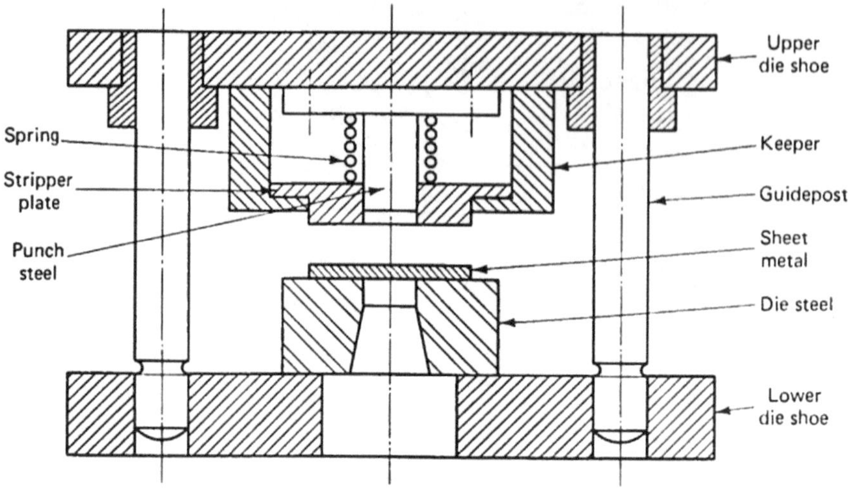

FIGURE 6.16 Die construction for simple drop-through blanking die.

Consequently, this die construction is applicable only if the bed of the press has a hole. On the other hand, if the diameter of the required blanks is too large, the use of a drop-through die may result in a defect called *dishing*. As shown in Figure 6.17, this defect involves slackening the middle of the blank in such a manner that it becomes curved and not flat. The answer to this problem lies in employing a *return-type die*. Figure 6.18 shows that in this type of die construction, the blank is supported throughout the operation by a spring-actuated block that finally pushes the blank upward above the surface of the strip, where it is automatically collected. A more complicated die construction, like that shown in Figure 6.19, can be used to perform two operations simultaneously. This is usually referred to as a *compound die*. As can be seen in Figure 6.19, the hollow blanking punch is

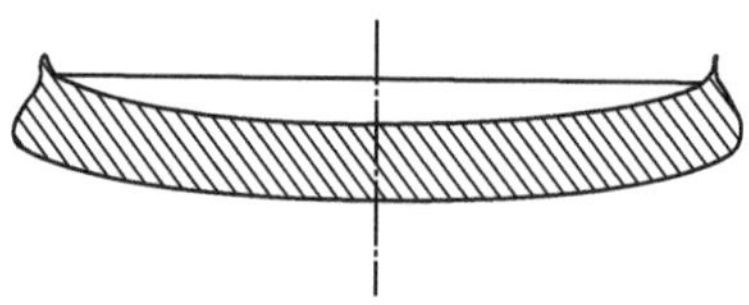

FIGURE 6.17 A vertical section through a blank with the dishing defect.

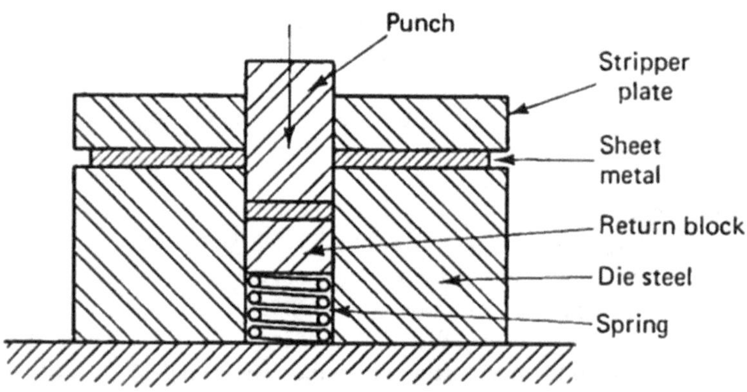

FIGURE 6.18 A return-type die.

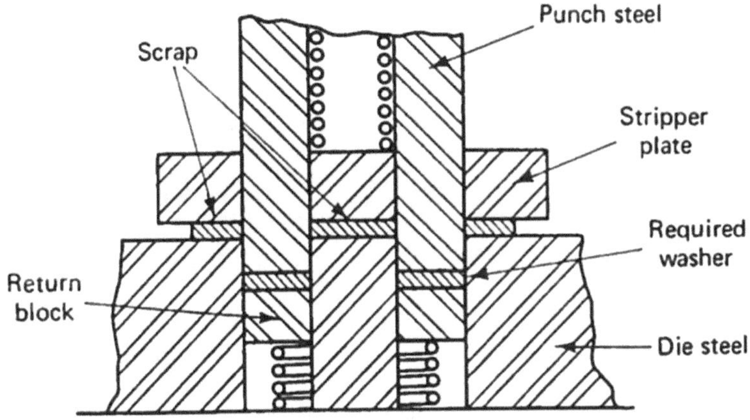

FIGURE 6.19 A compound die for producing a washer.

also a hole-punching die. This allows blanking and punching to be carried out simultaneously. The product, which is a washer, and the central scrap are removed by return blocks.

6.1.2 Bending Operations

Bending is the simplest operation of sheet metal working. It can, therefore, be carried out by employing simple hand tools. As opposed to cutting operations, there is always a clear displacement between the forces acting during a bending operation. The generated bending moment forces a part of the sheet to be bent with respect to the rest of it through local plastic deformation. Therefore, all straight unbent surfaces are not subjected to bending stresses and do not undergo any deformation. Figure 6.20 illustrates the most commonly used types of bending dies: the V-type, the wiping, and the channel (U-type) die. We can see that the displacement between forces is maximum in the case of the V-type die, and, therefore, lower forces are required to bend sheet metal when using this kind of die.

6.1.2.1 Mechanics of Bending

The bending of sheet metal resembles the case of a beam with a very high width-to-height ratio. When the load is applied, the bend zone undergoes elastic deformation; then plastic deformation occurs with a further increase in the applied load. During the elastic deformation phase, the external fibers in the bend zone are subjected to tension, whereas the internal fibers are subjected to compression. The distribution of stresses is shown in Figure 6.21a. Note that there is a neutral plane that is

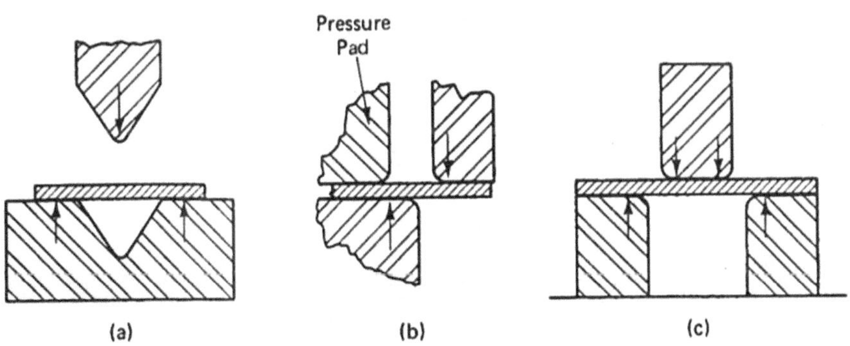

FIGURE 6.20 The three common types of bending dies: (a) V-type die, (b) wiping die, and (c) channel (U-type) die.

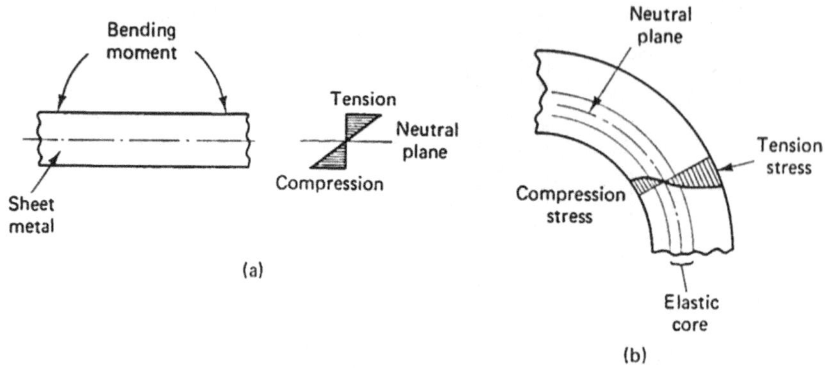

FIGURE 6.21 Distribution of stresses across the sheet thickness: (a) in the early stage of bending and (b) toward the end of a bending operation.

free of stresses in the middle of the thickness of the sheet. The length of the neutral axis remains constant and does not undergo either elongation or contraction. Next, when the plastic phase starts, the neutral plane approaches the inner surface of the bend, as can be seen in Figure 6.21b. The location of the neutral plane is dependent upon many factors, such as the thickness of the sheet metal, the radius, and the degree of bend. Nevertheless, the distance between the neutral plane and the inner surface of the bend is taken as equal to 40 percent of the thickness of the sheet metal as a first approximation for the calculations of blank development.

Let us now consider a very important phenomenon—namely, *springback*, which is an elastic recovery of the sheet metal after the removal of the bending load. As Figure 6.22 indicates, for bending by an angle of 90°, the springback amounts to a few degrees. Consequently, the obtained angle of bend is larger than the required one. Even toward the end of the bending operation, the zone around the neutral plane is subjected to elastic stresses and, therefore, undergoes elastic deformation (see Figure 6.21b). As a result, the elastic *core* tries to return to its initial flat position as soon as the load is removed. When doing so, the plastically deformed zones impede it. The final outcome is, therefore, an elastic recovery of just a few degrees. Consequently, the way to eliminate springback involves forcing this elastic core to undergo plastic deformation. This can be achieved through either of the techniques shown in Figure 6.23a,b. In the first case, the punch is made so that a projection squeezes the metal locally; in the second case, high tensile stress is superimposed upon bending. A third solution is overbending, as shown in Figure 6.23c. In this case, the amount of overbending should be equal to the springback so that the exact required angle is obtained after the elastic recovery.

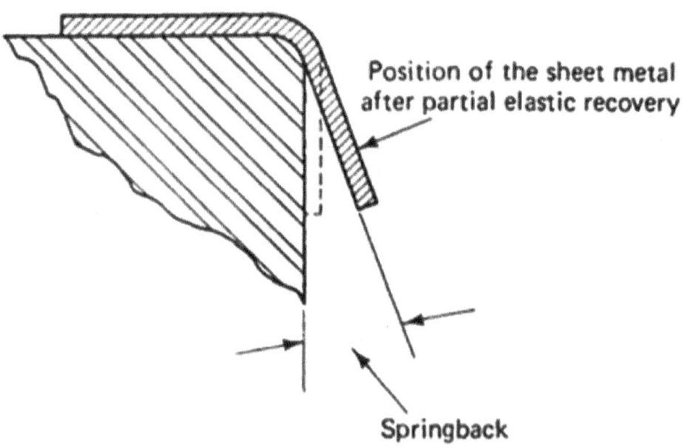

FIGURE 6.22 The springback phenomenon.

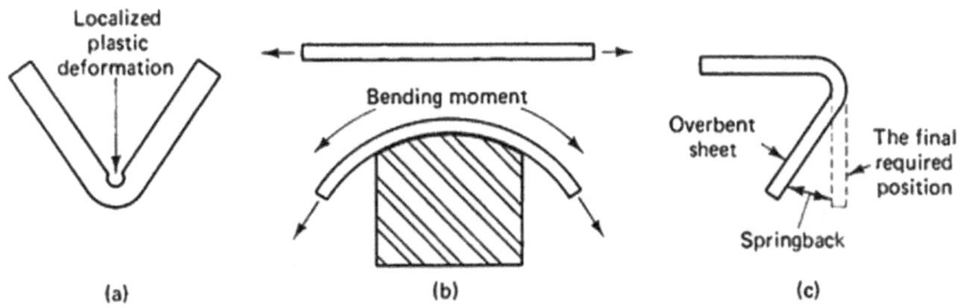

FIGURE 6.23 Methods used to eliminate springback: (a) bottoming, (b) stretch-forming, and (c) overbending.

6.1.2.2 Blank Development

We have previously referred to the fact that the neutral plane does not undergo any deformation during the bending operation and that its length, therefore, remains unchanged. Accordingly, the length of the blank before bending can be obtained by determining the length of the neutral plane within the final product. The lengths of the straight sections remain unchanged and are added together. The following equation can be applied to any general bending product, such as the one shown in Figure 6.24:

$$L = \text{Total length of blank before bending}$$

$$= l_1 + l_2 + l_3 + l_4 + \frac{\pi \times \alpha_1}{180} R_1 + \frac{\pi \times \alpha_2}{180} R_2 + \frac{\pi \times \alpha_3}{180} R_3 \tag{6.2}$$

where

R is equal to $r + 0.4t$
r is the inner radius of a bend
t is the thickness of the sheet metal
R is the radius of the neutral axis

6.1.2.3 Classification of Bending Operations

Various operations can be classified as bending, although each one has its own industrial name. They include, for example, conventional bending, flanging, hemming, wiring, and corrugating. The flanging operation is quite similar to conventional bending, except that the ratio of the lengths of the bent part to that of the sheet metal is small. Flanging is usually employed to avoid a sharp edge, thus eliminating the possibility of injury. It is also used to add stiffness to the edges of sheet metal and for assembly purposes.

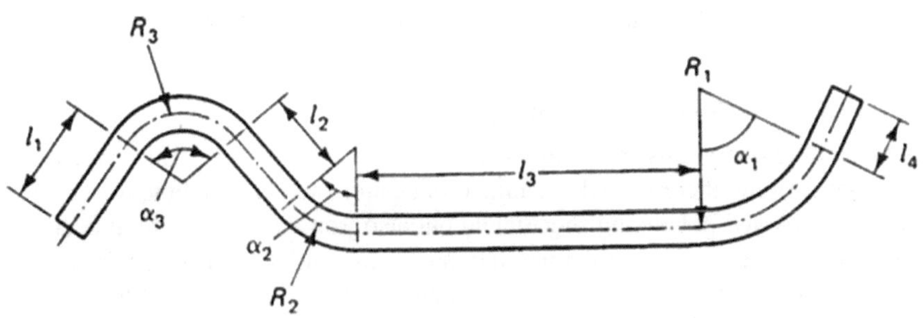

FIGURE 6.24 A bending product divided into straight and circular sections for blank development.

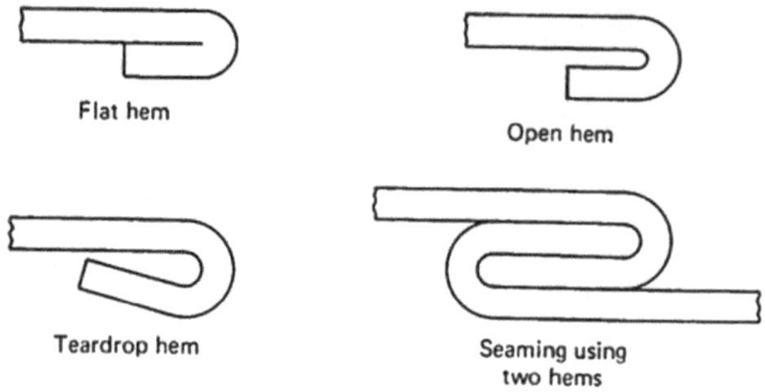

FIGURE 6.25 Different kinds of hems.

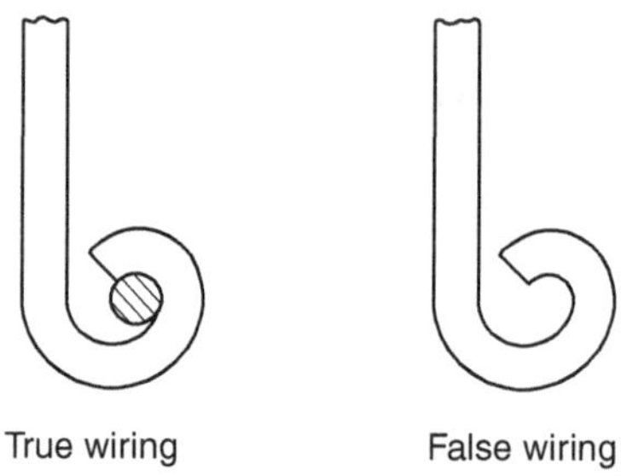

FIGURE 6.26 Wiring operation.

FIGURE 6.27 Different shapes of corrugated sheet metal.

Among the bending operations, *hemming* used to be a very important one, before the recent developments in welding and can-forming technologies. A hem is a flange that is bent by 180°; it is used now to get rid of a sharp edge and to add stiffness to sheet metal. A few decades ago, hems were widely employed for seaming sheet metals. Figure 6.25 shows four different kinds of hems. A similar operation is *wiring*, which is shown in Figure 6.26. *True* wiring involves bending the edge of the sheet metal around a wire. Sometimes, the operation is performed without a wire, and it is then referred to as *false* wiring.

Corrugating is another operation that involves bending sheet metal. Different shapes, like those shown in Figure 6.27, are obtained by this operation. These shapes possess better rigidity and can resist bending moments normal to the corrugated cross section mainly because of the increase in the moment of inertia of the section due to corrugation and because of the work-hardened zones resulting from bending.

6.1.2.4 Miscellaneous Bending Operations

Conventional bending operations are usually carried out on a press brake. However, with the developments in metal-forming theories and machine tool design and construction, new techniques have evolved that are employed in bending sheet metal but also iron angles, structural beams, and tubes. Figure 6.28 illustrates the working principles and the stages involved in roll bending. As can be seen in the figure, the rolls form a pyramid-type arrangement. Two rolls are used to feed the material, whereas the third (roll B) gradually bends it (see Figure 6.28a,b). The direction of the feed is then reversed, and roll A now gradually bends the beam (see Figure 6.28c,d).

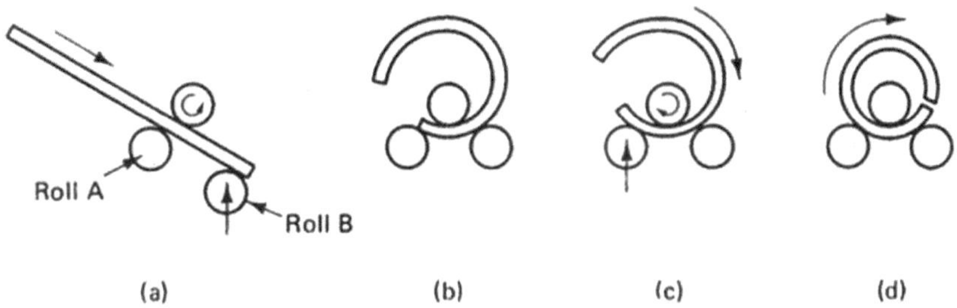

FIGURE 6.28 Stages involved in roll bending a structural beam: (a) feeding, (b) initial bending, (c) further bending, and (d) reversing the direction of feed.

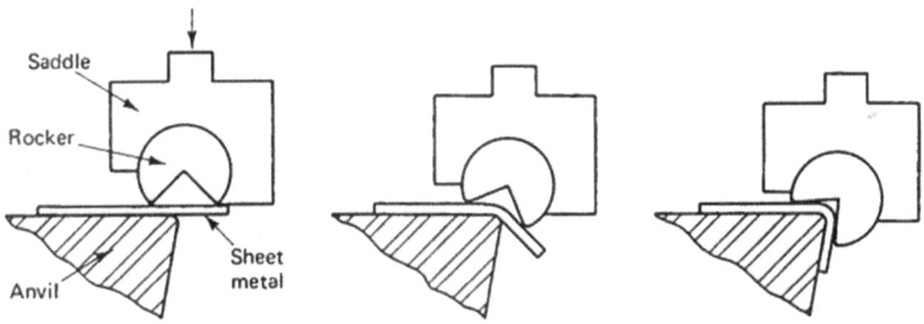

FIGURE 6.29 Working principles of rotary bending.

Another bending operation that recently emerged and that is gaining industrial application is *rotary bending*. Figure 6.29 illustrates the working principles of this operation. As can be seen, the rotary bender includes three main components: the saddle, the rocker, and the die anvil. The rocker is actually a cylinder with a V-notch along its length. The rocker is completely secured inside the saddle (i.e., the saddle acts like a housing) and can rotate but cannot fall out. The rotary bender can be mounted on a press brake. The rocker acts as both a pressure pad and a bending punch. Among the advantages claimed for rotary bending are the elimination of the pressure pad and its springs (or nitrogen cylinders), lower required tonnage, and the possibility of overbending without the need for any horizontal cams. This new method has been patented by the Accurate Manufacturing Association and is nicknamed by industrial personnel as the "Pac Man" bending operation.

A bending process that is usually mistakenly mentioned among the rolling processes is the manufacture of thin-walled welded pipes. Although rolls are the forming tools, the operation is actually a gradual and continuous bending of a strip that is not accompanied by any variation in the thickness of that strip. Figure 6.30 indicates the basic principles of this process. Notice that the width of the

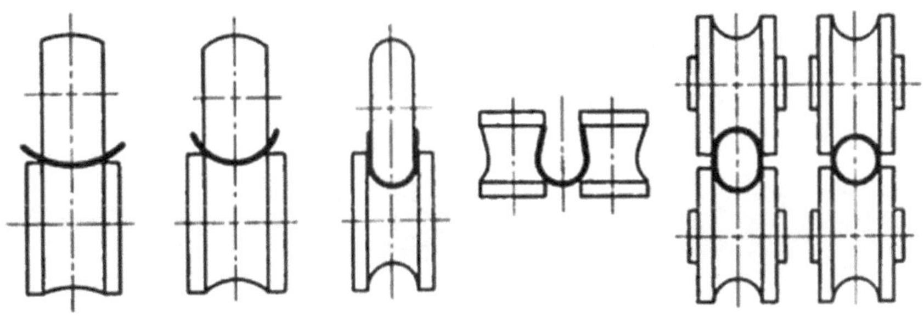

FIGURE 6.30 Roll bending as employed in the manufacture of seamed tubes.

strip is gradually bent to take the form of a circle. Strip edges must be descaled and mechanically processed before the process is performed to improve weldability. Either butt or high-frequency induction welding is employed to weld the edges together after the required circular cross section is obtained. This process is more economical and more productive than seamless tube rolling. Poor strength and corrosion resistance of seams are considered its main disadvantages.

6.1.3 Deep Drawing Operation

Deep drawing involves the manufacture of deep, cuplike products from thin sheet metal. As can be seen in Figure 6.31, the tooling basically involves a punch with a round corner and a die with a large edge radius. It can also be seen that the punch-die clearance is slightly larger than the thickness of the sheet metal. When a load is applied through the punch, the metal is forced to flow radially and sink into the die hole to form a cup. This is an oversimplification of a rather complex problem. For the proper design of deep-drawn products as well as the tooling required, we have to gain a deeper insight into the process and understand its mechanics.

6.1.3.1 Mechanics of Deep Drawing

Consider what happens during the early stages of applying the load. As Figure 6.32a shows, the blank is first bent onto the round edge of the die hole. With further increase in the applied load, the part of the blank that was bent is straightened in order to sink into the annular clearance between the punch and the die, thus forming a short, straight, vertical wall. Next, the rest of the blank starts to flow radially and to sink into the die hole, but because the lower surface of the blank is in contact with the upper flat surface of the die steel, frictional forces try to impede that flow. These forces are a result of static friction; their magnitude drops as the blank metal starts to move. Now consider what happens to a sector of the blank, such as that shown in Figure 6.32b, when its metal flows radially. It is clear that the width of the sector shrinks so that the large peripheral perimeter of the blank can fit into the smaller perimeter of the die hole. This is caused by circumferential compressive stresses acting within the plane of the blank. With a further increase in the applied load, most of the blank sinks into the die hole, forming a long vertical wall, while the remaining part of the blank takes the form of a small annular flange (see Figure 6.31). The vertical wall is subjected to uniaxial tension whose magnitude is increasing when going toward the bottom of the cup.

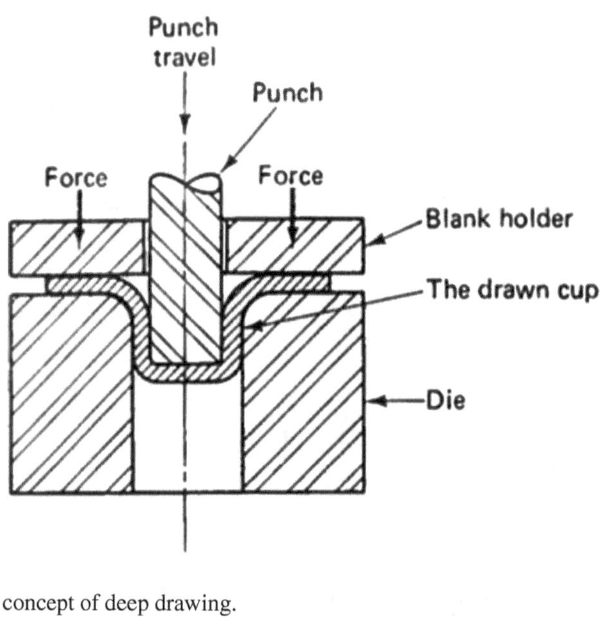

FIGURE 6.31 Basic concept of deep drawing.

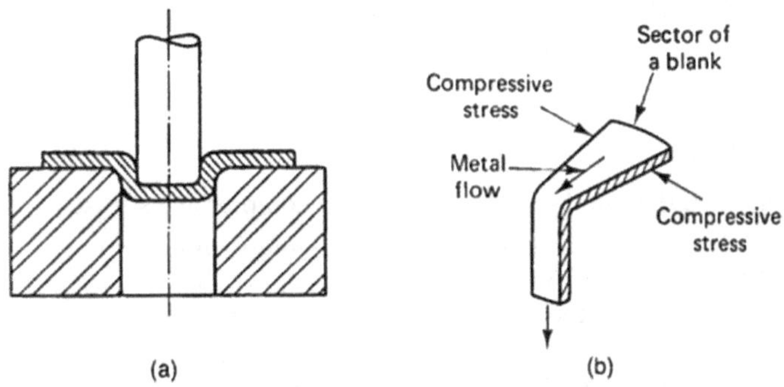

FIGURE 6.32 Mechanics of deep drawing: (a) first stage of deep drawing (i.e., bending) and (b) compression stage in deep drawing.

We can see from the preceding discussion that the deep drawing process involves five stages: bending, straightening, friction, compression, and tension. Different parts of the blank being drawn are subjected to different states of stress. As a result, the deformation is not even throughout the blank, as is clear in Figure 6.33, which shows an exaggerated longitudinal section of a drawn cup. While the flange gets thicker because of the circumferential compressive stress, the vertical wall gets thinner, and thinning is maximum at the lowest part of the wall adjacent to the bottom of the cup. Accordingly, if the cup is broken during the drawing process, failure is expected to occur at the location of maximum thinning. An upper bound for the maximum drawing force can, therefore, be given by the following equation:

$$F = \pi \times (d + t) t \sigma_T \tag{6.3}$$

where

F is the maximum required drawing force
d is the diameter of the punch

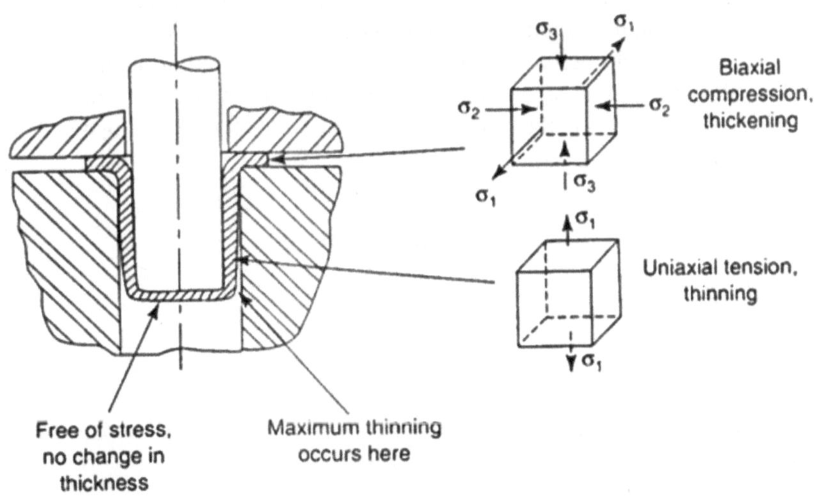

FIGURE 6.33 An exaggerated longitudinal section of a drawn cup, with the states of stress at different locations.

t is the thickness of the blank
σ_T is the UTS of the blank material

6.1.3.2 The Blank Holder

As previously mentioned, the thin blank is subjected to compressive stresses within its plane. This is similar to the case of a slender column subjected to compression, where buckling is expected to occur if the slenderness ratio (i.e., length/thickness) is higher than a certain value. Therefore, by virtue of similarity, if the ratio of the diameter of the blank to its thickness exceeds a certain value, buckling occurs. Actually, if $(D_0 - d)/t \geq 18$, where D is the blank diameter, d is the punch diameter, and t is the thickness, the annular flange will buckle and crimple. This is a product defect referred to as wrinkling.

One way to eliminate wrinkling (buckling) of the thin blank is to support it over its entire area. This is done by sandwiching the blank between the upper surface of the die steel and the lower surface of an annular ring that exerts pressure upon the blank, as shown in Figure 6.31. This supporting ring is called the *blank holder*, and the force exerted on it can be generated by die springs or a compressed gas like nitrogen. On the other hand, higher frictional forces will initiate at both the upper and lower surfaces of the blank as a result of the blank-holding force. For this reason, lubricants like soap in water, waxes, mineral oil, and graphite are applied to both surfaces of the blank. Moreover, the upper surface of the die steel, as well as the lower surface of the blank holder, must be very smooth (ground and lapped). As a rule of thumb, the blank-holding force is taken as 1/3 the force required for drawing.

6.1.3.3 Variables Affecting Deep Drawing

Now that we understand the mechanics of the process, we can identify and predict the effect of each of the process variables. For example, we can see that poor lubrication results in higher friction forces, and, accordingly, a higher drawing force is required. In fact, in most cases of poor lubrication, the cup cross section does not withstand the high-tensile force, and failure of the wall at the bottom takes place during the process. A small die corner radius would increase the bending and straightening forces, thus increasing the drawing force, and the final outcome would be a result similar to that caused by poor lubrication.

In addition to these process variables, the geometry of the blank has a marked effect not only on the process but also on the attributes of the final product. An appropriate quantitative way of expressing the geometry is the number indicating the thickness as a percentage of the diameter, or $(t/D) \times 100$. For smaller values of this percentage (e.g., 0.5), excessive wrinkling should be expected, unless a high blank-holding force is used. If the percentage is higher than 3, no wrinkling occurs, and a blank holder is not necessary. Another important variable is the drawing ratio, which is given by the following equation:

$$R = \frac{D}{d} \tag{6.4}$$

where

R is the drawing ratio
D is the diameter of the blank
d is the diameter of the punch

It has been experimentally found that the deep drawing operation does not yield a sound cup when the drawing ratio is higher than 2 (i.e., for successful drawing R must be less than 2).

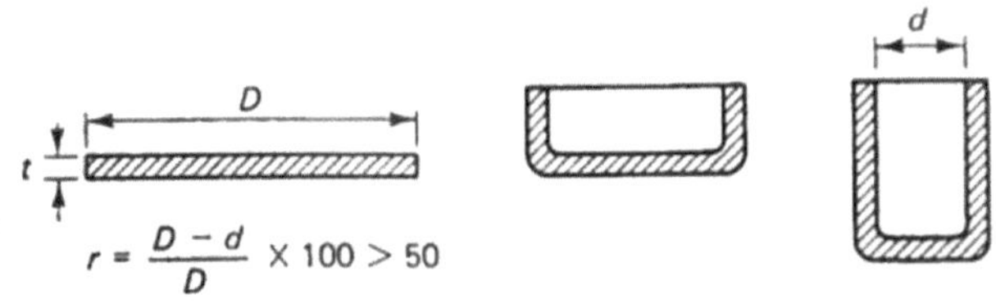

FIGURE 6.34 The use of an intermediate cup when the total required reduction ratio is high.

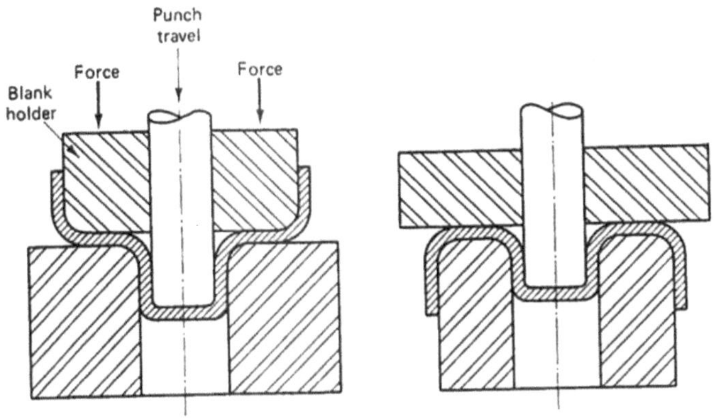

FIGURE 6.35 Redrawing an intermediate cup.

Another number that is commonly used to characterize deep drawing operations is the percentage reduction. It can be given by the following equation:

$$r = \frac{D-d}{D} \times 100 \quad (6.5)$$

where

r is the percentage reduction
D is the diameter of the blank
d is the diameter of the punch

It is a common industrial practice to take the value of r as less than 50 percent in order to have a sound product without any tearing. When the final product is long and necessitates a value of r higher than 50 *percent*, an intermediate cup must be obtained first, as shown in Figure 6.34. The intermediate cup must have dimensions that keep the percentage reduction below 50. It can then be redrawn, as illustrated in Figure 6.35, once or several times until the final required dimensions are achieved. The maximum permissible percentage reduction in the redrawing operations is always far less than 50 percent. It is usually taken as 30 percent, 20 percent, and 13 percent, in the first, second, and third redraws, respectively. If several redrawing operations are required, the product should then be annealed after every two operations in order to eliminate work hardening and thus avoid cracking and failure of the product.

6.1.3.4 Blank-Development Calculations

For the sake of simplicity, it is always assumed that the thickness of the blank remains unchanged after the drawing operation. Because the total volume of the metal is constant, it can then be

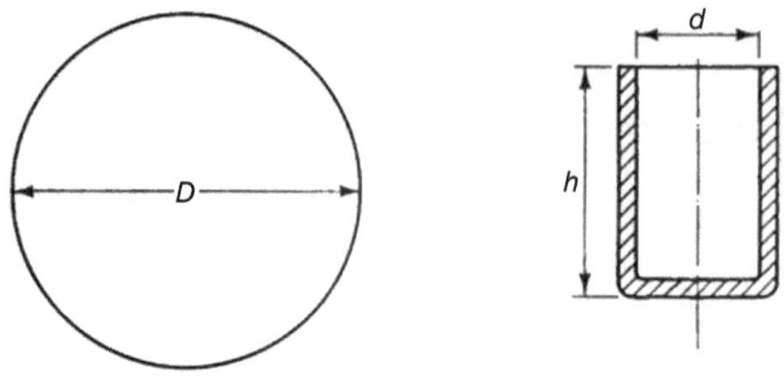

Surface area of blank $\frac{\pi}{4}D^2 = \frac{\pi}{4}d^2 + \pi dh$, i.e., surface area of the cup

FIGURE 6.36 A simple example of blank development.

concluded that the surface area of the final product is equal to the surface area of the original blank. This rule forms the basis for the blank-development calculations. Consider the simple example shown in Figure 6.36. The surface area of the cup is the area of its bottom plus the area of the wall:

$$\text{Surface area of cup} = \frac{\pi}{4} \times d^2 + \pi dh$$

This is equal to the surface area of the original blank; we can state that

$$\frac{\pi}{4} D^2 = \frac{\pi}{4} \times d^2 + \pi dh$$

or

$$D^2 = d^2 + 4dh$$

Therefore, the original diameter of the blank, which is unknown, can be given by the following equation:

$$D = \sqrt{d^2 + 4dh} \tag{6.6}$$

Equation (6.6) gives an approximate result because it assumes the cup has sharp corners, which is not the case in industrial practice. However, this equation can be modified to take round corners into account by adding the area of the surface of revolution resulting from the rotation of the round corner around the centerline of the cup, when equating the area of the product to that of the original blank. Note that the area of any surface of revolution can be determined by employing Pappas's first theorem, which gives that area as the product of the path of the center of gravity of the curve around the axis of rotation multiplied by the length of that curve.

6.1.3.5 Planning for Deep Drawing

The process engineer usually receives a blueprint of the required cup from the product designer. His or her job is to determine the dimensions of the blank and the number of drawing operations needed, together with the dimensions of intermediate cups, so that the tool designer can start designing the blanking and the deep drawing dies. That job requires experience, as well as close contact between the product designer and the process engineer. The following steps can be of great help to beginners:

- Allow for a small flange around the top of the cup after the operation is completed. This flange is trimmed at a later stage and is referred to as the *trimming allowance*. It is appropriate to take an allowance equal to 10–15 percent of the diameter of the cup.
- Calculate the total surface area of the product and the trimming allowance. Then, equate it to the area of the original blank with an unknown diameter. Next, solve for the diameter of the original blank.
- Calculate the thickness as a percentage of the diameter or $(t/D) \times 100$, in order to get a rough idea of the degree of wrinkling to be expected (see the preceding discussion on process variables).
- Calculate the required percentage reduction. If it is less than or equal to 50, then the required cup can be obtained in a single drawing. But if the required r is greater than 50, then a few redrawing operations are required; the procedure to be followed is given in the next steps.
- For the first draw, assume r to be equal to 50 and calculate the dimensions of the intermediate cup. Then, calculate r required for the first redraw. If $r = 30$, only a single redraw is required.
- If r is greater than 30 for the first redraw, take it as equal to 30 and calculate the dimensions of a second intermediate cup. The percentage reduction for the second redraw should be less than 20; otherwise, a third redraw is required, and so on.

6.1.3.6 Ironing

We can see from the mechanics of the deep drawing operation that there is reasonable variation in the thickness of the drawn cup. In most cases, such thickness variation does not have any negative effect on the proper functioning of the product and, therefore, the drawn cups are used as is. However, close control of the dimensions of the cups is sometimes necessary. In this case, cups are subjected to an ironing operation, in which the wall of the cup is squeezed in the annular space between a punch and its corresponding die. As can be seen in Figure 6.37, the punch-die clearance is smaller than the thickness of the cup and is equal to the final required thickness. Large reductions in thickness should be avoided in order to obtain a sound product. It is good industrial practice to take the value of the punch-die clearance in the range between 30 and 80 percent of the thickness of the cup. Also, the percentage reduction in thickness, which is given next, should fall between 40 and 60 in a single ironing operation. This is a safeguard against the fracture of the product during the operation. Following is the equation to be applied:

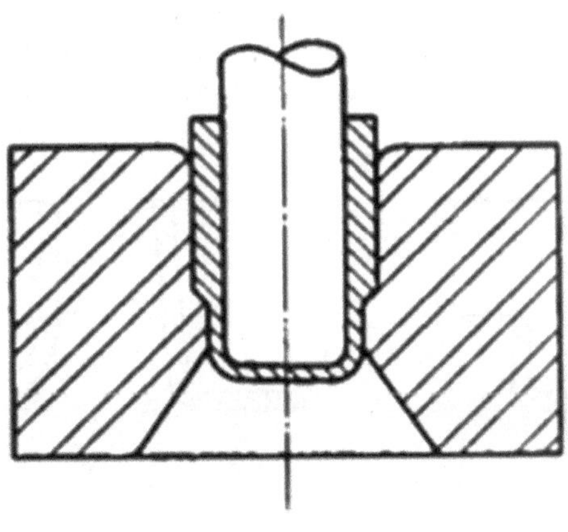

FIGURE 6.37 The ironing operation.

$$\text{Percentage reduction in thickness} = \frac{t_o - t_f}{t_o} \times 100 \quad (6.7)$$

where

t_o is the original thickness of the cup
t_f is the final thickness of the cup after ironing.

6.1.3.7 Drawing of Stepped, Conical, and Domed Cups

Stepped cups are those with two (or more) shell diameters (see Figure 6.38a). They are produced in two (or more) stages. First, a cup is drawn to have a large diameter, and, second, a redrawing operation is performed on only the lower portion of the cup. Tapered or conical cups (see Figure 6.38b) cannot be drawn directly. They first have to be made into stepped cups, which are then smoothed and stretched out to give the required tapered cups. A complex deep drawing operation is used for producing domed cups (see Figure 6.38c). So that the sheet metal stretches properly over the punch nose, higher blank-holding forces are required. Therefore, the process actually involves stretch forming, and its variables should be adjusted to eliminate either wrinkling or tearing.

6.1.3.8 Drawing of Box-Shaped Cups

When all press working operations of sheet metal are reviewed, there would be almost no doubt that the box-drawing process is the most complex and difficult to control. Nevertheless, in an attempt to simplify the problem, we can divide a box into four round corners and four straight sides. Each of these round corners represents 1/4 of a circular cup, and, therefore, the previous analysis holds true for it. On the other hand, no lateral compression is needed to allow the blank metal to flow toward the die edge at each of the straight sides. Accordingly, the process in these zones is not drawing at all; it is just bending and straightening. For this reason, the metal in these zones flows faster than in the round corners, and a square blank takes the form shown in Figure 6.39 after drawing. Note that there is excess metal at each of the four round corners, which impedes the drawing operations at those locations. It also results in localized higher stresses and tears almost always beginning at one (or more) of the corners during box drawing, as can be seen in Figure 6.40.

Several variables affect this complex operation as well as the quality of the products obtained. They include the die bending radius, the die corner radius, and the shape of the original blank. These process variables have been investigated by research workers, and it has been found that in order to obtain sound box-shaped cups, it is very important to ensure an easy, unobstructed flow of metal during the drawing operation. The absence of this condition results in the initiation of high tensile stresses in the vertical walls of the box, especially at the round corners, and results in considerable thinning, which is followed by fracture. Among the factors that can cause obstruction to the metal

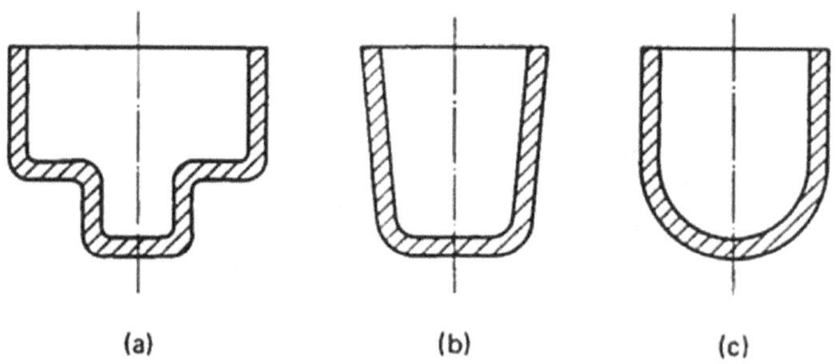

FIGURE 6.38 Deep-drawn cups: (a) stepped, (b) conical, and (c) domed.

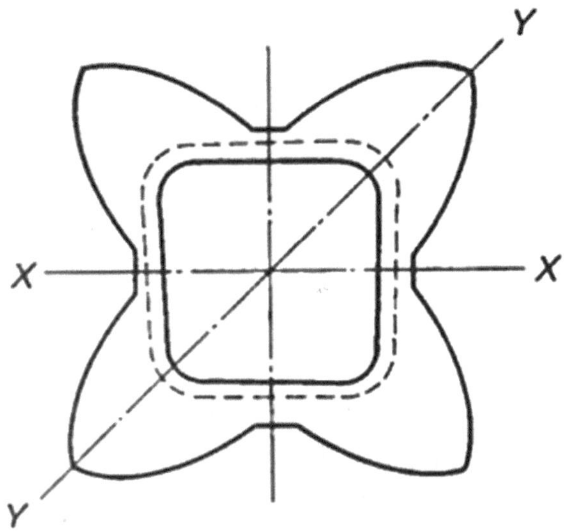

FIGURE 6.39 Final shape of a box-shaped cup, obtained by deep drawing a square blank.

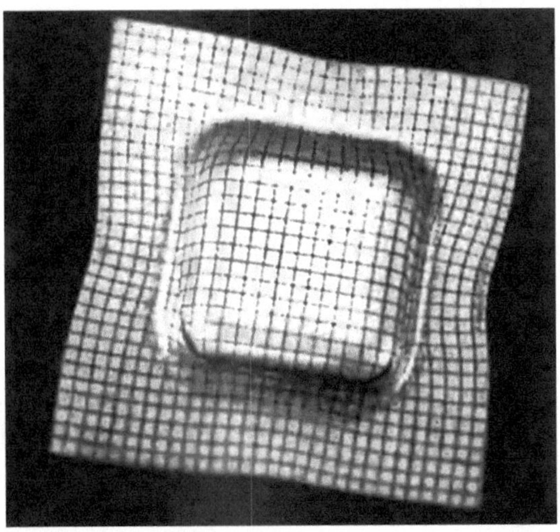

FIGURE 6.40 Tears occurring in a box drawing.

flow are smaller die radii, higher reduction ratios (at the corners), and poor lubrication. These are added to the presence of excess metal at the corners, which causes an appreciable increase in the transverse compressive stresses. Therefore, an optimum blank shape without excess metal at the corners is necessary for achieving successful drawing operations of box-shaped cups. A simple method for optimizing the shape of the blank is shown in Figure 6.41. It involves printing a square grid on the surface of the blank and determining the borders of the unreformed zone on the flanges at each corner (by observing the undistorted grid) so that it can be taken off the original blank. It has been found that the optimum shape is a circle with four cuts corresponding to the four corners. Also, the blank-holding force has been found to play a very important role. Better products are obtained by using a rubber-actuated blank holder that exerts low forces during the first third of the drawing stroke, followed by a marked increase in those forces during the rest of the drawing stroke to eliminate wrinkling and stretch out the product.

FIGURE 6.41 Optimized blank shape for drawing box-shaped cups.

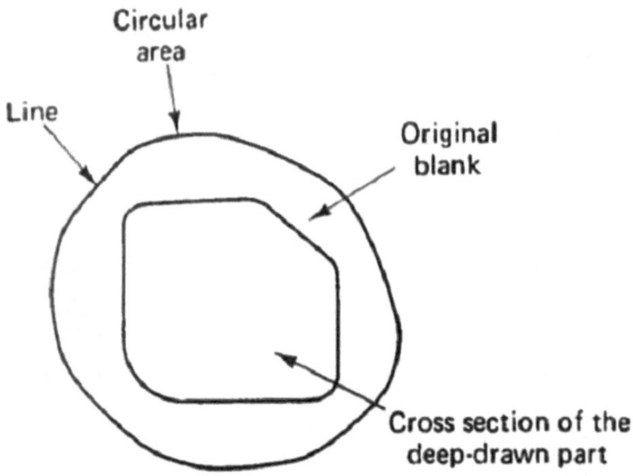

FIGURE 6.42 Optimized blank shape for drawing cups with an irregular cross section.

The preceding discussion can be generalized to include the drawing of a cup with an irregular cross section. This can be achieved by dividing the perimeter into straight sides and circular arcs. Professor Kurt Lange and his coworkers (Institute für Umformstechnik, Stuttgart University, Germany) have developed a technique for obtaining the optimum blank shape in this case by employing the slip-line field theory. The technique was included in an interactive computer expert system that is capable of giving direct answers to any drawing problem. An optimized blank shape obtained by that system is shown in Figure 6.42.

6.1.3.9 Recent Developments in Deep Drawing

A recent development in deep drawing involves cup drawing without a blank holder. Cupping of a thick blank has been accomplished by pushing the blank through a die having a special profile, as shown in Figure 6.43, without any need for a blank holder. This process has the advantages of reducing the number of processing stages, eliminating the blank holder, and using considerably simpler tool construction. A further advantage is that the operation can be performed on a single-acting press, resulting in an appreciable reduction in the initial capital cost required.

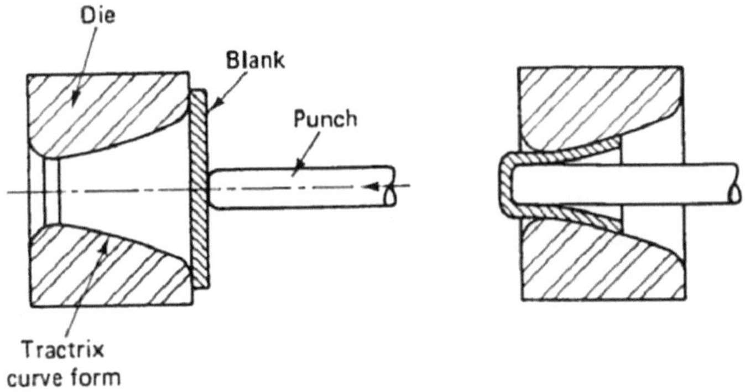

FIGURE 6.43 Drawing cups without a blank holder.

Another new development is the employment of ultrasonics to aid the deep drawing operation. The function of the ultrasonic waves is to enlarge the die bore and then leave it to return elastically to its original dimension in a pulsating manner. This reduces the friction forces appreciably, resulting in a marked reduction in the required drawing force and a clear improvement in the quality of the drawn cup. In many cases, the cup can be drawn by the force exerted by the human hand without the need for any mechanical force-generating device. It is, therefore, obvious that low-tonnage, high-production-rate presses can be used, which makes the process economically attractive.

6.1.3.10 Defects in Deep-Drawn Parts

These defects differ in shape and cause, depending upon the prevailing conditions and also on the initial dimensions of the blank. Following is a brief description of the most common defects, some of which are shown in Figure 6.44:

- *Wrinkling*. Wrinkling is the buckling of the undrawn part of the blank under compressive stresses; it may also occur in the vertical walls (see Figure 6.44a,b). If it takes place on the punch nose when drawing a domed cup, it is referred to as *puckering*.
- *Tearing*. Tearing, which always occurs in the vicinity of the radius connecting the cup bottom and the wall, is caused by high tensile stresses due to the obstruction of the flow of the metal in the flange.
- *Earing*. Earing is the formation of ears at the free edges of a deep-drawn cylindrical cup (see Figure 6.44c). It is caused by the anisotropy of the sheet metal. Ears are trimmed after a drawing operation, resulting in a waste of material.
- *Surface Irregularities*. Surface irregularities are caused by nonuniform yielding, like the orange-peel effect of Luder's lines.
- *Surface Marks*. Surface marks are caused by improper punch-die clearance or poor lubrication. These include draw marks, step rings, and burnishing.

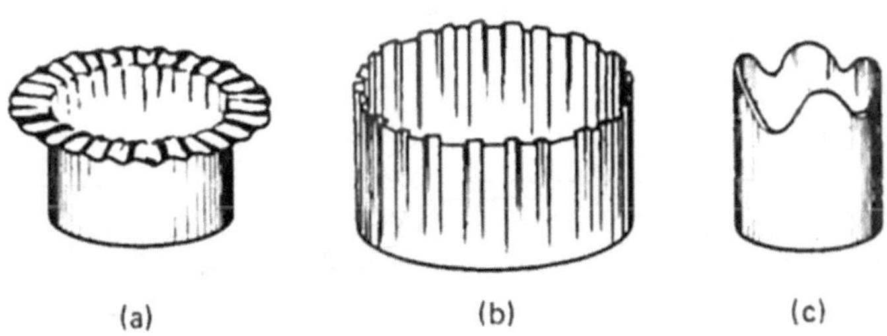

FIGURE 6.44 Some defects occurring in deep drawing operations: (a) wrinkling in the flange, (b) wrinkling in the wall, and (c) earing.

6.1.4 Forming Operations

In this section, we will discuss the various forming operations performed on sheet metals, not just flat sheets, but tubular sheets (i.e., thin-walled tubes) as well. Therefore, not only will operations like embossing and offsetting be discussed, but also tube bulging, expanding, and necking will be considered.

6.1.4.1 Forming of Sheets

True forming involves shaping the blank into a three-dimensional (or sculptured) surface by sandwiching it between a punch and a die. The strain is not uniform, and the operation is complex. The nonhomogeneity (or complexity) depends upon the nature and the unevenness of the required shape. Experience and trial and error were employed in the past to obtain an optimum blank shape and to avoid thinning of the blank or tearing. A printed grid on the original blank helps to detect the locations of overstraining where tearing is expected. It also helps in optimizing the shape of the original blank. With recent advances in computer graphics and simulation of metal deformation by the FEM, rational design of the blank can be performed by the computer, without any need for trial and error. In fact, a successful software package has been prepared by the Mechanical Engineering Department of Michigan Technological University.

6.1.4.2 Embossing Operations

Embossing operations involve localized deflection of a flat sheet to create depressions in the form of beads and offsets. This is sometimes called oil canning. Beads and offsets are usually employed to add stiffness to thin sheets, whether flat or tubular (e.g., barrels), as well as for other functional reasons. A typical example of a part that is subjected to embossing is the license plate of an automobile. The cross section of a bead can take different forms, such as those shown in Figure 6.45. Because this operation involves stretching the sheet, the achieved localized percentage elongation within the bead cross section must be lower than that allowable for the metal of the sheet. On the other hand, Figure 6.46 shows two kinds of offsets, where it is common practice to take the maximum permissible depth as three times the thickness of the sheet metal.

6.1.4.3 Rubber Forming of Flat Sheets

Rubber forming is not new and actually dates back to the nineteenth century when a technique for shearing and cutting paper and foil was patented by Adolph Delkescamp in 1872. Another rubber-forming technique, called the Guerin process, was widely used during World War II for forming aircraft panels. It involved employing a confined rubber pad on the upper platen of the press and a steel form block on the lower platen, as shown in Figure 6.47a. This method is still sometimes used. As can be seen in the figure, when a block of elastomer (usually incompressible artificial rubber) is

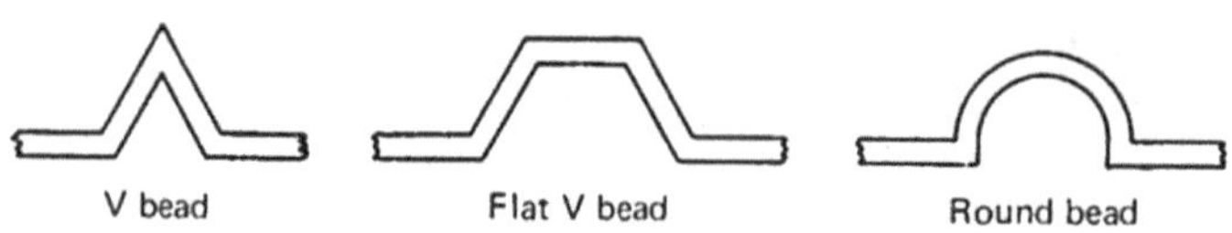

FIGURE 6.45 Different kinds of beads.

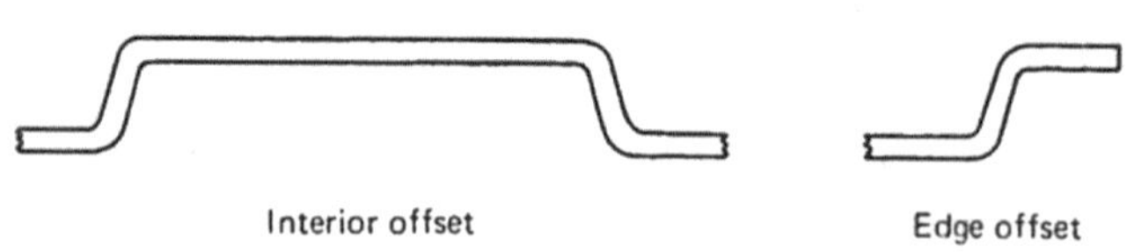

FIGURE 6.46 Offsetting operations.

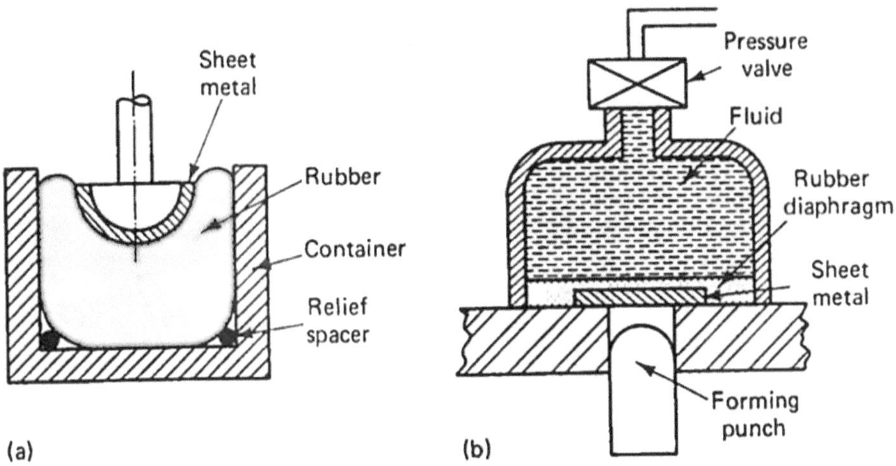

FIGURE 6.47 Rubber forming of flat sheets: (a) conventional rubber forming and (b) hydroform process.

confined in a rigid box, the only way it can flow when the punch sinks into it is up, thus forming the blank around the punch under uniform pressure over the whole surface. It is also common industrial practice to place spacers on the base of the metal box in order to provide a relief for the elastomer block, which, in turn, helps to avoid the initiation of high, localized strains in the blank area directly beneath the punch. Rubber forming has real potential when the number of parts required is relatively small and does not justify designing and constructing a forming die.

A modified version of this process, called the *hydroform process*, involves employing a pressurized fluid above the rubber membrane, as shown in Figure 6.47b. This is similar in effect to drawing the cup into a high-pressure container, as previously mentioned. Therefore, percentage reductions higher than those obtained in conventional deep drawing can be achieved.

6.1.4.4 Forming of Tubular Sheets

Figure 6.48a–d indicates tubular parts after they were subjected to beading, flattening, expanding, and necking operations, respectively. Tube bulging is another forming operation, in which the diameter of the tube, in its middle part, is expanded and then restrained by a split die and forced to conform to the details of the internal surface of the die. This can be achieved by internal hydraulic pressure or by employing an elastomer (polyurethane) rod as the pressure-transmitting medium, causing expansion of the tube. A schematic of this operation is given in Figure 6.49. At the beginning of the operation, the elastomer rod fits freely inside the tube and has the same length. Compressive forces are then applied to both the rod and the tube simultaneously so that the tube bulges outward in the middle and the frictional forces at the tube–rod interface draw more metal into the die space, thus decreasing the length of the tube. The method of using a polyurethane rod is simpler and cleaner, and there is no need for using oil seals or complicated tooling construction. A further advantage of rubber bulging is that it can be used for simultaneous forming, piercing, and shearing of thin tubular sheets.

6.2 HIGH-ENERGY-RATE FORMING (HERF)

In HERF, the energy of deformation is delivered within a very short period of time on the order of milliseconds or even microseconds. HERF methods include *explosive*, *electrohydraulic*, and *electromagnetic* forming techniques. These techniques are usually employed when short-run products or large parts are required. HERF is also recommended for manufacturing prototype components and new shapes in order to avoid the unjustifiable cost of dies. Rocket domes and other aerospace

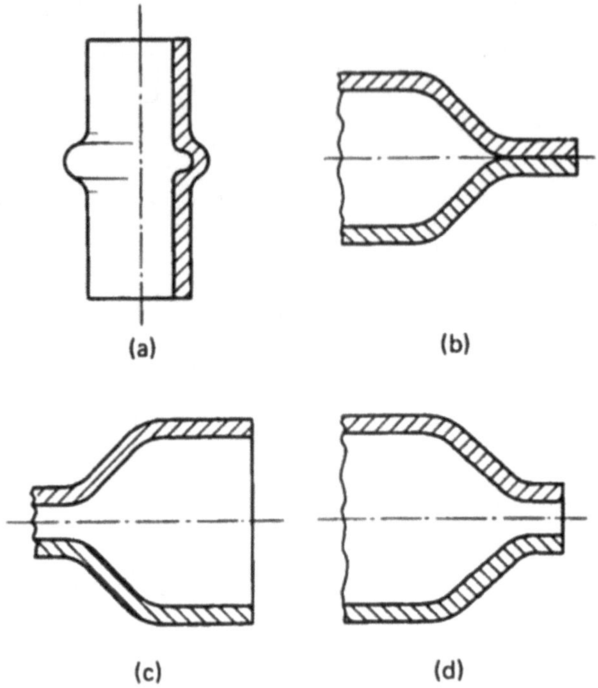

FIGURE 6.48 Different tubular parts after forming operations: (a) beading, (b) flattening, (c) expanding, and (d) necking.

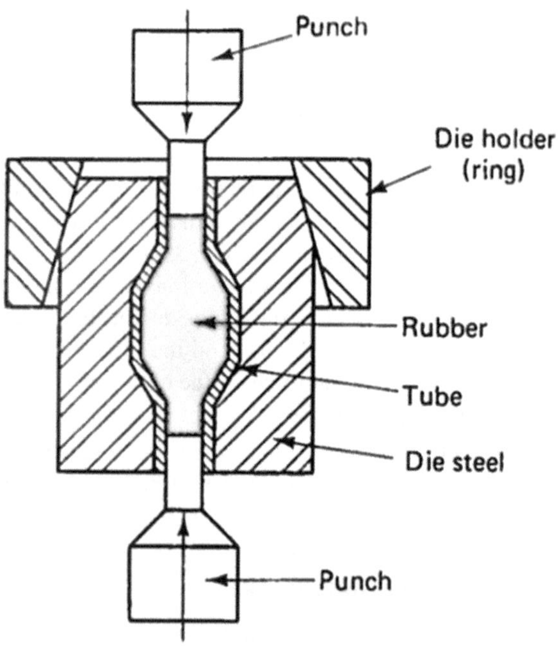

FIGURE 6.49 The tube-bulging operation with an elastomer rod.

structural panels are typical examples. During a HERF process, the sheet metal is given an extremely high acceleration in a very short period of time and is thus formed as a result of consuming its own kinetic energy to cause deformation.

6.2.1 Explosive Forming

The explosive forming of sheet metal received some attention during the past decade. The various explosive forming techniques fall under one or the other of two basic systems: confined and unconfined. In a *confined* system, which is shown in Figure 6.50a, a charge of low explosives is detonated and yields a large amount of high-pressure gas, thus forcing the sheet metal to take the desired shape. This system is mainly used for the bulging and flaring of small tubular parts. Its main disadvantage is the hazard of die failure because of the high pressure generated.

In an *unconfined*, or *standoff*, system, which is shown in Figure 6.50b, the charge is maintained at a distance from the sheet blank (the standoff distance), and both the blank and the charge are kept immersed in water. When the charge is detonated, shock waves are generated, thus forming a large blank into the desired shape. It is obvious that the efficiency of the standoff system is less than that of the confined system because only a portion of the surface over which the shock waves act is utilized (actually, shock waves act in all directions, forming a spherical front). However, the standoff system has the advantages of a lower noise level and of largely reducing the hazard of damaging the workpiece by particles resulting from the explosion. In a simple standoff system, the distance from the explosive charge to the water surface is usually taken as twice the standoff distance. The latter depends upon the size of the blank and is taken as equal to D (the blank diameter) for D less than 2 feet (60 cm) and is taken as equal to $0.5D$ for D greater than that. The best results are obtained when the blank is clamped lightly around its periphery and when a material with a low modulus of elasticity, like plastic, is used as a die material. This eliminates springback, thus obtaining closer tolerances. A modified version of this method is illustrated in Figure 6.51, where a reflector is used to collect and reflect the explosion energy that does not fall directly onto the blank surface. This leads to improved efficiency over the standoff system because a smaller amount of charge is needed for the same job.

6.2.2 Electrohydraulic Forming

The basic idea for the process of electrohydraulic forming, which has been known for some time, is based on discharging a large amount of electrical energy across a small gap between two electrodes

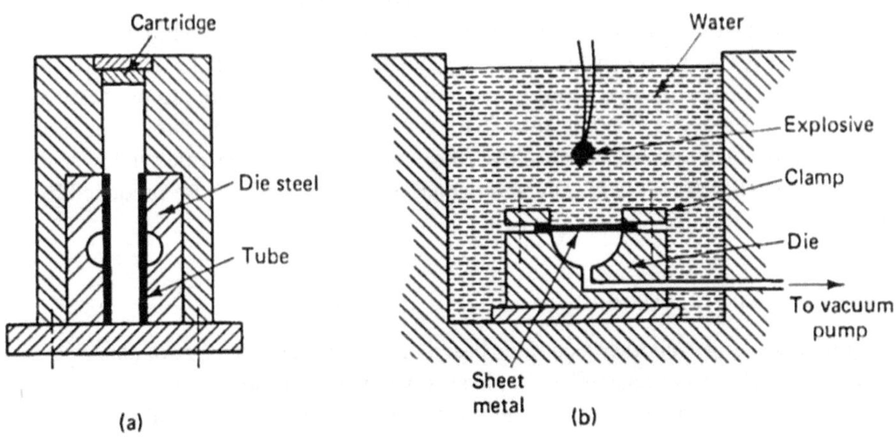

FIGURE 6.50 Explosive forming of sheet metal: (a) confined system and (b) standoff system.

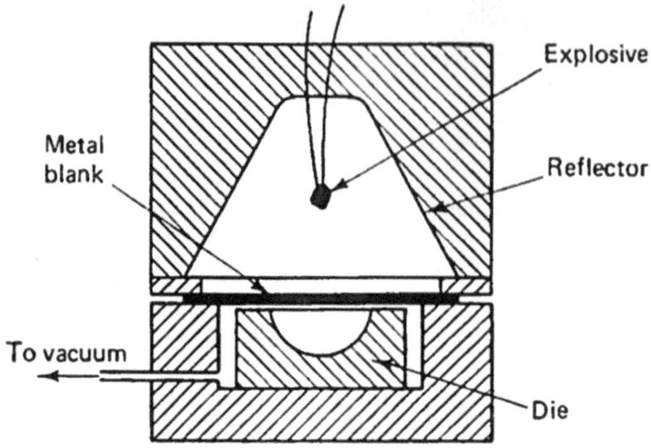

FIGURE 6.51 Increasing the efficiency of explosive forming by using a reflector.

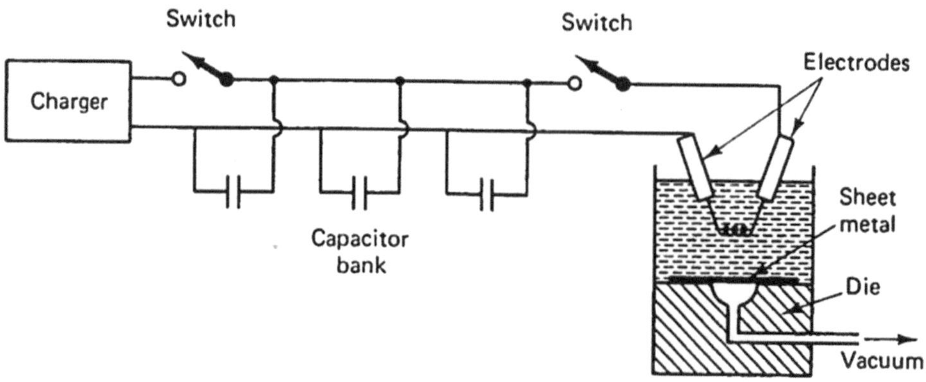

FIGURE 6.52 Electrohydraulic forming.

immersed in water, as shown in Figure 6.52. The high-amperage current resulting from suddenly discharging the electrical energy from the condensers melts the thin wire between the electrodes and generates a shock wave. The shock wave lasts for a few microseconds; it travels through water to hit the blank and forces it to take the shape of the die cavity. The use of a thin wire between the electrodes has the advantages of initiating and guiding the path of the spark, enabling the use of nonconductive liquids; also, the wire can be shaped to suit the geometry of the required product. The method is also safer than explosive forming and can be used for simultaneous operations like piercing and bulging. Nevertheless, it is not suitable for continuous production runs because the wire has to be replaced after each operation. Moreover, the level of energy generated is lower than that of explosive forming. Therefore, the products are generally smaller than those produced by explosive forming.

6.2.3 Electromagnetic Forming

Electromagnetic forming is another technique based on the sudden discharge of electrical energy. As we know from electricity and magnetism in physics, when an electric current passes through a coil, it initiates a magnetic field whose magnitude is a function of the current. We also know that when a magnetic field is interrupted by a conductive material (workpiece), a current is induced in that material that is proportional to the rate of change of the flux. This is called *eddy current* and produces its

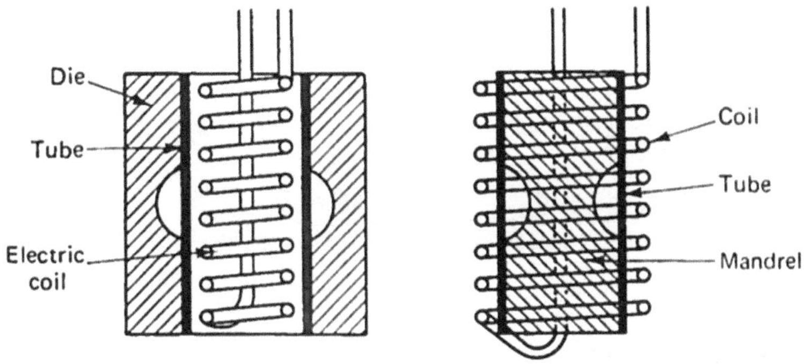

FIGURE 6.53 Examples of electromagnetic forming of tubes.

own magnetic field that opposes the initial one. As a result, repulsive forces between the coil and the workpiece force the workpiece to conform to the die cavity. This technique can be used to form flat as well as tubular sheets. As can be seen in Figure 6.53, it is employed in expanding as well as compressing tubes. It has proven to be very effective when forming relatively thin materials.

6.3 SPINNING OF SHEET METAL

Spinning is the forming of axisymmetric hollow shells over a rotating-form mandrel by using special rollers. Generally, the shapes produced by spinning can also be manufactured by drawing, compressing, or flanging. However, spinning is usually used for forming large parts that require very large drawing presses or when there is diversity in the products (i.e., when various shapes are needed but only a small number of each shape is required).

A schematic of the spinning operation is shown in Figure 6.54. In the beginning, the semifinished product (circular blank) is pushed by the tailstock against the front of the form mandrel (usually a wooden one) that is fixed on the rotating faceplate of the spinning machine (like a lathe). A pressing tool is pushed by the operator onto the external surface of the blank. The blank slips under the

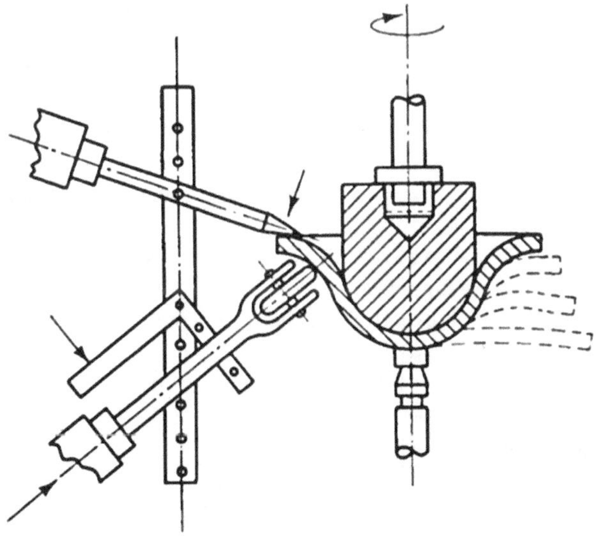

FIGURE 6.54 A schematic of the spinning operation.

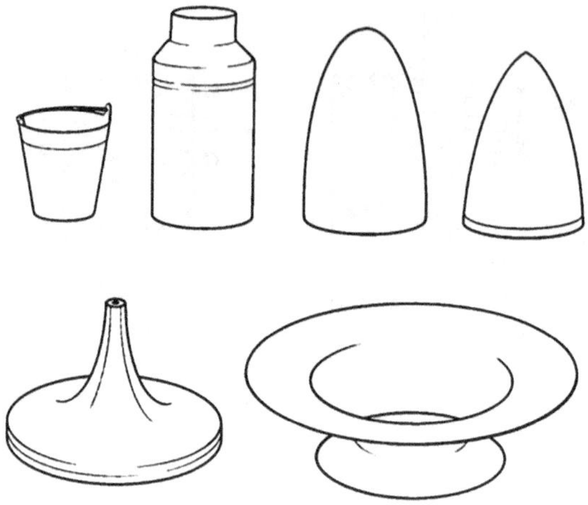

FIGURE 6.55 A group of parts produced by spinning.

pressing devices, which causes localized deformation. Finally, the blank takes the exact shape of the form mandrel. This technique can also be used to obtain hollow products with a diameter at the end (neck) smaller than that at the middle. In this case, it is necessary to use a collapsible-form mandrel, which is composed of individual smaller parts that can be extracted from the neck of the final product after the process is completed. Figure 6.55 shows a group of parts produced by spinning.

A modified version of this method involves replacing the operator with a numerically controlled (NC) tool. Auxiliary operations, like removing the excess metal, are also carried out on the same machine. Better surface quality and more uniform thickness are the advantages of NC spinning over conventional techniques.

REVIEW QUESTIONS

1. What main design feature characterizes sheet metal products?
2. List some of the advantages of press working sheet metals.
3. When are sheet metals formed in their hot state? Give examples.
4. What are the two main groups of press working operations?
5. What main condition must be fulfilled so that cutting of sheet metal (and not any other operation) takes place?
6. Use sketches to explain why the angle of inclination of the upper blade of guillotine shears must not exceed 15°.
7. Use sketches to differentiate between the following operations: shearing, cutoff, parting, blanking, and punching.
8. Why must attention be given to the careful layout of blanks on a sheet metal strip?
9. Describe a perforating operation.
10. What does an edge of a blank usually look like? Draw a sketch.
11. What is meant by the *percentage penetration?*
12. What does an edge of a blank look like when the punch-die clearance is too large?
13. What does an edge of a blank look like when the punch-die clearance is too tight?
14. When are punches sheared and why?
15. When are dies sheared and why?
16. In what aspect is fine blanking different from conventional blanking?
17. Use sketches to explain each of the following operations: shaving, piercing, and cropping.

18. Can a drop-through die be used on any press? Why not?
19. What is the function of a stripper plate?
20. List two types of die constructions for blanking operations.
21. How can a washer be produced in a single stroke?
22. What condition must be fulfilled so that the bending of sheet metal takes place?
23. Sketch the common types of bending dies.
24. Which die requires the minimum force for the same thickness of sheet metal?
25. Where is tearing expected to occur and where is wrinkling expected to occur when a sheet metal is subjected to bending?
26. What is springback? Why does it occur?
27. List three methods for eliminating the effects of springback.
28. On what assumption is blank development based?
29. List some operations that can be classified as bending. Use sketches and explain the design functions of the products.
30. How can structural angles be bent?
31. Explain rotary bending, using sketches, and list some of the advantages of this operation.
32. Explain how a seamed tube can be produced by continuous bending.
33. What are the disadvantages of seamed tubes?
34. Explain deep drawing, using sketches.
35. What are the stages involved in deep drawing a circular cup? Explain, using sketches.
36. Indicate the states of stress at different locations in a cup toward the end of a drawing operation.
37. Where is thickening expected to occur?
38. At what location is thinning maximum? To what would this lead?
39. Why is a blank holder sometimes needed?
40. List some of the variables affecting the deep drawing operation.
41. What is wrinkling? Why does it occur?
42. Describe an ironing operation.
43. Is there any limitation on ironing?
44. Why are conical cups not drawn directly?
45. What is actually taking place when drawing domed cups?
46. Is it feasible to take any blank shape for box-drawing operations and then trim the excess metal? Why?
47. What are the mechanics of deformation in the straight-sides areas?
48. What is the advantage of ultrasonic deep drawing?
49. How can plates be drawn without a blank holder?
50. List some of the defects experienced in deep-drawn products.
51. As a product designer, how can you make use of the embossing operation when designing sheet metal parts?
52. When would you recommend using rubber-forming techniques?
53. What is meant by high-energy-rate forming?
54. When would you recommend using explosive forming?
55. Should the dies used in explosive forming be made of a hard material, like alloy steel, or a softer one, like plastic? Why?
56. What happens if you make the hydraulic head very small in explosive forming?
57. What are the advantages of electrohydraulic forming? What are the disadvantages?
58. Use a sketch to explain the electromagnetic forming operation.
59. Describe spinning. When is it recommended?
60. Can spinning produce products with a diameter at the neck smaller than at the middle? How?

PROBLEMS

1. The blank shown in Figure 6.56 is to be produced from a sheet metal strip 0.0625 inches (1.6 mm) in thickness. The material is low-carbon steel AISI 1020. Estimate the required blanking force.

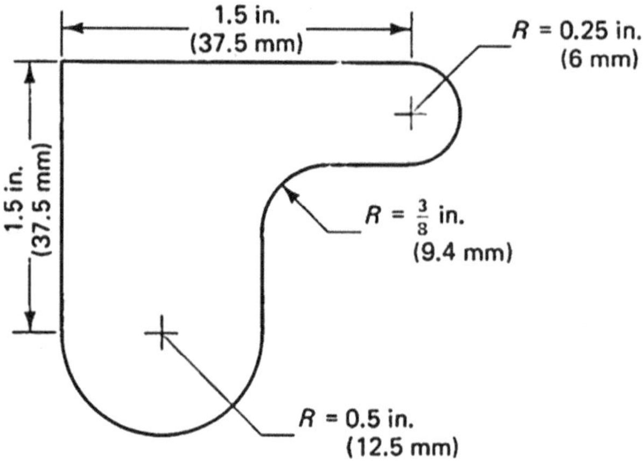

FIGURE 6.56 The blank shape required in problem 1.

2. The products shown in Figure 6.57a–c are produced by bending. Obtain the length of the original blank to the nearest 0.01 inches (0.25 mm). Take t as 0.0625 inches (1.6 mm).

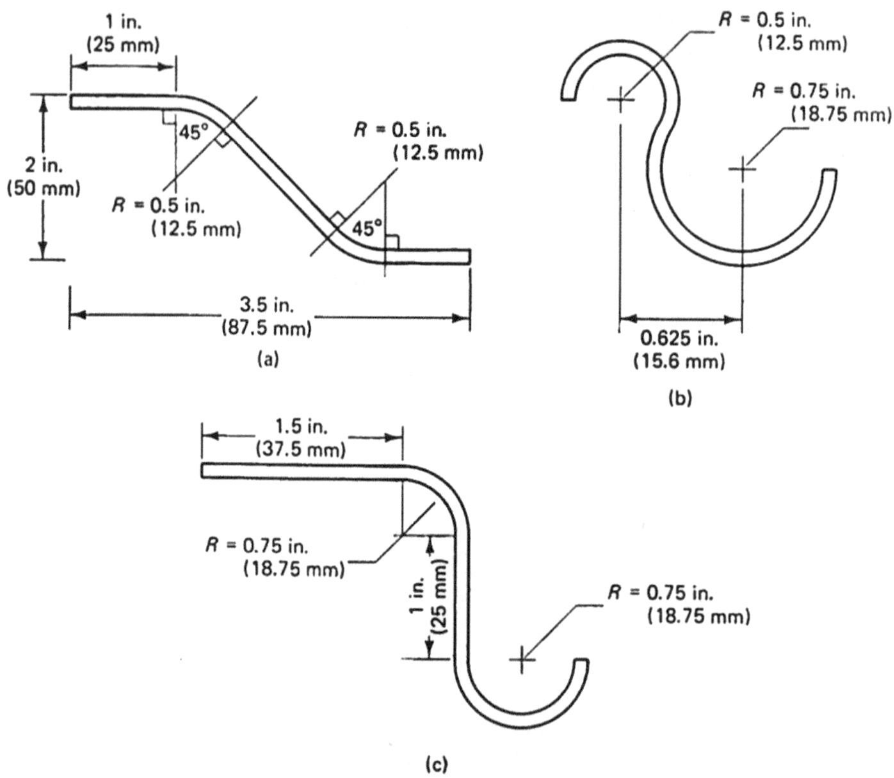

FIGURE 6.57 Products produced by bending in problem 2.

3. A cup is drawn from a sheet of 1020 steel. The thickness is 0.03 inch (0.8 mm), and the inner diameter is 1 inch (25 mm). Estimate the maximum force required for drawing. If the material is aluminum, what would the force be?
4. A cup with a height of 0.75 inches (18.75 mm) and an inner diameter of 1 inch (25 mm) is to be drawn from a steel strip 0.0625 inches (1.6 mm) in thickness. Plan for the drawing process by carrying out blank development, determining the number of drawings, and looking at the draw severity analysis.

DESIGN PROBLEM

Design a simple wrench to loosen (or tighten) a 1/2-inch (12.5-mm) nut (or bolt head). The 1/2 inch (12.5 mm) measures across the nut flats. The torque is 1 lb·ft (1.356 N·m), and 50,000 pieces are required annually. The wrench is to be produced by press working.

SOLUTION

A suitable method for production is fine blanking as there will be no need for any further machining operations. We cannot select steel that has a high carbon content because it will create problems during the fine-blanking operation. An appropriate choice is AISI 1035 CD steel. The dimensions of the wrench are the same as those given in the examples on forging and casting, although the tolerances can be kept much tighter. A detailed design is given in Figure 6.58.

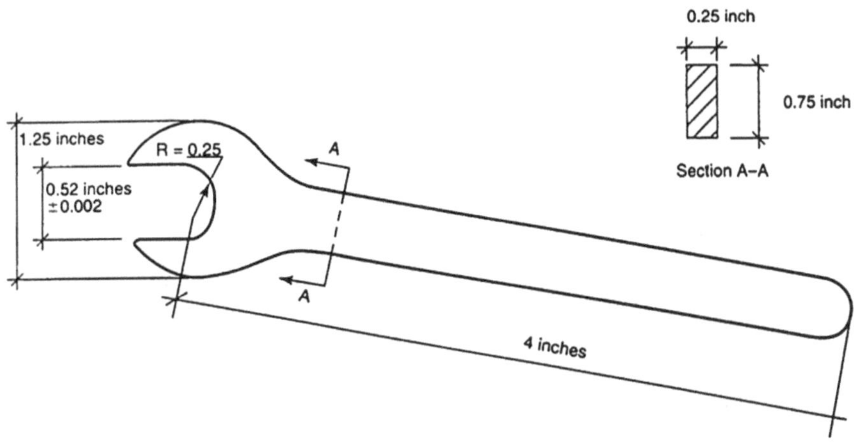

FIGURE 6.58 Detailed design of a wrench produced by stamping.

DESIGN PROJECTS

1. A pulley (for a V-belt) that has a 4-inch (100-mm) outer diameter and is mounted on a shaft that is 3/4 inch (19 mm) in diameter was manufactured by casting. The process was slow, and the rejects formed a noticeable percentage of the production. As a product designer, you are required to redesign this pulley so that it can be produced by sheet metal working and welding.
2. Design a connecting rod for a sewing machine so that it can be produced by sheet metal working, given that the diameter of each of the two holes is 0.5 inches (12.5 mm) and the distance between the centers of the holes is 4 inches (100 mm).

3. If a connecting rod four times smaller than the one of Design Project 2 is to be used in a little toy, how would the design change?
4. Design a table for the machine shop. The table should be 4 feet in height, with a surface area of 3 × 3 feet (900 × 900 mm), and should be able to carry a load of half a ton. Assume that 4000 pieces are required annually and that different parts will be produced by sheet metal working and then joined together by nuts and bolts.
5. A trash container having a capacity of 1 cubic foot (0.02833 m^3) is to be designed for manufacturing by sheet metal working. Assume that it is required to withstand an axial compression load of 200 pounds (890 N) and that the production rate is 50,000 pieces per year. Provide a detailed design for this trash container.
6. A connecting lever is produced by forging. The lever has two short bosses, each at one of its ends and each with a vertical hole that is 3/4 inch (19 mm) in diameter. The horizontal distance between the centers of the holes is 12 inches (300 mm), and the vertical distance is 3 inches (75 mm). The lever during functioning is subjected to a bending moment of 200 lb·ft (271 N·m). Because of the high percentage of rejects and low production rate, this connecting lever is to be produced by sheet metal working. Provide a detailed design so that it can be produced by this manufacturing method.

7 Powder Metallurgy

The basic powder metallurgy process is explained in detail, and then a survey of the various methods for consolidating metal powders is given. A discussion of the types and effects of the presence of pores is provided and an analytical approach is adopted. More important, the design considerations for powder parts are thoroughly covered. Examples of the important applications of powder metallurgy parts in various industries are also given. Finally, the recent developments in powder metallurgy are mentioned.

7.1 GENERAL

7.1.1 Definition

Powder metallurgy is the technology of producing useful components shaped from metal powders by pressing and simultaneous or subsequent heating to produce a coherent mass. The heating operation is usually performed in a controlled-atmosphere furnace and is referred to as *sintering*. The sintering temperature must be kept below the melting point of the powder material or the melting point of the major constituent if a mixture of metal powders is used. Therefore, sintering involves a solid-state diffusion process that allows the compacted powder particles to bond together without going through the molten state. This, in fact, is the fundamental principle of powder metallurgy.

7.1.2 Historical Background

Although powder metallurgy is becoming increasingly important in modern industry, the basic techniques of this process are very old indeed. The ancient Egyptians used a crude form of powder metallurgy as early as 3000 BC to manufacture iron implements. The technique involved reducing the ore with charcoal to obtain a spongy mass of metal that was formed by frequent heating and hammering to eject the slag and consolidate the iron particles together into a mass of wrought iron. This process was used because the primitive ovens then available were not capable of melting iron. The same technique was used later by smiths in India around AD 300 to manufacture the well-known Delhi pillar weighing 6.5 tons. This method was superseded when more advanced ovens capable of melting ferrous metals came into being.

At the beginning of the nineteenth century, powder metallurgy had its first truly scientific enunciation, in England, when Wallaston published details of the preparation of malleable platinum. As had happened in the past, Wallaston's technique was superseded by melting. However, the need for the powder metallurgy process arose again to satisfy the industrial demand for high-melting-point metals. An important application was the production of ductile tungsten in 1909 for manufacturing electric lamp filaments.

7.1.3 Why Powder Metallurgy?

As a result of the development of furnaces and melting techniques, the powder-consolidation process is now usually used when melting metal is undesirable or uneconomical. Fusion is not suitable when it is required to produce parts with controlled, unique structures, such as porous bearings, filters, metallic frictional materials, and cemented carbides. Also, it has been found that powder metallurgy can produce certain complicated shapes more economically and conveniently than other

DOI: 10.1201/9781003537106-7

known manufacturing processes. For this reason, the process currently enjoys widespread industrial application. As the price of labor and the cost of materials continue to rise, the powder-consolidation technique is becoming more and more economical because it eliminates the need for further machining operations, offers more efficient utilization of materials, and allows components to be produced in massive numbers with good surface finish and close tolerances.

7.2 METAL POWDERS

7.2.1 The Manufacture of Metal Powders

Different methods are used for producing metal powders. They include reduction of metal oxides, atomization of molten metals, electrolytic deposition, thermal decomposition of carbonyls, condensation of metal vapor, and mechanical processing of solid metals.

7.2.1.1 Reduction

In reduction, the raw material is usually an oxide that is subjected to a sequence of concentration and purification operations before it is reduced. Carbon, carbon monoxide, and hydrogen are used as reducing agents. Following is the chemical formula indicating the reaction between carbon and iron oxide:

$$2Fe_2O_3 + 3C \xrightarrow{\text{Heat}} 4Fe + 3CO_2 \uparrow \tag{7.1}$$

Because the reaction takes place at a high temperature, the resulting metal particles sinter together and form sponges that are subsequently crushed and milled to a powder suitable for consolidation. Such powders have low apparent densities and often contain impurities and inclusions, but they are cheap. Metal powders produced by this method include iron, cobalt, nickel, tungsten, and molybdenum.

7.2.1.2 Atomization

Atomization is frequently used for producing powders from low-melting-point metals such as tin, lead, zinc, aluminum, and cadmium. Iron powder can also be produced by atomization. The process involves forcing a molten metal through a small orifice to yield a stream that is disintegrated by a jet of high-pressure fluid. When compressed gas is used as the atomizing medium, the resulting powder particles will be spherical. The reason is that complete solidification takes a relatively long period, during which surface tension forces have the chance to spheroidize the molten metal droplets. However, when water is used, the droplets solidify very quickly and have a ragged or irregular form. Figure 7.1 illustrates the atomization technique.

7.2.1.3 Electrolytic Deposition

Electrolytic deposition involves obtaining metal powders from solutions by electrolysis. Process parameters such as current density and solution concentration are controlled to give a loose deposit instead of the coherent layer acquired in electroplating. The electrolytically deposited powders are then carefully washed, dried, and annealed. Such powders are relatively expensive, but their important advantage is their high purity and freedom from nonmetallic inclusions.

7.2.1.4 Thermal Decomposition of Carbonyls

Nickel and iron carbonyls are volatile liquids having low boiling points of 110°F and 227°F (43°C and 107°C), respectively. They decompose at temperatures below 572°F (300°C), and the metal is precipitated in the form of a very fine powder.

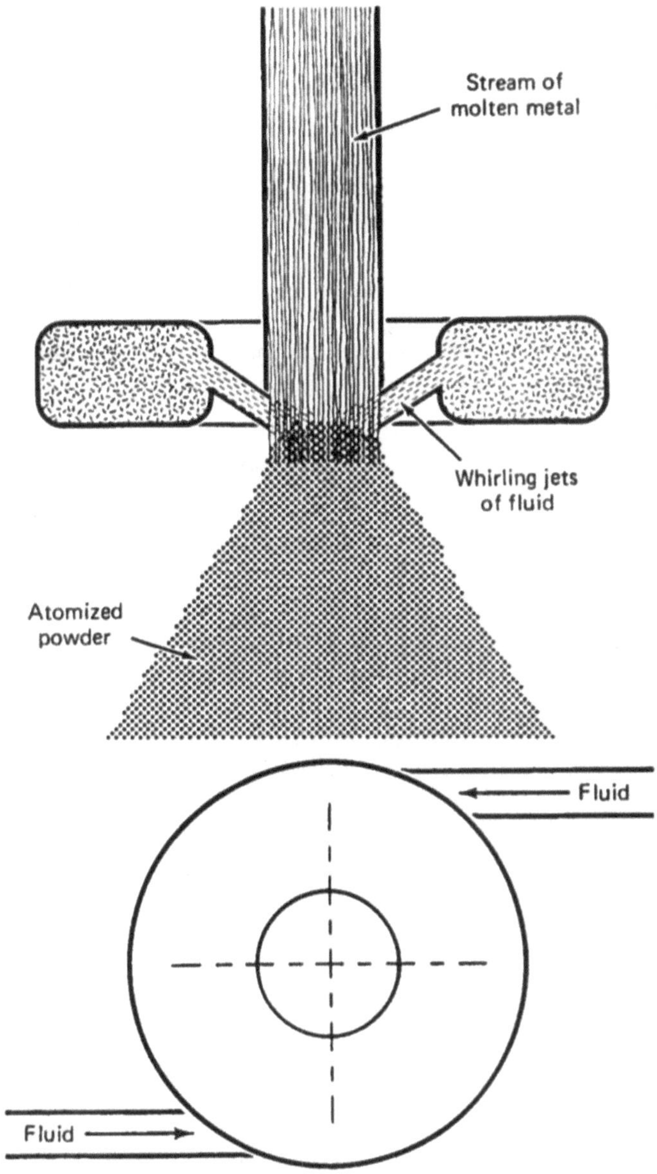

FIGURE 7.1 Production of metal powders by atomization.

7.2.1.5 Condensation of Metal Vapor

Condensation is employed only with some low-melting-point metals. For example, zinc powder can be obtained directly by condensation of the zinc vapor.

7.2.1.6 Mechanical Processing of Solid Metals

Production of metal powders by combination of solid metals is accomplished by machining, crushing, milling, or any combination of these. This method is limited to the production of beryllium and magnesium powders because of the expenses involved.

7.2.2 Properties of Metal Powders

The particular method used for producing a metal powder controls its particle and bulk properties, which, in turn, affect the processing characteristics of that powder. Therefore, comprehensive testing of all the physical and chemical properties of powders is essential prior to use in order to avoid variations in the final properties of the compacts. The following are the important characteristics of metal powders.

7.2.2.1 Chemical Composition

In order to determine the chemical composition, conventional chemical analysis is used in addition to some special tests that are applicable only to metal powders, such as weight loss after reduction in a stream of hydrogen, which is an indirect indication of the amount of oxide present. For example, in the case of iron powder, the following equation is used:

$$\%\text{iron oxide} = \%\text{weight loss} \times \frac{159.7}{48} \tag{7.2}$$

$$= \%\text{weight loss} \times 3.33 \tag{7.3}$$

In Equation (7.2), the fraction on the right-hand side is the ratio of the total weight of iron oxide to the weight of the combined oxygen in it, or $(Fe_2O_3)/(O_3)$, which can be calculated by summing up the atomic weights of each element in the numerator and denominator.

It is also important to mention that the percentages of nonmetallic inclusions will affect the maximum achievable density of the compacted powder (i.e., the full theoretical density). For example, if an iron powder (density of iron is 7.87 g/cm^3) consists of a percent Fe_2O_3, b percent carbon, and c percent sulfur, the following equation can be applied:

$$\text{Max.achievable density} = \frac{100}{\dfrac{100-(a+b+c)}{7.87} + \dfrac{a}{\rho_{\text{oxide}}} + \dfrac{b}{\rho_{\text{carbon}}} + \dfrac{c}{\rho_{\text{sulfur}}}} \tag{7.4}$$

Equation (7.4) can also be used in calculating the maximum achievable density for a mixture of powders.

7.2.2.2 Particle Shape

The particle shape is influenced by the method of powder production and significantly affects the apparent density of the powder, its pressing properties, and its sintering ability.

7.2.2.3 Particle Size

The flow properties and the apparent density of a metal powder are markedly influenced by the particle size, which can be directly determined by measurement on a microscope, by sieving, or by sedimentation tests.

7.2.2.4 Particle-Size Distribution

The particle-size distribution has a considerable effect on the physical properties of the powder. Sieve testing is the standard method used for the determination of the particle-size distribution in a quantitative manner. The apparatus used involves a shaking machine on which a series of standard sieves are stacked with the coarsest at the top and the finest at the bottom. The particle-size distribution is obtained from the percentage (by weight) of the sample that passes through one sieve but is retained on the next finer sieve. These sieves are defined by the mesh size, which indicates the number of apertures per linear inch. After the test is performed, the results are stated in a suitable form, such as a table of weight percentages, graphs of frequency distribution, or cumulative oversize and undersize curves where the cumulative size is the total weight percentage above or below a particular mesh size.

7.2.2.5 Specific Surface

Specific surface is the total surface area of the particles per unit weight of powder, usually expressed in square centimeters per gram (cm^2/g). The specific surface has a considerable influence on the sintering process. The higher the specific surface, the higher the activity during sintering because the driving force for bonding during the sintering operation is the excess energy due to the large area (high specific surface).

7.2.2.6 Flowability

Flowability is the ease with which a powder will flow under gravity through an orifice. A quantitative expression of the flowability of a powder is its flow rate, which is determined using a Hall *flowmeter*. As illustrated in Figure 7.2, this apparatus involves a polished conical funnel made of brass having a half-cone angle of 30° and an orifice of 0.125 inches (3.175 mm). The funnel is filled with 50 grams of the powder, and the time taken for the powder to flow from the funnel is determined, with the flow rate being expressed in seconds. The flow properties are dependent mainly upon the particle shape, particle size, and particle-size distribution. They are also affected by the presence of lubricants and moisture. Good flow properties are required if high production rates are to be achieved in pressing operations because the die is filled with powder flowing under gravity and because a shorter die-filling time necessitates a high powder-flow rate.

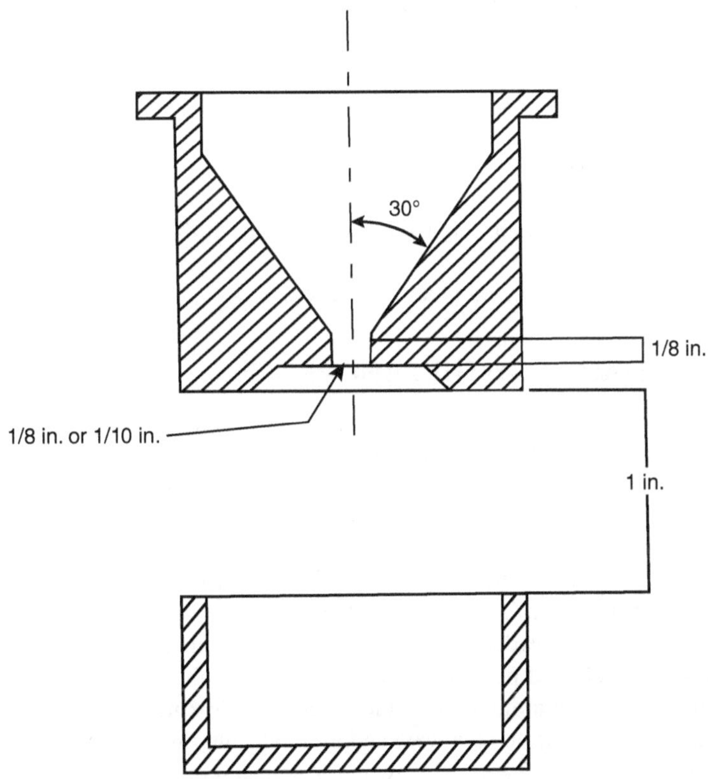

FIGURE 7.2 A sketch of the Hall flowmeter.

7.2.2.7 Bulk (or Apparent) Density

The bulk (or apparent) density is the density of the bulk of a powder mass. It can be easily determined by filling a container of known volume with the powder and then determining the weight of the powder. The bulk density is the quotient of the powder mass divided by its volume and is usually expressed in grams per cubic centimeter (g/cm^3). The apparent density is influenced by the same factors as the flowability—namely, the particle configuration and the particle-size distribution.

7.2.2.8 Compressibility and Compactibility

Compressibility and compactibility are very important terms that indicate and describe the behavior of a metal powder when compacted in a die. *Compressibility* indicates the densification ability of a powder, whereas *compactibility* is the structural stability of the produced as-pressed compact at a given pressure. A generalized interpretation of these terms involves graphs indicating the as-pressed density versus pressure (for compressibility) and the as-pressed strength versus pressure (for compactibility). It must be noted that these two terms are *not* interchangeable: a brittle powder may have good compressibility but usually has weak as-pressed compactibility.

7.2.2.9 Sintering Ability

Sintering ability is the ability of the adjacent surfaces of particles in an as-pressed compact to bond together when heated during the sintering operation. Sintering ability is influenced mainly by the specific surface of the powder used and is the factor responsible for imparting strength to the compact.

7.2.3 Factors Affecting the Selection of Metal Powders

Probably all metallic elements can be made in powderous form by the previously discussed manufacturing methods. However, the powder characteristics will differ in each case and will depend mainly upon the method of manufacture. The task of the manufacturing engineer is to select the type of powder appropriate for the required job. The decision generally depends upon the following factors:

- Economic considerations
- Purity demands
- Desired physical, electrical, or magnetic characteristics of the compact

These considerations will be discussed in a later section.

7.3 POWDER METALLURGY: THE BASIC PROCESS

The conventional powder metallurgy process normally consists of three operations: powder blending and mixing, powder pressing, and compact sintering.

7.3.1 Blending and Mixing

Blending and mixing the powders properly is essential for the uniformity of the finished product. Desired particle-size distribution is obtained by blending in advance the types of powders used. These can be either elemental powders, including alloying powders to produce a homogeneous mixture of ingredients, or prealloyed powders. In both cases, dry lubricants are added to the blending powders before mixing. The commonly used lubricants include zinc stearate, lithium stearate, calcium stearate, stearic acid, paraffin, acra wax, and molybdenum disulfide. The amount of lubricant added usually ranges between 0.5 and 1.0 percent of the metal powder by weight. The function of the lubricant is to minimize the die wear, to reduce the friction that is initiated between the die surface and powder particles during the compaction operation, and, hence, to obtain a more even density distribution throughout the compact. Nevertheless, it is not recommended that the just-mentioned limits of the percentage of lubricant be exceeded, as this will result in extruding the lubricant from the surfaces of the particles during compaction to fill the voids, preventing proper densification of the powder particles and impeding the compaction operation.

The time for mixing may vary from a few minutes to days, depending upon operator experience and the results desired. However, it is usually recommended that the powders be mixed for 45 minutes to an hour. Overmixing should always be avoided because it may decrease particle size and work-harden the particles.

7.3.2 Pressing

Pressing consists of filling a die cavity with a controlled amount of blended powder, applying the required pressure, and then ejecting the as-pressed compact, usually called the *green compact*, by the lower punch. The pressing operation is usually performed at room temperature, with pressures ranging from 10 tons/in.^{2} (138 MPa) to 60 tons/in.^{2} (828 MPa), depending upon the material, the characteristics of the powder used, and the density of the compact to be achieved.

Tooling is usually made of hardened, ground, and lapped tool steels. The final hardness of the die walls that will come in contact with the powder particles during compaction should be around 60 *Rc* in order to keep the die wear minimal. The die cavity is designed to allow a powder fill about three times the volume (or height) of the green compact. The ratio between the initial height of the loose powder fill and the final height of the green compact is called the *compression ratio* and can be determined from the following equation:

$$\text{Compression ratio} = \frac{\text{Height of loose powder fill}}{\text{Height of green compact}} = \frac{\text{Density of green compact}}{\text{Apparent density of loose powder}} \tag{7.5}$$

When pressure is first applied to metal powders, they will undergo repacking or restacking to reduce their bulk volume and to attain better packing density. The extent to which this occurs depends largely on the physical characteristics of the powder particles. The movement of the powder particles relative to one another will cause the oxide films covering their surfaces to be rubbed off. These oxide films will also collapse at the initial areas of contact between particles because these areas are small and the magnitude of the localized pressures is, therefore, extremely high. This leads to metal-to-metal contact and, consequently, to *cold-pressure welding* between the powder particles at the points of contact. When the pressure is further increased, interlocking and plastic deformation of the particles take place, extending the areas of contact between the individual particles and increasing the strength and density of the coherent compacted powder. The plasticity of the metal powder particles plays a major role during the second stage of the pressing operation. As the compaction pressure increases, further densification is increasingly retarded by work hardening of the particle material and by friction. Figure 7.3 shows a typical plot of the relationship between the achieved density and the compaction pressure. As can be seen, the density first goes up at a high rate, and then the rate of increase in density decreases with increasing pressure. Consequently, it is very difficult to achieve the full density because prohibitive pressure is required.

Frictional forces between the powder and the die wall always oppose the transmission of the applied pressure in its vicinity. Therefore, the applied pressure diminishes with depth in the case of single-ended pressing (i.e., when the compaction pressure is applied on only one side). This is accompanied by an uneven density distribution throughout the compact. The density always decreases with increasing distance from the pressing punch face. Figure 7.4 indicates the variation of pressure with depth along the compact as well as the resulting variation in density. It is always recommended that the value of the length-to-diameter ratio of the compact be kept lower than 2.0 in order to avoid considerable density variations.

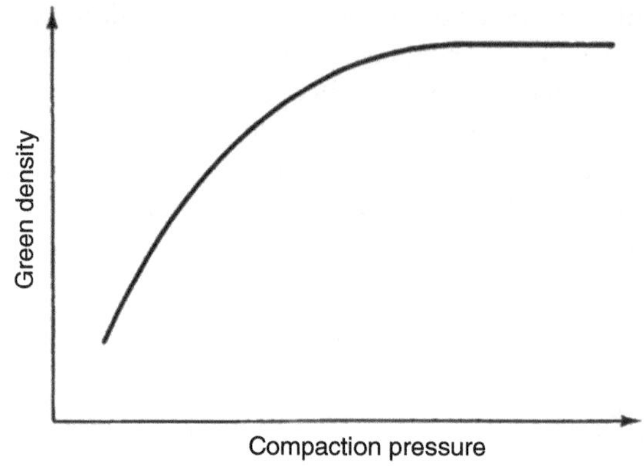

FIGURE 7.3 A typical plot of the relationship between the achieved density and the compaction pressure.

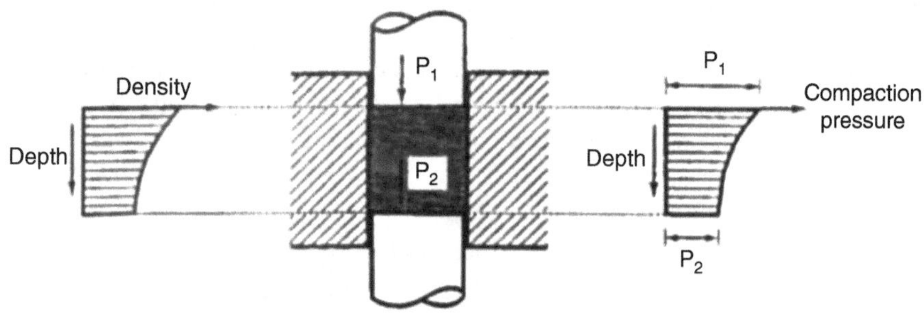

FIGURE 7.4 The variation of pressure with depth along the compact.

In order to improve pressure transmission and to obtain more even density distribution, lubricants are either admixed with the powder or applied to the die walls. Other techniques are also used to achieve uniform density distribution, such as compacting from both ends and suspending the die on springs or withdrawing it to reduce the effects of die-wall friction.

During the pressing of a metal powder in a die, elastic deformation of the die occurs in radial directions, leading to bulging of the die wall. Meanwhile, the compact deforms both elastically and plastically. When the compaction pressure is released, the elastic deformation tries to recover. However, because some of the compact expansion is due to plastic deformation, the die tightly grips the compact, which hinders the die from returning to its original shape. Accordingly, a definite load, called the *ejection load*, has to be exerted on the compact in order to push it out of the die. Figure 7.5 illustrates the sequence of steps in a pressing operation.

7.3.3 Sintering

Sintering involves heating the green compact in a controlled-atmosphere furnace to a temperature that is slightly below the melting point of the powder metal. When the compact is composed of mixed elemental powders (e.g., iron and copper), the sintering temperature will then have to be below the melting point of at least one major constituent. The sintering operation will result in the following:

- Strong bonding between powder particles
- Chemical, dimensional, or phase changes
- Alloying, in the case of mixed elemental powders

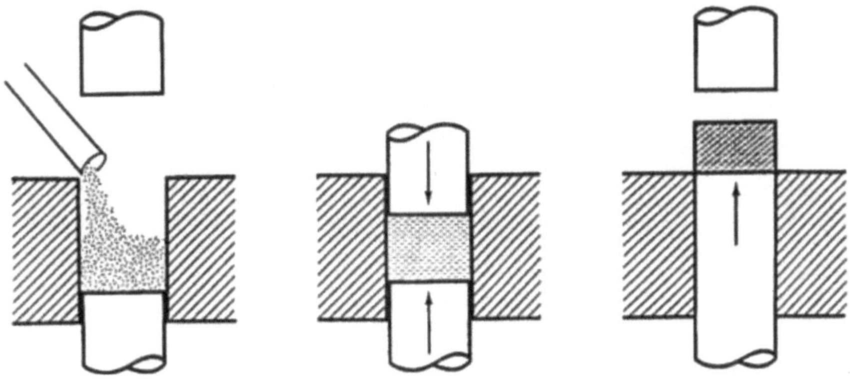

FIGURE 7.5 Sequence of steps in a pressing operation.

Such effects of the sintering operation are influenced by process variables such as sintering temperature, time, and atmosphere.

The amount, size, shape, and even nature of the pores are changed during sintering. There are two kinds of porosity: open, or interconnected, porosity (connected to the compact surface) and closed, or isolated, porosity. In a green compact, most of the porosity is interconnected and is characterized by extremely irregular pores. After sintering, interconnected porosity becomes isolated, and pore spheroidization takes place because of the surface tension forces. Also, the oxide films covering the particle surfaces of a green compact can be reduced by using the appropriate sintering atmosphere.

The most important atmospheres used in industrial sintering are carbon monoxide, hydrogen, and cracked ammonia. The latter is commonly used and is obtained by catalytic dissociation of ammonia, which gives a gas consisting of 25 percent nitrogen and 75 percent hydrogen by volume. Inert gases like argon and helium are occasionally used as sintering atmospheres, but the cost is a decisive factor in limiting their use. Vacuum sintering has also found some industrial applications in recent years; nevertheless, the production rate is the main limitation of this method.

There are two main types of sintering furnaces: continuous and batch-operated. In continuous furnaces, the charge is usually conveyed through the furnace on mesh belts. These furnaces are made in the form of tunnels or long tubes having a diameter of not more than 18 inches (45 cm). Heating elements are arranged to provide two heating zones: a relatively low-temperature zone, called a *dewaxing* zone, in which lubricants are removed so that they will not cause harmful reactions in the next zone, and a uniform heating zone, which has the required high temperature where sintering actually takes place. A third zone of the furnace tube is surrounded by cooling coils in order to cool the compacts to ambient temperature in the controlled atmosphere of the furnace, thus avoiding oxidation of the compacts. Flame curtains (burning gases like hydrogen) are provided at both ends of the furnace tube to prevent air from entering into the furnace. Figure 7.6 is a sketch of a continuous sintering furnace. This type of furnace is suitable for mass production because of its low sintering cost per piece and its ability to give more consistent products. When small quantities of compacts must be sintered, however, *batch-operated* furnaces are used. These furnaces (e.g., vacuum furnaces) are also more suitable when high-purity products are required.

The sintering time varies with the metal powder and ranges between 30 minutes and several hours. However, 40 minutes to an hour is the most commonly used sintering time in industry.

7.4 OPERATIONAL FLOWCHART

Because of the wide variety of powder metallurgy operations, it may be difficult for a person who is not familiar with this process to pursue the proper sequence of operations. The flowchart in Figure 7.7 is intended to clearly show the relationship between the various powder metallurgy operations

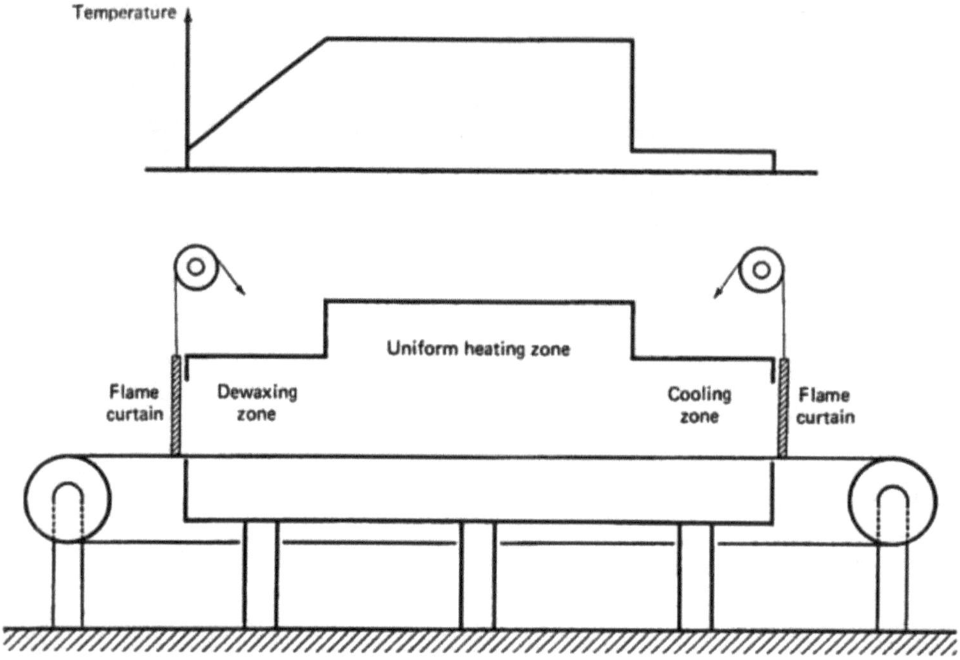

FIGURE 7.6 A sketch of a continuous sintering furnace.

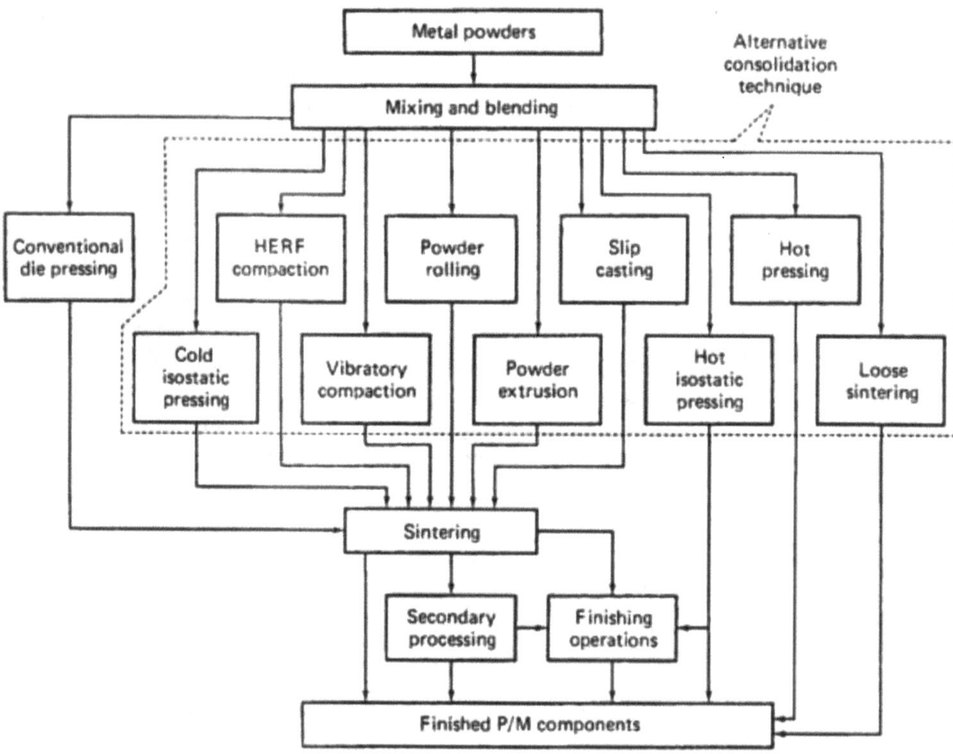

FIGURE 7.7 A flowchart showing the relationship between the various powder metallurgy operations.

(which will be discussed later) and to give a bird's-eye view of the flow of material to yield the final required product. Nevertheless, it must be remembered that there are exceptions and that some operations cannot be shown on the flowchart because they would make it overly detailed and complicated.

7.5 ALTERNATIVE CONSOLIDATION TECHNIQUES

There are many techniques for consolidating metal powders. They are classified, as shown in Figure 7.8, into two main groups: pressureless and pressure forming. The *pressureless* methods are those in which no external pressure is required. This group includes loose sintering, slip casting, and slurry casting. The *pressure-forming* methods include conventional compaction, vibratory compaction, powder extrusion, powder rolling, hot and *cold isostatic pressing* (CIP), explosive forming, and forming with binders. A detailed account of conventional powder metallurgy has been given; the following is a brief description of these other consolidation techniques.

7.5.1 Loose Sintering

Loose sintering is employed in manufacturing filters. It involves the sintering of loose metal powder in molds made of graphite or ceramic material. The temperature used is similar to that of conventional sintering, but the time involved is usually longer (2 days when manufacturing stainless steel filters).

7.5.2 Slip Casting

The application of *slip casting* is usually limited to the production of large, intricate components made from refractory metals and cermets (mixtures of metals and ceramics). The slip, which is a suspension of fine powder particles in a viscous liquid, is poured into an absorbent plaster-of-Paris mold. Both solid and hollow articles can be produced by this method. When making hollow objects, excess slip is poured out after a layer of metal has been formed on the mold surface.

7.5.3 Slurry Casting

Slurry casting is very similar to slip casting, except that the mixture takes the form of a slurry, and binders are usually added. Also, because the slurry contains less water, nonabsorbent molds can be used.

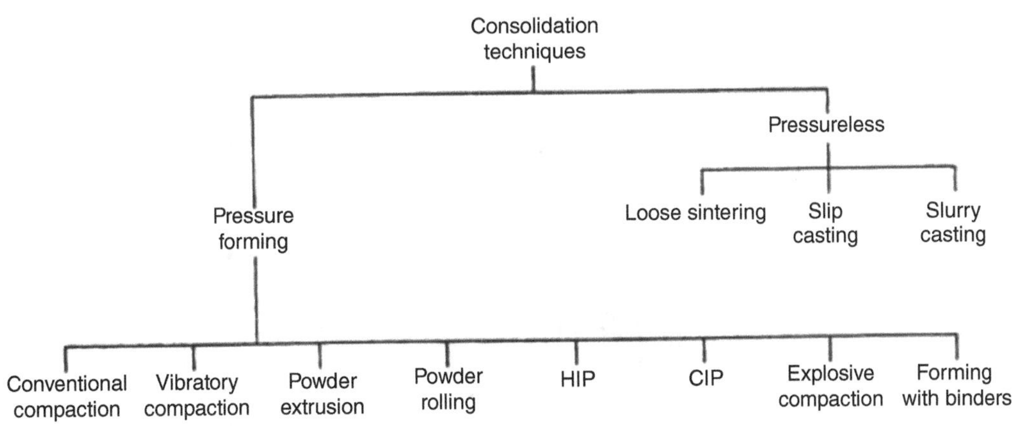

FIGURE 7.8 Classification of the techniques for consolidating metal powders.

7.5.4 Vibratory Compaction

Vibratory compaction involves superimposing mechanical vibration on the pressing load during the compaction operation. The advantages of this process include the considerable reduction in the pressure required and the ability to compact brittle particles that cannot be pressed by conventional techniques because the high compaction load required would result in fragmentation rather than consolidation of the powder particles. The main application involves the consolidation of stainless steel and uranium oxide powders for nuclear fuel elements.

7.5.5 Isostatic Pressing

In *isostatic pressing* (IP), equal all-around pressure is applied directly to the powder mass via a pressurized fluid. Accordingly, die-wall friction is completely eliminated, which explains the potential of the process to produce large, dense parts having uniform density distribution. The process can be performed at room temperature (CIP) or can be carried out at elevated temperatures (*hot isostatic pressing* or HIP).

In CIP, a flexible envelope (usually made of rubber or polymers) that has the required shape is filled with the packed powder. The envelope is then sealed and placed into a chamber that is, in turn, closed and pressurized to consolidate the powder. The lack of rigidity of the flexible envelope is countered by using a mesh or perforated container as a support (see Figure 7.9). The main disadvantage of this process is the low dimensional accuracy due to the flexibility of the mold.

In HIP, both isostatic pressing and sintering are combined. The powder is canned in order to separate it from the pressurized fluid, which is usually argon. The can is then heated in an autoclave, with pressure applied isostatically. Complete densification and particle bonding occur. The elevated temperature at which the powder is consolidated results in a softening of the particles. For this reason, the process is used to compact hard-to-work materials such as tool steels, beryllium, nickel-base

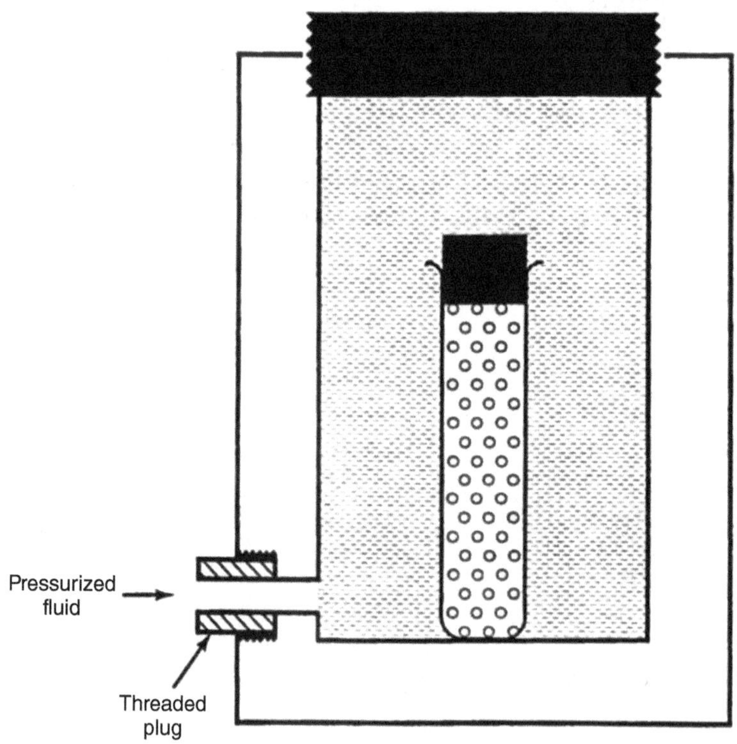

FIGURE 7.9 The isostatic pressing operation.

superalloys, and refractory metals. A good example is the manufacture of jet-engine turbine blades, where a near-net shape is made from nickel-base superalloys. A main disadvantage of this method is the long processing time.

7.5.6 Powder Extrusion

Powder extrusion is a continuous compaction process and can be performed hot or cold. It is employed in producing semifinished products having a high length-to-diameter ratio, a geometry that makes producing them by conventional powder metallurgy impossible. The conventional technique involves packing metal powder into a thin container that is, in turn, evacuated, sealed, and then extruded. An emerging technique involves the extrusion of suitable mixtures of metal (or ceramic) powders and binders such as dextrin and sugars. It has been successfully employed in the production of highly porous materials used as filters or fuel cells in batteries.

7.5.7 Powder Rolling

Direct *powder rolling*, or roll compacting, is another type of continuous compaction process. It is employed mainly for producing porous sheets of nonferrous powders like copper and nickel. This process involves feeding the metal powder into the gap between the two rolls of a simple mill, where it is squeezed and pushed forward to form a sheet that is sintered and further rolled to control its density and thickness.

7.5.8 High-Energy-Rate Compaction

The various *HERF compaction* techniques are based on the same principle, which is the application of the compaction energy within an extremely short period of time. Several methods were developed for compacting metal powders at high speeds. Examples are explosives, high-speed presses, and spark sintering. It is believed that explosive compaction is suitable only when the size of the compact and the density required cannot be achieved by the isostatic compaction process. Nevertheless, the danger of handling explosives and the low cycling times impose serious limitations on this technique in production.

The use of high-speed presses like the Dynapak (built by General Dynamics) and the Petro-Forge (built by the Mechanical Engineering Department, Birmingham University, England) for powder compaction is, in practicality, an extension of the die-pressing technique. These high-speed presses are particularly advantageous for pressing hard-to-compact powders and large components.

Some other powder-consolidation methods also can be classified as high-speed techniques. These include electrodynamic pressing, electromagnetic pressing, and spark sintering. Electrodynamic pressing involves utilizing the high pressure produced by the sudden discharge of electrical energy to compact powders at high speeds. Electromagnetic pressing is based on the phenomenon that a strong magnetic field is generated when an electric current is suddenly discharged through an inductance. This strong magnetic field is used for pressing a thin-walled metallic tube that contains the powder. Spark sintering involves the sudden discharge of electrical energy into the powder mass to puncture the oxide films that cover each individual powder particle and to build up pure metallic contacts between the particles. After about 10 seconds of impulsive discharging, the current is shut off, and a pressure of about 14,500 lb/in.^{2} (100 MPa) is applied to compact the powder to the final required form.

7.5.9 Injection Molding

Although *injection molding* is an emerging process, it can be considered a version of forming with binders, which is a rather old method. The process involves injection molding metal powders that are precoated with a thermoplastic polymer into a part similar in shape to the final required component but having larger dimensions. After removing the polymer with an organic solvent, the porous compact is then sintered for a long time in order to allow for volume shrinkage and, consequently, an increase in density. The main advantage of this process is that it offers promise in the forming of intricate shapes.

7.5.10 Hot Pressing

Hot pressing is a combination of both the compaction and the sintering operations. It is basically similar to the conventional powder metallurgy process, except that powders are induction heated during pressing, and, consequently, a protective atmosphere is necessary. For most metal powders, the temperatures used are moderate (above recrystallization temperature), and dies made of superalloys are used. The hot pressing of refractory metals (e.g., tungsten and beryllium), however, necessitates the use of graphite dies. The difficulties encountered in this technique limit its application to laboratory research.

7.6 SECONDARY CONSOLIDATION OPERATIONS

In most engineering applications, the physical and mechanical properties of the as-sintered compact are adequate enough to make it ready for use. However, secondary processing is sometimes required to increase the density and enhance the mechanical properties of the sintered component, thus making it suitable for heavy-duty engineering applications. The operations involved are similar to those used in forming fully dense metals, though certain precautions are required to account for the porous nature of the sintered compacts. Following is a survey of the common secondary operations.

7.6.1 Coining (Repressing)

Coining involves the pressing of a previously consolidated and sintered compact in order to increase its density. This operation is performed at room temperature, and considerable pressures are thus required. It is often possible to obtain significant improvement in strength not only because of the increased densification but also because of the work hardening that occurs during the operation. A further advantage of this process is that it can be employed to alter shape and dimensions slightly. *Repressing* is a special case of coining where no shape alteration is required.

7.6.2 Extrusion, Swaging, or Rolling of Compacts

Sintered powder compacts, whether in their cold or hot state, can be subjected to any forming operation (extrusion, swaging, or rolling). When processing at elevated temperatures, either a protective atmosphere or canning of the compacts has to be employed. Such techniques are applied to canned sintered compacts of refractory metals, beryllium, and composite materials.

7.6.3 Forging of Powder Preforms

Repressing and coining of sintered compacts cannot reduce porosity below 5 percent of the volume of the compact. Therefore, if porosity is to be completely eliminated, *hot forging* of powder preforms must be employed. Sintered powder compacts having medium densities (80–85 percent of the full theoretical density) are heated, lubricated, and fed into a die cavity. The preform is then forged with a single stroke, as opposed to conventional forging of fully dense materials, where several blows and manual transfer of a billet through a series of dies are required. This advantage is a consequence of using a preform that has a shape quite close to that of the final forged product.

(a)

FIGURE 7.10 Some forged powder metallurgy parts. (Courtesy of the Metal Powder Industries Federation, Princeton, NJ.)

The tooling used involves a precision flashless closed die; therefore, the trimming operation performed after conventional forging is eliminated. The forging of powder preforms combines the advantages of both the basic powder metallurgy and the conventional hot forging processes while eliminating their shortcomings. For this reason, the process is extensively used in the automotive industry in producing transmission and differential-gear components. Examples of some forged powder metallurgy parts are shown in Figure 7.10.

7.7 FINISHING OPERATIONS

Many powder metallurgy products are ready for use in their as-sintered state; however, finishing processes are frequently used to impart some physical properties or geometrical characteristics to them. Following are some examples of the finishing operations employed in the powder metallurgy industry.

7.7.1 Sizing

Sizing is the pressing of a sintered compact at room temperature to secure the desired shape and dimensions by correcting distortion and changes in dimensions that may have occurred during the sintering operation. Consequently, this operation involves only limited deformation and slight density changes and has almost no effect on the mechanical properties of the sintered compact.

7.7.2 Machining

Features like side holes, slots, or grooves cannot be formed during pressing, and, therefore, either one or two machining operations are required. Because cooling liquids can be retained in the pores, sintered components should be machined dry whenever possible. An air blast is usually used instead of coolants to remove chips and cool the tool.

7.7.3 Oil Impregnation

Oil impregnation serves to provide either protection against corrosion or a degree of self-lubrication or both. It is usually carried out by immersing the sintered porous compact in hot oil and then allowing the oil to cool. Oil impregnation is mainly used in the manufacturing of self-lubricating bearings made of bronze or iron.

7.7.4 Infiltration

Infiltration is the permeation of a porous metal skeleton with a molten metal of a lower melting point by capillary action. Infiltration is performed in order to fill the pores and give two-phase structures with better mechanical properties. The widely used application of this process is the infiltration of porous iron compacts with copper. The process is then referred to as *copper infiltration* and involves placing a green compact of copper under (or above) the sintered iron compact and heating them to a temperature above the melting point of copper.

7.7.5 Heat Treatment

Conventional heat treatment operations can be applied to sintered porous materials, provided that the inherent porosity is taken into consideration. Pores reduce the thermal conductivity of the porous parts and thus reduce their rate of cooling. For sintered porous steels, this means poorer hardenability. Also, cyanide salts, which are very poisonous and are used in heat treatment salt baths, are retained in the pores, resulting in extreme hazards when using such heat-treated compacts. Therefore, it is not advisable to use salt baths for surface treatment of porous materials.

7.7.6 Steam Oxidizing

A protective layer of magnetite (Fe_3O_4) can be achieved by heating the sintered ferrous parts and exposing them to superheated steam. This will increase the corrosion resistance of the powder metallurgy parts, especially if it is followed by oil impregnation.

7.7.7 Plating

Metallic coatings can be satisfactorily electroplated directly onto high-density and copper-infiltrated sintered compacts. For relatively low-density compacts, electroplating must be preceded by an operation to seal the pores and render the compacts suitable for electroplating.

7.8 POROSITY IN POWDER METALLURGY PARTS

The structure of a powder metallurgy part consists of a matrix material with a microstructure identical to that of a conventional fully dense metal and pores that are a unique and controllable feature of sintered porous materials. For this reason, powder metallurgy materials are grouped according to their porosity, which is quantitatively expressed as the percentage of voids in a part. Those materials having less than 10 percent porosity are considered high density; those with porosity more than 25 percent have low density. There is a relationship between porosity and density (both being expressed as fractions of the full theoretical density), and it can be expressed by the following equation:

$$\text{Porosity} = 1 - \text{density} \tag{7.6}$$

As previously explained, the theoretical density is not that of the fully dense pure metal but is the mean value of the densities of all constituents. These include not only alloying additives but also impurities. When considering green densities, the effect of lubricants must be taken into consideration.

Pores are classified with respect to their percentage, type, size, shape, and distribution. The type can be either interconnected or isolated. The volume of interconnected porosity can be determined by measuring the amount of a known liquid needed to saturate the porous powder metallurgy sample. The interconnected porosity is essential for successful oil impregnation and thus is very important for the proper functioning of self-lubricating bearings.

At this stage, it is appropriate to differentiate between the following three technical terms used to describe density:

$$\text{Bulk density} = \frac{\text{Mass of compact}}{\text{Bulk volume of compact}} \tag{7.7}$$

$$\text{Apparent density} = \frac{\text{Mass of compact}}{\text{Apparent volume}}$$
$$= \frac{\text{Mass of compact}}{\text{Bulk volume of compact} - \text{volume of open pores}} \tag{7.8}$$

$$\text{True density} = \frac{\text{Mass of compact}}{\text{True volume}}$$
$$= \frac{\text{Mass of compact}}{\text{Bulk volume of compact} - \begin{pmatrix} \text{volume of open pores} \\ +\text{volume of closed pores} \end{pmatrix}} \tag{7.9}$$

For a green compact produced by admixed lubrication, these densities are misleading and do not indicate the true state of densification due to the presence of lubricant within the space between metal particles. Therefore, the bulk density must be readjusted to give the true metal density (TMD) as follows:

$$\text{TMD} = \text{Actual bulk density} \times \frac{\%\text{of metal}}{100} \tag{7.10}$$

7.9 DESIGN CONSIDERATIONS FOR POWDER METALLURGY PARTS

The design of a powder metallurgy part and the design of the tooling required to produce it cannot be separated. A part design that needs long, thin tubular punches; tooling with sharp corners; or lateral movement of punches cannot be executed. For this reason, the design of powder metallurgy parts is often different from that of parts produced by machining, casting, or forging, and a component that is being produced by these methods has to be redesigned before being considered for manufacture by powder metallurgy. Following are various tooling and pressing considerations, some of which are illustrated in Figure 7.11.

7.9.1 Holes

Holes in the pressing direction can be produced by using core rods. In this case, there is almost no limitation on the general shape of the hole. However, side holes and side slots are very difficult to achieve during pressing and must be made by secondary machining operations (see Figure 7.11a).

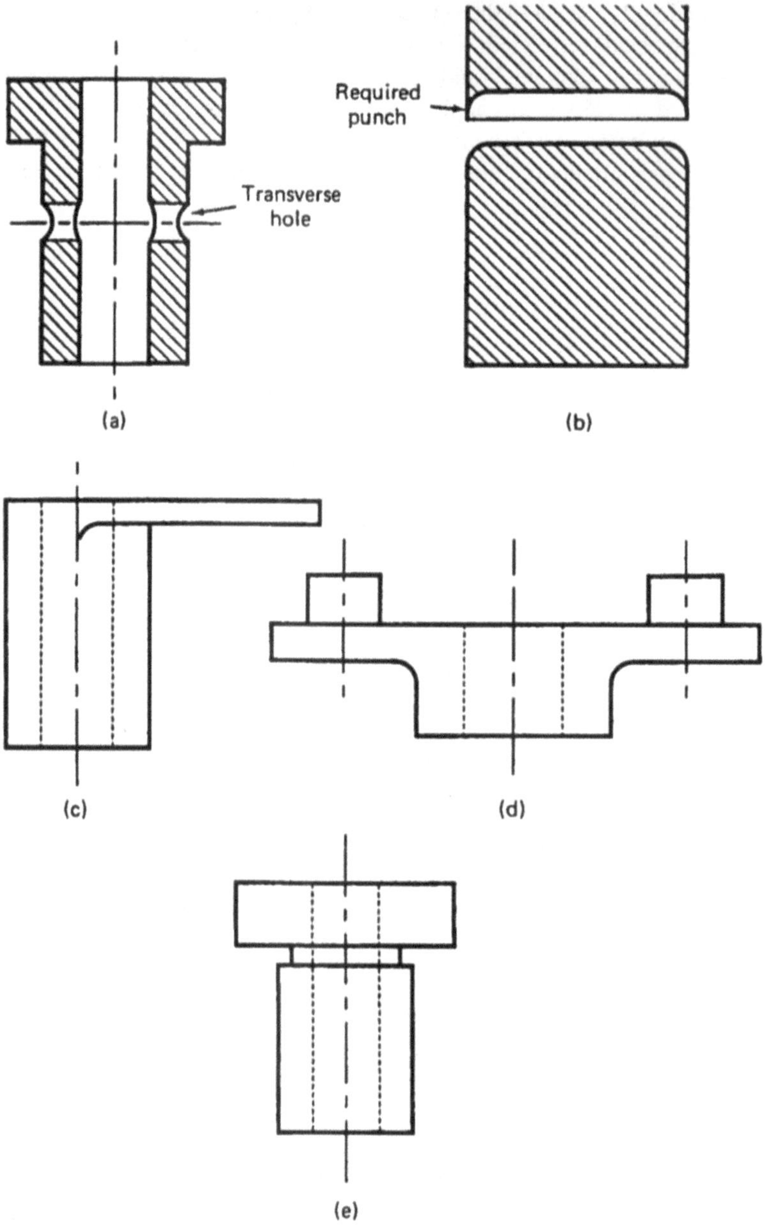

FIGURE 7.11 Design considerations for powder metallurgy parts: (a) holes, (b) fillets, (c) flanges, (d) bosses, and (e) undercuts.

7.9.2 Wall Thickness

It is not desirable to have a wall thickness of less than 1/16 inch (1.6 mm) because the punch required to produce the thickness will not be rigid enough to withstand the high stresses encountered during the pressing operation.

7.9.3 Fillets

It is recommended that sharp corners be avoided whenever possible. Fillets with generous radii are desirable, provided that they do not necessitate the use of punches with featherlike edges (see Figure 7.11b).

7.9.4 Tapers

Tapers are not always required. However, it is desirable to have them on flange-type sections and bosses to facilitate the ejection of the green compact.

7.9.5 Chamfers

As mentioned earlier, it is sometimes not desirable to use radii on part edges. Chamfers are the proper alternative for preventing burrs.

7.9.6 Flanges

A small flange, or overhang, can be easily produced. However, for a large overhang, ejection without breaking the flange is very difficult (see Figure 7.11c).

7.9.7 Bosses

Bosses can be made, provided that they are round in shape (or almost round) and that the height does not exceed 15 percent of the overall height of the component (see Figure 7.11d).

7.9.8 Undercuts

Undercuts that are perpendicular to the pressing direction cannot be made because they prevent the ejection of the part. If required, they can be produced by a secondary machining operation (see Figure 7.11e).

7.10 ADVANTAGES AND DISADVANTAGES OF POWDER METALLURGY

Like any other manufacturing process, powder metallurgy has advantages as well as disadvantages. The decision about whether to use this process or not must be based on these factors. The advantages of powder metallurgy are as follows:

- Components can be produced with good surface finish and close tolerances.
- There is usually no need for subsequent machining or finishing operations.
- The process offers a high efficiency of material utilization because it virtually eliminates scrap loss.
- Because all steps of the process are simple and can be automated, only a minimum of skilled labor is required.
- Massive numbers of components with intricate shapes can be produced at high rates.
- The possibility exists for producing controlled, unique structures that cannot be obtained by any other process.

The main disadvantages of the process are as follows:

- Powders are relatively high in cost compared with solid metals.
- Sintering furnaces and special presses, which are more complicated in principle and construction than conventional presses, are necessary.

- Tooling is very expensive as several punches or die movements are often used.
- High initial capital cost is involved, and the process is generally uneconomical unless very large numbers of components are to be manufactured.
- Powder metallurgy parts have inferior mechanical properties due to porosity (this does not apply to forged powder metallurgy parts), and the process is thus primarily suitable for the production of a large number of small, lightly stressed parts.

7.11 APPLICATIONS OF POWDER METALLURGY PARTS

The applications of powder metallurgy parts fall into two main groups. The first group consists of those applications in which the part is used as a structural component that can also be produced by alternative competing manufacturing methods, with powder metallurgy being used because of the low manufacturing cost and high production rate. The second group includes those applications in which the part usually has a controlled, unique structure and cannot be made by any other manufacturing method. Examples are porous bearings, filters, and composite materials. Following is a quick review of the various applications.

7.11.1 Structural Components

Powder metallurgy used to be limited to the production of small, lightly stressed parts. However, with the recent development in forging powder preforms, the process is commonly used in producing high-density components with superior mechanical properties. Cams, gears, and structural parts of the transmission system are some applications of the powder metallurgy process in the automotive, agricultural machinery, and domestic appliance industries.

The structural powder metallurgy components are usually made of iron-base powders, with or without additions of carbon, copper, and other alloying elements like nickel. Prealloyed powders are also employed, although they are less common than mixed elemental powders.

7.11.2 Self-Lubricating Bearings

Self-lubricating bearings are usually made by the conventional die-pressing technique, in which a porosity level between 20 and 40 percent is achieved. A sizing operation is performed for dimensional accuracy and in order to obtain smooth surfaces. The bearings are oil-impregnated either before or after sizing. Bronze powders are used in the manufacturing of porous bearings, but iron-base powders are also employed to give higher strength and hardness.

7.11.3 Filters

In manufacturing filters, the appropriate metal powder (e.g., bronze) is screened in order to obtain uniform particle size. The powder is then poured into a ceramic or graphite mold. The mold is put into a sintering furnace at the appropriate sintering temperature so that loose sintering can take place. The products must have generous tolerances, especially on their outer diameters, where 3 percent is typical.

7.11.4 Friction Materials

Clutch liners and brake bands are examples of friction materials. They are best manufactured by powder metallurgy. The composition includes copper as a matrix, with additions of tin, zinc, lead, and iron. Nonmetallic constituents like graphite, silica, emery, or asbestos are also added. The mixture is then formed to shape by cold pressing. After sintering, some finishing operations like bending, drilling, and cutting are usually required. It must be noted that friction materials are always joined to a solid plate, which gives adequate support to these weak parts.

7.11.5 Electrical Contact Materials

Electrical contact materials include two main kinds: sliding contacts and switching contacts. It is not possible to produce any of these contact materials except by powder metallurgy as both involve duplex structures.

Sliding contacts are components of electrical machinery employed when current is transferred between sliding parts (e.g., brushes in electric motors). The two main characteristics needed are a low coefficient of friction and good electrical conductivity. Compacts of mixtures of graphite and metal powder can fulfill such conditions. Powders of metals having high electrical conductivity, such as brass, copper, or silver, are used. These graphite–metal contacts are produced by conventional pressing and sintering processes.

Switching contacts are used in high-power circuit breakers. The three characteristics needed are good electrical conductivity, resistance to mechanical wear, and less tendency of the contact surfaces to weld together. A combination of copper, silver, and a refractory metal like tungsten provides the required characteristics. These contacts are produced either by conventional pressing and sintering or by infiltrating a porous refractory material with molten copper or silver.

7.11.6 Magnets

Magnets include soft magnets and permanent magnets. Soft magnets are used in DC motors or generators as armatures, as well as in measuring instruments. They are made of iron, iron–silicon, and iron–nickel alloys. Electrolytic iron powder is usually used because of its high purity and its good compressibility, which allows the high compact densities required for maximum permeability to be attained.

Permanent magnets produced by powder metallurgy have the commonly known name *Alnico.* This alloy consists mainly of nickel (30 percent), aluminum (12 percent), and iron (58 percent) and possesses outstanding permanent magnetic properties. Some other additives are often used, including cobalt, copper, titanium, and niobium.

7.11.7 Cores

The cores produced by powder metallurgy are used with AC high-frequency inductors in wireless communication systems. Such cores must possess high constant permeability for various frequencies as well as high electrical resistivity. Carbonyl iron powder is mixed with a binder containing insulators (to insulate the powder particles from one another and thus increase electrical resistivity) and then compacted using extremely high pressures, followed by sintering.

7.11.8 Powder Metallurgy Tool Steels

The production of tool steels by powder metallurgy eliminates the defects encountered in conventionally produced tool steels—namely, segregation and uneven distribution of carbides. Such defects create problems during tool fabrication and result in shorter tool life. The technique used involves compacting prealloyed tool-steel powders by HIP to obtain preforms that are further processed by hot working.

7.11.9 Superalloys

Superalloys are nickel- and cobalt-base alloys, which exhibit high strength at elevated temperatures. They are advantageous in manufacturing jet-engine parts like turbine blades. The techniques used in consolidating these powders include HIP, hot extrusion, and powder metallurgical forging.

7.11.10 Refractory Metals

The word *refractory* means "difficult to fuse." Therefore, metals with high melting points are considered refractory metals. These basically include four metals: tungsten, molybdenum, tantalum, and niobium. Some other metals can also be considered to belong to this group. Examples are platinum, zirconium, thorium, and titanium. Refractory metals, as well as their alloys, are best fabricated by powder metallurgy. The technique used usually involves pressing and sintering, followed by working at high temperatures. The applications are not limited to incandescent lamp filaments and heating elements; they also include space technology materials, the heavy metal used in radioactive shielding, and cores for armor-piercing projectiles. Titanium is gaining an expanding role in the aerospace industry because of its excellent strength-to-specific-weight ratio and its good fatigue and corrosion resistance.

7.11.11 Cemented Carbides

Cemented carbides are typical composite materials that possess the superior properties of both constituents. Cemented carbides consist of hard wear-resistant particles of tungsten or titanium carbides embedded in a tough strong matrix of cobalt or steel. They are mainly used as cutting and forming tools; however, there are other applications, including gauges, guides, rock drills, and armor-piercing projectiles. They possess excellent red hardness and have an extremely long service life as tools. Cemented carbides are manufactured by ball-milling carbides with fine cobalt (or iron) powder, followed by mixing with a lubricant and pressing. The green compact is then presintered at a low temperature, machined to the required shape, and sintered at an elevated temperature. A new dimension in cemented carbides is Ferro-Tic, involving titanium carbide particles embedded in a steel matrix. This material can be heat treated and thus can be easily machined or shaped.

7.12 RECENT DEVELOPMENTS IN POWDER METALLURGY

There are a few modern developments in the area of powder metallurgy. They can be summarized as follows:

- The most impressive recent development in powder metallurgy is the emerging additive manufacturing technology. At least five of the seven categories of additive manufacturing involve the processing of metal powders. Professional manufacturing engineers consider such a technology to constitute the fifth industrial revolution because of its characteristics and applications. For this reason, a whole chapter has been devoted to covering it in this textbook (Chapter 13).
- Some research work has been carried out at the University of Massachusetts Dartmouth (UMass Dartmouth) and elsewhere to fabricate and study the characteristics of powder metallurgy parts having porosity gradient, for different potential applications. At UMass Dartmouth, a specially designed novel indentor was used to "iron" the outermost layer and convert it to a fully dense one with zero porosity, while other researchers used laser to melt the outer layer and achieve that goal.
- Another new technology, developed at UMass Dartmouth, involves welding porous and solid materials together. The potential applications involve sound deadeners, vibration absorbers, and other applications in acoustics. This was achieved by optimization of the friction welding process. It was possible to weld pure iron to iron compacts having 10 percent porosity, as well as to stainless steel compacts with 40 percent porosity.

REVIEW QUESTIONS

1. Define each of the following technical terms: compressibility, compactibility, green density, impregnation, infiltration, flowability, and particle-size distribution.
2. List five advantages of the powder metallurgy process.
3. List four disadvantages of the powder metallurgy process.
4. What are the important characteristics of a metal powder?
5. Describe three methods for producing metal powders.
6. Explain briefly the mechanics of pressing.
7. Why are lubricants added to metal powders before pressing?
8. Is it possible to eliminate all voids by conventional die pressing? Why?
9. Explain briefly the mechanics of sintering.
10. Why is it necessary to have controlled atmospheres for sintering furnaces?
11. Explain why it is not possible to use the conventional pressing techniques as a substitute for each of the following operations: isostatic pressing, slip casting, and HERF compaction.
12. Differentiate between the following: coining, repressing, and sizing.
13. How is copper infiltration accomplished and what are its advantages?
14. Can powder metallurgical forging be replaced by conventional forging? Why?
15. How can machining of some powder metallurgy components be inevitable?
16. How is the plating of powder metallurgy components carried out?
17. Name five products that can only be produced by powder metallurgy.
18. Why are cemented carbides presintered?
19. Why is electrolytic iron powder used in manufacturing soft magnets?
20. Discuss four design limitations in connection with powder metallurgy components.

PROBLEMS

1. Following are the experimentally determined characteristics of three kinds of iron powder Plot the following for each powder:
   a. Cumulative oversize graph
   b. Cumulative undersize graph
   c. Frequency distribution curve (obtain median particle size for the powder)
   d. Histogram of particle-size distribution

Screen Analysis	Chemical	Composition	Apparent Density	Flow Rate
Sponge Iron Powder (I)				
+70.0%	H_2 loss	0.20%		
−70 to +100.1%	C	0.02%	2.4g/cm^3	35 s/50 g
−100 to +325.74%	SiO_2	0.20%		
−325.25%	P	0.02%		
	S	0.02%		
Sponge Iron Powder (II)				
+40.2%	H_2 loss	0.20%		
−40 to +60.40%	Cc	0.10%	2.4 g/cm^3	35 s/50 g
−70 to +100.30%	SiO_2	0.3%		
100 to +200.20%	S	0.02%		
−200.8%	P	0.02%		

(*Continued*)

Screen Analysis	Chemical	Composition	Apparent Density	Flow Rate	
Atomized Iron Powder					
+100	1	H_2 loss	0.20%		
+140	45	C	0.01%	2.9 g/cm³	24 s/50 g
+200	25				
+230	15				
+325	10				
−325	4				

The axis usually indicates the particle size in microns and not the mesh size. Use the following table:

1. Calculate the full theoretical density for a compact made of atomized iron powder, knowing that the density of carbon equals 2.2 g/cm³ and the density of iron oxide equals 2.9 g/cm³.
2. Determine the approximate height of the powder fill for each kind of iron powder given in problem 1, if the green density of the compact is 6.8 g/cm³ and its height is 2.1 cm.
3. Plot a graph indicating the maximum achievable green density versus the percentage of admixed zinc stearate for atomized iron powder (density of zinc stearate is 1.1 g/cm³). What can you deduce from the curve?
4. Which powder in problem 1 would fill the die cavity faster?
5. Calculate the maximum achievable green density of a mixture of atomized iron powder plus 1 percent zinc stearate and 10 percent pure copper (density of copper is 8.9 g/cm³).
6. Following is the relationship between density and pressure for atomized iron powder containing 1 percent zinc stearate. If it is required to manufacture a gear wheel having a green density of 6.8 g/cm³ using a press with a capacity of 1 MN, calculate the diameter of the largest gear wheel that can be manufactured. How can you produce a larger gear by modifying the design?

Sieve	40	60	70	100	140	200	230	325
Microns	420	250	210	149	105	74	63	44

1. A cylindrical compact of atomized iron powder plus 1 percent zinc stearate had a green bulk density of 7.0 g/cm³, a diameter equal to 2 cm, and a height equal to 3 cm. After sintering, its bulk density increased to 7.05 g/cm³. Calculate its new dimensions.
2. The sintered density of an atomized iron compact containing 10 percent copper was 7.2 g/cm³. What is the porosity?

Green Density, g/cm³	5.35	6.15	6.58	6.75	6.9
Pressure, MN/m²	157.5	315	472.5	629.9	787.4

DESIGN PROJECT

Figure 7.12 shows a part that is currently produced by forging and subsequent machining. Because the part is not subjected to high stresses during its actual service conditions, the producing company is considering the idea of manufacturing it by powder metallurgy in order to increase the production rate. Redesign this component so that it can be manufactured by the conventional die-pressing technique.

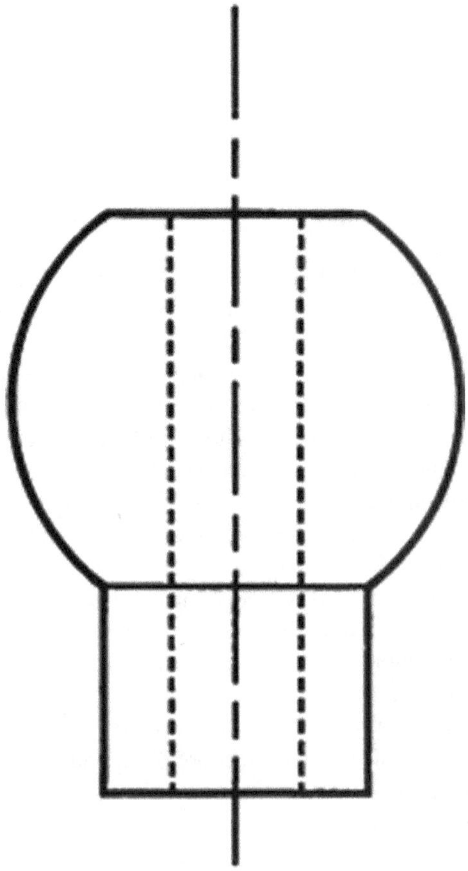

FIGURE 7.12 A part to be redesigned for production by powder metallurgy.

8 Plastics

8.1 INTRODUCTION

Plastics, which are more correctly called *polymers*, are products of macromolecular chemistry. In fact, the term *polymer* is composed of the two Greek words *poly* and *meres*, which mean "many parts." This is, indeed, an accurate description of the molecule of a polymer, which is made up of a number of identical smaller molecules that are repeatedly linked together to form a long chain. As an example, consider the commonly used polymer polyethylene, which is composed of many ethylene molecules (C_2H_4) that are joined together, as shown in Figure 8.1. These repeated molecules are always organic compounds, and, therefore, carbon usually forms the backbone of the chain. The organic compound whose molecules are linked together (like ethylene) is referred to as the *monomer*.

Now, let us examine why the molecules of a monomer tend to link together. We know from chemistry that carbon has a valence of 4. Therefore, each carbon atom in an ethylene molecule has an unsaturated valence bond. Consequently, if two ethylene molecules attach, each to one side of a third molecule, the valence bonds on the two carbon atoms of the center molecule will be satisfied (see Figure 8.1). In other words, the molecules of the monomer tend to attach to one another to satisfy the valence requirement of the carbon atoms.

The molecules of a monomer in a chain are strongly bonded together. Nevertheless, the long chains forming the polymer molecules tend to be more or less amorphous and are held together by weaker secondary forces that are known as the *van der Waals forces* (named after the Dutch physicist). Therefore, polymers are generally not as strong as metals or ceramics. It is also obvious that properties of a polymer such as strength, elasticity, and relaxation are dependent mainly upon the shape and size of the long chainlike molecules, as well as upon the mutual interaction between them.

A common, but not accurate, meaning of the term *polymer* involves synthetic organic materials that are capable of being molded. Actually, polymers form the building blocks of animal life; proteins, resins, shellac, and natural rubber are some examples of natural polymers that have been in use for a long time. On the other hand, synthetic, or manufactured, polymers have come into existence fairly recently. The first synthetic polymer, cellulose nitrate (celluloid), was prepared in 1869. It was followed in 1909 by the phenolics, which were used as insulating materials in light switches. The evolution of new polymers was accelerated during World War II due to the scarcity of natural materials. Today, thousands of polymers find application in all aspects of our lives.

8.2 CLASSIFICATION OF POLYMERS

There are, generally, two methods for classifying polymers. The first method involves grouping all polymers based on their elevated-temperature characteristics, which actually dictate the

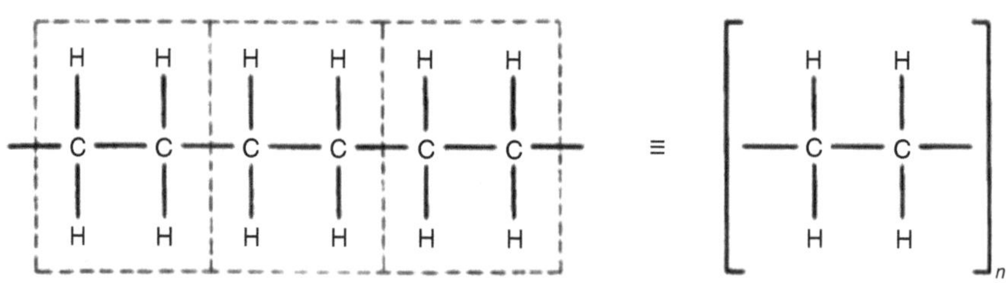

FIGURE 8.1 The molecular chain of polyethylene.

DOI: 10.1201/9781003537106-8

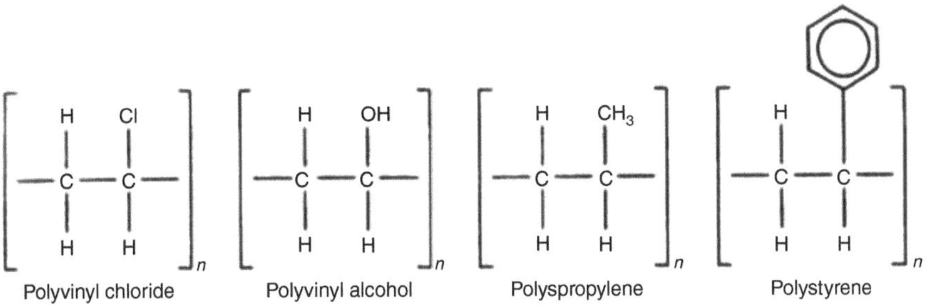

FIGURE 8.2 Structural formula of some polymers of the ethenic group.

manufacturing method to be used. The second method of classification groups polymers into *chemical families*, each of which has the same monomer. As an example, the ethenic family is based on ethylene as the monomer, and different polymers (members of this family such as polyvinyl alcohol or polystyrene) can be made by changing substituent groups on the basic monomer, as shown in Figure 8.2. As we will see later, this enables us to study most polymeric materials by covering just a limited number of families instead of considering thousands of polymers individually. However, before reviewing the commonly used chemical families of polymers, let us discuss in depth their elevated-temperature behavior. Based on this behavior, polymers can be split into two groups: *thermoplastics* and *thermosets*.

8.2.1 Thermoplastics

Thermoplastics generally have *linear* structures, meaning that their molecules look like linear chains having little breadth but significant length. This structure, as shown in Figure 8.3, is analogous to a bowl of spaghetti. Bonds between the various molecular chains are mainly of the van der Waals type (i.e., secondary forces). Therefore, this type of polymer softens by heating and can then flow viscously to take a desired shape because elevated temperatures tend to decrease the intermolecular coherence of the linear chains. When the solidified polymers are reheated and melted again, they can be given a different shape. This characteristic enables plastics fabricators to recycle thermoplastic scrap, thus increasing the efficiency of raw material utilization.

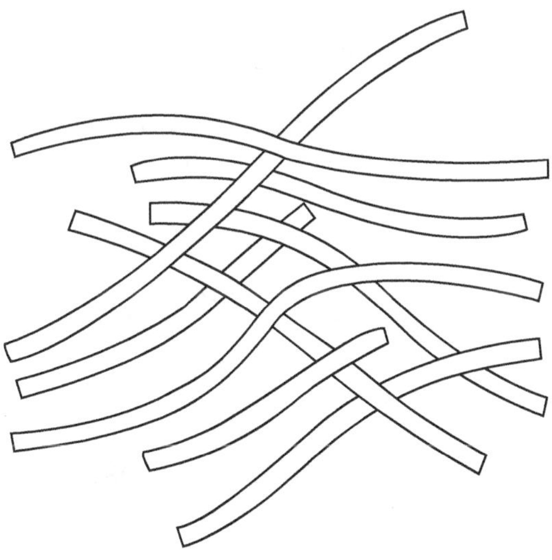

FIGURE 8.3 The molecular chains of a thermoplastic polymer.

Usually, a thermoplastic polymer consists of a mixture of molecular chains having different lengths. Therefore, each structure has a different melting point, and, consequently, the whole polymer melts, not at a definite temperature but within a range whose limits are referred to as the *softening point* and the *flow point*. It has been observed that when a thermoplastic is given a shape at a temperature between the softening and the flow points, the intermolecular tension is retained after the thermoplastic cools down. Therefore, if the part is reheated to a temperature above the softening point, it will return to its original shape because of this intermolecular tension. This phenomenon, which characterizes most thermoplastic polymers, is known as *shaping memory*. Many thermoplastic polymers are soluble in various solvents. Consequently, any one of these polymers can be given any desired shape by dissolving it into an appropriate solvent and then casting the viscous solution in molds. When the solvent completely evaporates, it leaves the rigid resin with the desired shape. Several chemical families of polymeric materials can be categorized as thermoplastic. These include the ethenics, the polyamides, the cellulosics, the acetals, and the polycarbonates. Their characteristics, methods of manufacture, and applications are discussed in detail later in this chapter.

8.2.2 Thermosets

The molecules of thermosets usually take the form of a three-dimensional network structure that is mostly *cross-linked*, as shown in Figure 8.4. When raw (uncured) thermosetting polymers are heated to elevated temperatures, they are set, cross-linked, or polymerized. If reheated after this curing operation, thermosets will not melt again but will char or burn. Therefore, for producing complex shapes of thermosetting polymers, powders (or grains) of the polymers are subjected to heat and pressure until they are cured as finished products. Such polymers are referred to as *heat-convertible resins*.

Some raw thermosets can take the form of liquids at room temperature. When required, they are converted into solids by curing as a result of heating and/or additives (hardeners). This characteristic enables fabricators to produce parts by casting mixtures of liquid polymers and hardeners into molds. Therefore, these polymers are referred to as *casting resins*.

The cured thermosets are insoluble in solvents and do not soften at high temperatures. Thus, products made of thermosets can retain their shape under combined load and high temperatures, conditions that thermoplastics cannot withstand.

8.3 PROPERTIES CHARACTERIZING PLASTICS AND THEIR EFFECT ON PRODUCT DESIGN

The properties of plastics differ significantly from those of metals, and they play a very important role in determining the form of the product. In other words, the form is dictated not only by the function but also by the properties of the material used and the method of manufacture, as we will see later. Following is a discussion of the effect of the properties characterizing plastics on the design of plastic products.

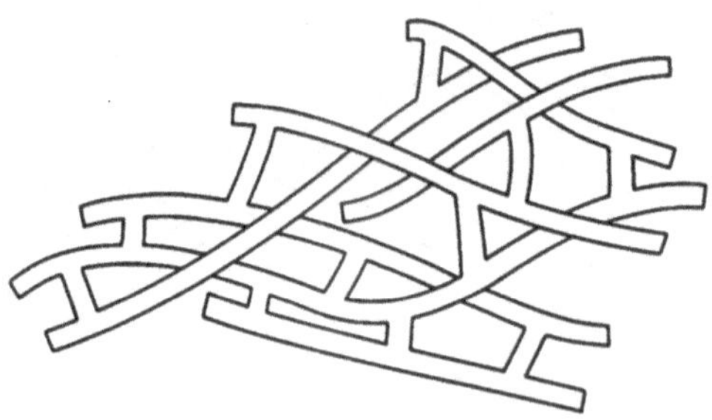

FIGURE 8.4 The molecular chains of a thermosetting polymer.

8.3.1 Mechanical Properties

The mechanical properties of polymers are significantly inferior to those of metals.

Strength and rigidity values for plastics are very low compared with the lowest values of these properties for metals. Therefore, larger sections must be provided for plastic products if they are to have a similar strength and/or rigidity as metal products. Unfortunately, these properties get even worse when plastic parts are heated above moderate temperatures. In addition, some plastics are extremely brittle and notch-sensitive. Accordingly, any stress raisers like sharp edges or threads must be avoided in such cases.

A further undesirable characteristic of plastics is that they tend to deform continually under mechanical load even at room temperature. This phenomenon is accelerated at higher temperatures. Consequently, structural components made of plastics should be designed based on their creep strength rather than on their yield strength. This dictates a temperature range in which only a plastic product can be used. It is obvious that such a range is dependent principally upon the kind of polymer employed.

Despite these limitations, the strength-to-weight ratio as well as the stiffness-to-weight ratio of plastics can generally meet the requirements for many engineering applications. In fact, the stiffness-to-weight ratio of reinforced polymers is comparable to that of metals like steel or aluminum.

8.3.2 Physical Properties

Three main physical properties detrimentally affect the widespread industrial application of polymers and are not shared by metals. First, plastics usually have a very high coefficient of thermal expansion, which is about ten times that of steel. This has to be taken into consideration when designing products involving a combination of plastics and metals. If plastics are tightly fastened to metals, severe distortion will occur whenever a significant temperature rise takes place. Second, some plastics are inflammable (i.e., not self-extinguishing) and keep burning even after the removal of the heat source. Third, some plastics have the ability to absorb large amounts of moisture from the surrounding atmosphere. This moisture absorption is unfortunately accompanied by a change in the size of the plastic part. Nylons are a typical example of this kind of polymer.

8.4 POLYMERIC SYSTEMS

This section surveys the commonly used polymeric materials and discusses their manufacturing properties and applications. Also discussed are the different additives that are used to impart certain properties to the various polymers.

8.4.1 Commonly Used Polymers

Following are some polymeric materials that are grouped into chemical families according to their common monomers.

8.4.1.1 Ethenic Group

The monomer is ethylene. This group includes the following polymers:

- *Polyethylene*. The properties of polyethylene depend upon factors like degree of crystallinity, density, molecular weight, and molecular weight distribution. This thermoplastic polymer is characterized by its chemical resistance to solvents, acids, and alkalis, as well as by its toughness and good wear properties. Polyethylenes also have the advantage of being adaptable to many processing techniques, such as injection molding, blow molding, pipe extrusion, wire and cable extrusion, and rotational molding. The applications of polyethylene are dependent upon the properties, which, in turn, depend upon the density and molecular weight. Low-density polyethylene is used in manufacturing films, coatings, trash bags, and throwaway products. High-density polyethylene (HDPE) is used for making injection-molded parts, tubes, sheets, and tanks that are used for keeping chemicals. The applications of ultra-high molecular weight (UHMW) polyethylene include wear plates and guide rails for filling and packaging equipment.

- *Polypropylene.* Polypropylene is a thermoplastic material. A molecule of this polymer has all substituent groups (i.e., CH_3) on only one of its sides. This promotes crystallinity and, therefore, leads to strength higher than that of polyethylene. The resistance of polypropylene to chemicals is also good. Polypropylene is mainly used for making consumer goods that are subjected to loads during their service life, such as ropes, bottles, and parts of appliances. This polymer is also used in tanks and conduits because of its superior resistance to chemicals.
- *Polybutylene.* Polybutylene is a thermoplastic polymer that has high tear, impact, and creep resistances. It also possesses good wear properties and is not affected by chemicals. Polybutylene resins are available in many grades, giving a wide range of properties and, therefore, applications. The properties of polybutylene have made it an appropriate material for piping applications. These pipes can be joined together by heat fusion welding or by mechanical compression. Some grades of polybutylene are used as high-performance films for food packaging and industrial sheeting.
- *Polyvinyl Chloride.* Polyvinyl chloride (PVC) is a thermoplastic polymer that can be processed by a variety of techniques like injection molding, extrusion, blow molding, and compression molding. It is fairly weak and extremely notch-sensitive but has excellent resistance to chemicals. When plasticized (i.e., additives are used to lubricate the molecules), it is capable of withstanding large strains. The applications of rigid PVC include low-cost piping, siding, and related profiles, toys, dinnerware, and credit cards. Plasticized PVC is used in upholstery, imitation leather for seat covers and rainwear, and as insulating coatings on wires.
- *Polyvinylidene Chloride.* Polyvinylidene chloride (PVDC) is nonpermeable to moisture and oxygen. It also possesses good creep properties. It is a preferred food-packaging material (e.g., saran wrap). Rigid grades are used for hot piping.
- *Polystyrene.* This thermoplastic polymer is known as "the cheap plastic." It has poor mechanical properties, can tolerate very little deflection, and breaks easily. Because of its cost, polystyrene is used for cheap toys and throwaway articles. It is also made in the form of foam (Styrofoam) for sound attenuation and thermal insulation.
- *Polymethyl Methacrylate (Plexiglas Acrylics).* This polymer has reasonably good toughness, good stiffness, and exceptional resistance to weather. In addition, it is very clear and has a white-light transmission equal to that of clear glass. Consequently, this polymer finds application in safety glazing and in the manufacture of guard and safety glasses. It is also used in making automotive and industrial lighting lenses. Some grades are used as coatings and lacquers on decorative parts.
- *Fluorocarbons Like Polytetrafluoroethylene (Teflon).* Teflon is characterized by its very low coefficient of friction and by the fact that even sticky substances cannot adhere to it easily. It is also the most chemically inert polymer. Nevertheless, it has some disadvantages, including low strength and poor processability. Because of its low coefficient of friction, Teflon is commonly used as a dry film lubricant. It is also used as a lining for chemical and food-processing containers and conduits.

8.4.1.2 Polycarbonate Group

These are actually polyesters. They are thermoplastic and have linear molecular chains. Polycarbonate exhibits good toughness, good creep resistance, and low moisture absorption. It also has good chemical resistance. It is widely used in automotive and medical and food packaging because of its cost-effectiveness. It is also considered a high-performance polymer and has found application in the form of solar collectors, helmets, and face shields.

8.4.1.3 Polyacetal Group

Included in this group is formaldehyde, with ending groups.

Formaldehyde is a thermoplastic polymer that can be easily processed by injection molding and extrusion. It has a tendency to be highly crystalline, and, as a result, this polymer possesses good mechanical properties. It also has good wear properties and a good resistance to moisture absorption.

Its applications involve parts that are made of nonferrous metals (like zinc, brass, or aluminum) by casting or stamping. These applications are exemplified by shower heads, shower-mixing valves, handles, good-quality toys, and lawn sprinklers.

8.4.1.4 Cellulosic Group

The monomer is cellulose. Cellulose itself is not a thermoplastic polymer. It can be produced by the viscous regeneration process to take the form of a fiber, as in rayon, or a thin film, as in cellophane. Cellophane applications involve mainly decoration. Nevertheless, cellulose can be chemically modified to produce the following thermoplastics:

Cellulose Nitrate. Good dimensional stability and low water absorption are the positive characteristics of this polymer. The major disadvantage that limits its widespread use is its inflammability. Cellulose nitrate is used in making table tennis balls, fashion accessories, and decorative articles. It is also used as a base for lacquer paints.

Cellulose Acetate. This polymer has good optical clarity, good dimensional stability, and resistance to moisture absorption. The uses of cellulose acetate include transparent sheets and films for graphical aids, visual aids, and a base for photographic films. It is also used in making domestic articles.

Cellulose Acetate Butyrate. This thermoplastic polymer is tough, has good surface quality and color stability, and can readily be vacuum formed. It finds popular use in laminating with thin aluminum foil.

Cellulose Acetate Propionate. This thermoplastic polymer has reasonably good mechanical properties and can be injection molded or vacuum formed. It is used for blister packages, lighting fixtures, brush handles, and other domestic articles.

8.4.1.5 Polyamide Group

This family includes high-performance melt-processable thermoplastics. One group of common polyamides is the nylons. These are characterized by their endurance and retention of their good mechanical properties even at relatively high temperatures. They also possess good lubricity and resistance to wear. The chief limitation is their tendency to absorb moisture and change size.

These polymers find use in virtually every market (e.g., automotive, electrical, wire, packaging, and appliances). Typical applications include structural components up to 10 pounds (4 kg), bushings, gears, earns, and the like.

8.4.1.6 ABS

The three monomers are acrylonitrile, butadiene, and styrene. Based on this three-monomer system (similar to an alloy in the case of metals), the properties of this group vary depending on the components. Fifteen different types are commercially used. They possess both good mechanical properties and processability. Applications of the ABS group include pipes and fittings, appliances and automotive uses, telephones, and components for the electronics industry.

8.4.1.7 Polyesters

These polymers result from a condensation reaction of an acid and an alcohol. The type and nature of the polymer obtained depend upon the acid and alcohol used. This multitude of polymers is mostly thermoplastic and can be injection molded and formed into films and fibers. Their uses include bases for coatings and paints, ropes, fabrics, outdoor applications, construction, appliances, and electrical and electronic components. Polyester is also used as a matrix resin for fiberglass to yield the composite fiber-reinforced polymer.

8.4.1.8 Phenolic Group

The monomer is phenol formaldehyde. As previously mentioned, phenolics are actually the oldest manufactured thermosetting polymers. They are processed by compression molding, where a product with a highly cross-linked chain structure is finally obtained. Phenolics are characterized by their high strength and their ability to tolerate temperatures far higher than their molding temperature.

Phenolics are recommended for use in hostile environments that cannot be tolerated by other polymers. They are used in electrical switch plates, electrical boxes, and similar applications. Nevertheless, the chief field of application is as bonding agents for laminates, plywood-grinding wheels, and friction materials for brake lining.

8.4.1.9 Polyimides

Polyimides are mostly thermosetting and have very complex structures. They are considered one of the most heat-resisting polymers. They do not melt and flow at elevated temperatures and are, therefore, manufactured by powder metallurgy techniques.

The polyimides are good substitutes for ceramics. Applications include jet engine and turbine parts, gears, coil bobbins, cages for ball bearings, bushings and bearings, and parts that require good electrical and thermal insulation.

8.4.1.10 Epoxies

Epoxies and epoxy resins are a group of polymers that become highly cross-linked by reaction with curing agents or hardeners. These polymers have low molecular weight and got their name from the epoxide group at the ends of the molecular chains. Epoxy resins are thermosetting and have good temperature resistance. They adhere very well to a variety of substrates. Another beneficial characteristic is their stability of dimensions upon curing.

The common application of epoxy resins is as adhesives. With the addition of fibers and reinforcements, laminates and fiber-reinforced epoxy resins can be obtained and used for structural applications.

8.4.1.11 Polyurethanes

Polyurethanes involve a wide spectrum of polymers ranging from soft thermoplastic elastomers to rigid thermosetting foams. While all polyurethanes are products of a chemical reaction of an isocyanate and an alcohol, different polymers are apparently obtained by different reacting materials.

Elastomers are used as die springs, forming-die pads, and elastomer-covered rolls. Some of these elastomers are castable at room temperature and find popular application in rubber dies for the forming of sheet metals. Flexible foam has actually replaced latex rubber in home and auto seating and interior padding. The rigid thermosetting foam is used as a good insulating material and for structural parts. Other applications of polyurethanes include coating, varnishes, and the like.

8.4.1.12 Silicones

In this group of polymers, silicon forms the backbone of the molecular chain and plays the same role as that of carbon in other polymers.

Silicones can be oils, elastomers, thermoplastics, or thermosets, depending on the molecular weight and the functional group. Nevertheless, they are all characterized by their ability to withstand elevated temperatures and their water-repellent properties. Silicones in all forms are mainly used for high-temperature applications. These include binders for high-temperature paints and ovens and good-handling tubing gaskets. Silicone oils are used as high-temperature lubricants, mold release agents, and damping or dielectric fluids.

8.4.1.13 Elastomers

These polymeric materials possess a percentage elongation of greater than 100 percent together with significantly high resilience. This rubberlike behavior is attributed to the branching of the molecular chains. Elastomers mainly include five polymers: natural rubber, neoprene, silicone rubber, polyurethane, and fluoroelastomers. Natural rubber is extracted as a thick, milky liquid from a tropical tree. Next, moisture is removed, additives (coloring, curing agents, and fillers) are blended with it, and the mixture is then vulcanized. The latter operation involves heating up to a temperature of 300°F (150°C) to start cross-linking and branching reactions.

The application of elastomers includes seals, gaskets, oil rings, and parts that possess rubberlike behavior such as tires, automotive and aircraft parts, and parts in forming dies.

8.4.2 Additives

Additives are materials that are compounded with polymers in order to impart and/or enhance certain physical, chemical, manufacturing, or mechanical properties. They are also sometimes added just for the sake of reducing the cost of products. Commonly used additives include fillers, plasticizers, lubricants, colorants, antioxidants, and stabilizers.

8.4.2.1 Fillers

Fillers involve wood flour, talc, calcium carbonate, silica, mica flour, cloth, and short fibers of glass or asbestos. Fillers have recently gained widespread industrial use as a result of the continued price increase and short supply of resin stocks. An expensive or unavailable polymer can sometimes be substituted by another filled polymer, provided that an appropriate filler material is chosen.

The addition of inorganic fillers usually tends to increase the strength because this kind of additive inhibits the mobility of the polymers' molecular chains. Nevertheless, if too much filler is added, it may create enclaves or weak spots and cause problems during processing, especially if injection molding is employed.

8.4.2.2 Plasticizers

Plasticizers are organic chemicals (high-boiling-temperature solvents) that are admixed with polymers in order to enhance resilience and flexibility. This is a result of facilitating the mobility of the molecular chains, thus enabling them to move easily relative to one another. On the other hand, plasticizers reduce the strength. Therefore, a polymer that meets requirements without the addition of plasticizers is the one to use.

8.4.2.3 Lubricants

Lubricants are chemical substances that are added in small quantities to the polymer to improve processing and flowability. They include fatty acids, fatty alcohols, fatty esters, metallic stearates, paraffin wax, and silicones. Lubricants are classified as external (applied externally to the polymer), internal, or internal–external. The last group includes most commercially used lubricants.

8.4.2.4 Colorants

Colorants may be either dyes or pigments. Dyes have smaller molecules and are transparent when dissolved. Pigment particles are relatively large (over 1 μm) and are, therefore, either translucent or opaque. Pigments are more widely used than dyes because dyes tend to extrude from the polymers.

8.4.2.5 Antioxidants

The use of antioxidants is aimed at enhancing the resistance to oxidation and degradation of polymers, thus extending their useful temperature range and service life. These substances retard the chemical reactions that are caused by the presence of oxygen.

8.4.2.6 Stabilizers

Stabilizers are substances that are added to polymers to prevent degradation as a result of heat or ultraviolet rays. The mechanism of inhibiting the degradation of polymers differs for different stabilizers. However, ultraviolet stabilizers usually function by absorbing ultraviolet radiation.

8.5 PROCESSING OF PLASTICS

A variety of processing methods can be employed in manufacturing plastic products. However, it must be kept in mind that no single processing method can successfully be employed in shaping all kinds of plastics. Each process has its own set of advantages and disadvantages that influence product design. Following is a survey of the common methods for plastic processing.

8.5.1 Casting

Casting is a fairly simple process that requires no external force or pressure. It is usually performed at room temperature and involves filling the mold cavity with monomers or partially polymerized syrups and then heating to cure. After amorphous solidification, the material becomes isotropic, with uniform properties in all directions. Nevertheless, a high degree of shrinkage is experienced during solidification and must be taken into consideration when designing the mold. Sheets, rods, and tubes can be manufactured by casting, although the typical application is in trial jigs and fixtures as well as in insulating electrical components. Acrylics, epoxies, polyesters, polypropylene, nylon, and PVC can be processed by casting. The casting method employed is sometimes modified to suit the kind of polymer to be processed. Whereas nylon is cast in its hot state after adding a suitable catalyst, PVC film is produced by solution casting. This process involves dissolving the PVC into an appropriate solvent, pouring the solution on a substrate, and allowing the solvent to evaporate in order to finally obtain the required film.

8.5.2 Blow Molding

Blow molding is a fast, efficient method for producing hollow containers of thermoplastic polymers. The hollow products manufactured by this method usually have thin walls and range in shape and size from small, fancy bottles to automobile fuel tanks.

Although there are different versions of the blow molding process, they basically involve blowing a tubular shape (parison) of heated polymer in a cavity of a split mold. As can be seen in Figure 8.5, air is injected through a needle into the parison, which expands in a fairly uniform thickness and finally conforms to the shape of the cavity.

8.5.3 Injection Molding and Guidelines for Good Design of Parts

Injection molding is the most commonly used method for mass production of plastic articles because of its high production rates and good control over the dimensions of the products. The process is used for producing thermoplastic articles, but it can also be applied to thermosets. The main limitation of injection molding is the required high initial capital cost, which is due to the expensive machines and molds employed in the process.

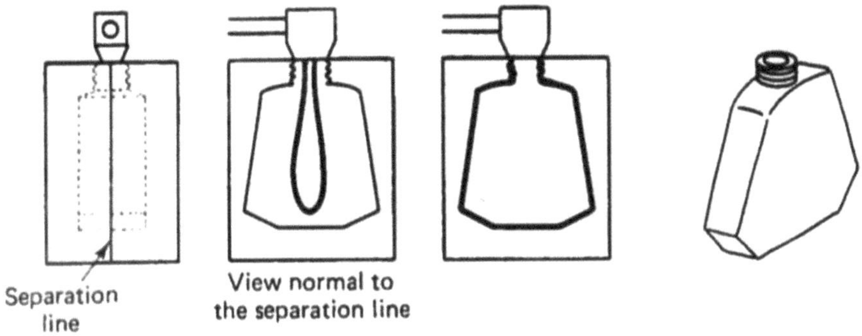

FIGURE 8.5 The blow molding process.

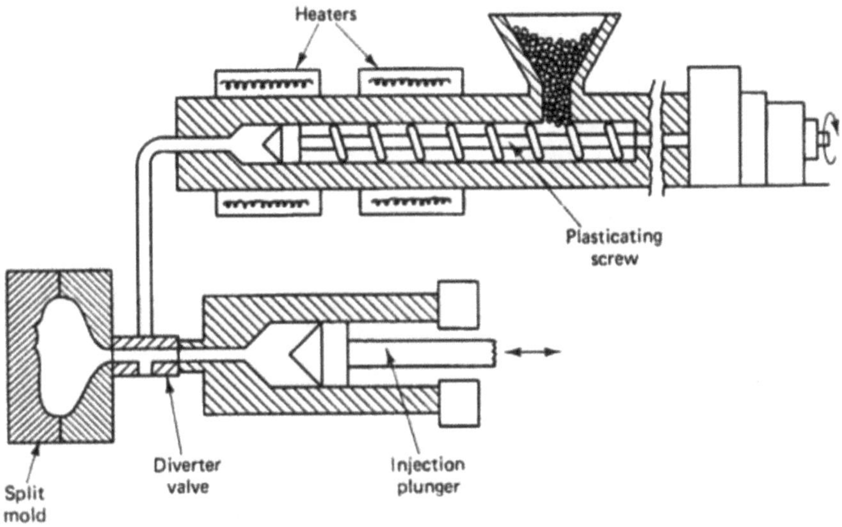

FIGURE 8.6 The injection molding process.

The process basically involves heating the polymer, which is fed from a hopper in granular pellet or powdered forms to a viscous melted state and then forcing it into a split-mold cavity, where it hardens under pressure. Next, the mold is opened, and the product is ejected by a special mechanism. Molds are usually made of tool steel and may have more than a single cavity.

Figure 8.6 shows a modem screw-preplasticator injection unit employed in injection molding of thermoplastics. As can be seen, the diverter valve allows the viscous polymer to flow either from the plasticating screw to the pressure cylinder or from the cylinder to the cooled mold. When thermosets are to be injection molded, a machine with a different design has to be used. Also, the molds must be hot so that the polymer can cure.

Once the decision has been made to manufacture a plastic product by injection molding, the product designer should make a design that facilitates and favors this process. Following are some design considerations and guidelines.

8.5.3.1 Make the Thickness of a Product Uniform and as Small as Possible

Injection molding of thermoplastics produces net-shaped parts by going from a liquid state to a solid state. (These net-shaped parts are used as manufactured; they do not require further processing or machining.) This requires time to allow the heat to dissipate so that the polymer melt can solidify. The thicker the walls of a product, the longer the product cycle, and the higher its cost. Consequently, a designer has to keep the thickness of a product to a minimum without jeopardizing the strength and stiffness considerations. Also, thickness must always be kept uniform; if the change in thickness is unavoidable, it should be made gradually. It is better to use ribs rather than increase the wall thickness of a product. Figure 8.7 shows examples of poor design and how they can be modified (by slight changes in constructional details) so that sound parts are produced.

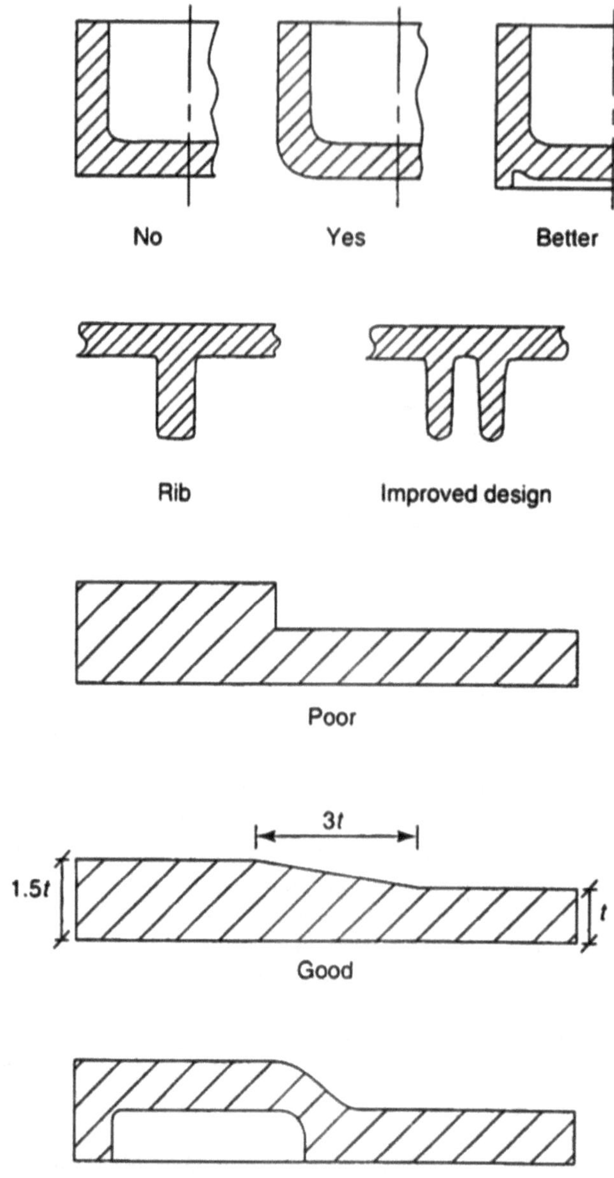

FIGURE 8.7 Examples of poor and good designs of walls of plastic products.

8.5.3.2 Provide Generous Fillet Radii

Plastics are generally notch-sensitive. The designer should, therefore, avoid sharp corners for fillets and provide generous radii instead. The ratio of the fillet radius to the thickness should be at least 1.4.

8.5.3.3 Ensure That Holes Will Not Require Complex Tooling

Holes are produced by using core pins. It is, therefore, clear that those through holes are easier to make than blind holes. Also, when blind holes are normal to the flow, they require retractable core pins or split tools, thus increasing the production cost.

8.5.3.4 Provide Appropriate Draft

As is the case with forging, it is important to provide a draft of 1° so that the product can be injected from the mold.

8.5.3.5 Avoid Heavy Sections When Designing Bosses

Heavy sections around bosses lead to wrappage and dimensional control problems. Figure 8.8 shows examples of poor and good designs of bosses.

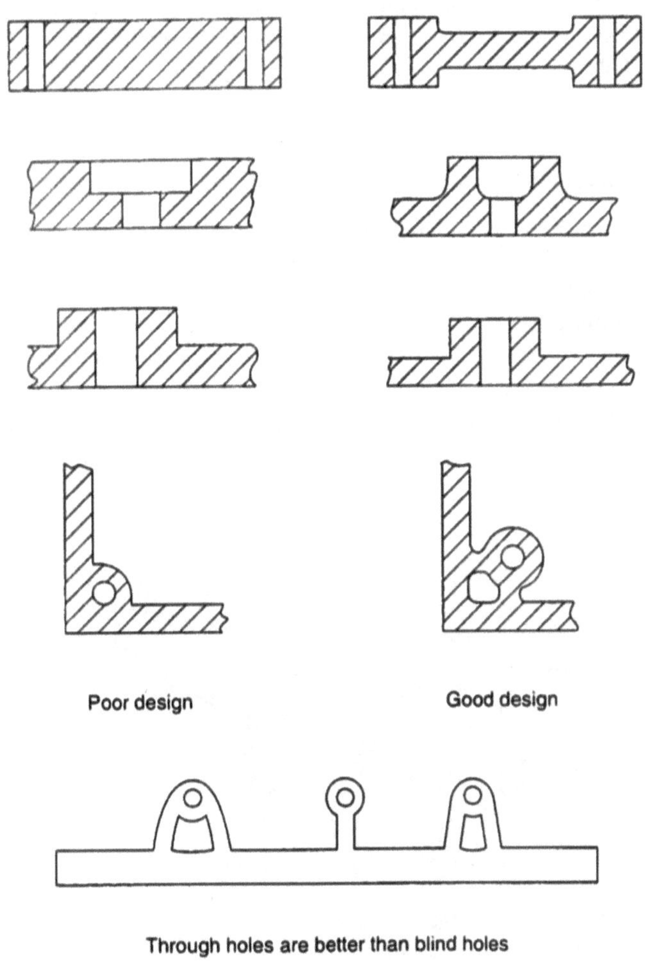

FIGURE 8.8 Examples of poor and good designs of bosses in injection-molded parts.

8.5.4 Compression Molding

Compression molding is used mainly for processing thermosetting polymers. The process involves enclosing a premeasured charge of polymer within a closed mold and then subjecting that charge to combined heat and pressure until it takes the shape of the mold cavity and cures. Figure 8.9 shows a part being produced by this process.

Although the cycle time for compression molding is very long when compared with that for injection molding, the process has several advantages. These include low capital cost (because the tooling and the equipment used are simpler and cheaper) and the elimination of the need for sprues or runners, thus reducing material waste. There are, however, limitations upon the shape and size of the products manufactured by this method. It is generally difficult to produce complex shapes or large parts as a result of the poor flowability and long curing times of the thermosetting polymers.

8.5.5 Transfer Molding

Transfer molding is a modified version of the compression molding process, and it is aimed at increasing productivity by accelerating the production rate. As can be seen in Figure 8.10, the process involves placing the charge in an open, separate "pot," where the thermosetting polymer is heated and forced through sprues and runners to fill several closed cavities. The surfaces of the sprues, runners, and cavities are kept at a temperature of 280°F to 300°F (140°C to 200°C) to promote curing of the polymer. Next, the entire shot (i.e., sprues, runners, product, and the excess polymer in the pot) is ejected.

8.5.6 Rotational Molding

Rotational molding is a process by which hollow objects can be manufactured from thermoplastics and sometimes thermosets. It is based upon placing a charge of solid or liquid polymer in a mold. The mold is heated while being rotated simultaneously around two perpendicular axes. As a result, the centrifugal force pushes the polymer against the walls of the mold, thus forming a homogeneous layer of uniform thickness that conforms to the shape of the mold, which is then cooled before the product is ejected. The process, which has a relatively long cycle time, has the advantage of offering almost unlimited product design freedom. Complex parts can be molded by employing low-cost machinery and tooling.

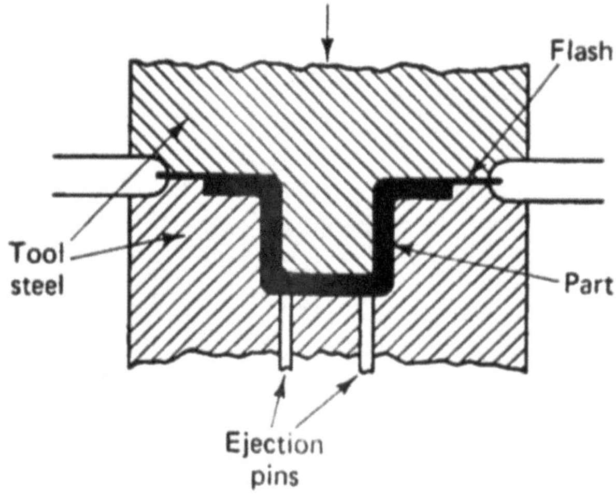

FIGURE 8.9 The compression molding process.

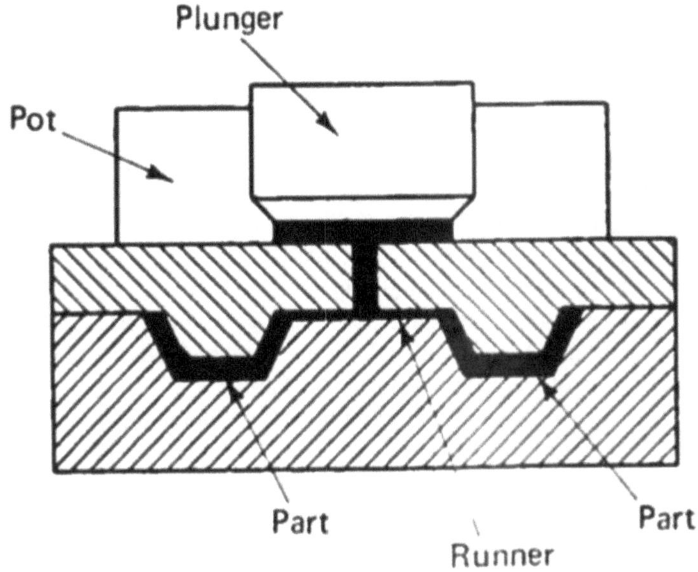

FIGURE 8.10 The transfer molding process.

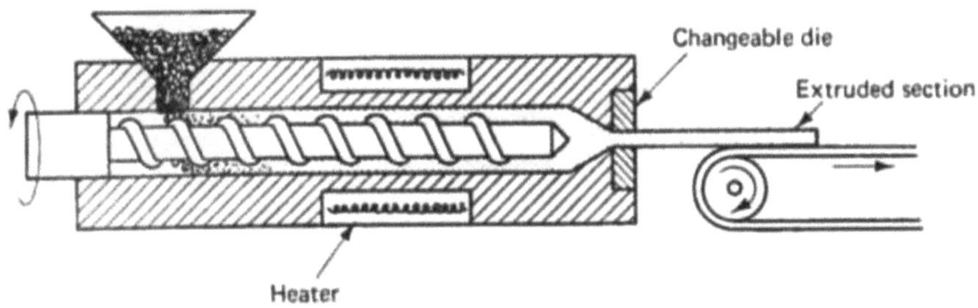

FIGURE 8.11 The extrusion process.

8.5.7 Extrusion

In extrusion, a thermoplastic polymer in powdered or granular form is fed from a hopper into a heated barrel, where the polymer melts and is then extruded out of a die. Figure 8.11 shows that plastics extrusion is a continuous process capable of forming an endless product that has to be cooled by spraying water and then cut to the desired lengths. The process is employed to produce a wide variety of structural shapes, such as profiles, channels, sheets, pipes, bars, angles, films, and fibers. Extrusions like bars, sheets, and pipes can also be further processed by other plastic manufacturing methods until the desired final product is obtained.

A modification of conventional extrusion is a process known as coextrusion. It involves extruding two or more different polymers simultaneously in such a manner that one polymer flows over and adheres to the other polymer. This process is used in industry to obtain combinations of polymers, each contributing some desired property. Examples of coextrusion include refrigerator liners, foamed-core solid sheaths, telephone wires, and profiles involving both dense material and foam, which are usually used as gaskets in automotive and appliance applications.

8.5.8 Thermoforming

Thermoforming involves a variety of processes that are employed to manufacture cuplike products from thermoplastic sheets by a sequence of heating, forming, cooling, and trimming. First, the sheet is clamped all around and heated to the appropriate temperature by electric heaters located above it. Next, the sheet is stretched under the action of pressure, vacuum, or male tooling and is forced to take the shape of a mold. The polymer is then cooled to retain the shape. This is followed by removing the part from the mold and trimming the web surrounding it. Figure 8.12a–d illustrates the different thermoforming processes.

Although thermoforming was originally developed for the low-volume production of containers, the process can be automated and made suitable for high-volume applications. In this case, molds are usually made of aluminum because of its high thermal conductivity. For low-volume or trial production, molds are made of wood or even plaster of Paris.

Examples of the parts produced by thermoforming include containers, panels, housings, machine guards, and the like. The only limitation on the shape of the product is that it should not contain holes. If holes are absolutely required, they should be made by machining at a later stage.

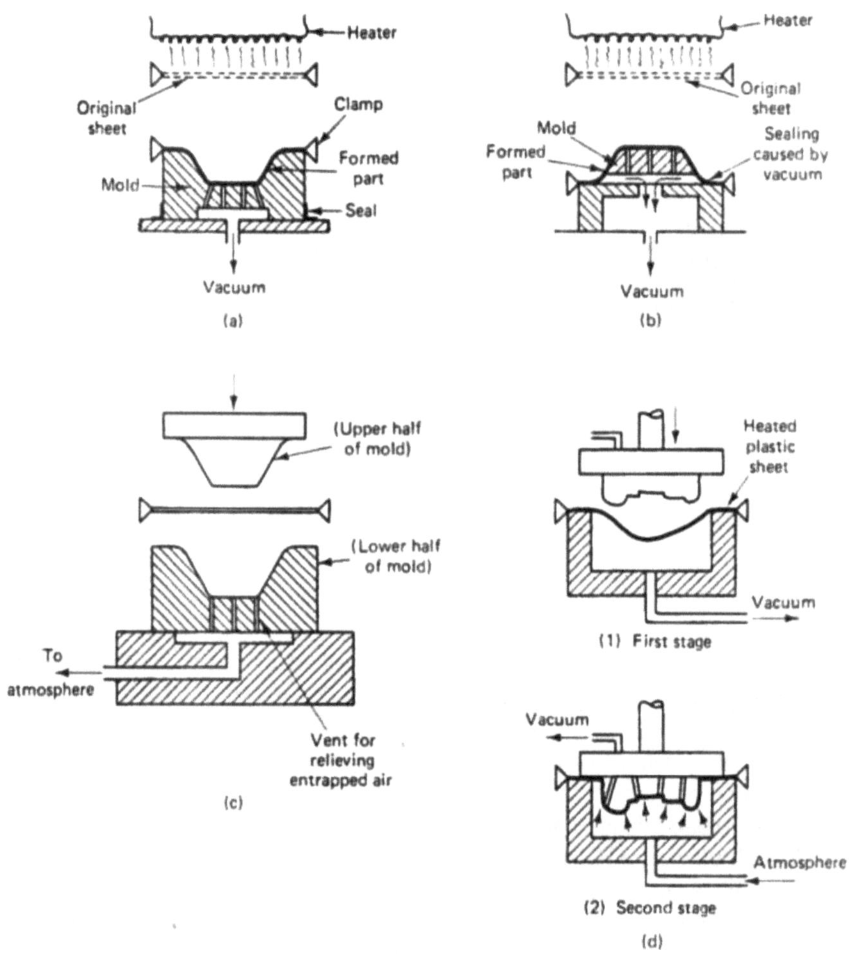

FIGURE 8.12 Different thermoforming processes: (a) straight vacuum forming, (b) drape forming, (c) matched-mold, and (d) vacuum snapback.

8.5.9 Calendering

Calendering is the process employed in manufacturing thermoplastic sheets and films. This process is similar to rolling with a four-high rolling mill, except that the rolls that squeeze the polymer are heated. The thermoplastic sheet is fed and metered in the first and second roll gaps, whereas the third roll gap is devoted to gaging and finishing.

Most of the calendering products are flexible or rubberlike sheets and films, although the process is sometimes applied to ABS and polyethylene. Figure 8.13 illustrates the calendering process.

8.5.10 Machining of Plastics

In some cases, thermoplastic and thermosetting polymers are subjected to machining operations like sawing, drilling, or turning. Some configurations and small lot sizes can be more economically achieved by machining than by any other plastic-molding method. Nevertheless, there are several problems associated with the machining of plastics. For instance, each type of plastic has its own unique machining characteristics, and they are very different from those of conventional metallic materials. A further problem is the excessive tool wear experienced when machining plastics, which results in the interruption of production as well as additional tooling costs. Although much research is needed to provide solutions for these problems, there are some general guidelines:

- Reduce friction at the tool–workpiece interface by using tools with honed or polished surfaces.
- Select tool geometry so as to generate continuous-type chips. Recent research has revealed that there exists a critical rake angle (see Chapter 10) that depends upon the polymer, depth of cut, and cutting speed.
- Use twist drills that have wide, polished flutes, low helix angles, and tool point angles (TPA) of about 70° and 120°.

Recently, lasers have been employed in cutting plastics. Because a laser acts as a materials eliminator, its logical application is cutting and hole drilling. High-pressure water jets also currently find some application in the cutting of polymers and composites.

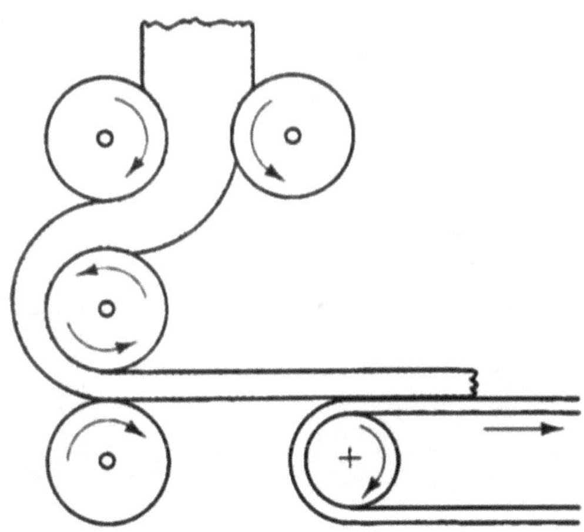

FIGURE 8.13 The calendering process.

8.5.11 Welding of Plastics

There are several ways to assemble plastic components. The commonly used methods include mechanical fastening, adhesive bonding, thermal bonding, and ultrasonic welding. Only thermal bonding and ultrasonic welding are discussed next because the first two operations are similar to those used with metals.

8.5.11.1 Thermal Bonding of Plastics

Thermal bonding, which is also known as fusion bonding, involves the melting of the weld spots in the two plastic parts to be joined and then pressing them together to form a strong joint. Figure 8.14 illustrates the steps involved in the widely used thermal bonding method known as hot-plate joining. As can be seen in the figure, a hot plate is inserted between the edges to be mated in order to melt the plastic parts; melting stops when the plate comes in contact with the holding fixture. Next, the plate is withdrawn, and the parts are pressed together and left to cool to yield a strong joint. The cycle time usually ranges from 15 to 20 seconds, depending upon the relationship between the melt time and the temperature (of the hot plate) for the type of plastic to be bonded. Also, this process is applied only to thermoplastics.

Figure 8.15 illustrates different types of joint design. The one to select is dependent upon both the desired strength and the appearance of the joint. The product designer must keep in mind that a small amount of material is displaced from each side to form the weld bead. This must be taken into account when dimensional tolerance is critical, such as when fusion-bonded parts are to be assembled together.

Another thermal bonding process, which is equivalent to riveting in the case of metals, is referred to as thermostaking. As can be seen in Figure 8.16, the process involves the softening of a plastic stud by a stream of hot air and then forming the softened stud and holding it while it cools down.

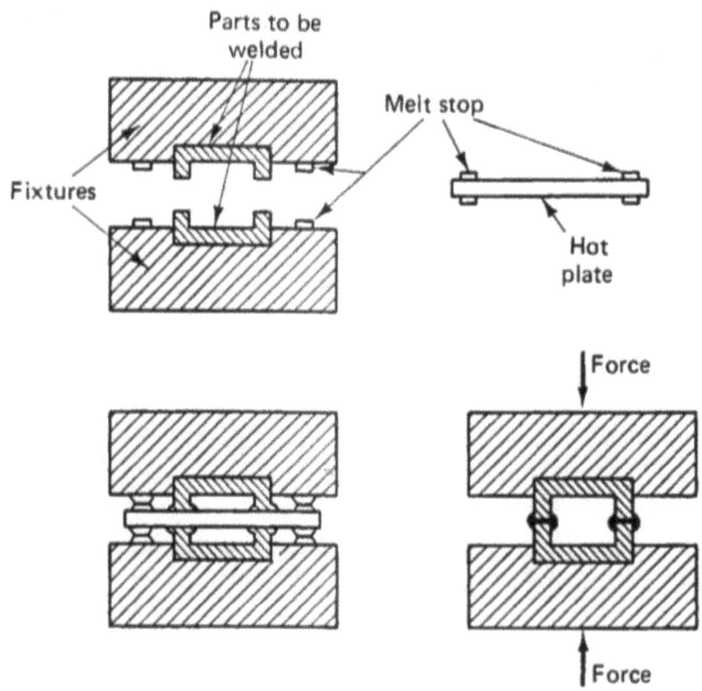

FIGURE 8.14 Steps involved in hot-plate joining.

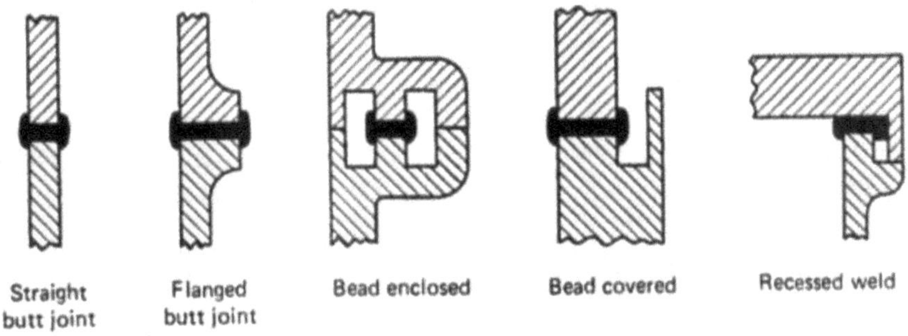

FIGURE 8.15 Different joint designs for fusion bonding.

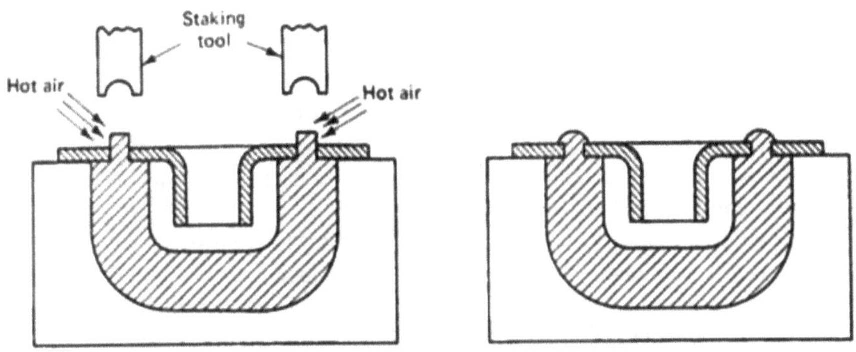

FIGURE 8.16 The thermostaking process.

Thermal bonding processes find widespread application in the automotive, appliance, battery, and medical industries.

8.5.11.2 Ultrasonic Welding of Plastics

Ultrasonic welding is gaining popularity in the industry because of its low cycle time of about 0.5 seconds and the strong, tight joints that are easily obtainable. The process is used for thermoplastics and involves the conversion of high-frequency electrical energy to high-frequency mechanical vibrations that are, in turn, employed to generate highly localized frictional heating at the interface of the mating parts. This frictional heat melts the thermoplastic polymer, allowing the two surfaces to be joined together.

The product designer must bear in mind that not all thermoplastics render themselves suitable for ultrasonic welding. Whereas amorphous thermoplastics are good candidates, crystalline polymers are not suitable for this process because they tend to attenuate the vibrations. Hydroscopic plastics (humidity-absorbing polymers, such as nylons) can also create problems and must, therefore, be dried before they are ultrasonically welded. In addition, the presence of external release agents or lubricants reduces the coefficient of friction, thus making ultrasonic welding more difficult.

The equipment used involves a power supply, a transducer, and a horn. The power supply converts the conventional 115-V, 60-Hz (or 220-V, 50-Hz) current into a high-frequency current (20,000 Hz). The transducer is usually a piezoelectric device that converts the electrical energy into high-frequency, axial-mechanical vibrations. The horn is the part of the system that is responsible for amplifying and transmitting the mechanical vibrations to the plastic workpiece. Horns may be made of aluminum, alloy steel, or titanium. The latter material possesses superior mechanical properties and is, therefore, used with heavy-duty systems. The horns amplify the mechanical vibration via a continuous decrease in the cross-sectional area and may take different forms to achieve that goal, as shown in Figure 8.17.

The task of joint design for ultrasonic welding is critical because it affects the design of the molded parts to be welded. Fortunately, there are a variety of joint designs, and each has its specific features, advantages, and limitations. The type of joint to be used should obviously depend upon the kind of plastic, the part geometry, the strength required, and the desired cosmetic appearance. Following is a discussion of the commonly used joint designs, which are illustrated in Figure 8.18.

Butt Joint with Energy Director. The butt joint (see Figure 8.18a) is the most commonly used joint design in ultrasonic welding. As can be seen in the figure, one of the mating parts has a triangular-shaped projection. This projection is known as an energy director because it helps to limit the initial contact to a very small area, thus increasing the intensity of energy at that spot. This causes the projection to melt and flow and cover the whole area of the joint. This type of joint is considered the easiest to produce because it is not difficult to mold into a part.

Step Joint with Energy Director. The step joint (see Figure 8.18b) is stronger than the butt joint and is recommended when a cosmetic appearance is desired.

Tongue-and-Groove Joint with Energy Director. The tongue-and-groove joint (see Figure 8.18c) promotes the self-locating of parts and prevents flash. It is stronger than both of the previously mentioned methods.

Interference Joint. The interference joint (see Figure 8.18d) is a high-strength joint and is usually recommended for square corners or rectangular-shaped parts.

Scarf Joint. The scarf joint (see Figure 8.18e) is another high-strength joint and is recommended for components with circular or oval shapes.

In addition to welding, ultrasonics are employed in inserting metallic parts into thermoplastic components. Figure 8.19 illustrates an arrangement for the ultrasonic installation of a metal insert into a plastic part.

Another useful application of these systems is ultrasonic staking, which is equivalent to riveting or heading. Figure 8.20 indicates the different types of stakes, as well as their recommended applications. Notice that these stakes can be flared, spherical, hollow, knurled, or flush.

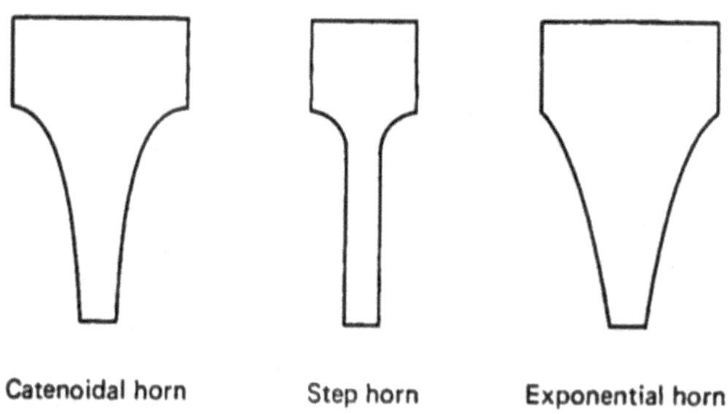

FIGURE 8.17 Different horn shapes employed in ultrasonic welding of plastics.

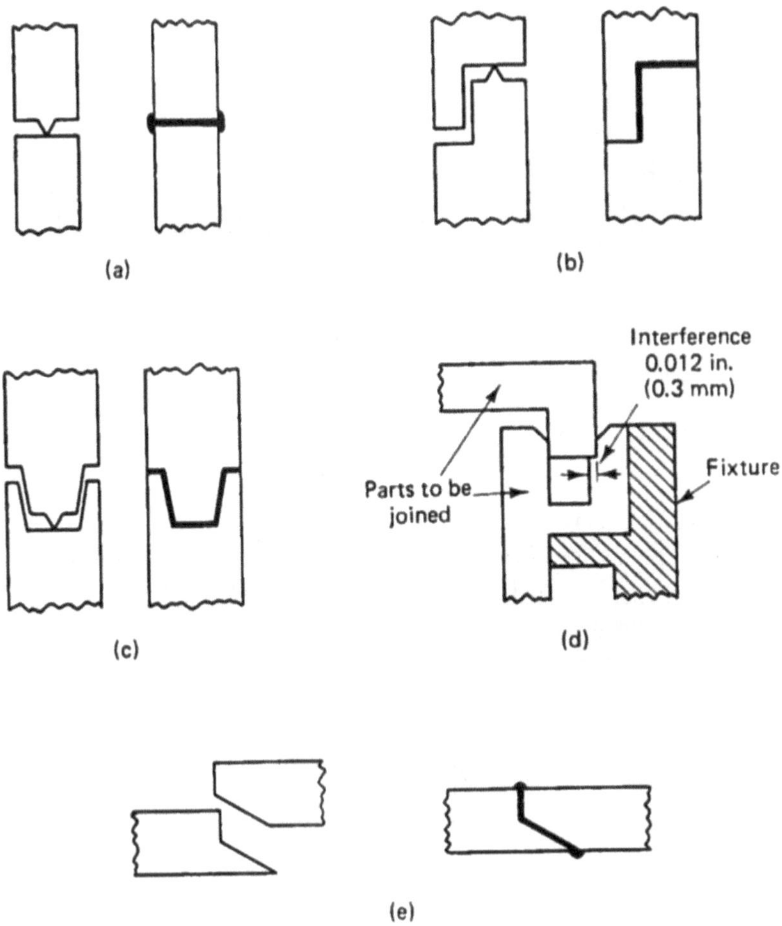

FIGURE 8.18 Different joint designs for ultrasonic welding: (a) butt joint, (b) step joint, (c) tongue-and-groove joint, (d) interference joint, and (e) scarf joint.

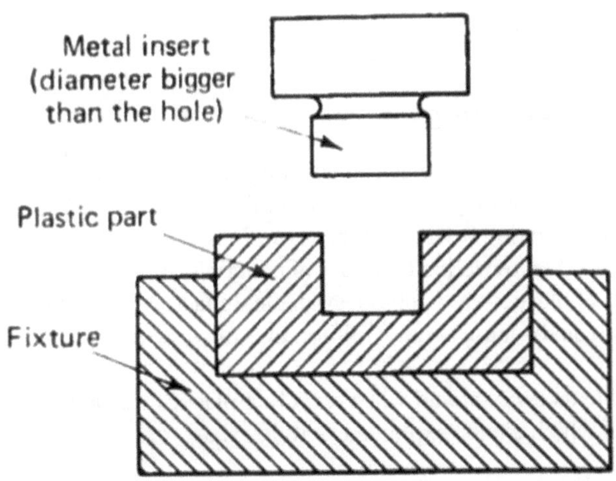

FIGURE 8.19 Ultrasonic installation of metal insert into plastic part.

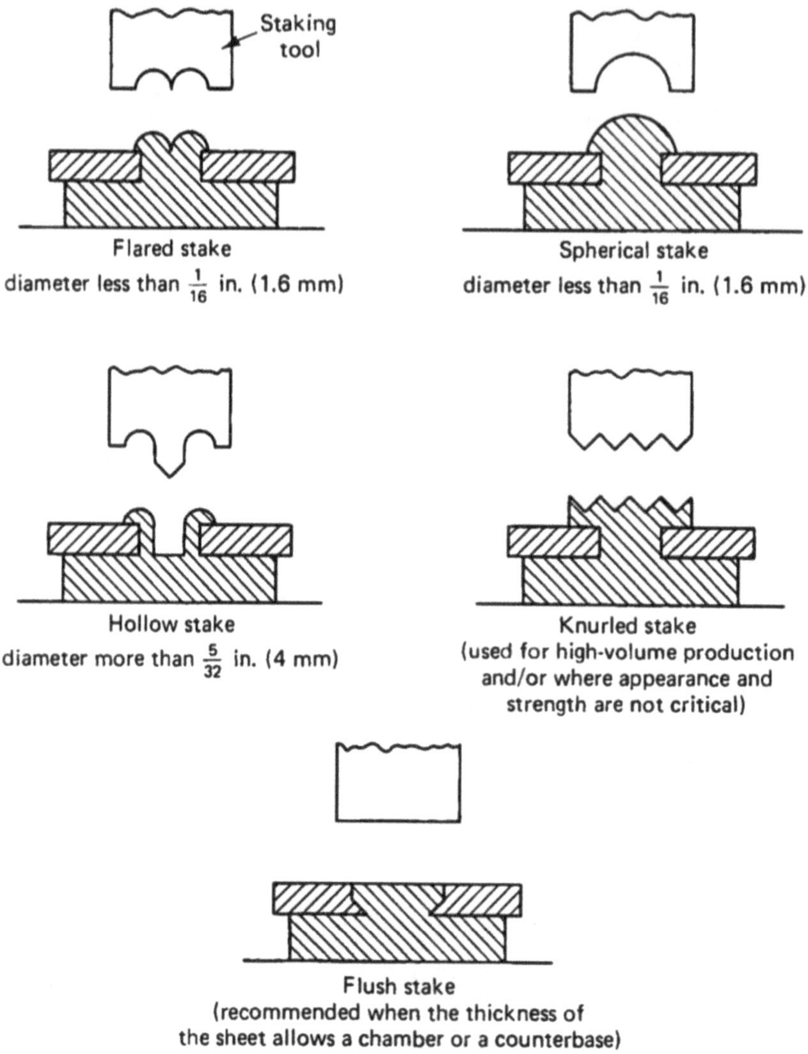

FIGURE 8.20 Ultrasonic staking.

8.6 RECYCLING OF PLASTICS

As previously explained in Chapter 1, plastics recycling is the use of plastics waste as a raw material for useful products that may or may not be similar to the original. Unlike the case of metals, recycling plastics is more challenging because of their low density, low value, and the numerous technical hurdles to overcome when recycling this class of materials. Nevertheless, it is an urgent necessity since the vast majority of plastics are nonbiodegradable. In fact, it is a part of global efforts to reduce plastics in the waste stream because about 8 million tons of waste plastics are dumped into the ocean every year.

The two main problems encountered when recycling plastics include contamination and mixing of the plastic material, which can be explained as follows:

Contamination. Recycling of plastics is usually classified as *primary recycling* and *secondary recycling*. Primary recycling is performed on the waste plastic that is collected in the factory, such as defective products or trimmed excess material, before it finds its way out of the factory and, therefore, before it gets contaminated with dirt or grease. It is usually clean enough to be used for producing the same original product. Secondary recycling is carried out on waste plastics obtained from recycling community collection centers, where they would have been subjected to various sources of contamination. Such materials need some initial cleaning or treatment in order to get rid of some of the contaminants before the recycling process is performed. It is almost always, however, that these plastics cannot be used in producing the original product, because their properties undergo degrading as a result of contamination. Consequently, other products for which demand in the market exists and which do not necessitate the use of high-quality raw plastic must be selected instead.

Mixing of Plastic Materials. This is a major problem indeed because different plastics having different chemical compositions and structural formulas look the same and cannot visually be identified. It is, therefore, easy for two kinds of plastics to be mixed by mistake and dealt with as only one and the same kind. However, unfortunately, when these

Plastic identification code	Type of plastic polymer	Properties	Common packaging applications
01 PET	Polyethylene terephthalate (PET, PETE)	Clarity, strength, toughness, barrier to gas and moisture	Soft drink, water and salad dressing bottles; peanut butter and jam jars; small consumer electronics
02 PE-HD	High-density polyethylene (HDPE)	Stiffness, strength, toughness, resistance to moisture, permeability to gas	Water pipes, hula hoop rings, five gallon buckets, milk, juice and water bottles; grocery bags, some shampoo/toiletry bottles
03 PVC	Polyvinyl chloride (PVC)	Versatility, ease of blending, strength, toughness	Blister packaging for non-food items; cling films for non-food use. May be used for food packaging with the addition of the plasticizers needed to make natively rigid PVC flexible. Non-packaging uses are electrical cable insulation; rigid piping; vinyl records
04 PE-LD	Low-density polyethylene (LDPE)	Ease of processing, strength, toughness, flexibility, ease of sealing, barrier to moisture	Frozen food bags; squeezable bottles, e.g. honey, mustard; cling films; flexible container lids
05 PP	Polypropylene (PP)	Strength, toughness, resistance to heat, chemicals, grease and oil, versatile, barrier to moisture	Reusable microwaveable ware; kitchenware; yogurt containers; margarine tubs; microwaveable disposable take-away containers; disposable cups; soft drink bottle caps; plates.
06 PS	Polystyrene (PS)	Versatility, clarity, easily formed	Egg cartons; packing peanuts; disposable cups, plates, trays and cutlery; disposable take-away containers
07 O	Other (often polycarbonate or ABS)	Dependent on polymers or combination of polymers	Beverage bottles; baby milk bottles. Non-packaging uses for polycarbonate: compact discs; "unbreakable" glazing; electronic apparatus housings; lenses including sunglasses, prescription glasses, automotive headlamps, riot shields, instrument panels

FIGURE 8.21 Plastic identification code (table).

two kinds of plastics are melted together, they tend to phase-separate, like oil and water, resulting in structural weakness in the plastic blend. Such a behavior is evidently observed when polypropylene and polyethylene are mixed together. For this reason, the PIC (Plastic Identification Code) was introduced by the Society of the Plastics Industry, Inc. to provide a uniform system for the identification of various polymer types and accordingly help separate various plastics for recycling. According to the PIC, each of the commonly used seven groups of plastics can be identified by a number and a letter abbreviation. The number appears inside a three-chasing-arrow recycling symbol indicating that the plastic can be recycled into new products. We can identify the plastic types based on the codes usually found at the base or at the side of the plastic products. The PICs for each of the widely used plastics, as well as the properties and applications, are shown in Figure 8.21.

In general, the procedure for recycling plastic involves separating the plastic waste by color before it is recycled. Next, the plastic recyclables are shredded into strips, and these strips are subsequently subjected to processes to eliminate contaminants and impurities such as paper labels. This cleaned material can be melted and extruded into pellets, which are then used to manufacture other products.

In some cases, further treatments such as melt filtering are used to produce food-contact-approved recycled polyethylene terephthalate (PET). There are some other applications for recycled PET, which include polyester fibers to create fabrics for use in the clothing industry, injection-molded engineering components, and building materials. Another plastic that is successfully recycled is HDPE. In fact, recycled plastic gained widespread acceptance and is usually in demand. It is used for manufacturing plastic lumber, tables, benches and outdoor furniture, trash receptacles, and other durable goods.

REVIEW QUESTIONS

1. What are plastics and why are they called *polymers?*
2. What is a *monomer?*
3. Are all polymers artificial? Give examples.
4. Why is the strength of polymers lower than that of metals?
5. Why is the electrical conductivity of polymers lower than that of metals?
6. When did polymers start to gain widespread application and why?
7. How are polymers classified based on their temperature characteristics?
8. What is meant by chemical families of polymers? Give examples.
9. What are the main characteristics of a thermoplastic polymer?
10. Does a thermoplastic polymer have a fixed melting temperature? Why?
11. What is meant by shaping memory?
12. What are the main characteristics of a thermosetting polymer?
13. How do molecules of a thermosetting polymer differ from those of a thermoplastic polymer?
14. Compare the properties of plastics with those of metals. How do the differences affect the design of plastic products?
15. How can we have different polymers starting from the same monomer?
16. List four polymers that belong to the ethenic group. Discuss their properties and applications.
17. What are the main applications of polyacetals?
18. What is cellophane and how is it produced?
19. What is the major disadvantage of cellulose nitrate?
20. What are the major applications of cellulose acetate?
21. What is the chief limitation of nylons?
22. What are the major characteristics of phenolics?
23. How are polyimides manufactured?

24. List the common applications for epoxies.
25. Discuss the properties of polyurethanes and list some of their applications.
26. What property characterizes silicones? Suggest suitable applications to make use of that property.
27. Explain how natural rubber is processed.
28. Why are additives compounded with polymers?
29. List some fillers. Why are they added to polymers?
30. What happens when too much filler is added?
31. How does the addition of plasticizers affect the properties of a polymer?
32. List some of the lubricants used when processing polymers.
33. What are the mechanisms for coloring polymers?
34. Are all polymers cast in the same manner?
35. What are the design features of parts produced by blow molding?
36. Using sketches, explain the injection molding process.
37. What is the chief limitation of injection molding?
38. What kinds of polymers are usually processed by compression molding?
39. List some advantages of the compression molding process.
40. What is the main difference between compression molding and transfer molding?
41. Explain briefly the operating principles of rotational molding.
42. List examples of plastic products that are manufactured by extrusion.
43. What is the coextrusion process? Why is it used in industry?
44. What are the design features of parts produced by thermoforming? Give examples.
45. What are the products of the calendering process?
46. What is the major problem experienced when machining plastics?
47. Using sketches, explain the process of hot-plate joining.
48. Describe thermal staking.
49. Explain how ultrasonics are employed in welding and assembling plastic parts.
50. Do all plastics render themselves suitable for ultrasonic welding? Explain.
51. What are the basic components of ultrasonic welding equipment?
52. Using sketches, show some designs of ultrasonic-welded joints. List the characteristics of each.

DESIGN PROJECTS

1. The current products of a company involve different fruit preserves in tin cans, each containing 8 ounces (about 250 g). The company uses 250,000 tin cans annually, and each cost 13 cents. Because their machines are almost obsolete and the cost of tin is rising every year, the company is considering replacing the tin cans with plastic containers. Design plastic containers to serve this goal, taking into account the plastic-processing method to be used. Also, make a feasibility study for the project.
2. Design a plastic cup that has a capacity of 8 ounces (about 250 g) of water. Assume the annual production volume is 20,000 pieces.
3. Design a high-quality plastic pitcher that has a capacity of 32 ounces (about 1 kg) of liquid. Assume the annual production volume is 15,000 pieces.
4. Design a wheel for a bicycle so that it can be produced by injection molding instead of sheet metal forming. The diameter is 24 inches (600 mm), and a load of 100 pounds (about 45 kg) is applied, through the axle, at its center. Assume the annual production volume is 100,000 wheels.
5. A trash container that has a capacity of 1 cubic foot (0.027 m^3) is made of sheet metal and can withstand an axial compressive load of 110 pounds (50 kg). Redesign it so that it can be made of plastic. Assume the annual production volume is 20,000 pieces.

9 Characteristics, Fabrication, and Design of Composites

9.1 OVERVIEW

A composite can be defined as a material made up of two (or more) identifiable materials (or phases), combined usually in an ordered fashion to provide specific properties different from and superior to those of the individual materials. Those two materials that constitute any composite are referred to as the matrix and the reinforcement. It is customary to classify composites based on the nature of their matrix materials. Based on that classification, there are three main groups: polymer-matrix, metal-matrix, and ceramic-matrix composites.

Most naturally occurring structural materials such as timber are, in fact, composites. Moreover, the merits of crude forms of composites have been known for centuries; for instance, straw-reinforced clay was reportedly used as a building material in Egypt in 1500 BC. It is only in the past 60 years that composites—and fiber-reinforced polymers in particular—have become important engineering materials. New synthetic high-strength, high-modulus fibers and new resins and matrix materials have elevated fiber-reinforced composites into the material of choice for innovative lightweight, high-strength engineered products. These developments, along with established engineering design criteria and special processing technology, have advanced fiber-reinforced composites close to the realm of a commodity material of construction. In the areas of automobile bodies, recreational boat hulls, bathtubs, and shower stalls, fiberglass-reinforced organic polymer resins have indeed become the material of choice. In more advanced applications, the first completely fiber-reinforced polymeric resin composite aircraft, the Boeing 787, came into existence in the 1980s and is now widely used by various airlines. In the 1990s, some other important applications emerged, such as sailboat masts, skiing gear, and wind turbine blades.

As the utilization of composite materials in functional engineering applications continues to grow, it is important for engineering students and professional engineers to gain a thorough understanding of the characteristics of these materials so as to consider them when selecting materials for various products. It is, therefore, the goal of this chapter to provide a profound knowledge of the characteristics, fabrication, and design of composites. Following is a coverage of the above-mentioned kinds of composites.

9.2 FIBER-REINFORCED POLYMERIC COMPOSITES

Let us now focus on the fiber/resin composite materials composed of higher-strength, higher-modulus fibers embedded in an organic polymer matrix. The integral combination of high-strength, high-modulus fibers, and relatively low-strength, low-rigidity polymer matrices results in some unique engineering materials. Fiber-reinforced polymeric composites (FRPCs) possess the material processing and fabrication properties of polymeric materials yet, due to their fiber reinforcement, can be designed to possess directional stiffness and strength properties comparable (or even superior) to those of metals. These mechanical properties can be achieved at a very light weight. This feature can be illustrated by establishing a map of the specific strength (tensile strength/density) versus the specific elastic modulus (tensile modulus of elasticity/density) for various FRPC materials as well as plastics and metals, as shown in Figure 9.1. As clearly evident, the commodity elastomers, plastics, and metals occupy only a very small portion of the lower left-hand side of this structural materials map, while reinforcement fibers and FRPCs occupy the outer regions. In fact,

DOI: 10.1201/9781003537106-9

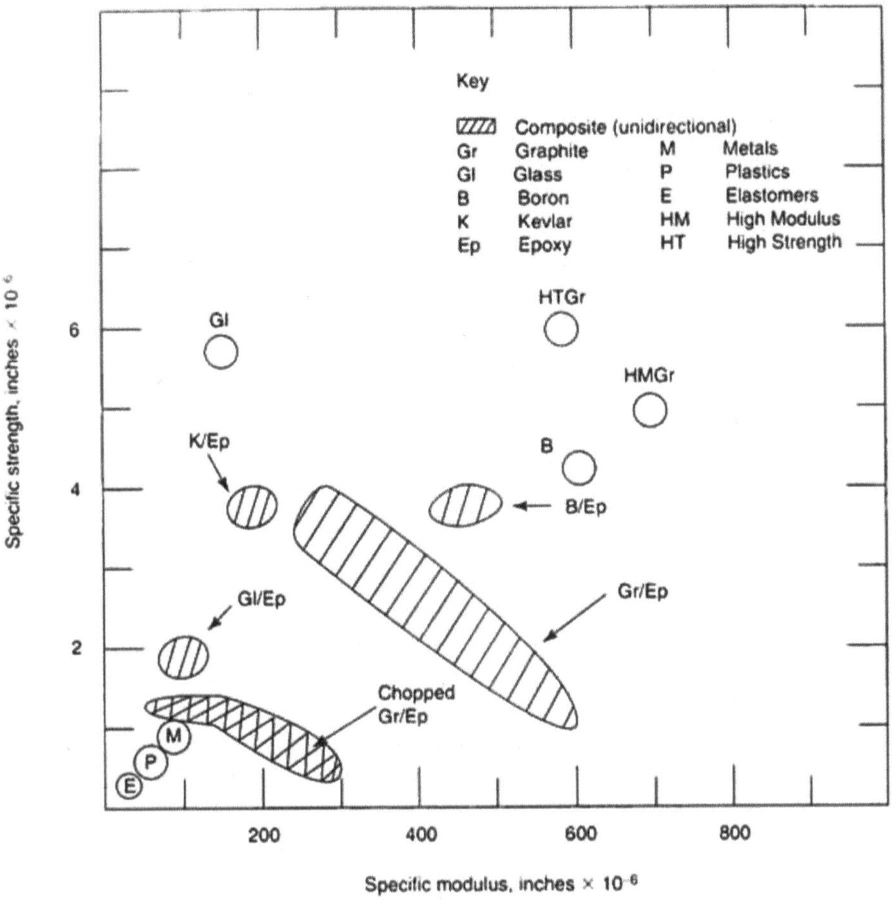

FIGURE 9.1 Specific strengths versus specific elastic moduli of materials.

FRPCs can have specific strengths and moduli up to six times those of common structural materials. In other words, for a given weight, FRPCs far outperform other engineering materials in their strength and stiffness. This property makes FRPCs unique among engineering structural materials and opens new horizons for novel engineering designs and applications. An example of an impressive achievement is the design, construction, and functional deployment of the U.S. Air Force graphite fiber-reinforced epoxy resin composite stealth reconnaissance aircraft.

9.2.1 Matrix Materials for FRPCs

Many types of polymers and resins can be reinforced by fibers to create FRPC materials. Polymer matrices can be classified into two basic categories: thermoplastic and thermosetting. Following is a detailed coverage of each of these two types of matrices.

9.2.1.1 Thermoplastics

Many of the polymers previously discussed can be reinforced with fibers to form composites. The most common types are chopped-fiber-reinforced thermoplastics. These materials can be processed in the same way as plastics that are not reinforced with fibers. Generally, chopped fibers are blended and mixed into a molten mass of the engineering thermoplastics, such as nylons, polycarbonates, and acetals, in a melt-extruder type of plastic-compounding machine. The fiber-containing plastic is extruded into a thin rod and then into molding chips or pellets. Next, these pellets are processed by injection molding or extrusion to produce engineering components, in the same way the unreinforced plastics are processed as explained in Chapter 8.

Continuous fibers of glass, carbon, and polyaramid have also been used with thermoplastic polymer matrices. The concept here is to first coat the thermoplastic polymer onto continuous-fiber yarn by hot melt or a polymer solution-solvent-dip process. These thermoplastic polymer-coated yarns can then be fabricated into shaped structures by a (hot press) matched-die compression molding technique or other techniques for affecting molten-polymer controlled consolidation. Nonetheless, discontinuous chopped-fiber thermoplastic composites are currently much more widely used than continuous fiber-reinforced composites.

The main advantage of thermoplastic-matrix composites is that they can be processed, for the most part, in conventional thermoplastic polymer fabrication equipment. Furthermore, any scrap or off-quality material can be recycled back into the injection molding or extrusion machine. Care must be taken, however, during this thermoplastic processing not to "overwork" these materials in the molten state. Excessive processing of the molten state severely shortens the overall reinforcing fiber length, which can diminish the reinforcement effect of the fiber in the polymer matrix.

9.2.1.2 Thermosetts

Fiber-reinforced composites are traditionally associated with thermosetting polymeric matrix materials such as unsaturated polyester and epoxy resins. In their cured state, thermosetting resins are composed of long polymer chains that are joined together through cross-bridges that link together all the molecules in the resin mass. The final, hardened, tough, and glassy state of the cured resin is the terminal condition of the polymer resin matrix. In this state, the resin serves the all-important role of structurally consolidating, supporting, and cohesively tying together the fiber reinforcement in the composite. However, during initial processing, it is important that the thermosetting resins undergo a gradual liquid-to-solid conversion. It is this feature that renders the thermosetting resins of unsaturated polyester and epoxy type most readily adaptable to fiber-reinforced composite fabrication.

At first, the resin is in a liquid state as it is received from the supplier. It may be more or less fluid depending upon its viscosity (from a consistency of a high-fluidity liquid to a high-viscosity syrup). At this stage, rheological thickeners to increase resin viscosity or reactive diluents to decrease resin viscosity may be added to the resin formulation. Frequently, the curative part of the resin system is much more fluid than the resin part. Here, the viscosity of the final mixture of the resin and the corrective system is low enough to accommodate proper flow in processing. Sometimes, the fluidity of the resin may be lowered by increasing the temperature of the resin upon its application to the fiber. In all, it is important that the viscosity of the liquid resin be adjusted so that it has the proper fluidity to wet, impregnate, and saturate the reinforcing fiber yarns, fabric, or mat. The next consideration is the need to chemically catalyze the resin so that it properly cross-links and cures under the prescribed conditions. It is also necessary to have the catalyzed resin react very slowly at ambient temperature so that it remains fluid while in the process of wetting the reinforcing fibers. This resin-system fluid time is referred to as the pot life or open time of the resin. The nature of the catalyst, the ambient temperature, and the bulk volume of the resin in the resin container control this fluid-time feature. Bear in mind that a bulk of resin being catalyzed is undergoing a heat-generating exothermic reaction. If the bulk volume were too large, heat would not be easily dissipated. The reaction in the fiber/resin dip tank would automatically accelerate, the resin would prematurely cure, or, worse, the heat of the reaction might cause serious fire or noxious fumes. Most often, however, the processing equipment would contain dual-component pumps and a mixing head that continuously meter and mix the two components of the resin system (resin and curative) at the appropriate moment and position to wet the fibers. Historically, phenolics emerged first as the thermosetting matrix for composites that found applications as circuit boards in the electrical industry. Epoxides, unsaturated polyester, and silicones then followed. Epoxides possess unique properties that make them the most important matrix material for high-performance structural FRPCs. In addition to its high strength,

TABLE 9.1
Properties of Important Polymeric Matrix Materials

	Density (g/cm^3)	Maximum Service Temperature, °F (°C)	Tensile Modulus of Elasticity, Ksi (GPa)	Tensile Strength, psi (MPa)	Elongation (%)
Material Thermoplastics					
Nylon	1.14	175 (80)	470 (3.24)	12,400 (85)	90
Polycarbonates	1.2	240 (125)	345 (2.38)	9500 (65)	135
Acetals	1.42	185 (120)	450 (3.1)	10,000 (70)	75
Thermosetts					
Phenolics	1.3	300 (140)	435–580 (3–4)	12,000 (83)	1.4
Epoxides	1.38	350 (176)	360–580 (2.5–4)	12,500 (85)	1.3
Polyester	1.1–1.4	150 (65)	290–580 (2–4)	4500 (30)	1.5
Silicones	1.6	600 (315)	290 (2)	4000 (28)	2
Polyimides	1.46	500 (260)	360 (2.5)	10,500 (72)	8

epoxy resin has low viscosity for wetting the reinforcing fibers, as well as superb dimensional stability during the cross-linking reaction (very low shrinkage), thus reducing the shear stress acting on the matrix/fiber bond. The unsaturated polyester resins are, however, by far the most commonly used thermosetting matrix material for general-purpose composite structures and components. They are used in fiberglass boat hulls, recreational vehicles, and countless commercial and military applications. This is mainly because of their low cost and the ease of using them to fabricate composites without the need for highly skilled labor. Nevertheless, polyester resins are not recommended for high-temperature applications, where silicones and polyimides are the recommended matrix materials. The important engineering properties of various matrix materials are provided in Table 9.1.

9.2.2 Fiber Reinforcement

Generally, reinforcement in FRPCs can be in the form of fibers, whiskers, or particles. In composite materials of the most commercial interest, fibers are the most important and have the most influence on composite properties. These reinforcing fibers, when used in composite fabrication, can take several forms, such as < 0.1-inch (3–4-mm) fiber "whiskers," 0.1–0.3-inch (3–10-mm) chopped fibers, 0.1–2-inch (3–50-mm) nonwoven matted fiber sheets, woven fabric (continuous) fiber with plain weave, and unidirectional/longitudinal (continuous) fiber ribbons. The various fiber reinforcement forms are illustrated in Figure 9.2. While we will discuss the pros, cons, and applications of each kind of commonly used fibers later in this section, let us first discuss why we use fibers and not bars, for example, as reinforcement. The first reason is the superior mechanical properties of fibers as compared with those of bars for the same material. This sounds very strange, but we can explain this phenomenon if we bear in mind the fundamentals of *fracture mechanics*. The smaller is the diameter of the fiber, the lower is the possibility of having microscopic cracks or imperfections in that small area, and the higher would be the strength of the material. This is known as the size effect. The second reason involves the high aspect ratio of a fiber (length-to-diameter ratio), a unique feature that enables any applied load to be transmitted via the matrix material to the fibers. Last but not least, fibers possess a high degree of "flexibility," meaning you can bend a fiber, even if it is made of a brittle material, to a reasonably small radius, something you cannot do with a bar. This last characteristic allows the use of a variety of techniques for fabricating composites. Now, let us try to quantify that concept of flexibility and identify the factors affecting it. As a start, let us consider the fiber to be an elastic beam subjected to a bending moment. Employing our knowledge of the *mechanics of materials*, we can apply the following equation:

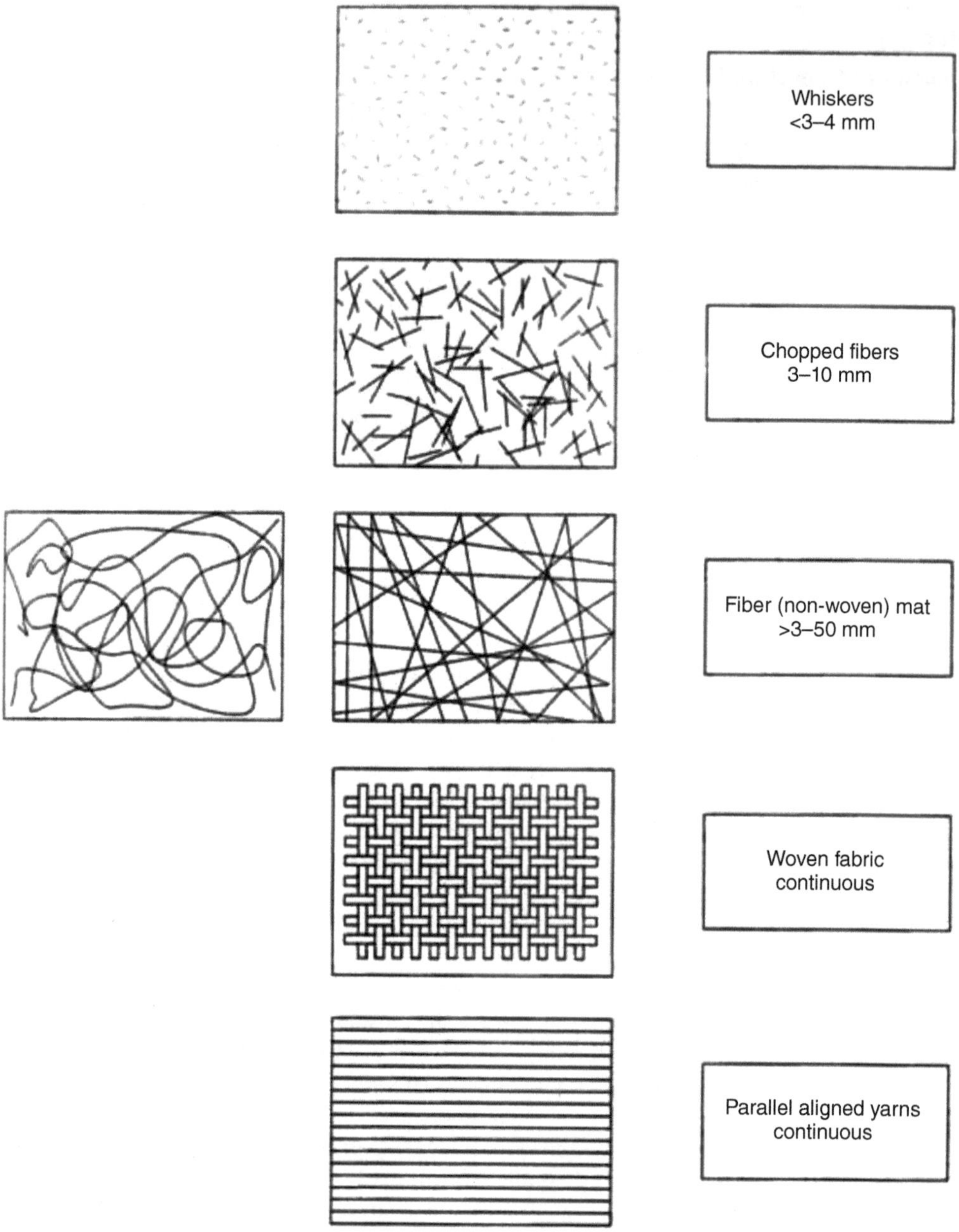

FIGURE 9.2 Comparison of fiber reinforcement forms.

$$\frac{M}{I} = \frac{E}{R}$$

where

M is the bending moment acting on the beam
R is the radius of curvature of the beam due to the bending moment
E is the modulus of elasticity of the material of the beam
I is the moment of inertia of the cross-sectional area of the beam

This equation can be rewritten as

$$MR = EI = E.\left(\pi d^4 / 64\right)$$

assuming the beam to be round (diameter *d*)

As well known, *EI* represents the "flexural rigidity" of the beam.

It is now logical to define flexibility as the inverse of the flexural rigidity and then can be expressed by

$$\text{Flexibility} = 1 / EI = \frac{64}{E.\pi.d^4}$$

It is clear that flexibility is a very sensitive inverse function of the diameter of the fiber. The smaller the diameter, the much higher the flexibility would be, and it can approach that of a metal or a polymer. As a consequence, it is possible to produce very flexible fibers out of brittle materials such as glass, provided that the diameter of the fiber is extremely small. Decreasing the diameter of the fibers has far more effect than selecting a fiber material with a low Young's modulus of elasticity.

Now, it is the time to discuss the different reinforcement materials.

9.2.2.1 Glass Fibers

There are many commercially available kinds of glass fibers, all being silica-based. The ASTM C162 standards provide designations for various kinds. Among these, three kinds having the designations E, C, and S are commonly used as reinforcement in FRPCs. The designation E stands for electrical resistivity with good strength, while C stands for corrosion resistance and S stands for high strength, particularly at high-temperature applications. The E glass fibers are, however, the favorite for reinforcing resins and fabricating composites. This is mainly because they possess good strength and a moderate modulus of elasticity coupled with low cost.

9.2.2.2 Carbon Fibers (Graphite Fibers)

Graphite is a crystalline allotrope of carbon that has a layered planar structure. The carbon atoms are arranged in each layer in a hexagonal lattice and are covalently bonded. This results in a very high packing factor in the plane of the layer, as well as high strength and an extremely high modulus of elasticity in that plane. This is not the case when we consider the direction normal to the plane of the layer, where both the strength and the modulus of elasticity are much lower. This is attributed to the weak secondary bond between the layers. In other words, graphite is highly anisotropic and has superior mechanical properties in the layer of the hexagonal lattice. Accordingly, it is always the goal when fabricating graphite fibers to make the axis of the fibers fall within the plane of that layer and not normal to it. The fabrication technique of graphite fibers involves carburization of organic precursor fibers, such as polyacrylonitrile (PAN), rayon, polyvinyl alcohol, and polyimides, followed by graphitization at high temperatures.

Graphite fibers are used nowadays for fabricating FRPCs that are gaining widespread industrial applications and are not limited just to the aerospace industry. This is due to the recent developments in the graphite fiber industry that enabled the production of graphite fibers with superior properties at a low cost. Examples of the new applications include wind turbine blades, high-pressure pipes for petroleum pipelines, oceangoing large yachts, and drums for the printing industry (particularly for printing on aluminum foils for food packaging). A printing drum made of Gr-epoxy composite material far outperforms that made of steel because it is much lighter, has a lower mass moment of inertia, and can thus be easily controlled to increase or decrease the rotational velocity.

9.2.2.3 Boron Fibers

Boron is a very brittle material, and, therefore, a boron fiber in itself is a composite material that is made by chemical vapor deposition of boron on a fine tungsten wire. While the specific gravity of boron is 2.34, it reaches 2.6 for boron fibers due to the tungsten substrate.

Extreme care must be taken when fabricating boron fibers in order to eliminate possible imperfections and microcracks resulting from the stress concentration at the boron–tungsten interface. This results in the high production cost of boron fibers, something that limits the spreading of their applications. They are used, however, for fabricating FRPCs for U.S. military aircraft and the space shuttle.

9.2.2.4 Aramid Fibers

Aramid fibers are organic, highly crystalline resin fibers. The most famous commercial name for aramid fibers is Kevlar®, which is a trademark of Du Pont Company. There are several Kevlar types of fibers; some have exceptionally high modulus of elasticity (such as K49 and K149), while others possess high strain to failure (K29 and K119). Kevlar 29 is usually used as reinforcement for cables and fabrics, as well as bulletproof vests, while Kevlar 49 is used for reinforcing polymeric matrices of epoxy and polyester for aerospace, marine, and other applications. They all, however, suffer from poor performance under compressive loading and sensitivity to ultraviolet light that causes their mechanical properties to deteriorate.

9.2.2.5 Whiskers

Whiskers are monocrystalline short fibers having extremely small diameters typically on the order of a few microns and a short length of a few millimeters. That can result in very high aspect ratios that can sometimes reach the value of 1000 but are usually around 100. This characteristic is very good for transmitting the load via the matrix to the reinforcement. Whiskers are usually made of silicon carbide (SiC) and can possess tensile strength and modulus of elasticity as high as 8.4 and 581 GPa, respectively.

Table 9.2 provides the important engineering properties of various reinforcement materials that are used in fabricating FRPCs.

TABLE 9.2
Properties of Important Reinforcement Fibers

Material	Density (g/cm^3)	Tensile Modulus of Elasticity, Ksi (GPa)	Tensile Strength, Ksi (GPa)	Elongation (%)
E glass	2.49	10,500 (72.3)	500 (3.445)	4.8
S-2 glass	2.49	12,600 (86.9)	624 (4.3)	4.8
PAN CT-300 carbon	1.76	33,060 (228)	464 (3.2)	1.4
Boron (fibers)	2.7	57,000 (393)	450 (3.1)	0.8
Kevlar 49	1.45	18,125 (125)	406 (2.8)	2.8

9.2.3 Forms of FRPCs and Fabrication Techniques

9.2.3.1 Discontinuous Fiber Reinforcement

The reaction injection molding (RIM) process involves bringing together two components of a thermosetting polymeric resin system in a mixing head and injecting the reacting mixture into a closed mold before the reaction is complete, as illustrated in Figure 9.3. The resin system then cures in the mold at a relatively low pressure of 50 psi (345 KPa). The timing of the curing reaction is very important because the reaction must occur at the moment the mold cavity is filled. Close process control is, therefore, required. Because the process involves low-viscosity intermediates, complex parts can be fabricated using the RIM method.

Reinforcement (glass fibers, or flakes) can be added to one of the resin components prior to mixing if increased flexural modulus, thermal stability, and, in some cases, a special surface finish are desired in the final molded product. This process, reinforced reaction injection molding (RRIM), is shown in Figure 9.4.

Structural reaction injection molding (SRIM) and resin transfer molding (RTM) are similar to RRIM, except that the reinforcement is placed directly into the mold prior to the injection of the resin. In SRIM, the reinforcement is typically a *preform* of reinforcement fibers or mats of nonwoven fibers. In RTM, as shown in Figure 9.5, a catalyzed resin is pumped directly into the mold cavity containing the reinforcement. The resin system is such that it cures without heat. The advantage of RTM is that, because no mixing head is involved, a relatively low investment is needed for equipment and tooling. Furthermore, large FRPC parts, as well as parts containing inserts and cores, can be fabricated using the RTM process. RIM, RRIM, SRIM, and RTM processes are widely used in the automotive and aerospace industries.

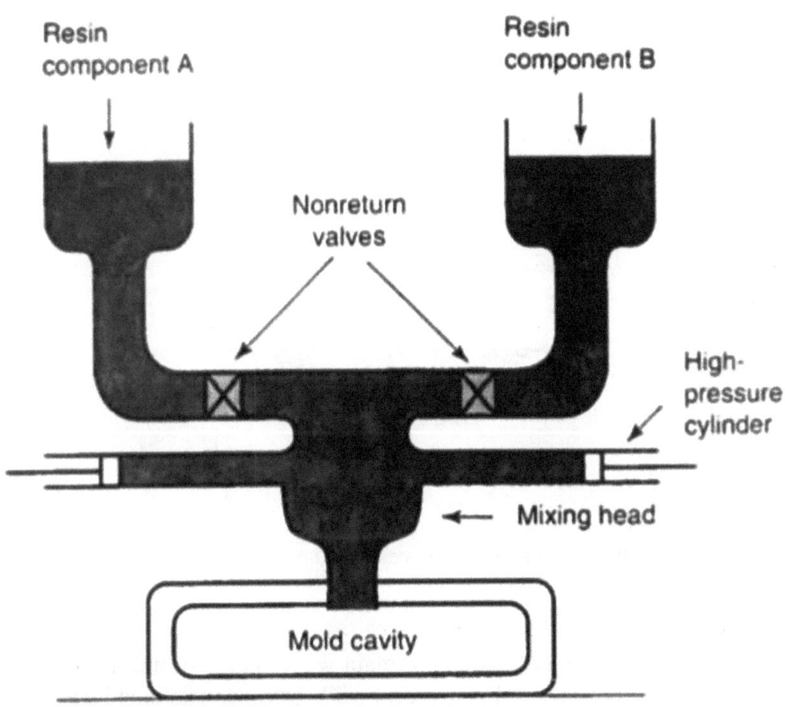

FIGURE 9.3 The reaction injection molding (RIM) process.

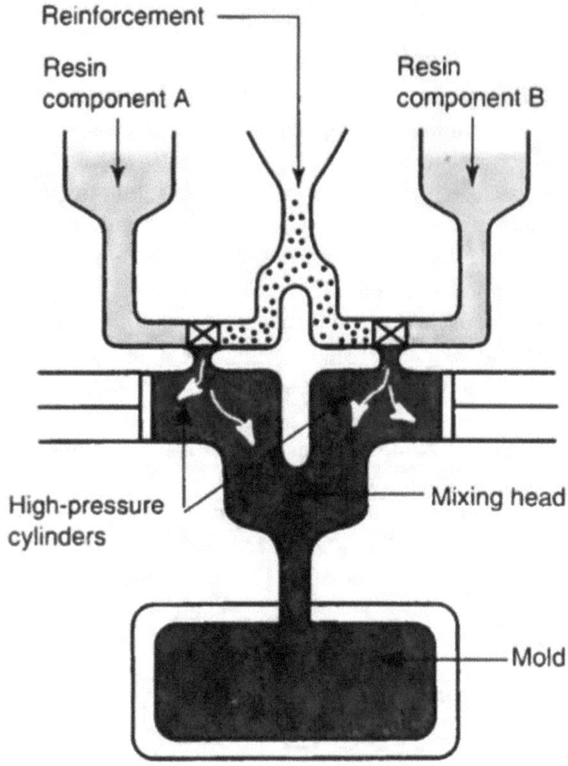

FIGURE 9.4 The reinforced reaction injection molding (RRIM) process.

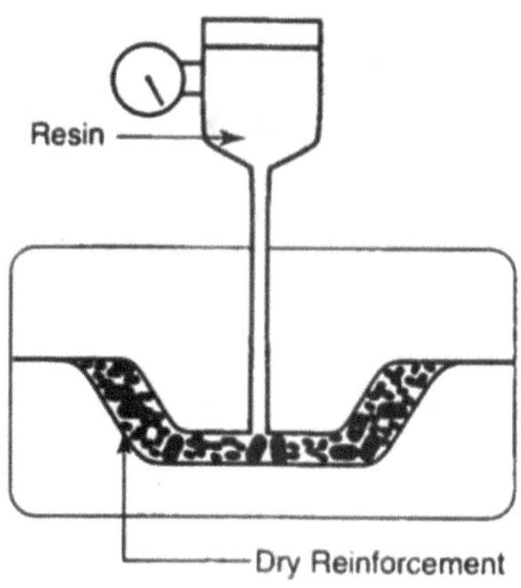

FIGURE 9.5 The resin transfer molding (RTM) process.

9.2.3.2 Wet Lay-Up and Vacuum Bagging

Embedding plies of glass, carbon, or polyaramid plain-weave fabric or fibrous mat into an uncured resin and allowing the liquid resin to solidify (cure) while being constrained by a mold or form is a common processing technique used in the pleasure boat building industry. A typical arrangement of the plies used in this technique, called the *wet lay-up* process, is shown in Figure 9.6. A recent automated version of this process is *automated fiber placement*, which involves placing the resin-impregnated fiber tows onto a mandrel using a numerically controlled head.

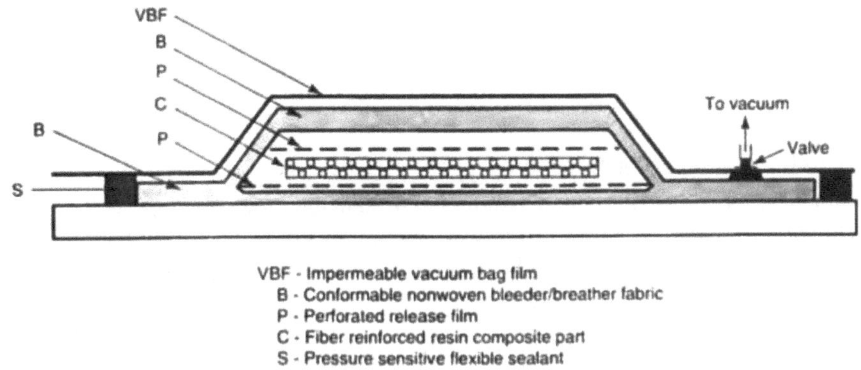

FIGURE 9.6 Arrangement of plies in wet lay-up assembly.

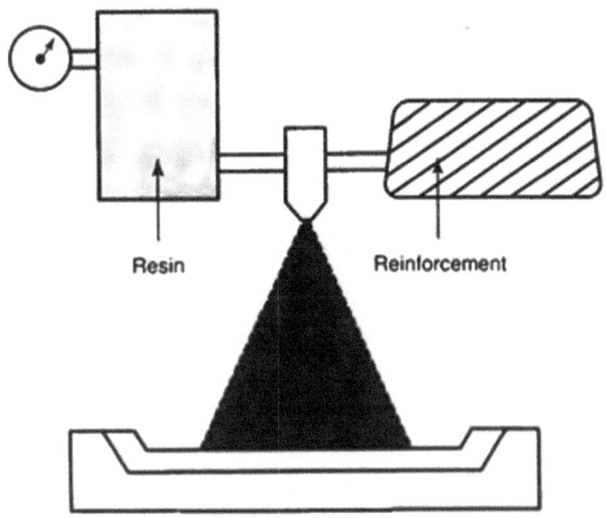

FIGURE 9.7 The spray-up process.

Related to this wet lay-up process is the *vacuum bagging* method of fabricating composite parts and shapes. The principle of vacuum bagging is quite simple. The shape to be fabricated is prepared by a room temperature wet lay-up procedure as just described. The parts to be fabricated are usually assembled over a form or shape of the desired (complex and/or contoured) part. The assembly, like the lay-up arrangement shown in Figure 9.7, is then placed in an airtight disposable plastic "bag" fitted with a vacuum tube fitting or stem. If the air is sealed off and then evacuated from it, the bag will automatically close in on the wet laid-up plies of fiber and liquid (uncured) resin and consolidate these plies by the action of atmospheric pressure. This composite assembly is then allowed to solidify (cure) at room or elevated temperature. After this cure time, the vacuum bag, bleeder ply, and resin-absorber material are removed from the assembly and discarded, leaving the fabricated composite part ready for subsequent finishing or treatment.

A variation of the wet lay-up method is the *spray-up* process, where a spray gun simultaneously sprays catalyzed resin and chops continuous glass yarn into specific lengths. As shown in Figure 9.7, chopped fibers enter the spray nozzle of the spray gun, and the materials are mixed together and sprayed onto an open-cavity mold. The mold is usually faced with a smooth coating of already cured resin called *gel-coat* or thermoplastic shell. This forms the outer surface of the structure being fabricated. When the sprayed-on fiber-reinforced resin cures, the part is removed from the mold. The laminar structure formed is composed of an aesthetically acceptable or otherwise finished outer skin. Adhered to and backing up this skin is the cured fiber-reinforced resin. Open-mold processing of this type is used extensively in bathtub and shower stall manufacturing.

9.2.3.3 Unidirectional-Fiber Resin Prepregs

Fiber-reinforced composite materials are commonly used in the form of a *prepreg*. Prepregs are typically side-by-side aligned fiber yarns that have been impregnated by a B-staged resin matrix (meaning that it has been deliberately partly cured). Unidirectional-fiber composite prepregs are commercially available in the form of rolls, tapes, and sheets. One drawback is that these prepregs must be kept frozen, below 32°F (0°C), during shipping and storage before use. They also have a relatively short shelf life. If not properly stored, the B-stage resin will cure slowly at room temperature, and their function will be destroyed.

Prepreg material is used to fabricate structures by plying together lay-ups of these resin-impregnated unidirectional fibers. The lay-ups can be designed to have different desired mechanical properties depending upon the geometrical arrangement or assembly of the reinforcing fibers in the cured lay-up. Some typical unidirectional-fiber ply arrangements are shown in Figure 9.8. These unidirectional (0°, 0°), cross-ply (0°, 90°), and quasi-isotropic (0°, +45°, 90°, −45°, 0°) plied laminates will

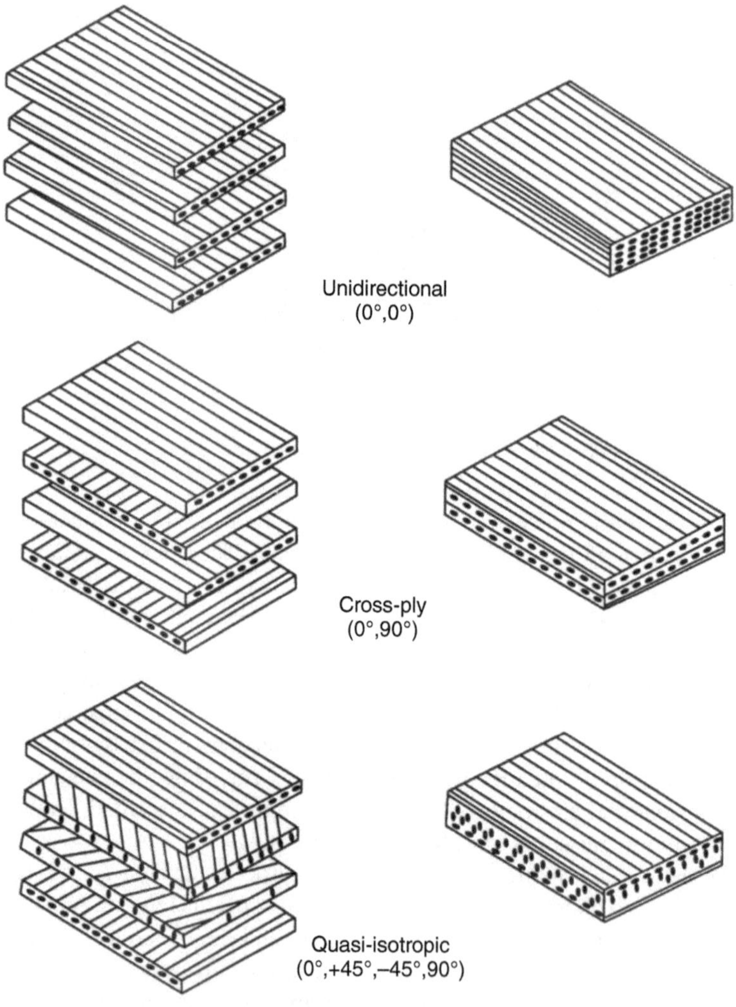

FIGURE 9.8 Various arrangements of unidirectional-fiber ply laminates.

have planar anisotropic properties. Their flexural stiffness will always be higher in the longitudinal direction of the fibers. Other forms of B-stage resin-impregnated fiber forms are commercially available (e.g., fabrics and fibrous mats). The numerous B-stage precomposite forms and types of fiber are all available to the composite materials design engineer in the construction of a fiber-reinforced composite structure.

9.2.3.4 Filament Winding

Filament winding is a fiber-reinforced composite processing procedure commonly used to fabricate tubular (hollow) and cylindrical tank or bottlelike structures. The apparatus used in the filament winding process is schematically shown in Figure 9.9a.

Basically, filamentary yarns are fed off a spool that is mounted on a creel. The yarn is immersed in a catalyzed, but still liquid, resin bath, where the yarn is impregnated with the resin. After squeezing out excess resin, the resin-impregnated yarn is wound onto a rotating mandrel in a controlled and directed manner. A system and control arm guide the yarn back and forth across the mandrel in a predetermined pattern. The computer controls the type of wind pattern and the number of layers of yarn filaments to be laid down on the mandrel surface. Two types of wind patterns are possible: circumferential and helical, as shown in Figure 9.9b. In the circumferential or hoop wind, the yarn is wound in a continuous manner in close proximity to itself. No crossover of the yarn occurs during

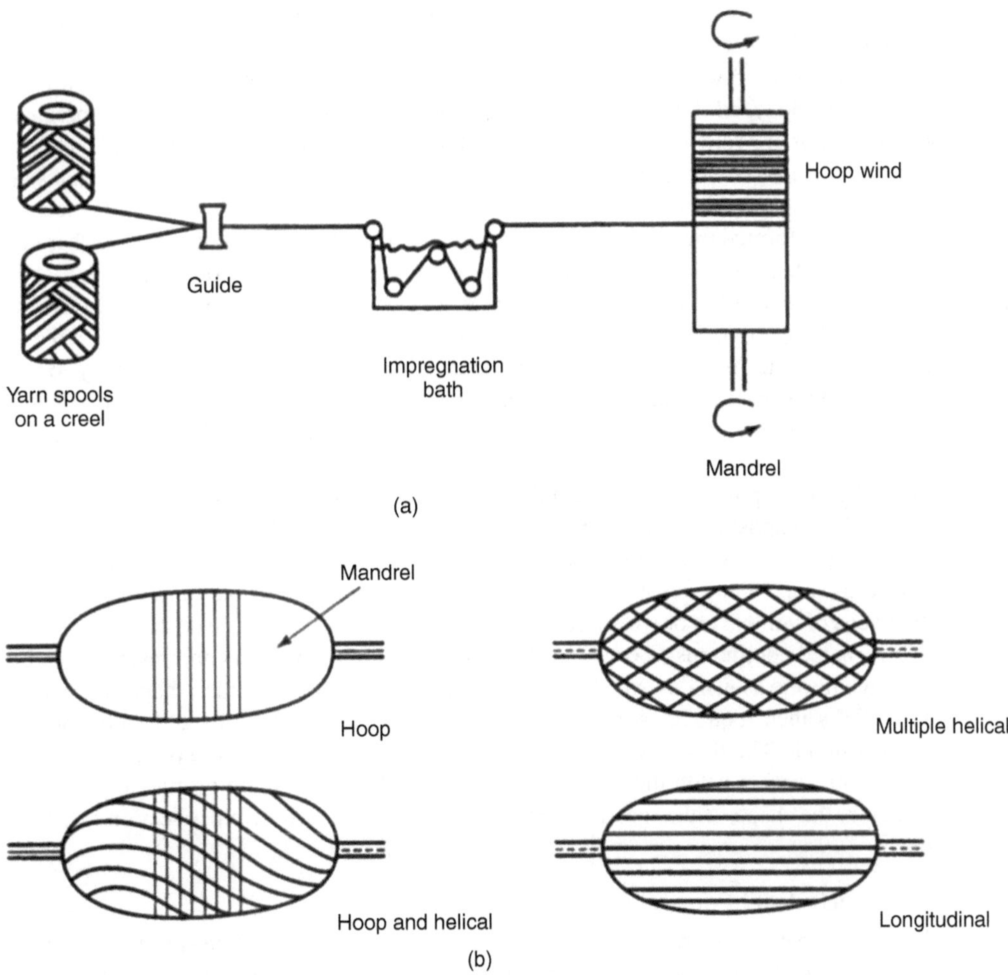

FIGURE 9.9 The filament winding process: (a) apparatus and (b) wind patterns.

the lay-down of a given layer and the pattern, and the lay-down pattern can thus be considered to be at a zero wind angle. The wind proceeds back and forth across the mandrel until the desired number of layers is accomplished. In the helical wind, the yarn is permitted to cross over itself and traverse the length of the mandrel at a prescribed angle (e.g., 0°, 30°, 45°). Again, the wind proceeds back and forth across the surface of the rotating mandrel until the desired number of layers is achieved.

In practice, combinations of hoop and helical winds are usually performed to fabricate a part. The desired lay-down sequence is programmed on the computer. While the desired (yarn) filament-wound resin composite is being formed on the mandrel, heating lamps can be focused on the resin/fiber mass to affect the partial cure of the resin during this lay-down step. Once the desired winding pattern is completed, the mandrel with its wound fiber/resin composite outer surface is left rotating. Rotation and heat-lamp curing continue until the resin material is in a rigid enough state that the rotation can stop and the cylindrical part and mandrel can be removed from the filament-winding machine. Postcuring of the wound composite and mandrel can then be accomplished by placing the assembly in an oven. After final curing, the mandrel is removed from the core of the assembly. To facilitate this, the mandrel form is generally made with a slight taper along its length so that the mandrel can easily be slipped out of an end, leaving the desired filamentary composite cylindrical "shell." The composite part can then be machined (if required) or posttreated to the desired condition.

9.2.3.5 Pultrusion Processing

Pultrusion is a fiber-reinforced resin-processing technique that is readily adaptable to the continuous manufacture of constant cross-sectional linear composite shapes. Rods, I-beams, angles, channel beams, and hollow tubes are commonly produced by pultrusion processing. Pultrusion is a linear-oriented processing method whereby yarns of reinforcing fiber are continuously immersed in and impregnated with a catalyzed fluid resin. As the term *pultrusion* indicates, these resin-impregnated continuous-fiber yarns are concurrently pulled through an elongated heated die designed so that the fiber/resin composite mass exiting the die is sufficiently cured and retains the cross-sectional shape of the die. The apparatus used in the pultrusion process is shown in Figure 9.10a. In practice, prescribed lengths of the formed piece can be cut using an in-line cutoff wheel. Pultrusion is, therefore, adaptable to low-cost, continuous production of constant cross-sectional composite shapes. The process of pultrusion is critically controlled by the resin system used (e.g., unsaturated polyester, epoxy, and vinyl ester resins), the temperature and temperature profile of the heated die, and the rate of pulling through the die. You may wonder why these sections would not be produced by forward extrusion, but when you bear in mind that fibers buckle under compression, it would then be easy to realize that it is not possible to push the fibers out of the die orifice while keeping them straight and parallel.

In the manufacture of pultruded shapes, such as those shown in Figure 9.10b, although the core cross section of the composite is linearly oriented, there is often a need to wrap the outer surface of the composite with a webbing (nonwoven or woven tape) of fibrous material. This serves to consolidate the pultruded shape and gives a much more durable outer surface to the finished part. In this instance, thin veils of nonwoven or woven fabric tapes are fed into the entrance of the die along with the resin-impregnated continuous-fiber yarns. This assembled mass of fibers and resin proceeds to be pulled through the die as just described. The manufacture of hollow pultruded shapes is common, and a special die is then required. A shaped insert or "torpedo" is fitted at the die entrance and extends partway into it. The fluid resin-impregnated fibers entering the die are now constrained by this center-core obstruction. With the proper pulling speed, die temperature profile, and catalyzed resin formulation, the shape of the insert is retained as the desired hollow cross section of the part exits the die.

9.2.3.6 Sandwich-Panel Construction

Structural sandwich-panel construction consists of face sheets made up of fiber-reinforced laminar composite material adhesively bonded to both sides of a core material. This concept is illustrated in

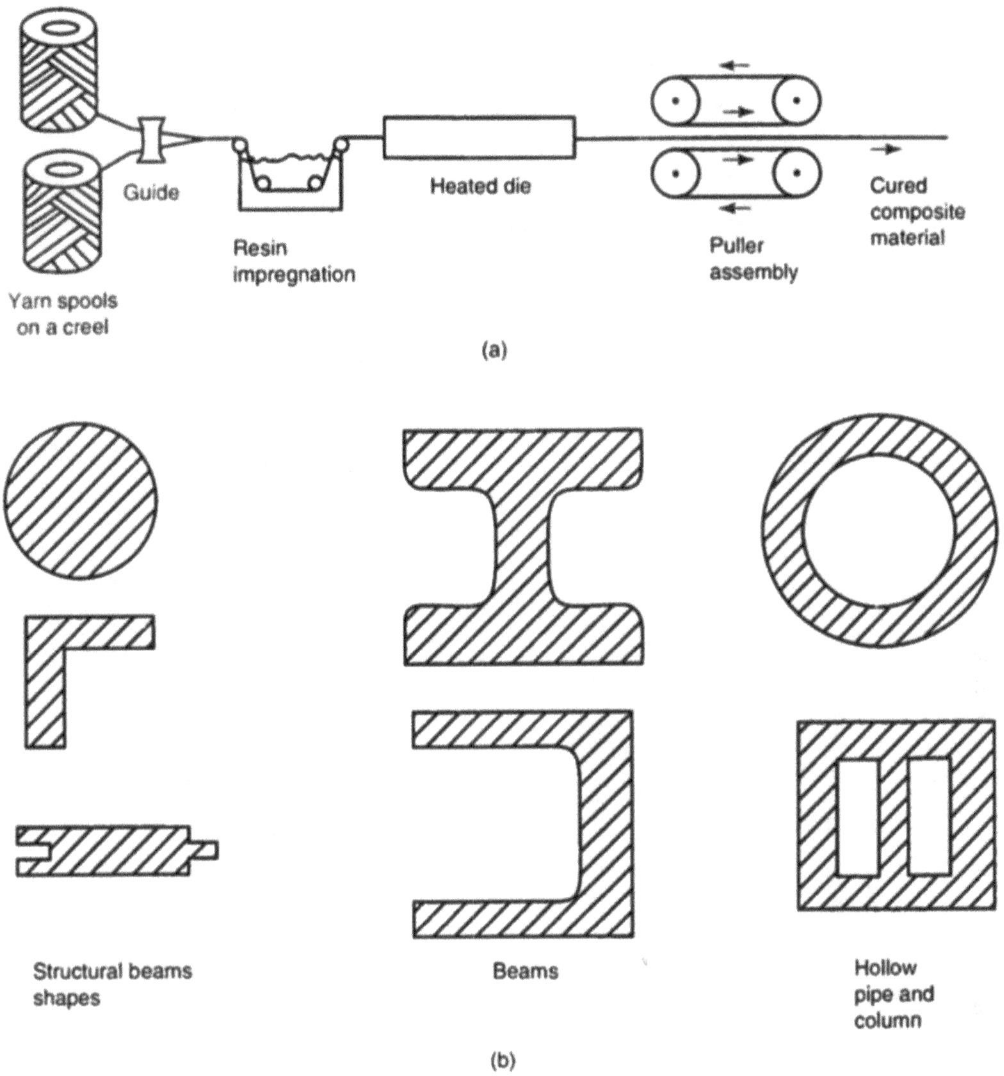

FIGURE 9.10 The pultrusion process: (a) apparatus and (b) cross-sectional designs.

Figure 9.11. The principle behind sandwich construction is that the core material spaces the facings away from the symmetric center of the panel. Therefore, in flexure, the faces or outer skin of the panel are in tension or compression. This construction leads to the reinforcement in the faces, which resists the bending of the panel. The columnar strength of the honeycomb core material then provides the shear and compression strength of this unique panel structure. Above all, the adhesive must be strong and have a high enough shear and peel strength to withstand these shear stresses.

Sandwich construction leads to the use of panels that give the highest stiffness-to-weight ratio of any material design. Sandwich-panel construction is used extensively in aircraft and aerospace applications, where the core materials are generally honeycombed in a geometric shape. Honeycomb cores can be made of thin metal (aluminum or titanium) or fiber-reinforced resin sheets (e.g., thin sheets of resin-impregnated glass, carbon, or polyaramid mat). The manufacture of the honeycomb core by the expansion process is shown in Figure 9.12. Manufacturing honeycomb core involves coating discrete strips of adhesive onto sheets of core material. The specially coated core material is then cured under compression to form a "log" or block of core material. The log must then be cut to

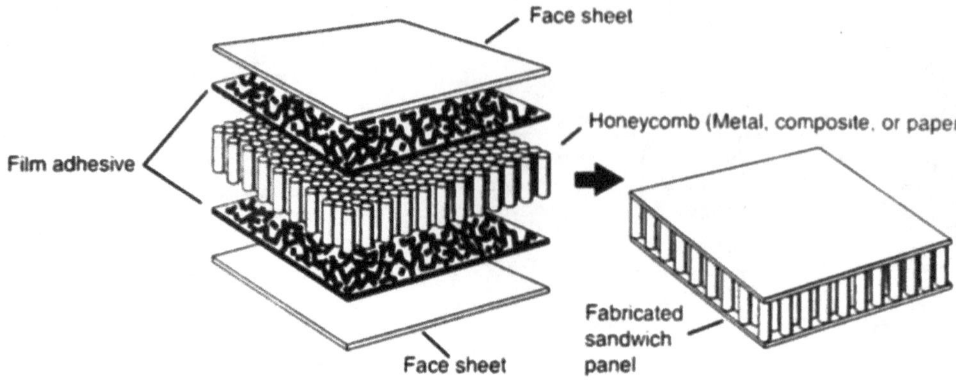

FIGURE 9.11 Structural sandwich-panel construction. (Courtesy: Strong, A. B. *Fundamentals of Composites Manufacturing: Materials, Methods, and Applications*. Dearborn, Michigan: Society of Manufacturing Engineers, 1989.)

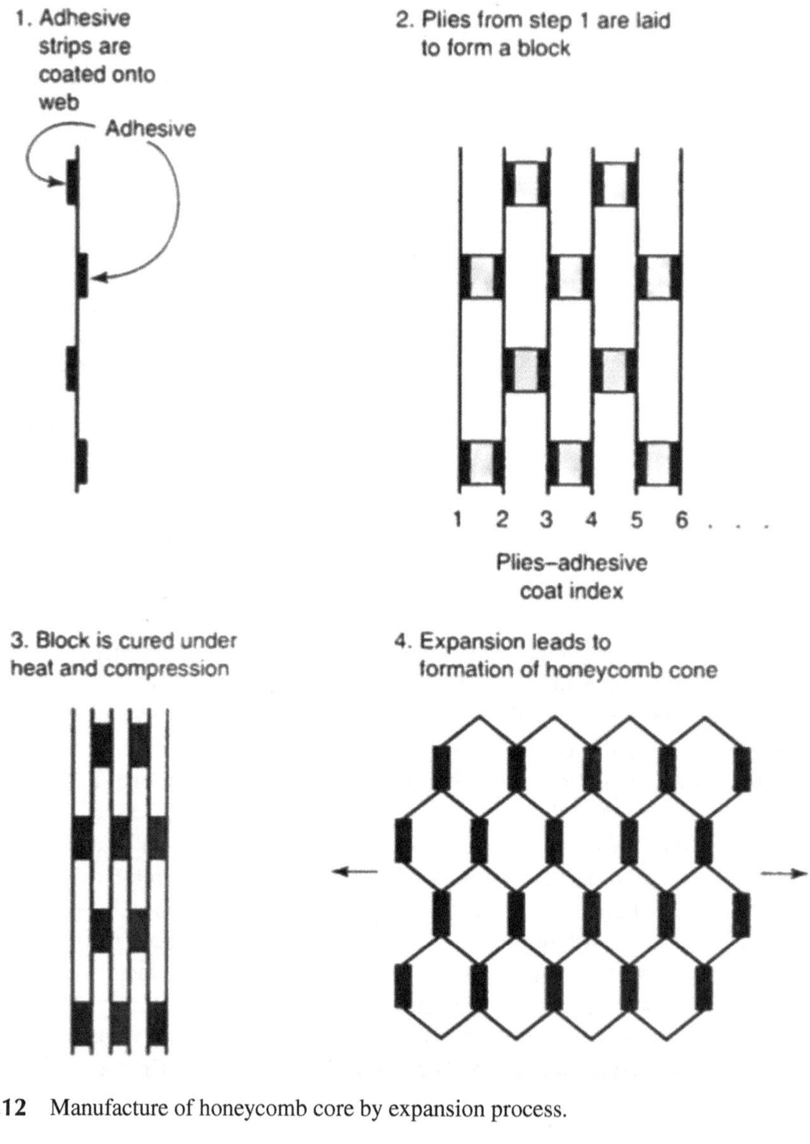

FIGURE 9.12 Manufacture of honeycomb core by expansion process.

the desired core height and subsequently expanded to form the final core material. In some instances, the core material is dipped into a resin solution so that the core structure can be consolidated or stiffened. Another method of making honeycomb is the direct corrugation process. In some less demanding stiffness and compression applications, a rigid foam core material can be used. Rigid foam and Kraft-paper-based honeycomb core panels are often used in truck cargo bed panels and door panels.

9.2.4 Post-Fabrication Processing

Although composite parts and structures are process molded to the near-finished state, machining, drilling, and trimming are often required as the final steps needed for the creation of a final ready-for-use product. There is always the possibility of damaging the composite material in these finishing posttreatments. Delamination, edge fraying, matrix cracking, and crazing leading to weak spots in the composite material structure are all possible. It is, therefore, important to take great care to maintain the structural integrity and surface quality of the composite part being machined.

Composites are machined, cut, and trimmed more easily using processes like grinding or abrasive cutting rather than conventional metal-cutting techniques. Also, the type of fiber reinforcement in the composite dictates the method used. Glass fiber, carbon fiber, and particularly polyaramid fiber composites all require their own procedures. For example, polyaramid-fiber-reinforced composite cannot be successfully drilled or machined using conventional processes. Water jet cutting, laser cutting, or diamond wire cutting are usually used to achieve acceptable quality of the final machined products. In the case of carbon fiber-reinforced composites, the thermal effects due to laser cutting result in extremely unacceptable quality of the finished part, because carbon fibers possess high thermal conductivity. A weakened, charred, heat-damaged zone often surrounds the laser-cut edge. Let us now discuss in some detail the two most common post-fabrication processes. Since Gr-epoxy (carbon-fiber-reinforced epoxy) composite is mainly the material of choice among composites for producing machine components, it will be the focus of our discussion.

9.2.4.1 Grinding of FRPCs

As mentioned before, there is a necessity for the use of precision machining, such as grinding, in order to produce structural machine components made of FRPCs. Successful examples of such applications include rollers for the processing of packaging film, rollers in the printing industry, and power transmission shafts of boats. Needless to say, in all these cases, the outer cylindrical surface of the shaft has to be extremely smooth and have a tight tolerance in order to enable the shaft to be supported in journal bearings. Evidently, that level of dimensional accuracy and surface quality of the finished component can only be achieved by employing the process of cylindrical grinding. A further advantage of that process is that it ensures a high degree of concentricity between the inner and outer surfaces. Unfortunately, the grinding of composites is far more difficult than the grinding of metals. Several problems are encountered when grinding FRPCs, particularly graphite-epoxy, which can be used to fabricate structural machine components. Such problems include charring, degradation, and thermal damage of the ground surface. Excessive noise, fumes, and vibrations occur during the grinding operation, and clogging of the grinding wheel may also take place. The reason is that the nature, physical characteristics, and mechanical properties of composites are different from those of metals. While metals, such as steel, are isotropic, graphite-epoxy is clearly anisotropic, with the value of any property being dependent upon the direction along which it is measured. More importantly, the thermal conductivity of graphite-epoxy is only one-tenth that of steel, while its coefficient of thermal expansion is four times higher than that of steel. As well known, thermal energy is generated when grinding any material, whether it is steel or graphite-epoxy. However, since the thermal conductivity of graphite-epoxy is very poor, that heat cannot be transferred quickly away from the spot being machined and is accordingly retained, resulting in a localized increase in the temperature of the surface of the graphite-epoxy workpiece during the grinding operation. Again, since the coefficient of thermal expansion of epoxy (actually FRPCs) is much higher than that of metals, appreciable radial expansion of the workpiece occurs toward the grinding wheel. The latter would result in the abovementioned problems.

Extensive research was carried out at the University of Massachusetts Dartmouth, with the aim of optimizing the process parameters and selecting an appropriate grinding wheel, in order to eliminate or at least minimize these problems and achieve a trouble-free cylindrical grinding operation. In cylindrical grinding, the relative velocity between the grinding wheel and the workpiece is an indication of the actual direction of cutting and can be determined by the vector addition of the axial feed rate and the peripheral velocity of the grinding wheel. It was found that, in order to minimize the heat generated and thus to eliminate the problems encountered during the process, the direction of cutting should be normal to the direction of the reinforcing fibers in the outer layer. The feed rate and the peripheral should accordingly be adjusted to achieve that goal. The type of grinding wheel used is another factor influencing the process. A wheel with a softer grade (such as AZ46I8V32A according to the American Standard Marking System) is superior to grinding wheels with harder grades. The reason is that, when the abrasive particles of the wheel plow into the composite during the grinding operation, they would be preferentially pulled off, rather than rupture the surface, whenever the graphite fibers come in their way. This is, in effect, a continuous process of self-dressing of the grinding wheel, resulting in better surface finish of the workpiece. Needless to say, the physical and mechanical properties that are dependent upon the fiber-to-matrix ratio have a marked effect on the quality of the ground surface. Best results have been obtained when the fiber content was 70 percent.

9.2.4.2 Drilling of FRPCs

FRPCs have gained widespread industrial applications because near-net-shape panels can conveniently be produced to meet specific needs. Nonetheless, further secondary processing is almost always indispensable for the production and assembly of final products. In this respect, drilling is the most commonly used machining process for composites, because of the need for holes for fastening mechanical components and subassemblies. Again, we have to bear in mind that the drilling of FRPCs and graphite-epoxy, in particular, is different from that of metals because of the nature and characteristics of the FRPCs. Some defects such as splintering, delamination, and burning of the composite material around the hole are frequently encountered. Those would lead to poor assembly tolerance, structural performance deterioration, and shorter service life. As was the case when grinding composites, the energy consumed in removing the material reappears as thermal energy and is retained in a narrow zone surrounding the drilled hole because of the low coefficient of thermal conductivity of the composite. This, in turn, results in a noticeable increase in the temperature localized around the hole that is undergoing machining. The resulting temperature is sometimes high enough to cause charring of the epoxy matrix of the composite material. Another detrimental effect of the high temperature generated is that the diameter of the drilled hole is always smaller than the nominal diameter of the twist drill used for producing it. A simple explanation is that the diameter of any drilled hole is definitely identical to that of the drilling tool, but only at the high temperature generated during the drilling operation. After the latter is completed and the drilling tool is withdrawn, the temperature of the localized zone surrounding the hole would then drop, resulting in the shrinking of the diameter of the drilled hole. Consequently, the higher the temperature generated during drilling, the more evident that drilling-induced defect becomes. It is, therefore, important to select the feeds and speeds used so as to minimize the heat generated and consequently the magnitude of that defect. A feed rate of 0.5 inch/min (0.212 mm/s) together with a cutting speed of 50 ft/min (0.254 m/s) is recommended. Flooding the workpiece with a coolant to flush the generated heat away is also advantageous.

In addition to the abovementioned thermal-induced defect, there is a different type of damage at the exit plane, which is usually referred to as thrust-induced damage. In order to understand it, we have to bear in mind that the action of the twist drill on the composite plate is actually similar to that of a pin exerting pressure on a shell, thus creating thrust force and bending moment. During the drilling operation, when the twist drill penetrates and reaches the lowest ply, the thrust coupled with the bending moment would result in delamination and fracture of that lowest ply.

9.2.4.3 Adhesive Bonding

Adhesives are the principal means of joining composite material to themselves as well as to other classes of materials such as metals and plastics. The reasons for this are numerous. More importantly, adhesive bonds are uniquely capable of distributing stress and can easily be joined into contoured shapes. In mechanical joining, hole drilling is necessary, which can result in delamination of the composite and stress concentration around the hole. The transfer of load from one material to another without creating noticeable stress concentration is the ultimate goal when joining two materials. This can be achieved better by adhesive bonding. Adhesives can often be incorporated into the structural laminar shape being fabricated as a one-step manufacturing process. Metal strips, layers, and/or fittings can easily be adhesively "molded" in the manufactured structure during the composite fabrication stage (e.g., wet lay-up, filament winding, RIM, printed injection molding [PRIM]). Adhesive bonding techniques lend themselves to the creation of integrally designed structures as previously described. The various adhesive joint designs were previously discussed in Chapter 4.

Structural adhesives are available in various forms and types. The most common are the *two-package* epoxy resins. These formulated products are very similar to the epoxy matrix resins used to create the fiber-reinforced composite materials themselves. Usually, these two-package products consist of part A, the epoxy resin prepolymer, and part B, the curative (such as amine or polyamide/amine). Fillers, thickeners, reactive diluents, tackifiers, and other processing aids such as silicone compounds to improve the moisture durability of the adhesive are added to the final formulation. These two-part adhesives are mixed just before being applied to the surfaces of the parts to be joined. The assembly is then placed in a compression mold, platen press, or vacuum bagging arrangement, where heat may be applied to consolidate the layers being joined and cure the adhesive. Some one-package paste adhesives also are formulated with a latent curative; the curative reacts only at a high temperature.

Another useful form of adhesive is the *film* adhesive. Film adhesives are used extensively in the aerospace industry. Here, adhesives take the form of sheets. These sheets are malleable and droppable and can be cut using shears to the desired size and shape. These films are then placed between the surfaces to be joined and are cured under consolidation pressure and elevated temperature. Like the one-package adhesives, these film adhesives are formulated with a high-temperature-reacting latent curative. Film adhesives, like the fiber-reinforced epoxy prepregs described earlier, must be stored at low temperatures and kept frozen until ready to use.

Acrylic adhesives are also used for bonding composite materials. Acrylic adhesives having different flexibilities are available. They cure at room temperature by a free-radical polymerization reaction. One feature of acrylic adhesives is that cure can be achieved by first coating the free-radical catalyst on the surfaces to be bonded. This "catalyst-primed" surface can then be stored until it is ready for bonding. An uncatalyzed acrylic adhesive is then coated onto the catalyst-primed surface. The surfaces to be joined are then mated under contact pressure and allowed to cure, undisturbed at room temperature. Acrylic adhesives can produce bonds that are very oil-resistant.

Finally, it is important that the surfaces to be joined are clean and free of oils, greases, and loose surface material layers. This is especially necessary when joining composite materials to metals. Vapor degreasing, followed by a chemically alkaline cleaning bath, is normally used for surface treatment of metals prior to adhesively bonding them to composites.

9.2.4.4 Painting and Coating

Standard coating methods can be used for painting or coating FRPC parts. In all cases, the surface of the composite must be thoroughly prepared before the final coating is applied. Surface cleaning, sanding, abrading, filling in surface blemishes, and a solvent wipe must be carried out before the paint sealer and final paint is applied. Paint sealers and the final paint coating must be dried/cured at temperatures below the cure temperature of the composite part. Drying with infrared heaters can be troublesome as the heat location and temperature cannot be properly controlled using this technique. Epoxy- and polyurethane-based surface coatings are especially useful in the painting of composite parts.

9.2.5 Micromechanics of Composites

Micromechanics enables us to estimate or more appropriately predict the properties of a composite if the properties of each component and how these components are arranged in it are known. Evidently, this is absolutely necessary for the successful design of components and products that are to be made of composites. Both physical and mechanical properties can be estimated with reasonable accuracy, although some restrictions have to be taken into consideration in the latter case. Examples of the procedures used to obtain various properties follow.

9.2.5.1 Density

Let us consider a composite piece that has a mass m_c and a volume v_c. It is easy to realize that the mass of the composite is the sum of the masses of the matrix and fiber reinforcement as follows:

$$m_c = m_m + m_f \tag{9.1}$$

On the other hand, the volume of the composite is the sum of the volumes of the components in the composite, assuming the absence of any voids in that piece. Therefore,

$$v_c = v_m + v_f \tag{9.2}$$

Dividing Equation (9.2) by v_c and denoting the volume fractions of the matrix and the fibers by V_m and V_f, respectively, we have

$$1 = V_m + V_f \tag{9.3}$$

$$\rho_c = \frac{m_c}{v_c} = \frac{\rho_m v_m + \rho_f v_f}{v_c} \text{ or } \rho_c = \rho_m V_m + \rho_f V_F \tag{9.4}$$

In other words, if the densities of the matrix and the fibers, as well as their percentages in the composite, are known, the density of the composite can be estimated. This equation is referred to as the *rule-of-mixtures* and forms the backbone of micromechanics.

EXAMPLE 9.1

An FRPC has 40% by weight glass fibers. If the density of the matrix is 1.2 g/cm^3, and the density of glass is 2.5 g/cm^3, compute the density of the composite assuming that no voids are present.

SOLUTION

The percentage of fiber reinforcement is usually given by volume, which is not the case in this example. Therefore, we have to obtain that percentage of fibers by volume.

Consider 100 g of the composite. Glass fibers would be 40 g and the polymeric matrix would be 60 g.

Now, we can calculate the volume of each component as follows:

Volume of fibers = 40/2.5 = 16 cm^3
Volume of matrix = 60/1.2 = 50 cm^3
Volume of composite = 16 + 50 = 66 cm^3
Volume fraction of the matrix = 50/66 = 0.758
Volume fraction of the fibers = 16/66 = 0.242
Now, apply Equation (9.4):
Density of composite = 0.242 × 2.5 + 0.758 × 1.2 = 1.51 g/cm^3

9.2.5.2 Mechanical Properties

Let us now see the method for predicting the elastic modulus of a composite if the elastic moduli of the components and the volume fractions are known.

Consider a unidirectional composite, as shown in Figure 9.8. Assume that we apply a force P_c in the direction of fibers and also that the fibers and the matrix adhere perfectly. Consequently, each component will undergo the same longitudinal elongation Δscript. Therefore, the strain in the composite and in each component is equal as follows:

$$\in_f = \in_m = \in_c = \frac{\Delta \ell}{\ell}$$

This is called the *isostrain* state. We can obtain the stress in each component, as well as that in the composite, as follows:

$$\sigma_f = E_f \in_c, \sigma_m = E_m \in_c, \text{and} \sigma_c = E_c \in_c$$

Let the total cross-sectional area of the composite be A_c and those of the fibers and the matrix be A_f and A_m, respectively. From simple statics, we have

$$P_c = P_f + P_m$$

Therefore,

$$\sigma_c A_c = E_f \in_c \cdot A_f + E_m \cdot \in_c \cdot A_m$$

Dividing by $A_c \in_c$, we obtain

$$E_c = \frac{\sigma_c}{\in_c} = E_f \cdot \frac{A_f}{A_c} + E_m \cdot \frac{A_m}{A_c}$$

For a given length of the composite piece

$$\frac{A_f}{A_c} = V_f \text{ and } \frac{A_m}{A_c} = V_m$$

then

$$E_c = E_f \cdot V_f + E_m \cdot V_m$$

which is again the rule of mixture Equation (9.4).

A similar expression can be derived for the UTS, for the isostrain state, and is given by

$$(U.T.S)_c = (U.T.S)_f \cdot V_f + (U.T.S)_m \cdot V_m$$

EXAMPLE 9.2

A unidirectional FRPC of glass fiber/epoxy has its volume fraction of fibers equal to 70 percent. The tensile strength of the glass fiber is 1 GPa, and its modulus of elasticity is 70 GPa. On the other hand, the tensile strength of the epoxy matrix is 60 MPa and its modulus of elasticity is 3 GPa. Estimate the strength of the composites in the longitudinal direction.

SOLUTION

First, calculate the failure strain for each component (since they are brittle, assume a linear relationship) as follows:

$$\epsilon_f = (U.T.S)_f / E_f = 1/70 = 0.014$$

$$\epsilon_m = (U.T.S)_m / E_m = 60/3000 = 0.02$$

Therefore, the fibers would fail first, and their failure strain must be the limiting strain for the matrix to calculate a modified strength.

$$\text{Modified } (U.T.S)_m = 0.014 \times 3 = 0.042\,\text{GPa}$$

Applying the rule of mixture, we have

$$(U.T.S)_c = 0.7 \times 1 + 0.3 \times 0.042 = 712.6\ \text{MPa}$$

We have to be careful, however, not to apply the rule of mixture automatically, without checking which component, fiber or matrix, has the lower failure strain. Sometimes, the fibers would fail at a lower strain than the matrix, leaving it to carry all the load alone. Accordingly, the failure strain of the fiber should be the limiting strain for the matrix in that case (and not its failure strain). The strength of the matrix should be modified based on that value of the failure strain of the fiber.

9.2.6 Macromechanics

As we previously mentioned, the individual unidirectional plies (the prepregs) are oriented in such a manner that the resulting composite body would have the desired mechanical characteristics in different directions. In other words, in this macromechanical analysis, we ignore the detailed microstructure nature of each ply and treat the plies as homogeneous orthotropic sheets stacked at specific orientations to get a composite structure with the desired characteristics. That requires the use of the theory of elasticity and the laws of stress and strain transformation, something that is above the level of this text.

9.2.7 Engineering Design with FRPCs

As previously discussed in Chapter 1, it is imperative for the engineer to consider the shape, the materials, and the manufacturing method as three interactive and interrelated factors affecting the design of the product. The advantages and disadvantages of the materials, as well as any limitation on the form resulting from the processing technique, must be recognized and considered. The designer should exploit the remarkable specific strengths and specific elasticity moduli of FRPCs. He or she can choose from a multitude of reinforcing fiber types and fiber geometry arrangements, as well as from a variety of matrix materials. The disadvantages, such as limited service temperature, sensitivity to moisture accompanied by swelling and distortion, and high cost, must be borne in mind in order to narrow down the number of alternative selections. Following are some guidelines:

- Some fiber-reinforced polymer-processing techniques are readily adaptable to the manufacture of specific shapes or forms and cannot therefore be used to produce other shapes. For instance, the pultrusion process can produce only linear composites having a constant cross section. Similarly, filament winding is limited to fabricating tubular and cylindrical tanklike or bottlelike structures, and plying of prepreg lay-ups is suited only for shell-like structures.
- It is always advantageous to make use of the commercially available B-stage FRPC sheets, rather than to create shapes starting with fibers and polymer, thus cutting the cost and reducing the production time.
- Sometimes it is essential to drastically reduce any deflection of the composite structure when subjected to loading during its service, in order for it to function properly. In such cases, the creative addition of stiffeners should be used. For example, it is not acceptable to allow deformation or deflection of the blade of a wind turbine during operation, and the airfoil cross section of the blade must keep its shape unchanged all the time. Figure 9.13 shows the appropriate stiffeners used to achieve that goal.
- Finally, it is sometimes impossible to obtain the desired shape directly. In such cases, a complicated part can be broken down into smaller simpler parts that are assembled together, at a later stage, by adhesive bonding, bolting, or even snapping together. Accordingly, each of these parts should be designed to facilitate assembly.
- Remember that slots and holes for attaching studs or metal subassemblies would be obtained by machining after the primary fabrication process is completed.

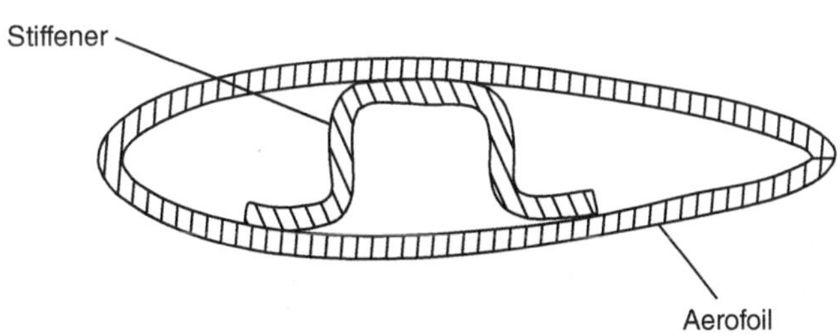

FIGURE 9.13 Appropriate stiffener for a wing turbine blade.

9.3 METAL-MATRIX COMPOSITES

We know from materials science that the strength of heat-treatable aluminum alloys (such as Al-Cu 5.7%) can be dramatically increased when these alloys are subjected to precipitation hardening (PH). This is because the microscopic hard particles that precipitate act as obstacles, which impede the movement of dislocation, thus obstructing deformation. Understandably, such strengthening is accompanied by a decrease in ductility but a negligible increase in the modulus of elasticity. Accordingly, for applications that require the metal alloy to have high stiffness, we can achieve a marked increase in the modulus of elasticity only by incorporating high-modulus fibers in the alloy. Also, since most of the reinforcement fibers are lighter than metal, incorporating them in the alloy would decrease the final density of the composite, and that is accompanied by a marked increase in the specific modulus and specific strength.

9.3.1 Reinforcement in Metal-Matrix Composites

As was the case with the resin-matrix composites, the reinforcement can be particles, short fibers/whiskers, or continuous fibers. Evidently, the use of particle reinforcement would not result in a sensible increase in the modulus of elasticity but would certainly yield an increase in the strength. The particle-reinforced metal-matrix (MM) composites have, however, become popular because they are fairly isotropic and are also inexpensive. The ceramic material of the reinforcing particles can be tungsten carbide, silicon carbide, alumina, titanium carbide, or boron nitride. As you remember, tungsten carbide is employed in the manufacture of cemented-carbide tips (referred to as Widia, from the German words *Wie Diamont*, meaning like diamond) that are used to tip various machining tools. On the other hand, the nature, properties, advantages, and disadvantages of whiskers have been previously covered when discussing the reinforcement of FRPCs. The third type of reinforcement involves continuous fibers, which can be made of carbon (graphite), boron, silicon carbide, or alumina. They have very high aspect ratios that can be as high as 1000, and the diameter of a single fiber may be as low as 3 microns. Such characteristics would certainly ensure the flexibility of the fibers as well as the adequate transfer of the load on an MM composite from the matrix to the fibers.

9.3.2 Metal-Matrix Alloys

Various alloys are used as matrix material in metal-matrix composites, most of them being lightweight alloys. Following is a brief discussion of these alloys, as well as their advantages and limitations.

9.3.2.1 Aluminum Alloys

As previously mentioned, aluminum alloys containing copper are very popular because they are heat-treatable by PH. Examples include Al-Cu-Mg and Al-Zn-Mg-Cu alloys. They are lightweight, possess good resistance to corrosion, and have excellent strength and toughness after heat treatment.

9.3.2.2 Magnesium Alloys

These alloys have very low densities, and that is a real advantage in some applications such as aircraft components. Nevertheless, remember that magnesium has a close-packed hexagonal lattice, which makes these alloys difficult to cold form.

9.3.2.3 Titanium Alloys

Titanium and its alloys are well-suited for the aerospace industry. Not only does titanium have an exceptionally high specific strength and specific modulus of elasticity, but it is also a refractory metal that retains its strength at elevated temperatures and has good corrosion resistance. Limitations for using these alloys include cost and the difficulties encountered in processing them.

9.3.2.4 Copper Alloys

Copper and its alloys have relatively high densities (higher than that of iron). Nevertheless, they have exceptionally high electrical and thermal conductivities.

The major application of copper alloys as matrix material in MM composites is in superconductors where a copper–niobium alloy is employed.

9.3.3 Fabrication Techniques of MM Composites

Some of the commonly used fabrication methods are as follows.

9.3.3.1 Casting

Casting involves blending the matrix material with ceramic particles, then melting them just above the liquidus temperature of the alloy before the composite blend is either cast in ingots or further processed by extrusion or rolling. When processing long continuous fibers, they are passed first through a liquid metal bath to be wet (as was the case in FRPCs), and that is followed by consolidation of bundles of wet fibers through extrusion.

9.3.3.2 Pressure Infiltration

This process involves placing a fibrous preform in a mold, and then forcing a liquid metal into it. Sometimes, slurry casting (see Chapter 7) is used, where the slurry contains short fibers or whiskers are poured into the mold, and then the composite part is further processed to get excess water out and complete consolidation.

9.3.3.3 Employing Powder Metallurgy

This method is suitable when the reinforcement takes the form of particles. A typical example is the fabrication of cemented carbides, where cobalt and tungsten carbide powders are blended (in a special way) and then compacted and subjected to liquid phase sintering.

9.3.3.4 Spray Forming

Ceramic particles are injected into the stream of a molten matrix alloy. A porous preform can be obtained and then subjected to a secondary consolidation process. Needless to say, this process is suitable only when particles are used as reinforcement.

9.3.4 Properties and Applications

The modulus of elasticity of fiber-reinforced metal-matrix composites can be estimated in the longitudinal direction using the rule of mixture. The prediction or estimation of the strength of a fiber-reinforced MM composite is more complicated because of the microstructural changes in the matrix metal, which lead to additional strengthening. We also know that when the reinforcement involves ceramic particles, the strength of the MM composite increases as the particle size of the ceramic particles decreases. The strength, however, increases with an increase in the volume fraction of the reinforcement, but up to a limit. However, a drop in toughness and ductility accompanies such an increase in strength.

Now, let us see what superior properties the MM composites have as compared with those of FRPCs. That would enable us to determine the appropriate applications for MM composites, where FRPCs cannot serve the purpose. The strength and modulus of elasticity are higher because the matrix is a metal alloy and not a polymer. More important, the service temperature of MM composites is higher (and sometimes much higher) than that of FRPCs. It is therefore evident that MM composites are more suitable for high-temperature applications, where FRPCs cannot function. Such applications include ballistic missiles and supersonic military aircraft, where the outer surface is subjected to extremely high temperatures. In the automobile industry, MM composites are used for the manufacture of piston crowns in diesel engines and cylinder liners for engine cylinders in some motorcar models.

Another fairly new application is the use of an alumina-reinforced aluminum composite for making high-voltage cables for overhead transmission lines. A cable would include a core made of such a composite, which is wrapped with aluminum zirconium wires. That replaced the old technology where the cable involved the use of steel wires as the core, which was then wrapped in aluminum wires. Evidently, the cable with the composite core would be much lighter, thus reducing the sag and, consequently, the height and the cost of the steel lattice towers of the transmission line.

9.4 CERAMIC-MATRIX COMPOSITES

As previously discussed, polymers and metals are reinforced with fibers in order to produce composites with strength and/or stiffness higher than those of the original matrix material. On the contrary, ceramics are not reinforced for that reason, but rather to eliminate an inherently major defect: brittleness. They pose superior properties such as high strength and high stiffness even at elevated temperatures, as well as having low density and being chemically inert. Nevertheless, they have very low ductility and accordingly low toughness, as a result of internal and surface flaws that make them susceptible to thermal shock and catastrophic failure during service. Evidently, incorporating fibers in the ceramic matrix can markedly reduce these latter shortcomings, without sacrificing the above-mentioned superior properties.

9.4.1 Fabrication Techniques of Ceramic-Matrix Composites

Following are the important commonly used processing methods of ceramic-matrix composites (CMCs).

9.4.1.1 Cold Pressing and Sintering

This is actually a special case of the conventional powered metallurgy for ceramics. It involves cold pressing of a mixture of the matrix powder and fibers, followed by high-temperature sintering. During the sintering operation, however, cracks may sometimes occur as a result of the considerable shrinkage that takes place.

9.4.1.2 Hot Pressing

As previously explained in Chapter 7, hot pressing involves simultaneous application of pressure and temperature that would produce a pore-free, fine-grained compact.

The fabrication of the composite in this case is carried out through two stages. First, the fiber tow or fiber preform sheet is impregnated with a slurry containing powder of the ceramic matrix (remember the slurry casting process in Chapter 7). This can be achieved by passing the tow or the preform through a slurry tank. The impregnated tow or preform sheet is then wound on a drum and dried. Next, sheets are cut to size, stacked, and finally consolidated in a hot press. Needless to say, these sheets can be arranged in any desired sequence, as was the case with the prepregs of the FRPCs.

9.4.1.3 Infiltration

This process involves infiltrating a preform made of the reinforcing fibers with the matrix material that is usually in the liquid state (it can also be in the solid or gaseous state). Infiltration by a molten intermetallic matrix material under pressure has been reported as successful when using the intermetallic compound Ti-Al.

9.4.2 Properties and Applications of CMCs

Despite the several advantages of ceramic-matrix composites, a main disadvantage, however, involves microcracking of the matrix material. In this respect, the ratio of the modulus of elasticity of the reinforcing fiber to that of the matrix material plays a very important role. Still, ceramic-matrix composites find applications in engineering components subjected to extremely high temperatures during their service. Examples include engine components of gas turbines, burner tubes, flame tubes, and other components for high-temperature furnaces.

9.5 CARBON–CARBON COMPOSITES

Carbon, like ceramics, can withstand extremely high temperatures, but only when the surrounding atmosphere must be inert or at least not an oxidizing one. For that reason, the carbon–carbon composite was originally developed for the space program and for use in military intercontinental ballistic missiles. There are now, however, quite a few high-temperature applications for the carbon–carbon composites in the domestic industry. Nevertheless, because of the high cost of these composites that only the military industry can afford, the new domestic applications are still limited. Examples of such applications include forging and hot-pressing dies and heating elements in furnaces.

9.5.1 Fabrication Techniques of Carbon–Carbon Composites

The fabrication of carbon–carbon composites is complicated, time-consuming, and, therefore, expensive. The reason is that, unlike other ceramic powders, carbon powder cannot be sintered to form the carbon matrix of the composite. Accordingly, other alternative approaches are necessary to create the carbon matrix. Following are some of the fabrication methods.

9.5.1.1 High-Pressure Impregnation Carbonization

Starting with a carbon-fiber preform, the polymeric matrix of graphite FRPC is obtained by impregnation of the fibers with a thermoplastic polymer (or a mixture of hydrocarbons). This process takes place under combined pressure and high temperature, and the polymeric matrix is then converted to carbon by pyrolysis. This procedure is repeated several times until an acceptable level of porosity is achieved.

9.5.1.2 Beginning with a Conventional Polymeric Matrix Composite Technique

The graphite FRPC is created using one of the conventional techniques for fabricating fiber-reinforced thermoset composites. The matrix is then converted to carbon by pyrolysis. Impregnations followed by pyrolysis are repeated until the desired density is achieved.

9.5.1.3 Chemical Vapor Deposition

In this process, a hydrocarbon gas is made to fill the interstices of the graphite-fiber preform. The temperature in then increased above 550°C, where the gas disintegrates and enables the deposition of carbon on the fibers to create the matrix.

9.5.2 Properties and Applications of Carbon–Carbon Composites

As was the case with other composites, the properties of the carbon–carbon composites depend upon the volume fraction of fibers, their distribution and their type, and the level of porosity in the composite. In addition to having good strength and toughness, they possess excellent thermal conductivity and good frictional properties. As previously mentioned, the main application of the carbon–carbon composite is as a high-temperature ablative material for the thermal protection of space vehicles, intercontinental ballistic missiles, and hypersonic aircraft.

REVIEW QUESTIONS

1. What is a composite material?
2. Mention the different types of composites.
3. Explain briefly the nature of FRPCs, and discuss their superior properties.
4. What are the forms of reinforcements in FRPCs?
5. List some of the fibers used as reinforcement in FRPCs.

6. Explain why boron fiber itself is considered a composite.
7. Briefly discuss the various matrix resins for FRPCs, indicating their advantages, disadvantages, and limitations.
8. List the different methods for fabricating FRPCs.
9. Explain the sequence of operations involved in open-mold processing of reinforced polymers.
10. What are the similarities and differences between the conventional extrusion of polymers and the pultrusion of FRPCs?
11. What are the design features of parts manufactured by PRIM?
12. What are the design features of parts manufactured by filament winding?
13. What are the design features of parts produced by pultrusion?
14. What are the design features of parts produced by the lay-up vacuum bagging?
15. Explain how sandwich-panel construction is produced, and list some of the advantages of that construction.
16. List some post-fabrication operations for producing an FRPC product.
17. What should we be careful about when using fiber resin prepegs?
18. How can we predict the properties of a composite? Provide a quantitative equation.
19. What are metal-matrix composites?
20. Since we can enhance the strength of metals by PH, why do we reinforce them with fibers to make metal-matrix composites?
21. Compare the advantages of FRPCs and metal-matrix composites, then suggest applications for the latter.
22. List some matrix materials used in metal-matrix composites.
23. List some reinforcements used in metal-matrix composites.
24. Discuss the advantages and disadvantages of the liquid metal processing method used in manufacturing metal-matrix composites. Compare them with those of other methods of fabricating metal-matrix composites.
25. Why do we reinforce ceramic materials with ceramic fibers to produce ceramic-matrix composites?
26. List some of the methods for the fabrication of ceramic-matrix composites.
27. What are the advantages of ceramic-matrix composites?
28. List some of the important applications of ceramic-matrix composites.
29. Why were the carbon–carbon composites developed?
30. Explain the different manufacturing methods of the carbon–carbon composites.
31. What is the main application for carbon–carbon composites?

PROBLEMS

1. A piece of Gr-epoxy composite was found in the lab, and the volume fraction of fibers was unknown. Using the data on the properties of graphite and epoxy given in the text, determine the volume fraction of the graphite fibers if the density of that piece was found to be 1.65 g/cm^3.
2. Draw a curve indicating the elastic modulus versus the volume fraction of fibers of Gr-epoxy FRPCs.
3. Draw a curve between the tensile strength of Gr-epoxy composites and the volume fraction of fibers. Use data in the text.
4. Estimate the percentage increase or decrease in the UTS, the modulus of elasticity along the fiber direction, and the specific strength and specific modulus when the volume fraction of fibers is increased from 0.5 to 0.7 in a Gr-epoxy composite.
5. An FRPC of fiber glass/epoxy has 60% fibers by volume. The tensile strength of the glass fiber is 3.5 GPa. The tensile strength of the epoxy matrix is 85 MPa. The elastic modulus of glass fibers and epoxy are 73 GPa and 2.5 GPa, respectively. Determine the tensile strength of the composite.

10 Physics of Metal Cutting

Metal cutting can be defined as a process during which the shape and dimensions of a workpiece are changed by removing some of its material in the form of chips. The chips are separated from the workpiece by means of a cutting tool that possesses a very high hardness compared with that of the workpiece, as well as certain geometrical characteristics that depend upon the conditions of the cutting operation. Among all of the manufacturing methods, metal cutting, commonly called *machining*, is perhaps the most important. Forgings and castings are subjected to subsequent machining operations to acquire the precise dimensions and surface finish required. Also, products can sometimes be manufactured by machining stock materials like bars, plates, or structural sections.

Machining comprises a group of operations that involve seven basic chip-producing processes: shaping, turning, milling, drilling, sawing, broaching, and grinding. Although one or more of these metal-removal processes are performed at some stage in the manufacture of the vast majority of industrial products, the basis for all these processes (i.e., the mechanics of metal cutting) is yet not fully or perfectly understood. This is certainly not due to the lack of research but rather is caused by the extreme complexity of the problem. A wide variety of factors contribute to this complexity, including the large plastic strains and high strain rates involved, the heat generated and high rise in temperature during machining, and, finally, the effect of variations in tool geometry and tool material. It seems, therefore, realistic to try to simplify the cutting operation by eliminating as many of the independent variables as possible and making appropriately implicit assumptions if insight into this complicated process is to be gained. In fact, we are going to take this approach in discussing the cutting tools and the mechanics of chip formation. We are going to consider two-dimensional cutting, in which a prismatic, wedge-shaped tool with a straight cutting edge is employed, as shown in Figure 10.1, and the direction of motion of the tool (relative to the workpiece) is perpendicular to its straight cutting edge. In reality, such conditions resemble the case of machining a plate or the edge of a thin tube and are referred to as *orthogonal cutting*.

10.1 CUTTING ANGLES

Figure 10.2 clearly illustrates that the lower surface of the tool, called *the flank*, makes an angle Ψ with the newly machined surface of the workpiece. This *clearance angle* is essential for the

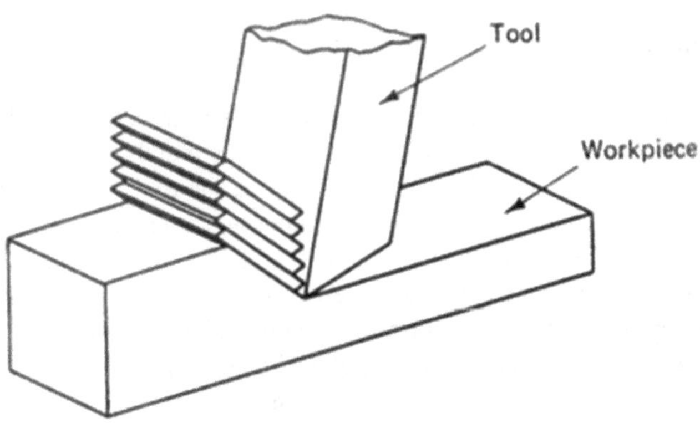

FIGURE 10.1 Two-dimensional cutting using a prismatic wedge-shaped tool.

DOI: 10.1201/9781003537106-10

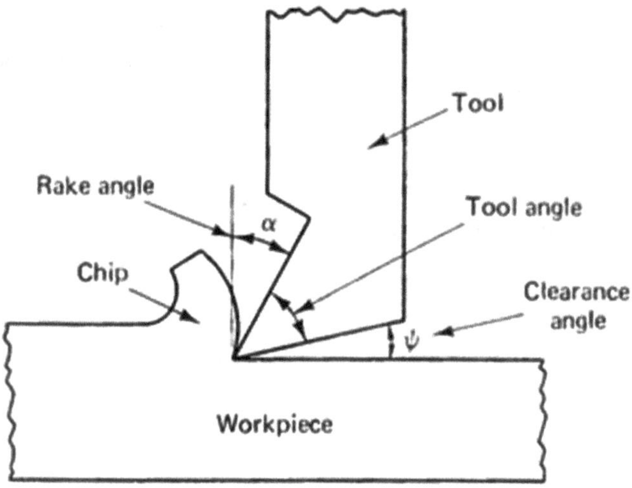

FIGURE 10.2 Tool angles in two-dimensional cutting.

elimination of friction between the flank and the newly machined surface. As can also be seen in Figure 10.2, there is an angle ∝ between the upper surface, or *face* of the tool along which chips flow and the plane perpendicular to the machined surface of the workpiece. It is easy to realize that the angle ∝ indirectly specifies the slope of the tool face. This angle is known as the *rake angle* and is necessary for shoveling the chips formed during machining operations. The resistance to the flow of the removed chips depends mainly upon the value of the rake angle. As a consequence, the quality of the machined surface also depends on the value of the rake angle. In addition to these two angles, there is the *tool angle* (or wedge angle), which is the angle confined between the face and the flank of the tool. Note that the algebraic sum of the rake, tool, and clearance angles is always equal to 90°. Therefore, it is sufficient to define only two of these three angles. In metal-cutting practice, the rake and clearance angles are the ones that are defined.

As you may expect, the recommended values for the rake and clearance angles are dependent upon the nature of the metal-cutting operation and the material of the workpiece to be machined. The choice of proper values for these two angles results in the following gains:

- Improved quality of the machined surface
- A decrease in the energy consumed during the machining operation (most of which is converted into heat)
- Longer tool life as a result of a decrease in the rate of tool wear because the elapsed heat is reduced to a minimum

Let us now consider how the mechanical properties of the workpiece material affect the optimum value of the rake and clearance angles. Generally, soft, ductile metals require tools with larger positive rake angles to allow easy flow of the removed chips on the tool face, as shown in Figure 10.3. In addition, the higher the ductility of the workpiece material, the larger the tool clearance angle that is needed in order to reduce the part of the tool that will sink into the workpiece (i.e., reduce the area of contact between the tool flank and the machined workpiece surface). On the other hand, hard, brittle materials require tools with smaller or even negative rake angles in order to increase the section of the tool subjected to the loading, thus enabling the tool to withstand the high cutting forces that result. Figure 10.4 illustrates tools having zero and negative rake angles required when machining hard, brittle alloys. In this case, the clearance angle is usually taken as smaller than that recommended when machining soft, ductile materials.

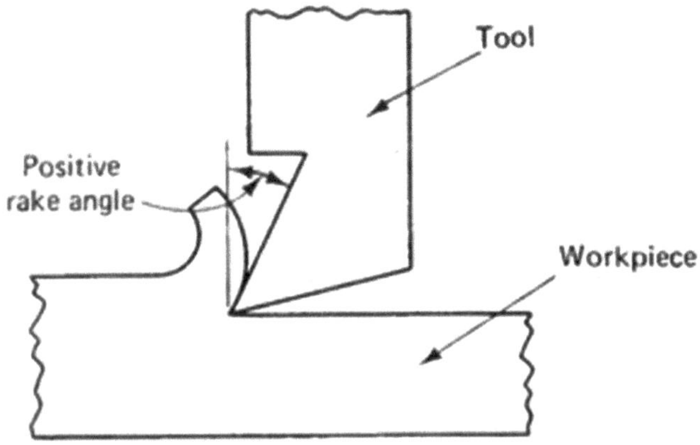

FIGURE 10.3 Positive rake angle required when machining soft, ductile metals.

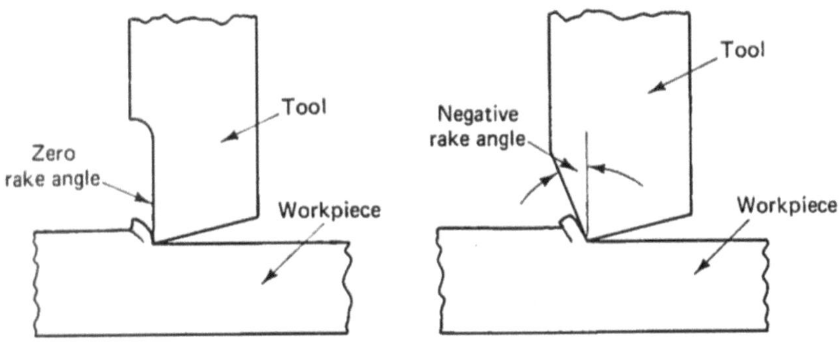

FIGURE 10.4 Zero and negative rake angles required when machining hard, brittle materials.

10.2 CHIP FORMATION

10.2.1 Mechanics of Chip Formation

There was an early attempt by Reuleaux at the beginning of the twentieth century to explain the mechanics of chip formation. He established a theory that gained popularity for many years; it was based on assuming that a crack would be initiated ahead of the cutting edge and would propagate in a fashion similar to that of the splitting of wood fibers, as shown in Figure 10.5. Thanks to modem research that employed high-speed photography and quick-stopping devices capable of freezing the cutting action, it was possible to gain a deeper insight into the process of chip formation. As a result, Reuleaux's theory collapsed and proved to be a misconception; it has been found that the operation of chip formation basically involves the shearing of the workpiece material. Let us see, step-by-step, how that operation takes place.

The stages involved in chip removal are shown in Figure 10.6. When the tool is set at a certain depth of cut (see Figure 10.6a) and is then pushed against the workpiece, the cutting edge of the tool and the face start to penetrate the workpiece material. The surface layer of the material is compressed; then pressure builds up and eventually exceeds the elastic limit of the material. As a result of the intense shear stress along the plane N-N, called the *shear plane*, plastic deformation takes place, and the material of the surface layer has no option but to flow along the face of the tool without being separated from the rest of the workpiece (see Figure 10.6b). With further pushing of the tool, the ultimate shear strength is exceeded, and a little piece of material (a chip) is separated from the workpiece by slipping along the shear plane (see Figure 10.6c). This sequence is repeated as

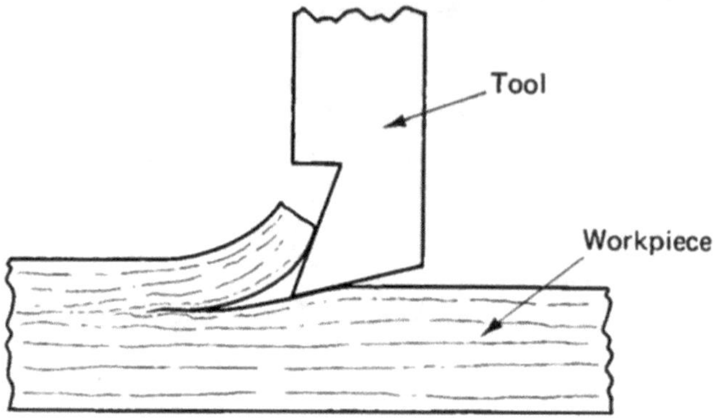

FIGURE 10.5 Reuleaux's misconception of the mechanics of chip removal.

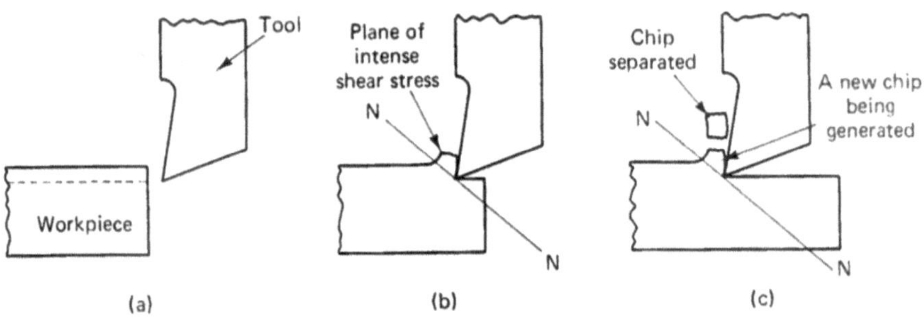

FIGURE 10.6 Stages in chip removal: (a) tool set at a certain depth of cut, (b) workpiece penetration, and (c) chip separation.

long as the tool continues to be pushed against the workpiece, and the second, third, and subsequent chips are accordingly separated.

10.2.2 Types of Chips

The type of chip produced during metal cutting depends upon the following factors:

- The mechanical properties (mainly ductility) of the material being machined
- The geometry of the cutting tool
- The cutting conditions used (e.g., cutting speed) and the cross-sectional area of the chip

Based on these factors, the generated chips may take one of the forms shown in Figure 10.7. Following is a discussion of each type of chip.

10.2.2.1 Continuous Chip

When machining soft, ductile metals such as low-carbon steel, copper, and aluminum at the recommended cutting speeds (which are high), plastic flow predominates over shearing (i.e., plastic flow continues, and shearing of the chip never takes place). Consequently, the chip takes the form of a continuous, twisted ribbon (see Figure 10.7a). Because the energy consumed in plastically deforming the metal is eventually converted into heat, coolants and lubricants must be used to remove the generated heat and to reduce friction between the tool face and the hot, soft chip.

FIGURE 10.7 Types of machining chips: (a) continuous, twisted ribbon; (b) discontinuous, irregular segments; and (c) sheared, short ribbon.

10.2.2.2 Discontinuous Chips

When machining hard, brittle materials such as cast iron or bronze, brittle failure takes place along the shear plane before any tangible plastic flow occurs. Consequently, the chips take the form of discontinuous segments with irregular shapes (see Figure 10.7b). As no plastic deformation is involved, there is no energy to be converted into heat. Also, the period of time during which a chip remains in contact with the face of the tool is short, and therefore, the heat generated due to friction is very small. As a result, the tool does not become hot, and lubricants and coolants are not required.

10.2.2.3 Sheared Chips

When machining semiductile materials with heavy cuts and at relatively low cutting speeds, the resulting sheared chips have a shape that is midway between the segmented and the continuous chips (see Figure 10.7c). They are usually short, twisted ribbons that break every now and then.

10.2.3 The Problem of the Built-Up Edge

When machining highly plastic, tough metals at high cutting speeds, the amount of heat generated as a result of plastic deformation and friction between the chip and the tool is large and results in the formation of a built-up edge, as shown in Figure 10.8. The combination of the resulting elevated temperature with the high pressure at the tool face causes localized welding of some of the chip material to the tool face (see Figure 10.8a). The welded material (chip segment) becomes an integral part of the cutting tool, thus changing the values of the cutting angles. This certainly increases

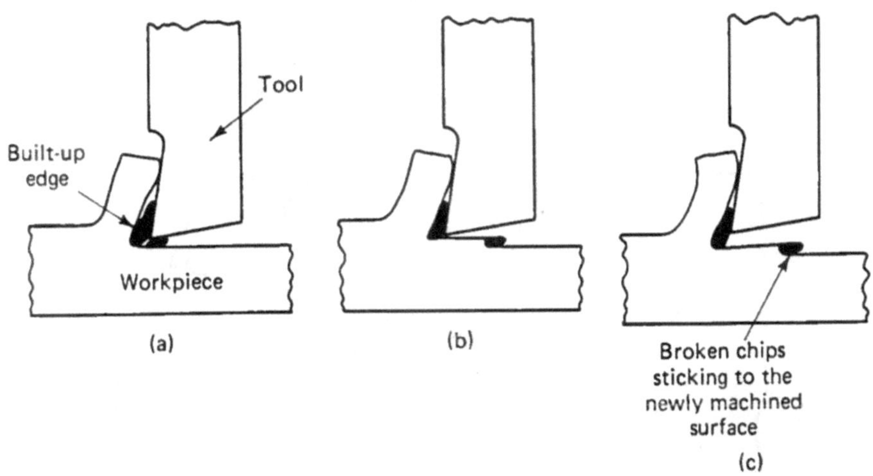

FIGURE 10.8 Stages in the formation of the built-up edge: (a) localized welding, (b) false cutting edge, and (c) flawed surface.

friction, leading to the buildup of layer upon layer of chip material. This newly formed false cutting edge (see Figure 10.8b) is referred to as the *built-up edge*. The cutting forces also increase, the built-up edge breaks down, and the fractured edges adhere to the machined surface (see Figure 10.8c). The harmful effects of the built-up edge are increased tool wear and a very poorly machined workpiece. The manufacturing engineer must choose the proper cutting conditions to avoid the formation of a continuous chip with a built-up edge.

10.2.4 The Cutting Ratio

As can be seen in Figure 10.9, during a cutting operation, the workpiece material just ahead of the tool is subjected to compression, and, therefore, the chip thickness becomes greater than the depth of the cut. The ratio of t_0/t is called the *cutting ratio* (r_c) and can be obtained as follows:

$$r_c = \frac{t_o}{t} = \frac{t_s \sin \varnothing}{t_s \cos(\varnothing - \propto)} = \frac{\sin \varnothing}{\cos(\varnothing - \propto)} \tag{10.1}$$

By employing trigonometry and carrying out simple mathematical manipulation, we can obtain the following equation:

$$\tan \phi = \frac{r_c \cos \alpha}{1 - r_c \sin \alpha} \tag{10.2}$$

Equation (10.2) is employed in obtaining the value of the *shear angle* ϕ when the rake angle $\propto$, the depth of cut, and the final thickness of the chips are known. In experimental work, the chip thickness is either measured directly with the help of a ball-ended micrometer or obtained from the weight of a known length of the chip (of course, the density and the width of the chip must also be known).

Let us now study the relationship between velocities. Considering the constancy of mass and assuming the width of the chip to remain constant, it is easy to see that

$$V \times t_o = V_C \times t$$

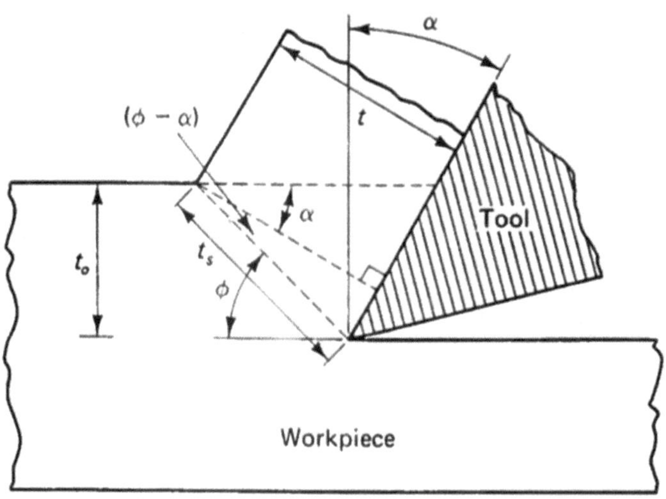

FIGURE 10.9 Geometry of a chip with respect to depth of cut.

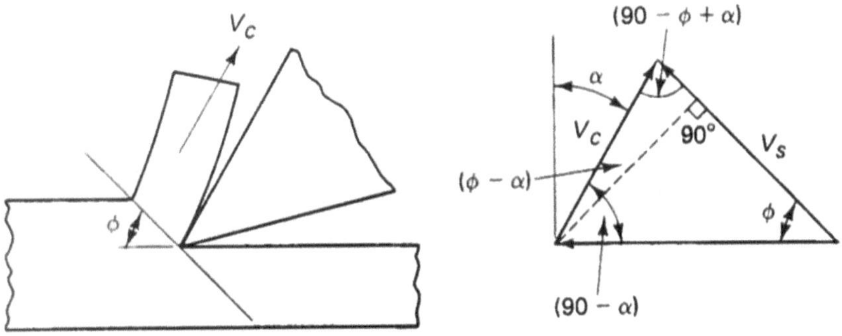

FIGURE 10.10 Velocity triangle and kinematics of the chip-removal process.

or

$$\frac{V_c}{V} = \frac{t_o}{t} = r_c$$

In other words,

$$V_C = Vr_c = \frac{V \sin \varnothing}{\cos(\varnothing - \propto)} \tag{10.3}$$

We can now draw the velocity triangle because we know the magnitudes and directions of two velocities, V and V_c. The shear velocity, V_s, which is the velocity with which the metal slides along the shear plane, can then be determined. Based on the velocity triangle shown in Figure 10.10 and applying the sine rule, the following can be stated:

$$\frac{V}{\sin(90 - \varnothing + \propto)} = \frac{V_s}{\sin(90 - \propto)} = \frac{V_c}{\sin \varnothing}$$

This equation can take the form

$$\frac{V}{\cos(\phi - \propto)} = \frac{V_s}{\cos \propto} = \frac{V_c}{\sin \phi}$$

Therefore,

$$V_s = V \frac{\cos \propto}{\cos(\varnothing - \propto)} \tag{10.4}$$

10.2.5 Shear Strain during Chip Formation

The value of the shear strain is an indication of the amount of deformation that the metal undergoes during the process of chip formation. As can be seen in Figure 10.11, the parallelogram *abda'* will take the shape *abed'* due to shearing. The shear strain can be expressed as follows:

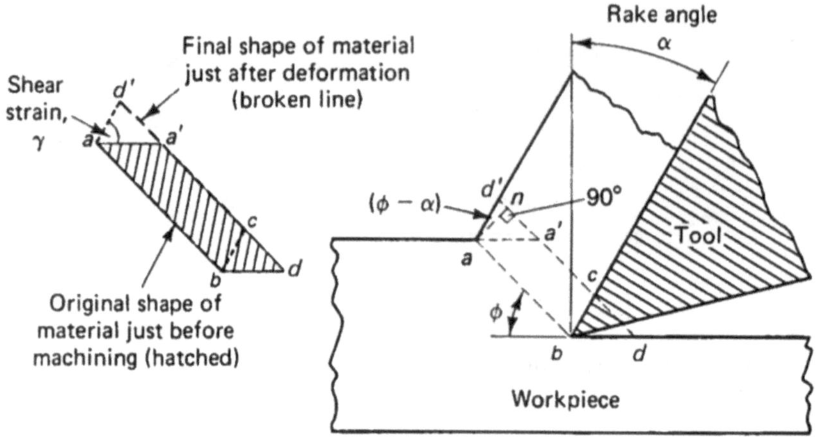

FIGURE 10.11 Shear strain during chip formation.

$$\gamma = \frac{a'n}{an} + \frac{d'n}{an} = \cot\phi + \tan(\phi - \alpha) \tag{10.5}$$

The shear strain rate can be obtained from Equation (10.5) as follows:

$$\gamma' = \frac{a'n}{an} \times \frac{1}{\Delta t} + \frac{d'n}{an} \times \frac{1}{\Delta t} = \frac{a'd'}{an} \times \frac{1}{\Delta t}$$

But

$$\frac{a'd'}{\Delta t} = V_s$$

Therefore,

$$\gamma' = \frac{V_s}{an}$$

where *an* is the thickness of the shear zone. Experimental results have indicated that the thickness of the shear zone is very small. Consequently, it can easily be concluded that the process of chip formation takes place at an extremely high strain rate. This finding is very important, especially for strain-rate-sensitive materials, where the strength and ductility of the material are markedly affected.

10.3 CUTTING FORCES

10.3.1 Theory of Ernst and Merchant

In order to simplify the problem, let us consider the two-dimensional, idealized cutting model of continuous chip formation. In this case, all the forces lie in the same plane and, therefore, form a coplanar system of forces. Walter Ernst and Eugene M. Merchant, both eminent American manufacturing scientists, based their analysis of this system of forces on the assumption that a chip acts as a rigid body in equilibrium under the forces acting across the chip–tool interface and the shear plane. As Figure 10.12 shows, the cutting edge exerts a certain force upon the workpiece. The magnitude of that force is dependent upon many factors, such as the workpiece material, the conditions of cutting, and the values of the cutting angles.

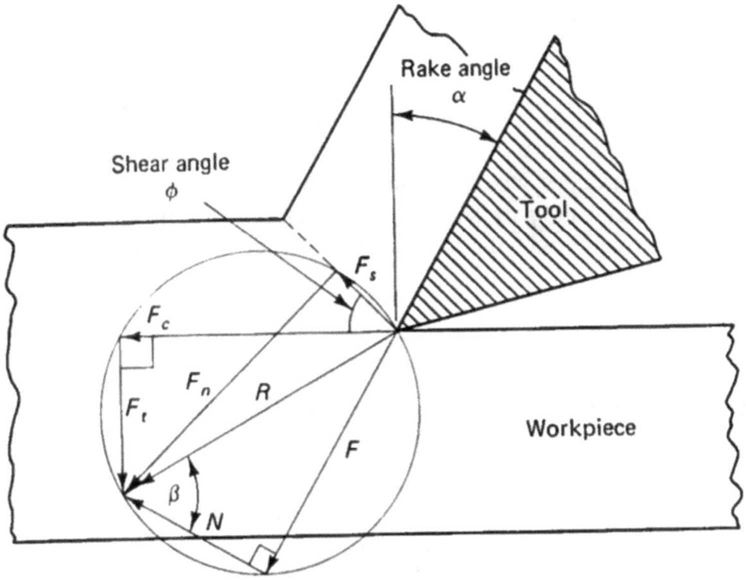

FIGURE 10.12 Cutting force diagram according to Ernst and Merchant.

By employing simple mechanics, the force can be resolved into two perpendicular components, F_c and F_t. As can be seen in Figure 10.12, F_c acts in the direction of tool travel and is referred to as the *cutting component*, whereas F_t acts normal to that direction and is known as the *thrust component*. The resultant tool force can alternatively be resolved into another two perpendicular components, F_s and F_n. The first component, F_s, acts along the shear plane and is referred to as the *shearing force*; the second component, F_n, acts normal to it and causes compressive stress to act on the shear plane. Again, at the chip–tool interface, the components of the resultant force that acts on the chip are F and N. Notice from the figure that F represents the friction force that resists the movement of the chip as it slides over the face of the tool, while N is the normal force. The ratio between F and N is actually the coefficient of friction at the chip–tool interface. Because each of the two components is perpendicular, it is clear from Euclidean geometry that the point of intersection of each two components must lie on the circumference of the circle that has the resultant force as a diameter. The cutting force diagram of Figure 10.12 lets us express F_s, F_n, F, and N in terms of F_C and F_t as follows:

$$F_s = F_C \cos\varnothing - F_t \sin\varnothing \tag{10.6}$$

$$F_n = F_c \sin\varnothing + F_t \cos\varnothing \tag{10.7}$$

$$F = F_C \sin \propto + F_t \cos \propto \tag{10.8}$$

$$N = F_C \cos \propto - F_t \sin \propto \tag{10.9}$$

The preceding equations can be used to determine different unknown parameters that affect the cutting operations. For instance, the coefficient of friction at the chip–tool interface can be obtained as follows:

$$\mu = \frac{F}{N} = \frac{F_C \sin \propto + F_t \cos \propto}{F_C \cos \propto - F_t \sin \propto} = \tan^{-1} \beta$$

Dividing both the numerator and denominator by cos ∝, we obtain

$$\mu = \frac{F_t + F_C \tan \propto}{F_C - F_t \tan \propto} \tag{10.10}$$

The shear force F_s is of particular importance as it is used for obtaining the magnitude of the mean shear strength of the material along the shear plane and during the cutting operation. This is equal to the mean shear stress acting through the shear plane and can be computed as follows:

$$\tau = \frac{F_s}{A_s}$$

where A_S, the area of the shear plane, equals A_{chip}, where A_{chip} is the cross-sectional area of the chip. Therefore,

$$\tau_s = \frac{\left(F_C \cos\phi - F_t \sin\phi\right)\sin\phi}{A_{chip}} \tag{10.11}$$

Experimental work has indicated that the mean shear stress, calculated from Equation (10.11), is constant for a given metal over a wide variation in the cutting conditions. This can be explained by the fact that the strain rate at which metal cutting occurs is sufficiently high to be the only factor that affects the shear strength for a given material. Therefore, the cutting speed, amount of strain, or temperature do not have any appreciable effect on the value of the mean shear stress of the metal being machined.

Ernst and Merchant extended their analysis and studied the relationship between the shear angle and the cutting conditions. They suggested that the shear angle always takes the value that reduces the total energy consumed in cutting to a minimum. Because the total work done in cutting is dependent upon and is a direct function of the component F_c of the cutting force, they developed an expression for F_c in terms of ϕ and the constant properties of the workpiece material. Next, that expression was differentiated with respect to ϕ and then equated to zero in order to obtain the value ϕ for which F_c and, therefore, the energy consumed in cutting is a minimum. Following is the mathematical treatment of this problem. From Figure 10.12, we can see that

$$F_s = R\cos\left(\phi + \beta - \propto\right) \tag{10.12}$$

Therefore,

$$R = \frac{F_s}{\cos\left(\phi + \beta - \propto\right)}$$

But,

$$F_s = \tau_s A_s = \tau_s \frac{A_{chip}}{\sin\phi}$$

Therefore,

$$R = \frac{\tau_s A_{chip}}{\sin\phi} \times \frac{1}{\cos\left(\phi + \beta - \propto\right)} \tag{10.13}$$

Again, it can be seen from Figure 10.12 that

$$F_c = R\cos(\beta - \propto) \tag{10.14}$$

Hence, from Equations (10.13) and (10.14),

$$F_c = \frac{\tau_s A_{\text{chip}}}{\sin\phi} \times \frac{\cos(\beta - \propto)}{\cos(\phi + \beta - \propto)} \tag{10.15}$$

Differentiating Equation (10.15) with respect to ϕ and equating the outcome to zero, we obtain the condition that will make F_c minimal. This condition is given by the following equation:

$$2\phi + \beta - \propto = \frac{\pi}{2} \tag{10.16}$$

It was found that the theoretical value of ϕ obtained from Equation (10.16) agreed well with the experimental results when cutting polymers, but this was not the case when machining aluminum, copper, or steels.

10.3.2 Theory of Lee and Shaffer

The theory of American manufacturing scientists E. Lee and Bernard W. Shaffer is based on applying the slip-line field theory to the two-dimensional metal-cutting problem. A further assumption is that the material behaves in a rigid, perfectly plastic manner and obeys the von Mises yield criterion and its associated flow rule. After constructing the slip-line field for that problem, it was not difficult for Lee and Shaffer to obtain the relationship between the cutting parameters and the shear angle. The result can be given by the following equation:

$$\phi + \beta - \propto = \frac{\pi}{4} \tag{10.17}$$

In fact, neither of the preceding theories quantitatively agrees with experimental results. However, the theories yield linear relationships between ϕ and $(\beta - \propto)$, which is qualitatively in agreement with the experimental results.

10.4 CUTTING ENERGY

We can see from the previous discussion that it is the component F_c that determines the energy consumed during machining because it acts along the direction of relative tool travel. The power consumption P_m (i.e., the rate of energy consumption during machining) can be obtained from the following equation:

$$P_m = F_c \times V \tag{10.18}$$

where V is the cutting speed.

The rate of metal removal during machining Z_m is also proportional to the cutting speed and can be given by

$$Z_m = A_o \times V \tag{10.19}$$

where A_o, the cross-sectional area of the uncut chip, equals t_o times the width of the chip. Now, the energy consumed in removing a unit volume of metal can be obtained from Equations (10.18) and (10.19) as follows:

$$P_C = \frac{P_m}{Z_m} = \frac{F_c \times V}{A_o \times V} = \frac{F_c}{A_o} \tag{10.20}$$

In Equation (10.20), P_c, a parameter that indicates the efficiency of the process, is commonly known as the *specific cutting energy* and also sometimes is called the *unit horsepower*. Unfortunately, the specific cutting energy for a given metal is not constant but rather varies considerably with the cutting conditions, as we will see later.

10.5 OBLIQUE VERSUS ORTHOGONAL CUTTING

Until now, we have simplified the metal-cutting process by considering only *orthogonal cutting*. In this type of cutting, the cutting edge of the tool is normal to the direction of relative tool movement, as shown in Figure 10.13a. It is actually a two-dimensional process in which each longitudinal section (i.e., parallel to the tool travel) of the tool and chip is identical to any other longitudinal section of the tool and chip. The cutting force is, therefore, also two-dimensional and can be resolved into two components, both lying within the plane of the drawing. Although this approach facilitated the analysis of chip formation and the mechanics of metal cutting, it is seldom used in practice because it applies only when turning the end face of a thin tube in a direction parallel to its axis.

The more common type (or model) of cutting used in the various machining operations is *oblique cutting*. In this case, the cutting edge of the tool is inclined to (i.e., not normal to) the relative tool travel, as can be seen in Figure 10.13b. It is a three-dimensional problem in which the cutting force can be resolved into three perpendicular components, as indicated in Figure 10.14. The magnitudes of these components can be measured by means of a special apparatus that is mounted either in the workholder or tool holder and is known as a *dynamometer*. As you may expect, the tool geometry is rather complicated and will be discussed later. For now, let us see the effect of each of the cutting force components on the oblique cutting operations.

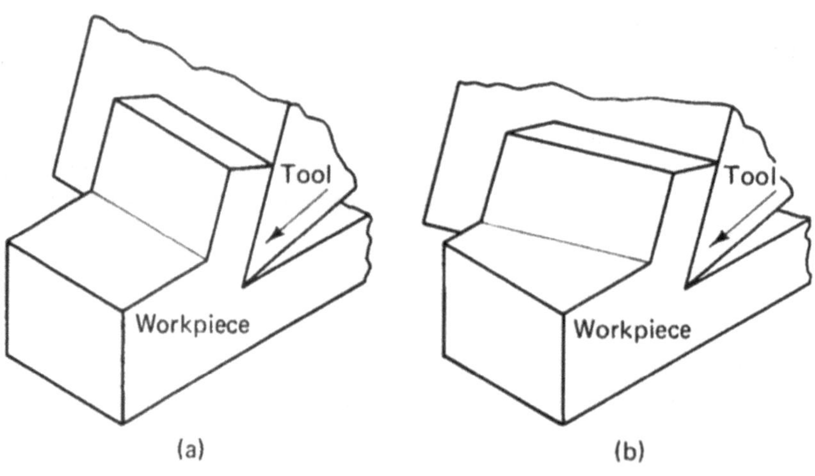

FIGURE 10.13 Types of cutting: (a) orthogonal and (b) oblique.

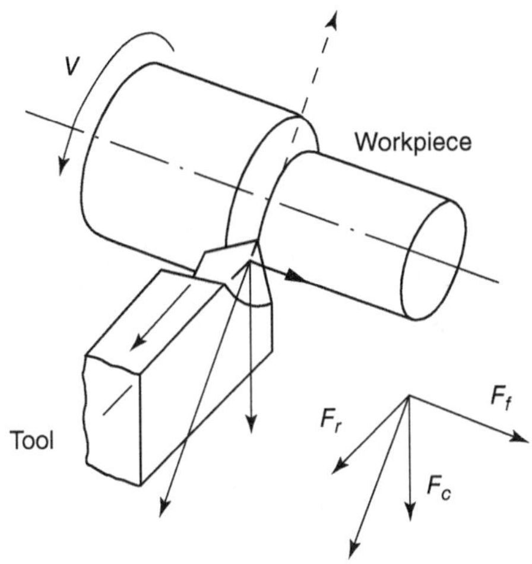

FIGURE 10.14 Components of the three-dimensional cutting force.

10.5.1 Forces in Oblique Cutting

Following is a discussion of the three components referred to as F_c, F_f, and F_r in Figure 10.14.

- F_C is the *cutting force* and acts in the direction where the cutting action takes place. It is the highest of the three components and results in 99 percent of the energy consumed during the process. The horsepower due to this force hp_c can be given by the following equation:

$$hp_c = \frac{F_c \times V_c}{550 \times 60} \tag{10.21}$$

  In Equation (10.21), F_c is in pounds and V_c is in feet per minute. Consequently, the appropriate conversion factors must be used if the horsepower is to be obtained in SI units.
- F_f is the *feed force* (or longitudinal force in turning). The term *feed* means the movement of the tool to regenerate the cutting path in order to obtain the machined surface. This force amounts to only about 40 percent of the cutting force. The horsepower required to feed the tool, hp_f, can be given as follows:

$$hp_f = \frac{F_f \times V_f}{550 \times 60} \tag{10.22}$$

  The horsepower given by Equation (10.22) amounts to only 1 percent of the total power consumed during cutting.
- F_r is the *thrust force* (or radial force in turning) and acts in the direction of the depth of cutting. This force is the smallest of the three components and amounts to only 20 percent of the cutting force or, in other words, 50 percent of the feed force. This component does not result in any power consumption, as there is no tool movement along the direction of the depth of cut.

These components of the cutting force are measured only in scientific metal-cutting research. The manufacturing engineer is, however, interested in determining beforehand the motor horsepower required to perform a certain job in order to be able to choose the right machine for that job. Therefore, use is made of the concept of *unit horsepower*, which was mentioned previously.

Experimentally obtained values of unit horsepower for various common materials are compiled in tables ready for use. The total cutting horsepower can be obtained from the following equation:

$$hp_c = unit\ hp \times \text{rate of metal removal} \times \text{correction factor} \tag{10.23}$$

where the rate of metal removal is in cubic inches per minute and the correction factor is introduced to account for the tool geometry and the variation in feed. Table 10.1 indicates the unit horsepower values for various ferrous metals and alloys having different hardness numbers. Table 10.2 provides the unit horsepower values for nonferrous metals and alloys. Figure 10.15a–c indicates the different correction factors for the unit horsepower to account for variations in the cutting conditions. The cutting horsepower is not of practical importance by itself. Its significance is that it is used in computing the motor horsepower. Obviously, the motor horsepower has to be higher than the cutting horsepower as some power is lost in overcoming friction and inertia of the moving parts. The following equation can be used for calculating the motor horsepower:

$$hp_m = hp_c \times \frac{1}{\eta} \tag{10.24}$$

where η is the machine efficiency, which can be taken from Table 10.3.

TABLE 10.1
Unit Horsepower Values for Ferrous Metals and Alloy

Ferrous Metals and Alloys			Brinell Hardness Number			
150–175	**176–200**	**201–250**	**251–300**	**301–350**	**351–400**	
ANSI						
1010–1025	0.58	0.67				
1030–1055	0.58	0.67	0.8	0.96		
1060–1095			0.75	0.88	1	
1112–1120	0.5					
1314–1340	0.42	0.46	0.5			
1330–1350		0.67	0.75	0.92	1.1	
2015–2115	0.67					
2315–2335	0.54	0.58	0.62	0.75	0.92	1
2340–2350		0.5	0.58	0.7	0.83	
2512–2515	0.5	0.58	0.67	0.8	0.92	
3115–3130	0.5	0.58	0.7	0.83	1	1
3160–3450		0.5	0.62	0.75	0.87	1
4130–4345		0.46	0.58	0.7	0.83	1
4615–4820	0.46	0.5	0.58	0.7	0.83	0.87
5120–5150	0.46	0.5	0.62	0.75	0.87	1
52100		0.58	0.67	0.83	1	
6115–6140	0.46	0.54	0.67	0.83	1	
6145–6195		0.7	0.83	1	1.2	1.3
Plain cast iron	0.3	0.33	0.42	0.5		
Alloy cast iron	0.3	0.42	0.54			
Malleable iron	0.42					
Cast steel	0.62	0.67	0.8			

Source: Turning Handbook of High-Efficiency Metal Cutting, 1980, courtesy Seco Tools, LLC.

TABLE 10.2
Unit Horsepower Values for Nonferrous Metals and Alloys

Nonferrous Metals and Alloys	Properties	Unit Horsepower
Brass	Medium	0.5
	Soft	0.33
	Free machining	0.25
Bronze	Hard	0.83
	Medium	0.5
	Soft	0.33
Copper	Pure	0.9
Aluminum	cast	0.25
	hard rolled	0.33
Monel	rolled	1
Zinc alloy	die cast	0.25

Source: Turning Handbook of High-Efficiency Metal Cutting, 1980, courtesy Seco Tools, LLC.

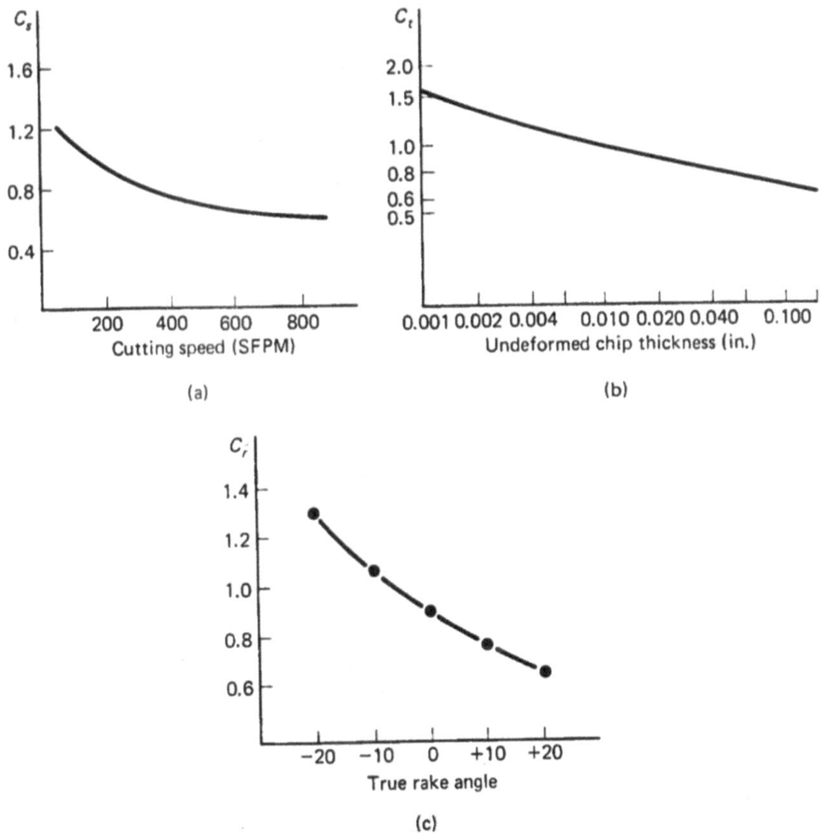

FIGURE 10.15 Different correction factors to account for variations in the cutting conditions: (a) cutting speed, (b) chip thickness, and (c) rake angle (*Source*: *Turning Handbook of High-Efficiency Metal Cutting*, 1980, courtesy Seco Tools, LLC.)

TABLE 10.3
Typical Overall Machine Tool Efficiencies (Except Milling Machines)

Type	Efficiency
Direct-spindle drive	90
One-belt drive	85
Two-belt drive	70
Geared head	70

Source: Turning Handbook High-Efficiency Metal Cutting, 1980, courtesy Seco Tools, LLC.

The cutting horsepower is used not only in calculating the motor horsepower but also for giving a fair estimate of the cutting force component F_c by using Equation (10.21). This force is very important when studying the vibrations associated with metal cutting, as we will see later. The following example illustrates how to estimate the cutting force component.

10.5.1.1 Example of Estimating Cutting Force Component

During a turning operation, the metal-removal rate (MRR) was found to be 3.6 cubic inches per minute. Following are other data of the process:

$$\text{Spindle hp} = \text{MRR} \times \text{unit hp} \times \text{correction factor}$$

Material	ANSI 1055, HB 250
Cutting speed	300 feet per minute
Unreformed chip thickness	0.01 inch
Tool character	0-7-7-7-15-15-1/32

Solution

The correction factor because of the cutting speed is 0.8, and the correction factor because of the undeformed chip thickness is 1. The true rake angle is

$$\tan \propto_{\text{true}} = \cos 15^\circ \tan 7^\circ + \sin 15^\circ \tan 0$$

$\propto_{\text{true}} = 6^\circ$. The correction factor because of the true rake angle is 0.83, and the unit horsepower is 0.8 from Table 10.1. Thus,

$$\text{Spindle hp} = 3.6 \times 0.8 \times 0.8 \times 1 \times 0.83 = 1.9\,\text{hp}$$

$$hp_c = \frac{F_C \times 300\,\text{ft/min}}{550 \times 60}$$

Therefore,

$$F_c = \frac{1.9 \times 550 \times 60}{300} = 209\,\text{pounds}$$

Note that the undeformed chip thickness equals feed (inches per revolution) times the cosine of the side cutting-edge angle (SCEA).

10.6 CUTTING TOOLS

10.6.1 Basic Geometry

In order for a tool to cut a material, it must have two important characteristics: first, it must be harder than that material, and, second, it must possess certain geometrical characteristics. The cutting tool geometry differs for different machining operations. Nevertheless, it is always a matter of rake and clearance angles. Therefore, we are going to limit our discussion, at the moment, to single-point tools for the sake of simplicity. Other types of tools will be considered when we cover the various machining operations.

As can be seen in Figure 10.16, the geometry of a single-point cutting tool can be adequately described by six cutting angles. These can be shown more clearly by projecting them on three perpendicular planes using orthogonal projection, as is done in Figure 10.17. Let us now consider the definition of each of the six angles.

10.6.1.1 Side Cutting-Edge Angle

The SCEA is usually referred to as the *lead angle*. It is the angle enclosed between the side cutting edge and the longitudinal direction of the tool. The value of this angle varies between 0° and 90°, depending upon the machinability, rigidity, and, sometimes, the shape of the workpiece (e.g., a 90° shoulder must be produced by a 0° SCEA). As this angle increases from 0° to 15°, the power consumption during cutting decreases. However, there is a limit to increasing the SCEA, beyond which excessive vibrations take place because of the large tool–workpiece interface. On the other hand, if the angle were taken as 0°, the full cutting edge would start to cut the workpiece at once, causing an initial shock. Usually, the recommended value for the lead angle should range between 15° and 30°.

10.6.1.2 End Cutting-Edge Angle

The end cutting-edge angle (ECEA) serves to eliminate rubbing between the end cutting edge and the machined surface of the workpiece. Although this angle takes values in the range of 5° to 30°, commonly recommended values are 8° to 15°.

10.6.1.3 Side Relief and End Relief Angles

Side and end relief angles serve to eliminate rubbing between the workpiece and the side and end flank, respectively. Usually, the value of each of these angles ranges between 5° and 15°.

10.6.1.4 Back and Side Rake Angles

Back and side rake angles determine the direction of flow of the chips onto the face of the tool. Rake angles can be positive, negative, or zero. It is the side rake angle that has the dominant influence on

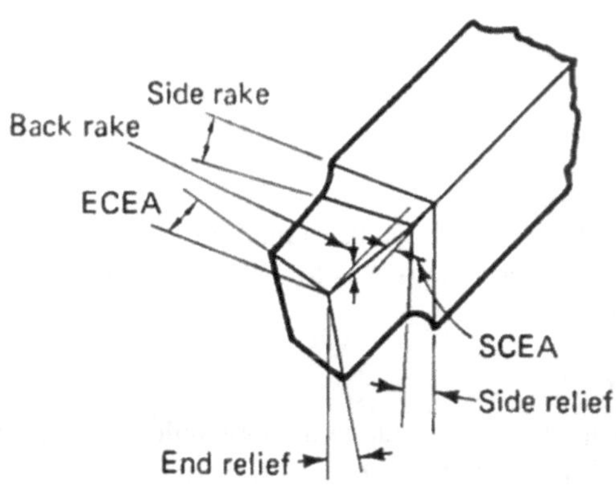

FIGURE 10.16 Geometry of a single-point cutting tool.

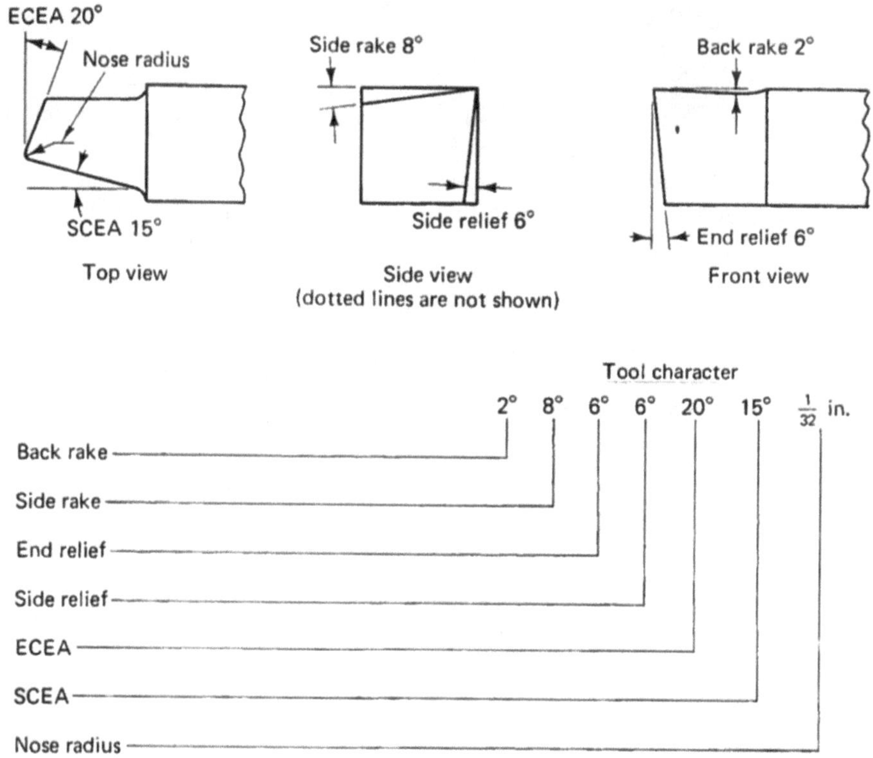

FIGURE 10.17 Orthogonal projection of the cutting angles of a single-point tool and tool character.

cutting. Its value usually varies between 0° and 15°, whereas the back rake angle is usually taken as 0°.

Another useful term in metal cutting is the *true rake angle*, which is confined between the line of major inclination within the face of the tool and a horizontal plane. It determines the actual flow of chips across the face of the tool and can be obtained from the following equation:

$$\text{True rake angle} = \tan^{-1}\left(\tan \propto \sin\lambda + \tan\beta \cos\lambda\right) \tag{10.25}$$

where

α is the back rake angle
β is the side rake angle
λ is the lead angle (SCEA)

As previously mentioned, the true rake angle has a marked effect on the unit horsepower for a given workpiece material, and a correction factor has to be used when calculating the power in order to account for variations in the true rake angle.

10.6.1.5 Tool Character

The tool angles are usually specified by a standard abbreviation system called the *tool character*, or the *tool signature*. As also illustrated in Figure 10.17, the tool angles are always given in a certain order: back rake, side rake, end relief, side relief, ECEA, and SCEA, followed by the nose radius of the tool.

10.6.2 Cutting Tool Materials

Cutting tools must possess certain mechanical properties in order to function adequately during the cutting operations. These properties include high hardness and the ability to retain it even at the elevated temperatures generated during cutting, as well as toughness, creep and abrasion resistance, and the ability to withstand high bearing pressures. Cutting materials differ in the degree to which they possess each of these mechanical properties. Therefore, a cutting material is selected to suit the cutting conditions (i.e., the workpiece material, cutting speed or production rate, coolants used, and so on). Following is a survey of the commonly used cutting tool materials.

10.6.2.1 Plain-Carbon Steel

Plain-carbon steel contains from 0.8 to 1.4 percent carbon, has no additives, and is subjected to heat treatment to increase its hardness. Plain-carbon steel is suitable only when making hand tools or when soft metals are machined at low cutting speeds as it cannot retain its hardness at temperatures above 600°F (300°C) due to tempering action.

10.6.2.2 Alloy Steel

The carbon content of alloy steel is similar to that of plain-carbon steel, but it contains alloying elements (in limited amounts). Tools made of alloy steel must be heat treated and are used only when machining is carried out at low cutting speeds. The temperature generated as a result of cutting should not exceed 600°F (300°C) to avoid any tempering action.

10.6.2.3 High-Speed Steel

High-speed steel (HSS) is a kind of alloy steel that contains a certain percentage of alloying elements, such as tungsten (18 percent), chromium (4 percent), molybdenum, vanadium, and cobalt. HSS is heat treated by heating (at two stages), cooling by employing a stream of air, and then tempering it. Tools made of HSS can retain their hardness at elevated temperatures up to 1100°F (600°C). These tools are used when relatively high cutting speeds are required. Single-point tools, twist drills, and milling cutters are generally made of HSS, except when these tools are required for high-production machining.

10.6.2.4 Cast Hard Alloys

Cast hard alloys can be either ferrous or nonferrous and contain about 3 percent carbon, which, in turn, reacts with the metals to form very hard carbides. The carbides retain their hardness even at a temperature of about 1650°F (900°C). Because such a material cannot be worked or machined, it is cast in ceramic molds to take the form of tips that are mounted onto holders by brazing or by being mechanically fastened.

10.6.2.5 Sintered Cemented-Carbide Tips

Sintered cemented carbide was developed to eliminate the main disadvantage of the hard cast alloys: brittleness. Originally, the composition of this material involved about 82 percent very hard tungsten carbide particles and 18 percent cobalt as a binder. Sintered cemented carbides are always molded to shape by the powder metallurgy technique (i.e., pressing and sintering, as was explained in Chapter 7). As it is impossible to manufacture the entire tool out of cemented carbide because of the strength consideration, only tips are made of this material: these tips are brazed or mechanically fastened to steel shanks that have the required cutting angles.

Cemented carbides used to be referred to as Widia, taken from the German expression *Wie Diamant*, meaning diamondlike, because they possess extremely high hardness, reaching about 90 R_C, and they retain such hardness even at temperatures of up to 1850°F (1000°C). Recent developments involve employing combinations of tungsten, titanium, and tantalum carbides with cobalt or nickel alloy as binders. The result is characterized by its low coefficient of friction and high abrasion resistance. Tools with cemented-carbide tips are recommended whenever the cutting speeds required or the feed rates are high and are, therefore, commonly used in mass production. Recently, carbide tips have been coated with nitrites or oxides to increase their wear resistance and service life.

10.6.2.6 Ceramic Tips

Ceramic tips consist basically of very fine alumina powder, Al_2O_3, which is molded by pressing and sintering. Ceramics have almost the same hardness as cemented carbides, but they can retain that hardness up to a temperature of 2200°F (1100°C) and have a very low coefficient of thermal conductivity. Such properties allow for cutting to be performed at speeds that range from two to three times the cutting speed used when carbide tips are employed. Ceramic tips are also characterized by their superior resistance to wear and to the formation of crater cavities. They require no coolants. Their toughness and bending strength are low, which must be added to their sensitivity to creep loading and vibration. Therefore, ceramic tips are recommended only for finishing operations (small depth of cut) at extremely high cutting speeds of up to 1800 feet per minute (600 m/min). Following are the three common types of ceramic tips:

- Oxide tips, consisting mainly of aluminum oxide, have a white color with some pink or yellow tint.
- Cermet tips, including alumina and some metals such as titanium or molybdenum, are dark gray in color.
- Tips that consist of both oxides as well as carbides are black in color.
- Ceramic tips should not be used for machining aluminum because of their affinity to oxygen.

10.6.2.7 Diamond

Diamond pieces are fixed to steel shanks and are used in precision cutting operations. They are recommended for machining aluminum, magnesium, titanium, bronze, rubber, and polymer. When machining metallic materials, a mirror finish can be obtained.

10.6.3 Tool Wear

There are two interrelated causes of tool wear: *mechanical abrasion* and *thermal erosion*. Although these two actions take place simultaneously, the role of each varies for various cutting conditions. Mechanical wear is dominant when low cutting speeds are used or when the workpiece possesses high machinability. Thermal wear prevails when high cutting speeds are used with workpieces having low machinability. Thermal wear is due to diffusion, oxidation, and the fact that the mechanical properties of the tool change as a result of the high temperature generated during the cutting operation.

The face of the cutting tool is subjected to friction caused by the fast relative motion of the generated chips onto its surface. Similarly, the flanks are also subjected to friction as a result of rubbing by the workpiece. Although the tool is harder than the workpiece, friction and wear will take place and will not be evenly distributed over the face of the tool. Wear is localized in the vicinity of the cutting edge and results in the formation of a crater. There are different kinds of tool wear:

- Flank wear
- Wear of the face that comes in contact with the removed chip
- Wear of the cutting edge itself
- Wear of the nose
- Wear and formation of a crater
- Cracks in the cutting edges occurring during interrupted machining operations such as milling

10.6.4 Tool Life

Tool life is defined as the length of actual machining time beginning at the moment when a just-ground tool is used and ending at the moment when the machining operation is stopped because of the poor performance of that tool. Different criteria can be used to judge the moment at which the machining operation should be stopped. It is common to consider the tool life as over when the flank wear reaches a certain amount (measured as the length along the surface generated due to abrasion starting from the tip). This maximum permissible flank wear is taken as 0.062 inches (1.58 mm) in the case of HSS tools and 0.03 inches (0.76 mm) for carbide tools.

The tool life is affected by several variables, the important ones being cutting speed, feed, and the coolants used. The effect of these variables can be determined experimentally and then represented graphically for practical use. It was found by Frederick W. Taylor (an American genius) that the relationship between tool life and cutting speed is exponential. It can, therefore, be plotted on a logarithmic scale so that it takes the form of a straight line, as shown in Figure 10.18. In fact, this was the basis for establishing an empirical formula that correlates tool life with cutting speed. A correction factor is also introduced into the formula to account for the effects of other variables. The original formula had the following form:

$$VT^n = C \tag{10.26}$$

where

n is a constant that depends upon the tool material (0.1 for HSS, 0.2 for carbides, and 0.5 for ceramic tools)
C is a constant that depends upon the cutting conditions (e.g., feed)
T is the tool life measured in minutes
V is the cutting speed in feet per minute

Equation (10.26) is very useful in obtaining the tool life for any cutting speed if the tool life is known at any other cutting speed.

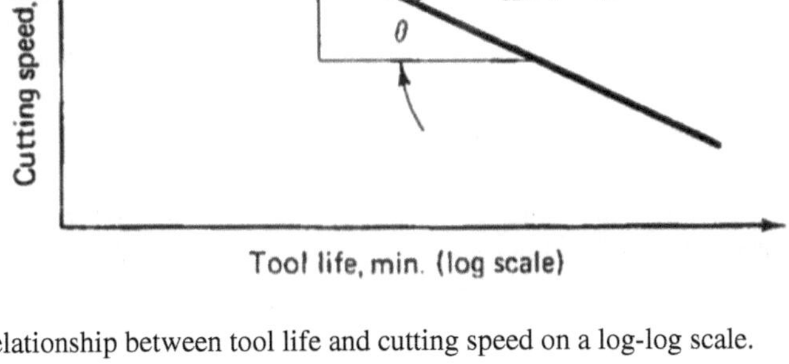

FIGURE 10.18 Relationship between tool life and cutting speed on a log-log scale.

10.7 MACHINABILITY

10.7.1 Machinability Defined

Machinability is a property characterizing the material of the workpiece: it is the ease with which that material can be machined. In order to express machinability in a quantitative manner, one of the following methods is used:

- The maximum possible rate of chip removal
- Surface finish of the machined workpiece
- Tool life
- Energy required to accomplish the cutting operation

It is clear that the tool life is the most important of these criteria as it plays an important role in maximizing production while minimizing the production cost. Moreover, criteria such as surface finish and machining precision depend upon many factors, such as the sharpness of the cutting edge, the rigidity of the tool, and the possibility of the formation of a built-up edge. As a consequence, it is the tool life that is most suitable as a criterion of machinability.

10.7.2 Machinability Index

Because machinability cannot be expressed in an absolute manner, it is appropriate to take a highly machinable metal as a reference and express the machinability of any other ferrous metal as a percentage of that of the reference metal. The reference metal chosen was steel SAE-AISI 1112 because of its superior machinability, which exceeds that of any other steel. Such steel is usually referred to as *free cutting steel.* The machinability index can now be given:

$$\text{Machinability index} = \frac{\text{Cutting speed of metal for tool life of 20 minutes}}{\text{Cutting of steel SAE 1112 for tool life of 20 minutes}} \times 100 \quad (10.27)$$

Table 10.4 indicates the machinability index for some commonly used metals and alloys.

TABLE 10.4
Machinability Index for Some Metals and Alloys (Using Carbide Tools)

Metal or Alloy	Machinability Index (%)
Steel SAE 1020 (annealed)	65
Steel SAE A2340	45
Cast iron	70–80
Stainless steel 18–8 (austenitic)	25
Tool steel (low tungsten, chrome, and carbon)	30
Copper	70
Brass	180
Aluminum alloy	300 and above

10.8 CUTTING FLUIDS

10.8.1 Necessary Characteristics

As previously mentioned, the process of metal cutting results in the generation of a large amount of heat and a localized increase in the temperature of the cutting tool. This effect is particularly evident when machining ductile metals. Accordingly, coolants are required to remove any generated heat, to lower the temperature of the cutting tool, and, consequently, to increase the tool service life. In order to fulfill such conditions and function properly, a cutting fluid must possess certain characteristics:

- The cutting fluid must possess suitable chemical properties (i.e., to be appropriate from the point of view of chemistry), must not react with the workpiece material or cause corrosion in any component of the machine tool, and should not promote the formation of rust or spoil the lubricating oil of the machine bearing and slides whenever it comes in contact with that oil.
- The cutting fluid must be chemically stable (i.e., must not change its properties with time).
- No poisonous gases or fumes should evolve during machining so that there is no possibility of problems regarding the safety or health of the workers.
- The lubricating and cooling properties of the cutting fluid must be superior.
- The fluid used should be cheap and should be recycled by a simple filtration process.

10.8.2 Types of Cutting Fluids

The following discussion involves the different kinds of cutting fluids that are used in industry to satisfy the preceding requirements.

10.8.2.1 Pure Oils

Mineral oils such as kerosene or polar organic oils such as sperm oil, linseed oil, or turpentine can be used as cutting fluids. The application of pure mineral oils is permissible only when machining metals with high machinability, such as free cutting steel, brass, and aluminum. This is a consequence of their poor lubricating and cooling properties. Although polar organic oils possess good lubricating and cooling properties, they are prone to oxidation, give off unpleasant odors, and tend to gum.

10.8.2.2 Mixed Oils

Mineral oils are mixed with polar organic oils to obtain the advantages of both constituents. In some cases, sulfur or chlorine is added to enable the lubricant to adhere to the tool face, giving a film of lubricant that is tougher and more stable. The oils are then referred to as sulfurized or *chlorinated* oils. The chlorinated oils have the disadvantage of the possible emission of chlorine gas during the machining operation.

10.8.2.3 Soluble Oils

Soluble oils are sometimes called *water-miscible fluids or emulsifiable oils*. By blending oil with water and some emulsifying agents, soapy or milky mixtures can be obtained. These liquids have superior cooling properties and are recommended for machining operations requiring high speeds and low pressures. Sometimes, extreme-pressure additives are blended with the mixtures to produce emulsions with superior lubricating properties.

10.8.2.4 Water Solutions

A solution of sodium nitrate and trinolamine in water can be employed as a cutting fluid. Caustic soda is also used, provided that the concentration does not exceed 5 percent. If the concentration of the solution exceeds this limit, the paint of the machine and the lubricating oil of the slides may be affected.

10.8.2.5 Synthetic Fluids

Synthetic fluids can be diluted with water to give a mixture that varies in appearance from clear to translucent. Extreme-pressure additives like sulfur or chlorine can be added to the mixture so that it can be used for difficult machining operations.

10.9 CHATTER PHENOMENON

When we feel cold in winter, our jaws and teeth may start to chatter. A similar phenomenon occurs when the cutting tool and workpiece are exposed to certain unfavorable cutting conditions and dynamic characteristics of the machine tool structure. The analysis of this chatter phenomenon is an extremely complex task. However, thanks to the work of the late Professor Stephen A. Tobias of the University of Birmingham in England, we are able to understand how vibrations of the cutting tool initiate and how they can be minimized. Left without remedy, these vibrations result in breakage of the cutting tool (especially if it is ceramic or carbide) and poor surface quality. They may also cause breakage of the entire machine tool. Two basic types of vibrations are generated during machining: forced vibrations and self-excited vibrations.

10.9.1 Forced Vibrations

Forced vibrations take place as a result of periodic force applied within the machine tool structure. This force can be due to an imbalance in any of the machine tool components or interrupted cutting action, such as milling, in which there is a periodic engagement and disengagement between the cutting edges and the workpiece. The frequency of these forced vibrations must not be allowed to come close to the natural frequency of the machine tool system or any of its components; otherwise, resonance (vibrations with extremely high amplitude) takes place. The remedy in this case is to try to identify any possible source for the imbalance of the machine tool components and eliminate it. In milling machines, the stiffness and damping characteristics of the machine tool are controlled so as to keep the forcing frequency away from the natural frequency of any component and/or the natural frequency of the system.

10.9.2 Self-Excited Vibrations, or Chatter

Self-excited vibrations, or *chatter*, occur when an unexpected disturbing force, such as a hard spot in the workpiece material or sticking friction at the chip–tool interface, causes the cutting tool to vibrate at a frequency near the natural frequency of the machine tool. As a result, resonance takes place, and a minimum excitation produces an extremely large amplitude. Such conditions drastically reduce tool life, resulting in poor surface quality, and may cause damage to the workpiece, machine tool, or both. This unfavorable condition can be eliminated, or at least reduced, by controlling the stiffness and the damping characteristics of the system. This is usually achieved by selecting the proper material for the machine bed (cast iron has better damping characteristics than steel), by employing dry-bolted joints as energy dissipators where the vibration energy is absorbed in friction, or by using external dampers or absorbers. Advanced research carried out at the University of Birmingham in England indicated the potentials of employing layers of composites as a means to safeguard against the occurrence of chatter.

10.10 ECONOMICS OF METAL CUTTING

Our goal now is to find out the operating conditions (mainly the cutting speed) that maximize the MRR or the tool life. These two variables are in opposition to each other; a higher MRR results in a shorter tool life. Therefore, some trade-off or balance must be made in order to achieve either minimum machining cost per piece or maximum production rate, whichever is necessitated by the production requirements.

Figure 10.19 indicates how to construct the relationship between the cutting speed and the total cost per piece for a simple turning operation. The total cost is composed of four components: machining cost, idle-time (nonproductive) cost, tool cost, and tool-change cost. An increase in cutting speed obviously results in a reduction in machining time and, therefore, lower machining cost. This is accompanied by a reduction in tool life, thus increasing tool and tool-change costs. As can be seen in Figure 10.19, the curve of the cost per piece versus the cutting speed has a minimum that corresponds to the optimum cutting speed for the minimum cost per piece.

The relationship between the production time per piece and the cutting speed can be constructed in the same manner, as shown in Figure 10.20. There is also a minimum for this curve that corresponds to the optimum cutting speed for maximum productivity (minimum time per piece). Usually, this value is higher than the maximum economy speed given in Figure 10.19. Obviously, a cutting speed between these two limits (and depending upon the goals to be achieved) is recommended.

We can also quantify and generalize our analysis of the economics of machining as follows.

The *components of tool cost* may be expressed in minutes and are the following:

t_1 is the time to change the tool in minutes

t_2 is the time to grind or sharpen the tool in minutes $= \dfrac{t_s}{N_2} \times \dfrac{R_s}{R_c}$

t_3 is the time equivalent to the depreciation cost of the tool in minutes $= \dfrac{C_T}{N_1 \times N_2} \times \dfrac{1}{R_C}$

where

t_s is the tool grinding time in minutes
R_s is the labor and overhead rate in \$/min
R_c is the labor and overhead rate applied to the metal-cutting operation in \$/min

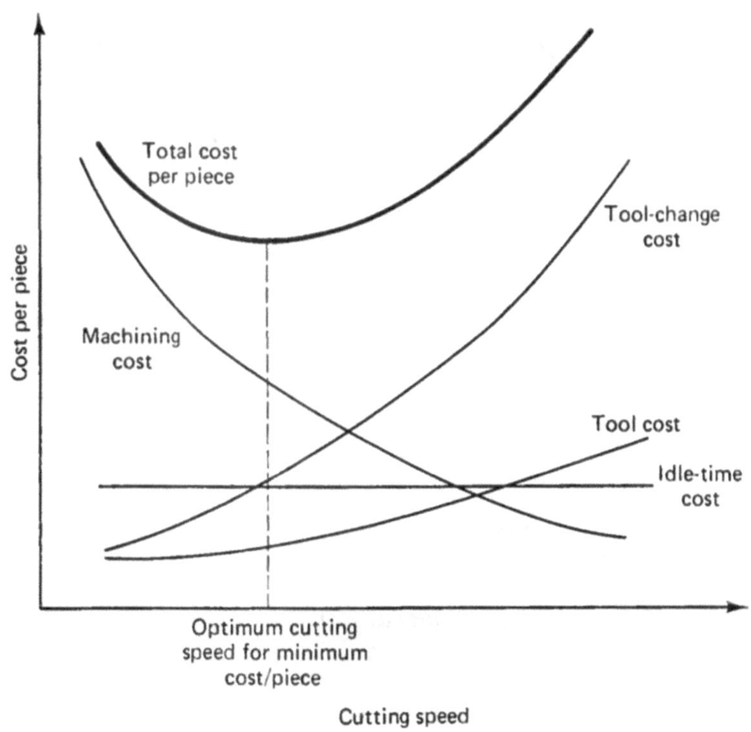

FIGURE 10.19 Relationship between cost per piece and cutting speed.

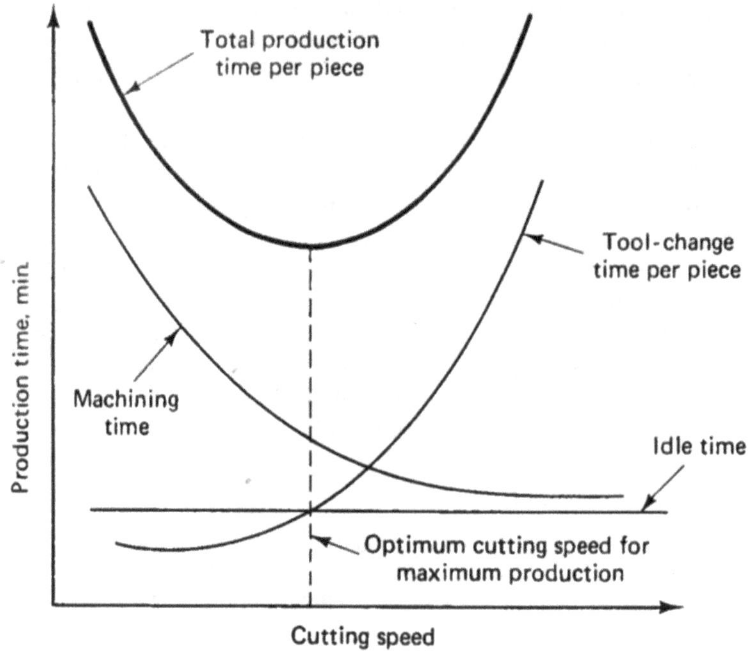

FIGURE 10.20 Relationship between production time per piece and cutting speed.

C_T is the original cost of the cutting tool in \$
N_1 is the number of times the tool can be ground
N_2 is the number of cutting edges obtained per grind

The total equivalent minutes required to replace the tool when it becomes dull:

$$T_e = t_1 + t_2 + t_3$$

The amount of metal Q in in.^{3} is cut during the life of one tool that is $(T + T_e)$, where T is the tool life. Thus, the rate of metal removal in in.^{3}/min is

$$R = \frac{Q}{(T+T_e)} = \frac{w \times d \times v_c \times T}{(T+T_e)} = \frac{T^{1-n} \times \text{constant}}{(T+T_e)}$$

where

w is the width of the chip
d is the depth of the cut

substitute for v_c by its value from Taylor's equation, then differentiate R with respect to T to obtain the max value of R, as follows:

$$dR / dT = 0$$

Therefore,

$$T_{\text{economical}} = \left(\frac{1}{n} - 1\right) \times T_e \tag{10.28}$$

The economical cutting speed that corresponds to the economical tool life can be found from Taylor's equation.

REVIEW QUESTIONS

1. How can the complex process of metal cutting be approached?
2. Define the *rake angle* and the *clearance angle* in two-dimensional cutting.
3. Why are the angles in question 2 required?
4. What is the upper surface of the tool called?
5. What is the lower surface of the tool called?
6. What are the cutting variables that affect the values of the rake and clearance angles?
7. List some drawbacks if the cutting angles are not properly chosen.
8. When should the rake angle be taken as a positive value?
9. When should the rake angle be taken as a negative value?
10. Can orthogonal cutting actually take place? Explain. Use sketches to explain the stages involved in the formation of chips during machining.
11. Use sketches to illustrate the different types of machining chips and explain when and why we can expect to have each of these types.
12. Explain the stages involved in the formation of the built-up edge.
13. Does the built-up edge have useful or harmful effects?
14. What is meant by the shear angle?
15. What is meant by the cutting ratio?
16. Derive an expression for the shear strain that takes place during orthogonal cutting.
17. Draw a sketch of the cutting force diagram proposed by Ernst and Merchant.
18. How can the relationship between the shear and rake angles be expressed according to Ernst and Merchant?
19. On what basis have Lee and Shaffer developed their theory?
20. Derive an expression for the specific energy during two-dimensional cutting.
21. Illustrate the difference between orthogonal and oblique cutting.
22. What are the components of the cutting force in oblique cutting? How do you compare their magnitudes with each other?
23. Define the *unit horsepower*.
24. Describe fully the geometry of single-point cutting tools.
25. Explain the effect of each of the cutting angles in oblique cutting on the mechanics of the process.
26. List the different cutting tool materials and enumerate the advantages, disadvantages, and applications of each.
27. What are the two main causes of tool wear?
28. List the different kinds of tool wear.
29. Define *tool life*.
30. What is the relationship between tool life and cutting speed?
31. Define *machinability* and explain how it is quantitatively expressed by the machinability index.
32. What are the necessary characteristics of cutting fluids?
33. List the different types of cutting fluids and provide the advantages and limitations of each.
34. What are the causes of forced vibrations during machining?
35. How can forced vibrations be minimized?
36. What is chatter and why does it occur?
37. How can we eliminate chatter?
38. What trouble can vibrations cause during machining?
39. Use sketches to explain how the value of the optimum cutting speed can be obtained for maximum economy and for maximum productivity.

PROBLEMS

1. In a turning operation, the diameter of the workpiece is 2 inches (50 mm), and it rotates at 360 revolutions per minute. How long would a carbide tool last ($n = 0.3$) under such conditions if an identical carbide tool lasted for 1 minute when used at 1000 feet per minute (305.0 m/min)?
2. Determine the increase in the tool life of a carbide tip as a result of a decrease in the cutting speed of 25, 50, and 75 percent.
3. When turning a thin tube at its edge, the following conditions were observed:
4. Calculate the following:
   a. Cutting ratio
   b. Shear angle
   c. Chip velocity
5. A geared-head lathe is employed for machining steel AISI 1055, BHN 250. The cutting speed is 400 feet per minute, and the rate of metal removal is 2.4 cubic inches per minute. If the tool used has the character 0-7-7-7-15-15-1/32, estimate the following:
   a. The energy consumed in machining per unit
   b. The power required at the motor
   c. The tangential component of the cutting force
6. Neglect the correction factor for the undeformed chip thickness.
7. A 5-hp, 2-V belt-driven lathe is to be used for machining brass under the following conditions:
8. Neglect the effect of chip thickness.
9. Does this lathe have enough power for the required job?

Cutting speed	600 ft/min
Rate of metal removal	7.2 in.^{3}/min
SCEA of the tool	30°

Depth of cut	0.125 inches
Chip thickness	0.15 inches
Back rake angle	8°

DESIGN PROJECT

Prepare a computer program that determines the optimum cutting speed that results in maximum productivity. The program should be interactive, the input being workpiece material, tool material, and depth of cut. Assume the time for changing the tool is 60 seconds and the time to return the tool to the beginning of the cut is 20 seconds. Take the workpiece material to be

1. Steel 1020
2. Brass
3. Aluminum
4. Stainless steel

11 Machining of Metals

11.1 INTRODUCTION

This chapter will focus on the technological aspects of the different machining operations, as well as the design features of the various machine tools employed to perform those operations. In addition, the different shapes and geometries produced by each operation, the tools used, and the workholding devices will be covered. Special attention will be given to the required workshop calculations that are aimed at estimating machining parameters such as cutting speeds and feeds, metal-removal rate, and machining time.

Machine tools are designed to drive the cutting tool in order to produce the desired machined surface. For such a goal to be accomplished, a machine tool must include appropriate elements and mechanisms capable of generating the following motions:

- A relative motion between the cutting tool and the workpiece in the direction of cutting
- A motion that enables the cutting tool to penetrate into the workpiece until the desired depth of cut is achieved
- A feed motion that repeats the cutting action every round or every stroke to ensure the continuation of the cutting operation

11.2 TURNING OPERATIONS

11.2.1 The Lathe and Its Construction

A *lathe* is a machine tool used for producing surfaces of revolution and flat edges. Based on their purpose, construction, number of tools that can simultaneously be mounted, and degree of automation, lathes, or, more accurately, lathe-type machine tools, can be classified as follows:

- Engine lathes
- Tool room lathes
- Turret lathes
- Vertical turning and boring mills
- Automatic lathes
- Special-purpose lathes

Despite the diversity of lathe-type machine tools, there are common features with respect to construction and principles of operation. These features can be illustrated by considering the commonly used representative type, the *engine lathe*, which is shown in Figure 11.1. Following is a description of each of the main elements of an engine lathe.

11.2.1.1 Lathe Bed

The *lathe bed* is the main frame, a horizontal beam on two vertical supports. It is usually made of gray or nodular cast iron to damp vibrations and is made by casting. It has guideways that allow the carriage to slide easily, lengthwise. The height of the lathe bed should be such that the technician can do his or her job easily and comfortably.

DOI: 10.1201/9781003537106-11

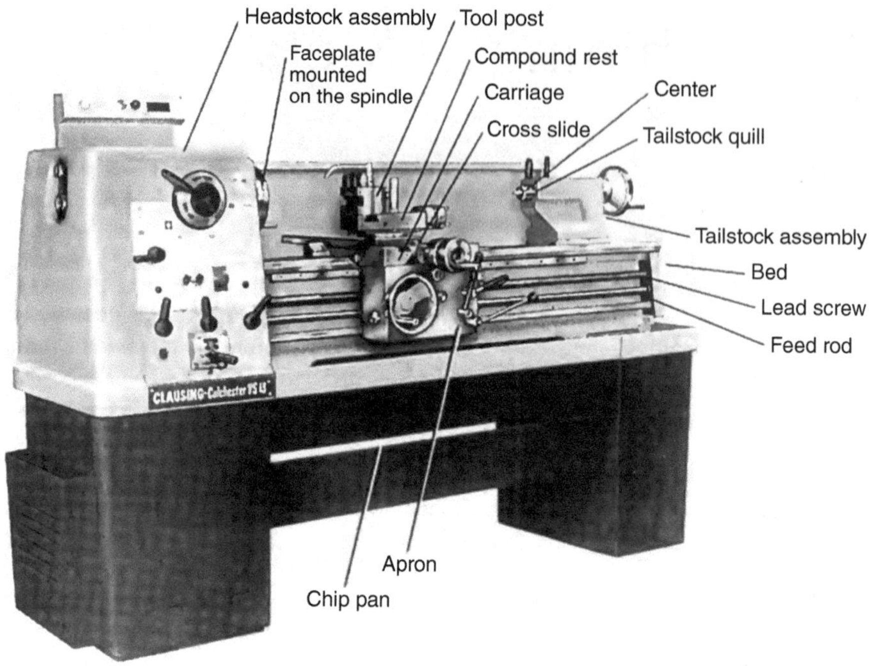

FIGURE 11.1 An engine lathe. (Courtesy of Clausing Industrial, Inc., Kalamazoo, Michigan.)

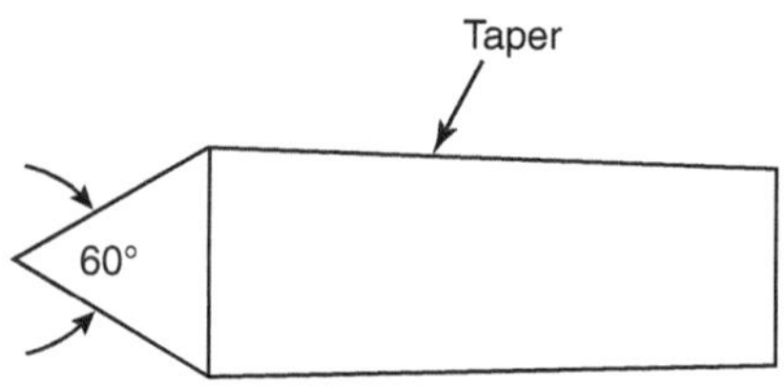

FIGURE 11.2 A plain lathe center.

11.2.1.2 Headstock

The *headstock* assembly is fixed at the left-hand side of the lathe bed and includes the *spindle*, whose axis is parallel to the guideways (the slide surface of the bed). The spindle is driven through the gearbox, which is housed within the headstock. The function of the gearbox is to provide a number of different spindle speeds (usually 6–18 speeds). Some modern lathes have headstocks with infinitely variable spindle speeds and that employ frictional, electrical, or hydraulic drives.

The spindle is always hollow (i.e., it has a through hole extending lengthwise). Bar stocks can be fed through the hole if continuous production is adopted. Also, the hole has a tapered surface to allow the mounting of a plain lathe center, such as the one shown in Figure 11.2. It is made of hardened tool steel. The part of the lathe center that fits into the spindle hole has a Morse taper, while the other part of the center is conical with a 60° apex angle. As explained later, lathe centers are used for mounting long workpieces. The outer surface of the spindle is threaded to allow the mounting of a chuck, a faceplate, or the like.

11.2.1.3 Tailstock

The *tailstock* assembly consists basically of three parts: its lower base, an intermediate part, and the quill. The lower base is a casting that can slide on the lathe bed along the guideways, and it has a clamping device so that the entire tailstock can be locked at any desired location, depending upon the length of the workpiece. The intermediate part is a casting that can be moved transversely so that the axis of the tailstock can be aligned with that of the headstock. The third part, called the *quill*, is a hardened steel tube that can be moved longitudinally in and out of the intermediate part as required. This is achieved through the use of a hand wheel and a screw, around which a nut fixed to the quill is engaged. The hole in the open side of the quill is tapered to allow the mounting of lathe centers or other tools like twist drills or boring bars. The quill can be locked at any point along its travel path by means of a clamping device.

11.2.1.4 Carriage

The main function of the *carriage* is to mount the cutting tools and generate longitudinal and/or cross-feeds. It is actually an H-shaped block that slides on the lathe bed between the headstock and tailstock while being guided by the V-shaped guideways of the bed. The carriage can be moved either manually or mechanically by means of the apron and either the feed rod or the lead screw.

The *apron* is attached to the saddle of the carriage and serves to convert the rotary motion of the feed rod (or lead screw) into linear longitudinal motion of the carriage and, accordingly, the cutting tool (i.e., it generates the axial feed). The apron also provides powered motion for the cross-slide located on the carriage. Usually, the tool post is mounted on the compound rest, which is, in turn, mounted on the cross-slide. The compound rest is pivoted around a vertical axis so that the tools can be set at any desired angle with respect to the axis of the lathe (and that of the workpiece). These various components of the carriage form a system that provides motion for the cutting tool in two perpendicular directions during turning operations.

When cutting screw threads, power is provided from the gearbox to the apron by the lead screw. In all other turning operations, it is the feed rod that drives the carriage. The lead screw goes through a pair of half nuts that are fixed to the rear of the apron. When actuating a certain lever, the half nuts are clamped together and engage with the rotating lead screw as a single nut that is fed, together with the carriage, along the bed. When the lever is disengaged, the half nuts are released and the carriage stops. On the other hand, when the feed rod is used, it supplies power to the apron through a worm gear. This gear is keyed to the feed rod and travels with the apron along the feed rod, which has a keyway extending along its entire length. A modern lathe usually has a quick-change gearbox located under the headstock and driven from the spindle through a train of gears. It is connected to both the feed rod and the lead screw so that a variety of feeds can easily and rapidly be selected by simply shifting the appropriate levers. The quick-change gearbox is employed in plain turning, facing, and thread-cutting operations. Because the gearbox is linked to the spindle, the distance that the apron (and the cutting tool) travels for each revolution of the spindle can be controlled and is referred to as the *feed*.

11.2.2 The Turret Lathe

A *turret lathe* is similar to an engine lathe, except that the conventional tool post is replaced with a hexagonal (or octagonal) turret that can be rotated around a vertical axis as required. Appropriate tools are mounted on the six (or eight) sides of the turret. The length of each tool is adjusted so that, by simply indexing the turret, any tool can be brought into the exact desired operating position. These cutting tools can, therefore, be employed successively without the need for dismounting the tool and mounting a new one each time, as is the case with conventional engine lathes. This results in an appreciable saving in the time required for setting up the tools. Also, on a turret lathe, a skilled machinist is required only initially to set up the tools. A laborer with limited training can operate the turret lathe thereafter and produce parts identical to those that can be manufactured when a skilled machinist operates the lathe. Figure 11.3 illustrates a top view of a hexagonal turret with six different tools mounted on its sides. Sometimes, the turret replaces the tailstock and can be either vertical (i.e., with a horizontal axis) or horizontal (i.e., with a vertical axis). In this case, four additional tools can be mounted on the square tool post, sometimes called a *square turret*, thus allowing 12 machining operations to be performed successively. Turret lathes always have work-holding devices with quick-release (and quick-tightening) mechanisms.

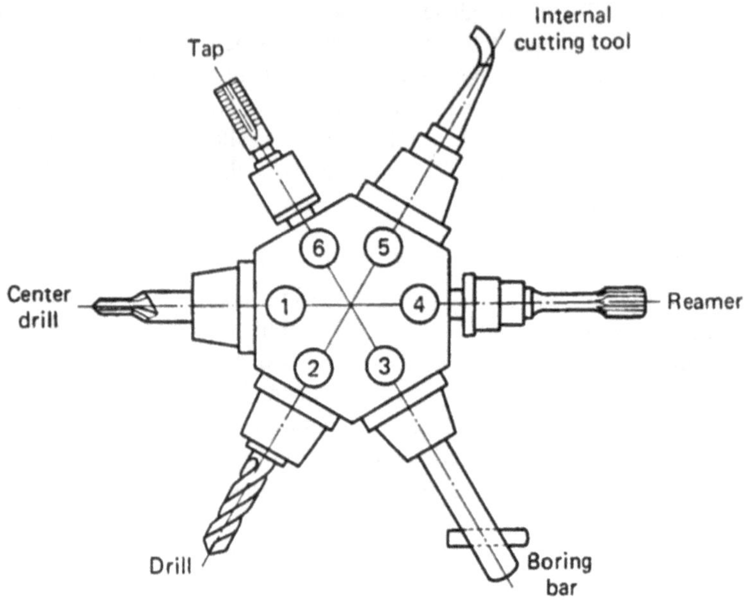

FIGURE 11.3 Top view of a hexagonal turret with six different tools.

11.2.3 Specifying a Lathe

It is important for a manufacturing engineer to be able to specify a lathe in order to place an order or to compare and examine contract bids. The specifications of a lathe should involve data that reveal the dimensions of the largest workpiece to be machined on that lathe. They also must include the power consumption, as well as information that is needed for shipping and handling. Table 11.1 indicates an example of how to specify a lathe.

TABLE 11.1
Example of Specification of a Lathe

Model	Example
Maximum swing over bed (largest diameter of workpiece)	12 in. (300 mm)
Maximum swing over carriage (largest diameter over carriage)	8 in. (200 mm)
Hole through spindle	0.75 in. (19 mm)
Height of centers	6 in. (150 mm)
Turning length	24 in. (600 mm)
Thread on spindle nose	
Taper in spindle and tailstock sleeves	3 Morse
21 spindle speeds	20–2000 rev/min
Metric threads	2–6 mm
Whitworth	4–28 teeth
Feeds per revolution	0.0002–0.0008 in. (0.05–0.2 mm)
Power required	1.6 kW
Net weight	1 ton
Floor space requirement (length × width × height)	64 × 36 × 56 in. (1600 × 900 × 1400 mm)

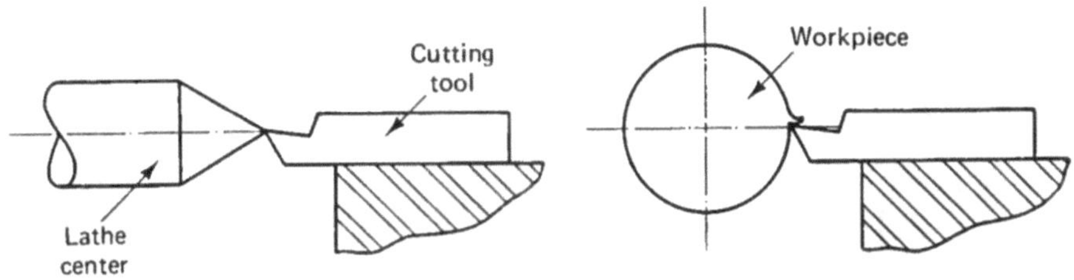

FIGURE 11.4 A simple method for tool setup.

11.2.4 Tool Holding

Tools for turning operations are mounted in a toolholder (tool post). On an engine lathe, it is located on the compound rest. More than one cutting tool (up to four) can be mounted in the toolholder in order to save the time required for changing and setting up each tool should only one tool be mounted at a time. In all turning operations, the following conditions for holding the tools must be fulfilled:

- The tip of the such as the one shown in with the level of the lathe axis. This can be achieved by using the pointed edge of the lathe center as a basis for adjustment, as shown in Figure 11.4. Failure to meet this condition results in a change in the values of the cutting angles from the desired ones.
- The centerline of the cutting tool must be horizontal.
- The tool must be fixed tightly along its length and not just on two points.
- A long tool overhang should be avoided in order to eliminate any possibility of elastic strains and vibrations.

11.2.5 Lathe Cutting Tools

The shape and geometry of lathe cutting tools depend upon the purpose for which they are employed. Turning tools can be classified into two main groups: *external* cutting tools and *internal* cutting tools.

Each of these groups includes the following types of tools:

- *Turning Tools*. Turning tools can be either *finishing* or *rough* turning tools. Rough turning tools have small nose radii and are employed when deep cuts are made. Finishing tools have larger nose radii and are used when shallower cuts are made in order to obtain the final required dimensions with a good surface finish. Rough turning tools can be a right-hand or left-hand tool, depending upon the direction of the feed. They can have straight, bent, or offset shanks. Figure 11.5 illustrates the different kinds of turning tools.
- *Facing Tools*. Facing tools are employed in facing operations for machining flat-side or end surfaces. As can be seen in Figure 11.6, there are tools for machining both left- and right-side surfaces. These side surfaces are generated through the use of cross-feed, contrary to turning operations, where longitudinal feed is used.
- *Cutoff Tools*. Cutoff tools, which are sometimes called *parting tools*, serve to separate the workpiece into parts and/or machine external annular grooves, as shown in Figure 11.7.
- *Thread-Cutting Tools*. Thread-cutting tools have triangular, square, or trapezoidal cutting edges, depending upon the cross section of the desired thread. Also, the plane angles of these tools must always be identical to those of the thread forms. Thread-cutting tools have straight shanks for external thread cutting and bent shanks for internal thread cutting. Figure 11.8 illustrates the different shapes of thread-cutting tools.
- *Form Tools*. As shown in Figure 11.9, form tools have edges specially manufactured to take a form that is opposite to the desired shape of the machined workpiece.

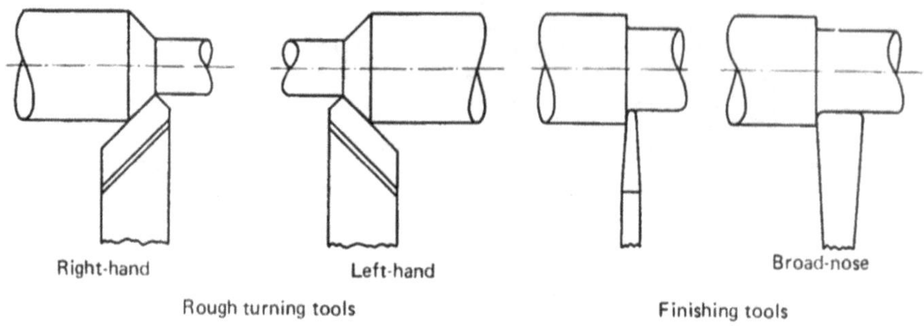

FIGURE 11.5 Different kinds of turning tools.

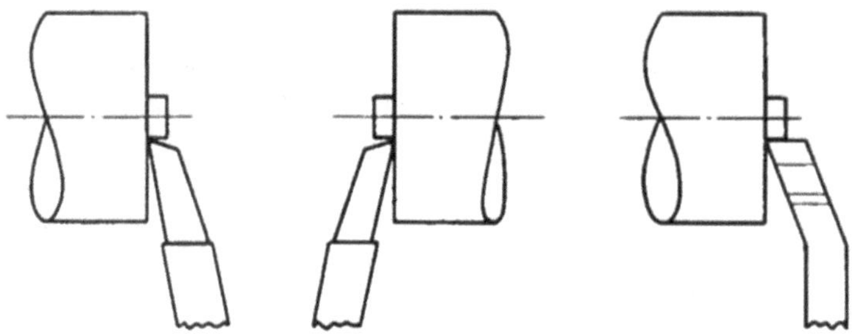

FIGURE 11.6 Different kinds of facing tools.

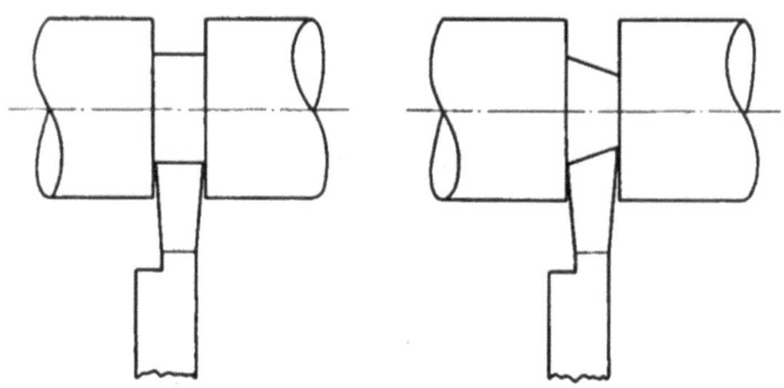

FIGURE 11.7 Cutoff tools.

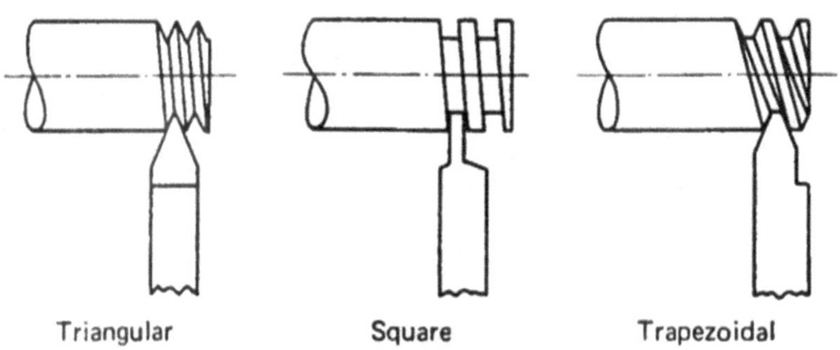

FIGURE 11.8 Different shapes of thread-cutting tools.

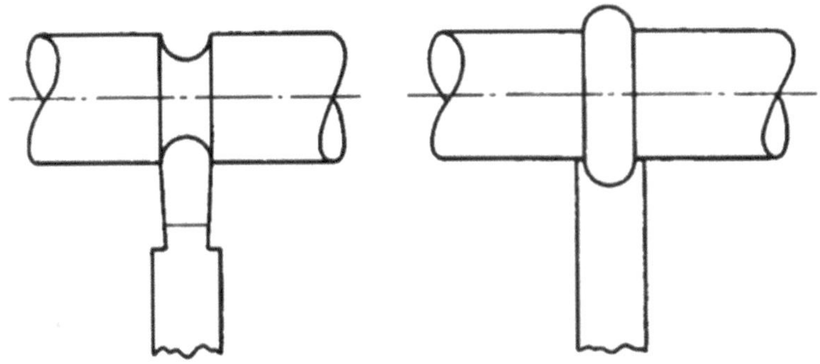

FIGURE 11.9 Form tools.

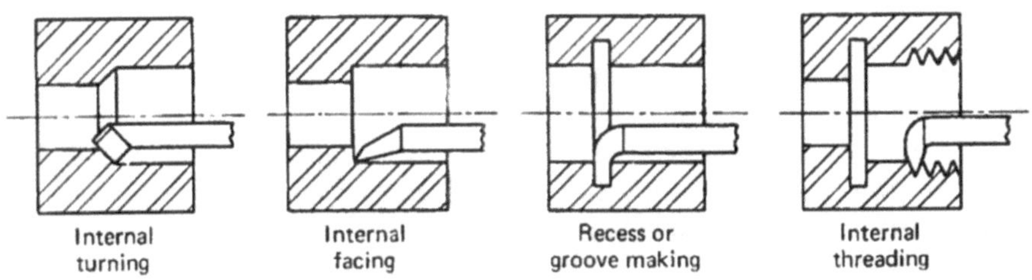

FIGURE 11.10 Different types of internal cutting tools.

11.2.5.1 Internal and External Tools

The types of internal cutting tools are similar to those of the external cutting tools. They include tools for rough turning, finish turning, thread cutting, and recess machining.

Figure 11.10 illustrates the different types of internal cutting tools.

11.2.5.2 Carbide Tips

As previously mentioned, an HSS tool is usually made in the form of a single piece, contrary to cemented carbides or ceramics, which are made in the form of tips. The tips are brazed or mechanically fastened to steel shanks. Figure 11.11 shows an arrangement that includes a carbide tip, a chip breaker, a seat, a clamping screw (with a washer and a nut), and a shank. As its name suggests, the function of a *chip breaker* is to break long chips every now and then, thus preventing the formation of very long, twisted ribbons that may cause problems during the machining operation. As shown in Figure 11.12, the carbide tips (or ceramic tips) have different shapes, depending upon the machining operations for which they are to be employed. The tips can either be solid or have a central through hole, depending upon whether brazing or mechanical clamping is employed for mounting the tip on the shank.

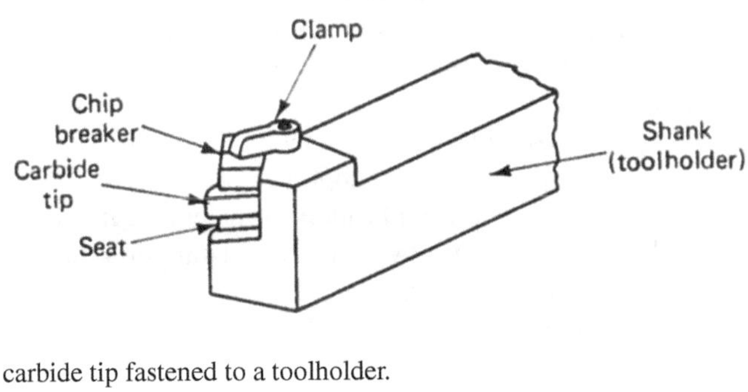

FIGURE 11.11 A carbide tip fastened to a toolholder.

FIGURE 11.12 Different shapes of carbide tips.

11.2.6 Methods of Supporting Workpieces in Lathe Operations

Some precautions must be taken when mounting workpieces on a lathe to ensure trouble-free machining. They can be summarized as follows:

- It is recommended that an appropriate gripping force that is neither too high nor too low be used. A high gripping force may result in distortion of the workpiece after the turning operation, whereas a low gripping force causes either vibration of the workpiece or slip between the workpiece and the spindle (i.e., the rotational speed, or rpm, of the workpiece will be lower than that of the spindle).
- The workpiece must be fully balanced, both statically and dynamically, by employing counterweights and the like if necessary.
- The cutting force should not affect the shape of the workpiece or cause any permanent deformation. A manufacturing engineer should calculate the cutting force using his or her knowledge of metal cutting (see Chapter 10) and then check whether or not such a force will cause permanent deformation by using stress analysis. Such calculations are very important when machining slender workpieces (i.e., those with high length-to-diameter ratios). Whenever it becomes evident that the cutting force will cause permanent deformation, the machining parameters must be changed to reduce the magnitude of the force (e.g., use a smaller depth of cut or lower feed). Following is a brief discussion of each of the work-holding methods employed in lathe operations.

11.2.6.1 Holding the Workpiece between Two Centers

The workpiece is held between two centers when turning long workpieces like shafts and axles having length-to-diameter ratios higher than 3 or 4. Before a workpiece is held, each of its flat ends must be prepared by drilling a 60° center hole. The pointed edges of the *live center* (mounted in the tailstock so that its conical part rotates freely with the workpiece) and the *dead center* (mounted in the spindle hole) are inserted in the previously drilled center holes.

As shown in Figure 11.13, a driving dog is clamped on the left end of the workpiece by means of a tightening screw. The tail of the lathe dog enters a slot in the driving-dog plate (or faceplate), which is screwed on the spindle nose. When very long workpieces having length-to-diameter ratios of 10 or more are turned between centers, rests must be used to provide support and prevent sagging of the workpiece at its middle. *Steady rests* are clamped on the lathe bed and thus do not move during the machining operation; *follower rests* are bolted to and travel with the carriage. A steady rest employs three adjustable fingers to support the workpiece. However, in high-speed turning, the steady rest should involve balls and rollers at the end of the fingers where the workpiece is supported. A follower rest has only two fingers and supports the workpiece against the cutting tool. A steady rest can be used as an alternative to the tailstock for supporting the right-hand end of the workpiece. Figure 11.14 illustrates a steady rest used to support a very long workpiece.

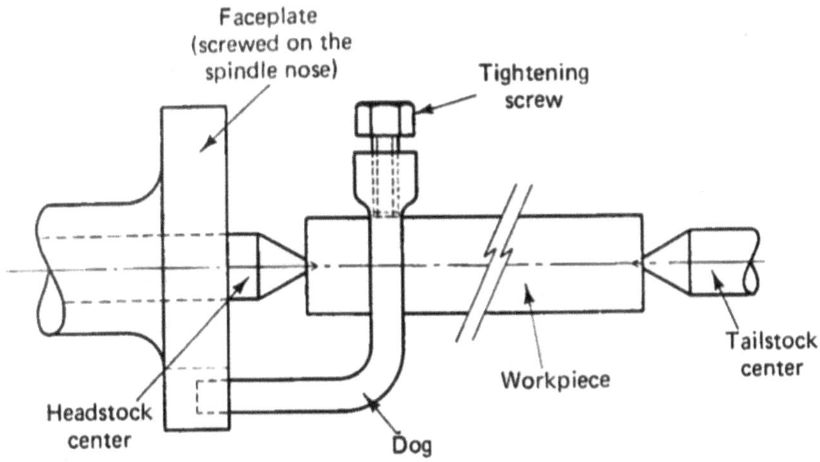

FIGURE 11.13 Holding the workpiece between two centers during turning.

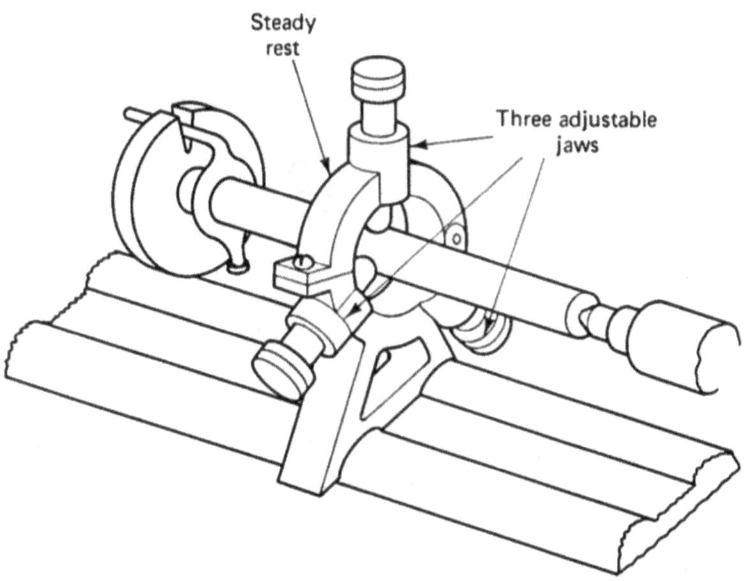

FIGURE 11.14 A steady rest used to support a very long workpiece.

11.2.6.2 Holding the Workpiece in a Chuck

When turning short workpieces and/or when performing facing operations, the workpiece is held in a *chuck*, which is screwed on the spindle nose. A universal, self-centering chuck has three jaws that can be moved separately or simultaneously in radial slots toward its center to grip the workpiece or away from its center to release the workpiece. This movement is achieved by inserting a chuck wrench into a square socket and then turning it as required. Four-jaw chucks are also employed; these are popular when turning complex workpieces and those having asymmetric shapes. Magnetic chucks (without jaws) are used to hold thin, flat workpieces for facing operations. There are also pneumatic and hydraulic chucks, and they are utilized for speeding up the processes of loading and unloading the workpieces. Figure 11.15 shows how a workpiece is held in a chuck.

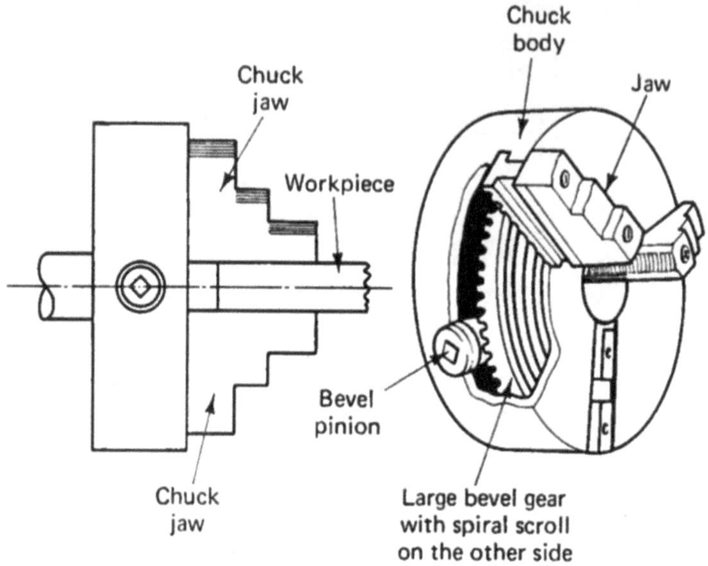

FIGURE 11.15 Holding the workpiece in a chuck.

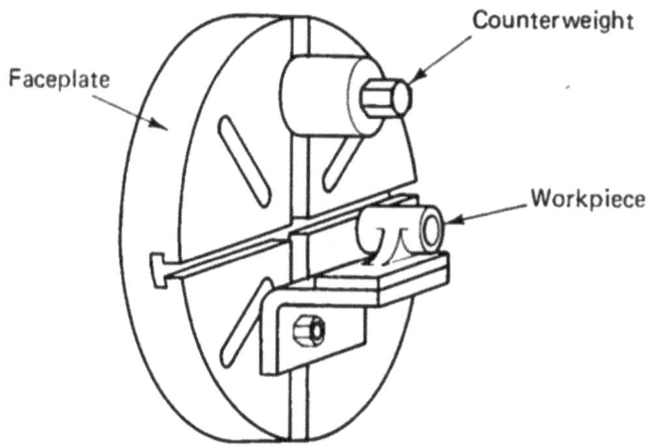

FIGURE 11.16 Mounting the workpiece on a faceplate.

11.2.6.3 Mounting the Workpiece on a Faceplate

A *faceplate* is a large circular disk with radial plain slots and T-slots in its face. The workpiece can be mounted on it with the help of bolts, T-nuts, and other means of clamping. The faceplate is usually employed when the workpiece to be gripped is large or noncircular or has an irregular shape and cannot, therefore, be held in a chuck. Before any machining operation, the faceplate and the workpiece must be balanced by a counterweight mounted opposite to the workpiece on the faceplate, as shown in Figure 11.16.

11.2.6.4 Using a Mandrel

Disklike workpieces or those that have to be machined on both ends are mounted on *mandrels*, which are held between the lathe centers. In this case, the mandrel acts like a fixture and can take different forms. As Figure 11.17 shows, a mandrel can be a truncated conical rod with an intangible

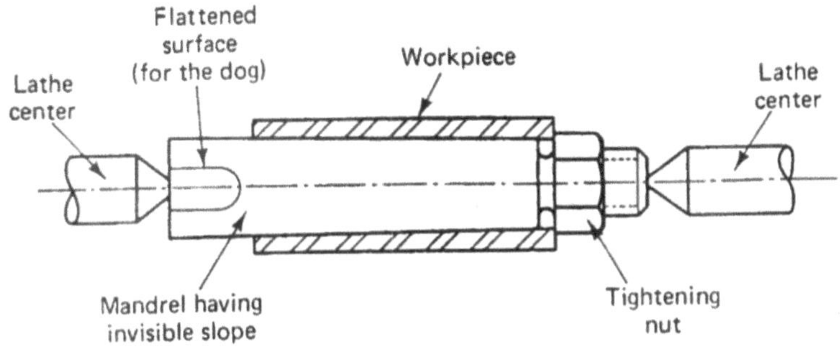

FIGURE 11.17 Mounting the workpiece on a mandrel.

slope on which the workpiece is held by the wedge action. A split sleeve that is forced against a conical rod is also employed. There are also some other designs for mandrels.

11.2.6.5 Holding the Workpiece in a Chuck Collet

A *chuck collet* consists of a three-segment split sleeve with an external tapered surface. The collet can grip a smooth bar placed between these segments when a collet sleeve, which is internally tapered, is pushed against the external tapered surface of the split sleeve, as shown in Figure 11.18.

11.2.7 Lathe Operations

The following sections focus on the various machining operations that can be performed on a conventional engine lathe. It must be borne in mind, however, that modern computerized numerically controlled (CNC) lathes have more capabilities and can do other operations, such as contouring, for example. The following are the conventional lathe operations.

11.2.7.1 Cylindrical Turning

Cylindrical turning is the simplest and the most common of all lathe operations. A single full turn of the workpiece generates a circle whose center falls on the lathe axis; this motion is then reproduced numerous times as a result of the axial feed motion of the tool. The resulting machining marks are, therefore, a helix having a very small pitch, which is equal to the feed. Consequently, the machined surface is always cylindrical.

The axial feed is provided by the carriage or compound rest, either manually or automatically, whereas the depth of cut is controlled by the cross-slide. In roughing cuts, it is recommended that

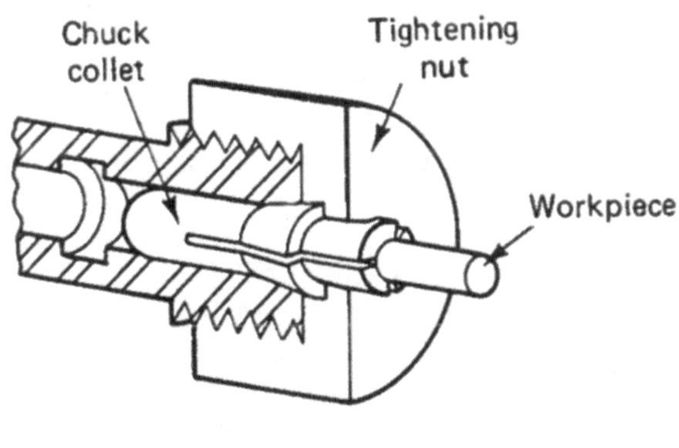

FIGURE 11.18 Holding the workpiece in a chuck collet.

Operation		Cutting Speed	Machining Time	Material-Removal Rate
Turning (external)	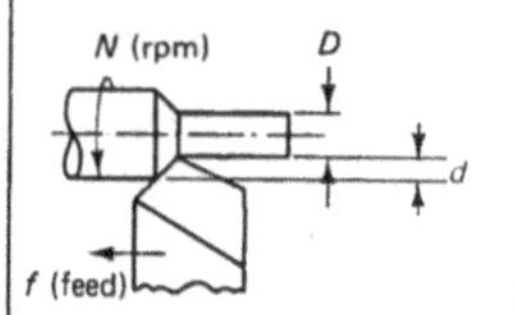	$V = \pi(D + 2d)N$	$T = \frac{L}{fN}$ where $L = L_{workpiece}$ + allowance i.e., length of the workpiece plus allowance	$MRR = \pi(D + d)N \cdot f \cdot d$
Boring	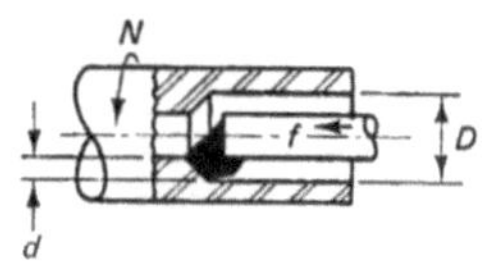	$V = \pi DN$	$T = \frac{L}{fN}$	$MRR = \pi(D - d)N \cdot f \cdot d$
Facing	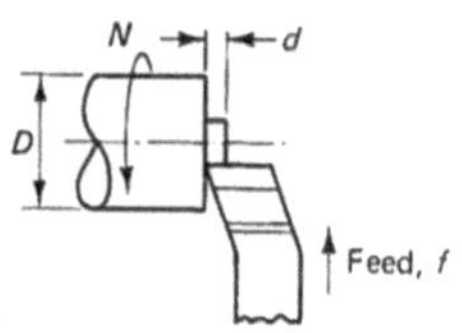	max. $V = \pi DN$ min. $V = 0$ mean $V = \frac{\pi DN}{2}$	$T = \frac{D + \text{allowance}}{2fN}$	max. $MRR = \pi DN \cdot f \cdot d$ mean $MRR = \frac{\pi DN \cdot f \cdot d}{2}$
Parting	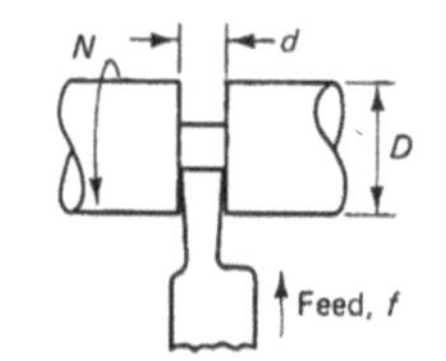	max. $V = \pi DN$ min. $V = 0$ mean $V = \frac{\pi DN}{2}$	$T = \frac{D + \text{allowance}}{2fN}$	max. $MRR = \pi DN \cdot f \cdot d$ mean $MRR = \frac{\pi DN \cdot f \cdot d}{2}$

FIGURE 11.19 Equations applicable to lathe operations.

large depths of cuts, up to 1/4 inch (6 mm) depending upon the workpiece material, and smaller feeds be used. On the other hand, very fine feeds, smaller depths of cut less than 0.05 inch (0.4 mm), and high cutting speeds are preferred for finishing cuts. Figure 11.19 indicates the equations used to estimate the different machining parameters in cylindrical turning.

11.2.7.2 Facing

The result of a *facing* operation is a flat surface that is either the entire end surface of the workpiece or an annular intermediate surface like a shoulder. During a facing operation, feed is provided by the cross-slide, whereas the depth of cut is controlled by the carriage or compound rest. Facing can be carried out either from the periphery inward or from the center of the workpiece outward. It is obvious that the machining marks in both cases take the form of a spiral. Usually, it is preferred to clamp the carriage during a facing operation as the cutting force tends to push the tool (and, of course, the whole carriage) away from the workpiece. In most facing operations, the workpiece is held in a chuck or on a faceplate. Figure 11.19 also indicates the equations applicable to facing operations.

11.2.7.3 Groove Cutting

In *cutoff* and *groove-cutting* operations, only cross-feed of the tool is employed. The cutoff and grooving tools that were previously discussed are employed.

11.2.7.4 Boring and Internal Turning

Boring and *internal turning* are performed on the internal surfaces by a boring bar or suitable internal cutting tool. If the initial workpiece is solid, a drilling operation must be performed first. The drilling tool is held in the tailstock, which is then fed against the workpiece.

11.2.7.5 Taper Turning

Taper turning is achieved by driving the tool in a direction that is not parallel to the lathe axis but inclined to it with an angle that is equal to the desired angle of the taper. Following are the different methods used in taper turning:

- One method is to rotate the disk of the compound rest with an angle equal to half the apex angle of the cone, as is shown in Figure 11.20. Feed is manually provided by cranking the handle of the compound rest. This method is recommended for the taper turning of external and internal surfaces when the taper angle is relatively large.
- Special form tools can be used for external, very short, conical surfaces, as shown in Figure 11.21 The width of the workpiece must be slightly smaller than that of the tool, and the workpiece is usually held in a chuck or clamped on a faceplate. In this case, only the cross-feed is used during the machining process, and the carriage is clamped to the machine bed.
- The method of offsetting the tailstock center, as shown in Figure 11.22, is employed for the external taper turning of long workpieces that are required to have small taper angles (less than 8°). The workpiece is mounted between the two centers and then the tailstock center

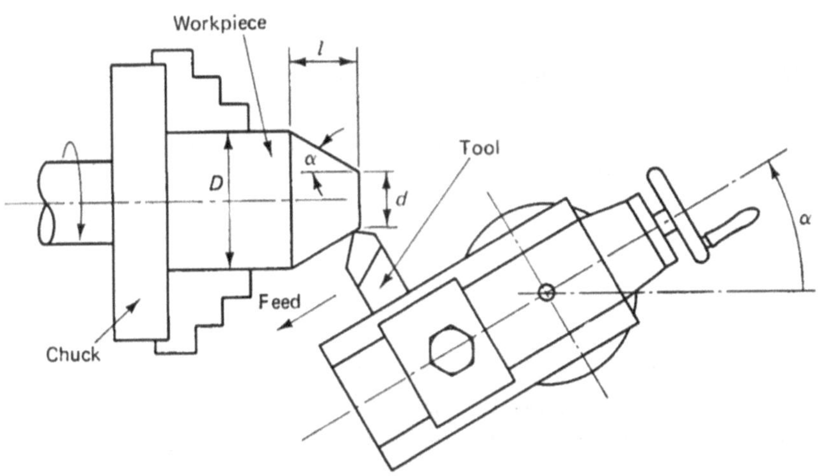

FIGURE 11.20 Taper turning by rotating the disk of the compound rest.

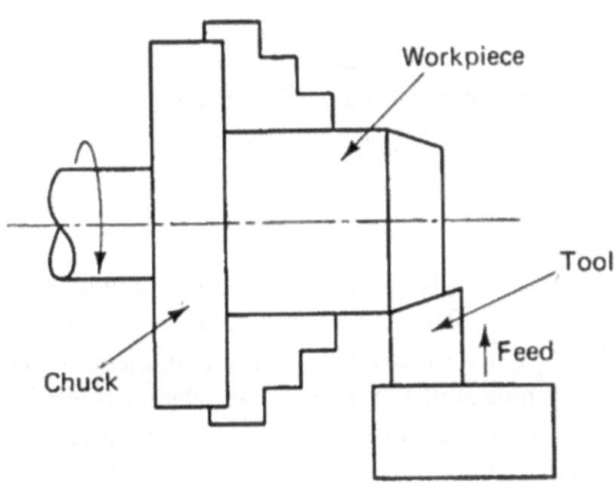

FIGURE 11.21 Taper turning by employing a form tool.

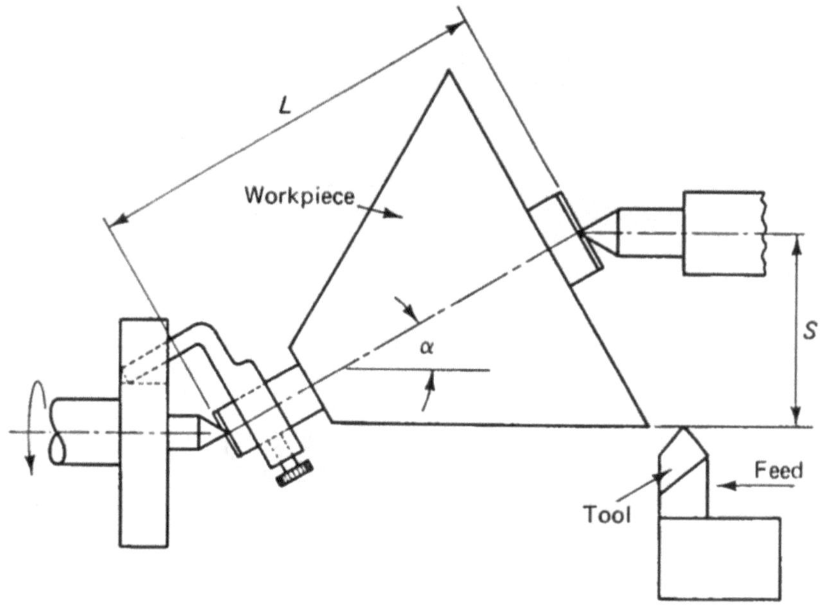

FIGURE 11.22 Taper turning by offsetting the tailstock center.

is shifted a distance S in the direction normal to the lathe axis. This distance can be obtained from the following equation:

$$S = \frac{L(D-d)}{2l} \tag{11.1}$$

where

- L is the full length of the workpiece
- D is the largest diameter of the workpiece
- d is the smallest diameter of the workpiece
- l is the length of the tapered surface

- A special taper-turning attachment, such as the one shown in Figure 11.23, is used for turning very long workpieces when the length is larger than the full stroke of the compound rest. The procedure followed in such cases involves complete disengagement of the cross-slide from the carriage, which is then guided by the taper-turning attachment. During this process, the automatic axial feed can be used as usual. This method is recommended for very long workpieces with a small cone angle (8° through 10°).

11.2.7.6 Thread Cutting

For *thread cutting*, the axial feed must be kept at a constant rate, which is dependent upon the rotational speed (rpm) of the workpiece. The relationship between both is determined primarily by the desired pitch of the thread to be cut. As previously mentioned, the axial feed is automatically generated when cutting a thread by means of the lead screw, which drives the carriage. When the lead screw rotates a single revolution, the carriage travels a distance equal to the pitch of the lead screw. Consequently, if the rotational speed of the lead screw is equal to that of the spindle (i.e., that of the workpiece), the pitch of the resulting cut thread is exactly equal to that of the lead screw. The pitch of the resulting thread being cut, therefore, always depends upon the ratio of the rotational speeds of the lead screw and the spindle.

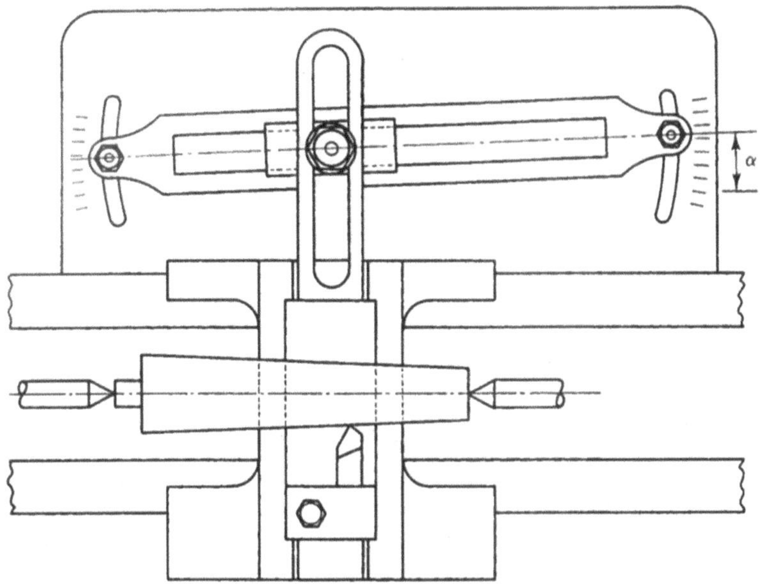

FIGURE 11.23 Taper turning by employing a special attachment.

$$\frac{\text{Pitch of lead screw}}{\text{Desired pitch of workpiece}} = \frac{\text{rpm of workpiece}}{\text{rpm of lead screw}} = \text{Spindle-to-carriage gearing ratio} \tag{11.2}$$

This equation is useful in determining the kinematic linkage between the lathe spindle and the lead screw and enables proper selection of the gear train between them. In thread-cutting operations, the workpiece can be either held in a chuck or mounted between two lathe centers for relatively long workpieces. The form of the tool used must exactly coincide with the profile of the thread to be cut (i.e., triangular tools must be used for triangular threads, and so on).

11.2.7.7 Knurling

Knurling is basically a forming operation in which no chips are produced. It involves pressing two hardened rolls with rough file-like surfaces against the rotating workpiece to cause plastic deformation of the workpiece metal, as shown in Figure 11.24. Knurling is carried out to produce rough, cylindrical (or conical) surfaces that are usually used as handles. Sometimes, surfaces are knurled just for the sake of decoration, in which case there are different knurl patterns to choose from.

11.2.8 Cutting Speeds and Feeds

The cutting speed, which is usually given in surface feet per minute (SFM), is the number of feet traveled in the circumferential direction by a given point on the surface (being cut) of the workpiece in 1 minute. The relationship between the surface speed and the rpm can be given by the following equation:

$$\text{SFM} = \pi DN$$

where

D is the diameter of the workpiece in feet
N is the rpm

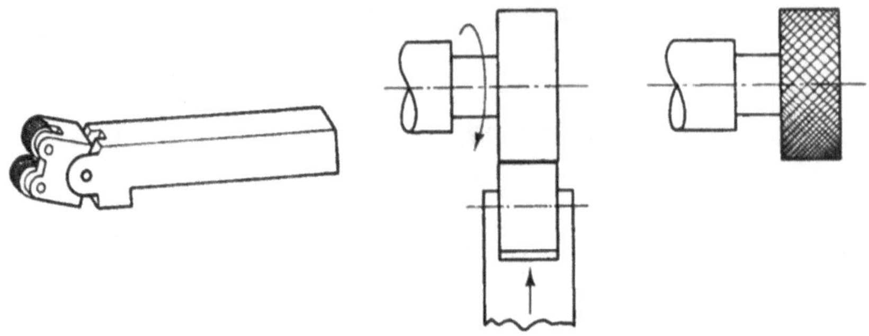

FIGURE 11.24 The knurling operations

The surface cutting speed is dependent upon the material being machined as well as the material of the cutting tool and can be obtained from handbooks and information provided by cutting tool manufacturers. Generally, the SFM is taken as 100 when machining cold-rolled or mild steel, as 50 when machining tougher metals, and as 200 when machining softer materials. For aluminum, the SFM is usually taken as 400 or above. Other variables also affect the optimal value of the surface cutting speed. These include the tool geometry, the type of lubricant or coolant, the feed, and the depth of cut. As soon as the cutting speed is decided upon, the rotational speed (rpm) of the spindle can be obtained as follows:

$$N = \frac{\text{SFM}}{\pi D} \tag{11.3}$$

The selection of a suitable feed depends upon many factors, such as the required surface finish, the depth of cut, and the geometry of the tool used. Finer feeds will produce a better surface finish, whereas higher feeds reduce the machining time during which the tool is in direct contact with the workpiece. Therefore, it is generally recommended to use high feeds for roughing operations and finer feeds for finishing operations. Again, recommended values for feeds, which can be taken as guidelines, are found in handbooks and in information booklets provided by cutting tool manufacturers.

11.2.9 Design Considerations for Turning

When designing parts to be produced by turning, the product designer must consider the possibilities and limitations of the turning operation as well as the machining cost. The cost increases with the quality of the surface finish, with the tightness of the tolerances, and with the area of the surface to be machined. Therefore, it is not recommended that high-quality surface finishes or tighter tolerances be used in the product design unless they are required for the proper functioning of the product. Figure 11.25 graphically depicts some design considerations for turning. Here are the guidelines to be followed:

- Try to reduce the area of the surfaces to be machined, especially when a large number of parts are required or when the surfaces are to mate with other parts (see Figure 11.25a).
- Try to reduce the number of operations required by appropriate changes in the design (see Figure 11.25b).
- Provide an allowance for tool clearance between different sections of a product (see Figure 11.25c).

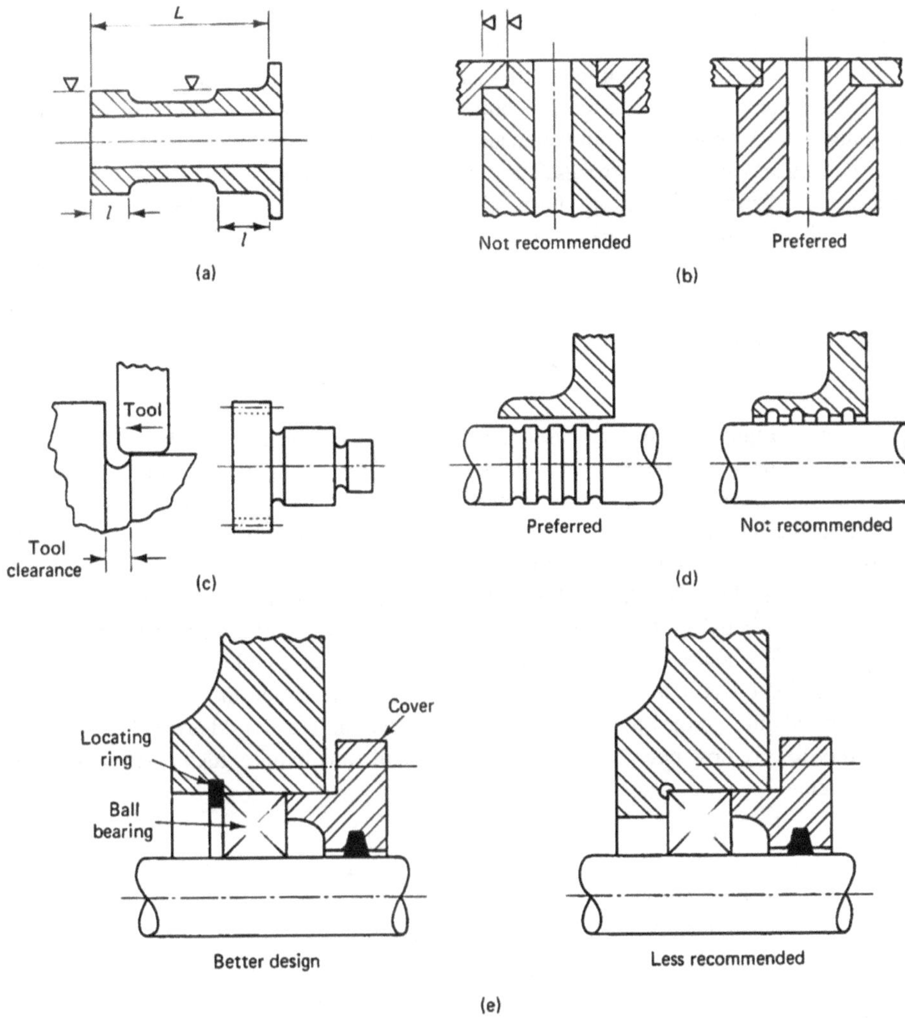

FIGURE 11.25 Design considerations for turning: (a) reduce the area of the surface to be machined, (b) reduce the number of operations required, (c) provide an allowance for tool clearance, (d) opt for machining external over internal surfaces, and (e) opt for through boring over alternatives.

- Always keep in mind that machining of exposed surfaces is easier and less expensive than machining of internal surfaces (see Figure 11.25d).
- Remember that through boring is easier and cheaper than other alternatives (see Figure 11.25e).

11.3 SHAPING AND PLANING OPERATIONS

Planing, *shaping*, and *slotting* are processes for machining horizontal, vertical, and inclined flat surfaces, slots, or grooves by means of a lathe-type cutting tool. In all these processes, the cutting action takes place along a straight line. In planing, the workpiece (and the machine bed) is reciprocated, and the tool is fed across the workpiece to reproduce another straight line, thus generating a flat surface. In shaping and slotting, the cutting tool is reciprocated, and the workpiece is fed normal to the direction of tool motion. The difference between the latter two processes is that the tool path is horizontal in shaping and it is vertical in slotting. Shapers and slotters can be employed in cutting external and internal keyways, gear racks, dovetails, and T-slots. Shapers and planers have become virtually obsolete because most shaping and planing operations have been replaced by more productive processes such as milling, broaching, and abrasive machining. The use of shapers and planers is now limited to the machining of large beds of machine tools and the like. In all three processes, there are successive alternating cutting and idle return strokes. The cutting speed is, therefore, the speed of the tool (or the workpiece) in the direction of cutting during the working stroke. The cutting speed may be either constant throughout the working stroke or variable, depending upon the design of the shaper or planer. Let us now discuss the construction and operation of the most common types of shapers and planers.

11.3.1 Horizontal Push-Cut Shaper

11.3.1.1 Construction

As can be seen in Figure 11.26, a *horizontal push-cut shaper* consists of a frame that houses the speed gearbox and the quick-return mechanism that transmits power from the motor to the ram and the table. The ram travel is the primary motion that produces a straight-line cut in the working stroke, whereas the intermittent cross-travel of the table is responsible for the cross-feed. The tool head is mounted at the front end of the ram and carries the clapper box toolholder. The toolholder is pivoted at its upper end to allow the tool to rise during the idle return stroke in order not to ruin the newly machined surface. The tool head can be swiveled to permit the machining of inclined surfaces. The workpiece can be either bolted directly to the machine table or held in a vise or other suitable fixture. The cross-feed of the table is generated by a ratchet and pawl mechanism that is driven through the quick-return mechanism (i.e., the crank and the slotted arm). The machine table can be raised or lowered by means of a power screw and a crank handle. It can also be swiveled in a universal shaper.

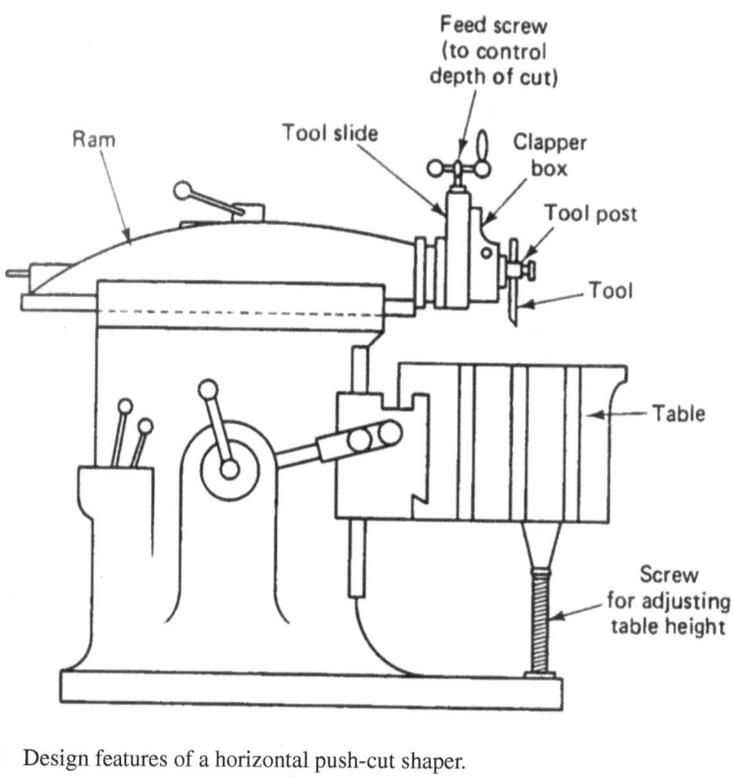

FIGURE 11.26 Design features of a horizontal push-cut shaper.

11.3.1.2 Quick-Return Mechanism

As can be seen in Figure 11.27, the *quick-return mechanism* involves a rotating crank that is driven at a uniform angular speed and an oscillating slotted arm that is connected to the crank by a sliding block. The working stroke takes up an angle (of the crank revolution) that is larger than that of the return stroke. Because the angular speed of the crank is constant, it is obvious that the time taken by the idle return stroke is less than that taken by the cutting stroke. In fact, it is the main function of the quick-return mechanism to reduce the idle time during the machining operation to a minimum. Now, let us consider the average speed (S) of the tool during the cutting stroke. It can be determined as a function of the length of the stroke and the number of strokes per minute as follows: it is also obvious that the total number of strokes required to machine a given surface can be given by the following equation:

$$S = \frac{2LN}{C} \text{in.ft / min} \left(\text{m / min}\right) \tag{11.4}$$

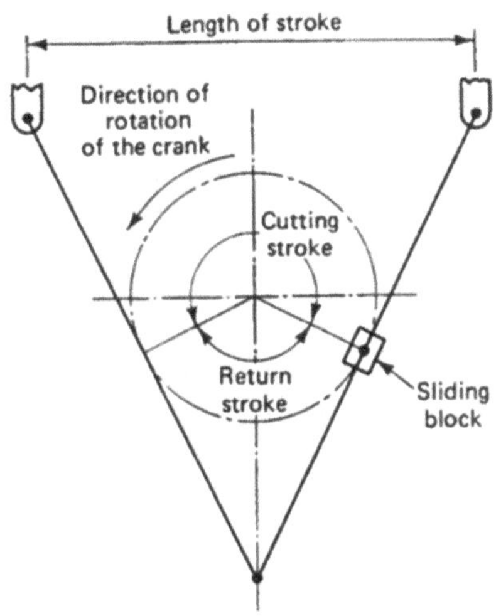

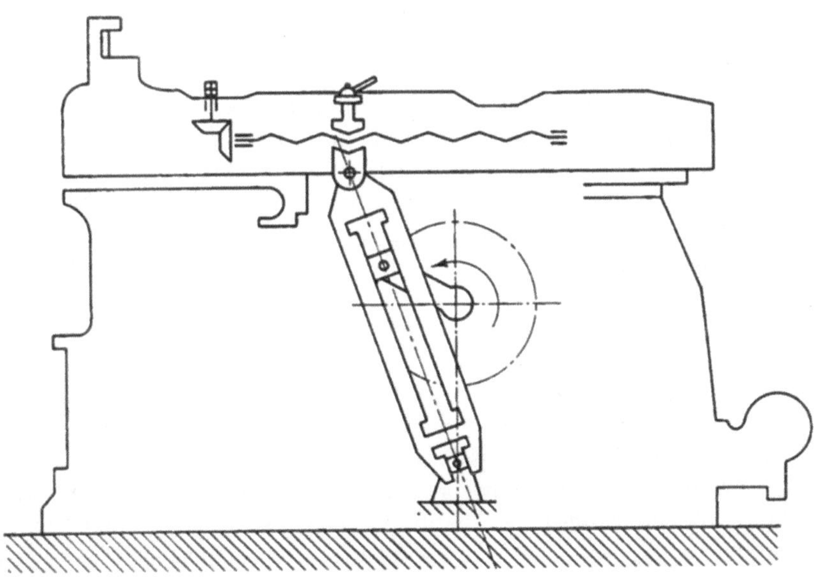

FIGURE 11.27 Details and working principles of the quick-return mechanism.

where

L is the length of the stroke in feet (m)
N is the number of strokes per minute
C is the cutting ratioNote that the cutting ratio is

$$C = \frac{\text{Cutting time}}{\text{Total time for one crank revolution}}$$
$$= \frac{\text{Angle corresponding to cutting stroke}}{2\pi}$$

It is also obvious that the total number of strokes required to machine a given surface can be given by the following equation:

$$n = \frac{W}{f} \tag{11.5}$$

where

W is the total width of the workpiece
f is the cross-feed (e.g., inches per stroke)

Therefore, the machining time T is n/N. After mathematical manipulation, it can be given as follows:

$$T = \frac{2LN}{S \times C} \tag{11.6}$$

$$\text{MRR} = T \times f \times L \times N\left(\text{in.}^3/\text{min}\right) \tag{11.7}$$

11.3.2 Vertical Shaper

The vertical shaper is similar in construction and operation to the push-cut shaper, the difference being that the ram and the tool head travel vertically instead of horizontally. Also, in this type of shaper, the workpiece is mounted on a round table that can have a rotary feed whenever desired to allow the machining of curved surfaces (e.g., spiral grooves). Vertical shapers, which are sometimes referred to as *slotters*, are used in internal cutting. Another type of vertical shaper is known as a *keyseater* because it is specially designed for cutting keyways in gears, cams, pulleys, and the like.

11.3.3 Planer

A *planer* is a machine tool that does the same work as the horizontal shaper but on workpieces that are much larger than those machined on a shaper. Although the designs of planers vary, the most common are the double-housing and open-side constructions. In a *double-housing* planer, two vertical housings are mounted at the sides of the long, heavy bed. A cross-rail that is supported at the top of these housings carries the cutting tools. The machine table (while in operation) reciprocates along the guideways of the bed and has T-slots in its upper surface for clamping the workpiece. In this type of planer, the table is powered by a variable-speed DC motor through a gear drive. The cross-rail can be raised or lowered as required, and the inclination of the tools can be adjusted as well. In an *open-side* planer, there is only one upright housing on one side of the bed. This construction provides more flexibility when wider workpieces are to be machined.

11.3.4 Planing and Shaping Tools

Planing and shaping processes employ single-point tools of the lathe type, but are heavier in construction. They are made of either HSS or carbon tool steel with carbide tips. In the latter case, the machine tool should be equipped with an automatic lifting device to keep the tool from rubbing the workpiece during the return stroke, thus eliminating the possibility of breaking or chipping the carbide tips.

The cutting angles for these tools depend upon the purpose for which the tool is to be used and the material being cut. The end relief angle does not usually exceed 4°, whereas the side relief varies between 6° and 14°. The side rake angle also varies between 5° (for cast iron) and 15° (for medium-carbon steel).

11.4 DRILLING OPERATIONS

Drilling involves producing through or blind holes in a workpiece by forcing a tool that rotates around its axis against the workpiece. Consequently, the range of cutting from this axis of rotation is equal to the radius of the required hole. In practice, two symmetrical cutting edges that rotate about the same axis are employed.

Drilling operations can be carried out by using either hand drills or drilling machines. The latter differ in size and construction. Nevertheless, the tool always rotates around its axis while the workpiece is kept firmly fixed. This is contrary to drilling on a lathe.

11.4.1 Cutting Tools for Drilling Operations

In drilling operations, a cylindrical rotary-end cutting tool, called a *drill*, is employed. The drill can have one or more cutting edges and corresponding flutes that are straight or helical. The function of the flutes is to provide outlet passages for the chips generated during the drilling operation and also to allow lubricants and coolants to reach the cutting edges and the surface being machined. Following is a survey of the commonly used types of drills.

11.4.1.1 Twist Drill

The *twist drill is* the most common type of drill. It has two cutting edges and two helical flutes that continue over the length of the drill body, as shown in Figure 11.28. The drill also consists of a neck and a shank that can be either straight or tapered. A tapered shank is fitted by the wedge action into the tapered socket of the spindle and has a tang that goes into a slot in the spindle socket, thus acting as a solid means for transmitting rotation. Straight-shank drills are held in a drill chuck that is, in turn, fitted into the spindle socket in the same way as tapered-shank drills.

As can be seen in Figure 11.28, the two cutting edges are referred to as the lips and are connected together by a *wedge*, which is a chisel-like edge. The twist drill also has two margins that allow the drill to be properly located and guided while it is in operation. The *tool point angle* (TPA) is formed by the two lips and is chosen based on the properties of the material to be cut. The usual TPA for commercial drills is 11 8°, which is appropriate for drilling low-carbon steels and cast irons. For harder and tougher metals, such as hardened steel, brass, and bronze, larger TPAs (130° or 140°) give better performance. The helix angle of the flutes of a twist drill ranges between 24° and 30°. When drilling copper or soft plastics, higher values for the helix angle are recommended (between 35° and 45°). Twist drills are usually made of HSS, although carbide-tipped drills are also available. The sizes of twist drills used in industrial practice range from 0.01 inches to 3.5 inches (0.25–80 mm).

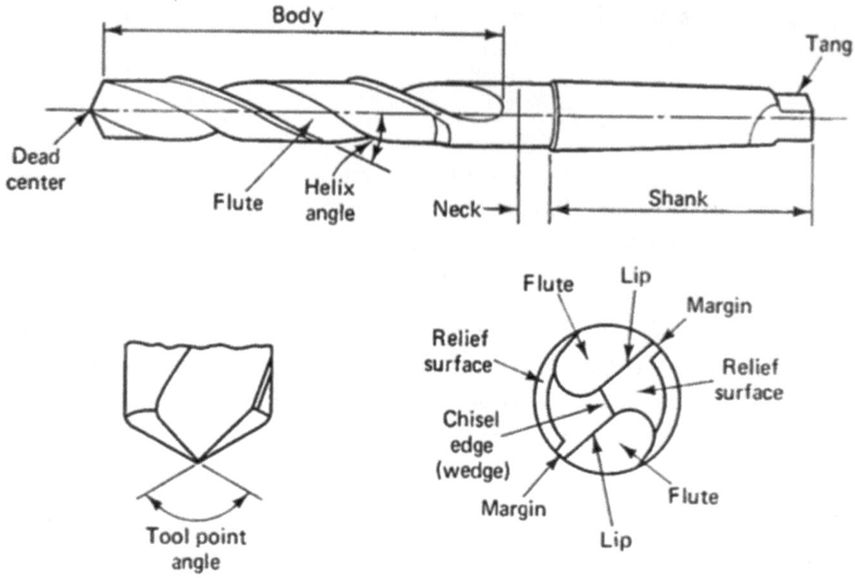

FIGURE 11.28 A twist drill.

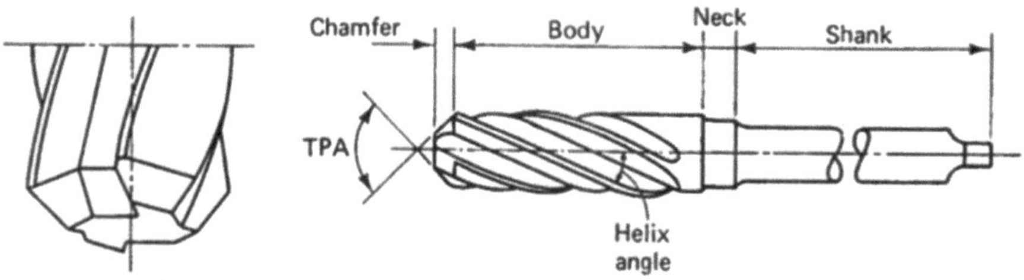

FIGURE 11.29 A core drill.

11.4.1.2 Core Drill

A *core drill* consists of the chamfer, body, neck, and shank, as shown in Figure 11.29. This type of drill may have three or four flutes and an equal number of margins, which ensures superior guidance, thus resulting in high machining accuracy. The figure also shows that a core drill has a flat end. The chamfer can have three or four cutting edges, or lips, and the lip angle may vary between 90° and 120°. Core drills are employed for enlarging previously made holes and not for originating holes. This type of drill promotes greater productivity, high machining accuracy, and superior quality of the drilled surfaces.

11.4.1.3 Gun Drill

A *gun drill* is used for drilling deep holes. All gun drills are straight-fluted, and each has a single cutting edge. A hole in the body acts as a conduit to transmit coolant under considerable pressure to the tip of the drill. As can be seen in Figure 11.30, there are two kinds of gun drills: the *center-cut* gun drill used for drilling blind holes and the *trepanning* drill. The latter has a cylindrical groove at its center, thus generating a solid core that guides the tool as it proceeds during the drilling operation.

11.4.1.4 Spade Drill

A *spade drill* is used for drilling large holes of 3.5 inches (90 mm) or more. The design of this type of drill results in a marked saving in tool cost as well as in a tangible reduction in tool weight that facilitates its ease of handling. Moreover, this drill is easy to grind. Figure 11.31 shows a spade drill.

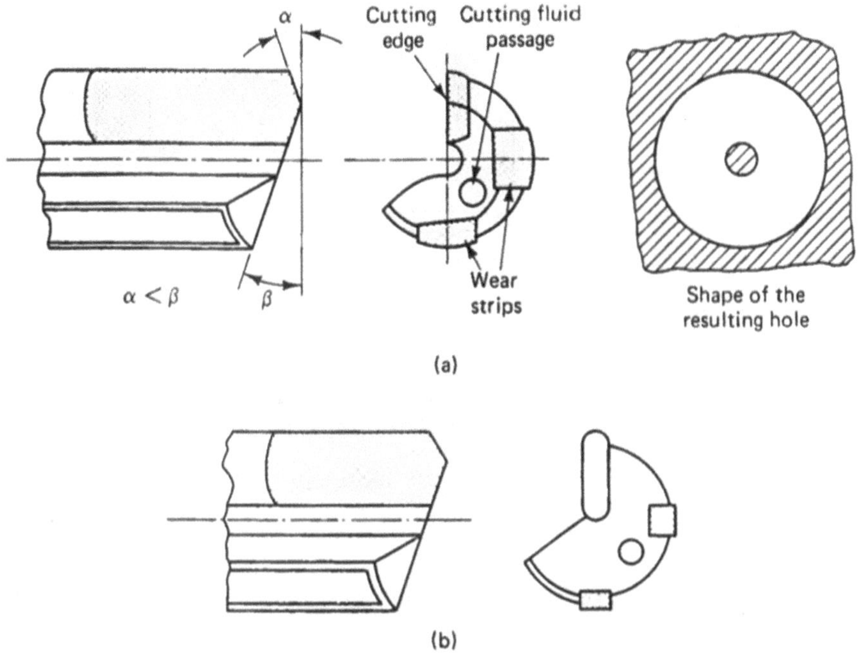

FIGURE 11.30 Gun drill: (a) trepanning gun drill and (b) center-cut gun drill.

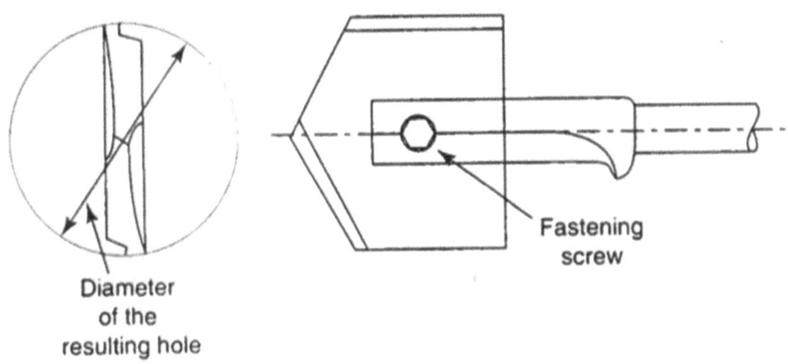

FIGURE 11.31 A spade drill.

11.4.1.5 Saw-Type Cutter

A *saw-type cutter*, like the one illustrated in Figure 11.32, is used for cutting large holes in thin metal.

11.4.1.6 Drills Made in Combination with Other Tools

An example is a tool that involves both a drill and a tap. Step drills and drill and countersink tools are also sometimes used in industrial practice.

11.4.2 Cutting Speeds and Feeds in Drilling

We can easily see that the cutting speed varies along the cutting edge. It is always maximum at the periphery of the tool and is equal to zero on the tool axis. Nevertheless, we consider the maximum speed because it is the one that affects the tool wear and the quality of the machined surface. The maximum speed must not exceed the permissible cutting speed, which depends upon the material of the workpiece as well as the material of the cutting tool. Data about permissible cutting speeds in

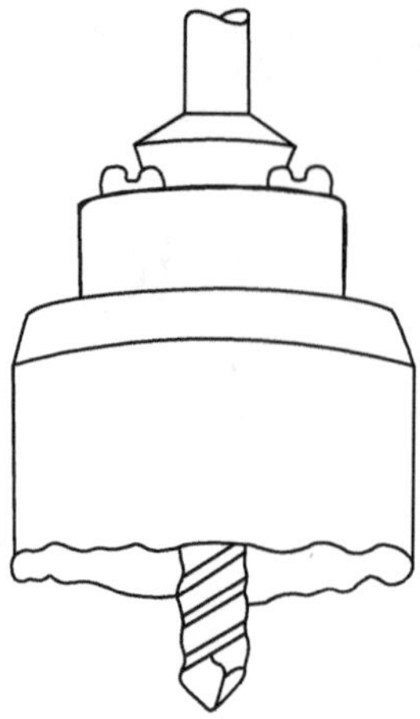

FIGURE 11.32 A saw-type cutter.

drilling operations can be found in handbooks. The rotational speed of the spindle can be determined from the following equation:

$$N = \frac{CS}{\pi D} \tag{11.8}$$

where

N is the rotational speed of the spindle (rpm)
D is the diameter of the drill in feet (m)
CS is the permissible cutting speed in ft/min (m/min)

In drilling operations, feeds are expressed in inches or millimeters per revolution. Again, the appropriate value of feed to be used depends upon the metal of the workpiece and drill material and can be found in handbooks. Whenever the production rate must be increased, it is advisable to use higher feeds rather than increase the cutting speed.

11.4.3 Other Types of Drilling Operations

In addition to conventional drilling, other operations are involved in the production of holes in industrial practice. Following is a brief description of each of these operations.

11.4.3.1 Boring

Boring involves enlarging a hole that has already been drilled. It is similar to internal turning and can, therefore, be performed on a lathe, as previously mentioned. There are also some specialized machine tools for carrying out boring operations. These include the vertical boring mill, the jig-boring machine, and the horizontal boring machine.

11.4.3.2 Counterboring

As a result of *counterboring*, only one end of a drilled hole is enlarged, as illustrated in Figure 11.33a. This enlarged hole provides a space in which to set a bolt head or a nut so that it will be entirely below the surface of the part.

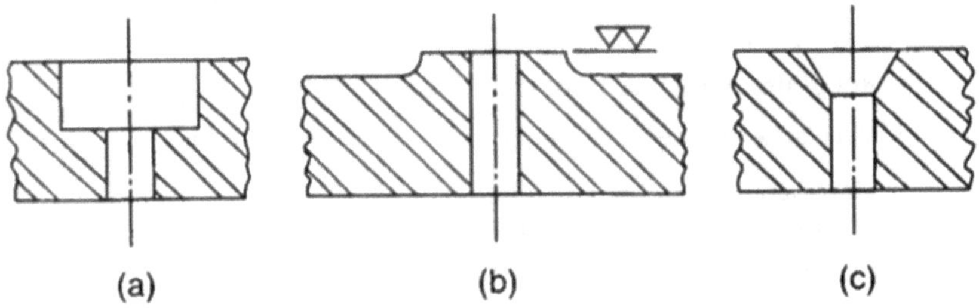

FIGURE 11.33 Operations related to drilling: (a) counterboring, (b) spot facing, and (c) countersinking.

11.4.3.3 Spot Facing

Spot facing is performed to finish off a small surface area around the opening of a hole. As can be seen in Figure 11.33b, this process involves removing a minimal depth of cut and is usually performed on castings or forgings.

11.4.3.4 Countersinking

As shown in Figure 11.33c, countersinking is done to accommodate the conical seat of a flat-head screw so that the screw does not appear above the surface of the part.

11.4.3.5 Reaming

Reaming is actually a "sizing" process, by which an already drilled hole is slightly enlarged to the desired size. As a result of a reaming operation, a hole has a very smooth surface. The cutting tool used in this operation is known as a *reamer*. As shown in Figure 11.34 a reamer has a fluted section, a neck, and a shank. The fluted section includes four zones: the chamfer, the starting taper, the sizing zone, and the back taper. The chamfer or bevel encloses an angle that depends upon the method of

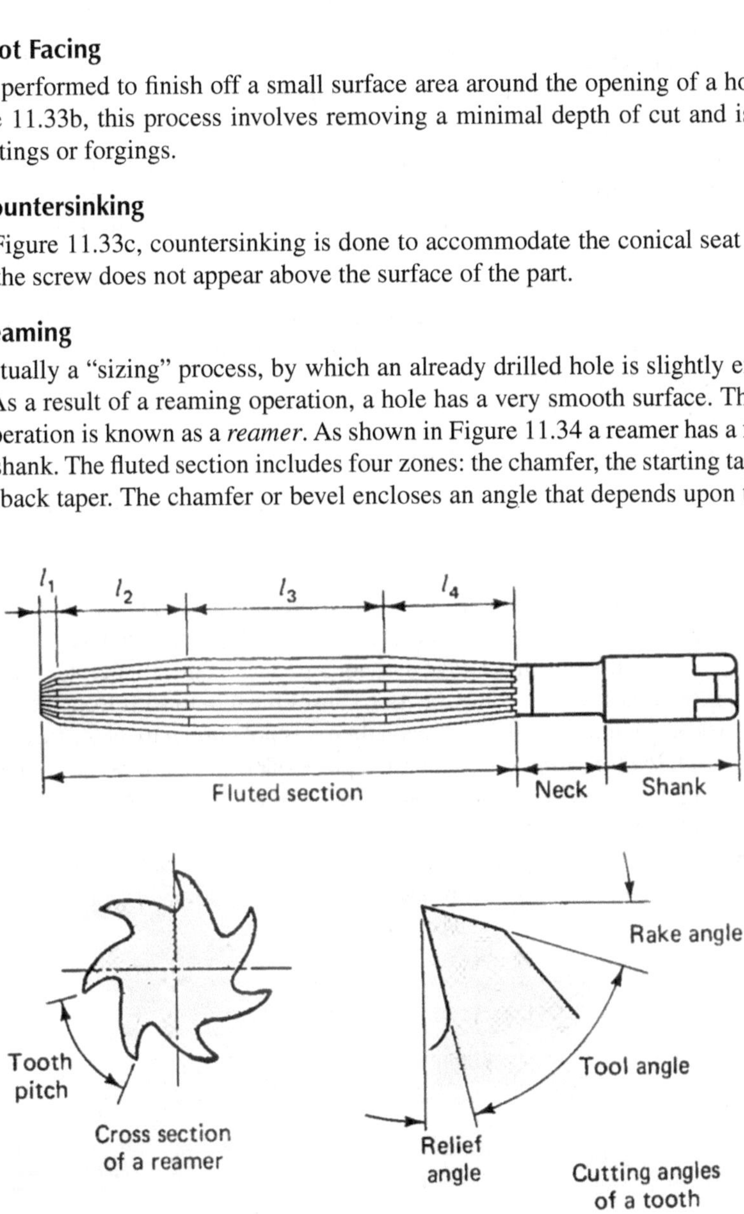

FIGURE 11.34 Details of a reamer.

TABLE 11.2
Recommended Values of the Chamfer Angles of Reamers

Metal to Be Removed	Steel	Cast Iron	Soft Metal
Manual reaming	1°–3°	1°–3°	1°–3°
Machining reaming	8°–10°	20°–30°	3°–5°

reaming and the material being cut. This is a consequence of the fact that this angle affects the magnitude of the axial reaming force. The larger the chamfer angle, the larger the required reaming force. Table 11.2 indicates some recommended values of the chamfer angle for different reaming conditions. The starting taper is the part of the reamer that actually removes chips. Figure 11.34 also shows that each tooth of that part of the reamer has a cutting edge as well as rake, relief, and tool (or lip) angles. The sizing zone guides the reamer and smooths the surface of the hole. Finally, the back taper serves to reduce friction between the reamer and the newly machined surface.

Reamers are usually made of hardened tool steel. Nevertheless, reamers that are used in mass production are tipped with cemented carbides in order to increase the tool life and the production rate.

11.4.3.6 Tapping

Tapping is the process of cutting internal threads. The tool used is called a *tap*. As shown in Figure 11.35, it has a boltlike shape with four longitudinal flutes. Made of hardened tool steel, taps can be used for either manual or machine cutting.

11.4.4 Design Considerations for Drilling

Figure 11.36 graphically depicts some design considerations for drilling. Here are the guidelines to be followed:

- Make sure the centerline of the hole to be drilled is normal to the surface of the part. This is to avoid bending and breaking the tool during the drilling operation. As previously mentioned, the twist drill has a chisel edge and not a pointed edge at its center. This, although it facilitates the process of grinding the tool, causes the tool to shift from the desired location and makes it liable to breakage, especially if it is not normal to the surface to be drilled. (See Figure 11.36a for examples of poor and proper design practices for drilled holes.)
- When tapping through holes, ensure that the tap will be in the clear when it appears from the other side of the part (see Figure 11.36b).
- Remember that it is impossible to tap the entire length of a blind or counterbored hole without providing a special tool allowance (see Figure 11.36c).

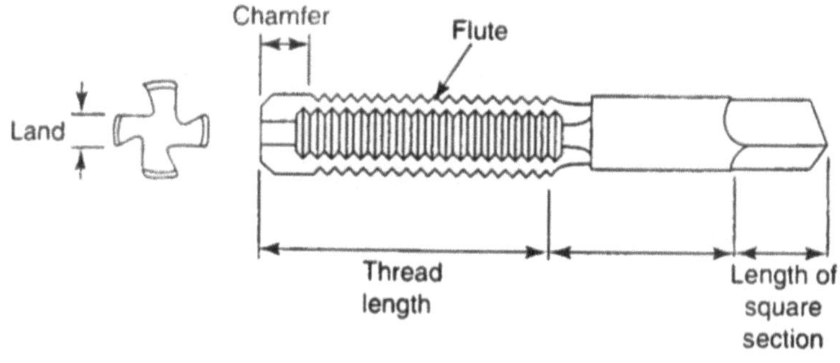

FIGURE 11.35 A tap.

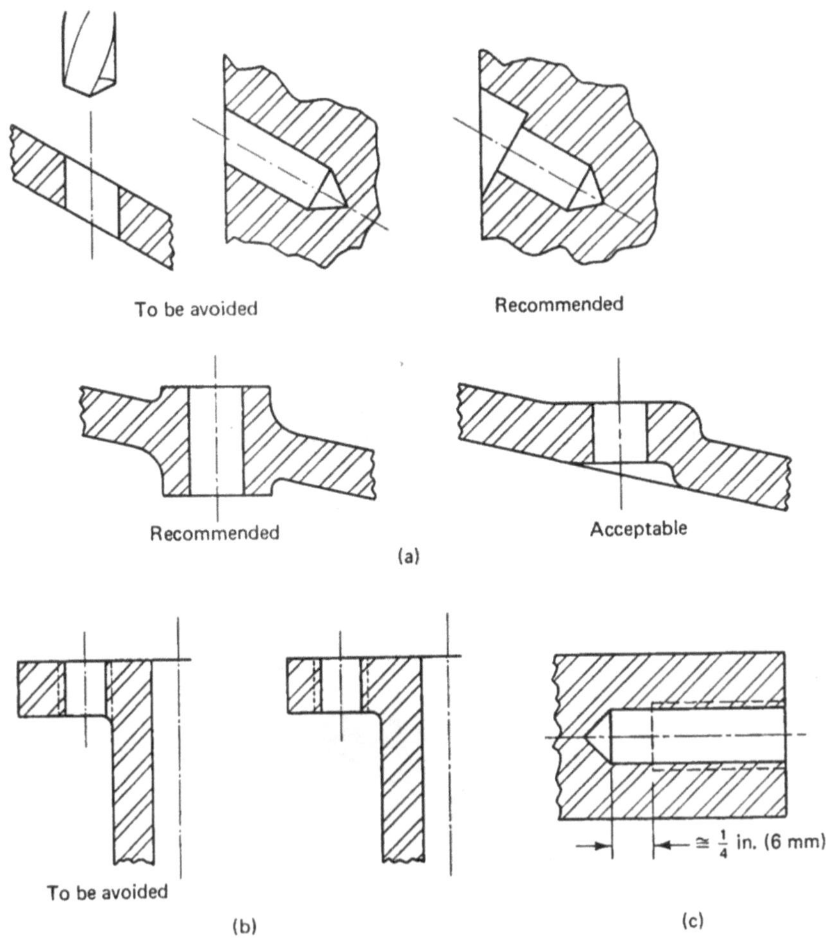

FIGURE 11.36 Design considerations for drilling: (a) set the centerline of the tool normal to the surface to be drilled, (b) ensure the tap is clear when it appears from the other side, and (c) provide allowance when tapping a blind hole.

11.4.5 Classification of Drilling Machines

Drilling operations can be carried out by employing small portable machines or by using the appropriate machine tools. These machine tools differ in shape and size, but they have common features. For instance, they all involve one or more twist drills, each rotating around its own axis while the workpiece is kept firmly fixed. This is contrary to the drilling operation on a lathe, where the workpiece is held in and rotates with the chuck. Following is a survey of the commonly used types of drilling machines.

11.4.5.1 Bench-Type Drilling Machines

Bench-type drilling machines are general-purpose, small machine tools that are usually placed on benches. This type of drilling machine includes an electric motor as the source of motion, which is transmitted via pulleys and belts to the spindle, where the tool is mounted. The feed is manually generated by lowering a lever handle that is designed to lower (or raise) the spindle. The spindle rotates freely inside a sleeve (which is actuated by the lever through a rack-and-pinion system) but does not rotate with the spindle. The workpiece is mounted on the machine table, although a special vise is sometimes used to hold the workpiece. The maximum height of a workpiece to be machined is limited by the maximum gap between the spindle and the machine table.

11.4.5.2 Upright Drilling Machines

Depending upon the size, *upright drilling* machines can be used for light, medium, and even relatively heavy jobs. A light-duty upright drilling machine is shown in Figure 11.37. It is basically similar to a bench-type machine, the main difference being a longer cylindrical column fixed to the base. Along the column is an additional sliding table for fixing the workpiece that can be locked in position at any desired height. The power required for this type of machine is greater than that for a bench-type drilling machine as this type is employed in performing medium-duty jobs. There are also large drilling machines of the upright type. In this case, the machine has a box column and a higher power to deal with large jobs. Moreover, gearboxes are employed to provide different rotational spindle speeds as well as axial feed motion, which can be preset at any desired rate.

11.4.5.3 Multispindle Drilling Machines

Multispindle drilling machines are sturdily constructed and require high power; each is capable of drilling many holes simultaneously. The positions of the different tools (spindles) can be adjusted as desired. Also, the entire head (which carries the spindles and the tools) can be tilted if necessary. This type of drilling machine is used mainly for mass production in jobs having many holes, such as cylinder blocks.

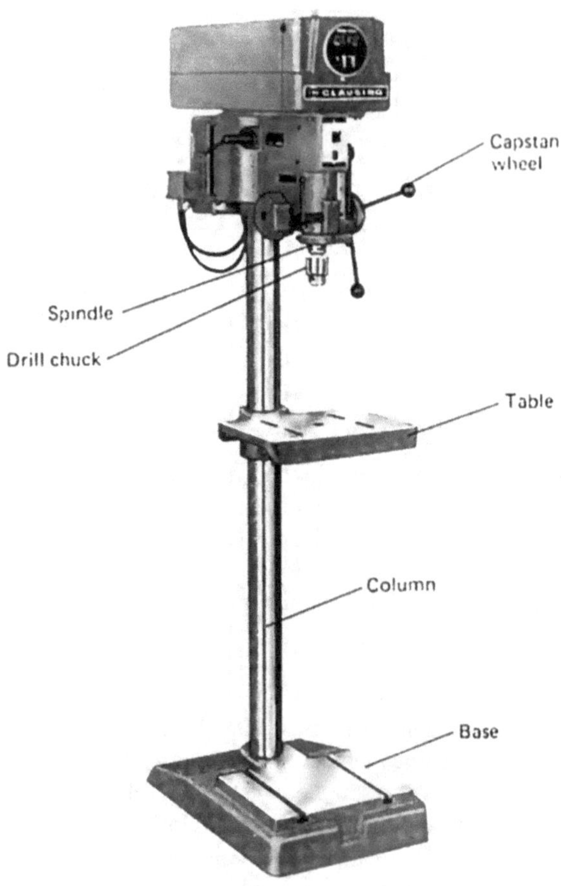

FIGURE 11.37 An upright drilling machine. (Courtesy of Clausing Industrial Inc., Kalamazoo, Michigan.)

11.4.5.4 Gang Drilling Machines

When several separate heads (each with a single spindle) are arranged on a single common table, the machine tool is then referred to as a gang drilling machine. This type of machine tool is particularly suitable where several operations are to be performed in succession.

11.4.5.5 Radial Drills

Radial drills are particularly suitable for drilling holes in large and heavy workpieces that are inconvenient to mount on the table of an upright drilling machine. As shown in Figure 11.38, a radial drilling machine has a main column that is fixed to the base. The cantilevered guide arm, which carries the drilling head spindle and tool, can be raised or lowered along the column and clamped at any desired position. The drilling head slides along the arm and provides rotary motion and axial feed motion. The cantilevered guide arm can be swung, thus allowing the tool to be moved in all directions according to a cylindrical coordinate system.

Radial drills are particularly suitable for drilling holes in large and heavy workpieces that are inconvenient to mount on the table of an upright drilling machine. As shown in Figure 11.38, a radial drilling machine has a main column that is fixed to the base. The cantilevered guide arm, which carries the drilling head spindle and tool, can be raised or lowered along the column and clamped at any desired position. The drilling head slides along the arm and provides rotary motion and axial feed motion. The cantilevered guide arm can be swung, thus allowing the tool to be moved in all directions according to a cylindrical coordinate system.

11.4.5.6 Turret Drilling Machines

Machine tools that belong in the *turret* drilling machine category are either semiautomatic or fully automatic. A common design feature is that the main spindle is replaced by a turret that carries several drilling, boring, reaming, and threading tools. Consequently, several successive operations can

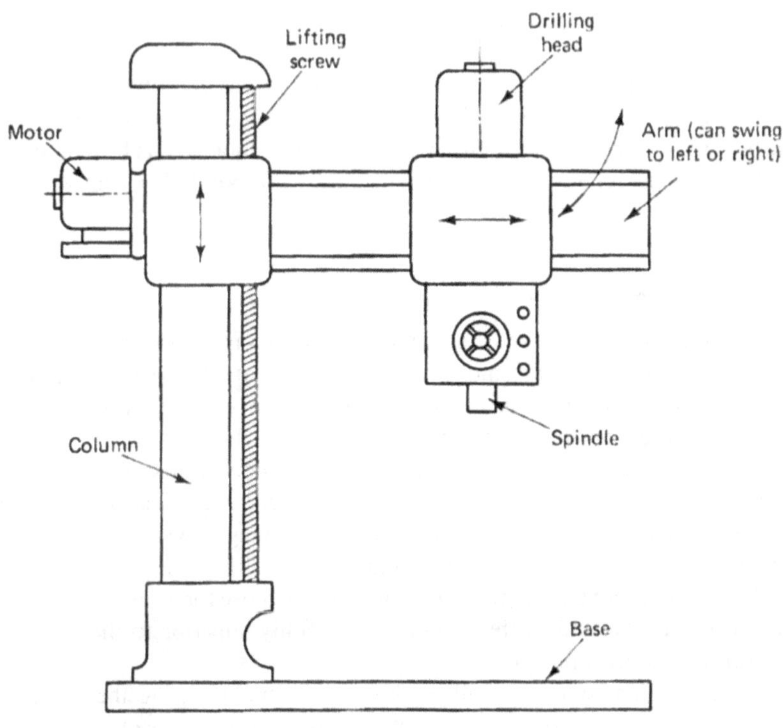

FIGURE 11.38 A radial drill.

be carried out with only a single initial setup and without the need for setting up the workpiece again between operations.

Automatic turret drilling machines that are operated by NC or CNC systems (see Chapter 14) are quite common. In this case, the human role is limited to the initial setup and monitoring. This type of machine tool has advantages over the gang-type drilling machine with respect to the space required (physical size of the machine tool) and the number of workpiece setups.

11.4.5.7 Deep-Hole Drilling Machines

Deep-hole drilling machines are special machines employed for drilling long holes like those of rifle barrels. Usually, gun-type drills are used and are fed slowly against the workpiece. In this type of machine tool, it is the workpiece that is rotated, while the drill is kept from rotary motion. A deep-hole drilling machine may have either a vertical or a horizontal construction. However, in both cases, the common feature is the precise guidance and positive support of the workpiece during the drilling operation.

11.4.5.8 Jig-Boring Machines

Jig-boring machines are specially designed to possess high precision and accuracy. A machine of this type not only drills the holes but also locates them because the table movements are monitored by electronic measuring devices. Jig-boring machines are usually employed in the manufacture of forming and molding dies, gauges, and work-holding devices like jigs and fixtures.

11.4.6 Work-Holding Devices in Drilling

During conventional drilling operations, the workpiece must be held firmly on the machine table. The type of work-holding device used depends upon the shape and size of the workpiece, the desired accuracy, and the production rate. For low production when the accuracy is not very important, conventional machine vises or vises with V-blocks (for round work) are used. For moderate production and when accuracy is of some importance, jigs are usually employed. A *jig* is a work-holding device that is designed to hold a particular workpiece (i.e., it cannot be used for workpieces having different shapes) and to guide the cutting tool during the drilling operation. This eliminates the need for laying out the workpiece prior to machining, thus saving the time spent in bluing and scribing when no jigs are employed. The design of jigs and fixtures is a separate topic and is beyond the scope of this text. Interested readers are referred to the books dealing with tool design and with jig and fixture design that are given at the end of this text.

11.5 MILLING OPERATIONS

Milling is a machining process that is carried out by means of a multiedge rotating tool known as a *milling cutter*. In this process, metal removal is achieved by simultaneously combining the rotary motion of the milling cutter and the linear motions of the workpiece. Milling operations are employed in producing flat, contoured, and helical surfaces, as well as for thread- and gear-cutting operations.

Each of the cutting edges of a milling cutter acts as an individual single-point cutter when it engages with the workpiece metal. Therefore, each of the cutting edges has the appropriate rake and relief angles. Because only a few of the cutting edges are engaged with the workpiece at a time, heavy cuts can be taken without adversely affecting the tool's life. In fact, the permissible cutting speeds and feeds for milling are three to four times higher than those for turning or drilling. Moreover, the quality of the surfaces machined by milling is generally superior to the quality of surfaces machined by turning, shaping, or drilling.

A wide variety of milling cutters is available in the industry. This, together with the fact that a milling machine is a very versatile machine tool, makes the milling machine the backbone of a machining workshop.

11.5.1 Milling Methods

As far as the direction of cutter rotation and workpiece feed are concerned, milling is performed by either of the following two methods.

11.5.1.1 Up Milling (Conventional Milling)

In *up milling*, the workpiece is fed against the direction of cutter rotation, as shown in Figure 11.39a. The depth of the cut (and, consequently, the load) gradually increases on the successively engaged cutting edges. Therefore, the machining process involves no impact loading, thus ensuring smoother operation of the machine tool and longer tool life. The quality of the machined surface obtained by up milling is not very high. Nevertheless, up milling is commonly used in industry, especially for rough cuts.

11.5.1.2 Down Milling (Climb Milling)

In *down milling*, the cutter rotation coincides with the direction of feed at the contact point between the tool and the workpiece, as shown in Figure 11.39b. The maximum depth of cut is achieved directly as the cutter engages with the workpiece. This results in a kind of impact, or sudden loading. Therefore, this method cannot be used unless the milling machine is equipped with a backlash eliminator on the feed screw. The advantages of this method include higher quality of the machined surface and easier clamping of workpieces as the cutting forces act downward.

11.5.2 Types of Milling Cutters

Milling cutters come in a wide variety of shapes, each designed to effectively perform a specific milling operation. Generally, a milling cutter can be described as a multiedge cutting tool having the shape of a solid of revolution, with the cutting teeth arranged on the periphery, on an end face, or on both. Following is a survey of the commonly used types of milling cutters.

11.5.2.1 Plain Milling Cutter

A *plain* milling cutter, as shown in Figure 11.40a, is a disk-shaped cutting tool that may have straight or helical teeth. This type of cutter is always mounted on horizontal milling machines and is used for machining flat surfaces.

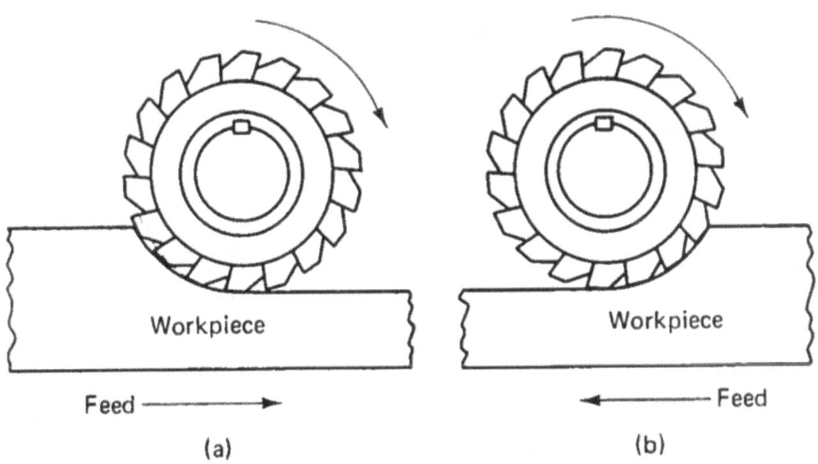

FIGURE 11.39 Milling methods: (a) up milling and (b) down milling.

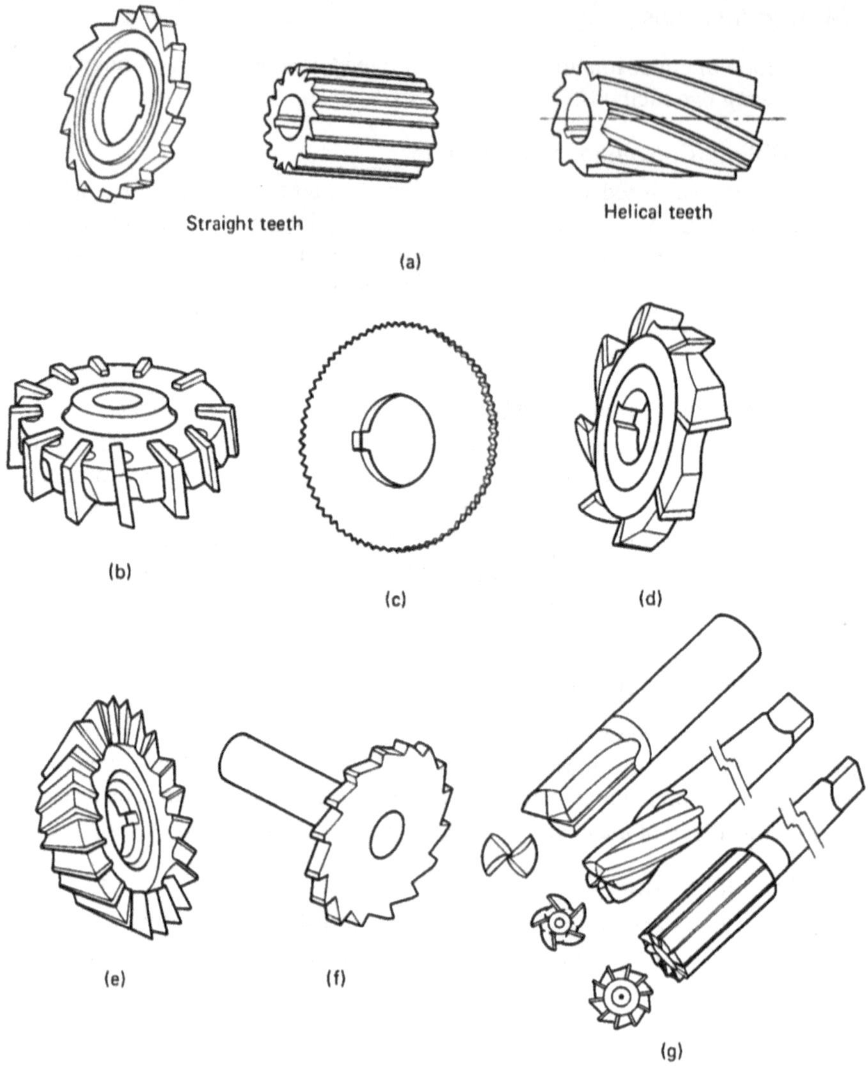

FIGURE 11.40 Types of milling cutters: (a) plain milling cutter, (b) face milling cutter with inserted teeth, (c) plain metal-slitting saw cutter, (d) side milling cutter, (e) angle milling cutter, (f) T-slot cutter, and (g) end mill cutter.

11.5.2.2 Face Milling Cutter

A *face* milling cutter, like the one in Figure 11.40b, is also used for machining flat surfaces. It is bolted at the end of a short arbor that is, in turn, mounted on a vertical milling machine.

11.5.2.3 Plain Metal-Slitting Saw

Figure 11.40c illustrates a *plain metal-slitting saw* cutter. Notice that it actually involves a very thin plain milling cutter.

11.5.2.4 Side Milling Cutter

A *side* milling cutter is used for cutting slots, grooves, and splines. As can be seen in Figure 11.40d, it is quite similar to the plain milling cutter, the difference being that this type has teeth on the sides. As is the case with the plain cutter, the cutting teeth can be straight or helical.

11.5.2.5 Angle Milling Cutter

An *angle* milling cutter is employed in cutting dovetail grooves, ratchet wheels, and the like. Figure 11.40e shows a milling cutter of this type.

11.5.2.6 T-Slot Cutter

As shown in Figure 11.40f, a T-*slot* cutter involves a plain milling cutter with an integral shaft normal to it. As the name suggests, this type of cutter is used for milling T-slots.

11.5.2.7 End Mill Cutter

An *end* mill cutter finds common application in cutting slots, grooves, flutes, splines, pocketing work, and the like. As Figure 11.40g indicates, an end mill cutter is always mounted on a vertical milling machine and can have two or four flutes, which may be straight or helical.

11.5.2.8 Form Milling Cutter

The teeth of a *form* milling cutter have a shape that is identical to the section of the metal to be removed during the milling operation. Examples of this type of cutter include gear cutters, gear hobs, and convex and concave cutters. Form milling cutters are mounted on horizontal milling machines, as is explained later when we discuss gear cutting.

11.5.3 Materials of Milling Cutters

The commonly used milling cutters are made of HSS, which is generally adequate for most jobs. Milling cutters tipped with sintered carbides or cast nonferrous alloys as cutting teeth are usually employed for mass production, where heavier cuts and/or high cutting speeds are required.

11.5.4 Cutting Speeds and Feeds in Milling

Figure 11.41 indicates methods of estimating the different machining parameters during milling operations. These parameters include the cutting speed, the feed, and the metal-removal rate. The cutting speed is the peripheral velocity at any point on the circumference of the cutter. The allowable value for the cutting speed in milling is dependent upon many factors, including the cutter material, material of the workpiece, diameter and life of the cutter, feed, depth of cut, width of cut, number of teeth on the cutter, and the type of coolant used. The feed in milling operations is the rate of movement of the cutter axis relative to the workpiece. It is expressed in inches (or mm) per revolution or inches (or mm) per minute. It can also be expressed in inches (or mm) per tooth, especially for plain and face milling cutters.

The depth of cut is the thickness of the metal layer that is to be removed in one cut. The maximum allowable depth of cut depends upon the material being machined and is commonly taken up to 5/16 inch (8 mm) in roughing operations and up to 1/16 inch (about 1.5 mm) in finishing operations. Another parameter that affects milling operations is the width of the cut, which is the width of the workpiece in contact with the cutter in a direction normal to the feed. The width of the cut should decrease with increasing depth of cut to keep the load and power requirement below those that can be met by the cutter and the machine tool, respectively.

11.5.5 Cutting Angles of Milling Cutters

As previously mentioned, the geometry of any tool is basically a matter of rake and relief angles. Figure 11.42 shows the cutting angles of a plain, straight-tooth milling cutter (for simplicity). The radial rake angle facilitates the removal of chips and ranges from 10° to 20°, depending upon the workpiece material to be cut. When machining hard metals with carbide-tipped cutters, a negative

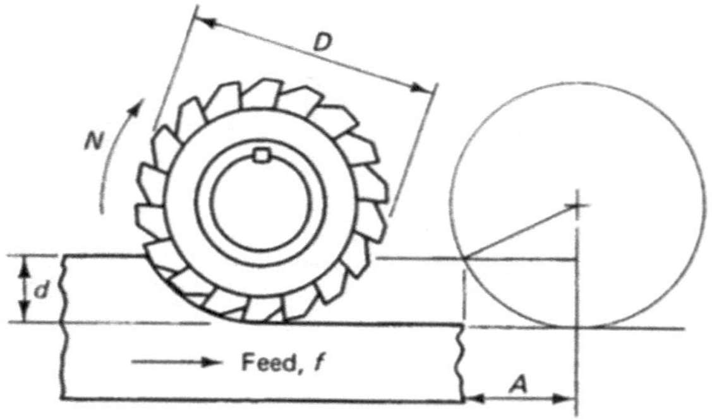

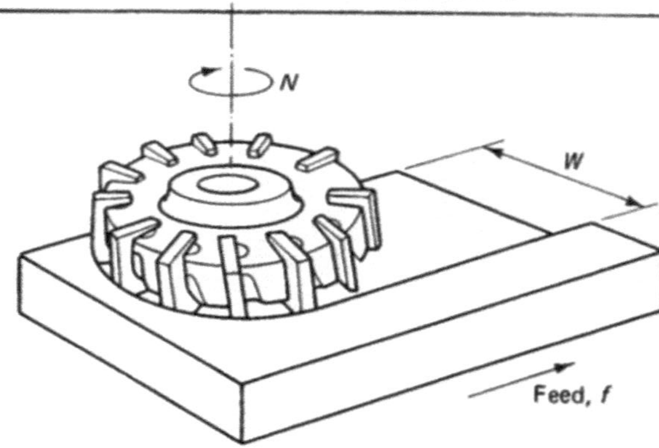

	Slab Milling (horizontal milling machine)	Face Milling (vertical milling machine)
Cutting speed	$V = \pi DN$	$V_{max} = \pi DN$ $V_{mean} = \frac{\pi DN}{2}$
Feed, f	$f = f_t Nn$ where f_t is the feed per tooth n is the number of teeth of the cutter	$f = f_t Nn$
Machining time	$T = \frac{L + 2A}{f}$ for each travel where L is the length of the workpiece $A = \sqrt{\frac{D^2}{4} - \left(\frac{D}{2} - d\right)^2}$ = distance between the point of approach and the edge of the workpiece	$T = \frac{L + 2A}{f}$ $A = \frac{D}{2}$ for $W = \frac{D}{2}$ up to D $A = \sqrt{W(D - W)}$ for $W < \frac{D}{2}$
Metal-removal rate	$MRR = W \cdot d \cdot f$ where W is the width of cut, i.e., either the width (or part of it) of the cutter engaged with the workpiece	$MRR = W \cdot d \cdot f$

FIGURE 11.41 Equations applicable to milling operations.

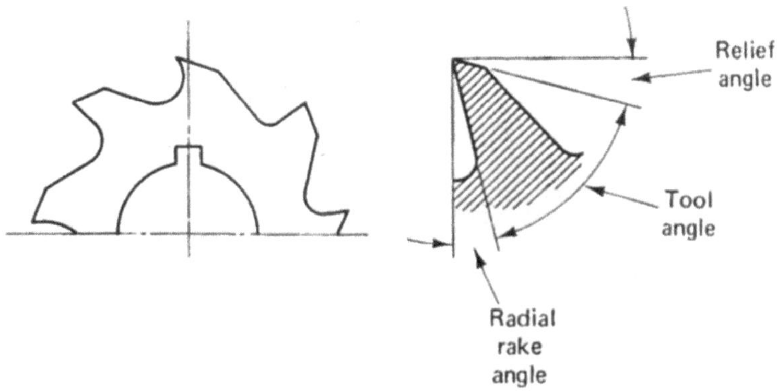

FIGURE 11.42 Cutter angles of a plain straight-tooth milling cutter.

rake angle of 10° is usually employed. The relief angle also depends upon the workpiece material and varies between 12° and 25°.

11.5.6 Types of Milling Machines

Several types of milling machines are employed in the industry. They are generally classified by their construction and design features. They vary from the common general-purpose types to duplicators and machining centers that involve a tool magazine and are capable of carrying out many machining operations with a single workpiece setup. Following is a survey of the types of milling machines commonly used in industry.

11.5.6.1 Plain Horizontal Milling Machine

The construction of the *plain horizontal milling machine* is very similar to that of a universal milling machine (see the discussion that follows), except that the machine table cannot be swiveled. Plain milling machines usually have a column and knee type of construction and three table motions (i.e., longitudinal, transverse, and vertical). The milling cutter is mounted on a short arbor that is, in turn, rigidly supported by the overarm of the milling machine.

11.5.6.2 Universal Milling Machine

The construction of a *universal milling machine* is similar to that of the plain milling machine, except that it is more accurate and has a sturdier frame and its table can be swiveled with an angle up to 50°. Universal milling machines are usually equipped with an index or dividing head that allows for the cutting of gears and cams, as is discussed later. Figure 11.43 shows a machine tool of this type.

11.5.6.3 Vertical Milling Machine

As the name *vertical milling machine* suggests, the axis of the spindle that holds the milling cutter is vertical. Table movements are generally similar to those of plain horizontal milling machines; however, an additional rotary motion is sometimes provided for the table when helical and circular grooves are to be machined. The cutters used with vertical milling machines are almost always of the end-mill type. Figure 11.44 shows a vertical milling machine.

11.5.6.4 Duplicator

A *duplicator* is sometimes referred to as a *copy milling machine* because it is capable of reproducing an exact replica of a model. The machine has a stylus that scans the model, at which time counterpoints on the part are successively machined. Duplicators were used for the production of large

FIGURE 11.43 A universal milling machine.

FIGURE 11.44 A vertical milling machine.

forming dies for the automotive industry, where models made of wood, plaster of Paris, or wax were employed. Duplicators are not commonly used in industry now because they have been superseded by CAD/CAM systems.

11.5.6.5 Machining Center

A *machining center* comprises a multipurpose CNC machine (see Chapter 14) that is capable of performing a number of different machining processes. A machining center has a tool magazine in which many tools are held. Tool changes are automatically carried out, and so are functions such as coolant turn-on/off. Machining centers are, therefore, highly versatile and can perform a number of machining operations on a workpiece with a single setup. Parts having intricate shapes can easily be produced with high accuracy and excellent repeatability.

11.5.6.6 Universal Dividing Head

The *head is* an attachment mounted on the worktable of a universal milling machine that is employed for cutting gears. The function of the dividing head is to index the gear blank through the desired angle each time the metal between two successive teeth is removed. Therefore, this attachment is sometimes known as an *index head.*

Figure 11.45 shows a universal dividing head, which consists of the body, the swivel block, the work spindle and its center, the index plate, and the index crank with a latch pin. The workpiece (with one of its ends supported by the center of the work spindle) is rotated through the desired angle by rotating the index crank through an angle that is dependent upon the desired angle. The index crank is fixed to a shaft that is, in turn, attached to a worm-gear reducer with a ratio of 40:1. Consequently, 40 turns of the index crank result in only one full turn of the workpiece. This index plate has six concentric circles of equally spaced holes to assist in measuring and controlling any fraction of revolution in order to crank the correct angle. The following equation is used to determine the angle through which the crank is to be rotated in gear cutting.

$$\text{Number of turns of index crank} = \frac{40}{\text{Number of teeth of desired gear}} \tag{11.9}$$

We can see from Equation (11.9) that if the gear to be cut has 20 teeth, the index crank should be rotated two full turns each time a tooth space is to be produced. As a consequence, the workpiece will be rotated each time through an angle equal to 18°. Similarly, if the desired gear has 32 teeth, the index crank must be rotated 1¼ turns each time.

FIGURE 11.45 A universal dividing head.

11.6 GRINDING OPERATIONS

Grinding is a manufacturing process that involves the removal of metal by employing a rotating abrasive wheel. The wheel simulates a milling cutter with an extremely large number of miniature cutting edges. Generally, grinding is considered to be a finishing process and is used for obtaining high dimensional accuracy and superior surface finish. Grinding can be performed on flat, cylindrical, or even internal surfaces by employing specialized machine tools, referred to as *grinding machines*. Obviously, grinding machines differ in construction as well as capabilities, and the type to be employed is determined mainly by the geometrical shape and nature of the surface to be ground (e.g., cylindrical surfaces are ground on cylindrical grinding machines).

11.6.1 Types of Grinding Operations

11.6.1.1 Surface Grinding

As the name *surface grinding* suggests, this operation involves the grinding of flat or plane surfaces. Figure 11.46 indicates the two possible variations: either a horizontal or a vertical machine spindle. With a horizontal spindle (see Figure 11.46a), the machine usually has a planer-type reciprocating table on which the workpiece is held. However, grinding machines with vertical spindles can have either a planer-type table like that of the horizontal-spindle machine or a rotating worktable. Also, the grinding action in this case is achieved by the end face of the grinding wheel (see Figure 11.46b), contrary to the case of horizontal-spindle machines, where the workpiece is ground by the periphery of the grinding wheel. Figure 11.46 also indicates the equations used to estimate the different parameters of the grinding operation, such as the machining time and the metal-removal rate. During the surface grinding operations, heavy workpieces are either held in fixtures or clamped on the machine table by strap clamps and the like, whereas smaller workpieces are usually held by magnetic chucks.

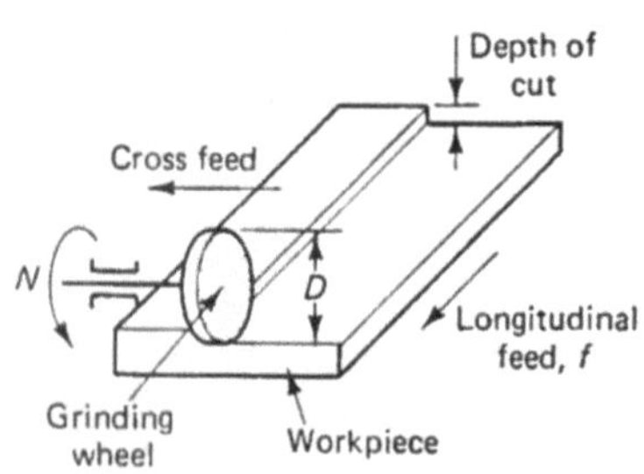

(a)

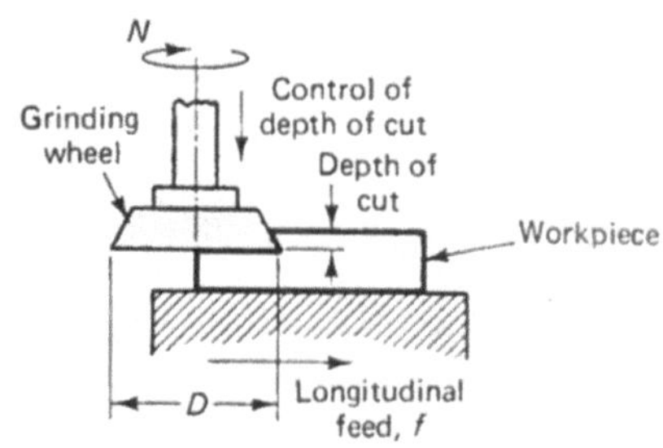

(b)

	(a)	(b)
Machining time	$T = \frac{L + 2A}{f}$ for each travel where: L is the length of the workpiece A = approach allowance $= \sqrt{\frac{D^2}{4} - \left(\frac{D}{2} - d\right)^2}$ where d is the depth of cut	$T = \frac{L + 2A}{f}$ $A = \frac{D}{2}$ for $W = \frac{D}{2}$ up to D $A = \sqrt{W(D - W)}$ for $W < \frac{D}{2}$ where W is the width of cut
Metal-removal rate	$MRR = W \cdot d \cdot f$ where W is the width of cut	$MRR = W \cdot d \cdot f$

FIGURE 11.46 Surface grinding: (a) horizontal spindle and (b) vertical spindle.

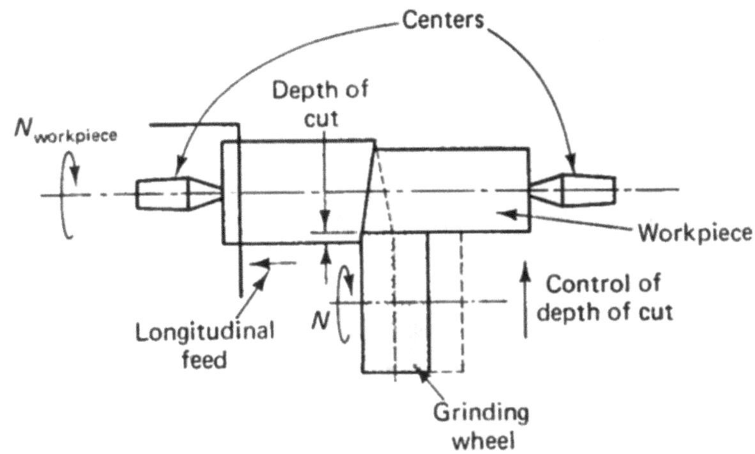

FIGURE 11.47 Cylindrical grinding.

11.6.1.2 Cylindrical Grinding

In *cylindrical grinding*, the workpiece is held between centers during the grinding operation, and the wheel rotation is the source and cause of the rotary cutting motion, as shown in Figure 11.47. Cylindrical grinding can be carried out by employing any of the following methods:

- In the *transverse* method, both the grinding wheel and the workpiece rotate, and longitudinal linear feed is applied so that the entire length can be ground. The depth of cut is adjusted by the cross-feed of the grinding wheel into the workpiece.
- In the *plunge-cut* method, grinding is achieved through the cross-feed of the grinding wheel, and no axial feed is applied. This method can be applied only when the surface to be ground is shorter than the width of the grinding wheel used.
- In the *full-depth* method, which is similar to the transverse method, the grinding allowance is removed in a single pass. This method is usually recommended when grinding short, rigid shafts.

11.6.1.3 Internal Grinding

Internal grinding is employed for grinding relatively short holes, as shown in Figure 11.48. The workpiece is held in a chuck or a special fixture. Both the grinding wheel and the workpiece rotate during the operation, and feed is applied in the longitudinal direction. Any desired depth of cut can be obtained by the cross-feed of the grinding wheel. A variation of this type of grinding is *planetary internal* grinding, and it is recommended for heavy workpieces that cannot be held in chucks. In this case, the grinding wheel not only spins around its own axis but also rotates around the centerline of the hole that is being ground.

11.6.1.4 Centerless Grinding

Centerless grinding involves passing a cylindrical workpiece, which is supported by a rest blade, between two wheels (i.e., the grinding wheel and the regulating or feed wheel). The grinding wheel does the actual grinding, while the regulating wheel is responsible for rotating the workpiece as well as generating the longitudinal feed. This is possible because of the frictional characteristics of this wheel, which is usually made of rubber-bonded abrasive. As can be seen in Figure 11.49, the axis of the regulating wheel is tilted at a slight angle with the axis of the grinding wheel. Consequently, the peripheral velocity of the regulating wheel can be resolved into two components: workpiece rotational speed and longitudinal feed. These can be given by the following equations:

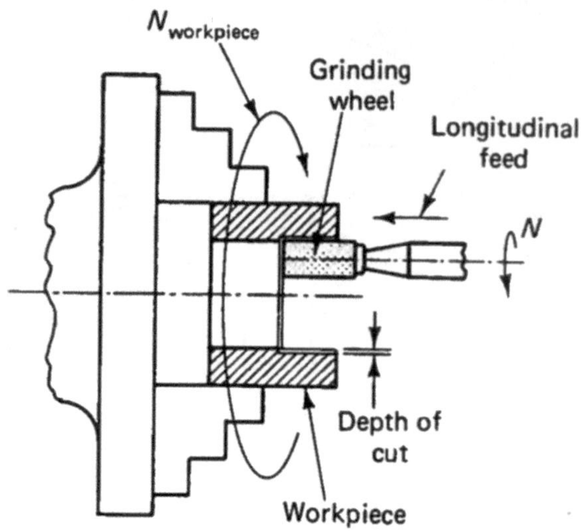

FIGURE 11.48 Internal grinding.

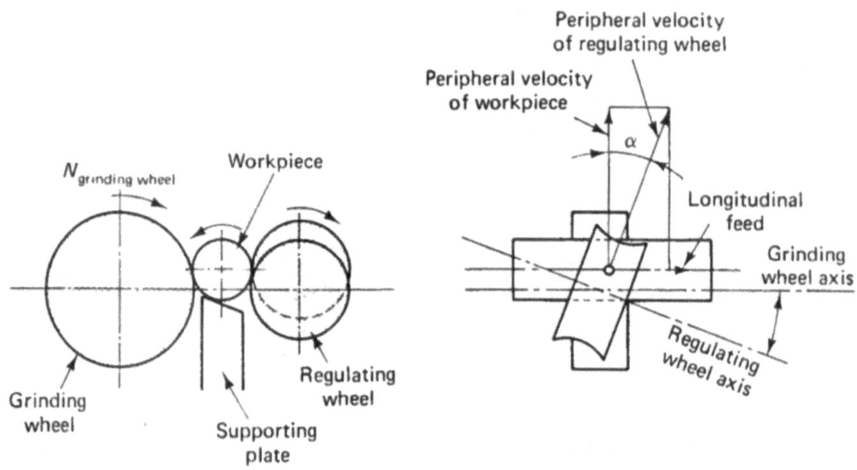

FIGURE 11.49 Centerless grinding.

$$V_{\text{workpiece}} = V_{\text{regulating wheel}} \times \cos \propto \quad (11.10)$$

$$\text{Axial feed} = V_{\text{regulating wheel}} \times C \times \sin \propto \quad (11.11)$$

Note that C is a constant coefficient that accounts for the slip between the workpiece and the regulating wheel ($C = 0.94$–0.98). The velocity of the regulating wheel is controllable and is used to achieve any desired rotational speed of the workpiece. The angle $\propto$ is usually taken from 1° to 5°; the larger the angle, the larger the longitudinal feed will be. When $\propto$ is taken as 0° (i.e., the two axes of the grinding and regulating wheels are parallel), there is no longitudinal feed of the workpiece. Such a setting is used for grinding short shoulders or heads of workpieces having such features.

11.6.2 Grinding Wheels

Grinding wheels are composed of abrasive grains having similar sizes and a binder. The actual grinding process is performed by the abrasive grains. Pores between the grains within the binder enable the grains to act like separate single-point cutting tools. These pores also provide space for the generated chips, thus preventing the wheel from clogging. In addition, pores assist the easy flow of coolants so that heat generated during the grinding process is efficiently and promptly removed. Grinding wheels are identified by their shape and size, kind of abrasive, grain size, binder, grade (hardness), and structure.

11.6.2.1 Shape and Size of Grinding Wheels

Grinding wheels differ in shape and size, depending upon the purpose for which they are to be used. Various shapes are shown in Figure 11.50 and include the following types:

- Straight wheels for surface, cylindrical, internal, and centerless grinding
- Beveled-face or tapered wheels for grinding threads, gear teeth, and the like
- Straight recessed wheels for cylindrical grinding and facing
- Abrasive disks for cutoff and slotting operations when the thickness is 0.02–0.2 inch (0.5–5 mm)
- Cylindrical, straight, and flaring cups for surface grinding with the end of the wheel

The main dimensions of a grinding wheel are the outside diameter D, the bore diameter d, and the height H. These dimensions vary widely, depending upon the grinding process for which the wheel is to be used.

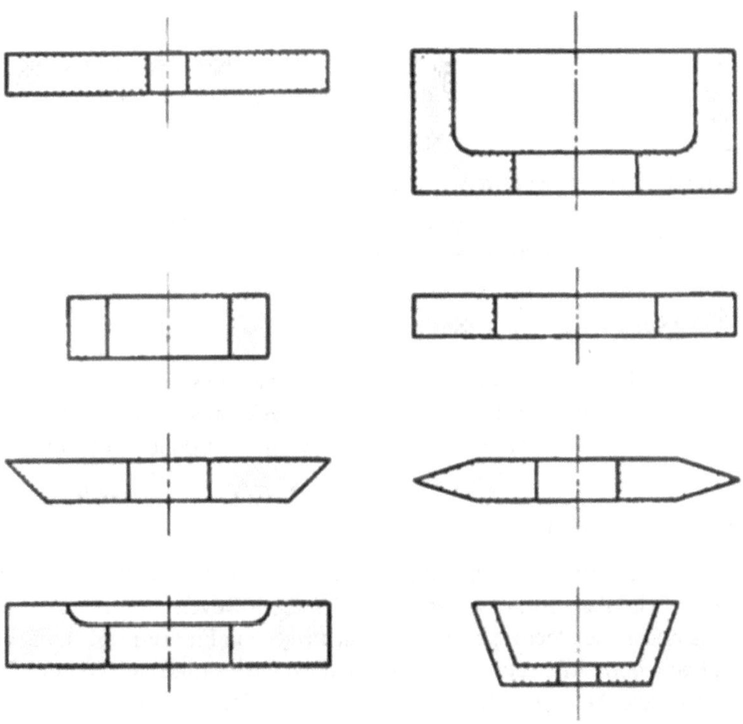

FIGURE 11.50 Various shapes of grinding wheels.

11.6.2.2 Kind of Abrasive

Grinding wheels can be made of natural abrasives such as quartz, emery, and corundum or of industrially prepared chemical compounds such as aluminum oxide or silicon carbide (carborundum). Generally, silicon-carbide grinding wheels are used when grinding low-tensile-strength materials like cast iron, whereas aluminum-oxide wheels are employed for grinding high-strength metals such as alloy steel and hardened steel.

11.6.2.3 Grain Size of Abrasive Used

As you may expect, the grain size of the abrasive particles of the wheel plays a fundamental role in determining the quality of the ground surface obtained. The finer the grains, the smoother the ground surface is. Therefore, coarse-grained grinding wheels are used for roughing operations, whereas fine-grained wheels are employed in final finishing operations.

11.6.2.4 Grade of the Bond

The grade of the bond is an indication of the resistance of the bond to the pulling off of the abrasive grains from the grinding wheel. Generally, wheels having hard grades are used for grinding soft materials and vice versa. If a hard-grade wheel were to be used for grinding a hard material, the dull grains would not be pulled off from the bond quickly enough, thus impeding the self-dressing process of the surface of the wheel and finally resulting in a clogged wheel and a burnt ground surface. The cutting properties of all grinding wheels must be restored periodically by dressing with a cemented-carbide roller or a diamond tool to give the wheel the exact desired shape and remove all worn abrasive grains.

11.6.2.5 Structure

Structure refers to the amount of void space between the abrasive grains. When grinding soft metals, large void spaces are needed to facilitate the flow of the removed chips.

11.6.2.6 Binder

Abrasive particles are bonded together in many different ways. These include the use of vitrified bond, silicate, rubber, resinoid, shellac, or oxychloride. The vitrified bond is the most commonly used binder.

11.6.2.7 Standard Marking System

The standard marking system shown in Figure 11.51 is employed for distinguishing grinding wheels by providing all the preceding parameters in a specific sequence.

11.7 SAWING OPERATIONS

Parting or cutoff operations can be performed on machine tools such as engine lathes, milling machines, and grinding machines. When cutting off is a basic operation in a large-volume production line, special sawing machines are required to cope with the production volume.

11.7.1 Types of Sawing Teeth

The cutting tool may take different forms, depending upon the type of sawing machine used. The tool can be a blade, a circular disk, or a continuous band. However, all these tools are multiedged with several cutting edges (i.e., teeth) per inch. As can be seen in Figure 11.52, teeth can be *straight*, *claw*, or *buttress*. Each tooth, irrespective of its form, must have a rake and a relief angle. Also, teeth are offset in order to make the kerf wider than the thickness of each individual tooth. This facilitates easy movement of the saw blade in the kerf, thus reducing the friction and the generated heat. The maximum thickness is usually referred to as the *saw set* and is equal to the width of the resulting kerf. When selecting the cutting speed and the number of teeth per inch, several factors have to be

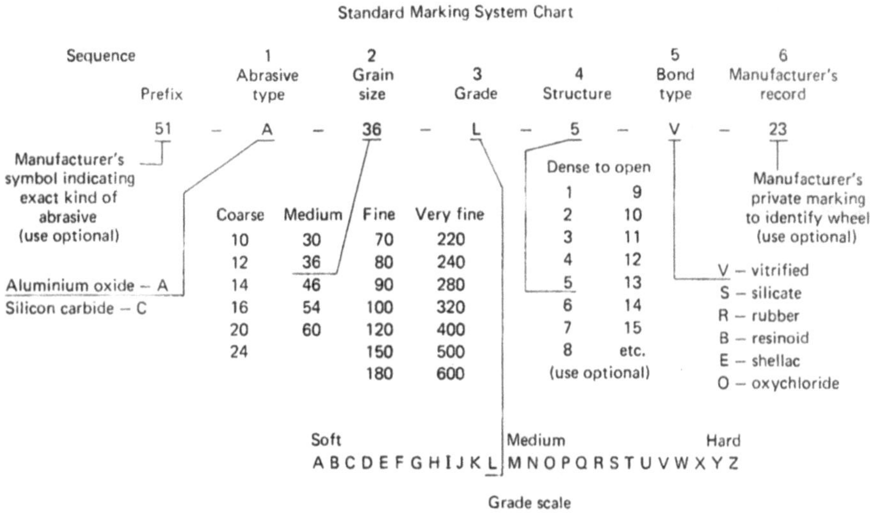

FIGURE 11.51 Standard marking system for grinding wheels.

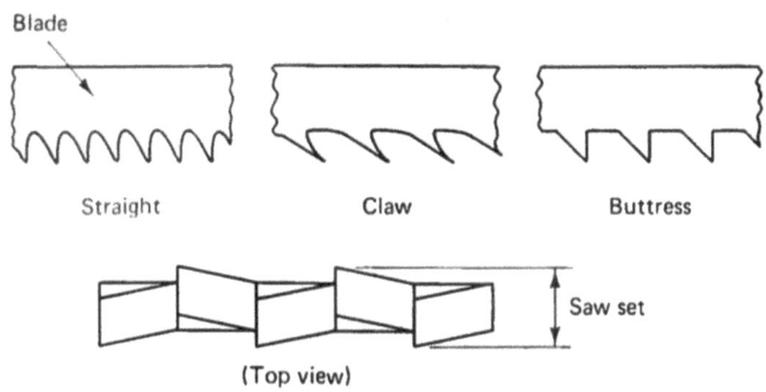

FIGURE 11.52 Types of sawing teeth.

taken into consideration, such as the cutting tool material, the material of the workpiece, the tooth form, and the type of lubricant (coolant) used.

11.7.2 Sawing Machines

Sawing machines differ in shape, size, and construction, depending upon the purpose for which they are to be used. They can be classified into three main groups.

11.7.2.1 Reciprocating Saw

In a *reciprocating saw*, a relatively large hacksaw blade is mechanically reciprocated. Depending upon the construction of the saw, the cutting blade may be either horizontal or vertical. Each cycle has a working (cutting) stroke as well as an idle stroke. Consequently, this type of saw is considered to be a low-productivity saw and is used only in small shops with low-to-moderate production volume.

11.7.2.2 Circular Saw

The cutting tool in a *circular saw* is a circular disk, with the cutting teeth uniformly arranged on its periphery. It looks like the slitting cutter used with milling machines. Although it is highly efficient, it can only be used for parting or cutoff operations of bar stocks or rolled sections.

11.7.2.3 Band Saw

The highly flexible and versatile *band saw* employs a continuous-band sawing blade. As can be seen in Figure 11.53, the band-saw blade is mounted on two pulleys, one of which is the source of power and rotation. Each machine has a flash-welding attachment that is used to weld the edges of the band-saw blade together after adjusting the length, thus forming a closed band. Band saws can be used for contouring and for large-volume cutoff operations. Loading and unloading of the bar stock as well as length adjustment are done automatically (by special attachments in the latter case).

11.8 BROACHING OPERATIONS

Broaching is a metal-removing operation in which a multiedge cutting tool, like that shown in Figure 11.54, is used. In this operation, only a thin layer or a limited amount of metal is removed. Broaching is commonly used to generate internal surfaces or slots, like those shown in Figure 11.55, that are very difficult to produce otherwise. However, it can also be used for producing intricate external surfaces that require tight tolerances.

11.8.1 Broaching Machines

A broaching machine simply comprises a sturdy frame (or bed), a device for locating and clamping the workpiece, the cutting tool, and a means for moving the cutting tool (or the workpiece). The commonly used types of broaching machines are as follows.

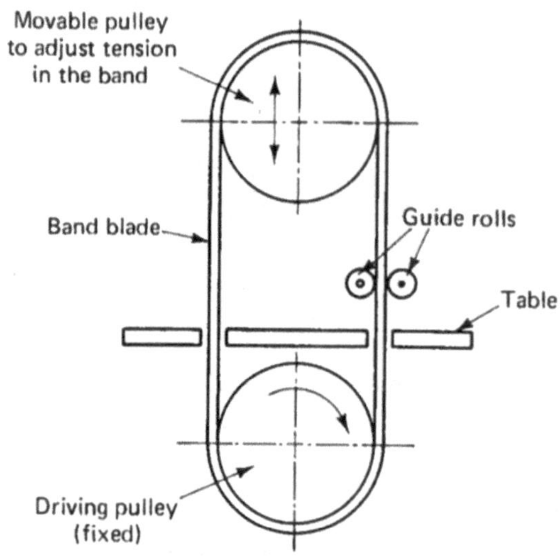

FIGURE 11.53 Basic idea of a band saw.

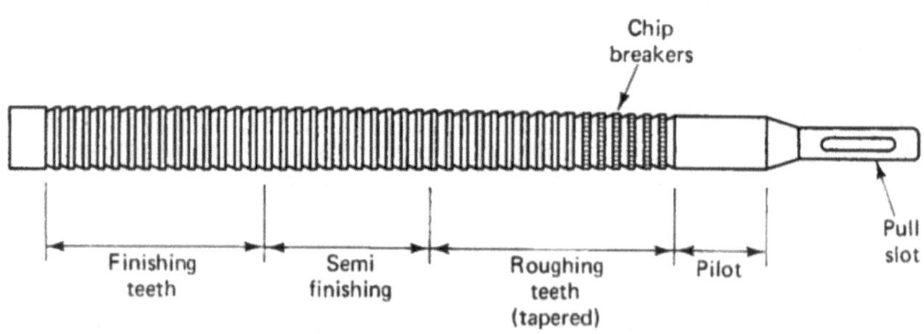

FIGURE 11.54 A broaching tool.

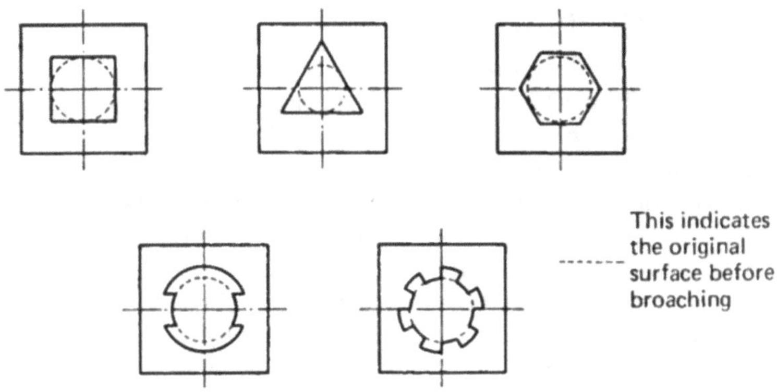

FIGURE 11.55 Different shapes produced by broaching.

11.8.1.1 Pull-Type Machines

In this type of machine, the broaching tool is withdrawn through the initial hole in the tightly clamped workpiece.

11.8.1.2 Push-Type Machines

In this type of machine, the broaching tool is pushed to generate the required surface.

11.8.1.3 Surface-Broaching Machines

In this type of machine, either the tool or the workpiece moves to generate the desired surface.

11.8.1.4 Continuous-Broaching Machines

In this type of machine, the workpiece moves continually over a fixed broaching tool in a straight or circular path.

11.8.2 Advantages and Limitations of Broaching Operations

It is important to know the advantages and limitations of broaching in order to make full use of the potential of this operation. The advantages include the high cutting speed and high cycling time, the close tolerances and superior surface quality that can easily be achieved, and the fact that both roughing and finishing are combined in the same stroke of the broaching tool. Nevertheless, this operation can be performed only on through holes or external surfaces and cannot be carried out on blind holes, for example. Also, broaching involves only light cuts and, therefore, renders itself unsuitable for operations where the amount of metal to be removed is relatively large. Finally, the high cost of broaching tools and machines, together with the expensive fixturing, make this operation economically unjustifiable unless a large number of products are required.

11.9 NONTRADITIONAL MACHINING OPERATIONS

The need for nontraditional machining processes came as a result of the shortcomings and limitations of the conventional, mechanical, chip-generating processes. Whereas conventional processes can be applied only to soft and medium-hard materials, very fine features of extremely hard materials can be produced using nontraditional machining processes. There are a variety of nontraditional processes, and each has its own set of advantages and fields of application. Following is a brief discussion of each of the nontraditional processes commonly used in industry.

11.9.1 Ultrasonic Machining

The mechanism of the vibration ultrasonic machining is actually microchipping or erosion of the surface of the workpiece. The machining tool does not come in contact with the workpiece because a constant stream of abrasive slurry flows between them and flushes away the debris from the cutting area. The slurry consists of very fine abrasive particles of boron carbide, aluminum oxide, or silicon carbide in a suspension of water (20% to 60% by volume). High-frequency (between 18 and 40 kHz), low-amplitude mechanical vibrations impart kinetic energy through a tool called the sonotrode to the slurry (and the abrasive particles), causing it to impact the surface of the workpiece. The material is removed from the workpiece by abrasion where the slurry contacts it, and the result of machining is to cut a perfect negative of the sonotrode's profile into the workpiece. Ultrasonic vibration machining allows extremely complex and nonuniform shapes to be cut into the workpiece with extremely high precision.

Machining time depends on many factors such as the strength and hardness of the workpiece, the material and particle size of the particles in the slurry, and the amplitude of the applied ultrasonic mechanical vibrations. On the other hand, the surface finish of the machined workpiece depends heavily on its hardness and strength; the softer and weaker the materials, the smoother the surface finish will be.

Ultrasonic machining is capable of manufacturing brittle materials that are not electrically conductive and that cannot be machined by alternative methods such as electrical discharge machining. Examples include glass, ceramics, and precious stones. The process enables producing parts with tight tolerances and without residual stresses. The main disadvantages include that the material removal rate is relatively slow and wear can take place at the tip of the vibrating tool.

11.9.2 Abrasive-Jet Machining

In abrasive-jet machining, liquids in which abrasive particles are suspended in a fluid (water or gas) are pumped under extremely high pressure out from a nozzle. The resulting jet stream is then employed in processes like deburring, drilling, and cutting of thin sheets and sections of hard, brittle materials. Since the cutting action is cool, the process is also very successful when cutting materials that are susceptible to thermal damage. Therefore, it is particularly advantageous when cutting glass, ceramics, and sheets of FRP composites. Material is removed by impingement of the fine abrasive particles. As may be expected, the nozzle is usually made of tungsten carbide to resist the abrasive action of these particles while moving fast and rubbing it. The shortcomings of this process involve the problems associated with using high-pressure pumps and the relatively low feed rate employed. Water jets (without abrasive particles) can also be used for cutting leather, textile, and paper.

11.9.3 Chemical Machining

This process is employed in fabricating thin sheet metal components. It involves the application of chemically resistant material on the areas not to be removed (this stage is called masking) and then applying an etchant by spraying or dipping. The etchant reacts with the unprotected selected areas of the sheet metal, essentially corroding it away fairly quickly. After neutralizing and rinsing, the remaining resist is removed and the sheet (or parts) is cleaned and dried. This process can produce highly complex parts with very fine details accurately and economically. There are two types of chemical machining:

- Chemical blanking, a process that produces details (contours of the blank) that penetrate the sheet metal entirely
- Chemical milling, a process that involves eroding the metal to produce blind details, pockets, or channels

11.9.3.1 Etchants

Many chemicals are commercially available as etchants. These include aqueous solution of ferric chloride ($FeCl_3$), chromic acid, aqueous solution of ferrous nitrate ($FeNO_3$), hydrofluoric (HF) acid, and nitric acid (HNO_3). The selection of an etchant for a certain job depends upon many factors, but generally, $FeCl_3$ is recommended for aluminum, copper, and nickel alloys. For tool steel, HNO_3 is usually employed, while HF acid is recommended for titanium.

11.9.3.2 Masking Methods

Masking involves covering the surfaces of the workpiece that are not to be machined with a layer or film of chemical-resistant material that totally prevents etching of those areas. There are three classifications of maskants and resist: cut and hand-peel maskants, photographic resists, and screen resists. Following is a brief description of each:

- The cut and hand-peel method is basically used for chemical milling of structural parts in the aviation industry, in order to reduce the weight of components and increase the strength-to-weight ratio. The masking materials include neoprene, vinyl-base, and butyl-base materials. The coatings are applied by flow, dip, or spray, and the dry-film thickness of these materials can go to 0.015 inch (approximately 0.4 mm), and the etching depths can reach 0.5 inch (12.5 mm). Because scribing and peeling are carried out by hand, it is difficult to keep a good level of accuracy.
- Photoetching employing photographic resist involves covering the workpiece with an extremely thin film (by dip, spray, flow, or laminating) of a photoresist (enamel) that is etchant resistant but sensitive to ultraviolet light. The shape of the part is printed onto optically clear and dimensionally stable photographic film, which is used later as a "phototool." Now, you have to bear in mind that if minute fine details are required, a magnified drawing showing all those details can be photographically reduced to produce the phototool. The next step involves placing the phototool on the coated sheet metal and ensuring intimate contact by drawing a vacuum. The sheet metal (with the film) is then exposed to UV light that allows the areas of resistance that are in the clear sections of the film to be hardened. After exposure, the plate is "developed," washing away the unexposed resist and leaving the areas to be etched unprotected. Next, the etchant is applied to machine the shape and then neutralized followed by rinsing and drying. Thin-gauge sheet metal under 0.050 inch (1.3 mm) in a broad range of alloys are candidates for this type of chemical blanking. Metals include aluminum, brass, copper, Inconel, manganese, nickel, silver, steel, stainless steel, zinc, and titanium. This process has economical advantages over stamping, punching, laser and water jet cutting, or wire electron discharge machining (EDM) for thin-gauge precision parts. This is particularly evident in prototyping or small-batch production because the tooling is inexpensive and quickly produced. It maintains dimensional tolerances and does not create burrs or sharp edges. It can make a part in hours after receiving the drawing. Industrial applications include fine screens and meshes, motor and transformer laminations, metal gaskets, and retainers.
- Screen printing by employing screen resists is actually an application that draws on silk-screen printing technology. Screen resists are applied through a polyester or stainless steel mesh that has an image stencil on it. While the chemical of the coating is higher than that of photographic resists, the accuracy of this method is clearly lower than that of photographic resist, though higher than that of the cut and hand-peel of maskants. It is used for a large number of parts with acceptable accuracy.

11.9.4 Electrochemical Machining

11.9.4.1 The Basic Process

Electrochemical machining (ECM) is the process of removing metal by an electrochemical reaction. The mechanism with which ECM takes place is reciprocal to that of the electroplating process. In fact, similar equipment is used in both cases. For this reason, ECM is referred to as "reverse electroplating," in that it removes material instead of adding it. In ECM, a DC power supply is used. Therefore, the commonly available AC current has to be converted to DC by a rectifier. The workpiece is connected to the anode, while the cathode is connected to the machining tool that is usually made of copper, although brass, graphite, and copper–tungsten are also used. That ECM tool is guided along the desired path and advanced into the workpiece but without touching it. Low-voltage, high-amperage direct current is used, and a pressurized electrolyte (sodium chloride in water) is pumped into the gap between the workpiece and the tool. That gap is very small and varies between 0.003 and 0.03 inch (80–800 μm). Also, it has to be maintained at that level by keeping the feed rate of the ECM tool the same as the rate of corrosion of the workpiece material. As is always the case in a corrosion process, when the electrons cross the gap, the workpiece material dissolves and the tool gradually forms the desired shape in the workpiece. The electrolyte flushes away the products of the reaction from the gap.

11.9.4.2 Process Parameters

As is the case in electroplating, the amperage plays an important role in determining the rate of metal removal from the anode during the ECM process. Moreover, it is advantageous to keep the density of the current flow constant. Another process parameter is the uniformity of the gap (known as the overcut) between the tool and the workpiece, which is in turn affected by the electrolyte flow (requires a substantial pressure of 200 lb/in.^{2} or 1.38 MPa) as well as the electric current density. The accuracy and the surface finish of the tool (cathode) are also important considerations. In fact, the surface finish of the tool is reproduced on the surface of the machined part. The voltages used range between 5 and 20 volts, but the amperage varies between 100 and 40,000 amperes depending upon the size of the workpiece and the rate of material removal required.

11.9.4.3 Advantages and Limitations

ECM can cut intricate contours having fine details, or cavities, in hard and exotic metals, such as Inconel, and high nickel and cobalt alloys. Both external and internal geometries can be machined. It is, therefore, usually used for mass production of parts made of extremely hard materials or materials that are difficult to machine using conventional methods. Complicated shapes such as turbine blades are produced with good surface finish in difficult-to-machine materials. Further advantages of the ECM machining process include producing parts with excellent surface finish and no residual stresses, while the tool wear is zero, thus allowing the tool to be used an infinite number of times. Nonetheless, the use of the ECM process is limited to only electrically conductive materials, and the specific energy consumption is high.

11.9.4.4 Applications

ECM is used in die sinking operations and multiple hole drilling. Its most important application, however, is in machining both steam and gas turbine blades.

It is also widely and effectively used as a deburring process. The process is fast and often more convenient than the conventional methods of deburring by hand or nontraditional machining processes.

11.9.5 Electrical Discharge Machining

Electrical discharge machining or EDM was discovered by Russians and Americans, more or less at the same time, toward the middle of the past century. The concept of the process involves the rapid and recurring discharge of electrical current between an electrode and a workpiece, which are separated by a dielectric liquid. The dielectric fluid acts as an electrical insulator until the applied voltage becomes high enough to bring it to its ionization point when it becomes an electrical conductor. The resulting sparks gradually erode the workpiece to form the desired shape by removing minute chips that take the form of hollow spheres. Other names for this process include spark machining and arc machining. There are three types of EDM machines; all operate on the same principle of erosion by electrical discharge. They include die sinker (ram EDM), wire EDM (cheese cutter), and hole-drilling EDM (hole popper).

11.9.5.1 Die Sinker EDM

Die sinker EDM is used to create complex cavity shapes in tool and die applications, such as forging and cold-forming dies, and plastic injection molds. This is mainly because it provides an answer to such high accuracy, demanding machining applications where conventional metal-removal processes are not possible. EDM can cut hardened materials and exotic alloys while also providing excellent surface finishes. Any workpiece that is going to be machined with EDM has to be electrically conductive, but the process is not successful for heterogeneous metallic materials, especially those with impurities. It cannot be used for plastics or ceramics.

- *Construction and Operation.* As can be seen in Figure 11.56, an EDM system comprises a work tank connected to a dielectric reservoir with a pump and a filter, as well as a pulse generator. The electronic components also include a control system for the Z-axis servo (not shown in the figure). The workpiece is placed in the work tank, which is filled with a hydrocarbon dielectric (such as kerosene), and a small gap of about 0.02 inch (0.5 mm) is maintained between the workpiece and the electrode that is a "positive" of the desired cavity. When voltage (usually on the order of 100 V) is applied, minute particles of metal in the form of hollow spheres are removed by melting and vaporization. Those tiny spheres are composed of material from both the electrode as well as the workpiece. Evidently, unless those spheres are removed from the cutting zone, it would be difficult for the machine to maintain a stable electrical arc inside the sparking gap. Therefore, they need to be removed from the cutting zone, which is accomplished by continually pumping clean dielectric fluid as the flushing medium through the sparking gap to whip away the conductive chips. It also serves as a coolant, cooling the heated material to form the EDM chips.
- *Electrode Material.* The electrode is usually made of a material that is electrically conductive and that can be easily shaped. The electrode must be given a shape that fits exactly into the desired final cavity. The selection of the electrode material depends upon factors such as the cost and machinability of the material, the surface finish of the machined workpiece,

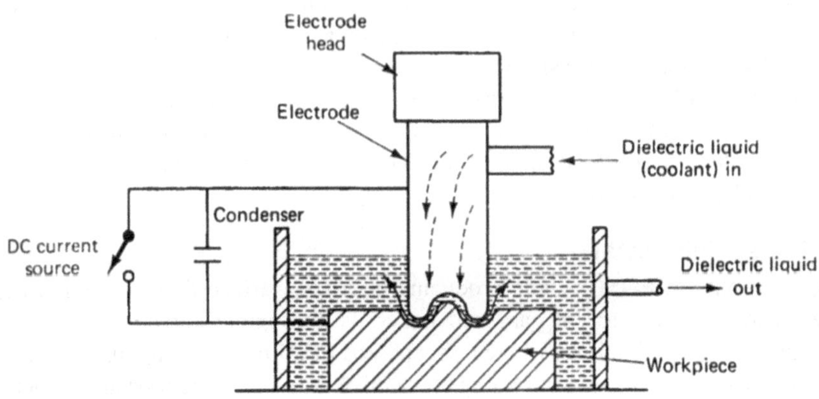

FIGURE 11.56 The EDM process.

and the desired metal-removal rate during the process. The commonly used materials include graphite, copper, brass, copper tungsten, silver tungsten, and zinc alloys. Each material is available in different grades to meet specific needs. Graphite is the predominant electrode material because it is easy to machine, gives a good metal-removal rate per ampere, and is relatively inexpensive. It is usually used for machining steel. Nonetheless, since it is a sintered material and has porosity, care must be taken when a fine surface finish of the machined workpiece is required. Graphite copper mixtures are also used as electrode material and work well for machining tungsten carbides in both roughing and finishing operations. Copper electrodes are often used when the finest surface finish of the workpiece is required. Copper tungsten is recommended as electrode material when fine-detailed, high-precision products are required. It has to be borne in mind, however, that it belongs to the expensive class of electrode materials.

- *Rough versus Finish Machining*. Roughing EDM machining operations require higher power levels in order to achieve greater metal-removal rates. While the number of sparks per minute is lower in the case of roughing operations, the energy level in each spark is much higher, with the final outcome being larger amounts of EDM chips. The roughing processes produce a rougher surface finish and lower dimensional accuracy. On the contrary, finishing processes require a higher frequency of sparks, but the energy level in each spark is reduced. This results in a lower metal-removal rate but a finer surface finish and higher accuracy. Finishing processes always follow roughing ones, in order to minimize the amount of material that finishing operations must remove in order to achieve an efficient total overall cycle time.
- *Advantages and Limitations*. EDM provides an answer to a very difficult problem—namely, machining metals with high hardness. More important, that process is more predictable, accurate, and repeatable than conventional metal-removal processes. Other benefits include that all EDM machining is performed unattended. Thus, the direct labor cost and, accordingly, the production cost are lower than other manufacturing methods. In addition, EDM is practically a stress-free machining process that would not cause any distortion or permanent deformation of the workpiece as a result of machining. The main disadvantage is that it is much slower than other machining methods, where finishing a job may take hours or even days. Moreover, since it is a thermal process, it may affect the surface integrity if any molten metal not flushed during the process is resolidified to form a hard skin. This may cause hair cracks to form at the grain boundaries.

11.9.5.2 Wire EDM

This "cheese cutter" EDM works well, but it has an important limitation: it can essentially produce a two-dimensional cut in a three-dimensional part. The concept of that process is shown in Figure 11.57, wherein the conventional electrode is replaced by a tensioned wire of copper or tungsten that is guided by a CNC system to trace any desired contour. This process has revolutionized the tool and die-making industry. Whereas sharp corners are avoided in tools manufactured by conventional processes to prevent breakage or cracking during subsequent heat treatment, wire EDM can cut heat-treated steels directly to the desired shape. Therefore, large dies having intricate shapes and sharp corners can be produced by this technique.

11.9.5.3 Hole-Drilling EDM

The principal benefits of EDM include producing excellent surface finish, minimal heat-affected zone (HAZ), and its ability to cut hardened materials and exotic alloys. Those advantages are exploited in blind hole applications, where a specialized EDM hole-making machine called a "hole popper" is used. That machine uses a rotating conductive tube as its electrode and a continuous flow of deionized water as the dielectric fluid to flush the EDM chips.

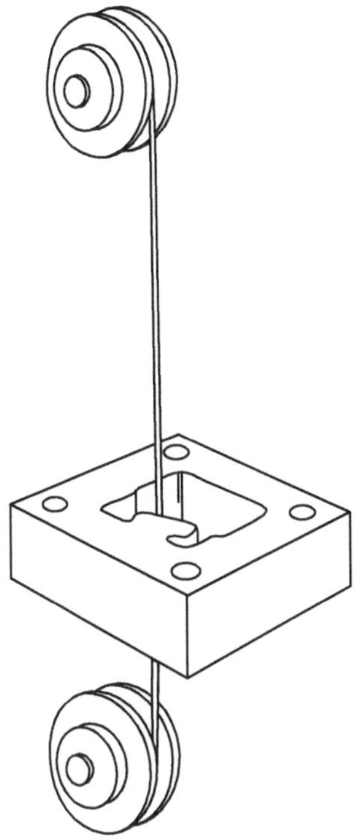

FIGURE 11.57 The concept of wire EDM.

REVIEW QUESTIONS

1. What are the three motions necessary to generate a surface during machining operations?
2. List six different types of lathes.
3. What are the main elements of an engine lathe?
4. Why is the spindle of an engine lathe hollow and why does it have a Morse taper?
5. Discuss briefly the construction of the tailstock.
6. What is the main function of the carriage?
7. How does the carriage receive its motion?
8. Use sketches to explain the difference between a turret lathe and an engine lathe.
9. How do you specify a lathe?
10. What are the conditions for proper tool holding?
11. Use sketches to illustrate the following: turning tools, facing tools, cutoff tools, thread-cutting tools, and form tools.
12. What precautions should be taken when supporting a workpiece during lathe operations?
13. When should a workpiece be held between two centers?
14. When is it necessary to hold a workpiece in a chuck?
15. When would a workpiece be mounted on a faceplate?
16. How would you hold a disklike workpiece that has to be machined on both sides?
17. What do the machining marks look like in cylindrical turning?
18. How is the axial feed provided and how is the depth of cut controlled in cylindrical turning?
19. What do the machining marks look like in facing operations?
20. What are the suitable work-holding devices for facing operations?
21. List three methods that can be employed to generate a tapered surface.
22. What provides the feed during thread cutting?
23. Describe knurling.

24. What are the variables that affect the optimal value of the surface cutting speed?
25. Discuss the considerations that must be taken into account when designing turned parts.
26. What is the difference between shaping and planing?
27. What kind of surfaces can be produced by shaping and planing operations?
29. Explain the working principles of the quick-return mechanism.
30. List the commonly used types of drills and discuss the applications of each.
31. What is meant by the TPA? How does the workpiece material affect the optimal value for this angle?
32. List some other hole-making operations and discuss the applications of each.
33. Discuss the considerations that must be taken into account when designing drilled parts.
34. List the various types of drilling machines and discuss the characteristics and fields of application of each.
35. What is a jig? When is the use of jigs recommended?
36. Define *milling*.
37. Why are the permissible cutting speeds in milling four times higher than those for turning?
38. Differentiate between up milling and down milling.
39. List the various types of milling cutters and discuss the applications of each.
40. List the various types of milling machines and discuss the applications of each.
41. What is the function of a universal dividing head?
42. Define *grinding*.
43. List the types of grinding operations and discuss the applications of each.
44. Of what are grinding wheels composed?
45. How can grinding wheels be identified?
46. List the different types of sawing machines and discuss the constructional features as well as the fields of application of each.
47. Use sketches to illustrate the types of teeth of saw blades.
48. Define *broaching*. When is the use of this process recommended? Discuss the advantages and limitations of broaching operations.
49. Why did the need arise for nontraditional machining operations?
50. How is ultrasonic energy employed on machine surfaces?
51. Explain the working principles of abrasive-jet machining.

PROBLEMS

1. It is required to maintain a cutting speed of 120 feet per minute (37 m/min) in a turning operation. If the initial workpiece diameter is 3.25 inches (82 mm) and the depth of cut is 0.1 inch (2.5 mm), calculate the rpm of the spindle during the third and sixth passes.
2. A 24-inch-diameter (600-mm) part with a 6-inch-diameter (150-mm) hole in the center is to be faced starting at the outside. The rotational speed of the spindle is 7 revolutions per second, the depth of cut is 0.15 inch (3.75 mm), and the feed is 0.01 inch per revolution (0.25 mm/rev). Calculate the machining time as well as the maximum and minimum rate of metal removal.
3. Two thousand bars that are 3.25 inches (81 mm) in diameter and 12 inches (300 mm) long are to be turned down to 2.75-inch (69-mm) diameters. Heavy cuts followed by a light-finishing cut are to be used. For finishing, the feed is 0.005 inch (0.1 25 mm), the cutting speed is 300 feet per minute (90 m/min), and the depth of cut is 0.07 inch (1.75 mm). Two roughing cuts (two passes) are required, where the cutting speed is only 200 feet per minute (60 m/min) and the feed is 0.01 inch (0.25 mm). Calculate the overall production time when the time taken to return the tool to the beginning of the cut is 15 seconds and the load/unload time is 2 minutes.

4. A bronze bushing is 2 inches (50 mm) in diameter, 3 inches (75 mm) long, and has a central hole of 1.25 inches (31.25 mm). It is to be produced on a lathe, starting with a solid bar stock having a 2-inch (50-mm) outer diameter. Estimate the production time per piece. Take the feed as 0.01 inch per revolution (0.25 mm/rev) and the cutting speed as 200 feet per minute (60 m/min). Assume any missing data.
5. A workpiece having a length of 3 inches (75 mm) is to be taper-turned by offsetting the tailstock. If the maximum diameter of the workpiece is 1⅛ inches (28.125 mm) and the minimum diameter is 1.0 inch (25 mm), calculate the amount of offset.
6. A 10-inch-diameter (250-mm) part having a 4-inch-diameter (100-mm) hole is to be bored for 4 inches (100 mm) of its length to a diameter of 4.4 inches (110 mm). A depth of cut of 0.08 inch (2 mm) is to be used with a feed of 0.004 inch (0.1 mm) and a cutting speed of 330 feet per minute (100 m/min). If it takes 15 seconds to return the tool to the starting point and set the depth of cut, calculate the time required to complete this job.
7. A part is to be tapered in such a manner as to have the following dimensions:
8. Calculate the tailstock offset.
9. How far must the tailstock be offset to cut a 0.5-inch-per-foot (41.7-mm/m) taper on an 8-inch-long (200-mm) workpiece?
10. In a drilling operation, the desired depth of the hole is 1 inch (25 mm), the drill size is 0.4 inch (10 mm), the rpm is 100, and the feed is 0.01 inch (0.25 mm). Calculate the cutting speed and estimate the drilling time.
11. A standard twist drill is used to drill a number of 3/8-inch (9.5-mm) through holes in a 5/8-inch-thick (16-mm) SAE 1020 steel plate. The cutting speed is 60 feet per minute (18.3 m/min), and the feed is 0.004 inch (0.1 mm). Calculate the time required for drilling each hole and the metal-removal rate.
12. It is required to drill a 1-inch-deep (25-mm) hole in each of 75,000 cast-iron blocks. If the rotational speed is 600 rpm and the feed is 0.002 inch (0.05 mm), estimate the required working hours. Assume that it takes 30 seconds to load and unload the part and that 15 seconds must be allowed each time the drill bit is changed. Take the number of bit changes as 10.
13. In a drilling operation, the feed rate is 1 inch per minute (25 mm/min), the cutting speed is 37.2 feet per minute (12 m/min), and the diameter of the hole to be drilled is 0.6 inch (15 mm). What is the feed?
14. In milling a step 1/8 × 1/8 inch (3.18 × 3. 18 mm) in a 2-inch-long (50-mm) workpiece, a two-fluted 1/2-inch-diameter (12.5-mm) end mill is used. The rotational speed is 700 rpm, and the feed is 0.006 inch per tooth (0.15 mm per tooth). Estimate the metal-removal rate and the milling time.
15. In a face milling operation, the depth of cut is 1/4 inch (6 mm), and the table moves at 0.2 inch per second (10 mm/s). The width of the workpiece is 2.0 inches (50 mm), and the cutter has a diameter of 3.25 inches (81 mm). The rotational speed of the cutter is 120 rpm, and the feed is 0.01 inch per tooth (0.25 mm per tooth). If the length of the workpiece is 10 inches (250 mm), calculate the number of teeth of the cutter, the metal-removal rate, and the milling time.
16. An 18-tooth, 1-inch-wide (25-mm) HSS cutter having a diameter of 4 inches (100 mm) is to be used in slot milling a 10-inch-long (250-mm) workpiece. If the desired depth of the slot is 0.24 inch (6 mm), the cutting speed is 93 feet per minute (30 m/min), and the feed is 0.005 inch per tooth (0.125 mm per tooth), estimate the milling time.
17. Estimate the machining time in face milling given the following data:
18. The recommended feed for milling this kind of steel is 0.01 inch per tooth (0.25 mm per tooth).

19. When using a helical milling cutter with 20 teeth, if the cutting speed is 70 feet per minute (22.5 m/min) and the cutter diameter is 4 inches (100 mm), calculate the feed rate of the table.
20. When gear-cutting processes are to be performed on a universal milling machine by using the indexing head, explain the procedure in each case when the gear has the following number of teeth:
    a. 20 teeth
    b. 32 teeth
    c. 22 teeth
    d. 15 teeth

Cutter	20 teeth and 4 inches (100 mm) in diameter
Rotational speed	300 rpm
Depth of cut	0.25 inch (6 mm)
Feed	0.001 inch per tooth (0.025 mm per tooth)
Length of workpiece	20 inches (500 mm)
Cutter width	Larger than that of the workpiece
Total length	6 inches (150 mm)
Tapered length	1½ inches (62.5 mm)
Large diameter	1.0 inch (25 mm)
Small diameter	0.625 inch (16 mm)

12 Design for Assembly

12.1 INTRODUCTION

Modern societies are undergoing continuous development, which necessitates large-scale use of sophisticated products like appliances, automobiles, and health-care equipment. Each of these products involves a large number of individual components, assemblies, and subassemblies that must be brought together and assembled into a final product during the last step of the manufacturing sequence. A rational design should, therefore, be concerned with the ease and cost of assembly, especially when given the fact that 70–80 percent of the cost of manufacturing a product is determined during the design phase. It is for this reason that the concept of design for assembly (DFA) emerged. It is simply a process for improving the product design for easy and low-cost assembly. In other words, this assembly-conscious design approach not only focuses on functionality but also concurrently considers *assemblability*.

Although the use of the term *design for assembly* is fairly recent, several companies can claim, in good faith, that they have developed and have been using guidelines for assembly-conscious product design for a long time. For instance, the General Electric Company published in 1960, for internal use only, the *Manufacturing Producibility Handbook*. It included compiled manufacturing data that provided designers in the company with information necessary for sound and cost-efficient design. Later, in the 1970s, research institutions and research groups started to become more and more interested in the subject when the Conference Internationale pour le Recherches de Production (CIRP) established a subcommittee for that purpose and Professor Geoffrey Boothroyd began his pioneering research at the University of Massachusetts Amherst.

The traditional approach for DFA has been to reduce the number of individual components in an assembly and to ensure an easy assembly for the remaining parts through design modifications. When a design is altered in such a manner that two components are replaced by just a single one, the logical consequence is the elimination of one operation in manual assembly or a whole station of an automatic assembly machine. Accordingly, many benefits have been credited to the DFA methodology, including simplification of products, lower assembly costs, reduced assembly (and manufacture) time, and reduced overheads. Recently, the DFA concept has been extended to incorporate process capacity and product mix considerations so that products can be designed to assist in balancing assembly flow, thus eliminating the problem of stressing one process too heavily while underutilizing others. Many people now are calling for extending DFA over the whole product life cycle, in which case environmental concerns would be addressed and designs would be developed that facilitate disassembly for service as well as for recycling at the end of the life cycle.

A first step toward a rational design for easy and low-cost assembly is the selection of the most appropriate method for assembling the product under consideration. The design guidelines for the selected method can then be applied to an assembly-conscious design for that product. The next step is the use of a quantitative measure to evaluate the design in terms of ease of assembly and to pinpoint the sources of problems so that the design can then be subjected to improvement. Many iterations of this evaluation/improvement process, as necessary, can be done in order to achieve an optimal design. It is, therefore, essential for us now to discuss the different assembly methods currently available.

12.2 TYPES AND CHARACTERISTICS OF ASSEMBLY METHODS

As you may expect, there is no single method that is always "better" than other methods under all conditions. In other words, each method has its own domain or range within which it can most

DOI: 10.1201/9781003537106-12

successfully and economically be applied. Factors like the number of products assembled per year and the number of individual components in an assembly play a major role in determining the range of economical performance of an assembly system. Following is a description of the different assembly methods, as well as the characteristics of each.

12.2.1 Manual Assembly

In manual assembly, the operations are carried out manually with or without the aid of simple, general-purpose tools like screwdrivers and pliers. Individual components are transferred to the workbench either manually or by employing mechanical equipment such as parts feeds or transfer lines and then are manually assembled. This assembly method is characterized by its flexibility and adaptability—a direct consequence of the very nature of the key element of the system, the human brain. The assembly cost per product, however, is virtually constant and is independent of the production volume. There is an upper limit to the production volume, above which the practicality and feasibility of the manual assembly method are, to say the least, questionable. This upper limit depends upon the number of individual components in an assembly and the number of different products assembled. Nevertheless, it is important to remember that the capital investment required for this type of assembly system is close to zero.

12.2.2 Automatic Assembly Using Special-Purpose Machines

In the type of assembly system referred to as fixed automation or the Detroit type, either synchronous indexing machines and automatic feeders or nonsynchronous machines where parts are handled by a free-transfer device are used. The system, in both cases, should be built to assemble only one specific product. Such is the case with the automotive assembly lines in Detroit, where each one is dedicated to the production of a specific model of car (and hence the reason this name is given to this type of assembly system). There is an inherent rigidity in this method of assembly, meaning that these systems lack any flexibility to accommodate tangible changes in the design of the product. Moreover, a system of this type requires a large-scale capital investment, as well as considerable time and engineering work, before actual production can be started. Also, the individual components must be subjected to strict quality-control inspection before they can be assembled because any downtime due to defective parts will result in considerable production delays and, therefore, cash losses. Nevertheless, a real advantage of this assembly system is the decreasing assembly cost per product for increasing production volume. Naturally, when the production volume increases, the share of each product from the capital investment becomes smaller, which makes this assembly system particularly appropriate for mass production. It is worth mentioning that an underutilized assembly system will simply result in an increase in the assembly cost per product because the cost of equipment has to be divided between a smaller number of products, thus increasing the cost share of each product.

In order to come up with a more flexible version of the automatic assembly system that can tolerate some minor changes in the design of the product being assembled, the nonsynchronous machines are fitted with programmable workheads and parts magazines. Thus, the assembly sequence and characteristics can be tailored to match the attributes of the modified design. Although this system provides some flexibility, it is still considered most appropriate for mass production.

12.2.3 Automatic Assembly Using Robots

In robotic assembly, the production volume is higher than that of a manual assembly system but lower than that of an automatic assembly system that incorporates special-purpose machines. It, therefore, fills a gap in production volume between these other two assembly systems. Robotic assembly systems may take one or more of the following forms:

- A one-arm, general-purpose robot operating at a single workstation that includes parts feeders, magazines, and so on. The end effector of the arm is tailored to suit the specific operation performed.
- Two robotic arms operating at a single workstation. A programmable controller (PLC) is employed to simultaneously control and synchronize the motions of the two arms. This setup is referred to as a robotic assembly cell and is, in fact, very similar to a flexible manufacturing cell. Other supporting equipment like fixtures and feeders is also included in the cell.
- Multistation robotic assembly system. This system is capable of performing several assembly operations simultaneously. It can also perform different assembly operations at each station. Accordingly, this robotic assembly system possesses extremely high flexibility and adaptability to design changes. On the other hand, a production volume that is quite close to that of the automatic assembly mass production system can be achieved using this type of system.

12.3 COMPARISON OF ASSEMBLY METHODS

Clearly, manual assembly requires the least capital investment followed by the two simplest forms of robotic assembly. On the other hand, compared to the automatic system with special-purpose machines, the multistation robotic assembly system requires more capital investment for a large production volume but less capital investment for a moderate production volume. A better way of illustrating this comparison is to plot a graph indicating the relationship between the assembly cost per product and the annual production volume for the three assembly methods. As shown in Figure 12.1, the assembly cost per product is constant for manual assembly and decreases linearly with increasing production volume for automatic assembly using special-purpose machines. In the case

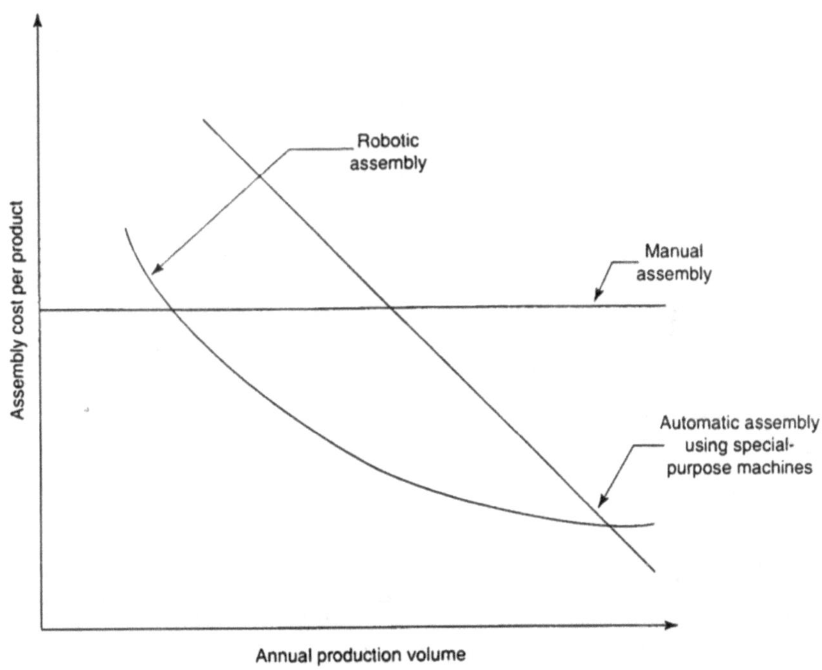

FIGURE 12.1 Assembly cost per product versus annual production volume for three assembly methods.

of robotic assembly, the assembly cost per product also decreases with increasing production volume but not linearly because the type of system used and its physical size depend upon the production volume as well. Figure 12.1 also helps to determine the range of production volume within which each of the assembly methods is cost-effective. Consequently, such a graph is a valuable tool for selecting the appropriate assembly method for a specific project.

12.4 SELECTION OF ASSEMBLY METHOD

Several factors must be taken into consideration by the product designer and the manufacturer when selecting an assembly method. These factors include the cost of assembly, the annual production volume (or production rate), the number of individual components to be assembled in a product, the number of different versions of a product or products, the availability of labor at a reasonable cost, and, last but not least, the payback period. The factors are interactive, and it is impossible to have a single mathematical relationship or a single graph that incorporates them all and indicates an appropriate range or domain for each assembly method. Usually, a two-variable chart is constructed based on fixed specific values for the other variables.

Figure 12.2 indicates the appropriate ranges of application for each of the various assembly methods when there is only one type (or version) of the product to be assembled. As can be seen, the two variables, which are pivotal in most cases, are the annual production volume and the number of individual components in an assembly. Notice that the manual assembly method is suitable for low production volumes and a limited number of individual components per assembly. Robotic assembly is recommended for moderate production, with the one-arm robot being more appropriate for assemblies that have less than eight individual components. When a large number of assemblies are to be produced, the use of assembly systems with special-purpose machines becomes a must.

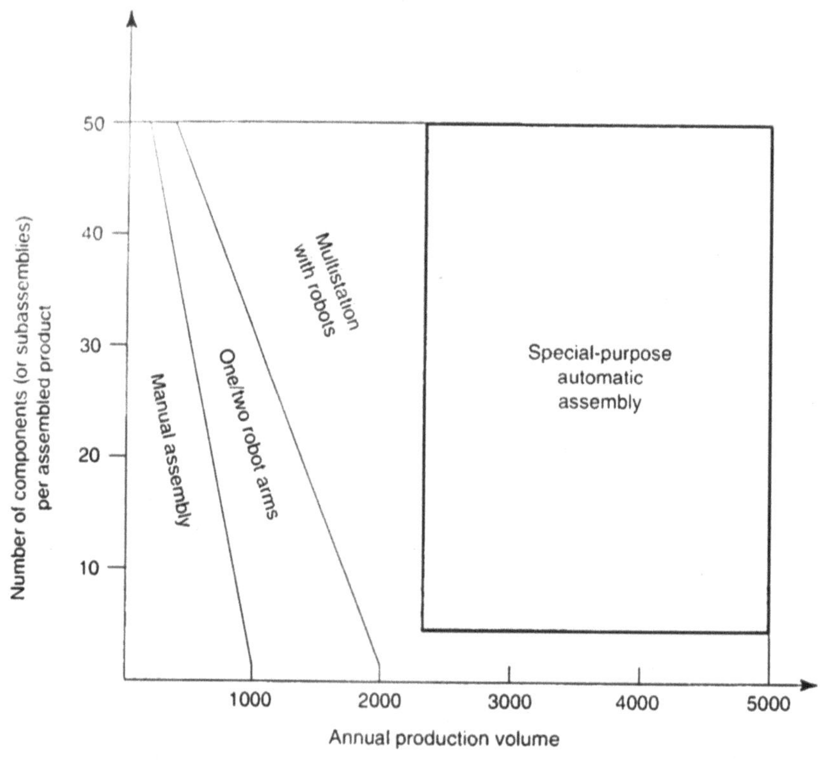

FIGURE 12.2 Appropriate ranges of application for various assembly methods.

Remember that with an increasing number of different types or versions of assemblies, the recommended ranges of application for each assembly method will differ from those shown in Figure 12.2. For instance, a multistation robotic assembly system would be more appropriate than an automatic assembly system with special-purpose machines for relatively high production volumes. The most important point here is that the assembly rate of the selected assembly method should not result in any bottleneck but rather should ensure trouble-free production. Also, it is always advisable to estimate the cost of assembly whenever more than one assembly method is under consideration. Assuming that all other factors are comparable, the method that gives the lowest assembly cost is the one to select.

12.5 PRODUCT DESIGN FOR MANUAL ASSEMBLY

We are now in a position to discuss the rules and guidelines to be followed when designing components for manual assembly. It is important here to emphasize that blind adherence to these rules is not recommended. In fact, this approach can result in very complex components that are difficult and expensive to manufacture. The use of good engineering sense, rational thinking, and accumulated knowledge will ensure that these rules are wisely applied. The strategy to adopt when designing products for manual assembly is to strive to reduce both the assembly time and the skills required of assembly workers. Here are the guidelines for product design for manual assembly:

- Eliminate the need for any decision making by the assembly worker, including his or her having to make any final adjustments. Remember that assembly workers are usually unskilled and are paid at or close to the minimum wage and it is, therefore, not logical or fair to rely on them to make these adjustments.
- Ensure accessibility and visibility. It is not logical or fair to require the worker, for example, to insert and tighten a bolt in a hole that is not visible or easily accessible.
- Eliminate the need for assembly tools or special gauges by designing the individual components to be self-aligning and self-locating. Parts that fit and snap together eliminate the need for fasteners, thus resulting in an appreciable reduction in both the assembly time and cost. Also, features like lips and chamfers can greatly aid in making parts self-locating, as is clearly demonstrated in Figure 12.3, where two pins, one having a chamfer and the other

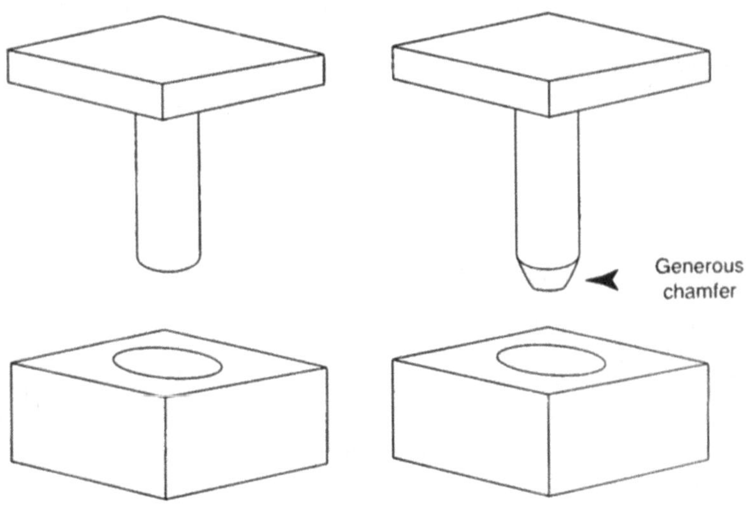

FIGURE 12.3 Using a chamfer to make a part self-locating.

without, are being inserted into two identical holes during an assembly operation. Obviously, it is far easier and takes less time to insert the pin with the chamfer.

- Minimize the types of parts by adopting the concept of standardization as a design philosophy. Expand the use of standard parts as well as multifunction and multipurpose components. Although more material may be consumed to manufacture multipurpose parts, the gains in reducing assembly time and cost will exceed that waste.
- Minimize the number of individual parts in an assembly by eliminating excess parts and, whenever possible, integrating two or more parts together. Certainly, handling one part is far easier than handling two or more. The criteria for reducing the parts count per assembly, established by G. Boothroyd and P. Dewhurst (see the references at the end of this book), involve negative answers to the following questions:
  1. Does the part move relative to all other parts already assembled?
  2. Must the part be of a different material or be isolated from other parts already assembled? (Only fundamental reasons concerned with material properties are acceptable.)
  3. Must the part be separate from all other parts already assembled because otherwise necessary assembly or disassembly of other parts would be impossible?

  If the answer to each of these questions is no, then the part can be integrated or combined with another neighboring part in the assembly. When applying this rule, however, remember that combining two or more parts into a complicated one may result in making the part difficult to manufacture.
- Avoid or minimize reorienting the parts during assembly. Try to make all motions simple by, for example, eliminating multimotion insertions. Avoid rotating or reorienting the assembly as well as releasing and regripping individual components. These are wasteful motions and result in increased assembly time and cost. The best time to eliminate them is during the design phase. The use of vertical insertion (along the Z-axis) is ideal, especially when you take advantage of gravity.
- Ensure ease of handling of parts from the bulk by eliminating the possibility of nesting or tangling them. This is achieved by simple modifications in the design. In addition, avoid the use of fragile or brittle materials, as well as flexible parts like cords and cables.
- Design parts having maximum symmetry in order to facilitate easy orientation and handling during assembly. If symmetry is not achievable, the alternative is to design for asymmetry that is easily recognizable by the assembly worker. Failure to observe the preceding rules may result in serious problems during assembly in terms of higher assembly costs or jams and delays. Consequently, many companies avoid manual assembly and sell their products unassembled. Examples in the United States include grills, furniture, and toys. As you may have experienced, some of these products are not properly designed for easy assembly, and it takes customers an extremely long time to assemble them. It is no surprise that such faulty designs do not pay off as they adversely affect the sales of unassembled products.

12.6 PRODUCT DESIGN FOR AUTOMATIC ASSEMBLY

Parts that are designed to be assembled by automatic special-purpose machines must possess different geometric characteristics from those of parts to be assembled manually. Automatic assembly requires parts that are uniform, are of high quality, and have tighter geometric tolerances than those of manually assembled parts. These requirements are dictated by the need to eliminate any downtime of the assembly system due to parts mismatch or manufacturing defects. As a consequence, problems related to locating and inserting parts, though they need to be addressed, are not of primary importance. These problems require design changes to ease assembly; by revising the product design, each assembly operation becomes simple enough to be performed by a machine rather than by a human being. The most important concerns to address involve the orientation, handling, and feeding of parts to the assembly machine. The efficiency of performing these tasks has a considerable effect on the efficiency and output of the assembly system and, of course, on the assembly cost. This approach is referred to as design for ease of automation. Here are the guidelines for product design for automatic assembly:

- Reduce the number of different components in an assembly by using the three questions listed previously in the design guidelines for manual assembly. An appropriate approach is to use value analysis in identifying the required functions performed by each part and finding out the simplest and easiest way to achieve those functions. An example is shown in Figure 12.4, where two products are contrasted, one designed to facilitate assembly through a reduction in the parts count and the other designed without ease of assembly being taken into consideration.
- With the new developments in casting and plastics injection-molding technologies, complex components can replace entire subassemblies. Nevertheless, the designer has to be very careful when combining parts so as not to adversely affect the manufacturing cost. In fact, in order to reduce the parts count in assemblies, subcontractors and suppliers of electronics manufacturers have been continually asked to fabricate extremely complex parts. In short, the rule of reducing the number of parts should not be applied blindly because, in many cases, more efficient manufacturing can be achieved by breaking a single component into two or more parts, as shown in Figure 12.5, which indicates two methods for manufacturing a 2-foot axle shaft and flange.
- Use self-aligning and self-locating features in parts to facilitate the process of their assembly. Considerable improvement can be achieved by using chamfers, guide pins, dimples, molded-in locators, and certain types of screws (e.g., cone and oval screws). Figure 12.6 is an example of how to facilitate assembly through simple design modifications, while Figure 12.7 shows the types of screws that are suitable for assembly operations.

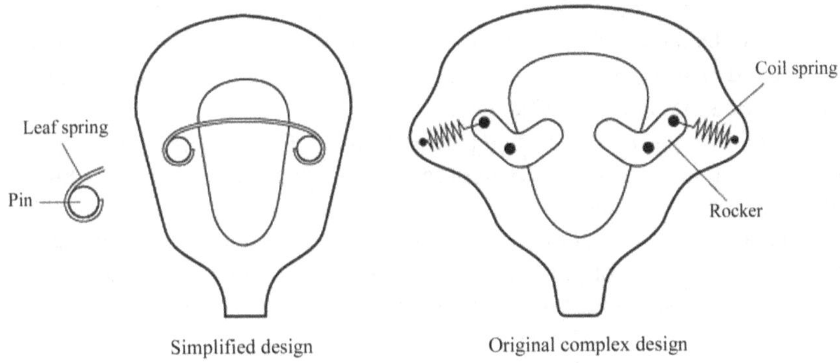

FIGURE 12.4 Facilitating assembly through a reduction in parts count.

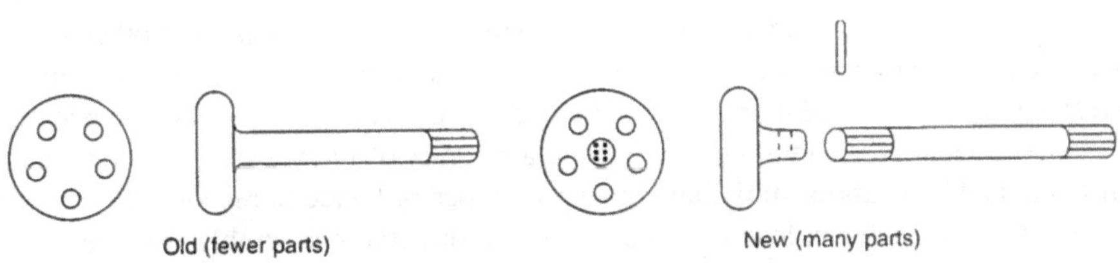

FIGURE 12.5 Two methods for manufacturing an axle shaft and flange. (Redrawn after Lane, J. D., ed. *Automated Assembly*, 2nd ed., Society of Manufacturing Engineers, 1986. Used by permission.)

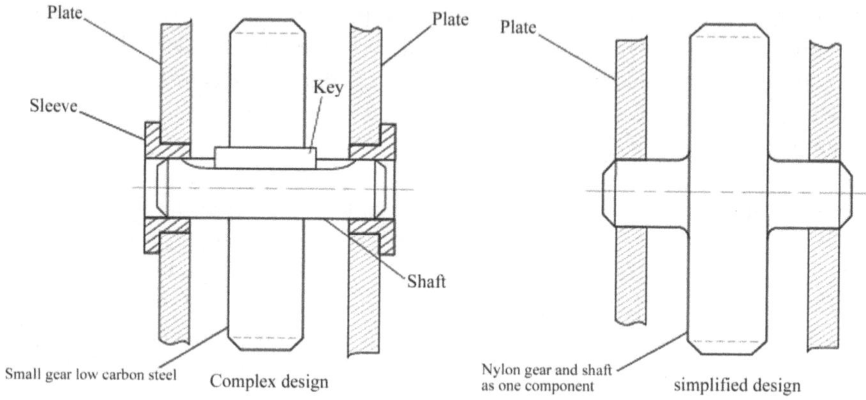

FIGURE 12.6 Facilitating assembly through simplification of design.

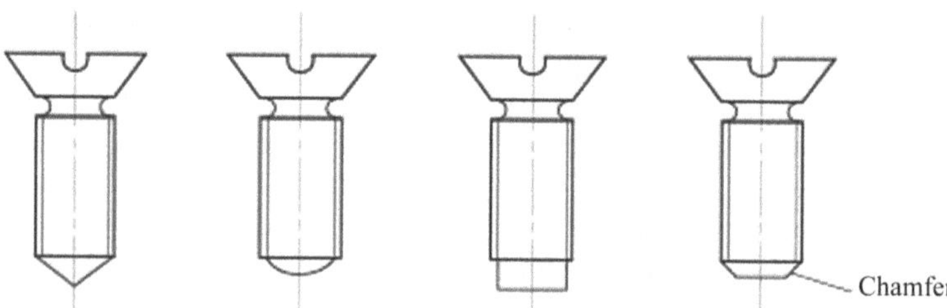

FIGURE 12.7 Types of screws suitable for assembly operations.

- Avoid, whenever possible, fastening by screws because that process is both expensive and time-consuming. It is, therefore, recommended to design parts that will snap together or be joined together by a press fit. Tighter tolerances are then required, and problems may also be encountered in disassembly for maintenance repair, or recycling. If screws must be used, then unify their types and head shapes.
- Make use of the largest and most rigid part of the assembly as a base or fixture where other parts are stack-assembled vertically in order to take advantage of gravity. This will eliminate the need for employing an assembly fixture, thus saving time and cost. Also, remember that the best assembly operation is one that is performed in a sandwich-like or layered fashion. If this is difficult or impossible to do, the alternative is to divide the assembly into a number of smaller subassemblies, apply the rule stated herein to each separately, and then plug all the subassemblies together.
- Actively seek the use of standard components and/or materials. There should be a commitment, at all levels, to the goal of using a high percentage of standard parts in any new design. A very useful concept to be adopted in order to achieve this goal is group technology. Standardization should begin with fasteners, washers, springs, and other individual components. This translates into standardization of assembly motions and procedures. The next step is to use standard modules that are assembled separately and then plugged together as a final product. Each module can include a number of individual components that are self-contained in a subassembly having a specific performance in response to one or more inputs. This approach can lead to a considerable reduction in assembly costs, as well as in manufacturing and inventory costs.
- Avoid the possibility of parts tangling, nesting, or shingling during feeding. A few changes in the geometric features may eliminate these problems without affecting the proper

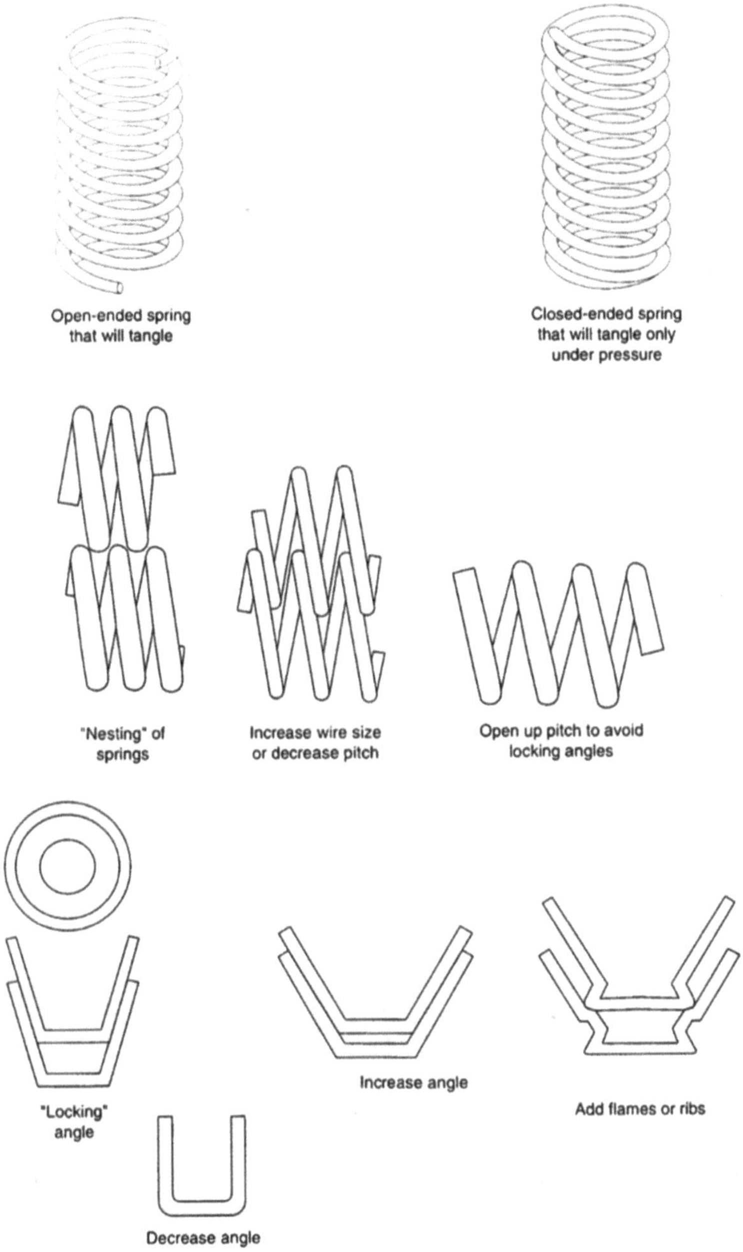

FIGURE 12.8 Parts that tend to nest during feeding and design modifications that eliminate the problem. (Redrawn after Lane 1986. Used by permission of SME.)

functioning of the component. Figure 12.8 shows some parts that tend to nest during feeding and the design modifications that eliminate this problem.

- Avoid flexible, fragile, and abrasive parts and ensure that the parts have sufficient strength and rigidity to withstand the forces exerted on them during feeding and assembly.
- Avoid reorienting assemblies because each reorientation may require a separate station or a machine, both of which cause an appreciable increase in cost.
- Design parts to ease automation by presenting or admitting the parts to the assembly machine in the right orientation after the minimum possible time in the feeder. The process

Number of Orientations	Types of Parts	Required Number of Parts/Hour (out of the feeder)	Minimum Required Rate of Feeding Parts/Hour (into the feeder)
1	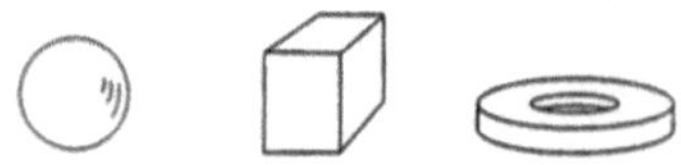 Sphere; Symmetrical cube; Symmetrical flat washer	600	600
2	 Tapered washer; Parts that naturally fall in one of two possible positions	600	1200
4	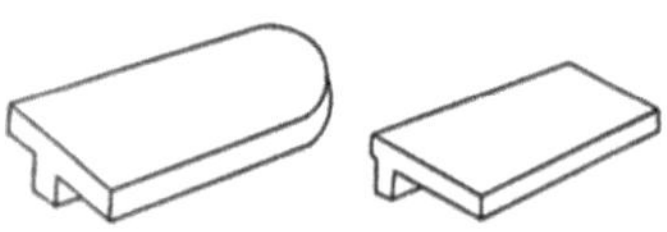 Parts having four natural positions	600	2400

FIGURE 12.9 Effect of the possible number of orientations on the efficiency of feeding. (Redrawn after Lane 1986. Used by permission of SME.)

in the feeder consists of rejecting parts resting in any position but the one desired. Consequently, reducing the number of possible orientations of a part actually increases the odds of that part going out of the feeder on its first try. Figure 12.9 shows the effect of the possible number of orientations on the efficiency of feeding. According to W. V. Tipping, two types of parts can easily be oriented: parts that are symmetrical in shape (e.g., a sphere or cube) and parts with clear asymmetry (preferably with marked polar properties either in shape or weight). Symmetrical parts are easily oriented and handled. Therefore, try to make parts symmetrical by adding nonfunctional design features like a hole or a projection.

- Figure 12.10 shows some small changes in the design of parts that result in full symmetry. Generally, it is easy to achieve symmetry with sheet metal and injection-molded parts because the manufacturing cost of adding a feature is relatively low.
- If it is too difficult or too expensive to achieve symmetry, nonfunctional features must then be added to make identification and grasping easier. This approach is also employed for parts for which orientation is based on hard-to-detect features like internal holes. In addition, components having similar shapes and dimensions are difficult to identify and orient, and changes in dimensions or additions of design features must be made. Recent research work has come up with a concept, called *feedability*, that involves quantitative estimation of the odds of feeding a part having certain geometric characteristics to the assembly

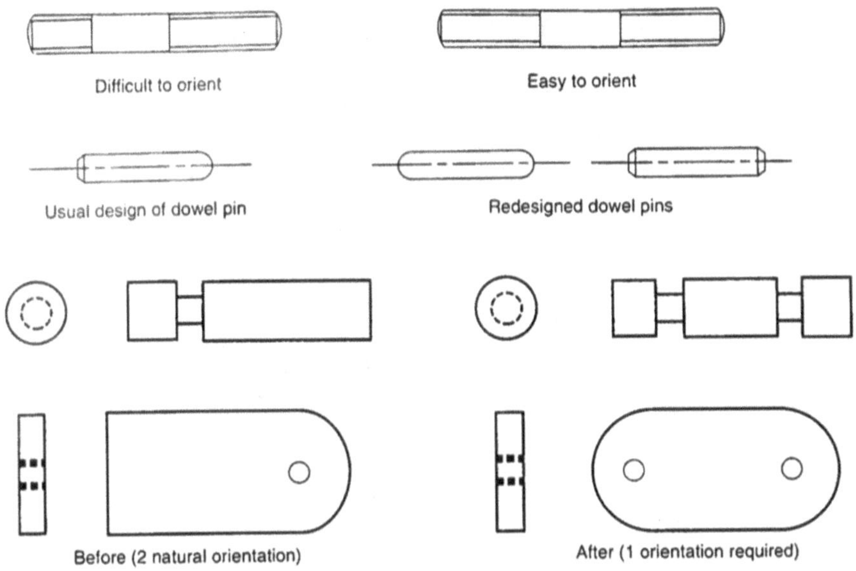

FIGURE 12.10 Examples of design changes that give full symmetry. (Redrawn after Lane 1986. Used by permission of SME.)

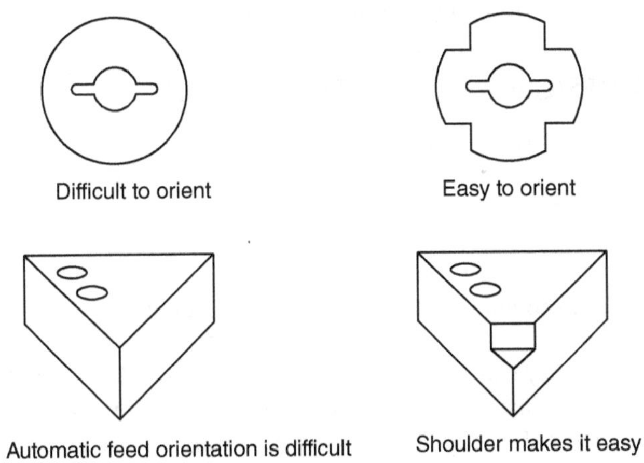

FIGURE 12.11 Examples of design changes that exaggerate asymmetry or indicate hidden features.

station in a specific orientation. Figure 12.11 shows some design changes that exaggerate asymmetry or indicate hidden features, while Figure 12.12 shows the effect of changing geometric features on the calculated values of feedability.

- Try to design parts with a low center of gravity (i.e., it should not be far above the base). This gives the part a natural tendency to be fed in one particular orientation. Also, when such a part is transferred on a conveyor belt, it will not tip or be disoriented due to the force of inertia.

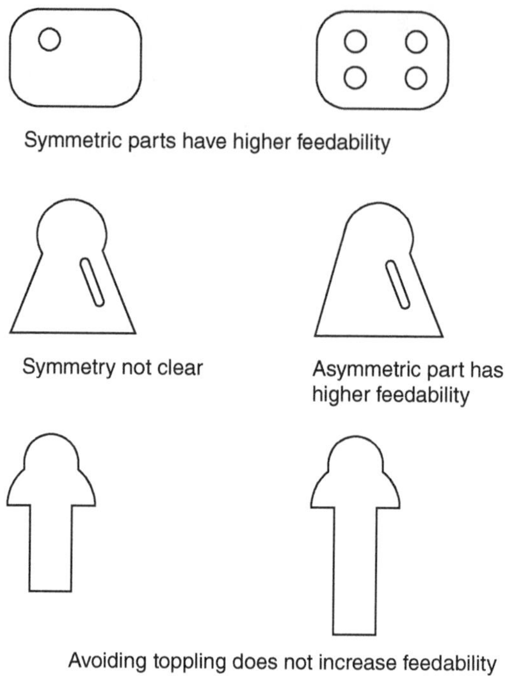

FIGURE 12.12 Effect of changing geometric features on calculated values of feedability.

12.7 PRODUCT DESIGN FOR ROBOTIC ASSEMBLY

The product design rules for robotic assembly are basically the same as those for manual and/or automatic assembly. However, two very important and crucial considerations have to be taken into account when designing components for robotic assembly. They can be summed up as follows:

- Design a component so that it can be grasped, oriented, and inserted by that robot's end effector. Failure to do so will result in the need for an additional robot and, consequently, higher assembly costs.
- Design parts so that they can be presented to the robot's arm in an orientation appropriate for grasping. Also, eliminate the need for reorienting assemblies (or subassemblies) during the assembly operation. Ignoring this rule will cause an increase in assembly time by consuming the robot's time for no valid reason. It also will cause an increase in the assembly cost per unit.

12.8 METHODS FOR EVALUATING AND IMPROVING PRODUCT DFA

At this point, let us review some of the methods currently used in industry, in America and abroad, for evaluating and improving product DFA. Because so many methods, systems, and software packages have recently been developed, the survey here will be limited to the most commonly known and used methods, for which substantial information and details have been published. There is no bias here for or against any method that has or has not been covered.

As you will soon see, most of the methods are based on measuring the ease or difficulty with which parts can be handled and assembled together into a given product. This does not mean that the components are physically brought together but rather that an analytical procedure is followed where the problems associated with the component design are detected and quantitatively assessed.

The right answer or optimal design comes from you, the engineer, when you use a particular DFA method as a tool in evaluating and comparing alternative design solutions. Following is a survey of each method.

12.8.1 The Boothroyd–Dewhurst DFA Method

The Boothroyd–Dewhurst DFA method was developed in the late 1970s by Professor Geoffrey Boothroyd, a pioneer in the area of DFA, at the University of Massachusetts Amherst in cooperation with Salford University of England. First, the appropriate assembly method is selected by means of charts. Then, the analytical procedure corresponding to the assembly method selected is used (i.e., there is a separate, though similar, procedure for each of the assembly methods). Figure 12.13 is a diagram of the stages of the Boothroyd–Dewhurst DFA method.

As an example, let us now examine the analytical procedure for manual assembly, as the DFA analysis procedures for the other assembly methods are not much different. Note that the analysis cannot be employed to create a design from nothing but rather is used to evaluate and refine an existing design. In other words, the starting point is an assembly drawing of the product (or a prototype or an actual product). The first step in the analysis is to determine the assembly sequence (i.e., the part that is to be placed first and the parts that are to follow it in the order to be used for attaching them together). Boothroyd and Dewhurst proposed the worksheet shown in Figure 12.14 for effective bookkeeping of the assembly time and cost.

When more than one part is to be used in an operation, the assembly time for that operation is obtained by multiplying the assembly time for one part by the number of parts (see Figure 12.14). It is also required that nonassembly operations must also be included in the sequence. Each time the unfinished assembly is reoriented during the assembly process, the reorientation operation is entered into the worksheet and a time is allocated for it. The assembly time for each component part is then obtained by adding the handling time of that part to its insertion time. These two times are extracted from charts that include assembly data. The data were compiled by Boothroyd, Dewhurst, and their coworkers based on practical observation over long periods of time and on research. In order to use the handling-time chart, a two-digit handling code must first be determined for each part based on

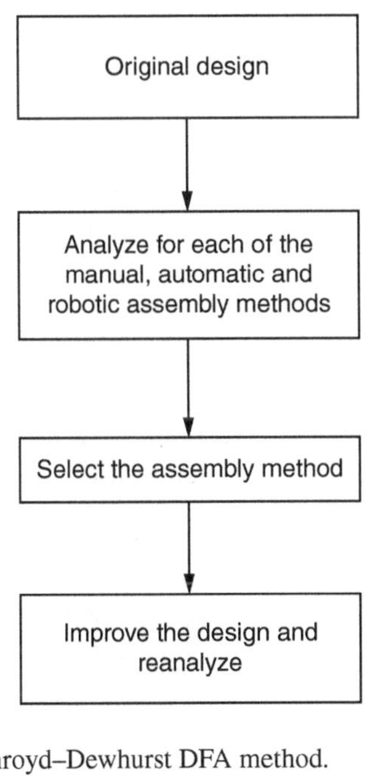

FIGURE 12.13 Stages of the Boothroyd–Dewhurst DFA method.

1	2	3	4	5	6	7	8	9*	Name of Assembly ↓
Part (component) ID number	Number of times the operation is carried out consecutively	Two-digit manual handling code (obtained from charts)	Manual handling time per part	Two-digit manual insertion code (obtained from charts)	Manual insertion time per part	Operation time [(2) × (4) + (6)]	Operation cost	Figure for estimating the theoretical minimum number of parts	
4	1							1	
2	1							0	
3	2							1	
1	1							1	
Record totals here →						T_M	C_M	N_M	

* In column 9, if part is not essential, put 0; if required, put 1.

FIGURE 12.14 The Boothroyd–Dewhurst bookkeeping worksheet.

its size, weight, and geometric attributes. A two-digit insertion code (and thus time) must also be obtained for each part based on accessibility, vision restriction, and resistance to insertion. Once the components and the assembly time for each are known, it is easy to estimate the total assembly time and assembly cost for the existing design.

The next step is aimed at reducing the parts count by totally eliminating some parts or combining them with neighboring parts. This is achieved by answering the previously listed three questions about the movement of the part relative to adjacent parts, its materials, and the need to have it separate for assembly and/or disassembly. Candidates for elimination can be identified and subtracted from the total number of parts to obtain the number of "theoretically needed" parts. Assuming that an ideal assembly operation of a component takes 3 seconds (1.5 seconds for handling and 1.5 seconds for inserting), the total ideal assembly time is given by the following equation:

$$\text{Total ideal assembly time} = 3N_M \tag{12.1}$$

where N_M is the theoretical minimum number of parts.

Boothroyd and Dewhurst used a design efficiency index to evaluate the improvement in design in a quantitative manner. This index can be given by the following equation:

$$\text{Design efficiency} = \frac{3N_M}{\text{Calculated total assembly time}} \tag{12.2}$$

The mechanism for improving the design, according to this method, involves a review of the worksheet in order to pinpoint components that can be eliminated and that have relatively high handling and insertion times. The number of components or parts must then be reduced by eliminating some or most of the components so identified. This process is repeated until an optimal design (i.e., one having a design efficiency much higher than that of the initial design) is obtained.

Because it is rather time-consuming to perform the Boothroyd–Dewhurst procedure manually, a software package for DFA analysis based on their structured analysis has been developed. The latest commercially available version is very user-friendly and runs in a Windows environment. Again, note that the system does not make any decisions for the designer; it is the designer who, with rational thinking and good engineering sense, ultimately decides what is right and appropriate.

One final note here: Although this DFA analysis would certainly decrease the parts count, it can often result in the manufacture and use of complex components. Bearing in mind that the assembly cost is only about 5 percent of the total production cost, the finalized "optimal design" may be easy to assemble but expensive (or difficult) to manufacture. In fact, the absence of a manufacturing-knowledge-based supporting system was the main shortcoming of the initial DFA techniques. Realizing that fact, Boothroyd and Dewhurst supplemented their DFA software with what they called *design for manufacture* software. This software is actually a product cost estimator for a few selected manufacturing processes and is used to estimate the manufacturing cost of the different alternative designs. The optimal design can then be selected based on both the assembly and the manufacturing costs.

12.8.2 The Hitachi Assembly Evaluation Method

Another method with a proven record of success is the Hitachi assembly evaluation method (AEM). It was employed to refine the designs of tape recorder mechanisms in order to develop an automatic assembly system for producing those subassemblies. That pioneering and original work by S. Hashizure (a research engineer at Hitachi) and his coworkers was awarded the Okochi Memorial Prize in 1980. Although this method does not explicitly distinguish between manual and automatic assembly, this difference is accounted for implicitly within the structured analysis. Also, the method was subjected to refinement in 1986 with improvements to its methodology, and a computer-based version is now available.

The Hitachi AEM approach is based on assessing the assemblability of a design by virtue of the following two indices:

- An assemblability evaluation score (E) is used to assess the design quality or difficulty of assembly operations. The procedure to compute E is based on considering the simple downward motion for inserting a part as the "ideal reference." For more complicated operations, penalty scores that depend upon the complexity and nature of each operation are

Elemental Operation	Assembly Evaluation Symbol	Penalty Points
single downward movement (assisted by gravity)	$\downarrow$	0
soldering	S	
tightening a screw		
adhesive bonding		

FIGURE 12.15 Examples of Hitachi method symbols and penalty points.

assigned. The Hitachi method uses symbols to represent operations, and there are about 20 of them covering operations like the straight downward movement for part insertion and the operation of soldering, as shown in Figure 12.15.

After completing a worksheet in the same order as the anticipated assembly sequence, the penalty score for each part is manipulated to give the assemblability evaluation score for that part. The "E" values for all parts are then combined to produce an assemblability evaluation score for the whole assembly. Because a penalty score of zero corresponds to an E value of 100 percent, the higher the E score for an assembly, the lower the assembly time and cost. Accordingly, if each part of an assembly is to be added by a simple downward motion, the E score for each part and, therefore, for the whole assembly will be 100 percent. The E score is employed in simplifying the various operations and not explicitly in reducing the parts count.

- An estimated assembly cost ratio (K) is an indication of the assembly cost improvements. As the name suggests, K is the ratio between the assembly cost of the new (modified) design divided by the assembly cost of the initial and/or standard design. It is clear that when K is 0.7, there is a 30 percent saving in the assembly cost as a result of modifying the design. The method of estimating the time (and cost) of an operation involves breaking it into its elemental components and allocating time for each elemental motion based on compiled practical observations. Any saving in the assembly cost can be achieved by reducing the parts count in a product and/or simplifying the assembly operations.

12.8.3 The Lucas DFA Method

The Lucas DFA method was developed in the 1980s as a result of collaborative work between the Lucas Corporation and the University of Hull (both in England). The motivation for developing this method, as stated by its creators, B. L. Miles and K. G. Swift, was to have the best features of the commercially available DFA software packages within a simple system and to aim for its application at an early stage of the design process. Unlike the previous two methods, the Lucas DFA evaluation is not based on monetary costs but on three indices that give a relative measure of assembling difficulty. The goal of reducing the parts count and the analysis of the insertion operations based on

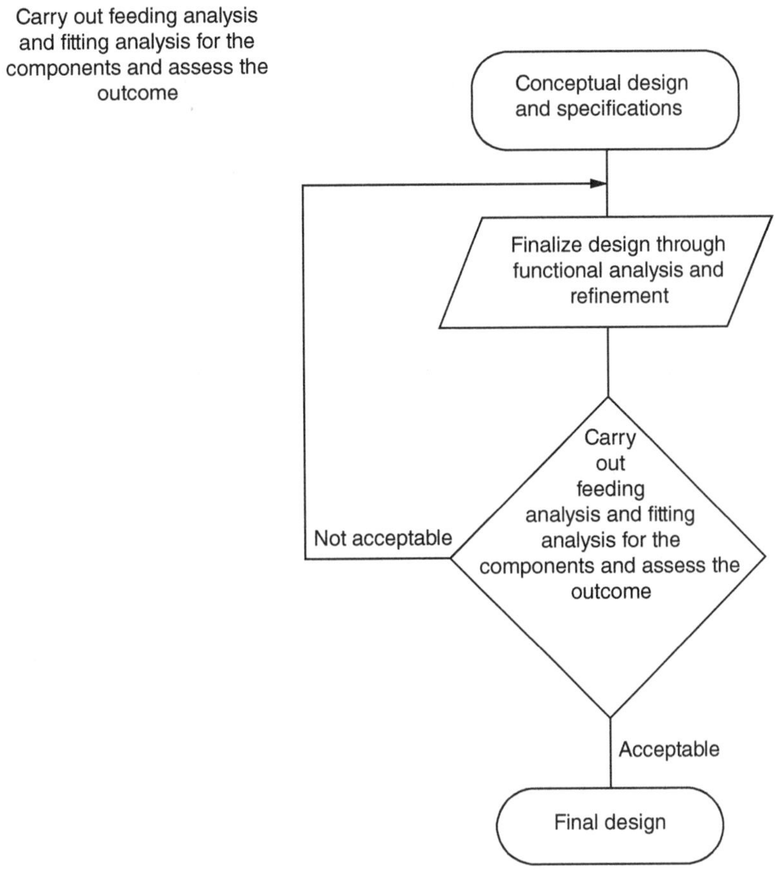

FIGURE 12.16 The Lucas DFA assembly sequence flowchart.

an encoded classification system, however, are shared with the previous two methods. Also, an easy-to-use computer version of this method is now commercially available.

Figure 12.16 shows an assembly sequence flowchart (ASF) of the Lucas DFA procedure. As can be seen, the analysis is carried out in three sequential stages: the functional, feeding (or handling), and fitting analyses. It can also be seen that the existence of a well-defined product design specification (PDS) is a must for carrying out the first stage of the DFA analysis.

12.8.3.1 Functional Analysis

In the functional analysis, components are divided into two main groups. The first group includes components that perform a primary function and, therefore, exist for fundamental reasons. These components are considered to be essential, or "A," parts. The second group involves nonessential, or "B," components that perform only secondary functions like fastening and locating. The design efficiency is the product of dividing the number of essential parts by the total number of parts and can be given by the following equation:

$$\text{Design efficiency} = \frac{A}{(A+B)} \times 100 \quad (12.3)$$

According to the flowchart (see Figure 12.16), if the design efficiency is low, it should be improved through design modifications aimed at eliminating most of the nonessential parts. A clear advantage of the Lucas DFA method is that performing the functional analysis separately, before the other two analyses, acts as an initial "screening mechanism" that returns back poor designs before further effort is encountered in the detailed analysis. For this initial stage, the target objective is to achieve a design efficiency of 60 percent.

12.8.3.2 Feeding Analysis

The feeding analysis is concerned with the problems associated with handling components (and subassemblies) until they are admitted to the assembly system. By answering a group of questions about the size, weight, handling difficulties, and orientation of a part, its feeding/handling index can be calculated. Next, the feeding/handling ratio can be calculated by using the following equation:

$$\text{Feeding / handling ratio} = \frac{\text{Feeding / handling index}}{\text{Number of essential components}} \tag{12.4}$$

An ideal value for this ratio and one that is often taken as a target goal is 2.5.

12.8.3.3 Fitting Analysis

The fitting analysis is divided into a number of subsystems, including gripping, insertion, and fixing analyses. An index is given to each part based on its fixturing requirements, resistance to insertion, and whether or not there will be restricted vision during assembly. High individual values and/or a high total value of these indices mean costly fitting operations; in such cases, the product should be redesigned with the goal of eliminating or at least reducing these operations. The fitting index is manipulated to yield the fitting ratio as given by the following equation:

$$\text{Fitting ratio} = \text{fitting index / number of essential components} \tag{12.5}$$

Again, for the design to be acceptable, the value of the fitting ratio should be around 2.5.

Note that while the feeding/handling and fitting ratios can certainly be used as "measures of performance" to indicate the effectiveness of the design quality with respect to assembly, the absence of a mechanism to evaluate the effect of design changes on the manufacturing cost is a clear shortcoming of this method.

12.8.4 Other Methods

In addition to the DFA methods previously discussed, some software packages are commercially available and/or being applied in-house in large corporations. These include packages by Fujitsu, AT&T, Sony, and Sapphire. Unfortunately, technical details are not available, so coverage of these methods was not possible in this text.

REVIEW QUESTIONS

1. What options are available when selecting an assembly method?
2. What are the major characteristics of each of the available assembly methods?
3. Discuss some of the factors that affect the selection of an appropriate assembly method.
4. Define the term *design for assembly*.
5. What are the benefits of applying the concept of design for assembly?
6. What has always been the traditional approach for DFA?
7. What three questions form the criteria for eliminating a part from an assembly or combining it with its neighboring part?
8. List the guidelines for product design for manual assembly.

9. What is the ideal insertion motion? Why?
10. Why should you try to avoid reorienting parts during assembly?
11. What effect does the concept of standardization of parts have on the assembly process?
12. Does nesting (or tangling) of parts while in the bulk have any effect on the assembly operation?
13. Why should parts be symmetrical?
14. What is your advice if you cannot get the parts to be symmetrical?
15. List the guidelines for product design for automatic assembly.
16. How can the use of self-aligning and self-locating features facilitate automatic assembly?
17. Can screws be considered as essential parts in a product? Why?
18. What is the ideal fixturing method in automatic assembly?
19. Discuss the concept of feedability.
20. Why should parts with a low center of gravity be favored in automatic assembly?
21. List two rules for product design for robotic assembly.
22. What are the methods for performing DFA analysis?
23. List some of the advantages, characteristics, features, and limitations of each DFA analysis method.

DESIGN PROJECT

Choose a fairly simple product (e.g., a shower handle or coffeemaker), disassemble it, and make an assembly drawing or an exploded view of its parts. Next, study the function and material of each part, as well as the assembly sequence. Then, use the three questions (elimination criteria) of Boothroyd and Dewhurst to identify parts that are candidates for elimination or combination with other parts. Finally, modify your design in order to reduce the parts count and provide an assembly drawing of the new design, as well as a workshop drawing for each part.

13 Additive Manufacturing

13.1 INTRODUCTION

When Mr. Parsons of Traverse City (Michigan) developed his crude form of a numerically controlled machine tool, it was indeed the beginning of a new era in which manufacturing has undergone, and is still undergoing, a radical change. That revolutionary methodology paved the way for computerized numerical control (CNC), direct numerical control (DNC), computer-aided manufacturing (CAM), and eventually computer-aided drafting/manufacturing (CAD/CAM). Digital manufacturing became a reality, where one or a few of the same products can be produced by a CNC machine tool without the need for a skilled machinist if a digital blueprint is readily available for that product. It does not matter whether the source of that blueprint is located next to the machine tool or hundreds, possibly thousands, of miles far from it. In fact, according to Mr. Joe Kaiser, the CEO of Siemens A.G., the future of manufacturing would be characterized by mass customization and mass digitization. That is indeed what additive manufacturing would help achieve.

According to the American Society for Testing and Materials (ASTM), additive manufacturing (AM) is defined as "the process of joining materials to make objects from 3D model data, usually layer upon layer, as opposed to subtractive manufacturing technologies, such as traditional machining." Understandably, this definition is quite broad and does not specify either the material of the objects or the method of creating the layers. The reason is that there are several versions of additive manufacturing; each has a different method for creating the layers and producing the objects, and each possesses unique advantages when processing some materials. In fact, there is a long list of commercial variations of this process technology, and it would be challenging educationally to explain each version included in that list. Therefore, a need arose to have categories for grouping AM technologies so that one can discuss a category of machines rather than a list of variations. For this reason, ASTM International and the Society of Manufacturing Engineers collaborated and produced the ASTM Standards F2792. Following are the seven categories of AM technologies, with a brief definition and description of the terms of each:

- *Binder Jetting*. An additive manufacturing process in which a liquid bonding agent is selectively deposited to join powder materials.
- *Direct Energy Deposition*. An additive manufacturing process in which focused thermal energy is used to fuse materials by melting them as they are being deposited.
- *Material Extrusion*. An additive manufacturing process in which material is selectively dispensed through a nozzle or orifice.
- *Material Jetting*. An additive manufacturing process in which droplets of build materials are selectively deposited.
- *Powder Bed Fusion*. An additive manufacturing process in which thermal energy selectively fuses regions of a powder bed.
- *Sheet Lamination*. An additive manufacturing process in which sheets of material are bonded to form an object.
- *Vat Photopolymerization*. An additive manufacturing process in which liquid photopolymer in a vat is selectively cured by light-activated polymerization.

As we can see, some of these groups actually emerged in the 1980s under other names such as rapid prototyping, stereolithography, and 3D printing, but now they are considered by the ASTM to belong

DOI: 10.1201/9781003537106-13

to additive manufacturing because of their additive nature. In the next section, we are going to review in depth each category, including process details and parameters, advantages, limitations, and applications in various industries.

13.2 THE VARIOUS ADDITIVE MANUFACTURING CATEGORIES

13.2.1 Material Jetting (3D Printing)

Material jetting is among the oldest additive manufacturing processes that was originally called 3D printing. In this process, the desired object is created in a simple way, which involves depositing drops of a liquid polymer mix selectively to build layer upon layer of the object. Because of such simplicity, university students in the two African nations of Togo and Tanzania have recently built 3D printers from e-waste (old printers, computers, and scanners). Recent developments have been in three areas—namely, software application packages, the material of the object, and the design features of the material jetting mechanism. Current 3D printers have gone far beyond just prototyping components for demonstration or validation of ideas. Nevertheless, the process is still quite simple, where a liquid, paste, or gel is dispensed from a material cartridge through a needle tip. The latter is the end part of a three-axis system that selectively dispenses the droplet preplanned by the computer to create the 3D object. The essential requirement is that the dispensed material must easily and quickly solidify through a polymerization reaction initiated physically (e.g., ultraviolet radiation) or chemically (hardening agent). A large variety of materials can be processed by this method, including different polymers, metals, ceramics, and living cells. Products can be quite large in size (5800 in.^{3} or 0.09 m^3), while still having a high accuracy level (sometimes down to 30 microns) coupled with an impressive surface finish that eliminates any need for postprocessing by machining. The size (or build envelope) of the machine, the build material, and the accuracy level depend upon the required application and in which industry it would be used.

13.2.1.1 Applications of 3D Printing Technology

Let us now discuss the applications of this technology in various industries and see its impact and how it would revolutionize industry, as follows:

- Naturally, the first application we should expect is building end-use prototypes for design verification.
- Another important application is the production of 3D print patterns for direct investment casting or lost-wax casting. While this application originally originated in the jewelry industry, it is currently used in other industries such as the automotive industry where large cast parts such as engine mount blocks are produced using 3D print patterns. In the dental sector, crowns, bridges, and partial frameworks are cast with high precision using 3D-printed patterns. The build material that fills the cartridge in all these cases is a blend of liquid wax and wax in the form of nanopowder.
- Food and Drug Administration (FDA)—approved, biocompatible material is processed by 3D printing for producing all types of denture bases.
- In the medical sector, a medical-grade polycaprolactone (PCL), which is a biodegradable thermoplastic polyester that is processed by 3D printing at high temperatures, is used for applications such as bone regeneration, cartilage regeneration, and drug release scaffolds. Also, virtual planning of surgery and guidance can be performed using 3D-printed iconic models of the human body.
- In bioengineering research, 3D printing is extensively used to build parts and components from silicones, thermoplastics, ceramic and metal pastes, hydrogels, or even living cells.
- The newest and most impressive application is additively manufacturing woven fiber composites preimpregnated with thermoplastics. This revolutionary technology can be used to

manufacture an automobile exterior body, customized according to the client's description and extremely fast. In fact, a company in China and another in England started doing just that.

13.2.1.2 Advantages of 3D Printing Technology

As we can see, some characteristics common to all the above-mentioned processes and applications are as follows:

- *Speed of Building Objects*. Whether producing a prototype, a pattern for investment casting, or a mold, 3D printing does the job much faster than any competing process.
- *Extremely Lightweight Designs*. The process allows for the production of very thin walls of products, thus saving material and making the overall weight of the product much lighter.
- *Design Complexity and Freedom*. The 3D printing process enables the production of parts having very complex geometries, thus giving the designer more freedom during the design process.
- *Customization*. The ease and speed with which a product is built enable the manufacturing engineer to tailor the attributes of the product to the requirements of the client. In fact, automobile customization will become common in the next decade. An interesting story came from Turkey, where a hawk lost a leg and was lucky to get a replacement, which was made of plastic and produced by 3D printing.
- *Reduces or Even Eliminates the Need for Storage*. Instead of moving raw materials for factories far away, production can take place just next to the source of materials, thus eliminating transportation and storage and making the supply chain management (SCM) much simpler.

13.2.1.3 Disadvantages/Limitations of 3D Printing

Despite the several advantages mentioned above, some limitations, have to be taken into consideration before selecting this process for building our product. They are as follows:

- *Size Limitation*. There is a limit to the maximum size of the product manufactured by this process. This is particularly important when considering it for making the exterior body of an automobile, for example.
- Unless care is taken when selecting the build material layer thickness, "stairsteps" would occur. This happens when the thickness of the layer is too large, making the edge appear as a step in a staircase.
- The process is basically the most advantageous for polymers (including wax mix). While it can be used to produce metallic and ceramic objects, disposing of a paste of metal (or ceramic) powder in a viscous polymer, the product still has to be sintered first in order to burn off the polymer and make the metal powder particles bond together. However, even in that case, porosity would be an issue.

13.2.2 Vat Photopolymerization (Stereolithography)

This process was patented in 1986 under the name *stereolithography* (SL). In this process, a light source (laser) is used to selectively cure a photopolymer resin inside a vat or tray and harden it by triggering a polymerization reaction, which finally yields a thermosetting material that holds its shape. Figure 13.1 indicates a schematic sketch of the process. As can be seen in that figure, a UV laser beam literally draws (lithography means "to write") out a part into a vat of resin. Wherever the laser beam hits the resin, curing and hardening occur and take place on the upper layer of resin in

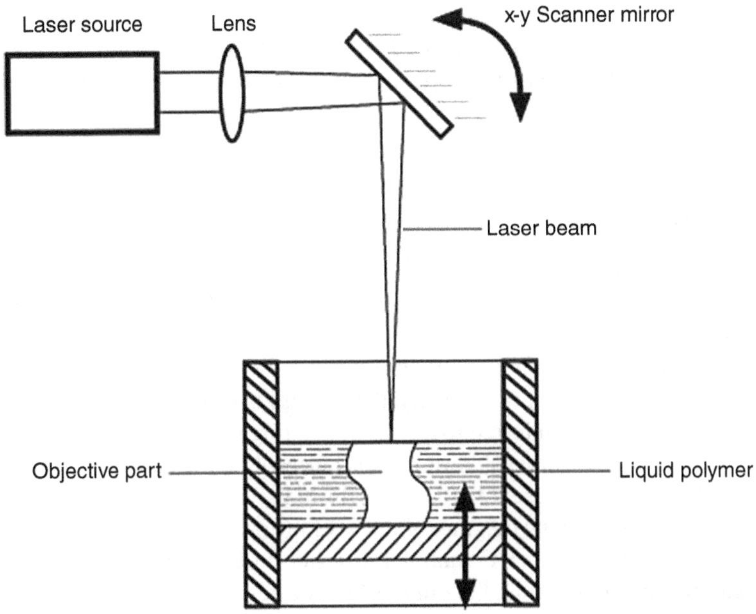

FIGURE 13.1 A schematic sketch of the stereolithography process.

the vat or tray. After curing each layer, the platform is lowered along the Z-axis by a distance equal to the thickness of the layer, and the next layer can then be cured onto the part. The process continues until the whole part is built.

13.2.2.1 Applications of Stereolithography

The major application for this process has always been and still is rapid prototyping for quick design verification. Nevertheless, with recent advancements in synthesizing new materials, the process has found new applications in jewelry, consumer goods, medical devices, and manufacturing industries. Now high- and low-wax materials can be used to build patterns with fine details and good surface finish, along with a clean burnout for investment casting. Also, ceramic-filled and glass-filled photopolymer for building plastic injection molds.

13.2.2.2 Advantages of Stereolithography

Advantages like low weight and complex designs of parts, as well as the ability to customize a product, are common to the various additive manufacturing processes, including stereolithography. A further advantage is the low cost of the required capital equipment, thus making the process feasible for small and medium projects.

13.2.2.3 Limitations of Stereolithography

- Since its inception, stereolithography has had one large and troubling limitation: speed. Using a fine laser beam, the system draws out parts in photosensitive polymer, layer by layer. Even with software advancements and years of development, the process is slow because of the need to limit the thickness of each layer and therefore the necessity to draw an extremely large number of layers. If the thickness of the layer is taken too large, the surface of the product suffers from the stairsteps defects, and the functionality of the product is detrimentally affected.

- The laser beam produces highly accurate curing "only" in the centermost spots of the build envelope. Evidently, if the build envelope is wide (i.e., large cross section of the object), the laser beam loses focus at the furthest spots of the build envelope, resulting in reduced accuracy and poor surface finish.

13.2.3 Powder Bed Fusion

Basically, two different types of processes fall under this category, depending upon the source of thermal energy that causes fusion—namely, the laser (SLS, SLM, DMLS commercial names) and electron beam processes. In both cases, net-shape parts with high geometric complexity can be produced directly from powders, without passing through any intermediate state, by selectively melting the particles at the surface, layer by layer. Since the process is usually performed in a chamber, the process is sometimes mistakably referred to as "laser sintering or LS" in the case when a laser beam is used. We know from Chapter 7 that sintering takes place while the main phase is in its solid state, which is not the case in laser beam additive manufacturing.

Figure 13.2 indicates a schematic diagram of the working principles of laser beam additive manufacturing (developed in 1995 by the Fraunhofer Institute in Germany). It is basically similar to electron beam additive manufacturing, the only difference being that an electron beam replaces the laser beam as the source of fusion energy. First, the desired powder is poured from a hopper to fill the cavity above the base plate, thus creating the powder bed for one layer only. The surface of the powder bed is then flattened and smoothened using the rollers. Next, as can be seen in the figure, a laser beam is traced across the metal powder bed, according to 3D data fed from a computer to the machine. The laser fuses the powder particles to form a solid. When each layer is completed, the powder bed base plate drops incrementally, thus allowing a new layer to be added to the previous one, and the process continues until the whole part is created. We know, however, that some problems arise during laser-melting additive manufacturing. Because of the high temperature, the gas in the pores expands, causing flashing and powder scatter. The remedy is to carry out the process in a

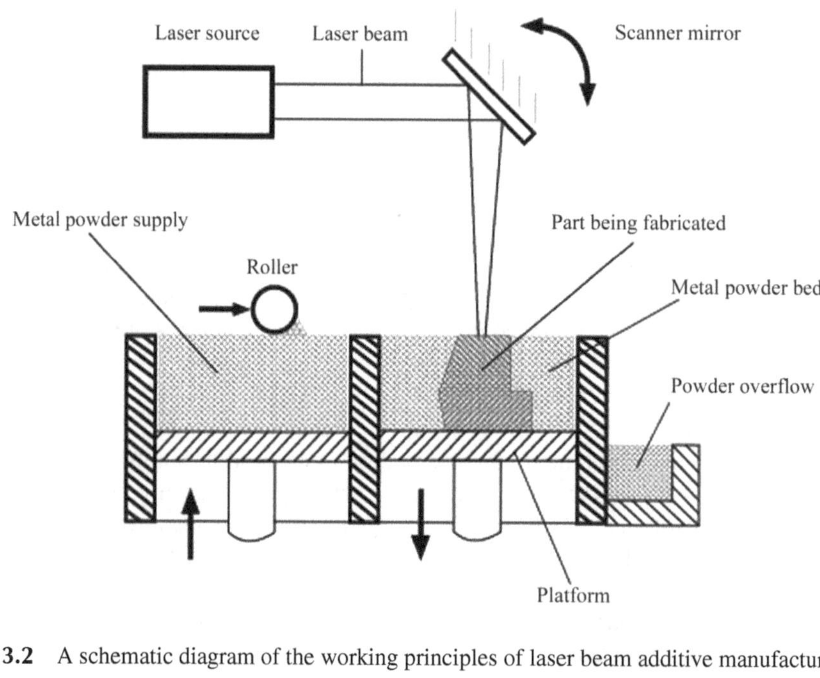

FIGURE 13.2 A schematic diagram of the working principles of laser beam additive manufacturing.

vacuum so as to evacuate the pores from gas and eliminate the problem. In the case of electron beam additive manufacturing, the charge of the electron would cause problems, and the powder bed should be given a positive charge in order to avoid the problem of delamination between layers. Alternatively, the powder bed should first be preheated to form a cohesive mass by a low-energy-level electron beam, before the final melting of the layer.

Evidently, in order to get a sound, defect-free product, the parameters of the process have to be optimized. These include laser power, scan speed, and scan pattern, as well as the characteristics of the powder used such as particle size, apparent density, and so on. Such optimization would materialize either through experimentation and trial and error or through rational modeling of the process and employing the rules and equations of the science of heat transfer. Because of the complexity of the problem, current models have an accuracy of around 50 percent, which is unacceptable. On the other hand, the manufacturing corporations that managed to determine these optimum conditions through lengthy experimentation unfortunately kept the valuable results proprietary.

13.2.3.1 Applications of Powder Bed Fusion (Metal Additive Manufacturing)

This process is among a few that are usually called metal additive manufacturing (MAM). Its main application is manufacturing parts that would be subjected to a heavy loading condition coupled with very high temperatures, or components that need to be very light and must withstand corrosive environments. Such application necessitates the use of raw materials that are very difficult to manufacture by conventional processes. Titanium and its alloys are the typical examples because they are difficult to cast, forge, or machine. Titanium producers therefore have to find a new application for the MAM process. A lot of research to find potential applications is currently taking place in Australia, a country whose production of titanium accounts for more than half of the world's production of that metal. Following are some applications in the various industrial sectors:

- In the aerospace industry, MAM is used to process high-performance specialty alloys to produce aerospace components that require total failure-free operation. Examples include jet-engine fuel nozzles, rocket thrust chambers, turbo and turbine blades, and other parts of the gas turbine. Moreover, MAM is proving to be an ideal method for producing spare parts for aircraft. In fact, both the Israeli Air Force and the Royal Australian Air Force are beginning to produce spare parts for their aircraft using MAM. A recent promising application has been the development of a helicopter engine with 16 parts instead of 900, through integrating components together and employing additive manufacturing technology.
- The medical sector is another promising area for employing MAM. That technology has been successfully used to make medical devices and tools that are used by surgeons in the operating room. More importantly, it has also been used to manufacture patient-matched implants that fit one unique individual. Currently, the surgeon goes to the operating room with three sets of implants that he or she believes are close enough to the patient's knee, then selects the one that is close enough to the actual knee of the patient during the surgery. That is contrary to what happens when employing MAM, where the actual computed tomography or magnetic resonance imaging digital files of the patient's knee can be processed to create a 3D digital model that is used to create a replica of the patient's knee by employing an additive manufacturing machine. The product is, indeed, a truly customized implant that fits one unique individual. In addition, facial reconstruction surgeries have been carried out with great success using additively manufactured parts.
- In the automotive industry, additive manufacturing is also employed in fabricating spare parts. The process has also been used to manufacture components for the development and manufacture of Formula 1 cars, where those components need to be sturdy and lightweight.
- In the marine industry, military propeller blades each weighing 300 kg (660 lb) were successfully produced by additive manufacturing.

13.2.3.2 Advantages of Powder Bed Fusion

Many professional manufacturing engineers consider this revolutionary process to constitute the "Fifth Industrial Revolution" because of its potential and various advantages. It will reshape the manufacturing technology in the next decade. Following are some of its advantages:

- The additive manufacturing process enables the processing of metals and alloys that are very difficult to cast, forge, or machine. Examples include, but are not limited to, nickel superalloys, Inconel 738, Inconel 718, Inconel 625, Hastelloy, titanium and its alloys, stainless steel, tool steels, and cobalt alloys.
- Very lightweight structures that have complex geometries can be manufactured by using thin sections, thus minimizing the weight, without sacrificing structural integrity, functionality, or performance. Figure 13.3 indicates a structure made of a titanium alloy by MAM technology. It is also worth mentioning that a group of scientists at the California Institute of Technology (Caltech) have reportedly developed a new process that enables the MAM of structures smaller than a human hair and invisible to the naked eye.
- A complex assembly of components can be made in one piece, thus eliminating the time and cost of assembly.
- The need for laborious tooling development, which is required in other conventional manufacturing technology and results in long lead times, is eliminated.
- The product can be customized to an individual client. We saw that when we discussed the truly customized knee implant that fits one unique individual. An interesting story in this respect came from Turkey, where a boat propeller cut and removed the beak of a turtle. The injured sea reptile could not eat and was about to die until manufacturing engineers created a graphical model for the lost beak, then produced it from a titanium alloy by additive manufacturing. The life of the turtle was saved.
- Easier and faster production coupled with shorter lead time would enhance the efficiency of the supply chain, thus yielding numerous benefits. Also, it would change how a product gets to the client.
- The last important advantage is that a product can be manufactured at the place of consumption, thus eliminating transportation and warehousing.

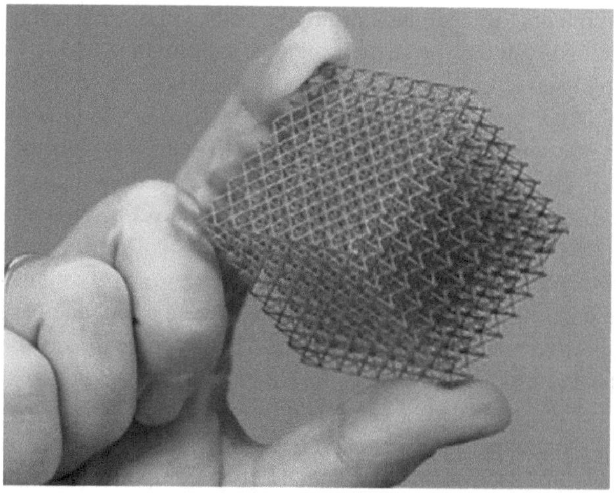

FIGURE 13.3 A structure made of a titanium alloy by MAM technology. (Courtesy of "Picture": Siemens A.G., by permission.)

13.2.3.3 Disadvantages (Limitations) of Powder Bed Fusion

Following are some issues that may be considered as limitations to the use of this MAM process:

- There is always a limit to the size of the components, which can be manufactured by this process. That limit is dictated by the size of the chamber in which the process is carried out. Large components cannot, therefore, be manufactured by this process or other MAM, and conventional methods should be considered whenever it is required to produce large parts.
- In this process, fusion and solidification of each spot on the path of the laser (or electron) beam take place in a very short time. Therefore, a mixture of elemental powders cannot be used because there would not be enough time for diffusion between the individual elements to occur. This is evidently not the case in conventional powder metallurgy, where a blend of elemental powders is compressed and sintered for a period of time that is long enough to allow diffusion to take place. Accordingly, special prealloyed powders of any desired alloy must always be used to fabricate a part made of that alloy. Therefore, many of the famous metal powder producers are currently involved in developing and producing special prealloyed powders that are recommended for use with the MAM technology.

13.2.3.4 Design for Additive Manufactured Parts

When considering additive manufacturing for production applications, parts should be designed (or actually redesigned) specifically for MAM by taking advantage of the unique possibilities offered by this technology. Employing additive manufacturing to fabricate a product that was designed to be manufactured by conventional processes can result in parts that are not commercially viable due to cost. Following are some guidelines:

- Try to save material by using thinner sections without jeopardizing the structural integrity, thus reducing the weight and the overall cost of the product. Weight reduction is vital in applications such as those in the aerospace and automotive industries.
- See whether it is justifiable costwise to use other materials such as titanium and its alloys, which are difficult to process by conventional methods but have superior needed properties.
- Try to make a complex multipart assembly as one single piece, thus eliminating the cost of assembly and reducing the production time.
- The rational selection of the orientation of an AM part during the process with respect to the buildup plane is of utmost importance and has to be considered in the early design stage. It affects the direction of the dendrites in the microstructure of the part and therefore influences its overall properties. A known fact is that the microstructure of a casting and an AM part is quite similar, although the latter has finer grains.
- Avoid horizontal overhang surfaces (known as down skins). The reason is that during the process, the powder below the first layer of the overhang surface would be sintered and attached to the bottom surface of the overhang because of the heat in that layer. Postprocessing may be required to remove that powder, resulting in added cost and longer processing time. Optimal orientation of the AM part may also help eliminate the shortcomings of overhang surfaces (making an overhang vertical or inclined instead of horizontal).
- As much as possible, avoid down-facings, which are surfaces pointing downward opposite to the buildup direction. As can be seen in Figure 13.4a,b, this can be achieved through rethinking the orientation of the AM part or by changing the design. Additionally, a "sacrificial" support structure can be added to the part design and fabricated with it, then removed by postprocessing. It would provide a needed scaffold under the down-facing surface. Let us see what happens if that rule is ignored. In this AM process, as each layer is melted, it

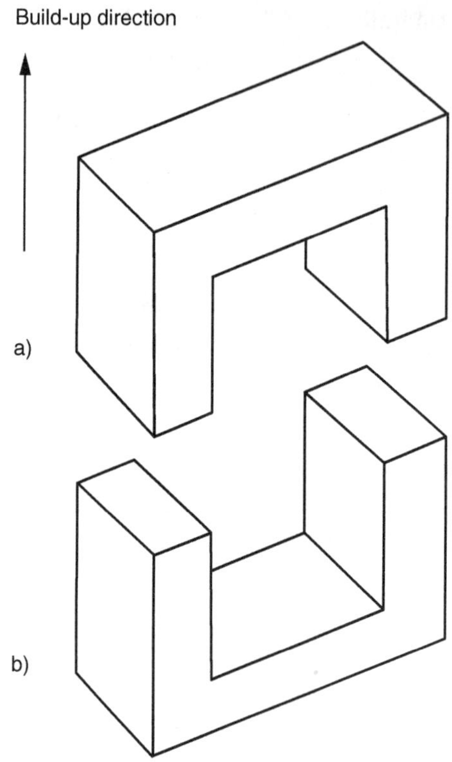

FIGURE 13.4 Avoid the down-facings by rethinking the orientation of the AM part: (a) bad orientation and (b) proper orientation.

relies upon the layer just below it to provide physical support and to act as a heat sink, taking the heat away from the molten layer. When the laser (or electron) beam is melting powder in an area where the layer below is solid metal, the latter acts as a supporting structure and a heat sink. But when there is only powder below the layer being melted, the latter would not find physical support or a path for heat to flow away. The molten layer would take longer to solidify and would not be supported during that time. The final outcome would be the "sagging" of that layer, creating a concave surface.

- Remember that the accuracy of the additive manufacturing machine, as well as the parts it manufactures, is not the same in all directions. It is excellent along the X- and Y-directions and far less accurate along the Z-direction (the buildup direction).
- Minimize residual stresses by eliminating the design features that cause them to occur. Residual stresses are destructive and may result in cracking of the product or distortion and curl-up at the edges. We have to understand what causes residual stress, and then see how we can eliminate it. The top powder layer melts, creating a molten-metal pool when the laser beam scans it, and then cools through heat conduction into the solid metal layer below it. Once it solidifies, it contracts, thus producing shear stress between it and the layer below, because the new layer is constrained by the solid layer below it. Therefore, it is recommended to avoid large uninterrupted areas of molten metal and to avoid abrupt changes in cross sections. Also, use the other guidelines for castings, which we discussed in Chapter 3, such as generous radii or chamfers whenever there is a change in the thickness of the part.

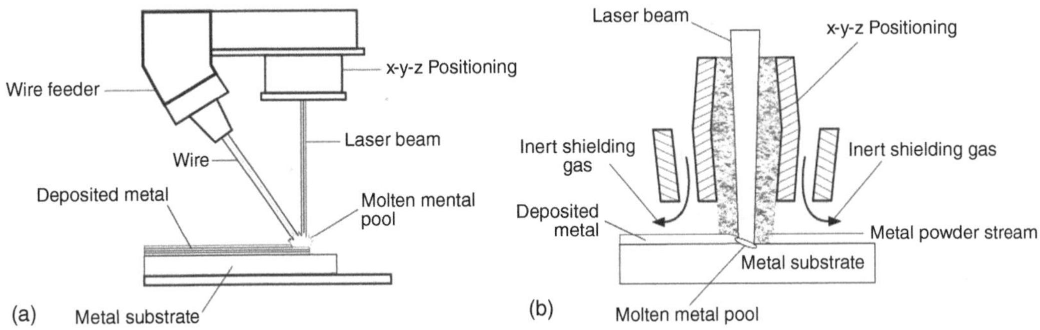

FIGURE 13.5 The working principle of the direct energy deposition process: (a) the material being deposited is a wire, and (b) the material being deposited is a powder.

13.2.4 Direct Energy Deposition

This process involves using an energy source such as plasma arc, laser beam, or electron beam, which is focused to fuse and melt the metal as it is being deposited on a build surface. As shown in Figure 13.5a,b, the material being deposited can be in the form of either a wire or a powder. The basic process involves the free deposition of a wire using a plasma arc to melt the wire and produce a part, and there is no need for it to be carried on in a chamber. In fact, it initially appeared as an advancement from laser welding, when companies started exploring the use of CNC welding equipment to build near-net shape parts having complex geometries. But when the source of thermal energy is a laser or an electron beam, the process has to be carried out inside a vacuum chamber. When the metal being deposited is in the form of a powder, the process also has to be performed in a chamber using a laser beam. In this latter case, it is referred to as direct powder deposition. When a plasma arc is employed to melt and deposit a wire from a spool, the rate of metal fusion and deposition is very high, and the process can be used to build very large parts that can be in some cases longer than 82 feet (5 m). Nevertheless, parts are typically very rough as they come straight out of the machine and are often subjected to postprocessing by a CNC milling machine to mill them down to the right size and improve their surface finish. On the other hand, when it is required to produce accurate near-net-shape parts, metal powder is fused and deposited using a laser beam and in a chamber. The modern apparatus for this direct energy deposition consists of a nozzle for feeding the metal powder (or wire) with a high-power laser beam passing through it (or electron beam). That assembly is mounted on a four- or five-axis robotic arm and can therefore melt and project the fed metal supply onto the target surface from any angle, thus enabling complex geometries to be achieved.

13.2.4.1 Applications of Direct Energy Deposition

The applications of this process are based on its capabilities and characteristics and include, but are not limited to, the following:

- It has already been used for manufacturing the structural parts for satellites and military aircraft. It is also being considered for manufacturing commercial aircraft structures and will soon find similar applications in the automotive industry and shipbuilding.
- It is uniquely and extensively used in repairing high-value equipment and components such as turbine blades and titanium parts of military aircraft.
- The plasma arc version of the process is used for fabricating very large metallic structures. It has been reported that a stainless steel bridge was fabricated by using this particular version of MAM.
- It is uniquely used for adding "grow-outs" to pipes that are then used in machines and equipment.

13.2.4.2 Advantages of Direct Energy Deposition

- There are no constraints on the size of the part to be produced, thus enabling the fabrication of very large parts.
- The speed with which the supply wire is fused and deposited is very high. This is added to the lower cost of the wire, thus making the process economically attractive.
- The ability to add features to existing parts is a further advantage of this process.
- The resulting microstructure is fully dense, indicating the absence of porosity.

13.2.4.3 Disadvantages of Direct Energy Deposition

Despite the several advantages provided by this process, CNC milling of the parts is almost always required, adding cost and making production time longer.

13.2.5 Binder Jetting

Binder jetting is a 3D printing process that is employed in producing functional metal parts, full-color prototypes, porcelains, and large sand-casting cores and molds. Figure 13.6 shows a diagrammatic sketch for a 3D binder jetting machine. In this process, an object is built by bonding powder particles together with an adhesive layer by layer. As can be seen in that figure, the powder-supplying hopper first spreads a thin layer of powder over the bed. Next, a print head equipped with an inkjet nozzle (similar to that used in desktop 2D printers) selectively deposits an adhesive agent, thus binding the powder particles together at the desired spots and forming a layer. The build platform then moves down an increment of one layer, and then this sequence of operations is repeated until the part is complete. The as-printed part is very fragile because it is in its green state (remember Chapter 7), and it must be subjected to either infiltration or sintering in order to gain strength before it can be used. However, the unbound powder must first be removed from the surface of the part using brushes or a compressed air stream.

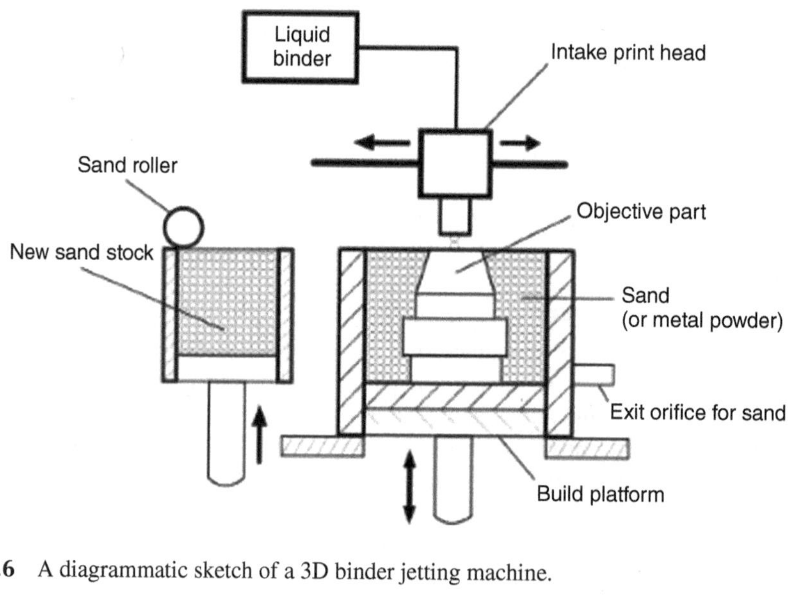

FIGURE 13.6 A diagrammatic sketch of a 3D binder jetting machine.

Infiltration involves placing the part in a furnace at a high temperature for a period of time until the binder is burned out, leaving pores. The part is then infiltrated with bronze by employing capillarity, thus filling the pores and increasing the density and strength of the part. On the other hand, in sintering, the part remains at a much higher temperature and for a longer time until the powder particles sinter together (see Chapter 7). The part becomes strong and porosity drops to about 3 percent. Again, as previously explained, both the infiltration and the sintering processes are accompanied by shrinkage (i.e., a decrease in the dimensions), which must be taken into account when designing the green part.

13.2.5.1 Applications of the Binder Jetting Process

The binder jetting process has diverse capabilities and a variety of applications, as follows:

- The production of functional metal parts having complex shapes that would be very expensive and difficult to produce traditionally and from a range of metals and alloys. The process is also used for small-batch production of noncritical mechanical components that are not subjected to heavy loading conditions. The metals fabricated by this process include stainless steel and the Inconel alloy.
- The process is used for manufacturing wear-resistant silicon carbide tools used in metalworking, petroleum, and construction industries.
- A typical application for this process is the production of full-color concept models and prototypes using sandstone powder, as well as porcelains using ceramic powder.
- This particular 3D printing process found popularity in the foundry industry. An updated version, a robotic arm with a special print head assembly, has an open tray at its backside. The tray is filled with sand from a hopper sand delivery system. The robot positions the print head, drops sand, and spreads it and prints, all in one pass. To generate the next layer, the robot rises in height by the layer thickness to add vertical depth. It is easy to build molds within just a few hours and without any need for patterns. Cores can be produced in the same manner.

13.2.5.2 Advantages of the Binder Jetting Process

- The process enables flexibility in design since the parts are printed at low temperatures, thus eliminating defects such as warping that are experienced in processes like SLS or DMLS. The high temperatures at which these processes are carried out cause such defects.
- There is no need for a support structure as the parts are fully compressed and, therefore, supported by the surrounding powder.
- The process is simple and cost-effective, particularly for making silica sand molds for casting, since silica sand is abundant and not expensive.

13.2.5.3 Disadvantages of the Binder Jetting Process

- The produced parts are very fragile in their green state, and thin sections should accordingly be avoided since they may break while handling the part in its green state.
- As a remedy for this problem, the dimensions of thin sections should be increased. Alternatively, a sacrificial structure is added particularly to support those thin sections and then removed later.
- Metal parts produced by the binder jetting process still have some porosity and are therefore weak. They are not suitable for applications that require carrying higher loads.

13.2.5.4 Design for Parts Produced by Binder Jetting

As previously mentioned, parts produced by this method are very fragile in their green state. This creates restrictions on the designs that can be manufactured by the binder jetting process in order to avoid breakage of weak sections while the parts are still in their green state. Manufacturers of binder jetting machines have published guidelines regarding the various design features and the minimum allowable dimension in each case. Beginner designers should consult with these publications before starting the design process of a part that is to be produced by this method.

13.2.6 Additive Manufacturing Material Extrusion

Material extrusion is the name given in the ASTM standards. Nevertheless, a few other names are used in industry that depend upon the material being extruded and include fused deposition modeling (FDM), fused filament fabrication (FFF), and additive manufacturing of concrete (AMoC). In fact, FDM is just a legal term that was employed to get a patent that widely covers all possible FFF processes. The basic process is simple and the machine used is cheap, thus making it the most popular process for those who want to pursue 3D printing as a hobby. It involves an extrusion nozzle that is attached to a moving, heated printer head. A continuous filament of a thermoplastic material is fed from a spool through the head to melt and is then forced out of the nozzle and deposited selectively to create the product layer by layer. Actually, the printer head moves in two dimensions using CNC technology to trace any desired contour and create one layer at a time. Also, the nozzle has an on-off mechanism, to interrupt the deposition of the molten material whenever needed, and then goes on again in order to continue completing an interrupted layer. Next, the head moves incrementally upward (or the build platform moves downward) to create a new layer, and the process continues until the part is completely created by the extruded flattened strings of molten material. The latter immediately hardens after extrusion from the nozzle.

A wide variety of materials are processed by the FDM method. They include thermoplastics like acrylonitrile butadiene styrene (ABS), high-impact polystyrene (HIPS), thermoplastic polyurethane, and nylon. In addition, paste-like ceramics and metals can be 3D printed by this method.

The other version of this 3D printing process—namely, AMoC—is gaining acceptance in the construction industry, as it has the potential to improve on current construction methods that are labor-intensive and also create health hazards to the workers. The Occupational Health and Safety Administration of the U.S. Department of Labor lists these hazards as eye, skin, and respiratory tract irritation from exposure to cement dust. In the AMoC process, these hazards are minimized or totally eliminated; concrete in the form of a wet paste is extruded in the same manner explained above, to additively create buildings. Evidently, no heating is needed in this latter case. It has been recently reported that a five-story apartment building in Suzhou (China) was constructed using this technology.

13.2.6.1 Applications of Additive Manufacturing Material Extrusion

Initially, the FFF method was dominantly used as a 3D printing hobby just for leisure. Now, it is not the case as the motor of the world's first electric automobile was produced by employing this method, where highly viscous metallic and ceramic pastes were extruded through a nozzle to build the body of the part in layers, before sintering to gain strength and hardness. On the other hand, AMoC is gaining ground in the construction industry, particularly in China. Building on-demand shelters for the military is another emerging application. An on-site concrete printer, operated by two people, could build a 2400-ft^2 (216-m^2) shelter capable of accommodating a C-130 aircraft in only 24 hours.

13.2.6.2 Advantages of Additive Manufacturing Material Extrusion

- The main advantage of the FFF method is the cost. Equipment is cheap and easy to use. Also, the process can be used to manufacture a variety of materials, including polymers, ceramics, and metals.
- The main advantages of AMoC are the speed of construction and the elimination of industrial hazards to the worker, and this can be accomplished by a very small number of workers, thus cutting construction costs.

13.2.6.3 Disadvantages of Additive Manufacturing Material Extrusion

There are no real disadvantages for either FFF or AMoC. The application of the FFF process is, however, limited to the production of lightly stressed parts.

13.2.6.4 Design for Parts Produced by the FFF Process

The design considerations involve more or less the same rules that were previously given, namely, avoiding overhang surfaces and down-facings.

13.2.7 Sheet Lamination

This additive manufacturing process was among the early technologies that were developed in the mid-1980s and began commercialization in the early 1990s. The commonly used name for the sheet lamination process has been *laminated object manufacturing* or simply (LOM). This process, as well as early processes, was developed to enable faster product development by creating 3D models of the form and function of the product being developed. As expected, a graphical 3D model of the object first has to be created by the software, then "sliced" into layers, in order to obtain the contour of each layer. Each of these digital contours is then used to drive a knife cutter by a computer to cut

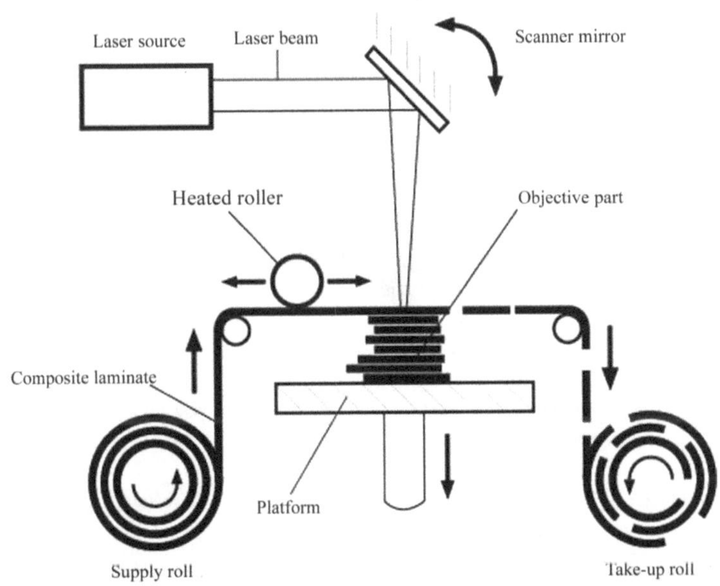

FIGURE 13.7 A schematic illustration of the sheet lamination process.

an adhesive-coated paper to shape. Next, the layers are stacked in the right order and glued together to produce the desired object.

Later, the process was developed further to fabricate objects from plastic sheets, metal laminates, and finally fiber-reinforced polymeric composites. A schematic illustration of the process is shown in Figure 13.7, where the knife cutter has been replaced by a computer-guided laser beam. Actually, this is currently the most widely used application of the sheet lamination AM process.

13.2.7.1 Applications of Sheet Lamination

As mentioned above, the main application is in the manufacture of fiber-reinforced polymeric composites.

13.2.7.2 Advantages of Sheet Lamination

These were previously discussed in Chapter 9 (composites), as well as the limitations.

TABLE 13.1
Summary of the Different Additive Manufacturing Processes

Process Type, Names	Description	Materials	Applications
Powder bed fusion, DMLS, LS, SLS	A process used to produce objects from powdered materials using a laser beam or electron beam, to selectively fuse or melt the particles at the surface, layer by layer, in an enclosed chamber	Metal powders	Functional metallic components
Directed energy deposition	Focused thermal energy is used to fuse materials by melting as they are being deposited	Metal wires, metal powders	Large metallic components
Material extrusion, FDM, FFF, AMoC	A process used to make thermoplastic parts through heated extrusion and deposition of materials layer by layer	Polymers, pastes of metals and ceramics, concrete	Polymers, metals and ceramic parts, concrete
Vat photopolymerization, SL	A process in which liquid photopolymer in a vat is selectively cured by light-activated polymerization	Polymers	Prototypes
Binder jetting	A liquid bonding agent is selectively deposited to join powder materials	Polymers, metal powders, silica sand	Casting molds, metallic components
Material jetting, 3D printing	A process in which droplets of build material are selectively deposited	Polymers, pastes of metals, biomaterials	Casting patterns, prototypes
Sheet lamination, LOM	A process in which sheets of material are bonded to form an object	Papers, metals, FRPC composites	FRPC components

13.3 SUMMARY OF THE VARIOUS ADDITIVE MANUFACTURING PROCESSES

Table 13.1 indicates the different additive manufacturing processes with their commercial names, short descriptions, fabricated materials, and applications. It is meant to be a guide for beginners in that field.

13.4 IMPACT ON INDUSTRY

Let us give an example of the impact of additive manufacturing on industry. A good example is the medical devices industry. Evidently, additive manufacturing has the potential for how medical devices are made, distributed, sold, and used. It can enhance device functionality and doctor and patient satisfaction. This comes as a result of bringing new types of products to the market, more quickly and cheaply, without the need for massive infrastructure, through large-scale customization. It would also affect how devices get to the patient. Moreover, it would enable making customized implants (e.g., for an individual patient) instead of manufacturing generic ones that would be adjusted to suit the patient. In fact, parts can be produced in situ, and hospitals can become manufacturers, making the required devices at the points of care, thus eliminating warehouses and simplifying the SCM by eliminating logistics.

REVIEW QUESTIONS

1. Define additive manufacturing.
2. Why is additive manufacturing considered by professional manufacturing engineers to constitute the fifth industrial revolution?
3. What are the two most important capabilities that form the bases for modern manufacturing, which all categories of additive manufacturing provide?
4. List the seven categories/groups of additive manufacturing according to ASTM/SME terminology.
5. What are the common features in all those categories?
6. Explain the vat photopolymerization process. Why is it considered an additive manufacturing process?
7. List some of the advantages and limitations of the vat photopolymerization process.
8. Explain briefly the working principles of the powder bed fusion process.
9. Why do we consider the name "selective laser sintering" to be slightly misleading? How will you correct it?
10. List some of the advantages and shortcomings of the SLS process.
11. List a few applications of the SLS process.
12. Explain using sketches the direct energy deposition process, and differentiate between its two versions. Which version can be used to produce large metallic structures?
13. The additive manufacturing material extrusion category includes two processes. What are they, and which material do they extrude?
14. List some applications of material extrusion additive manufacturing.
15. What is the commonly used name for material jetting?
16. Explain the working principles of a material jetting machine.
17. List the advantages and limitations of the material jetting machine.
18. List some applications of the material jetting process.
19. How does the binder jetting process differ from the material jetting process?
20. Explain how the binder jetting process has been used to automate the foundry industry and make it faster and more efficient.
21. What is the current main application of sheet lamination additive manufacturing?

DESIGN PROJECT

This is a multidisciplinary project that can be a good senior design capstone project.

Using e-waste such as old printers and computers, design and build a binder jetting machine.

Tip: use a mix of fine silica sand and starch as the material to build the object and an aqueous solution with appropriate chemical as the binder.

14 Computer-Aided Manufacturing

14.1 INTRODUCTION

Computer-aided manufacturing (CAM) has been defined by Computer-Aided Manufacturing International (CAM-I) as "the effective utilization of computer technology in the management, control, and operations of the manufacturing facility through either direct or indirect computer interface with the physical and human resources of the company." Although this definition of CAM is broad and flexible and covers a variety of tasks, the dominant application of CAM is still in numerical control (NC) part programming. For this reason, many people subscribe to a narrower concept of CAM that involves mainly computer-assisted NC part programming. This concept of CAM may also stem from the fact that CAM has its roots in NC systems. Consequently, a logical step in studying CAM is to discuss NC systems as well as manual (unassisted by computer) NC part programming and then proceed to extensions of NC and the different methods of computer-assisted part programming.

14.2 NUMERICAL CONTROL (NC)

14.2.1 Overview

Before discussing NC systems, how they began, and how they evolved to their present status, let us consider a simple definition that will be a logical entry to the subject. Numerical control (NC) can be defined as control of the operation of machine tools (or other sheet working and welding machines) by a series of coded instructions called a program, which consists mainly of alphanumeric characters (numbers and letters). It is obvious from this definition that the sequence of events is both preplanned and predictable. In other words, any desired sequence of events can be obtained by coding the appropriate instructions and can also be changed by changing those coded instructions. Therefore, NC systems are considered to be the typical form of programmable automation.

14.2.2 Historical Background

The basic concept of numerical control is not new at all and dates back to the early years of the Industrial Revolution, when Joseph Jacquard developed a method to control textile looms by using punched cards. When he applied for a patent for his invention, however, the queen of England denied him that right because she believed that it would displace poor workers (notice the similarity with the use of robots today). In fact, this old invention, together with the player piano, which is operated by a roll of punched paper tape, can be considered as simple, crude forms of mechanical NC. A modern version of NC emerged in 1947 at the Parsons Engineering Company of Traverse City, Michigan, as a result of the need of John C. Parsons (the owner of the company) to manufacture helicopter rotor blades fast enough to meet his contracts. Later, Parsons Engineering was awarded a study contract by the U.S. Air Force Material Command to speed up production and develop continuous-path machining, with the subcontractor being the Massachusetts Institute of Technology (MIT). The job was later given in its entirety to MIT, and the machine they developed was successfully demonstrated in 1952. Between the years 1953 and 1960, the rate of building and selling NC

DOI: 10.1201/9781003537106-14

machines in the United States was very slow. This type of machine tool later gained widespread industrial application because of the need for consistency of dimensions and tighter tolerances.

14.2.3 Simplified Idea of Numerical Control

To understand the basic idea behind numerical control, let us assume that a hole has to be drilled in a plate using a drill press and that, according to the blueprint, the hole is 8 inches to the right of the left edge and 5 inches above the lower edge. We start by clamping the workpiece on a positioning table on the drill press; we then crank the two hand wheels of the two perpendicular slides to locate the corner of the plate (the point where the left and the lower edges meet) exactly under the center of the spindle. Now, if a single turn of the hand wheel causes the table of the drill press to move by 0.10 inch, it is obvious that we need to move the table to the left 8 inches and downward 5 inches by cranking the appropriate hand wheels 80 and 50 turns, respectively. In order to automate this operation, we replace the hand wheels with electric servomotors that are operated by push buttons. Let us say for convenience that a single quick push of a button (like a dot in the telegram code) causes the attached servomotor to turn by 1/100 revolution. Consequently, we need to push the button of the first servomotor 8000 times and that of the second servomotor 5000 times in order to position the center of the spindle exactly above the desired location of the hole. After doing so, we will be in a position to perform the drilling operation and obtain a part that conforms precisely to the blueprint. A closer look at this example shows that the machine is driven by numerical values (a number of pulses or button pushes for each direction) and responds by converting these values to meaningful physical quantities. This is, in fact, what is meant by numerical control, no matter how these numerical values are fed into the machine. We can, therefore, say that an NC system is a system that readily converts numerical values into physical quantities such as dimensions.

14.2.4 Advantages of Numerical Control

The advantages of NC machine tools are felt not only on the factory floor but also in many other departments of a business corporation. The following are some advantages that can be used as justification for employing NC machine tools:

- NC machine tools ensure positioning accuracy and repeatability. In other words, if the same program is employed to produce a number of parts, they will have identical dimensions.
- NC machine tools can produce complex-shaped components automatically with closer tolerances and very high degrees of reliability, which provides the designer with a great degree of flexibility and freedom.
- Because NC machine tools have high dimensional accuracy and repeatability, parts can be manufactured that require a long series of operations. Such parts are difficult to produce by conventional methods because accumulated errors result in completely unacceptable results.
- Because, after being programmed, NC machine tools can perform any desired task (within their capability) without the need for a human operator on the shop floor, they can be employed to carry out operations in hostile environments, such as the machining of polymeric materials that emit poisonous gases.
- NC systems transfer a substantial portion of the planning for the processing operation from the shop floor to the engineering offices, where specialists prepare NC part programs in comfortable surroundings, and production is then directly monitored and controlled by management.
- NC machine tools have the capability of performing more than one machining operation by automatically changing the tool used without changing the location of the workpiece. In

other words, a sequence of machining operations can be performed in a single setup, which reduces the number of transfers of a workpiece between different machine tools or machining departments. This capability is considered one of the major advantages of NC machine tools because the nonproductive time used in setups and workpiece transfer amounts to a high percentage of the total production time, as evidenced by statistical data. In this respect, it is worth mentioning that actual machining time involves only 5 percent of the production life cycle of a typical component.

- As a result of the preceding advantages and because of the minimal idle time involved, the use of NC machine tools is always accompanied by increased productivity.
- The high dimensional accuracy and repeatability of NC machine tools provide a profound basis for the interchangeability of work between different production plants.
- NC systems reduce part scrapping due to machining errors and lower the inspection and assembly costs as a result of the uniformity and reliability of the products produced by this technology.

14.2.5 Elements of an NC System

This section will focus on the elements of a tape-operated NC system. Although very few if any of these systems exist in the United States now, an understanding of this basic NC system will provide an adequate basis for comprehending the advanced NC systems currently used in industry. A sketch of the basic elements of an NC system that controls a machine table along a single direction is shown in Figure 14.1. Two systems like this one have to be used when the motion of the machine table must be controlled along two directions. Following is a brief discussion of the elements of an NC system.

Tape Reader. As previously mentioned, the desired sequence of events is converted into a series of coded instructions (i.e., the program). The program is then recorded onto a tape. Next, the coded tape is read by the *tape reader* (a device, located in the machine control console, that has the function of winding and reading the tape). There are different types of tape readers (electromechanical, electronic, and optical); each has a different method of operation.

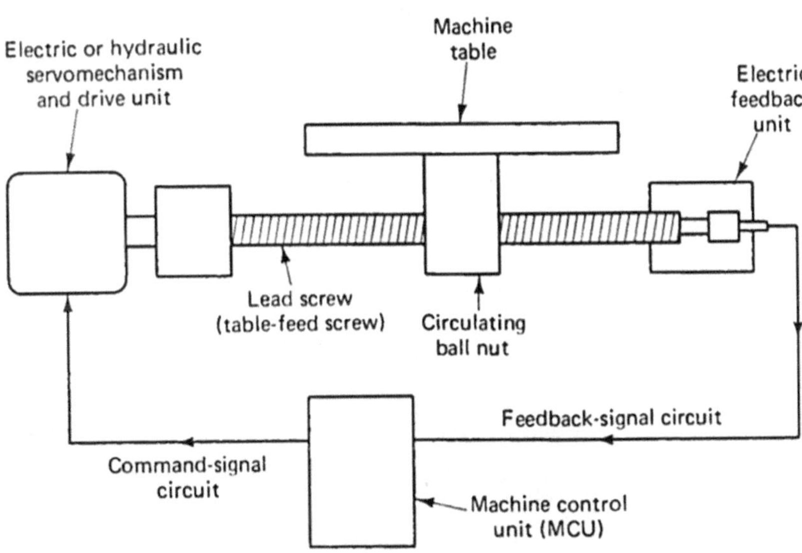

FIGURE 14.1 The basic elements of an NC system.

Machine Control Unit. The *machine control unit* (MCU) receives the coded instructions from the tape reader, decodes them by converting them into signals representing the preplanned commands, and then transmits the signals to the servomotors to generate the machine movements. In the early days of NC, the MCU was hardwired; today, it consists mainly of a microcomputer.

Servomechanism. The function of the *servomechanism* is to amplify the signals received from the MCU and to provide power to produce the required tool (or machine table) movements. These signals generally take the form of pulses, whereas the servomotor is often a DC electric motor that drives the tool (or machine table) through a lead screw. Hydraulic systems are also in use.

Controlled Element. A *controlled element* is any part (of the machine tool) that is numerically controlled. It can be a tool, a turret for an NC lathe, or the machine table for an NC drill press.

Feedback Unit. The function of the *feedback unit* is to record the achieved movement of the tool (or machine bed) and then send a feedback signal to the MCU. The MCU compares the achieved position with the required or programmed one and automatically compensates for any discrepancy. Systems with feedback units are usually referred to as *closed-loop* systems.

14.2.6 The Coordinate System and Dimensioning Modes

As with any engineering application, NC programming is based on the Cartesian coordinate system (sometimes on the polar coordinate system as well). According to the Cartesian system, any point within a plane can be defined by its distances from the X-axis and the Y-axis (i.e., the X- and the Y-coordinates, respectively). Also, the point of intersection of these two perpendicular axes is called the *origin*, or *zero point*. The coordinates of a point can be both positive, both negative, or one negative and the other positive, depending upon the location of that point. The two perpendicular intersecting axes divide the plane into four quadrants, which are numbered counterclockwise, as shown in Figure 14.2. Notice that all values of X and Y are positive in the first quadrant and negative in the third quadrant. In the second quadrant, all values of X are negative, and all values of Y are positive;

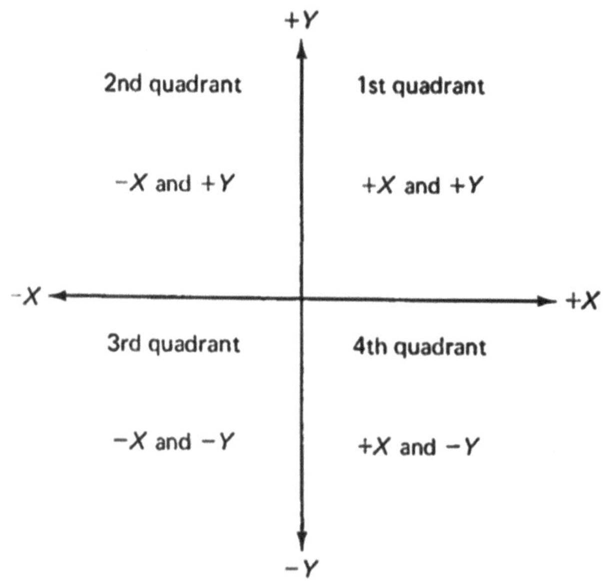

FIGURE 14.2 The coordinate system and quadrant notation.

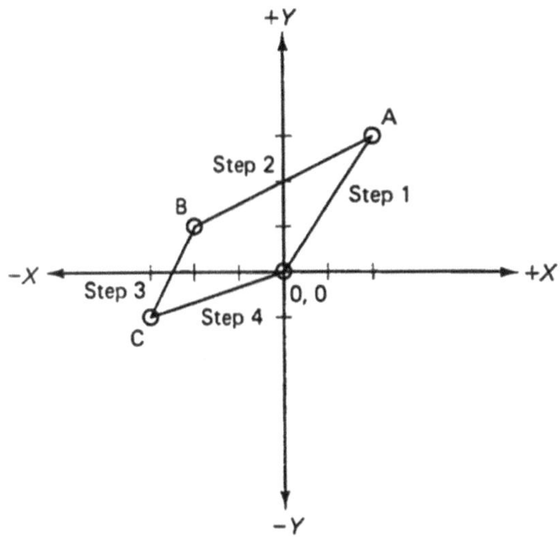

FIGURE 14.3 Absolute and incremental dimensioning modes.

in the fourth quadrant, it is the other way around. In other words, when a point falls to the right of the Y-axis, its X-coordinate is positive, but when it is to the left of that axis, its X-coordinate is negative. Similarly, the Y-coordinate of a point is positive when the point is above the X-axis, but it is negative when the point is below that axis.

The Cartesian coordinate system can be extended to describe a point in space by adding a third dimension along the Z-axis, which is perpendicular to the plane of the X- and Y-axes. As we will see later, this third dimension will enable us to deal with more complicated work.

We are now in a position to discuss the dimensioning modes, which include two types: *absolute* and *incremental*. The absolute mode is similar to the Cartesian coordinate system, where all coordinates of points are always given in reference to the origin or the machine zero point. When the incremental mode is used, reference is made to the latest position, which is actually equivalent to considering each location to be the zero point for the next location. In other words, incremental programming involves an *increment* from the present position to the new one, together with an associated sign indicating the direction. The following example will clarify the difference between the two programming modes. As can be seen in Figure 14.3, the centerline of the tool (e.g., a drill) coincides with the zero point, and we are required to write statements to move it first to point A, then to points B and C, and finally back to the origin. Here are the desired statements in both absolute and incremental modes.

Although absolute programming is more logical, it requires all drawings (i.e., blueprints of parts) to be done with all dimensions referenced to a single point in order to make the programmer's job easier. Incremental programming is advantageous for positioning work. NC programs are, therefore, usually a blend of both programming modes so that use can be made of the as-drafted dimensions of the part.

14.2.7 NC Machine Motions

The Electronic Industries Association (EIA) lists in its RS-267A standards the various types of NC machine motions or axes designations, whereas RS-267 indicates some 25 different NC machines. The single-spindle-drilling machine is the simplest of all. It is generally a two-axis NC machine tool because it can be program-controlled on two axes: the X- and Y-axes. The Z motion of the spindle (raising and lowering) is controlled manually or by using a system of cams. In some NC machines, a tape command calls a preset depth, but this cannot be considered an axis of motion. A true *axis of*

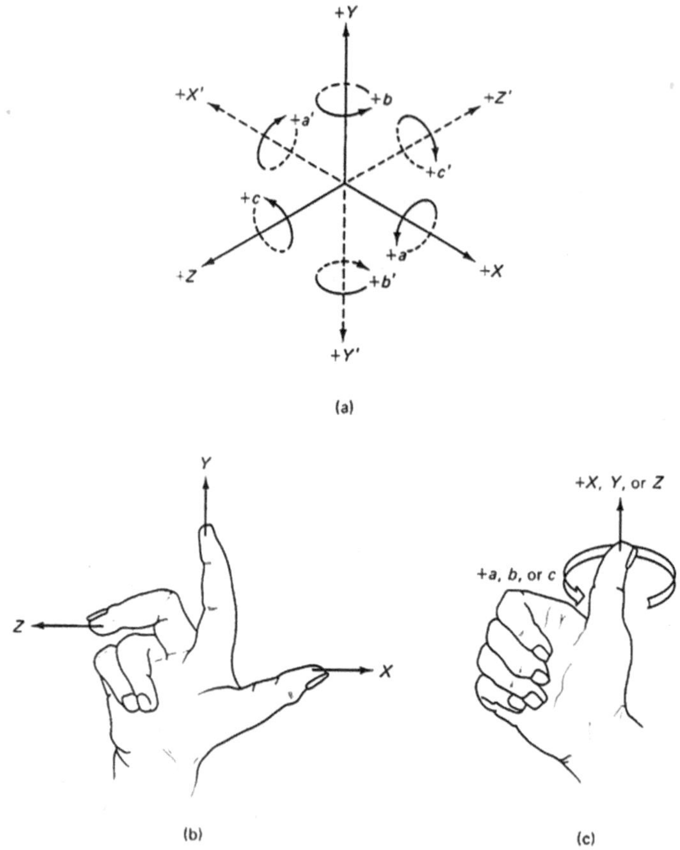

FIGURE 14.4 The right-hand rule for the relative location of coordinate axes.

motion is one along which an infinite number of locations for the tool (or machine bed) can be obtained. An axis of motion may be either *linear* or *rotational*. According to the EIA standards, the X-, Y-, and Z-axes are the linear axes, whereas the rotational axes are the *a*-, *b*-, and *c*-axes, which are used to indicate the rotary motion around the X-, Y-, and Z-axes, respectively, as shown in Figure 14.4a. The positive direction of any of the rotary motions can be obtained by employing the *right-hand rule*. As can be seen in Figure 14.4b, this rule involves using three fingers of the right hand to indicate the linear axes and then using the thumb pointing out in the positive direction of the linear axis (the one that forms the center of rotation of the rotary motion under investigation) with the other fingers curved to indicate the positive direction or the rotary motion.

It is important to remember that the X-, Y-, and Z-axes are neither arbitrary nor interchangeable. It has been agreed upon by the ElA and NC machine tool builders that the Z-axis is always a line parallel to the spindle of the machine tool. Consequently, the Z-axis can be either vertical or horizontal, depending upon the kind of machine tool. In the case of an NC lathe, the Z-axis is horizontal, whereas it is vertical for a vertical milling machine or a drill press. The function of the spindle differs, depending upon the kind of machine tool. The spindle is employed for rotating the tool on milling and drilling machines, whereas it is the workpiece-rotating means on engine lathes and similar machines.

In addition to the primary linear axes X, Y, and Z, sometimes secondary linear axes are used that are parallel to the primary axes and are designated U, V, and W, respectively. NC systems that have both primary and secondary linear axes provide more flexibility when a program is to consist of both absolute and incremental dimensioning modes. In such a case, it is common to devote the X-, Y-, and Z-coordinates to absolute dimensions and employ U, V, and W to indicate incremental motions.

It is also common to have the controlled element be the machine table (or, in other words, the workpiece) and not the cutting tool. The controlled element then responds to the tape command in an opposite direction; any movement of the workpiece in the established positive direction for the tool is considered negative. This is, in fact, equivalent to saying that it is the relative movement of the tool with respect to the workpiece that is actually considered.

NC machines can have two, three, four, or even five axes. In this respect, the word *axis* means any direction of linear or angular motion that is truly and fully controlled by the NC system. As previously mentioned, indexing or calling a preset dimension does not fall under the definition of a true NC axis of motion. Following is a brief survey of each type of NC machine.

- *Two-Axis NC Machines*. In two-axis NC machines, motions along only two axes (usually X and Y or X and Z) are fully controlled by tape commands. Figures 14.5 and 14.6 show an NC turret lathe and an NC drill press, respectively, that belongs to this type of machine. Notice that, for the turret lathe, the positive direction of X is going away from the workpiece and the positive direction of Z is going away from the headstock. It is also important to note the difference between the positive direction of the axes and the machine table movements in Figure 14.6.
- *Three-Axis NC Machines*. Vertical knee mills, drilling machines, and jig borers are examples of three-axis NC machines. In this type of machine, the motion is controlled along the Z-axis as well as the X- and Y-axes.
- *Four-Axis NC Machines*. In four-axis NC machines, in addition to motion along the X-, Y-, and Z-axes, the machine table is rotated by command at a controlled rate during the machining operation. Again, note the difference between NC controlling and indexing (even when the latter involves a very large number of indexed positions).
- *Five-Axis NC Machines*. Five-axis NC machines are used for producing sculptured surfaces because the machine head can swivel at a controlled rate (in addition to the previously mentioned four axes of motion). The tool can, therefore, be brought perpendicular to the desired surface. An example of this type of machine is the five-axis profile and contour mill with a tilting head.

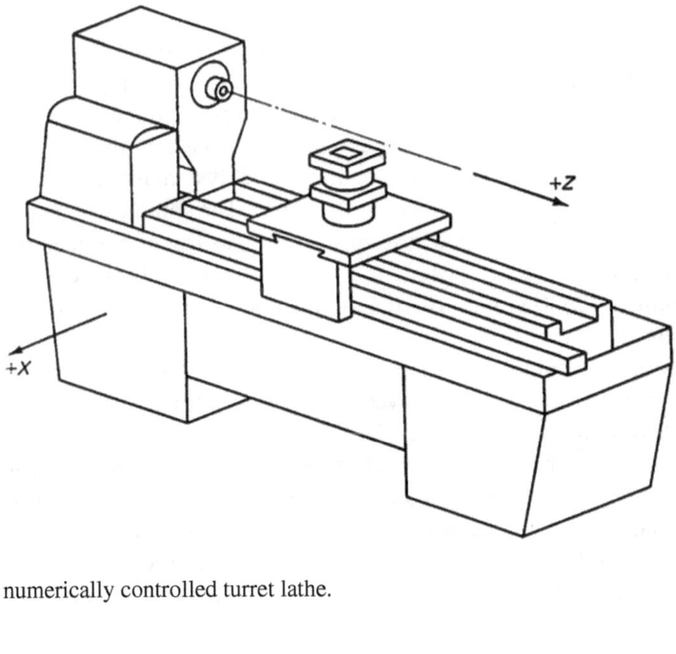

FIGURE 14.5 A numerically controlled turret lathe.

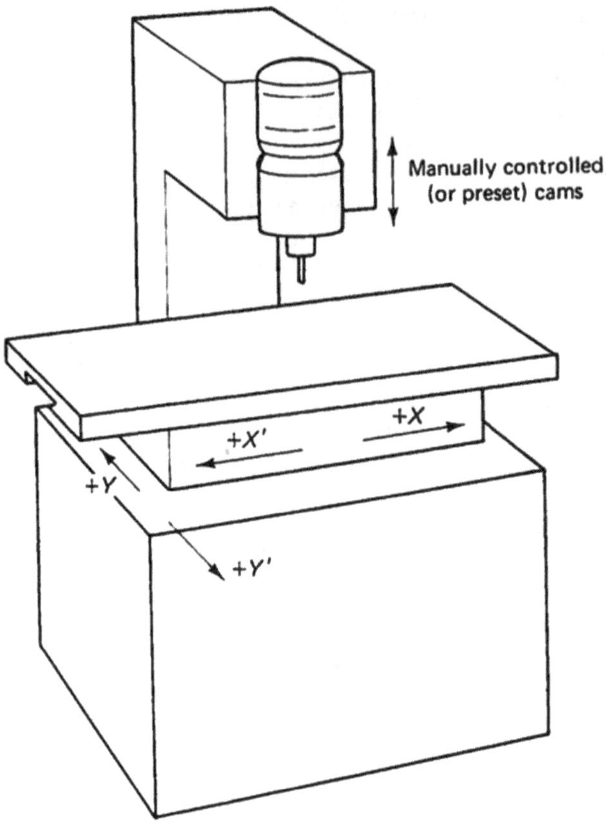

FIGURE 14.6 A numerically controlled drill press.

14.2.8 Types of NC Systems

There are three basic types of control systems for NC machine tools: point-to-point, straight-cut, and contouring.

- *Point-to-Point System.* The point-to-point system is also referred to as numerical positioning control (NPC) and is usually used in NC drilling machines that are employed in drilling precise patterns of holes. The function of the NC system is, therefore, to move the spindle (or machine table) to the exact location, as given by a tape command, so that a hole can be drilled. As soon as the desired hole is drilled, the NPC system moves the spindle to the next programmed location to drill another hole, and so on. The spindle (or machine table) movement from one hole location to the next must be done as fast as possible to bring to a minimum the nonproductive time spent in movement. Accordingly, speeds of more than 100 inches per minute (2500 mm/min) are quite common. Provided that our main concern is positioning the spindle in each of the desired locations, it is of no importance to control the path along which the spindle moves from one location to the next. In fact, that path is not necessarily a straight line as it just covers the shortest amount of time. It usually involves two intersecting straight lines, as is explained later.
- *Straight-Cut System.* The straight-cut system is quite similar to the point-to-point system, except that the feed rate of the spindle along each machine axis is controlled so as to be suitable for machining (e.g., a milling operation on a vertical mill). Again, the spindle cannot be controlled so that it moves along a line inclined to the X- and Y-axes of the machine; the motion along one axis is independent of the motion along the other axis because it is

controlled by a separate NC circuit (or subsystem). Nevertheless, motions along lines coinciding with or parallel to either the X- or Y-axis can be accurately controlled. The spindle can move forward, to the right, backward, and to the left in a rectangular path, and, for this reason, the system is sometimes referred to as a picture-frame system. The sequence of motions may not necessarily yield a rectangular path. Figure 14.7 indicates some tool paths that can be produced by this NC system and that are employed in machining rectangular configurations, in face milling, and in pocketing. NC machines fitted with straight-cut control are also capable of performing point-to-point positioning at very high speeds. They are, therefore, more versatile than NPC machines. However, their cost is also higher and must be justified by the kind of products required.

- *Contouring System.* In order to make angular cuts on a workpiece, the two driving servomotors (one for X-axis motion and the other for Y-axis motion) have to run at unequal speeds. In fact, the rate of travel in the Y direction divided by the rate of travel in the X direction must be equal to tanθ, where θ is the angle that the angular cut makes with the X direction. The capability of a control system to regulate the rate of spindle (or table) travel along two axes of motion at the same time is called linear interpolation.

In addition to point-to-point positioning and picture-frame cutting, a contouring control system can produce curves to very close tolerances. Therefore, it is sometimes referred to as the *continuous-path* system. The method of employing linear interpolation to produce curves involves breaking down a curve or an arc into a large number of straight lines in such a manner that the end of each line is the beginning of the next one (tip-to-tail fashion). Each and every line segment must, therefore, be programmed in order for the path to conform to the desired curve. Obviously, the larger the number of segments taken, the smaller each of the line segments becomes and the smoother the machined curve becomes. This concept is illustrated in Figure 14.8. Breaking down a curve into

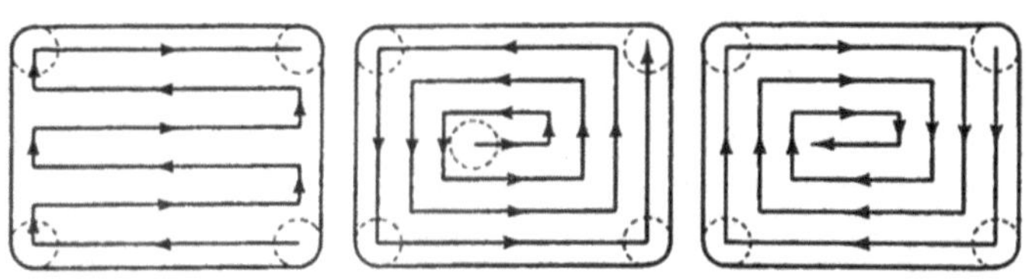

FIGURE 14.7 Some tool paths that can be produced by a straight-cut NC system.

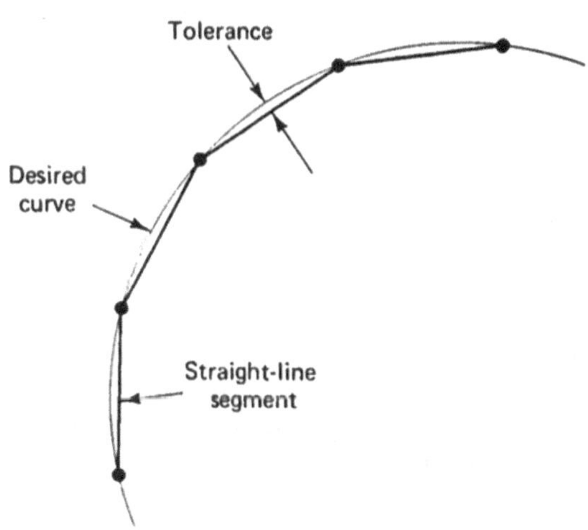

FIGURE 14.8 Approximation of a curve using straight-line segments.

hundreds of straight lines and programming each line is very complex. This process, if carried out manually, would take much time and effort. Therefore, it is always performed using a computer, as we will see later when discussing *computerized numerical control* (CNC).

14.2.9 Punched Tape and Tape Coding

14.2.9.1 The NC Tape

The NC tape is the oldest type of input media used for storing all the data (i.e., the NC program) needed to generate a desired part. When the tape is run on an NC machine, the prepared program is simply read, and the desired part can thus be machined. The same part can be produced several times by running the prepared tape as many times as required. The identical part can also be made years after the tape is prepared, as long as the tape is kept in good condition. It is very important that the MCU and the tape always be compatible. In other words, they both must be based on the same coding system and the same coding format. Although punched tapes are not as commonly used as they used to be, a discussion of the coding and format of a punched tape will, nevertheless, provide an adequate and clear picture of how instructions are fed into NC machines.

The punched tape used for NC systems is standardized by the EIA to have a width of 1.000 ± 0.003 inches (25.4 ± 0.076 mm) and a thickness of 0.004 ± 0.0003 inches (0.1 ± 0.008 mm). The tape can be made of paper, a paper Mylar sandwich, or an aluminum–Mylar laminate. Paper tapes are cheap and easy to damage, so their use is limited to short runs. For high production and frequent use, aluminum–Mylar tapes are more suitable because of their durability, but they are more expensive. NC tapes are purchased in the form of rolls that are 8 inches (200 mm) in diameter, each having a tape length of up to 2000 feet (600 m). The tape is divided into eight main channels, or *tracks* (i.e., parallel to the edges), where holes can be punched. There is also a track of smaller holes to the right of the third main track. These smaller holes fit the tape-feeding sprocket in order to ensure a positive drive of the tape. Letters of the English alphabet, digits from 0 to 9, and symbolic signals to the MCU each have a specific arrangement of punched holes in a line, or *row*, perpendicular to the edges of the NC tape. A single instruction given to the MCU usually consists of a set of letters and numbers; a set of rows of punched holes is referred to as a *word*. There are, however, some words that take only a single row. A number of words that are grouped together form a *data block*. The block is the smallest unit of a program that provides the NC system with complete information for an operation (or tool motion).

14.2.9.2 Punching the Tape

First, the programmer prepares the manuscript of the part program, which is commonly called the *program sheet*. It involves a list of detailed instructions that describe the step-by-step operation of the NC system. The information on the program sheet is then transferred to a blank tape by punching holes into it that stand for the required codes. This is done by typing on a Flexowriter or similar tape-punching piece of equipment. The result is not only the punched tape but also a printout of the program sheet that can be used to check for errors and make corrections.

14.2.9.3 Tape Codes

NC tapes are coded in a *binary-coded decimal* (BCD) system, which is a further development of the binary coding system. This system is based on considering the presence of a hole as *on* and its absence as *off*; each is called a *bit*. The presence of a hole in the first track means 2^0 (i.e., two to the power zero), or 1, whereas its absence means 0. The second, third, and fourth tracks mean the number 2 raised to powers 1, 2, and 3, respectively. In other words, the presence of a hole in the second track is equivalent to 2, the presence of a hole in the third track means 4, and the presence of a hole in the fourth track means 8. Thus, any digit (i.e., a number from 0 to 9) can be represented in one row of the tape by an arrangement of holes in the first four tracks (from left to right). Some examples will clarify this coding system. The digit 3 is designated by a combination of two holes, one in the first track and the other in the second track. The digit 7 is a combination of holes in the first, second, and third tracks.

In fact, each numerical digit, letter, or symbolic signal has its own designated combination of holes in a single row. When the tape reader reads a numerical value such as 4732, it reads a single digit (one row) at a time and, through its electronic circuit, places that digit in its proper decimal position with respect to the preceding and succeeding digits. In this way, the decimal value of any digit is determined by its relative position in a set of rows representing a numerical value.

Tape coding has to be standardized in order to facilitate the interchangeability of tapes and communications between industrial firms. Two tape codes are commonly used: the EIA code and the ASCII (American Standard Code for Information Interchange) sponsored by the American National Standards Institute (ANSI). Figure 14.9 indicates both standard tape codes. As can be seen, the EIA

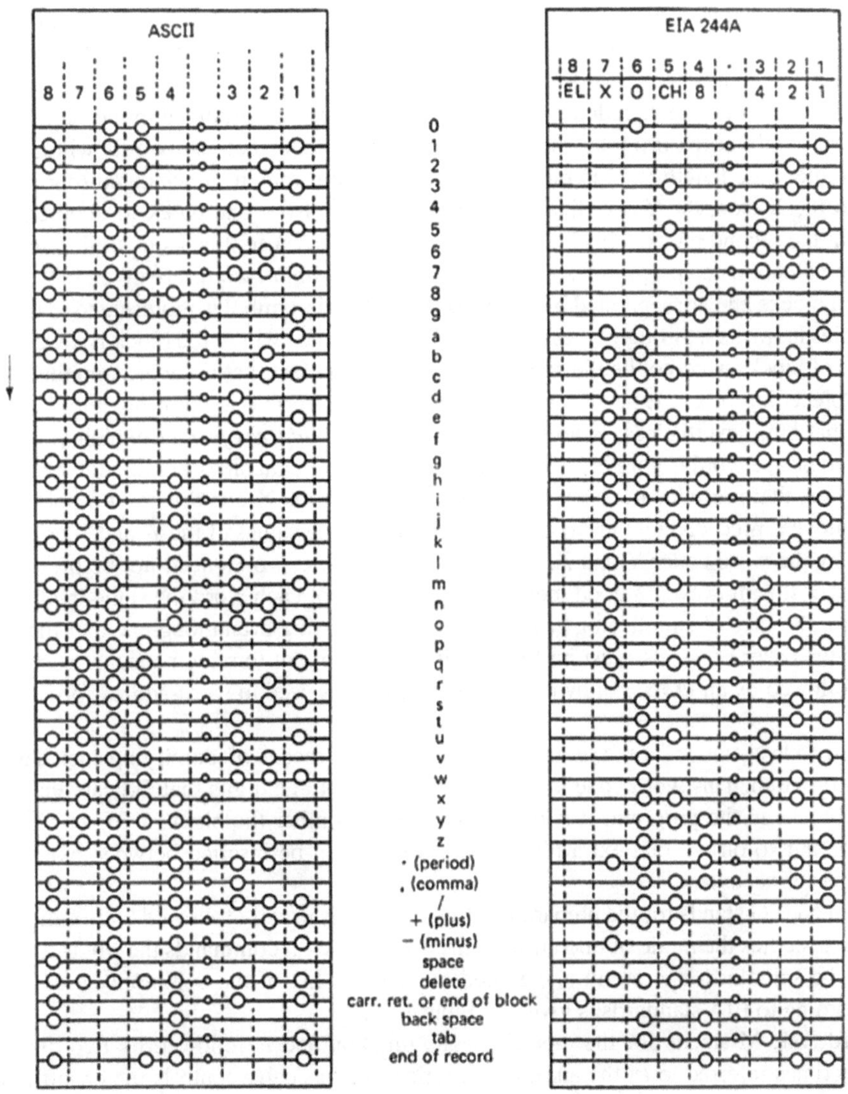

FIGURE 14.9 ASCII and EIA tape codes.

code uses only six tracks of the eight available on a tape and always has an odd (uneven) number of holes in any row. The fifth track contains a hole whenever the number of holes representing a character is even. This method of detecting errors in a punched tape is called a *parity check*. An EIA-coded tape must have odd parity as an indication that no punching mistakes have been made. Also, as previously mentioned, each operation or movement is represented by a set of rows on the tape (i.e., a data block). Each block of information must be separated from the following one by a special character called the *end of block* (EOB). It is represented by a hole in the eighth track of the tape. The EIA code provides 63 different combinations of holes, which is both logical and sufficient for NC applications. The ASCII code was introduced to more appropriately meet the needs of computer organizations, the government, and the communications industry. It utilizes all eight tracks of a tape and, therefore, provides 128 characters (i.e., possible combinations of holes). It has even parity, contrary to the EIA code.

14.2.9.4 Tape Formats

The term *format* refers to the way NC words are arranged in a data block. Although three formats gained some common use, two of them are not used now, and it is only the third one, the word-address format, that has gained widespread application in industry.

In the first type of format, called *fixed-block format*, words are arranged in the same sequence in all blocks throughout the program. In addition, not only blocks but also words within each block must have the same length (i.e., number of rows). If a word remains unchanged (e.g., in a motion parallel to the X-axis, the Y-coordinate remains the same) from one block to another, it must be repeated in the second block. This format is not used now because it lacks flexibility and results in lengthy and complicated programming. In *TAB sequential format*, words are given in the same fixed order in all blocks and are separated by the TAB symbol. This symbol is represented in the EIA code by five holes in tracks 2 through 6. Although the order of words within the block is always the same, the length of blocks need not be the same. This is due to the fact that if a word remains unchanged, as in the preceding block, it need not be given again. This format is more flexible than the fixed-block format but, nevertheless, is not used today. In *word-address format*, which is currently the format used now, a word is not identified by its location in the data block but rather by a single letter, or *word code*, that precedes it. As an example, the value of the X-coordinate of a point to which a tool is to move is preceded by the word code X. Similarly, the values of the Y- and Z-coordinates are preceded by the word codes Y and Z, respectively. Therefore, words need not be presented in any special order. It is, however, a good idea to keep the order of each word the same in all blocks of a program for the sake of simplifying programming and checking even though neither the length of a block nor the order of words must be fixed. It is also important to remember that each data block must be followed by an EOB. Following are the word codes used in word-address programming:

- Word code N stands for the sequence number and is a means of identifying each data block in a program. This word code is usually followed by three digits that indicate the order of the blocks in the program. It is quite common to number the blocks by fives (e.g., N05, N10, and so on) so that extra blocks (operations) can be inserted between existing ones whenever necessary.
- Word code G stands for the preparatory function and is usually followed by two digits. This code specifies the mode of operation of the control (i.e., it commands the MCU, thus causing the spindle to operate in a specified manner). A list of the commonly used G codes according to EIA standards is given in Table 14.1.
- Word code M stands for the miscellaneous function and is followed by two digits. It is sometimes referred to as the auxiliary function and basically controls the on-off machine operations such as coolant on, tool change, and the like. A list of the commonly used M codes according to EIA standards is given in Table 14.2.

TABLE 14.1
Commonly Used G Codes

G00	Rapid traverse, linear interpolation in Cartesian coordinates for tool positioning at a speed of 200 inches/minute (5 m/minute)
G01	Linear interpolation in Cartesian coordinates, between the current location and the next programmed one at a specified feed rate
G02	Circular interpolation in Cartesian coordinates, clockwise
G03	Circular interpolation in Cartesian coordinates, counterclockwise
G10	Rapid traverse, linear interpolation in polar coordinates
G11	Linear interpolation in polar coordinates at a feed rate
G12	Circular interpolation in polar coordinates, clockwise
G13	Circular interpolation in polar coordinates, counterclockwise
G17	XY-plane designation, tool axis Z
G40	No tool compensation
G41	Tool radius compensation to contour, offset left
G42	Tool radius compensation to contour, offset right
G54	Datum shift to a specified point
G73	Rotate the X- and Y-coordinates by a specified angle
G74	Slot milling
G79	Cycle call
G83	Pecking cycle
G84	Tapping
G70	Dimensions in inches
G71	Dimensions in millimeters
G90	Absolute dimensioning
G91	Incremental dimensioning
G98	Assign label number
G99	Tool definition

TABLE 14.2
Commonly Used M Codes

M00	Program run stop, spindle stop, coolant off
M02	Program run stop, spindle stop, coolant off
M03	Spindle on clockwise
M04	Spindle on counterclockwise
M05	Spindle stop
M06	Tool change
M08	Coolant on
M09	Coolant off
M30	Same as M02

- Word code X is for the X-coordinate dimension.
- Word code Y is for the Y-coordinate dimension.
- Word code Z is for the Z-coordinate dimension.

14.2.10 Manual Part Programming

As previously mentioned, the job of the NC programmer involves manually (unassisted by computer) preparing step-by-step detailed instructions on a program sheet. This task requires that he or she be familiar with the NC machine on which the part is to be processed. The programmer should know, for example, the location of the *setup point* with respect to the machine's *zero point*. Let us define these terms before we discuss manual part programming.

14.2.10.1 Zero Point

The zero point is the point where all coordinate axes meet. Therefore, at the zero point location, each of the X, Y, and Z values is equal to zero. Also, as is the case in analytical geometry, the coordinate dimensions of any point are measured from that origin or zero point.

Some NC machines have the zero point at a specific point (on the machine table) that cannot be changed. This is referred to as the *fixed zero point*. On the other hand, some MCUs allow the zero point to be established at any convenient spot selected by the programmer. This is referred to as the *floating zero point*. In this case, it is necessary to let the MCU know where the tool is located with reference to the selected zero point. This must be the first piece of information on the tape, directly after the units and dimensioning mode (i.e., inches or millimeters and absolute or incremental). It is usually achieved with the preparatory function code G92, followed by the coordinates of the location of the tool at the home position with reference to the selected program zero.

14.2.10.2 Setup Point

Consider the simple case of an NC machine with a fixed zero point. It is not difficult to see that the programmer must know where the workpiece is to be located on the machine table with reference to the zero point so that he or she can refer all dimensions to that zero point and thus be able to write the program. There is, therefore, a need for an actual point on the workpiece whose location in relation to the fixed zero point must be known beforehand. This point is called the *setup point*, and it can be the intersection of two straight edges of the workpiece or a machined hole in the workpiece. It is also obvious that the setup point can be a defined point on the fixture holding the workpiece. Figure 14.10 shows how the absolute coordinates of any point on the workpiece can be obtained if the coordinates of the setup point are known.

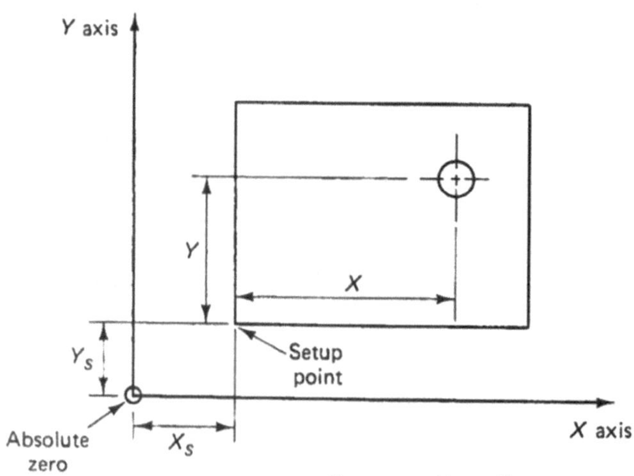

$$X_{\text{absolute}} = X_S + X$$
$$Y_{\text{absolute}} = Y_S + Y$$

FIGURE 14.10 Using the coordinates of the setup point to obtain absolute coordinates of any point on the workpiece.

14.2.10.3 Program Preparation

In NC programming, it is not enough for the program to be capable only of producing the required part. The goal of the programmer should also be to reduce the time spent by the workpiece on the table of the NC machine. The task of eliminating wasteful and unnecessary movements of the tools as well as reducing the setup time is not easy; it requires a lot of experience and skill. Here are some guidelines that the beginning programmer should follow:

1. Check the dimensions on the part blueprint to see whether they can be given in a way that makes programming easier.
2. Study the part blueprint and prepare process sheets indicating the details and sequence of operations required to complete the job. Also, check the number of setups needed and try to divide the number of tools to be used between the setups.
3. Determine the most suitable fixturing method by studying the part configuration and correlate basic dimensions between the blueprint and the machine layout.
4. Prepare a tool layout, including the sizes and lengths of all tools to be used, in order to facilitate the replacement of broken tools and to simplify setup.
5. Prepare the program sheet using the information gained in the preceding four planning steps.
6. Have a typist prepare the tape using a Flexowriter. The printout of the program for that tape should be checked to make sure that there are no errors. (This step is not valid for modern CNC machine tools.)
7. Test the tape on the machine while operating in a single-block mode to ensure against collisions and to eliminate wasteful motion. Run the entire tape, bypassing errors but keeping a record of them for subsequent corrections.
8. Inspect the obtained part to make sure that its attributes fall within an acceptable range.
9. Correct the tape.
10. Prepare a folder containing the blueprint of the part, the fixture design, the program sheet, and a copy of an actual punched tape (or a magnetic tape) for the part program. Always write down the NC machine tool that can be used (that is not needed nowadays).

14.2.11 Computerized Numerical Control (CNC)

In 1970, a new era for NC systems began with the emergence of CNC technology, which involves replacing the hardwired MCU of a conventional NC system with a microcomputer that, together with its software, accomplishes all the functions of a standard MCU. In addition to data decoding, feed rate control, buffering, and position loop control, a CNC system has many new features that are possible simply because a digital computer is used and improves the usefulness of the MCU. The following are the important features:

- *Ability to Store Programs*. There is no need for the frequent use of a tape reader. Once a tape is run, it can be stored in the memory of the computer. It can be recalled later, as many times as required, directly from computer memory and without the need to rerun the tape. A tape reader fitted on a CNC system thus requires less maintenance than one used with a conventional NC system. Also, the computer is much faster in obtaining information (by retrieval from its memory) than the tape reader. Therefore, the use of CNC systems results in an appreciable saving in time.
- *Editing*. It is very seldom that a satisfactory part program is obtained on the first attempt. Even experienced programmers need to make corrections, modifications, and improvements after running a program. The editing feature of CNC systems enables the programmer to make changes right on the factory floor. Also, all changes made go directly to

computer memory without any reference to or use of the original punched tape. Consequently, a data-input device is needed in addition to a means of editing the program. The program is edited on a cathode-ray tube (CRT) that is similar to but smaller than that of a computer. The *manual data input* (MDI) device provides a means of entering programs into computer memory without any need for a tape reader.

- *Ability to Produce Tapes*. After all necessary changes and improvements in the part program are made, a corrected punched (or magnetic) tape can easily be obtained by using an appropriate device that is plugged into the machine controller. (This device is not used nowadays.)
- *Expanded Tool Offsets*. In CNC systems, the tool offsets (i.e., deviations in the lengths of the different tools from a reference value) are stored in the memory of the computer. Large numbers of offsets can be stored, which is not the case in conventional NC systems.
- *Expanded Control of Machine-Sequence Operations*. CNC software usually handles machine operations for tool changes or control of the spindle or turret, thus making programming and operation of the machine much easier.
- *Digitizing*. Digitizing is usually provided as an option at extra cost. This feature allows a part program to be obtained directly from a model or an existing part. This is achieved by employing a stylus to scan the model while the CNC system monitors the movements and records the signals indicating the coordinate dimensions of points on the surface of the model. Again, a punched (or magnetic) tape can be obtained if needed. This feature eliminates the time-consuming and cumbersome operation of manual program preparation.
- *Circular Interpolation*. Very smooth arcs can be obtained because the computer of the CNC system has the capability to divide an arc into a large number of very small chord segments, calculate the coordinates of the endpoint of each segment, and establish a subprogram to generate the desired arc. This is achieved in programming by using either G02 or G03 as appropriate, together with the coordinates of the endpoint of the arc (say, X and Y) and the offsets of the center of the arc from its starting point along the X- and Y-axes, which are referred to as I and J, respectively. Following is a block of information that can drive a tool of a CNC lathe along the path required to machine an arc:
- N30 G02 X 3410 Z 1606 I 0400 K 0325
- *Parametric Programming*. Parametric programming provides flexibility to a program by allowing several different-sized components having similar shapes to be machined using the same program. This is easily accomplished by changing the value of a few parameters, or *program variables*. Parametric programming can also be employed to obtain very smooth curves or sculptured surfaces, provided that their mathematical equations are known.
- *Do Loops*. A do loop is useful in cutting the program short when it involves an operation that is to be repeated several times through incremental steps.
- *Roughing to a Defined Shape*. An example of roughing is using the code G73, which involves a cycle to rough out a bar to a defined shape when cutting along the Z-axis (turning). The shape is defined in a series of blocks called up in the G73 block, together with a parameter that defines the incremental depth of the cut.
- *Subroutines*. Subroutines allow the programmer to program, store, and repeat a pattern at different locations on the workpiece. Examples of applications include patterns of holes (bolt-hole circles), series of standard grooves on a shaft, and canned cycles that are created by the user.
- *Diagnostic Capability*. Diagnostic capability refers to the ability to detect faults when the CNC system goes down. This feature is a major advantage of CNC systems; it is due mainly to the diagnostic capabilities inherent in computer systems in general. In most cases, error messages are displayed to the operator. Also, a special diagnostic tape may be supplied with the CNC system, or the system may be connected through the telephone to the computer of the manufacturer's service department.

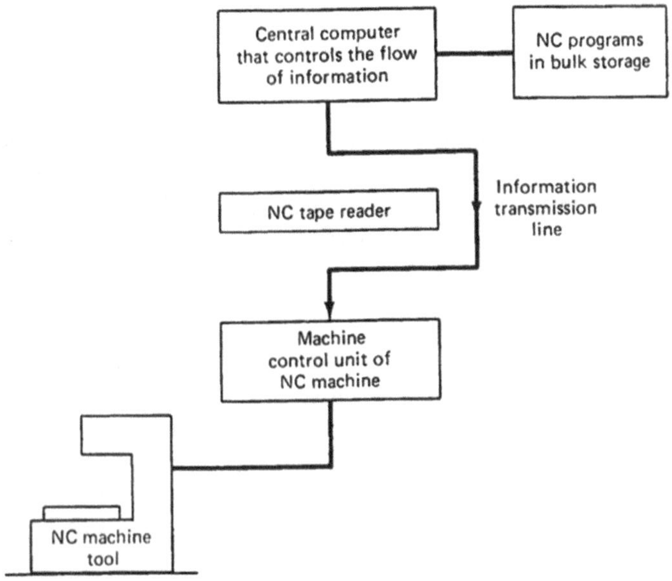

FIGURE 14.11 Structure and operation of a DNC system.

14.2.12 Direct Numerical Control (DNC)

As is the case with CNC, DNC is a hybrid of both NC and computer technologies. It involves bypassing the weakest and least reliable link in the system, the tape reader, by supplying part program data directly from the bulk storage device to the controller of the machine tool through telecommunication lines. Figure 14.11 shows how a DNC system functions by employing an additional hardware module as a connection between the mainframe computer and the machine tool controller. This piece of hardware is referred to as a *behind-the-tape-reader* (BTR). It is actually an additional source of data that does not depend on or make any use of the tape reader in its functioning. The tape reader is used, however, as a practical backup when the DNC system breaks down (actually, it has totally disappeared nowadays).

The main elements of a DNC system include a large, remote computer, a bulk storage device, telecommunication lines, and a number (as many as 100) of NC machine tools. The operation of a DNC system is based upon the continuous flow of information from the bulk storage device to the NC machine tools and vice versa. This takes place on a time-sharing basis and in real time. As a result, information is transmitted to the NC machine tool almost instantaneously when a signal from the machine indicating a need for instructions reaches the computer. This DNC configuration forms the backbone of today's flexible manufacturing systems.

DNC has several advantages. For instance, a 10–20 percent increase in productivity has been reported when DNC is employed. This is due to the monitoring abilities of the system, together with the increased machine runtime. In addition to the elimination of the problems of the tape reader and the cost of tapes, DNC offers a step-by-step approach for establishing an integrated system that starts with a few NC machine tools and expands as required.

14.3 PROGRAM PREPARATION AND CODING

14.3.1 Programming for CNC Milling Machines

As previously mentioned, a CNC milling machine is a three-axis machine, namely, X, Y, and Z. In order to write an executable program, you should first prepare an accurate process sheet together with a list of the tools that would be used and their characteristics. Next, you convert the steps included in the process sheet into corresponding coded blocks of information that form a CNC program. A program is prepared to be related to or based on a point, that is, the datum point (or setup point). The latter has to be known to the CNC machine in relation to the absolute zero of the machine. The machine operator carries out this task during the setup process. It involves touching the workpiece with an edge finder (or a tool) that is mounted on the quill in the X, Y, and Z directions and then letting the machine recognize those coordinates. Let us first become familiar with the various G codes and M codes before preparing a simple program. Table 14.1 shows the commonly used G codes according to the ISO (International Standardization Organization) and the function of each, while Table 14.2 shows the M codes.

14.3.1.1 Introductory Programming

Now, let us learn how to create a simple program, and the best way to go is evidently to have an example. We will engrave the letter "D," with a height of 2.0 inches, radius of curve 1.0 inch, and depth of cut 0.1 inch in an aluminum block 3 × 4 inches. Figure 14.12 indicates a sketch of the tool path with some points on it identified by giving each a reference number. Notice that we will explain every block in plain English under "Remarks," in order to be able to review and possibly correct our program later when we "walk through it." The program is given in Table 14.3.

Let us now have some tips to reduce the typing effort, without affecting the logic or the information provided in each block. Try to make use of these tips when writing a program. They are as follows:

- G00 is "modal" (i.e., it will be valid in the subsequent blocks, even if you do not mention it). In other words, you do not have to repeat it.
- G01 and the feed rate value that follows "F" are also modal.
- If either X, Y, or Z value in the block you are writing is not different from that in the preceding block, you do not have to repeat it.
- The symbol (*) denotes EOB.
- Notice that the blocks are numbered 5, 10 in steps of five, in order to enable adding and inserting blocks in between.

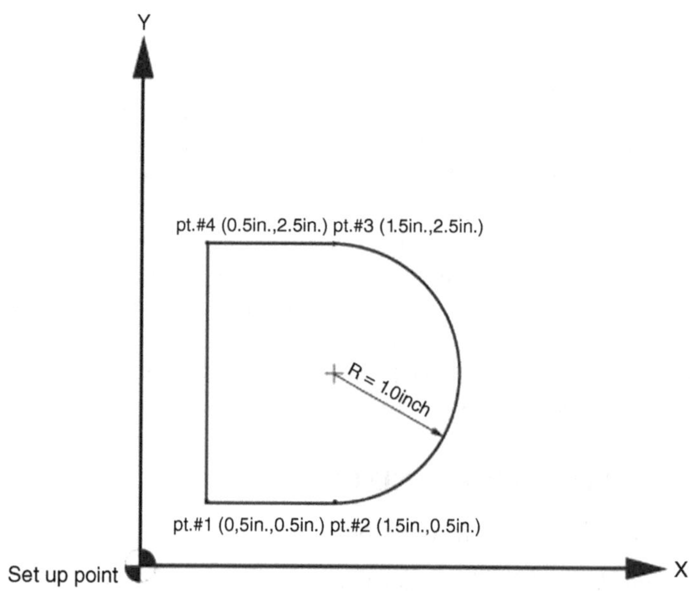

FIGURE 14.12 A sketch of the tool path for engraving the letter "D."

TABLE 14.3
A Program to Engrave the Letter "D" in an Aluminum Block

	Coded Statement	Remarks
	EOR% # 7 G70	End of record, program 7, dimensions in inches.
N05	To G17 G00 G40 G90 Z + 2.0*	No tool assigned, XY plane, rapid traverse, no tool compensation, absolute programming, retract tool 2.0 inches above the workpiece.
N10	X-1.5 Y-1.5*	Move the tool quickly to the home position, 1.5 inches to the left and 1.5 inches below the lower LHS corner of the workpiece.
N14	G99 T1 L + 0 R + 0.0625 M00*	Define tool 1, tool offset 0, tool radius 0.0625, stop the spindle.
N20	T1 S 2000,00*	Call tool 1, select a constant cutting speed of 200 feet/minute.
N25	X + 0.5 Y + 0.5*	Go quickly above the starting point 1 with X + 0.5 and Y + 0.5.
N30	Z + 0.2 M03*	The tool goes down very fast to only 0.2 inches above point 1 on the workpiece. Start the spindle counterclockwise.
N35	G01 Z − 0.2 F5*	Penetrate into the workpiece to a depth of 0.2 inches, at a feed rate of 0.5 inches/minute. The tool is now at point 1 on the tool path.
N40	X + 1.5*	Cut until the tool reaches point 2.
N45	G03 I + 1.5 J + 1.5 X + 1.5 Y + 2.5*	The tool goes cutting until point 3. X and Y are the coordinates of point 3, and I and J are the coordinates of the center of the arc from 2 to 3.
N50	G01 X + 0.5*	The tool goes cutting until point 4.
N55	G01 Y + 0.5*	The tool goes cutting until point 1, thus finishing the engraving job.
N60	G00 Z + 2.0*	Retract the tool 2.0 inches above the surface of the workpiece.
N65	X − 1.5 Y − 1.5 M02*	The tool goes quickly to the home position and program end.

14.3.1.2 Canned Cycles

A block containing a canned cycle would provide the same information as that included in several blocks, and it is evidently, therefore, used as a substitute for those blocks. Accordingly, the use of a canned cycle command would cut the length of the program short without jeopardizing the tasks to be performed. There are two types of canned cycles: machining cycles and cycles for coordinate transformations. The machining canned cycles include the pecking cycle G83, the tapping cycle G84, the slot milling cycle G74, and the pocket milling cycles (G77 and G78 round pocketing, and G75 and G76 rectangular pocketing). On the other hand, the coordinate transformation cycles include the linear datum shift cycle G54 and the angular rotation of the coordinate axes cycle G73. We will provide some examples of the canned cycle commands later in the sections to follow in this chapter.

14.3.1.3 Do Loops and Subroutines

You are able to repeat any task (involving a sequence of command blocks) a finite number of times by initiating a "do loop" and then indicating the number of times you want it to be repeated. The do loop is initiated within a program by the command G98 followed by the label address "L" and a number you assign to identify that particular do loop. The sequence of command blocks then follows, and finally, the number of repetitions is indicated following the label "L." Remember that the sequence of blocks would be executed a number of times equal to the indicated number of repetitions plus one, in order to account for the first time it is executed before repetitions.

Do loops can be nested (i.e., a do loop can be inserted within another do loop). Now, the best way to understand do loops is to have an example. We will write a "do loop" as a part of a program to drill a set of four equally spaced holes as shown in Figure 14.13 (for now, the lower row only). The diameter of each hole is 0.5 inches, the depth is 1.0 inch, and the distance between each two successive centers is 1.0 inch. The coordinates of the center of the first hole are X = +1 and Y = + 1. The example do loop is given in Table 14.4.

We can drill the second row of holes, 1.0 inch above the first one, by nesting another "do loop" as shown in Table 14.5. You can see that loop L1 was used to create the columns, while L2 was used

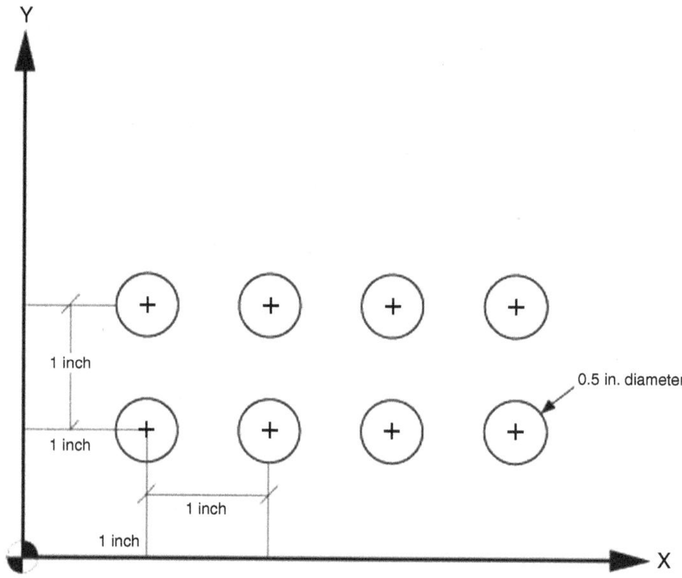

FIGURE 14.13 A set of holes to be drilled using do loops.

TABLE 14.4
Application of the Do Loop

N25	G00	X + 1	Y + 1	M03*	Rapid traverse to above the first point of X + 1 and Y + 1 then turn the spindle on.
N30	G98	L1	G90*		Initiate a do loop and label it as number 1, absolute programming.
N35	G83	P01–0.2	P02–1.0	P03–0.5 P04 0 P05 0.5*	This is a pecking cycle, P01 clearance above the workpiece, P02 total drilling depth, P03 pecking depth, P04 dwell time, and P05 feed rate.
N40	G79				This is a cycle call, and it always comes after any canned cycle to execute it. The first hole is drilled and tool up again.
N45	G91	G00	X + 1*		While Y is constant, move the tool an incremental step of +1 inch, just above the location of the next hole.
N50	L1, 3*				Repeat the label L1 three times additionally, i.e., blocks N30, N35, N40, and N45.
N55	G00	G90	X – 1.5	Y – 1.5 M02*	Go to the home position and end the program.

TABLE 14.5
Application of Nested Do Loops

N27	G98 L2	G90*				Initiate a do loop and label it 2, absolute dimensioning.
N51	G00	G90	X + 1.0	G91	Y + 1.0*	Rapid traverse to X + 1 absolute and increment Y by +1.0.
N52	L2, 1*					Repeat loop L2 once.

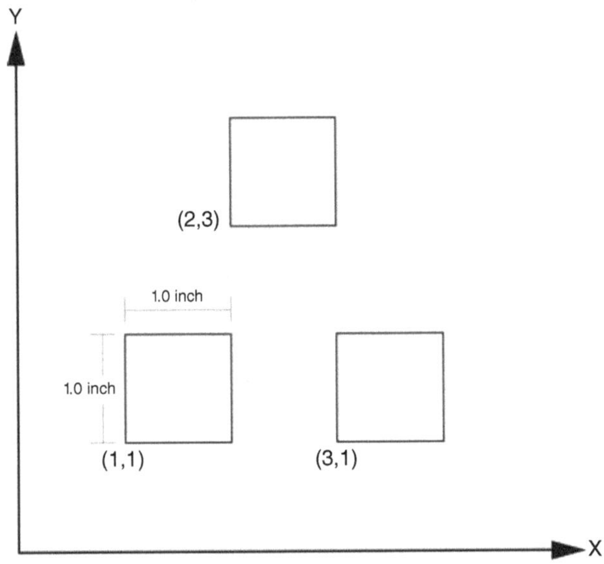

FIGURE 14.14 A workpiece with a square to be engraved three times using a subroutine.

to create the rows. You can apparently apply the same approach to create say 10 columns and 7 rows, using a small number of command blocks in a short program.

On the other hand, a subroutine (also called subprogram) is also initiated within a program by the command G98 followed by the label address "L" and a number you assign to identify that particular subroutine. The sequence of command blocks then follows, and finally, the subroutine is ended with the command G98 and the label L0. Remember that the subroutine will neither be executed nor can it be repeated. You are actually keeping that group or sequence of command blocks in the memory unused until you call the subroutine by its label and number, and then it would be executed just once. When it is required to machine a feature a few (or several) times at different locations, a subroutine would be very helpful. You just write it once and then call it at each location as needed. Let us have a simple example for clarification. Figure 14.14 shows a block of aluminum with a square (each side 1.0 inch) engraved three times at points (1,1), (3,1), and (2,3). The solution is given in Table 14.6.

14.3.1.4 Parametric Programming

In this section, we will cover the basics of the most advanced feature of CNC programming that is parametric programming (also known as programming with macros). It is a very powerful tool because it allows for designing a representative model that can be described by several related variables, and when a specific size or form of the model is desired, the appropriate numeric values are then put into the program. It is evident that it is very easy to edit this type of program to produce multiple sizes of the same part. Another advantage of parametric programming is the ability to use

TABLE 14.6
Application of the Subroutine

N45	G98 L3*			Create a subroutine and label it 3.
N50	G00 G90	Z + 0.2	M03*	The tool goes very quickly down to only 0.2 inches above the workpiece. Absolute dimensioning.
N55	G01	Z − 0.2	F5*	Penetrate into the workpiece to a depth of 0.2 inches at a feed rate of 0.5 inches/minute.
N60	G91	X + 1*		The tool goes an increment of 1 inch in the X direction. Increment of 1.0 inch in the X direction.
N65		Y + 1*		The tool goes an increment of 1.0 inch in the Y direction.
N70		X − 1*		The Tool goes an increment of −1.0 inch in the X direction.
N75		Y − 1*		The tool goes an increment of −1 inch in the Y direction.
N80	G90 G00	Z + 0.2*		Go above the workpiece by 0.2 inches. Absolute dimensioning.
N85	G98 L0*			End the subroutine that has the label 3.
N120	G00	X + 1	Y + 1*	Go above point (1,1).
N125	L3*			Call subroutine labeled L3.
N130	G00	X + 3	Y + 1*	Go above point (3,1).
N135	L3*			Call subroutine labeled L3.
N140	G00	X + 2	Y + 3*	Go above point (2,3).
N145	L3*			Call subroutine labeled L3.
N150	G00 G90	X − 1.5	X − 1.5 M02	Go to home position and end program.

conditional statements that are used in loop programming. This allows for the number of times a particular loop is executed to be variable based on a check of the value of a certain parameter or parameters, instead of a fixed number of times. A further valuable aspect of parameterization is the ability to increment and decrement a variable, which is useful in creating user-defined canned cycles. In other words, parametric programming adds computer-related features such as arithmetic, variables, and logic to the known CNC-related features. As a result of the above-mentioned features and advantages, parametric programming found some important industrial applications, which can be indicated as follows:

- Can be used to produce families of parts, with one program for all members with minimum changes every time a member is to be made
- Can be used to create user-tailored canned cycles
- Can be used to produce highly complex geometric shapes such as spheres, pyramids, ellipses, and splines

Let us now learn about the format of the command blocks in a parametric program. Unfortunately, the syntax of a parametric program differs for different manufacturers of CNC machine controllers. Nonetheless, they all follow the same logic, and if you learn one, you would be able to understand and learn any other one easily. We are going to explain the syntax of the programming language of the Heidenhain controller, because of its ease and similarity with the assembler computer language. After the block sequence number, the operation command is given, which can be a parameter definition, arithmetic computations, trigonometric or algebraic functions, or logical operations. Next comes the parameter under consideration, and since there are several parameters in a program of this kind, each is designated by the letter Q followed by two numerals ranging between 1 and 99. You should always bear in mind that a parameter is a temporary storage location for a variable. The initial

TABLE 14.7
Different Operation Commands, Block Format, and Syntax

Command	Description	Block Format	Explanation
D00	Definition of a parameter	Nxx D 00 Qxx P01 xxx*	P01 is the parameter address and what follows can be a numeric value or another parameter provided that it has been defined before.
D01	Addition	Nxx D01 Q3 P01 Q1 P02 Q2*	Add Q1 to Q2, store the outcome in Q3.
D02	Subtraction	Nxx D02 Q3 P01 Q1 P02 Q02*	Subtract Q1 from Q2, store the outcome in Q3.
D03	Multiplication	Nxx D03 Q3 P01 Q1 P02 Q2*	Multiply Q1 by Q2, store the outcome in Q3.
D04	Division	Nxx D04 Q3 P01 Q1 P02 Q2*	Divide Q1 by Q2, store the outcome in Q3.
D05	Square root	Nxx D05 Q3 P01 Q1*	Take the square root of Q1 and put it under Q3*.
D06	Sine	Nxx D06 Q2 P01 Q1*	Take the sine of Q1 and put it under Q2.
D07	Cosine	Nxx D07 Q5 P01 Q3*	Take the cosine of Q3 and put it under Q5.
D09	If equal jump	Nxx D09 P01 Q7 P02 Q8 P03 1*	If the value of Q8 becomes equal to Q7, go to label 1.
D12	If less than jump	Nxx D12 P01 Q7 P02 Q9 P03 3*	If the value of Q7 is less than Q9, then go to label 3.

value of that variable then comes after P01, P02, or P03 (depending on the operation command) and can actually be a numeric value or another previously defined parameter. There are several operation commands as mentioned before, and each of them has a certain format that must be followed for proper execution. Also, each operation command must be given in a separate block. Table 14.7 shows the different operation commands, block format, and syntax for the Heidenhain programming language.

The best way to understand and be able to write a parametric program is to see an example as follows:

EXAMPLE 14.1

Write a program to cut an elliptical hole with a major axis of 3.25 inches and a minor axis of 2.25 inches in a steel block AISI 1020 that has a thickness of 0.25 inch. The block is 9 × 8 inches.

SOLUTION

The equation of an ellipse would be a general second-degree equation, which is rather complicated if we take the datum point as the LHS corner of the workpiece as we always do. The equation will become far simpler if we take the point of intersection of the major and minor axes as our datum point. While we can handle it that way, it would be even better if we express each of X and Y in terms of a third variable and change that variable to get the corresponding values of X and Y. In other words, we use the parametric equations of an ellipse with θ being the independent variable and both X and Y the dependent variables as follows:

$$X = a\cos\theta, \text{and } Y = b\sin\theta$$

where a = major axis and b = minor axis.

We have, however, to account for the tool radius when programming by subtracting it from each of the major and minor axes. Assuming that we use a 0.5-inch cutter, then a becomes 3.0 inches and b will be 2 inches. We should now develop a strategy for machining the hole, in the form of a flowchart of a pseudo-code, as shown in Figure 14.15.

Now, it is fairly easy to code this logic and cut the hole. The parametric program is given in Table 14.8.

Let us now have another example of producing half a sphere using the CNC vertical milling machine and employing parametric programming.

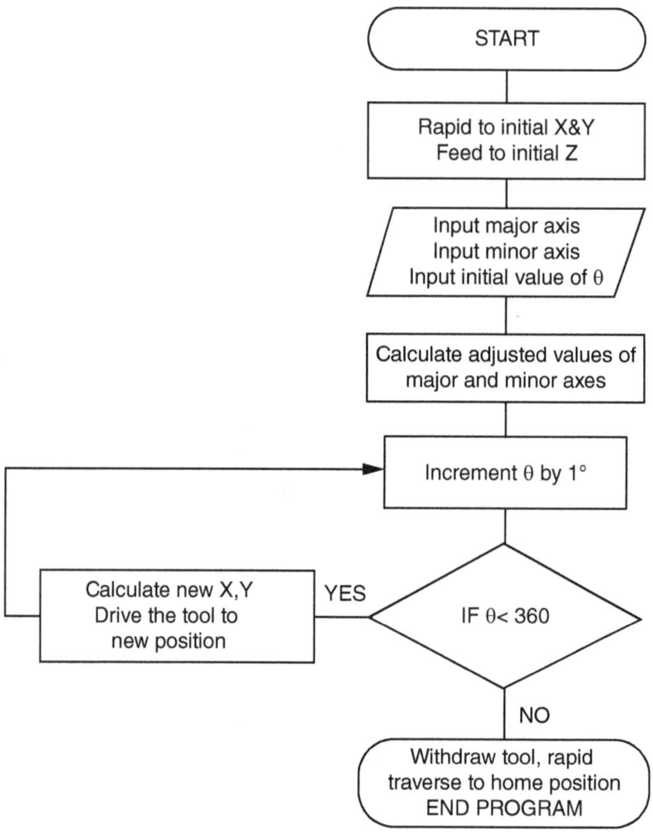

FIGURE 14.15 A flowchart of a program for cutting an ellipse.

TABLE 14.8
A Parametric Program to Cut an Ellipse

	% 77 G70*	Program 77, Dimensions in Inches
N05	T0 G17 G00 G40 G90 Z + 2*	No tool, XY plane, rapid traverse, no tool compensation, absolute dimensions.
N10	X − 2 Y − 2 M05*	Rapid traverse to home position, spindle stop.
N14	G99 T1 L + 0 R + 0.25 M00*	Define tool 1, L is the reference, R = 0.25, spindle stop.
N20	T1 G17 S2000*	Call tool 1, XY plane, and cutting speed 200 ft/minute.
N25	D00 Q1 P01 + 3*	Define parameter Q1 = 3 inches = major axis of the ellipse.
N30	D00 Q2 P01 + 2*	Define parameter Q2 = 2 inches = b = minor axis of the ellipse.
N35	D00 Q3 P01 + 0*	Define parameter Q3 = θ.
N40	G00 X + 7.5 Y + 4*	Rapid traverse to above the end of the major axis.
N45	Z + 0.2 M03*	Tool to 0.2 inches above the surface, start the spindle.
N50	G01 Z −0.4 F5*	The tool penetrates at 0.5 inches/minute into the workpiece.
N55	G54 X + 4.5 Y + 4*	Shift the datum to the center of the ellipse.
N60	G98 L1 G90*	Initiate the macro loop 1, absolute dimensions.
N65	D01 Q3 P01 Q3 P02 + 1*	Put θ = θ + 1, i.e., increment θ by 1 degree.
N70	D07 Q4 P01 Q3*	Parameter Q4 = cos θ (new).
N75	D03 Q5 P01 Q4 P02 Q1*	Parameter Q5 = a cos θ (new) = X (new).
N80	D06 Q6 P01 Q3*	Parameter Q6 = sin θ (new).
N85	D03 Q7 P01 Q6 P02 Q2*	Parameter Q7 = b sin θ (new) = Y (new).
N90	G01 X + Q5 Y + Q7 F5*	Cut to the next position on the ellipse, new X, new Y.
N95	D12 P01 Q3 P02 + 360 P03 1*	If the new value of θ is less than 360, go to the L1 loop.
N100	G00 G90 Z + 2*	When θ = 360, the ellipse is cut, the tool goes up 2 inches above the surface of the workpiece.
N105	G54 X − 4.5 Y − 4*	Shift datum back to its original position.
N110	X − 2 Y − 2 M02*	Retreat to home position, end the program.

EXAMPLE 14.2

Write a parametric program to cut half a sphere having a radius of 1.5 inches in a block of steel AISI 1020 with dimensions 4 × 4 × 2 inches.

SOLUTION

Figure 14.16 shows a cross section through the block indicating the final shape of the half sphere. Our strategy will be to move the datum to the center of the upper face of the block, then increment θ to increase both the depth of penetration as well as the radius of the circle corresponding to it. Remember two things: we have to account for the tool radius, and we have to make the tool produce a full circle at each level because we are producing a 3D surface of revolution. From Figure 14.16, we can obtain the following relations:

Depth of cut $= -Z = -(R - R\cos\theta) = R(\cos\theta - 1)$
Radius of circle "r" $= R\sin\theta$.
We will use a 0.5-inch ball-end milling cutter.
The program is given in Table 14.9.

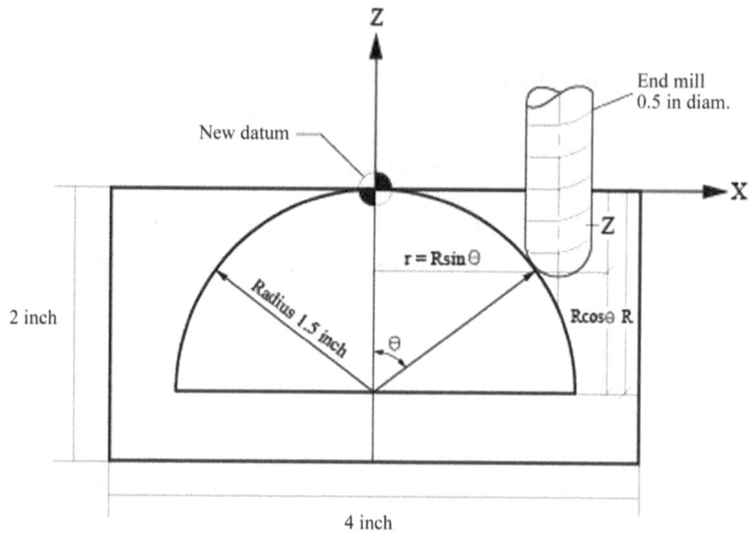

FIGURE 14.16 A cross section of the workpiece showing the final shape of the half sphere.

TABLE 14.9
A Parametric Program to Cut Half a Sphere

	% 97 G70*	**Program 97, Dimensions in Inches**
N05	T0 G17 G00 G40 G90 Z + 0.2*	Rapid traverse to 0.2 inches above the surface.
N10	X − 1.5 Y − 1.5 M05*	Rapid traverse to home position spindle off.
N14	G99 T1 L + 0 R.25 M00*	Define tool 1, standard, radius 0.25 inch, spindle off.
N20	T1 G17 S2000*	Call tool 1, define cutting speed.
N25	G00 X + 2 Y + 2*	The tool is above the center of the upper face.
N30	G54 X + 2 Y + 2*	Shift datum to the center of the face.
N35	G01 Z + 0 F5*	The tool goes at a feed rate and touches the surface at the center.
N40	D00 Q1 P01 + 0*	Define parameter Q1 = 0 = Initial θ.
N45	G98 L1 G90*	Initiate macro loop 1.
N50	D01 Q1 P01 Q1 P02 + 1*	Increment θ by 1 degree.
N55	D06 Q2 P01 Q1*	Define parameter Q2 = sin θ.
N60	D03 Q3 P01 Q2 P02 + 1.5*	Set Q3 = R sin θ = 1.5 sin θ.
N65	D07 Q4 P01 Q1*	Define parameter Q4 = cos θ.
N70	D03 Q5 P01 Q4 P02 + 1.5*	Set parameter Q5 = R cos θ = 1.5 cos θ.
N75	D02 Q6 P01 Q5 P02 + 1.5*	Set parameter Q6 = Q5 − R = R cos θ − 1,5 = −Z.
N80	D01 Q7 P01 Q3 P02 0.25*	Set Q7 = Q3 + tool radius = Q3 + 0.25.
N85	G01 X Q7 F5*	Move a horizontal increment.
N90	Z Q6*	Penetrate a vertical increment.
N95	G11 R Q7 H 360 F10*	Cut a full circle having this radius.
N100	D12 P01 Q1 P02 90 P03 1*	If theta is less than 90, go to loop 1.
N105	G00 Z + 0.2*	Tool 0.2 inches above the surface of the workpiece.
N110	G54 X − 2 Y − 2*	Return to original datum.
N114	X − 1.5 Y − 1.5 M02*	Rapid traverse to home position, end program.

14.3.2 Programming for CNC Lathe

Based on the right-hand rule and the EIA standards previously discussed, a CNC lathe has only two control axes, namely, the X-axis and the Z-axis. The latter is horizontal and coincides with the axis of the lathe, pointing out from the quill. There is no need for the Y-axis since the tool will always be in the XZ plane. The location of the tool post or turret affects the positive direction of the X-axis. If the tool post is to the left of the Z-axis, then the X-axis will be pointing toward it and away from the operator. On the contrary, if the tool post or turret is to the right relative to the positive direction of the Z-axis, then the positive direction of X would be toward the operator as shown in Figure 14.17. This would evidently affect the direction of the Y-axis, making it either point upward or downward, which would in turn affect the direction of the rotary axis (i.e., G02 and G03), as shown in Figure 14.17.

There are two distinct features of CNC lathe programming. The first one is the use of the diameter value rather than the radius in indicating the X-coordinate, which indeed facilitates the actual diameters of parts being machined for the purpose of quality assurance. The other feature is the "floating zero," where the programmer chooses the location of the zero point to facilitate writing the program. In other words, the programmer locates the floating zero at a convenient point, usually the center of the end face of the workpiece. A special code (G92) is then entered followed by $Z = 0$. Alternatively, the center of the other end face can be used, provided that the length of the piece L is known, and in that case G92 would be followed by $Z = L$. This is illustrated in Figure 14.18. You have to bear in

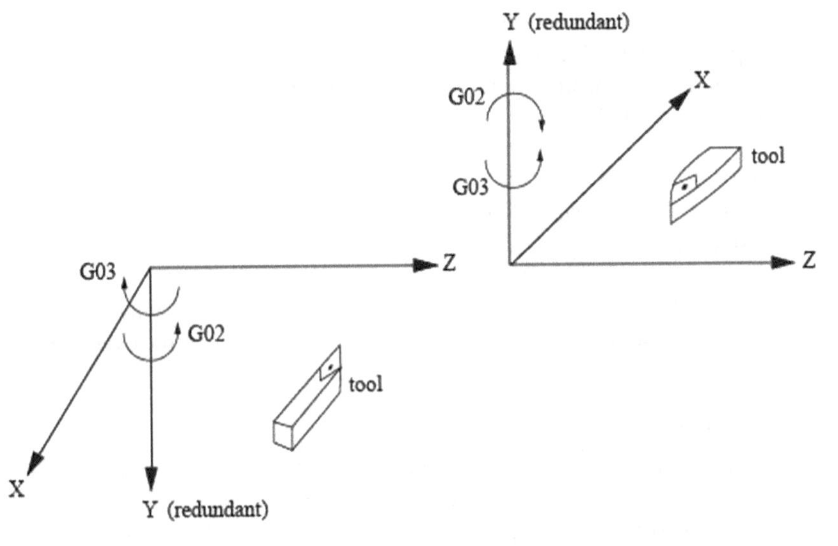

FIGURE 14.17 Direction of rotary axes in a CNC lathe.

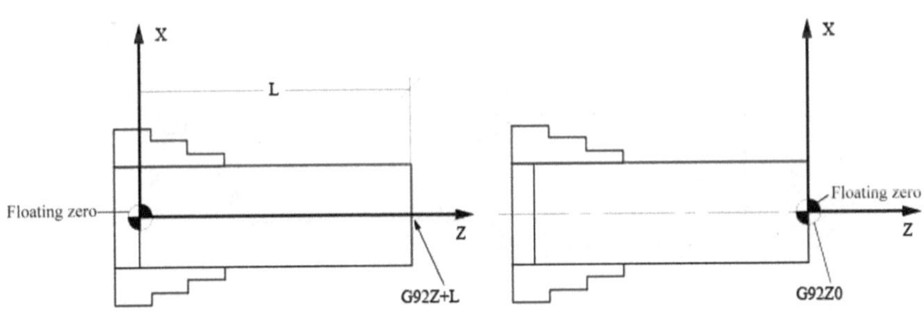

FIGURE 14.18 Selection of the floating zero in a CNC lathe.

TABLE 14.10
The G Codes and M Codes That Are Used with a CNC Lathe Only

Code	Description
G20	Roughing cycle, single
G21	Threading cycle, single
G24	Facing cycle, single
G28	Approach reference point
G33	Thread cutting
G72	Finishing cycle
G73	Longitudinal turning cycle
G74	Facing cycle
G75	Pattern repetition
G96	Constant cutting speed, i.e., rpm increases automatically when the stock diameter decreases
M25	Release clamping device
M26	Close clamping device
M30	Main program end

mind, however, that all the values of the Z-coordinate in the first case would accordingly be negative. Meanwhile, the G codes and M codes are generally the same for both milling and turning, except for some codes that apply to the CNC lathe only. Sometimes, however, some controllers use G0 and G1 instead of G00 and G01, so bear that in mind. Table 14.10 indicates the G codes and M codes that are used only with the CNC lathe.

14.3.2.1 Introductory Programming

Let us have a simple example to show CNC lathe programming. We have a stock of 0.5-inch AISI 1040 cold-rolled steel, and we want to reduce the diameter to only 0.3 inches and make the end as a slope as shown in Figure 14.19. First of all, it is better to designate a letter or a number on the tool path and then prepare a table indicating the coordinates of each of those points.

Since this is a simple case, there is no need for the table. Following is the example program given in Table 14.11.

14.3.2.2 Fixed Cycles

As you can see from Table 14.11, a fairly long program was needed just to reduce the stock diameter from 0.5 inches to 0.3 inches. Accordingly, as we used the canned cycles when programming the CNC milling machines, we can use equivalent cycles specifically for the CNC lathe, which are called *fixed cycles*, in order to make the programs simpler and shorter. The various fixed cycles are shown in Table 14.10, and we can have an example in order to understand them better, as follows:

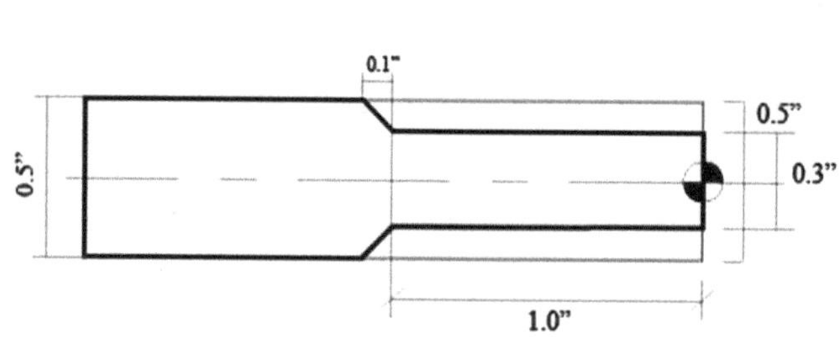

FIGURE 14.19 A workpiece for introductory CNC lathe programming.

TABLE 14.11
A Simple Example of a CNC Lathe Program

N05	(Stock 3X 0.5 Aluminum) G92 set floating zero as shown.	
N10	G40 G70 G90*	No tool compensation, dimensions in inches, absolute mode.
N14	G95 G96 G98*	Feed inch/rev, constant cutting speed, return to start plane.
N20	G0 Z 2.0*	Safe move to Z 2.0 inches away from the end face, rapidly.
N25	T0202 S1000 M4*	Right-hand tool, speed call out, rpm spindle on counterclockwise.
N30	G0 X 0.4 Z 0.1*	Rapid traverse from the safe point to the start point A.
N35	G1 Z – 1.0 F0.004*	Go and do the first cut.
N40	G0 X 0.5*	Rapid traverse away from the workpiece.
N45	Z 0.1*	Rapid ready for the second cut.
N50	X 0.3*	The start point for the second cut.
N55	G1 Z – 1.0 F0.004*	Finish the second cut.
N60	X 0.5 Z – 1.1*	Cut the slope.
N65	G0 Z 2.0 M30*	Rapid traverse to the safe point and end program.

EXAMPLE 14.3

Figure 14.20 indicates the final shape of a part to be cut from aluminum stock. Employ a fixed cycle to produce a short program for machining this part.

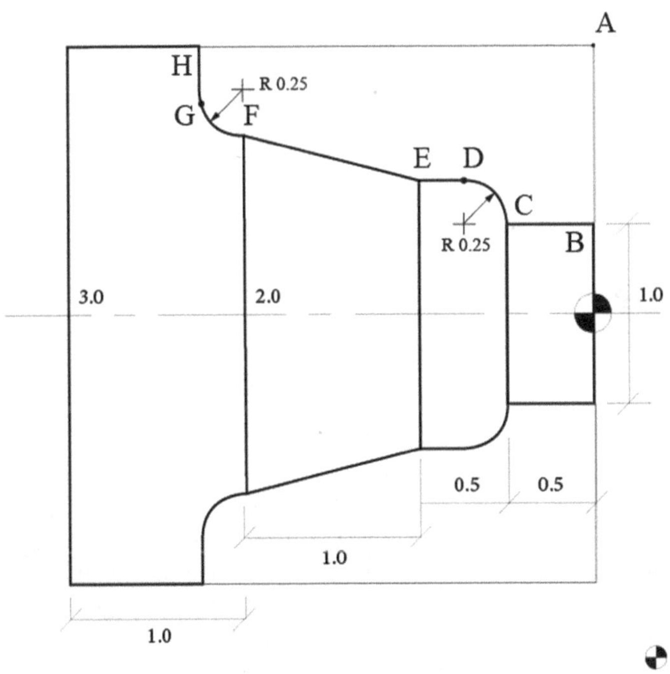

FIGURE 14.20 The final shape of a part to be cut from aluminum stock by employing fixed cycles.

Solution

The program with an explanation of each block is shown in Table 14.12.

TABLE 14.12
CNC Program Utilizing Fixed Cycles

N05	(Stock 5X 3 Aluminum) G92 set floating zero as shown.	
N10	G40 G70 G90*	No tool compensation, dimensions in inches, absolute mode.
N14	G95 G96 G98*	Feed inch/rev, constant cutting speed, return to the start plane.
N20	G0 Z 2*	Safe move away from the end face.
N25	T0202 S1000 M4*	Right-hand tool, speed call out 1000 rpm, spindle on counterclockwise.
N30	G00 X 3.0 Z0.1	Start point of a cycle.
N35	G73 U.04 R.02	Start a roughing cycle, U depth of cut, retract distance.
N40	G73 P45 Q75 U.01 W.005 F.004	Continue cycle, first block of contour 45, last block of contour 75, feed .004 inch/rev.
N45	G0 G42 X 1.0	Tool compensation, X 1, but 0.1 inches away from the end face.
N50	G1 Z – 0.5*	To point C on the contour.
N55	G3 X 1.5 Z – 0.75 R0.25*	To point D.
N60	G1 Z – 1.0	To point E.
N65	X 2.0 Z – 2.0	To point F.
N70	G2 X 2.5 Z – 2.25 R0.25	To point G.
N75	G1 X 3.0	To point H.
N80	G0 X 3.3 Z.2 T0100	To the tool change, get a finishing tool.
N85	S1400 F.002	Speed 1400 rpm, feed .002 inches/rev.
N90	G72 P45 Q75	Finishing cycle
N95	G0 X 3 Z 2	Rapid traverse to X 3, Z 2 ready for the next piece to be cut.
N100	M30	End program.

14.4 COMPUTER-AIDED PART PROGRAMMING

There are generally two methods of regenerating an NC part program: by *manual programming* or with *computer assistance*. In manual part programming, the programmer prepares a set of detailed instructions by which the desired sequence of operations is performed. He or she has to calculate manually the coordinates of all the various points along the required tool path before writing the program directly in a coded form and arranged in a format that can be understood by the MCU of the NC system. In other words, the programmer writes the part program in a *machine language* that can be directly read and processed by the NC system. Consequently, unless the part configuration is relatively simple and only a few different types of NC machines are employed, the task of manual part programming becomes time-consuming and cumbersome. When programs are to be prepared for complex parts requiring contouring or having complicated patterns of holes, it is almost impossible, or at least impractical, to do all the required geometric and trigonometric calculations by hand, and use must be made of that magic data-processing tool, the computer. Computer-aided part programming becomes a necessity when programming three-, four-, or five-axis NC machines that are used for generating sculptured surfaces or when complicated contouring is required. Also, when the plant includes several different types of NC machines, each having its own programming codes and format, computer-aided part programming is the right solution. As can be expected from this discussion, computer-aided part programming is much easier and faster than manual part programming.

All a programmer has to do is define the desired operations using the English-like words of the NC computer language that he or she is using. Once the part program is loaded into the computer, the computer takes care of all calculations and converts the input statements into a machine language compatible with the particular NC machine to be used. Typing errors can be corrected by means of the editing routines before a program is compiled. Programming errors are detected when compiling a program, thanks to the diagnostic error messages given by the computer. Nevertheless, the program may still contain some undetected errors that, if left uncorrected, will result in a part configuration different from the desired one. Therefore, the graphics capabilities of the NC computer system are always used to obtain a plot of the part geometry and the tool path in order to verify that the program will indeed produce the desired part. One of the most important advantages of computer-aided part programming is program verification because it results in fewer scrapped parts and saves most of the time used in debugging the program at the machine tool. This method of programming also has the advantages of simplicity, reducing the time needed for programming, and accuracy due to the elimination of any cumulative errors in calculations.

14.4.1 Internal Computer Operation

NC computer software systems can be divided into two distinct parts: the *general processor* and the *postprocessor*. The postprocessor is, in turn, composed of the postprocessor control system and the postprocessor machine segment. Let us now see what happens when a part program is loaded into the computer.

As mentioned, a part program consists of English-like statements that are used to define the desired operations. Therefore, the first step in processing a part program involves translating the input file (written in a general-purpose programming language such as APT and consisting of English-like words) into an equivalent computer machine-language program. This process, referred to as *compilation*, is necessary because the computer can understand only its own machine language. During the compilation process, if any syntax error is detected, further processing of the program is promptly stopped. Once the compilation is complete, the processor handles all geometry and motion commands in the machine-language version of the program and carries out all the necessary calculations in the arithmetic-logic unit (ALU) of the central processing unit (CPU) of the computer. The output of these calculations adequately defines the tool path or cutter location (CL). The tool path indicates the center of the cutting tool and not the boundaries of the workpiece. Therefore, some of the calculations are concerned with offsetting the tool path from the desired part outline by a distance equal to the radius of the cutter (in milling operations).

Different NC systems use different control-tape coding and formats. Also, NC machine tools have different characteristics, depending upon the builder. Therefore, the output from the processor (which is in the form of CL data) must be reprocessed so that the precise output codes and format required for the given NC system and machine tool can be obtained. This is referred to as *postprocessing*, and it is carried out by a subprogram called the *postprocessor*. The output from the postprocessor is a tape image that can be converted to a punched (or magnetic) tape that can be employed to operate the given machine tool.

The postprocessor is a very specialized program that can reprocess data (the output from the processor) to operate only a particular combination of machine tools and NC controller. Because there are a large number of such combinations in industry, the postprocessor is divided into two distinct parts in order to facilitate the task of tailoring a complete postprocessor suitable for a given

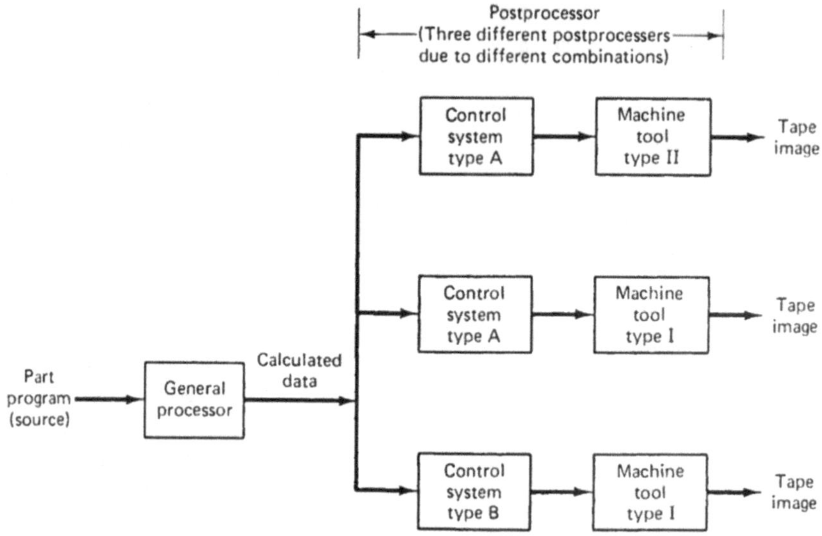

FIGURE 14.21 A computer-aided programming system involving various combinations of postprocessor components.

combination of machine tools and controller. The *control-system part* of the postprocessor is developed mainly to format the numerical data in a way to be received and understood by the NC controller. The *machine-segment part* of the postprocessor is in charge of processing statements dealing with coolant control (e.g., ON, OFF, FLOOD, MIST), automatic tool changing, spindle-speed selection, and the like. This part of the postprocessor is machine-dependent (i.e., it depends upon the features of the individual machine tool and changes from machine to machine). In order to better understand the relationship between the different parts of a computer-aided programming system, let us consider the block diagram shown in Figure 14.21. As can be seen in the figure, two identical types of controllers operating two different machines require two similar but not identical postprocessors. Also, two different NC controllers running two identical machines require two different postprocessors. We have to remember, however, that the output from the postprocessor in each case will always be exactly the same as the coded manual program prepared for the given machine tool.

14.4.2 NC Programming Languages

Since the emergence of computer-aided part programming in the 1970s, numerous NC programming languages have been developed. Most of them have found limited use, and only a few are commonly used in industry. Therefore, the survey of NC programming languages is limited to the more or less general-purpose languages.

- *APT*: APT stands for *Automatically Programmed Tools*. The language, which was originally developed at the MIT, is the most widely used and most comprehensive language. An APT part program is written in English-like words and consists of a series of statements that define the part geometry, the desired operations, and the machine (and tool) characteristics. Each statement consists of a major word followed by a slash and some modifier words. There are 80 major words and 180 modifiers plus punctuation in the APT language. APT, although capable of producing sculptured surfaces, requires a large mainframe, which has limited the use of the language in the past. APT is the parent of two other NC programming languages that eliminate the need for a large computer. Recently, a simplified PC version of APT that is capable of driving a two-axis NC machine has become commercially available.

- *ADAPT*: ADAPT stands for *Air Material Command-Developed APT*. This language is a simplified version of the APT language and can be run on a much smaller computer. ADAPT has 160 words (major and modifiers) plus punctuation, and it is limited mainly to applications that require plane contouring with a third axis of linear control.
- *UNIAPT*: UNIAPT is another modified compatible version of the APT language. Although it can be used for programming three-axis and most of the four- and five-axis NC machines, UNIAPT was specifically developed to be run on minicomputers.
- *SPLIT*: SPLIT is the acronym for *Sundstrand Processing Language Internally Translated*. SPLIT was developed to be used with Sundstrand machine tools. Therefore, its processor was dedicated and machine-dependent, and there was no need for a separate postprocessor. Accordingly, SPLIT could not be used with any type of machine tool other than Sundstrand, which had markedly limited its industrial use.
- *ACTION*: ACTION is considered a child of SPLIT. It is a modified version that has a general-purpose processor and a machine-dependent postprocessor.
- *COMPACT II*: COMPACT II is a child of ACTION and a grandchild of SPLIT. It has a general-purpose processor and a machine-dependent postprocessor. The COMPACT II language has the advantages of being simple and easy to learn, and it satisfies the vast majority of programming requirements. In addition to availability on a time-sharing basis from M.D.S.I./APPLICON in Ann Arbor, Michigan, there is now a version that can run on a VAX 785/11, as well as another new version that has been specially developed to run on a microcomputer. Most of the microcomputer CAM systems now available are similar to COMPACT II. Because microcomputers have found their way into manufacturing and computer-aided part programming, let us now study the COMPACT II language in more detail, as a representative example of the NC computer languages. The information provided here is published with special permission from M.D.S.I./APPLICON, Ann Arbor, Michigan. The cooperation of that firm is greatly appreciated.

14.4.3 Details of the COMPACT II Language

A COMPACT II program consists of a series of statements, each providing information or instructions to the system. A statement must always begin with a major word, followed by a set of associated minor words. Major words indicate the operation to be performed by the system, whereas minor words provide details about the location and manner in which that operation is to be accomplished. Minor words are also used to indicate tool description and to define cutting speeds, feeds, and the like.

14.4.3.1 Guidelines

In a COMPACT II program, statements must form a logical sequence of events. Following are some guidelines for building a program:

- The initialization statements must come first in the program.
- If the BASE (similar to the setup point in conventional NC) is to be used as a reference for the coordinate system, it must be defined prior to its use.
- Points, lines, and circles must first be defined before being used in defining the part geometry.
- A tool-change statement should precede the motion statements in which that tool is used.
- The termination statement END must be the last statement in the program.

Syntax. Like any computer language, the COMPACT II language has a certain syntax that must be followed. Here are the rules for punctuation and arithmetic operations:

- The comma is used to separate the units of information that form a statement.
- The slash operator is usually used to modify the parameters associated with a geometric element.
- Parentheses are employed to combine a set of information into a single unit. They must also be used to enclose a division operation.
- The semicolon allows the programmer to avoid repeating the same major word in successive statements by acting as a substitute for the major word after the first statement.
- The percent sign (%) is used to specify the opposite input mode for dimension. If the input mode is inches, a dimension followed by % is in millimeters.
- The dollar sign is used in pairs to enclose comments that may be continued up to three lines.

14.4.3.2 Structure

The structure of a COMPACT II program involves five different groups of instructions, each group consisting of one or more statements. The five groups of instructions serve different purposes. The statements in the first group, which are always given at the beginning of a program, are the initialization statements. Other groups are used for defining the geometry of the workpiece, giving tool-change commands, initiating and defining tool motion, and terminating the program. A tool-change statement can be given as many times as required, depending upon the number of tools needed. In addition, each time a tool-change statement is given, the corresponding tool-motion statements must follow it. Following is a discussion of each of these kinds of statements.

1. *Initialization Sequence*. The initialization sequence usually includes four statements:
   - The MACHIN statement is always the first statement of a COMPACT II program. It provides the name of the machine tool and consists the major word MACHIN, followed by the name of the machine tool link, for example,

     MACHIN, MILL
   - The IDENT statement is the second statement in the program. It is used for identifying the program (or the machine-control tape) and consists of the major word IDENT, followed by the part name or number or any alphanumeric combination, for example,

     IDENT, TEST 2 PROGRAM
   - The SETUP statement is used mainly to specify the home position of the gauge-length reference point (GLRP), relative to the absolute zero of the machine tool, at both the beginning and the end of the program. The GLRP is actually the point from which the tool gauge lengths are measured (e.g., for a milling machine tool, it is the center point at the surface of the quill). This statement is used to specify the program zero when a floating-zero machine is used. For a milling machine tool, this statement takes the following form:

     SETUP, 3LX, 4.5LY, 10 LZ

     where 3 is the dimension from the absolute zero along the X-axis to the GLRP, 4.5 is the dimension from the absolute zero along the Y-axis to the GLRP, and 10 is the dimension from the absolute zero along the Z-axis to the GLRP. The given numbers (3, 4.5, and 10) are arbitrary and differ from program to program. You should always keep in mind that the home position is, at the same time, the load–unload position. Therefore, when using a lathe, for example, the home position of the turret should be selected such that the longest tool is clear of the maximum outer diameter of the workpiece. Sometimes, the SETUP statement is also used to establish the travel limits of the machine tool by specifying the LIMIT parameter, as follows:

     SETUP, 10 X, 20 Z, LIMIT (X 0/14, Z 0/30)

     For a lathe, the SETUP statement takes the following form:

     SETUP, 5. 75 X, 7. 5 Z

where 5.75 is the dimension from the spindle centerline to the GRLP and 7.5 is similar to SETUP X but along the Z-axis.

The BASE statement is used to define a secondary coordinate system shifted from but parallel to the original coordinate system, with the aim of facilitating the programming task. Here, BASE is a datum point located on the part blueprint and from which the part blueprint has been dimensioned (similar to the setup point in conventional NC). It is always advantageous to reference the BASE to the absolute zero, although it can also be referenced to other defined points. Following is a BASE statement where A means absolute (as opposed to XB, meaning with reference to the BASE):

BASE, 3XA, 4YA, 2ZA

2. *Geometry Definition*. The shape of a workpiece can be precisely defined by defining its geometric elements (i.e., points, lines, circles, and planes):
   a. A *point* is defined by using the major word DPT (define point). Any associated minor words describe how that point is specified. As is the case in analytical geometry, a point can be defined by providing a set of coordinate dimensions from absolute zero (or BASE) or by specifying the location of a point as lying at the intersection of two lines, a line and a circle, or two circles. In the latter two cases, a selector is required because the intersection will yield two points. There are also other methods for defining a point in COMPACT II. Following are some examples of statements used in defining points:

      DPT1, 5XB, 3YB, 6ZB
      DPT5, LN1, LN2, 5ZB
      DPT6, LN2, CIR2, X L

      In the last statement, XL is the selector, meaning that the point that has a larger coordinate dimension along the X-axis is the required one.
   b. A line is defined by using the major word DLN (define line). Any associated minor words describe how that line is specified. A line can be defined as passing through a point and making a certain angle with a reference axis, as passing through two defined points, or as an implied line perpendicular to one of the coordinate system axes. Following are some examples of statements used in defining lines:

      DLN1, PT1, 30 CW
      DLN2, PT1, PT2
      DLN3, 4XB
   c. A circle is defined by using the major word DCIR (define circle), followed by any associated minor words. As is the case when defining points and lines, the methods used are adopted from analytical geometry. A circle can be defined by its center and radius, by three points through which it passes, by being concentric with an existing circle, or by being tangential to two existing lines. Following are some examples of statements used in defining circles:

      DCIR1, PT1, 1.5 R
      DCIR2, LN2, LN3, 2.5R
      DCIR3, PT1, PT2, PT3
      DCIR4, CIR3/5R
   d. A plane is defined by using the major word DPLN (define plane), followed by any associated minor words. One of the methods for defining a plane involves specifying three points through which that plane passes. A plane can also be defined as perpendicular to an axis by programming its axis intercepts. Following are some examples of statements used in defining plane:

      DPLN1, PT1, PT2, PT3
      DPLN2, 10ZA

3. *Tool-Change Statements*. The tool-change cycle is started by using one of the two major words ATCHG or MTCHG, followed by any associated minor words specifying the tool configuration, feed rate, spindle speed, and so on. The ATCHG (automatic tool-change) statement causes the spindle to stop and move to the tool-change position, where the current tool is returned to the magazine and replaced by the new tool whose number is given in the statement. The MTCHG (manual tool-change) statement serves to stop the machine function so that the operator can perform the tool-change operations manually. Following is a typical tool-change statement used in milling applications:

   - ATCHG, TOOL4, GL6, 0.5TD, 300 RPM, 0.01 IPR
   - The minor words in this statement have the following meanings:
   - TOOL4 is a command to get the tool in pocket 4 in the magazine and mount it in the spindle.
   - GL6 means that tool 4 has a gauge length (i.e., the length appearing beyond the GLRP) of 6 inches.
   - STD means that tool 4 has a tool diameter of 0.5 inches.
   - RPM indicates the rpm of the spindle after tool 4 is mounted. Sometimes, it is replaced by the cutting speed in feet per minute (e.g., 80 FPM). In this case, the system automatically calculates the rpm using the tool diameter and the cutting speed. 0.01 IPR indicates the feed in inches per revolution. Sometimes, the feed rate in inches per minute is used instead (e.g., 1.5 IPM).
   - In lathe applications, the minor words in a tool-change statement are slightly different from the preceding ones. For one thing, the tool gauge length has to be specified along both the X- and the Z-axes (i.e., the distances of the tool tip from the reference point at the center of the turret along the X- and Z-axes). Also, the radius of the tool nose has to be given. Following is a typical tool-change statement used in lathe applications:
     ATCHG, 3 GLX, 6GLZ, Tool2, 0.05 TLR, 100 FPM, 0.015 IPR

4. *Motion Statements*. Major words are used to identify either linear or circular motion; these are followed by minor words that specify and terminate the path of the tool. The major and minor words used for linear motion are different from those used for circular motion.
   - The two major words that generate linear motion are MOVE and CUT. MOVE generates rapid traverse motion and is used to position the tool prior to a cutting operation. (A clearance must be left between the final position of the tool and the workpiece surface to avoid accidents.) CUT generates feed rate motion for machining with the tool in contact with the part. Following are some examples of statements used to generate linear motion:

     MOVE, OFFLN1/ 0. 3 XS, OFFLN2/ YS
     CUT, PARLN1, OFFLN3/XL
     CUT, PARLN2, PASTLN4
     CUT, PARLN8, TOLN7
     CUT, PARLN1, ONLN2

     In the preceding statements, TOLN and PASTLN are determined relative to the present tool position, and OFFLN must be followed by a modifier (XL, XS, YL, YS), which is sometimes accompanied by a tool offset (OFFLN1/XS).
   - The three major words that generate circular motion are CONT, ICON, and OCON. They indicate the location of the path of the tool center with respect to the circular arc to be obtained after machining. As can be seen in Figure 14.22, CONT is used when the tool center always falls on the arc, while OCON (outside contour) and ICON (inside contour) are used to produce convex and concave surfaces, respectively. In all cases, the major words are followed by minor words indicating the direction of motion and its start

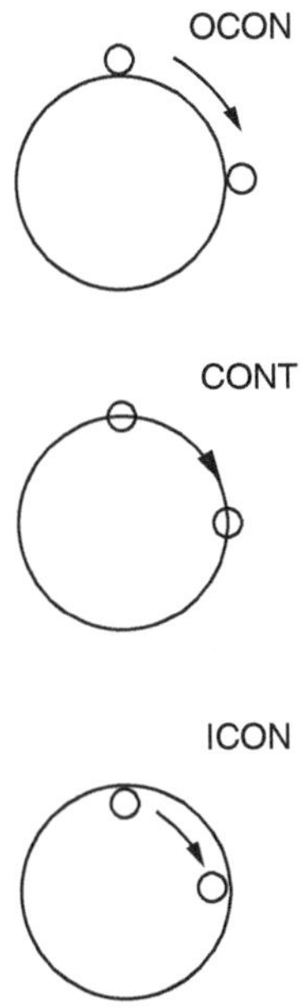

FIGURE 14.22 Major words for circular motion in COMPACT II.

and finish locations. An interesting feature of the COMPACT II language is that the linear motion from the current tool location to the start location of an arc need not be programmed and is automatically included in all circular motion statements. Following are some examples of statements used to generate circular motion:

ICON, CIR2, CCW, S(TANLN3), F(TANLN4)
OCON, CIR4, CW, S(TANLN4), F(TANLN6)
ICON, CIR3, CCW, S(TANLN5), F(90)

5. *Program Termination*. The END statement is always the last statement of a COMPACT II program, and it contains only the major word END. As soon as this command is given, the resulting tape image will have blocks of information that return the machine tool axes to the home position, reset the control, and rewind the tape.

14.4.4 Graphics NC Systems

14.4.4.1 Features

In graphics NC systems, which have been developed for machining centers and lathes programming these CNC systems, the programmer inputs data and communicates with the system in a user-friendly language. A message and a menu corresponding to each position of the cursor are displayed in the lower part of the CRT screen. Depending upon the message, the programmer inputs data by pressing either the appropriate menu key or one or more numeric keys. Generally, the operational procedure for machine tools with graphics NC systems takes the following steps:

1. Register tools (input tools in the tool file picture).
2. Prepare the desired program.
3. Allocate tools in the different pockets of the magazine or in the different stations of a turret in the case of a lathe (tool layout picture).
4. Input tool data, such as actual diameter and lengths. A tool gauge length can be accurately measured using a special unit attached to the machine.

5. Input coordinate dimensions for the program zero (this is similar to the BASE point in COMPACT II).
6. Check the geometry of the part and the tool path on the graphic display (the CRT screen). If the desired part configuration and tool path are obtained, start automatic machining. However, if an error is observed in the product shape or in the tool path, modify the program until the error is eliminated.

14.4.4.2 Advantages of Graphics NC Systems

Following are some advantages of the graphics NC systems:

1. Graphics NC systems involve a user-friendly NC data input method with respect to operation, routines, capabilities, and efficiency.
2. These systems have increased machine performance and high productivity.
3. They provide automatic programming routines, such as automatic optimum tool selection, automatic determination of tool paths, calculation of cutting conditions, insertion of chamfers and rounding corners, and calculation of the points of intersection of the geometric elements.
4. Programming and editing are possible while the computer is controlling the machine tool. The prepared program can be checked on the graphic display of the controller.
5. The tool path that is to be followed during machining can be checked on the CRT screen at a higher speed without the need for actually running the machine tool. This eliminates the danger of collision between the tools and any obstacle. It also allows the actual machining time to be obtained and then displayed on the CRT screen.
6. The systems have the capability of automatic machining-accuracy compensation for the tool wear that occurs during machining and that depends upon the workpiece machining time and the number of workpieces machined.
7. The systems are highly reliable as a result of the full adoption of the latest microelectronics technology.

14.5 CAD/CAM SYSTEMS

The recent trend to establish a direct link, through electronic channeling, between the product design and manufacturing departments is aimed at eliminating the duplication of efforts by design and manufacturing personnel. When using such systems, interactive graphics software is employed to establish the geometry, dimensions, and tolerances for the desired part design, which can be displayed on the CRT screen. The geometry of the part design can be stored in the memory of the computer in the form of digital data that can, in turn, be adopted as a database for preparing an NC part program. By entering the tool data, the cutter location file can be generated and then postprocessed in order to obtain a tape image for the part program. These CAD/CAM systems are particularly advantageous when the shop includes different types of NC machine tools.

14.6 OTHER APPLICATIONS OF COMPUTER-AIDED MANUFACTURING

The discussion of CAM applications has thus far been limited to the use of computers to drive tools in order to machine parts. Following is a brief discussion of some other applications of CAM.

14.6.1 Computerized Cost Estimation

The task of determining the cost of a new product is usually both time-consuming and cumbersome because it involves analysis of indirect expenses as well as overheads. Because the computer is efficient at information handling and processing, it is widely used to accurately estimate the cost of new products in the shortest possible time.

14.6.2 Computer-Aided Process Planning

Computer-aided process planning involves employing the computer to determine the optimal sequence of operations that should be employed to manufacture a desired part and also keep the production time and cost to a minimum. This application of CAM has recently been used in computerized automated manufacturing systems.

14.6.3 Computerized Machinability Data Systems

In computerized machinability data systems, the role of the computer is to provide the feed and speed that should be used to machine a given workpiece material by a given tool material. This is achieved either through retrieving the recommended values from a database created by experienced people from experimental observations or by employing mathematical modeling and Taylor's equation.

14.6.4 Computer-Aided Monitoring and Control of Manufacturing Processes

Computer-aided monitoring involves a variety of applications, ranging from data acquisition and computer process control to CNC and adaptive control. Adaptive control has special important applications in modern automated manufacturing systems, so let us discuss it here in some detail.

We previously came to the conclusion that although NC machines result in an appreciable reduction in the overall production time, the actual machining time remains virtually identical to that for conventional machine tools. This is a consequence of the fact that an NC system provides *preplanned control* without any feedback mechanism to account for real-time variations in the process parameters. For this reason, Bendix Research Laboratories developed an *adaptive control* system in the early 1960s to operate the machining process more efficiently. The main function of a practical adaptive control system is the real-time optimization of a performance index. This index is taken as the metal-removal rate or the cost per volume of metal removed in machining operations. The adaptive control system performs its function by detecting any variability in the condition of the workpiece being machined and adjusts the feed and the cutting speed to account for that variability and to maximize the performance index. The variability can take different forms, such as hard spots, which require a reduction in feeds and cutting speeds, or the presence of air gaps in the workpiece, where the feed should be doubled or even tripled to minimize the idle time during which the tool travels across the air gap. Figure 14.23 shows a typical adaptive control system used in industry for controlling NC machine tools. In this system, the controlled variable is the feed, and the system monitors the spindle deflection (or the horsepower consumed) and keeps it below a certain predetermined value by controlling the feed. This type of adaptive control system is practical and is most commonly used in industry. It is referred to as *adaptive control constraint* because it limits, or constrains, the measured variable (e.g., spindle deflection or horsepower consumed) below a desired value.

In the 1970s, adaptive control was not widely used because it reduced only the in-process time, which usually accounted for less than 5 percent of the production cycle of a part. The system was used either when the actual machining time amounted to a high percentage of the total production time or when there were significant sources of variability in the workpiece. Now, adaptive control is gaining wider industrial application, especially in computer-integrated manufacturing (CIM) systems, where human intervention to compensate for variability is not required. Also, adaptive control has found application in chipless manufacturing processes like sheet metal working and welding.

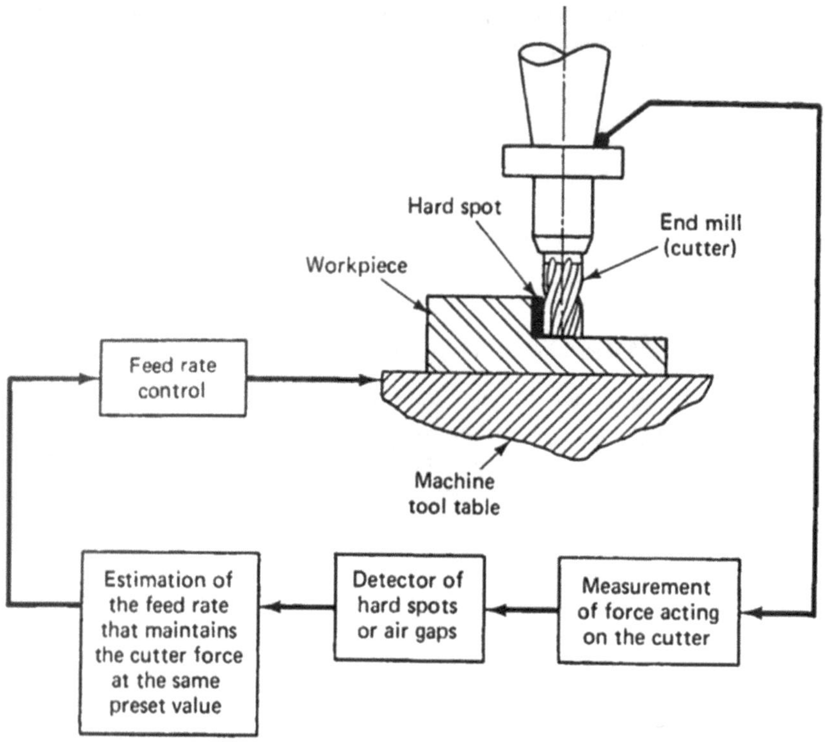

FIGURE 14.23 A typical adaptive control system used in industry for controlling NC machine tools.

REVIEW QUESTIONS

1. Define *CAM*.
2. What is the dominant application of CAM?
3. What is meant by numerical control?
4. When and from where did numerical control emerge in modern times?
5. Discuss some of the main advantages of NC machine tools.
6. What are the main elements of an NC system? Discuss the functions of each briefly.
7. What is the difference between the absolute and incremental dimensioning modes?
8. Explain the NC machine axes.
9. How many axes does an NC machine have?
10. An NC machine can be indexed in eight positions by tape command. Can this be considered a machine axis?
11. How do you identify the Z-axis of an NC machine tool?
12. What kind of NC machine tools can be employed for producing sculptured surfaces?
13. Define *NPC*. What is the main application for this kind of system?
14. What is the difference between the NPC and picture-frame systems?
15. What additional features must an NC system have so that it can perform contouring jobs? What do we call such an NC system?
16. What is the function of a punched tape?
17. What materials are NC tapes made of? Give an application for each of these materials.
18. What is meant by the binary-coded decimal system? Explain how numbers are coded on NC tapes.
19. Define the following terms: *row*, *bit*, *word*, *track*, and *block*.
20. How many tracks does an NC tape have?
21. What is a parity check? How is it performed in the EIA and the ASCII codes?

22. Why is the word *EOB* used?
23. Why is the word *EOR* used?
24. Describe some tape formats.
25. Explain the basic concept of word-address programming.
26. Explain the meaning of the N, G, and M codes.
27. Describe manual part programming.
28. What is the absolute zero?
29. Differentiate between the fixed zero and floating zero points.
30. What is the setup point?
31. List the steps included in a programming task.
32. Define *CNC*.
32. List some of the features of CNC systems and discuss each briefly.
33. Define *DNC*.
34. Explain what a BTR is.
35. What are the advantages of DNC?
36. In what way can a computer assist in preparing a part program?
37. What are the advantages of computer-assisted part programming? Explain each briefly.
38. Discuss briefly what you feed into a computer and what happens within it so that a tape image can be obtained to drive a machine tool.
39. What are the functions of the processor and the postprocessor?
40. Of what is a postprocessor usually composed?
41. List some NC computer programming languages.
42. Briefly describe each of the languages you listed in question 42.
43. What is the relationship between SPLIT, ACTION, and COMPACT II?
44. Explain the function of each of the following in COMPACT II: comma, slash, parentheses, semicolon, percent sign, and dollar sign.
45. What kinds of statements form a COMPACT II program? Explain each briefly.
46. Explain the meaning of BASE in COMPACT II.
47. How can we define the geometry of a part in COMPACT II?
48. What are ATCHG and MTCHG in COMPACT II?
49. What data does a tool-change statement in COMPACT II include?
50. What is the GLRP?
51. Describe the home position.
52. What statement in COMPACT II follows the tool-change statement? What major words can be used in that statement?
53. What is the last statement in a COMPACT II program?
54. What is graphics NC?
55. Describe the general procedure of preparing a graphics NC program.
56. List and discuss the main advantages of graphics NC systems.
57. Why is programming a graphics NC system considered very easy?
58. Explain briefly the structure of an NC graphics program.
59. Differentiate between the process data and sequence data in an NC graphics program.
60. What are the advantages of CAD/CAM systems? When would you recommend these systems?
61. What are the other applications of computer-aided manufacturing?
62. What is adaptive control and why is it now gaining industrial application?

DESIGN PROJECT

Design a chess piece (king, queen, pawn, or knight) and then select the appropriate tools, plan the tool path, and write a program to machine it using the CNC lathe.

15 Automated Manufacturing Systems

15.1 INTRODUCTION

Since the Industrial Revolution, U.S. industry has been undergoing a process of continuous development. During the nineteenth century, new machines expanded the productivity of workers; at the beginning of the twentieth century, new automation technology in the form of mass production through transfer and assembly lines emerged. Factory manufacturing grew to include more and more new sectors of work and finally became a composite of different departments, each separate from the others and dedicated to a certain task. Although automation and computerization of these isolated islands of work have been increasing steadily since the early 1950s, barriers to understanding methods of working have also been growing. Although the number of blue-collar workers on the shop floor has continually decreased, large numbers of white-collar personnel, ranging from managers to clerks, have been needed to handle paperwork and to transfer information among the different departments in order to tie them together. It is, therefore, obvious that automating isolated tasks in the process of product development, although relatively cost-effective, cannot alone achieve either significant savings in lead time taken to develop a product or gains in productivity. It is also clear that these goals can be accomplished only by automating the flow of information in the business organization and by optimizing the process of product development as a whole through the adoption of a system's approach.

The solution is a complete implementation of CIM. Figure 15.1a shows how different departments in a corporation can be electronically channeled so that each department has immediate access to all other departments as well as to the mainframe database. This arrangement ensures efficient control of the corporation and, accordingly, optimization of the whole system. On the other hand, Figure 15.1b illustrates how many companies function today, with isolated automated work

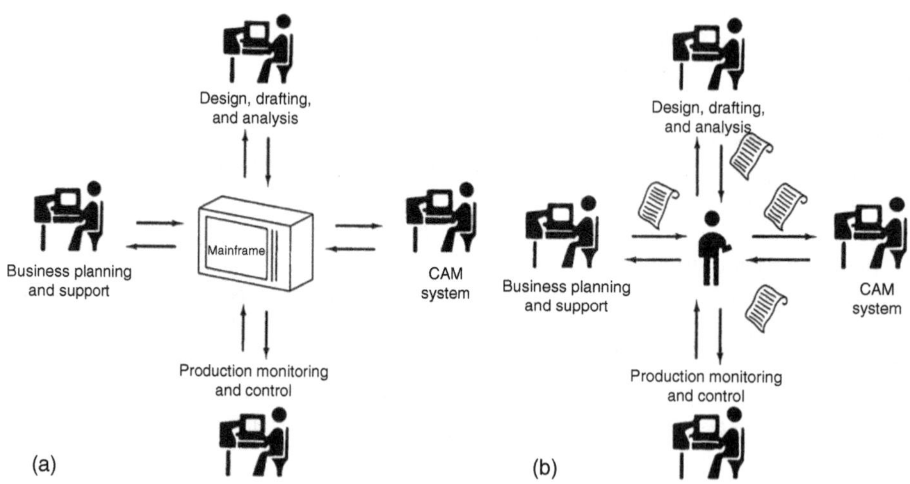

FIGURE 15.1 Types of information transfer between computerized departments of a plant: (a) departments electronically channeled (CIM) and (b) white-collar workers passing documented information between departments.

DOI: 10.1201/9781003537106-15

islands and white-collar workers doing paperwork to pass information from one department to another. A comparison of these figures will reveal to us the anticipated benefits of CIM. But first, let us consider some definitions of CIM.

15.2 COMPUTER-INTEGRATED MANUFACTURING (CIM)

An interesting definition of CIM was given by Eugene Merchant (the father of the theory of metal cutting):

> [CIM is] a closed-loop feedback system whose prime inputs are product requirements and product concepts and whose prime outputs are finished products. It comprises a combination of software and hardware: product design, production planning, production control, production equipment, and production processes.

Another definition, given by Richard G. Abraham in his paper presented at the AUTOFACT III Conference, was adopted and published by the Computer and Automated Systems Association (CASA) of the Society of Manufacturing Engineers (SME). It stated the following:

> A timely integrated CAD/CAM or CIM system provides computer assistance to all business functions from marketing to product shipment. It embraces what historically have been classified as "business systems" applications, including order entry, bill of material processing, inventory control, and material-requirements planning; design automation, including drafting. design, and simulation: manufacturing planning, including process planning. routing and rating, tool design, and parts programming; and shop floor applications such as numerical control, assembly automation, testing, and process automation.

A succinct definition of the business planning and execution functions in a CIM system was also provided. It included economic simulations, long-term business forecasting, customer-order servicing, and finished-goods inventory management. It is important to note that both definitions apply not only to engineering activities but also to management and business activities.

Thus, as a result of CIM, it appears that the role (as well as the quality) of engineers involved in design, manufacturing, and production planning will change. Product design not only will be dictated by the desired functions but also will be affected by manufacturing considerations. It is, therefore, anticipated that the boundaries now existing between the design and the manufacturing phases will disappear, or at least fade. Consequently, the engineers who carry out the process of product design and development will need to be thoroughly knowledgeable about the details of manufacturing. Similarly, an increase in the skill levels of other personnel will also be required. For instance, workers will become responsible for maintenance and initial setup work instead of just the simple, repetitious, and boring tasks involved in transfer or assembly lines. Managers will also need more technical backgrounds in order to be able to make appropriate decisions. In all cases, it is obvious that knowledge of computer systems is an absolute necessity.

In summary, the proper implementation of CIM must be based not only on the mechanization, optimization, and computerization of the various processes but also on achieving these goals in synchronization with the automation of information flow in order to deal with real-time planning and control of the business organization as a whole, from order entry to shipment of the finished product. In addition to a well-structured database, some fundamental methodologies are employed to achieve this goal. Among the methodological tools used for integrating design, manufacturing, and production control are group technology (GT), CAPP, and MRP.

15.2.1 Benefits of CIM

Despite the contributions of computer-aided design drafting (CADD) and computer-aided manufacturing (CAM) in increasing productivity by reducing the total product development costs and time, industrial automation experts predict the greatest gains coming from the implementation of CIM and replacing white-collar personnel and their paperwork with computer control. Following are some of the anticipated benefits of CIM.

15.2.1.1 Improved Product Quality

Improved product quality is a result of the CIM concept of developing the product within the computer, thus basing the product development process on rational and profound analysis instead of today's common philosophy of "build and test." Prototypes will still be built, but not to find out how a product performs. Instead, they will be used to verify the results of the analysis carried out by the computer. Other factors that contribute to improving product quality and consistency include a lower probability of human error and ensured uniformity of product attributes because of the use of online computer-aided inspection and quality control.

15.2.1.2 Improved Labor Productivity

Automating the information flow will result in a decrease in indirect labor, while at the same time increasing the efficiency of information transfer and eliminating redundant data collection. Any decrease in indirect labor will lead to a reasonable decrease in unit cost because the total costs for a typical highly automated production plant (in the United States today) are composed of 0–6 percent for direct labor, 18–22 percent for overhead and indirect labor, and about 75 percent for materials and machines. (You can see that paperwork costs more than the manufacturing work.) Also, increasing the efficiency of information transfer will result in more effective management.

15.2.1.3 Improved Equipment Productivity

Equipment productivity is improved because of the better utilization of machines when CIM is implemented. Factors like the programmability of equipment and computerized monitoring and control of the entire manufacturing facility will largely improve the efficiency of machine utilization.

15.2.1.4 Lower Costs of Products

Higher labor and equipment productivity will certainly result in lower product costs. This is added to the advantages of designing the product with the required manufacturing processes in mind (i.e., design for manufacturing [DFM]), which can easily be achieved through the integration of CAD and CAM. The use of DFM will ensure the production of a part through the easiest and cheapest methods, thus reducing its cost. In fact, it has been found that designs that take into account only the product and its functions generally create the need for special manufacturing equipment, leading to a noticeable increase in production cost.

15.2.1.5 Increased Market Share and More Profit

By its very nature, CIM increases the flexibility of the manufacturing facility, thus enabling it to react quickly to fast-changing market demands. The reason is that much less lead time is taken to develop a product in a corporation where CIM is implemented (lead time is the time from the moment at which design work begins until the moment the product is shipped out of the factory). Also, less lead time means lower manufacturing costs of the products, which translates into greater manufacturing flexibility.

15.2.2 Implementation of CIM

It is clear that the development of a totally CIM system is a difficult task that requires a long-term commitment. However, the implementation of CIM can be accomplished gradually because a computer-integrated business organization consists of computerized modular subsystems that are interconnected through a network. The nature and basic functions of each of these modular subsystems depend primarily upon the products and the activities of business corporations and differ from one company to another. It is, therefore, impossible to purchase a turnkey CIM system from a vendor. Of course, components of a CIM system can be purchased, but the backbone of a CIM system has to be established within the corporation or at least tailored to suit the particular conditions of that corporation.

The first step that management should take in order to implement CIM is to study and review the activities to be performed in each subsystem, the type and amount of information flowing to and from that subsystem, and the interaction between the different subsystems in the business organization. Second, the rational development of a long-term plan involving the architecture of the CIM system has to be carried out.

Third, a feasibility study and/or economic justification of each subsystem should be performed to establish the order of designing and implementing the subsystems during the gradual phase-in. Each subsystem is then integrated into the system directly after it is implemented, according to the main plan, with the end result being a fully integrated CAM system.

15.2.3 The CIM Database

As mentioned, the heart of integration is a well-structured database that enables designers, manufacturing engineers, production managers, marketing and purchasing personnel, and partners and subcontractors to have access to the same set of factual data about products, production, and inventory. The typical architecture of a CIM system should be based upon a group of computer systems (CAD, CAM, and others), a neutral data format, a number of processors linked to a computer system to convert local data into neutral data (free format) and vice versa, and, finally, a database management system (DBMS). The major purpose of a DBMS is to see that all users or elements (such as the terminal or machine controller) automatically receive the updated version of data when any user alters data in his or her local system or element. For example, if for some reason, a manufacturing engineer in the CAM department changes the design to facilitate the manufacturing phase, all modifications made will be transmitted directly to the database and to the CAD department.

15.2.3.1 Classes of CIM Database

For easier and more logical management of a CIM database, it is appropriate to group the data in that database into four classes: product data, manufacturing data, production control and operational data, and business data. Product data involve the attributes and geometrics of the objects to be manufactured. Manufacturing data deal with how parts are to be manufactured, and operational data involve lot sizes, schedules, routes, and the like. These three classes are mostly technical, contrary to business data which deal with resources, people, and money.

15.2.3.2 Logical and Physical Databases

The logical database relates to algorithms and programming methodologies or, in other words, what the user is concerned with. On the other hand, the physical database relates to what hardware personnel see. However, these two concepts, though distinct, are interrelated, and, therefore, it is advisable to separate them, making each more flexible regarding its own functions and demands in order to have a successful CIM database that is responsive to fast-changing technology and users' needs. Because the logical and physical structures are interrelated, it is necessary to define a concept of the CIM database in such a manner that it is completely independent of any specific logical or physical requirement. This is done through the use of data standards that control the meanings of data and are defined using a data-modeling tool (like IDE FIX from USAF ICAM). They describe data in terms of entities and attributes of entities and are stored and maintained in a data dictionary. A data dictionary defined by Daniel S. Appleton in his original paper entitled "The CIM Database" was adopted and published by CASA as "a Rosetta stone for providing access to users in a physical database environment."

15.2.4 Communication Networks

A CIM system is a complex system involving communication networks (i.e., a large number of separate but interconnected computers that must communicate with one another through many computer networks). These can be different from the viewpoint of computer system structure or configuration and can incorporate different computers, operating systems, and interfaces. However, the computers must be able to communicate with one another on a real-time basis and must also be able to process one another's data and execute one another's programs. In a conventional network, the distance between the user and his or her data may be several hundred miles. This is not the case with local-area networks (LANs) used in CIM, in which the system consists of many processors (computers, machine controllers, etc.) located close together. The individual LANs are connected through internetworking to form the CIM system. The structure of a network can take different forms, each having its own set of advantages and disadvantages. Nevertheless, almost all communication networks and LANs consist of two components: switching elements (specialized computers or microprocessors [MPs]) and transmission lines (circuits).

15.2.4.1 Network Structures

Network structures involve either point-to-point channels or broadcast channels. In the first type of structure, the network contains numerous channels, each one connecting a pair of nodes or interface message processors (IMPs), which are the switching elements. Figure 15.2 shows some topologies of network structure. The star configuration (Figure 15.2a) was used in the early days of networking. But because star-type LANs suffer from the disadvantage of not being decentralized, ring topology (Figure 15.2b) became more common because of its decentralized structure and decentralized access techniques. Nevertheless, because this topology necessitates passing messages unidirectional from node to node until they reach their destinations, any failure at a single node can bring the network down unless a bypass is provided. On the other hand, the second type of network structure (i.e., broadcast channels) involves a single communication channel shared by all IMPs (nodes). In this type of structure, messages sent by any IMP are simultaneously received by all other IMPs (nodes). This type is usually referred to as bus topology (Figure 15.2c) when it is used in LANs. Each node knows to whom the message is being sent and can, therefore, ignore any message not intended for itself. This is achieved by providing a piece of information (in the message itself) that specifies who the message is for. In fact, the bus topology has an important advantage in that it has the ability to insert splitters and create branches, thus facilitating network reconfiguration.

15.2.4.2 Network Architectures

Because of the complexity of the system and the different communication needs of nodes, CIM networks are organized as a series of layers or levels. Each layer is built upon its predecessor and is designed to provide certain services to the higher-level layers without involving them in the details of how those services are implemented. According to this architecture, a layer with a certain level (order) on one machine communicates only with a layer having the same order on another machine. The set of rules and conventions stating how these two layers should interact during a conversation is known as the protocol. In reality, physical communication between two machines takes place only at the lowest layer, whereas communication between higher layers on two machines is only virtual. What happens is that each layer passes data to the layer directly below it until it reaches the lowest layer, where physical communication is possible. This idea is illustrated in Figure 15.3, which also indicates the need for an interface between each two successive layers to define the primitive operations and services that are offered by the lower layer to the upper one. It has been agreed to call the set of layers and their protocols the network architecture. In the ISO standard, as well as in the Manufacturing Automation Protocol (MAP), the network architecture comprises seven layers: physical, data link, internetwork, transport, session, presentation, and applications, as illustrated in Figure 15.4.

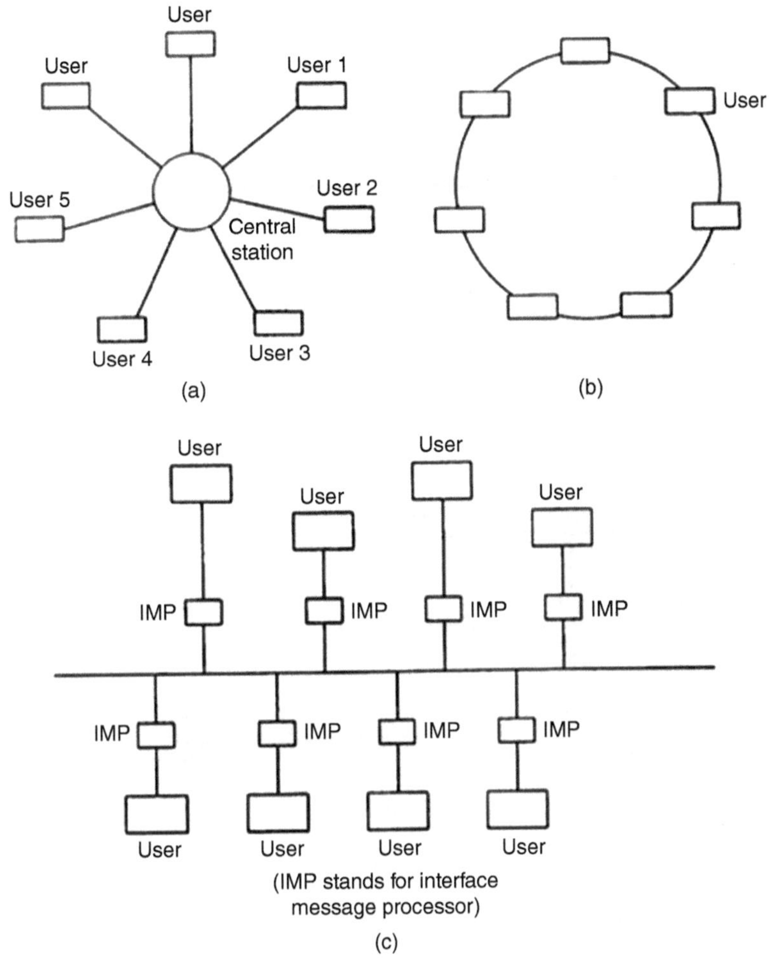

FIGURE 15.2 Some topologies of network structure: (a) star type, (b) ring type, and (c) bus type.

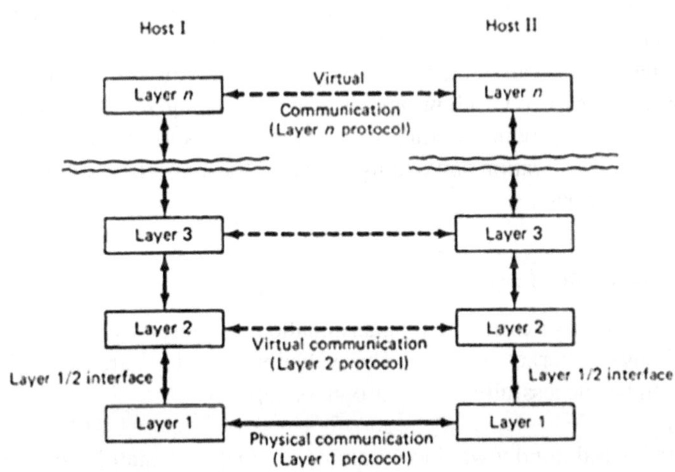

FIGURE 15.3 Layers involved in network architecture.

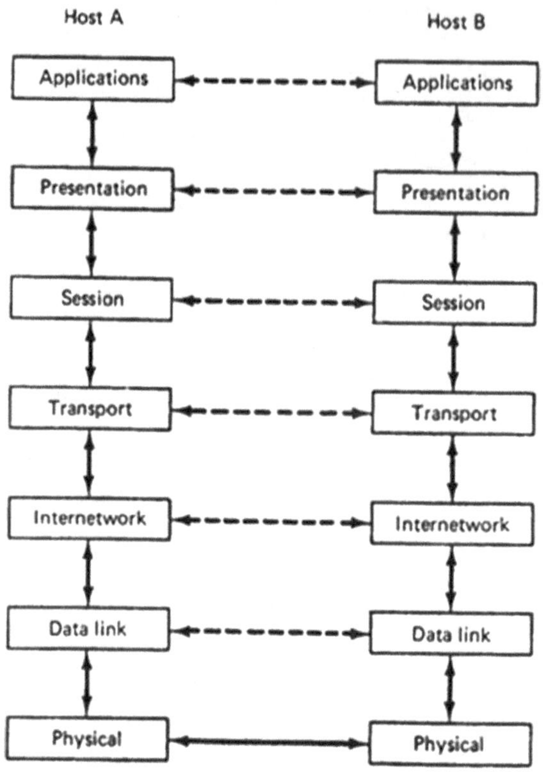

FIGURE 15.4 Network architecture when MAP is used.

15.3 GROUP TECHNOLOGY (GT)

Group technology (GT) is a manufacturing philosophy that involves identifying and grouping components having similar or related attributes in order to take advantage of their similarities in the design and/or manufacturing phases of the production cycle. Historically, this novel idea came into being in the United States in 1920, when Frederick Taylor supported the concept of grouping parts that required special operations. He was followed by the Jones and Lamson Machine Company in the early 1920s, which used some crude form of group technology to build machine tools. Their manufacturing approach involved such principles as departmentalization by product rather than by process and minimizing the product routing paths. Today, GT is implemented through the application of well-structured classification and coding systems and supporting software to take advantage of the similarities of components.

15.3.1 Reasons for Adopting GT

Modern manufacturing industries are facing many challenges caused by growing international competition and fast-changing market demands. These challenges, which are exemplified in the following list, have and can be successfully met by group technology:

- There is an industrial trend toward low-volume production (small lot sizes) of a wide variety of products in order to meet the rising demand for specially ordered products in today's affluent societies. In other words, the share of batch-type production in industry is growing day by day, and it is anticipated that 75 percent of all manufactured parts will be produced in small lot sizes.

- As a result of the trend toward batch-type production, conventional shop organization (i.e., departmentalization by process) is becoming inefficient and obsolete, with wasteful routing paths of products between the various machine tool departments.
- There is a need to integrate the design and manufacturing phases in order to cut short the lead time, thus achieving competitiveness in the international market.

15.3.2 Benefits of GT

15.3.2.1 Benefits of Product Design

In product design, the principal benefit of GT is that it enables product designers to avoid reinventing the wheel or duplicating engineering efforts. In other words, it eliminates the possibility of designing a product that was previously designed because it facilitates storage and easy retrieval of engineering designs. When an order of a part is released, the part is first coded, and then existing designs that match that code are retrieved from the company's library of designs stored in the memory of a computer, thus saving a large amount of time in design work. If the exact part design is not included in the company's computerized files, a design close to the required one can be retrieved and modified in order to satisfy the requirements. A further advantage of GT is that it promotes standardization of design features, such as corner radii and chamfers, thus leading to the standardization of production tools and work-holding devices.

15.3.2.2 Standardization of Tooling and Setup

Because parts are grouped into families, a flexible design for a work-holding device (jig or fixture) can be made for each family in such a manner that it can accommodate every member of that family, thus reducing the cost of fixturing by reducing the number of fixtures required. Also, a machine setup can be made once for the whole family (because of the similarities between the parts of a family) instead of several times for each of the individual parts.

15.3.2.3 More Efficient Material Handling

When the plant layout is based on GT principles, such as dividing the plant into cells, each consisting of a group of different machine tools and wholly devoted to the production of a family of parts, material handling is more efficient because of the minimal routing paths of parts between machine tools. This is in contrast to the "messy" flow lines in the conventional departmentalization-by-process layout. For comparison, both layouts are clearly illustrated in Figure 15.5a,b.

15.3.2.4 Improved Economies of Batch-Type Production

Usually, batch-type production involves a wide variety of nonstandard parts, seemingly with nothing in common. Therefore, grouping parts (and processes) in families allows economies that are obtainable only in mass production to be achieved.

15.3.2.5 Easier Scheduling

Grouping the parts into families facilitates the task of scheduling because this work will be done for each family instead of for each part.

15.3.2.6 Reduced Work-in-Process and Lead Time

Reduced work-in-process (WIP) and lead time result directly from reduced setup and material-handling time. Parts are not repetitively transferred between machining departments because material handling is carried out efficiently within each of the individual cells. This is in contrast to the production in a typical plant with a process-type layout, where a piece that requires only a few minutes of machining may spend days on the shop floor. This situation involves increased WIP, which adversely affects inventory turnover and the cash-flow cycle. Also, the lead time for a product manufactured in a plant designed according to GT principles is far shorter than that of a product manufactured in a plant with a process-type layout.

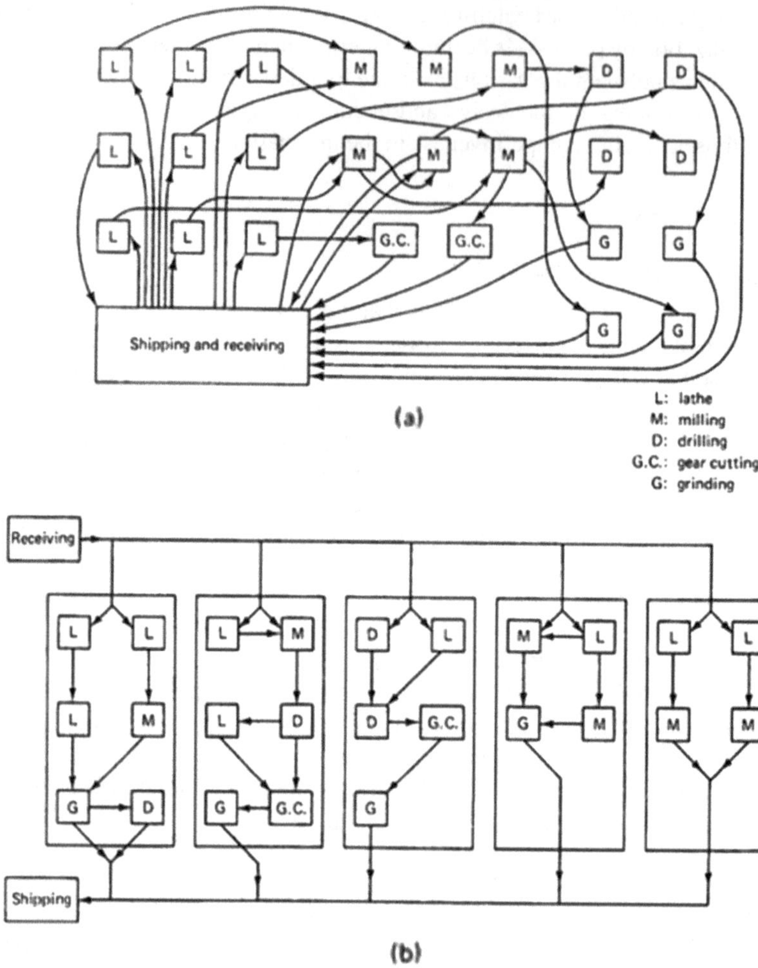

FIGURE 15.5 Flow of parts processed in a plant: (a) when departmentalization by process and (b) when group technology rules are applied.

15.3.2.7 Faster and More Rational Process Planning

Group technology paves the way for automated process planning. This can be achieved through an appropriate parts classification and coding system, where a detailed process plan for each part is stored under its code and thus can be easily retrieved.

15.3.3 Factors Preventing Widespread Application of GT

15.3.3.1 Problems Associated with Rearrangement of Physical Equipment

As mentioned, GT is always associated with the concept of cellular manufacturing. The formation of cells necessitates rearrangement of the existing physical equipment (machine tools) and involves costly and cumbersome work that is sometimes difficult to justify.

15.3.3.2 Need for Large Amount of Upfront Work

In order to implement GT, it would seem that every single part in the inventory of the industrial corporation must be coded so that parts families can be established. This appears to be a huge task that in itself would create a barrier to any tendency toward the implementation of GT. However, a manufacturing corporation already deals with ordered and similar groups of parts because of the area of specialty of that corporation and/or its product line (i.e., range of products).

Accordingly, an appropriate approach for solving the problem of the upfront work required is to do it gradually by coding just the blueprints released to the workshop. As a result, the number of truly new (uncoded) parts released to the workshop tends to level out after a short period of time.

15.3.3.3 Natural Resistance to Anything New

As human beings, we naturally shy away from anything risky or unknown. For this reason, many managers and administrators avoid the adoption of new concepts and philosophies such as group technology.

15.3.4 Classification and Coding of Parts

15.3.4.1 Implementation of a Classification and Coding System

In order to implement a classification and coding system based on GT principles, parts must be classified according to suitable features, and then a meaningful code must be assigned to reflect those features. The process of retrieving or grouping parts with similar features is rather simple. As an example, consider ZIP codes as representing the basic features of a classification and coding scheme. A ZIP code indicates a geographic location by progressively classifying it into subdivisions, starting with the state and proceeding to county, city, neighborhood, and street. Codes that are numerically close indicate locations that are, in reality, geographically close. It is this particular feature that similarly enables the formation of a family of parts based on codes, without the need for physically examining the parts or their drawing.

Although many classification and coding systems have been developed all over the world, none of them has yet become universally standard. The reason is that a system must meet the specific needs of the organization for which it has been developed. The right approach, therefore, is to develop a GT classification and coding system based on the specific needs of the client or to tailor an existing turnkey system to meet those needs.

While there are many benefits for group technology, they fall within two main areas of application: design and manufacturing. Although it is always the ultimate goal to combine the advantages in both areas, it is usually very difficult to do so, and the result is either a design-oriented or a manufacturing-oriented system.

15.3.4.2 Construction of a Coding System

A coding system can be based only on numbers or only on letters, or it can be alphanumeric. When letter codes are used, each position (or digital location) has 26 different alternatives; when number codes are used, position values are limited to 10. Consequently, letters are employed to widen the scope of a coding scheme and make it more flexible.

There are basically two types of code construction: monocodes and polycodes. A monocode, also referred to as a hierarchical or tree-structure code, is based on the approach that each digit amplifies the information given in the preceding digit. It is, therefore, obvious that the meaning of each digit (or what a digit indicates) is dependent upon the digits preceding it. Monocodes tend to be short and are shape-oriented. However, they do not directly indicate the attributes of components because of their hierarchical structure. Consequently, they are usually used for design storage and retrieval and are not successful for manufacturing applications.

In contrast, the meaning of each digit in a polycode is completely independent of any other digit and provides information that can be directly recognized from its code. For this reason, a polycode is sometimes referred to as an attribute code. Figure 15.6 indicates how a polycode is structured. We can easily see that a polycode is generally manufacturing-oriented because the easily identifiable attributes help the manufacturing engineer determine the processing requirements of parts. Moreover, a polycode involves a string of features, a structure that makes it particularly suitable for computer analysis. Nevertheless, polycodes tend to be long, and a digit location must be reserved whether or

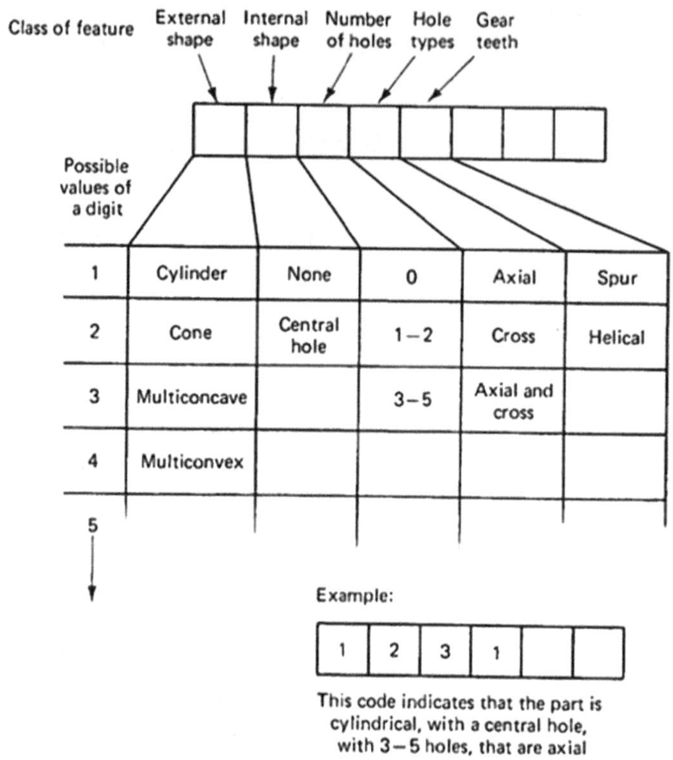

FIGURE 15.6 Structure of a polycode.

not that particular feature applies to a part of a family being coded. Common industrial practice, therefore, uses a hybrid construction that combines the advantages of each of the two basic types of codes while eliminating their disadvantages. In a combination code, the first digit divides the whole group of parts into subgroups, where shorter polycodes are employed. Also, in order to eliminate completely the possibility of error when coding a part, an interactive conversational computer program is employed, where the computer asks questions and automatically assigns a code for the part based on answers provided by the user at the computer terminal. An example of this kind of automated coding is the Metal Institute Classification System (MICLASS system), which was developed by the Netherlands Organization for Applied Scientific Research and has gained industrial application in the United States during the past decade.

Now, let us examine a real classification and coding system that is popular in industry and see how we can make use of it. A good choice would be the Opitz classification and coding system that was developed by Professor Opitz of Aachen Technical University for the Machine Tool Builders Association in Germany. The complete coding system is too complex to provide a comprehensive description here. We will, however, explain its features as well as its main part.

The Opitz coding system includes the *basic code*, which is divided into two parts: the *form code*, which consists of five digits describing the primary design attribute of the part, and the supplementary code, which consists of four digits describing some attributes that can be of use in manufacturing (e.g., workpiece material, accuracy). An additional four digits, which are called the secondary code, are to be designed by the firm to serve its own particular needs. The details of the form code are indicated in Figure 15.7. We can see that a code consisting of five zeros (i.e., 00000) represents a very short smooth cylindrical part that can possibly be a stamped circular blank. Comparing the two codes 10000 and 2000, we can promptly visualize the shape of each and how to machine it. The first (10000) represents a relatively short smooth cylinder that can be chucked on a lathe, while

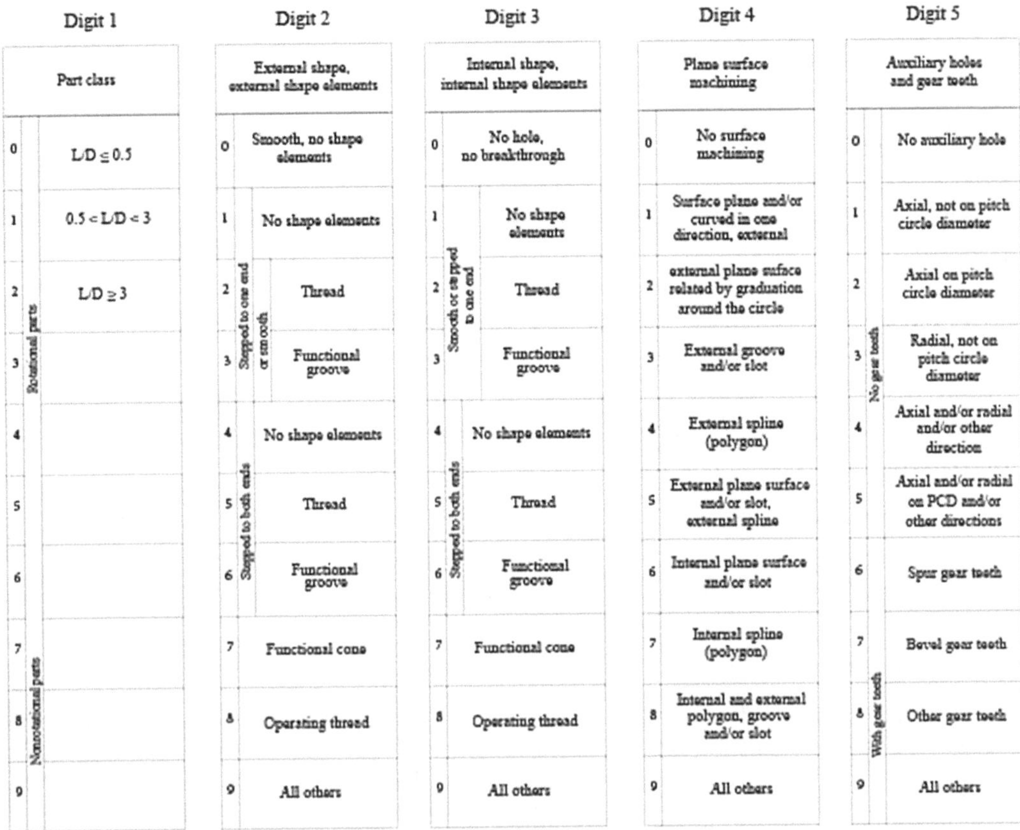

FIGURE 15.7 Details of the form code of the Opitz classification and coding system.

the second represents a long smooth cylindrical that should be suspended between two centers during machining. A simple washer is represented by the code 00100. It is easy to realize that we could put the shape of any part in a digital form.

15.3.5 Design of Production Cells

Before we see how a production cell can be designed, we need to understand a new concept, that of the composite part. This is a hypothetical part that has all of the processing attributes possessed by all of the individual parts of a family. Consequently, the processes required to manufacture the parts of a family will all be employed to produce the composite part representing that family. Any part that is a member of that family can then be obtained by deleting, as appropriate, some of the operations required for producing the composite part. Figure 15.8 illustrates the concept of a composite part.

The next step is to design the machining cell to provide all machining capabilities based on the processing attributes of the composite part for the family of parts that is to be manufactured in that machining cell. The number of each kind of machine tool in a manufacturing cell depends upon how frequently that machining operation is needed. In other words, the number of each kind of machine tool in a machining cell is not necessarily the same for all the different kinds of machine tools in the cell. After determining the kinds and numbers of machine tools in the cell, the layout of machines within that cell is planned to achieve an efficient flow of workpieces through the cell. Of course, the cells are also arranged to guarantee an easy flow of raw stock into the cells and finished products out from them.

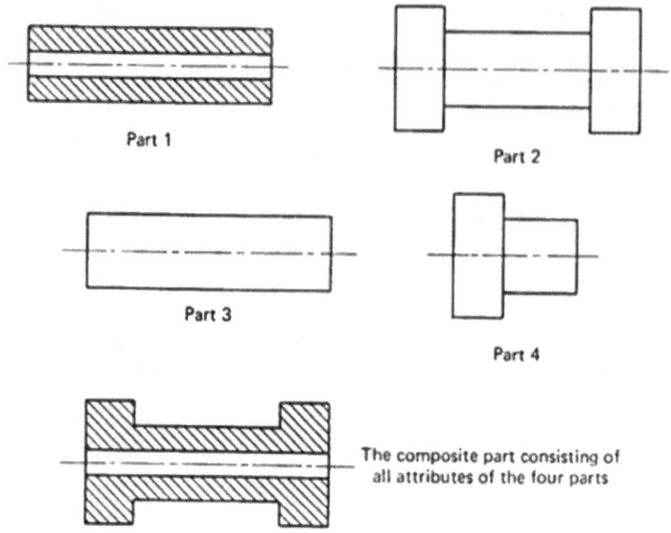

FIGURE 15.8 The concept of a composite part.

15.3.6 Production-Flow Analysis (PFA)

Production-flow analysis (PFA) is a method in which parts families are identified and machine tools are grouped based on the analysis of the sequences of operations for the various products manufactured in the plant. Parts that may not be similar in shape but that require identical or similar sequences of operations are grouped together to form a family. The resulting families can then be used to design or establish machine cells. Although PFA is simple and readily applicable, it does very little to improve and optimize the machine routing being used. It is, therefore, used just as a first step toward the application of the group technology concept in an industrial firm.

After collecting the required data (i.e., the part number and machine routing for every product), the computer is employed to sort out these products into groups, each of which contains parts that require identical process routings and is called a pack. Each pack is then given an identification number, and, by means of graphical illustration, packs having similar routing are grouped together. Next, zoning is used to identify the machine tools that form rational machine cells. Figure 15.9a, b shows how machine cells are established using PFA.

15.4 COMPUTER-AIDED PROCESS PLANNING (CAPP)

Process planning is defined in the *Tool and Manufacturing Engineer's Handbook* as "the systematic determination of the methods by which a product is to be manufactured, economically and competitively." It is the manufacturing engineer's task to set up a process plan for each new product design released from the design office. This process plan is an important stage linking design and manufacturing in an industrial organization. Any process plan involves the sequence as well as the details of the operations required to manufacture a part whose design is given. That sequence of detailed operations is documented on a route sheet. The routing must be optimal to bring the production cost of a part to a minimum, thus making the product economically competitive. Unfortunately, this is not the case today in conventional process planning, where the plan is determined primarily by the experience and the opinion of the planner. In other words, if ten process planners are asked separately to develop a process plan for a given part, they will probably come up with ten different plans. Moreover, it is possible that none of these plans is optimum or close to the optimal plan. In fact, it

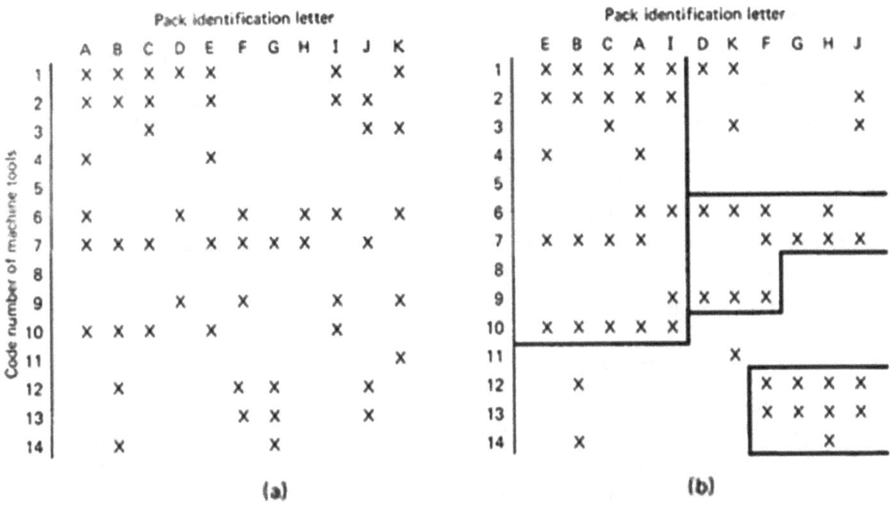

FIGURE 15.9 Establishing machine cells using PFA: (a) initial PFA chart with codes of machine tools for each pack and (b) PFA chart after arrangement with possible cells of machine tools.

has been proven by experimental evidence that a conventional process plan usually reflects a stubborn commitment to the personal preference of the planner and his or her industrial and educational backgrounds.

In order to rationally design a process plan for manufacturing a product economically and competitively, a planner has to study all possible alternatives (i.e., different combinations and/or sequences of operations, machine tools available, process parameters, and the like). Obviously, such a task requires thousands of lengthy calculations. Consequently, a computer must be used if that task is to be accomplished within a reasonable period of time. Also, time is indeed a decisive factor because optimal process plans do not usually remain static but change with changing conditions, such as lot sizes, available equipment, and any emerging new technology. In short, a process plan that is developed for a part today may be different from the plan that would be developed after a year for the same part in the same manufacturing facility. It is, therefore, a major advantage of CAPP that it accounts for any variation in the manufacturing parameters.

15.4.1 Benefits of CAPP

CAPP, together with group technology (which helps pave the way for it), bridges the gap between engineering design and manufacturing and is a key factor in integrating the activities in a manufacturing organization. Some of the benefits of CAPP are as follows.

15.4.1.1 Improved Productivity

Improved productivity is due to the more efficient utilization of resources such as machines, tooling, stock material, and labor, which is based on accurate and lengthy computations, as opposed to conventional process planning, where personal experience, preference, and sometimes even prejudice are the determining factors.

15.4.1.2 Lower Costs of Products

Improved productivity leads to cost savings in direct labor, tooling, and material and ultimately results in lower product costs.

15.4.1.3 Consistency of Process Plans

Because the plans are based on the same analytical logic, with every planner having access to the same updated database, planners will come up with the same plan for the same part. This also occurs because of the elimination of human error during the lengthy computations performed with a computer.

15.4.1.4 Reduction in Time Required to Develop a Process Plan

As a result of computerizing the work, a job that used to take several days can now be done in about 20 minutes. Consequently, increased volumes of work can be easily handled with CAPP, which is not the case with conventional manual process planning.

15.4.1.5 Faster Response to Changes in the Production Parameters

The fact that the logic is stored in the memory of the computer makes CAPP more responsive to any changes in the production parameters than the manual method of process planning. When CAPP is used, it takes the process planner a few minutes to input updated data, such as lot size, machines available, and cost of raw material, and to obtain a modified optimal version of the process plan.

15.4.1.6 Less Clerical Effort and Paperwork

The paperwork involved is far less than in manual process planning because its routing through the system (between different specialized planners) is greatly reduced or even eliminated. This is in agreement with and promotes the goals of CIM in reducing clerical white-collar jobs and thus ending up with a paperless factory.

15.4.2 Types of CAPP

The two types of CAPP, variant and generative, are based on two completely different approaches.

15.4.2.1 Variant Type

The variant approach involves preparing beforehand a group of standard process plans, one for each of the parts families produced by the plant. The parts families should be identified based on GT principles through an appropriate parts classification and coding system. The standard process plans are filed and stored in the memory of a computer, and each one can be retrieved and edited by inputting its associated GT code. When a new part is given to the process planner, he or she can relate it to an appropriate family. Next, the standard plan for the parts family is edited and modified by the deletion or addition of some operations to suit the given part (remember our discussion about the composite part concept). The computer is employed in the initial analysis and computations required to develop the standard process plans because it is available in the organization. However, during the actual running of a variant CAPP system, the role of the computer is limited to that of a word processor. Thus, the main problem with a variant CAPP system would appear to be the large amount of upfront work required to establish the group of standard process plans (which may include hundreds of plans). In addition, these standard process plans need to be revised and modified on occasion to support any changes in the production requirements or parameters. It is primarily for these reasons that the generative approach of CAPP has been developed.

15.4.2.2 Generative Type

In the generative approach, the computer is used to synthesize each individual plan automatically and without reference to any prior plan. What is stored in the memory of the computer are rationales and algorithms that permit the appropriate technological decision to be made. The human role involved in running the system is minimal and includes inputting the GT code of the given part design and monitoring the function; it is the computer that determines the sequences of operations and the manufacturing parameters. It is, therefore, the computer's role to select the processes and the machines, to determine the sequences of operations, and, finally, to sum these up into the optimal process plan.

Despite the attractive features of a generative CAPP system, it is important to note that the decision logic often varies from company to company and that, therefore, the theoretical process parameters have to be adjusted to conform to the practical manufacturing conditions. For this reason, it is very difficult to establish a truly generative system. Nevertheless, some attempts are quite close to that goal. Generative process planning (GENPLAN), which was developed by Lockheed Corporation, Georgia, is among the successful attempts. The manufacturing logic included in GENPLAN's database was based on a comprehensive analytical study of process plans that were developed over the course of 25 years. As soon as a code is assigned to a part design by the process planner, the software quickly accesses the different alternatives and makes the appropriate decisions to establish an optimal process plan for a generative CAPP.

Another successful attempt is the DCLASS system, which was developed by Brigham Young University of Utah. The DCLASS approach is based on using a hierarchical tree of orderly classified processes as a common reference, to which process parameters and part-attributes information trees are compared and evaluated. These trees take the form of keywords and GT codes. Despite these efforts, much research and development (R&D) still has to be carried out in this area. Moreover, use has to be made of emerging concepts in computer science, such as artificial intelligence, in order to achieve the goal of a complete and truly generative CAPP system.

15.5 MATERIAL-REQUIREMENT PLANNING

Let us now look at what MRP is and why it is important. Consider a plant that is well operated by an active, qualified manufacturing staff and that is suddenly faced with the fact that there is no stock material. Production will stop, although machine tools and manufacturing personnel are available because there is simply nothing that can be manufactured. It is, therefore, of supreme importance to manage inventories to ensure a continuous supply of stock material if production is to run uninterrupted. MRP is a computerized method for ensuring this. The function of MRP software is to continually update all information about inventory, to check whether the necessary raw materials and purchased parts will be available when required, and to issue purchase orders whenever necessary.

MRP software packages can be obtained from various vendors, but they have to be tailored to the specific needs of the client. Major computer producers also provide MRP packages as part of the software developed for their machines. Software houses and consulting bureaus offer MRP packages that can be run on different types and makes of computers. It is even possible, in some cases, to obtain MRP on a time-share basis.

A recent trend in MRP development calls for linking inventory planning with the financial system of the company in order to achieve a total business plan. In this case, MRP is usually referred to as material-resource planning (MRP II). Whether MRP or MRP II is employed, the gains have been impressive. Elimination of late orders and delays, increased productivity, and reduced WIP inventories are among the benefits of MRP systems. A further benefit is that MRP promotes the integration of manufacturing systems.

15.6 THE POTENTIAL OF ARTIFICIAL INTELLIGENCE IN MANUFACTURING

The Latin root of the word *intelligence* is "intelligere," which literally means to gather, to assemble, and then to form an impression. The word thus has a meaning that involves understanding and perceiving, followed by the skilled use of reason. If we assume that an artifact gathers knowledge, chooses among facts based on sound reasoning, understands, and perceives, then what we have is indeed an artificial intelligence (AI). The desired artifact or reasoning machine must have the special capability of processing knowledge information at very high speeds. In fact, this was the goal of the

Japanese who, in 1982, established the Institute for New Generation Computer Technology (ICOT) to guide a 10-year research program aimed at developing hardware and software for the fifth generation of computers, or the knowledge information processing systems (KIPS).

Let us now look at the differences between fifth-generation and fourth-generation computers. The design of all fourth-generation computers is generally the same, and they are called von Neumann machines. Each consists of a central processing unit CPU (a memory, an arithmetic unit, and a controller) and input/output devices. A characteristic feature of all von Neumann machines is that they operate in an orderly sequence (serial fashion). The difference is not, therefore, in the working principles but in the building unit of the hardware or central technology. Whereas first-generation computers were composed of vacuum tubes, the second generation was transistorized, the third involved integrated circuits, and the fourth was based on very large-scale integrated circuits (VLSICs). Fifth-generation computers, however, involve new computer architectures, new memory organizations, and new programming languages. These computers are designed to handle and process symbols and not just numbers, like the von Neumann machines. Also, programs can be run in any order (i.e., not necessarily in a serial fashion). This requires a special computer architecture involving parallel processors (more than one computer working together at the same time). There are special AI programming languages; the most commonly used are PROLOG and LISP. PROLOG has become a favorite in Europe and Japan and was also selected for the Japanese ICOT R&D project. LISP has been widely used in the United States and has become the dominant language. So, what role can artificial intelligence play in the future of advanced manufacturing systems? In order to answer this question, we should keep in mind that AI programs are mainly concerned with symbolic reasoning and problem-solving. It is, therefore, anticipated that fields of AI technology such as expert systems, artificial vision, and intelligent robots will have great potential in manufacturing. Following is a brief discussion of each of these AI technology applications in manufacturing.

15.6.1 Expert Systems

15.6.1.1 Definition

An expert system can be defined as an intelligent computer program that has enough knowledge and capability to allow it to operate at the expert's level (i.e., it can solve extremely difficult problems that can be solved only by employing the expertise of the best practitioners in the field). The person who develops an expert system is called a knowledge engineer, and his or her job involves acquiring knowledge and structuring it into an AI program. The program usually consists of a large number of if-then rules, perhaps as many as a few thousand, in addition to a knowledge base.

15.6.1.2 Applications in Manufacturing

It is expected that expert systems will have widespread application in CIM systems. They can be employed in maintaining the systems and quickly detecting any faults or disturbances that may occur. Another emerging application involves intelligent control of machine tools. As previously discussed, CNC, DNC, and computer-assisted part programming are different forms of preplanned computerized control of machine tools. In all cases, the tool path has to be established beforehand through a program. The person who prepares the program employs his or her experience in order to bring the processing time to a minimum without causing any damage or distortion to the workpiece. This is, in many cases, a difficult problem involving many factors, alternatives, and constraints. It is exactly where an expert system is needed.

15.6.2 Artificial Vision

Artificial vision is currently the most exciting emerging AI technology in manufacturing. Although artificial vision is commercially available today, a lot of research is being carried out on more advanced aspects of this technology. Generally, it can be stated that artificial vision is aimed at locating, identifying, and guiding the manipulation of industrial components. As research continues in computer science to develop a superior pattern-recognition methodology, we expect to see more and more commercial and industrial applications of artificial vision.

15.6.3 Intelligent Robots

Intelligent robots have always been a dream of manufacturing engineers' intents of making the fully automated factory of the future attainable. It is artificial intelligence that will make this dream come true. By definition, an intelligent robot is one that is able to think, sense, and perform so as to cope with a changing environment and learn from experience. Because thinking is a brain function, it is obvious that it would fall within the domain of artificial intelligence if it were to be performed by a computer. Integration between reasoning, sensing, and performing would unify artificial intelligence and robotics, with the final outcome being an intelligent robot.

15.7 FLEXIBLE MANUFACTURING SYSTEMS (FMS)

Flexible manufacturing is one of the most exciting emerging automation concepts. In order to understand this concept, we must first discuss automation in general. As we know, mechanization started with the Industrial Revolution, when machines performed some of the functions of labor. Automation included mechanization and went beyond it. With the twentieth century came the new concept of mass production. The automation technology then available (i.e., the assembly line) markedly increased the productivity of workers, and the United States grew to lead all other nations both in productivity and prosperity. Both the assembly line and the transfer line belong to the domain of fixed automation, in which the sequence of processing operations remains unchanged throughout the production run because the sequence of operations is determined mainly by the nature and arrangement of the physical equipment in the production line, which, once established, cannot be changed (except with great difficulty). It can, therefore, be seen that fixed automation, which is based on integrating and coordinating simple operations, has three characterizing features. First, it requires a very high initial capital cost; second, it is suitable only for high production volume (to pay back for its high capital cost); and third, it is inflexible and cannot handle product changeover or even major changes in the product design.

In contrast to fixed automation, programmable automation is flexible and can handle a large variety of parts. A good example of this kind of automation is a stand-alone CNC machine tool, where new part designs are accommodated simply by preparing new part programs, without the need for changing the machine. Programmable automation is characterized by its low production volume, its flexibility and ability to tolerate product changeovers, and the fact that its initial capital cost is not as high as that of fixed automation.

An interesting way of illustrating the difference between these two types of automation is to plot the number of different shapes (or designs) of workpieces that the automated system can tolerate versus the annual production volume. This is clearly indicated in Figure 15.10, which also shows the recommended successful domain for each of the two automated systems. Notice that a gap exists between the high-production transfer lines (fixed automation) and the low-production, though flexible, individual CNC machines. This gap involves medium production volume accompanied by a limited variety of part designs. In fact, it is this kind of production that is required most today because of the need for insurance against unexpected circumstances, which calls for flexibility and tailored products demanded by customers in quantities that are not suitable for mass production. A reasonable solution to this problem is to develop a hybrid of fixed and programmable automations that combines the best features of both. This new production system should be responsive to the changing needs of manufacturing, yet highly automated. It is, therefore, appropriately called a FMS.

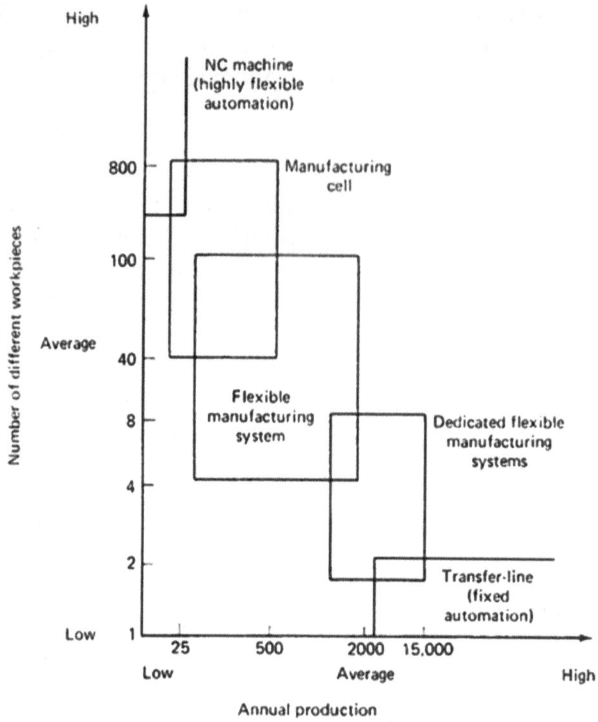

FIGURE 15.10 Areas of application of the different automated manufacturing systems.

15.7.1 Elements of a Flexible Manufacturing System

Although FMSs may differ slightly in features, depending upon a system's manufacturer, they are all based on the same idea of incorporating several individual automation technologies into a unified, highly responsive production system. We can easily see that this new method of automation is actually based on three main elements: machine tools, an automatic material-handling system, and a computer-control system that coordinates the functions of the other two elements in order to achieve flexibility. Figure 15.11 is a chart of these basic FMS elements.

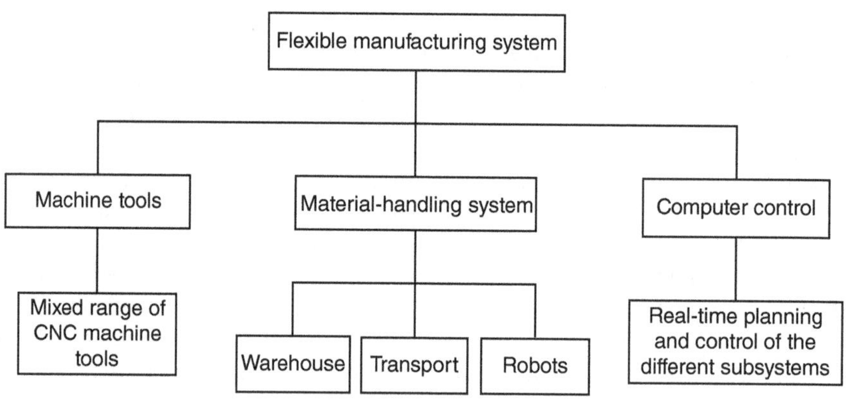

FIGURE 15.11 Elements of an FMS.

15.7.1.1 Machine Tools

An FMS consists of a mixed range of CNC machine tools. The characterizing features of these machine tools depend upon the configuration of the components to be manufactured as well as the desired degree of flexibility of the system. Therefore, some manufacturing experts tend to divide FMSs into two categories: dedicated and random FMSs. Dedicated FMSs possess a low degree of flexibility and are built to meet long-term, well-defined manufacturing requirements. Such systems can, therefore, tolerate only a limited variety of processing needs. Special machine tools form the backbone of this type of system. Random FMSs are designed to machine (or process) a wide range of different parts in random sequence. Therefore, the machine tools used must be highly flexible. New part designs can easily be handled by random FMSs. Four- and five-axis CNC machine tools are commonly used with this type of system.

Although the machining processes involved differ from one FMS to another, they often include operations like turning, milling, drilling, tapping, cylindrical grinding, spline hobbing, and broaching. Batches usually fall within the range of 25–100 and can be completed, in most cases, in 3 days or less. The system and the machine tools are designed and planned so that they operate without human supervision during unfavorable work hours.

15.7.1.2 Material-Handling System

The material-handling system involves robots that are used for the automatic loading and unloading of machine tools and a means of transportation that links the various machine tools together and moves parts around (e.g., power roller conveyors, shuttle carts on fixed railroads, vehicles that float and move on an air or a fluid cushion, industrial robots that travel on the factory floor). All movements of parts (or pallets) in the system must be randomly independent (i.e., parts can move from one station to another without interference). These movements are directly controlled by the main computer of the FMS. Each pallet is uniquely coded so that the computer can track it. Codes (which are usually binary) are affixed to a coding tag that is mounted onto a pallet identification strip. This enables the computer to use conveniently located sensors to identify and accurately locate pallets. Industrial robots are then used to load pallets and/or parts in the machine and also return pallets loaded with machined parts to the transport system. Temporary storage is provided at each machining station so that a workpiece can be promptly loaded in the machine as soon as a machined part is removed from it, thus increasing machine utilization by saving time. Another function of the robots is to select components for inspection under computer control.

15.7.1.3 Computer-Control System

An FMS cannot be operated without a digital computer because of the complexity of the system and the need for real-time planning and control of the different subsystems. The function of the computer is not limited to the storage and distribution of the NC part programs, the operation of load–unload robots, and the control of shuttle-cart traffic between stations. It also covers tool monitoring (following each tool throughout the system and monitoring its life) and production control (employing a data entry unit in the load–unload station to act as an interface between the operator and the computer).

15.8 FLEXIBLE MANUFACTURING CELLS

A FMC usually consists of two CNC machining centers (or one machining center and one or two special machines) that are linked with each other as well as with their tool storage area through an industrial robot. Because of the flexibility inherent in the design of an FMC, its productivity is higher than that of separate stand-alone machines, and it can also handle a larger variety of parts configurations. FMCs are, therefore, used when the production volume does not justify the purchase of an FMS yet cannot be met by stand-alone machines. FMCs are a kind of automation that fills the gap between FMSs and individual stand-alone CNC machine tools. The domain of successful

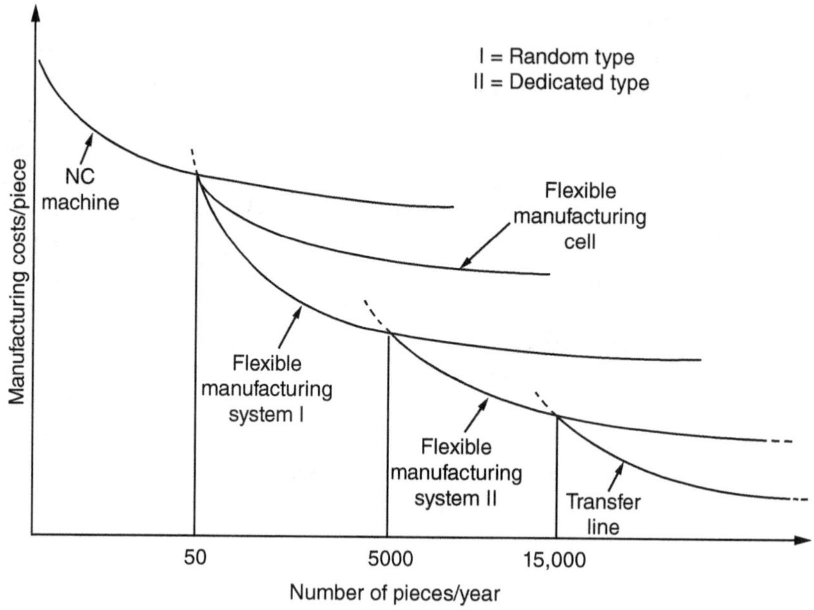

FIGURE 15.12 Cost per piece versus production volume for different automated manufacturing systems.

application for FMCs, therefore, involves a production volume between 20 and 500 pieces and a variation in configuration ranging between 30 and 500 different shapes.

The economic advantage of using FMSs and FMCs can easily be realized by comparing each system with individual machine tools in relation to the production cost per component. We must not, however, forget that the production volume plays a very important role in determining the cost per piece. Figure 15.12 shows the cost per piece versus the production volume for each of the different kinds of automated manufacturing systems.

15.9 INTERNET OF THINGS (IoT)

The IoT can be defined as a system of physical objects, digital equipment, computers, and mechanical machines that are assigned unique identifiers (IP addresses) and provided with the ability to exchange data and communicate with one another over a network without requiring human-to-human or human-to-computer interaction. Evidently, that would help tear down the walls between operational technology and information technology, yielding opportunities to use data from the devices to improve efficiency and productivity. Furthermore, IoT devices can be scattered all over the world and still be connected together, enabling that convergence on a global scale and resulting in substantial economic benefits.

While there were some crude attempts to bring about the idea of a network of smart devices connected together, the concept was truly defined by Reza Raji in *IEEE Spectrum* (1994) as "moving small packets of data to a large set of nodes, so as to integrate and automate everything from home appliances to entire factories." The term *Internet of Things* was later forged by Kevin Ashton, cofounder and executive director of the MIT Auto-ID Center in 1999 (it was then called the *Internet for Things*). He considered radiofrequency identification (RFID) to be the essential element to allow computers to control all individual things.

Evidently, IoT devices have extensive applications in consumer, business, and infrastructure; we are concerned here with its applications in manufacturing. IoT incorporates CIM and by far surpasses it, as it brings together and automates the various distant islands of intelligent systems and is therefore rightfully considered to be the backbone of smart manufacturing. Network control is not limited to manufacturing aspects but also includes assets, business and service information, and SCM. Industrial *Big Data* move at an extremely high speed between a large number networked sensors, thus enabling the real-time exchange of needed information. Accordingly, enterprises can have access to more data about their own products and their own internal systems and, consequently, a greater ability to make prompt changes as needed. As you can see, IoT devices provide dynamic responses to product demands and real-time optimization of manufacturing and business systems. Occasionally, industrial manufacturing professionals refer to this industrial subset of applications as the *Industrial Internet of Things*, or IIoT, and consider it to be the Fourth Industrial Revolution (4IR). It, therefore, earned its commonly used nickname of "Industry 4.0."

15.10 SUPPLY CHAIN MANAGEMENT (SCM)

As we saw in Chapter 1, when discussing the production turn, the manufacturing and business activities are entangled together, with marketing playing a vital role in ensuring the continuity of the cycle. With globalization dominating the world economy and the emergence of the IoT, there is a necessity for business and manufacturing activities to come even closer in order to achieve alignment and communication between all the entities. The final goal is to create a competitive advantage and ensure that the consumer receives the best-quality goods, at the best possible price. This requires an endeavor to adopt strategic planning and synchronizing supply with demand, which is certainly what SCM encompasses.

The Council of Supply Chain Management Professionals (CSCMP) provides the following definition for *SCM*:

> Supply chain management encompasses the planning and management of all activities involved in sourcing and procurement, conversion, and all logistics management activities. Importantly, it also includes coordination and collaboration with channel partners, which can be suppliers, intermediaries, third party service providers, and customers. In essence, supply chain management integrates supply and demand management within and across companies.[1]

We will discuss and explain this definition and how SCM interacts with and affects our various manufacturing activities. In order to do that, let us first explain some business terms, which the engineering student might not be familiar with the following:

- *Sourcing* means tracking down and locating the necessary raw materials, components, or subassemblies that meet quality requirements and that are needed for manufacturing the products. It is aimed at reevaluating and improving the purchasing activities.
- *Procurement* involves finding the best prices, agreeing on terms, and acquiring goods (raw materials, etc.) from an external source, usually through tenders or a competitive bidding process.
- *Logistics management activities* was defined by the CSCMP as

> that part of supply chain management that plans, implements, and controls the efficient, effective forward and reverses flow and storage of goods, services and related information between the point of origin and the point of consumption in order to meet customers' 'requirements.'[2]

It is often confused with SCM or used synonymously, although it is just a component. It manages activities such as packaging, transportation, distribution, warehousing, and delivery, as well as the associated IT.

We can clearly see that SCM has a marked effect on production operations because it is entangled with MRP, which has been explained previously. This convergence ensures the availability of raw materials that meet the required quality, just in time and at the anticipated location, in order to avoid any interruption of production.

Furthermore, SCM plays an essential role in product development and continuous improvement by establishing the attributes and characteristics of an expected or potential product. This is achieved through the statistical analysis of the clientele demands, returns for excess, and defective products.

In brief, SCM is an integrating function with primary responsibility for linking major business functions and business processes within and across companies into a cohesive and high-performing business model. It includes all of the logistics management activities noted above, as well as manufacturing operations, and it drives the coordination of processes and activities with and across marketing, sales, product design, finance, and information technology.

15.11 INDUSTRY 4.0

Industry 4.0, also known as the Fourth Industrial Revolution (4IR) is a pivotal force shaping the future of industries globally. It constitutes the full digitization and the integration of intelligent digital technologies into manufacturing and industrial processes. The latter can be exemplified by, but not limited to the following:

- Artificial intelligence (AI)
- Big data
- Industrial IoT networks
- Robotics
- Automation
- Machine learning
- Augmented reality

Industry 4.0 enhances productivity and efficiency in manufacturing activities and supply chain operations. It also supports more rational and data-driven decision-making.

The benefits of Industry 4.0 are the following:

- Increased labor productivity
- Reduced machine downtime
- Reduced inventory
- Improved forecasting
- Reduced human exposure to hazardous environments
- Remote monitoring
- Enhanced safety through predictive maintenance
- Creation of smart factories where advanced technologies seamlessly collaborate

The driving forces behind Industry 4.0 include artificial intelligence, robotics, and 3D printing. Its fundamental requirements include distributed computing and control, pervasive sensing, and authenticated and secure systems.

Implementing Industry 4.0 requires open software and communication standards that enable sensors, controllers, people, machines, equipment, logistics systems, and products to interact and collaborate effectively. However, challenges include the following:

- High implementation costs
- Cybersecurity risks
- Potential job displacement due to increased demand for skills in programming, operating, and maintaining technologies like industrial robots, automated systems, and big data analytics tools

15.11.1 Societal Impact of Industry 4.0

The influence of Industry 4.0 extends beyond industries to address critical global challenges such as resource and energy efficiency, urban production, and demographic changes.

It facilitates the supply chain by emphasizing a global network of machines in smart factories that autonomously exchange information and coordinate operations.

15.11.2 Conclusion

Industry 4.0 is poised to create a new digital era, transforming how products are designed, manufactured, and delivered. This revolution is not just reshaping industries but also contributing to solving some of the world's pressing problems while paving the way for a more efficient and connected future.

REVIEW QUESTIONS

1. What does CIM stand for and what does that mean?
2. Why is the implementation of CIM necessary? Can CAD, CAM, and other advanced technologies be a substitute? Why?
3. Describe the role (and the quality) of the engineers who will operate a CIM system.
4. What steps should be taken in order to implement CIM properly?
5. List the benefits of CIM. Briefly describe each.
6. Describe the typical architecture of a CIM system.
7. What is the major purpose of a DBMS?
8. What are the classes of CIM? Briefly describe each.
9. In a CIM system, what is meant by communication networks? What is their function?
10. List and use sketches to illustrate some LAN topologies. What are the advantages and disadvantages of each?
11. What is meant by network architecture?
12. Explain and use sketches to illustrate the layers included in the network architecture involving MAP.
13. What is group technology?
14. List the reasons for adopting GT.
15. List the benefits of GT. Briefly describe each.
16. What factors prevent the widespread application of GT?
17. Describe the classification and coding of parts. What important feature does this system have?
18. Is there a universally standard classification and coding system? Why?
19. How can a coding system be constructed?
20. What are the basic types of code construction? Explain briefly the characteristics of each, as well as the advantages and disadvantages.
21. What is meant by cellular manufacturing?
22. What is the difference between a plant design based on GT principles and another one based on old-fashioned concepts?
23. Explain the meaning of a composite part.
24. What is the right approach to take when designing a production cell?
25. What is meant by computer-aided process planning? How is it linked with group technology?
26. How do you compare CAPP with the conventional process planning performed today?
27. List the benefits of CAPP. Briefly describe each.
28. What are the main types of CAPP?
29. Give some examples of commercially available CAPP systems.
30. What is meant by material-requirement planning? What is its function?

31. What is the recent trend in MRP development? How does this serve the goal of implementing CIM?
32. What is artificial intelligence?
33. What are the differences between fifth-generation and fourth-generation computers?
34. Differentiate between the first four generations of computers.
35. What computer languages are suitable for artificial intelligence?
36. What does the term *expert system* mean?
37. What application will expert systems have in manufacturing?
38. What is meant by artificial vision? What is its main function?
39. What is the difference between an intelligent robot and a conventional industrial robot? Explain.
40. What are the two basic types of automation? List the main characteristic features of each.
41. Are the two basic types of automation enough to meet today's market demands?
42. Why is there a need for flexible manufacturing systems?
43. What are the main elements of an FMS? Briefly describe each.
44. What are the two kinds of FMSs?
45. Are these two kinds of FMSs enough to fill the gap that exists between the two basic types of automation? If not, how can this gap be filled? Describe a similar automated manufacturing system that can fill this gap.

DESIGN PROJECT

Following are form codes of parts according to the Opitz system:

11106, 11002, 10110, 11100, 10002, 11000

You are required to draw a neat sketch of each part (orthogonal projection) and then synthesize the composite part. Finally, make a preliminary design of a manufacturing cell to produce this family of parts. Assume the number of parts is equally divided among the different codes.

NOTES

1 Council of Supply Chain Management Professionals, "CSCMP Supply Chain Management Definitions and Glossary," https://cscmp.org/CSCMP/Educate/SCM_Definitions_and_Glossary_of_Terms/CSCMP/Educate/SCM_Definitions_and_Glossary_of_Terms.aspx.
2 CSCMP, "Definitions and Glossary."

16 Fabrication of Microchips

16.1 BEGINNING OF THE ELECTRONICS AGE

After the experiments of Professor Rutherford of New Zealand, scientists understood the nature of electricity as a flow of negatively charged particles, or electrons, in wires. This understanding paved the way for the electronic age, which began in 1904 with the invention of the diode by John Ambrose Fleming.

A diode resembles an incandescent electric bulb but has a single pole with two elements, which are insulated by a vacuum or inert gas: one connected to the anode and the other to the cathode. When the cathode is electronically heated and the anode is given a positive charge, electrons flow from the cathode to the anode, creating an electric current that powers the circuit.

A diode acts as a one-way valve for electricity, allowing current to flow in one direction while blocking it in the opposite direction. It quickly found industrial applications in electric rectifiers, converting AC current to DC current.

Soon after, in 1906, the triode was invented. Like the diode, it is a vacuum tube but differs by having three electrodes inside the evacuated glass bulb: a heated filament or cathode, a plate or anode, and a grid in between that controls the flow of electrons by adjusting its charge. If the grid is given a negative charge, no electrons can pass, and the current stops flowing. However, if the grid is given a positive charge, it accelerates and increases the flow of electrons. This means that a small amount of power applied to the grid can be amplified into a much larger output. Triodes were used as amplifiers and to generate high-frequency currents.

One of the biggest challenges in the early decades of the twentieth century was finding a mechanism that could switch electrical signals on and off using electrical signals themselves. This goal was achieved with the invention of the triode. Information could then be manipulated by controlling the flow of electricity. When Guglielmo Marconi invented the wireless telegraph, vacuum tubes found important applications in that invention. Five hundred vacuum tubes were employed in an apparatus that sent the first trans-oceanic wireless signal. Triodes were also used in consumer electronic devices, such as radio sets. These radios were bulky and filled with tubes and wiring. You might see one in a movie from the 1940s or 1950s, and you would be surprised at how large it is compared to a modern mobile phone, which includes radio functionality.

16.2 COMPUTING MACHINES

In the 1920s and 1930s, the word "computer" had a different meaning than it does today. It referred to a human being with a ream of papers and a pencil who carried out calculations manually. Small companies hired many human computers to perform all necessary arithmetic calculations. This expensive and time-consuming method led some electrical engineers to consider employing Boolean algebra to build a machine using triodes to perform arithmetic calculations. Boolean algebra involved a system where logical statements could be expressed in mathematical terms, which could then be executed using mathematical operations on symbols. Only 0 and 1 were used, aligning well with the performance of the triode, where 0 meant off and 1 meant on (see Chapter 14 for a detailed explanation of this system).

The next step involved connecting a few vacuum tubes, copper wires, and other rudimentary electrical components to perform basic logic functions. These were the building blocks of the

DOI: 10.1201/9781003537106-16

electronic computer and were referred to as logic circuits. Vacuum tube calculating machines began replacing calculators that used gears and levers.

The first attempt to build an electronic computer took place in 1938 at Iowa State College (now Iowa State University). Funded by the U.S. Army Ordnance Corps in 1943, John Mauchly and J. Presper Eckert constructed the first-generation computer, ENIAC, at the University of Pennsylvania. By the time it was taken out of operation, the ENIAC had 20,000 vacuum tubes. Each digit in an algorithm required 36 tubes. This technology, however, suffered from significant problems, such as being power-hungry and generating excessive heat, requiring the tubes to be spaced to avoid overheating, which necessitated a large building to accommodate the electronic equipment. There was a need for new technology to eliminate these shortcomings, leading to the development of transistors.

16.3 THE ERA OF TRANSISTORS

The era of transistors began when it was discovered that a crystal of Germanium, a semiconductor element, would allow electric current to flow in one direction only. This led to the development of better transistors, known as bipolar transistors, which replaced vacuum tubes. Vacuum tubes were bulky and required a lot of power, whereas transistors were more compact and consumed less power. Eventually, silicon replaced germanium as the preferred material for transistors due to its availability.

The transition to transistors resulted in significant advancements in electronic equipment. As mentioned above, transistors were smaller, consumed less power, and required less cooling than vacuum tubes, leading to a reduction in the size of electronic devices. This allowed for innovations such as portable radio sets powered by batteries, which could be carried in a small bag to the beach.

The second generation of computers utilized transistors instead of vacuum tubes for logic circuits. In 1953, the University of Manchester created the Transistor Computer, marking a significant milestone in computing history.

Bipolar Junction Transistors (BJT) became common electronic devices with three terminals as shown in Figure 16.1: the emitter (source), the collector (sink), and the base (gate). One type of BJT is the NPN transistor, which consists of two N-type (negative) semiconductor materials separated by a thin layer of P-type (positive) semiconductor material. The flow of electrons from the emitter to the collector is controlled by the current flow in the base terminal. A small current at the base terminal causes a larger current to flow from the emitter to the collector.

When the base of a transistor is ON (or has a value of 1), electricity flows from the emitter to the collector, and the transistor is said to be ON. Conversely, when the base is OFF (or has a value of 0), electricity does not flow from the emitter to the collector, and the transistor is said to be OFF.

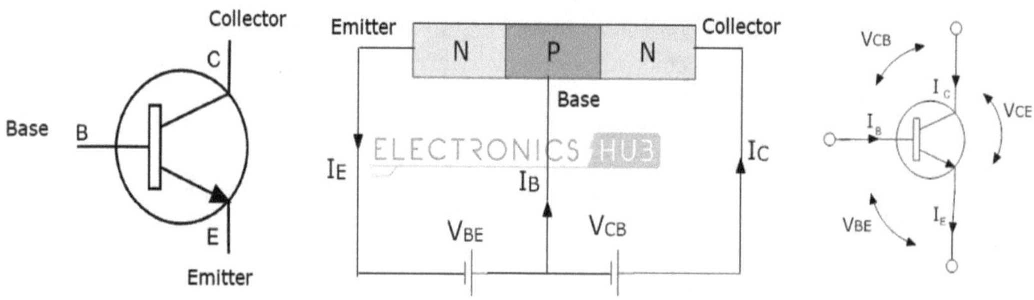

FIGURE 16.1 BJT transistor.

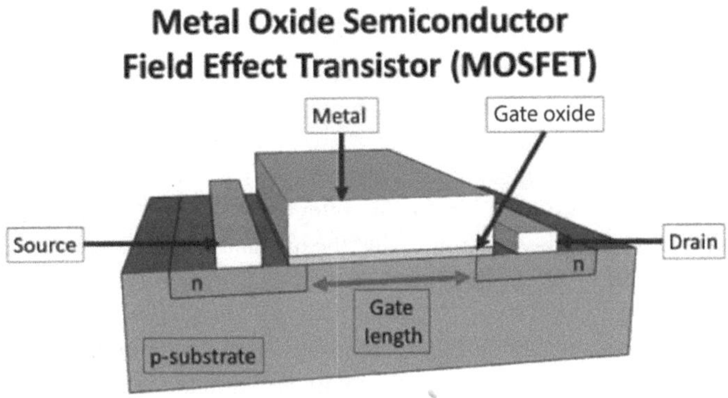

FIGURE 16.2 The MOSFET transistor.

16.4 THE ROAD TO THE MICROCHIP

The achievement of utilizing transistors instead of vacuum tubes made computers (as well as other electronic equipment) smaller in size. However, the computing power and capacity were still quite limited due to the size of the transistor valves. This necessitated the development of new technology. In 1957, Egyptian American physicist and electrical engineer Mohamed Atalla and his Korean American assistant Dawon Kahng achieved this breakthrough at Bell Laboratories in New Jersey.

While conducting research on the passivation of surfaces, Atalla invented the MOSFET (metal-oxide-silicon field-effect transistor), also known as the MOS transistor. MOSFETs offered rapid switching speeds, enabling fast response times and efficient operation in digital electronic equipment. This new method of creating semiconductors involved coating a silicon wafer with an insulating layer of silicon oxide, allowing electricity to overcome the surface state and reliably penetrate into the semiconducting silicon below. A diagrammatic sketch of a MOSFET is shown in Figure 16.2. This type of semiconductor includes a thin metallic layer acting as the gate (G), insulated from the P-type substrate terminal by a very thin layer of silicon oxide. The source (S) and the drain (D) are heavily doped N+, with the supporting base usually connected to the source terminal. The current flows as a result of the flow of electrons, while the gate voltage controls the electrons in the channel between the source and the drain, thus controlling the flow of electrons.

Based on this new technology, Atalla and Kahng were able to miniaturize the MOSFET. They created numerous valves on the same wafer, eliminating the need for a motherboard on which traditional transistor valves are mounted in integrated circuits. They produced a 0.5-inch × 0.5-inch wafer with 10,000 transistor valves, marking the creation of the first VLSIC (very large-scale integrated circuits). Although primitive compared to today's microchips, each containing billions of microscopic transistor valves, their work paved the way for modern microchip technology.

16.5 MANUFACTURING OF VLSIC

After becoming familiar with the components and function of the MOSFET transistor, we need to mimic it on a nanoscopic scale in a way that allows for mass production. The microchip, or VLSIC, is a small but sophisticated device composed of two main parts: a tiny and fragile silicon chip (referred to as the "die") and packaging material designed to protect the chip and enable practical, safe, and easy handling.

The production sequence of a microchip includes two main parts. First is the fabrication of the tiny silicon chip, a highly sophisticated and intricate sequence of the processing referred to in the industry as the "Front End." Second is the packaging of the die, a very precise automated process referred to as the "Back End." Evidently, the Front End comes after the initial preparation of the design and the production of the raw stock for processing.

16.5.1 Initial Phase

In the predesign phase, the goal is to create billions of integrated and connected transistors on a semiconductor surface that is extremely small, making the design process highly complex. This phase involves defining the microchip's functions, simulating its technical and physical properties, testing its functionality, and working out the connections between individual transistors. Special design tools are used to draw up the plan for the integrated circuits, develop a three-dimensional geometric pattern for the various layers comprising the die, and detail the patterns in each layer. This process enables the creation of photo images of the circuits, which will be used later for chemical machining operations.

It is reasonable to ask how the silicon-based chip is produced and prepared for further processing to yield the die. After the Front-End sequence is completed, the die is cut from a thin silicon wafer using a diamond tool. This means that the processed wafer has numerous dies on it, resembling postage stamps on a sheet until the Front-End process is completed (as shown in Figure 16.3).

The first step in fabricating a microchip is to prepare a very thin silicon wafer. Silicon is the second most abundant element in the Earth's crust, but it is always found combined with oxygen to form silicon dioxide, or silica, which is pure sand, like that found in Silicon Valley, California. To obtain silicon, silica must undergo a chemical reaction to extract silicon by reduction. For example, if you mix sand with magnesium in a test tube and heat it, you will end up with silicon and magnesium oxide. In industry, sand is blended with carbon and heated to an extremely high temperature to extract silicon. Complex physical and chemical processes are performed to ensure the silicon's purity, so the wafer meets the high standards required for microchip fabrication.

Next, pure silicon undergoes specialized processing to create a monocrystalline ingot, commonly known as "the bull." This process achieves an exceptionally high level of purity, with only one impurity atom per 10 million silicon atoms. These ingots are produced in various diameters, typically 6 inches (150 mm), 8 inches (200 mm), and 12 inches (300 mm). From these ingots, extremely thin wafers are sliced, each measuring 0.650 mm (650 microns) in thickness. A special sawing machine is used to ensure precision in cutting. After slicing, the wafers undergo a thorough cleaning process to remove any contaminants. They are then transferred to a clean room, where the Front-End stage

FIGURE 16.3 The wafer with finished dies unseparated.

of subsequent processing begins. This clean environment is crucial for maintaining the purity and integrity of the wafers for semiconductor manufacturing.

A clean room should not be mistaken for an ordinary residential or commercial room. These specialized environments can be as large as eight football fields and are meticulously maintained to be dust-free, with stable levels of temperature and humidity. The presence of even the tiniest dust particles can significantly impact the wiring pattern on a microchip during processing. Therefore, the cleanliness of a clean room is a million times better than the air around us and maybe a hundred times cleaner than the air in a heart-transplant operating room.

The cleanliness of the environment in which fabrication takes place is measured by the size and number of dust particles in one cubic foot of air. In some areas of a clean room, the level is restricted to a maximum of only one particle, with a diameter not exceeding one micron (typically 0.5 microns), per cubic foot of air.

To maintain this exceptional level of cleanliness, very sophisticated ventilation and filtration systems operate 24/7. Millions of cubic meters of air are circulated every hour, ensuring a constant airflow and preventing any contamination.

16.5.2 The Front-End Part of Fabrication

After slicing the silicon ingot to produce raw wafers, they are transferred to the clean room known as the "Fabo." Returning to the MOSFET transistor of Mr. Attallah, we must create it on a microscopic or, more accurately, nanoscopic scale. This requires forming the P-type and N-type regions with the semiconductor material in between. The silicon wafer naturally provides the semiconductor material, but the P and N regions are created through a process called "doping."

Doping involves introducing atoms of a dopant into the crystal lattice of silicon using a high-energy electron beam. This process alters the electrical properties at specific locations to form N-type (negative) and P-type (positive) carriers. The most commonly used dopants are phosphorus and arsenic for N-type regions, and boron for P-type regions. This process, known as "Ion Implantation," is widely used in the industry because it allows precise control over the depth of doping, although it is relatively expensive.

Historically, doping was achieved through diffusion, where the silicon wafer was placed in a high-temperature furnace, and doping gases were introduced onto its surface.

Ion Implantation, while highly effective, causes some damage to the silicon wafer's crystal lattice. Therefore, an annealing process is necessary to repair this damage.

Doping is a crucial step in the process of creating nanoscopic transistor valves, which function similarly to the vacuum tubes used in radios from the 1940s. Once the fabrication process is complete, the transistors are located on the surface of the chip, beneath a complex 3D maze of wires distributed across more than a dozen layers.

The first step in the process involves subjecting the wafer's surface to oxidation in a high-temperature furnace operating at around 1000 degrees Celsius. This creates a non-conductive thin layer of silicon dioxide (SiO_2) on the wafer's surface. After this, the wafer is prepared to receive the pattern that defines the boundaries of the active regions of the integrated circuit, where transistors, capacitors, diffused resistors, and interconnects will be formed as part of a complex, multi-layered integrated circuit.

To create space for the pattern, the thin silicon dioxide layer must be selectively removed. This is accomplished through a process called photolithography, followed by chemical etching. In this process, a photoresist material is uniformly applied to the oxide layer using centrifugal force, creating a light-sensitive coating on the wafer.

As illustrated in Figure 16.4. A photomask is placed over the photoresist-coated wafer, and the wafer is exposed to ultraviolet light through the photomask in a specialized exposure machine known as a "stepper." Chip templates, known as reticles (as shown in the figure), are used to transfer the intricate geometric pattern of the circuit design onto each of the microchips on the silicon wafer.

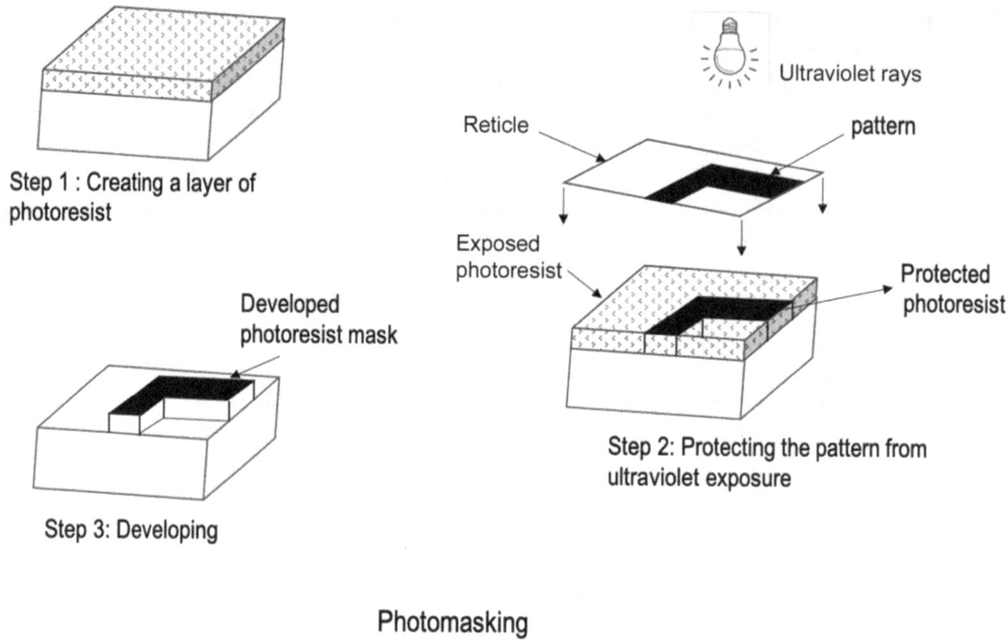

FIGURE 16.4 The masking process.

The exposed areas of the pattern are then developed, revealing the underlying silicon dioxide layer, while the unexposed areas remain covered by the photoresist, protecting the oxide layer beneath.

Next, the exposed oxide layer is etched away (Figure 16.5) in the developed areas using wet or plasma etching, creating vertical openings through the silicon dioxide down to the bare silicon beneath. After etching, the photoresist residue is stripped away, and the wafer is cleaned.

The wafer is then moved to the sputtering (shown in Figure 16.5) or vapor deposition tool, where a sequence of metals is deposited to fill the exposed pattern, coating the wafer in a very thin layer of metal. Finally, the wafer undergoes planarization, during which the metal is polished down until a layer of SiO_2 and conductive copper (or aluminum) interconnects are formed, completing a single layer of the integrated circuit. Each pair of successive layers must be separated by a dielectric material. Where a connection between two layers is required, a miniature hole is laser-drilled and filled with tungsten, which acts as a connector between the circuits in those layers.

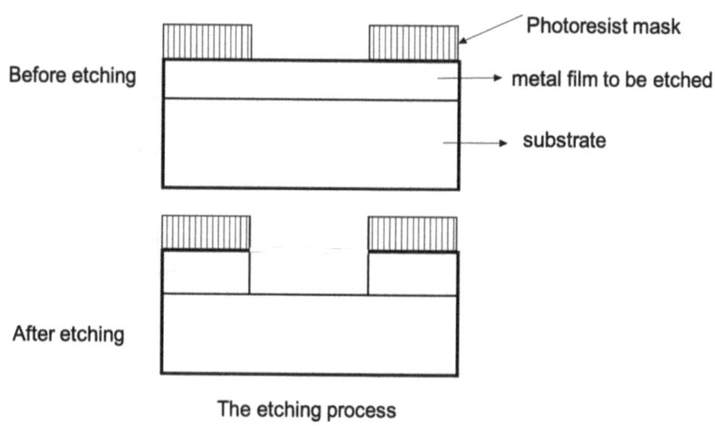

FIGURE 16.5 Chemical machining.

It is crucial to remember that microchip pattern dimensions are measured in tenths of a micron, which significantly influences the technologies used in photolithography and quality control (QC) operations. In photolithography, extremely short-wavelength ultraviolet light is necessary to reveal the intricate details of the patterns. This technology is employed by machines produced by a single company in the Netherlands. For QC operations, special Scanning Electron Microscopes (SEMs) with nanometer-level resolution are used to capture images of the wafer's surface. These images help detect defects, such as imperfectly patterned layers or particles on the surface.

Given that the production sequence in the Front-End is long, time-consuming, and expensive, it is logical to perform QC both during and after the Front-End process. This approach helps identify and eliminate defective chips at an early stage, thus saving time and resources. The QC operation conducted after the Front-End is commonly referred to in the industry as "probing." Naturally, the final QC operation, which occurs after the device is fully manufactured, is the final test—an area of expertise for electronics engineers.

It's important to note that the discussion above covers only the main processes of the Front-End; the actual number and nature of operations can vary between companies.

16.5.3 The Back-End Part of Fabrication

The first step in the Back-End part of fabrication is the separation of individual chips from the wafer in a process known as "die cutting." Once separated, the die is placed on a "lead frame," where the leads form the microchip legs that will later be soldered onto or inserted into a socket on a printed circuit board. At this stage, the device is fully functional, containing millions of interconnected electronic components capable of executing complex electronic functions.

However, handling the device safely poses a significant challenge, as any scratch or impact could compromise its reliability and potentially lead to failure. To prevent such damage, the die must be enclosed in a plastic or ceramic package to provide protection. Additionally, to connect the microchip to external devices or components, several thin wires are attached to the chip.

Like any product, a traceability code is engraved on the chip to indicate its function and the date of fabrication. Finally, at the end of the assembly process, the chip undergoes automated testing to ensure it performs its intended function before being packed and shipped.

16.6 A FINAL WORD

We are seeing big improvements in the Internet of Things (IoT), smart devices, and more. But to understand where these innovations are taking us and what to expect, we first need to learn the basic technologies behind them.

Semiconductor technology is a key part of this progress, making the IoT and AI grow faster than ever. For the past three decades, advances in semiconductors have been a major factor in boosting computing power. In fact, semiconductors make up about 50% of the cost of computing hardware. Thanks to these advancements, AI-powered devices will become more smoothly integrated into our everyday lives.

REVIEW QUESTIONS

1. What role does a diode play in an electric circuit?
2. In a hydraulic circuit, which element performs a function similar to that of a diode in an electric circuit?
3. What additional characteristics did the triode introduce compared to the diode?
4. Is it feasible to build a computer using only triodes?
5. What advantages did transistors bring to electronics?
6. Is it feasible to construct a computer entirely using transistors? What are the drawbacks?

7. How are logic circuits constructed?
8. Explain the working principle of a BJT transistor.
9. What advancements did Mr. Atalla's MOSFET transistor offer to the world of electronics?
10. How did the introduction of Atalla's transistor impact the construction of modern computers?
11. What does VLSI stand for?
12. What are the two main components of a microchip?
13. What material is a microchip made from?
14. What is a wafer, and how is it made?
15. How can a nanoscale transistor be created on the surface of a wafer?
16. What is ion implantation?
17. How is the wiring between nanoscale transistors fabricated?
18. Why are frequent quality control (QC) operations necessary during the Front-End part of fabrication?
19. Why is it essential to perform the Front-End part of fabrication in a clean room?
20. If the microchip is functional after the Front-End part of fabrication, why is packaging still necessary?

Appendix

A.1 INTRODUCTION

This appendix is aimed at those students who have not taken any materials science or materials engineering courses. It provides them with the fundamentals of this important subject so that they can understand the various manufacturing processes and the concept of design for manufacturing. It is not meant as a substitute for a textbook for those who want to study materials engineering more comprehensively. Industrial engineering students will find it helpful to study this appendix thoroughly before reading the current text. The coverage here is concise and limited to those areas that provide the background necessary to learn material processing.

A.2 TYPES OF MATERIALS

There are three basic classes of engineering materials: metals, ceramics, and polymers. How the atoms of a certain material are bonded together and how they are arranged have the ultimate influence on the nature and the properties of that material. Let us now scratch the surface of each of the three basic classes of materials to obtain a view on the atomic scale or level.

A.2.1 Metals

Metals can be elements or alloys (combinations of elements). As you studied in chemistry, the atoms of a metal are relatively large and heavy. Therefore, the attraction forces that keep the electrons circling in the outer orbits are not strong enough to sustain that dynamic equilibrium. As a consequence, the electrons are free to move throughout the piece of metal, forming an electron mist. On the other hand, the atoms that lose these electrons have positive charges. It is, therefore, clear that a solid metal is composed of atoms (that have positive charges) held together by a matrix of electrons (that have negative charges). This basic nature of metals is responsible for their useful properties. For instance, good electric conductivity is a result of the abundance of free electrons that flow when a wire is subjected to a magnetic field. Other properties of metals include their ability to undergo major permanent deformation and their opacity.

We can see from the preceding discussion that metals take the solid form when a certain bond (positive atoms with positive charges in an electron mist) exists between the component atoms. This type of bonding is referred to as a metallic bond. The atoms of a metal are arranged in a repetitive three-dimensional pattern to form tiny (usually) microscopic crystallites or grains. The smallest arrangement of atoms that, when repeated in all six directions in space, produces a grain is referred to as the unit cell. Although there are 14 different types of unit cells, only three types are common, and they are shown in Figure A.1. Each type has a different name: body-centered cubic (BCC), face-centered cubic (FCC), and hexagonal close-packed (HCP). Each metal (element) has one of these types of atomic arrangements; for example, iron and chromium are BCC, copper and aluminum are FCC, and magnesium and zinc are HCP. As we will see later on, the properties of a solid metal are affected by the type of crystal structure it has.

In a few cases, metals have two or more crystal structures (e.g., one at room temperature and one at high, or at low, temperature). A specific example is iron; it is BCC at room temperature but suddenly changes to FCC at 1674°F (912°C). This phenomenon is called allotropy and change from one type of unit cell to another is called allotropic transformation. The properties of the metal (such as its ability to dissolve other elements) also change before and after the allotropic transformation.

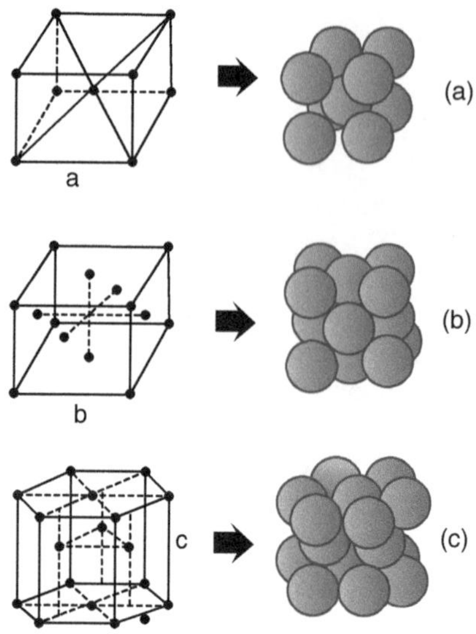

FIGURE A.1 The most common types of unit cells. (Reprinted from *Physical Metallurgy: Principles and Design*, by G. N. Haidemenopoulos, CRC Press, by permission.)

When metals undergo permanent deformation as a result of applying loads, atoms slide over one another along planes and directions where the atom population density is maximum. This mechanism of deformation is known as slip. On the other hand, if the crystal lattice has a defect in the form of impurities or distortion, which is referred to as dislocation, slip, and, therefore, deformation will be more difficult. In fact, the distortion in the crystal lattice that impedes slip can be a result of deformation in the first place. In other words, the more deformation a specimen of metal undergoes, the more difficult it is to produce more deformation. This is usually referred to as work hardening and is encountered when forming metals in their cold state. Work hardening can, however, be eliminated or inhibited if the temperature of the metal is raised. The reason is that the movements and activity of atoms at elevated temperatures eliminate dislocations and create new grains or crystals that are distortion-free. The process of creating new crystals is known as recrystallization, and it always takes place above a certain temperature, called the recrystallization temperature, which differs for different metals.

A.2.2 Ceramics

A ceramic can be defined as a chemical compound including one or more metals with a nonmetallic element. The atoms comprising a ceramic body are bonded together with either ionic or covalent bonds. As you studied in chemistry, these types of atomic bonding are very strong and rigid compared with the metallic bonds that exist in metals. Therefore, the mechanism of failure in ceramics is not slip, as in the case of metals but rather the separation or cleavage failure that characterizes brittle materials. That is to say, tensile loading tends to result in cleavage failure in ceramics and is accompanied by very little or no deformation.

Because of the absence of any ability to undergo plastic deformation and the strong bonds between atoms, ceramics possess high hardness. Also, because the electrons are tied up in that strong type of bonding, ceramics are chemically inert (i.e., do not react with other materials) and are good electrical insulators. Examples of ceramics include aluminum oxide, boron nitride, and silicon carbide.

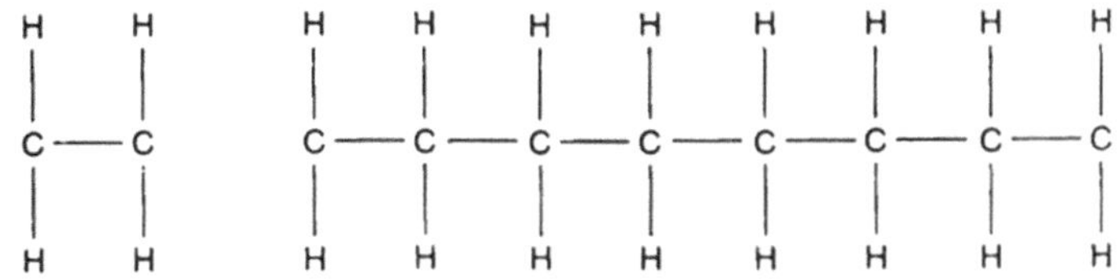

FIGURE A.2 Structural formula of ethylene and polyethylene.

A.2.3 Polymers

Polymers is the scientific name for the engineering materials commonly known as plastics. The word *polymer* stems from two Greek words: "poly," which means many, and "meras," which means parts. Polymers are composed of long chains, each being a giant molecule that, in turn, includes many small molecules linked together. For instance, if many molecules of ethylene (C_2H_2) are linked together, the result will be a long molecular chain of the polymer polyethylene, which has many applications in our everyday life. As can be seen in Figure A.2, the carbon atoms in ethylene have unsaturated valence bonds (carbon has a valence of 4, and hydrogen has a valence of 1). Therefore, if other ethylene molecules attach to each side of that molecule, the valence bonds on the carbon atoms in the molecule will be satisfied. In other words, for the carbon-to-carbon bond to satisfy valence requirements, numerous ethylene molecules must attach to one another and form long chains. When the chains grow longer, they get tangled, and a sample of polyethylene is analogous to a bowl of spaghetti.

Polymers are classified into two groups based on their manufacturing properties: thermoplastic and thermosetting polymers. In thermoplastic polymers such as polyethylene, the only force that binds these molecular chains together is the van der Waals attraction bond, which is rather weak. Consequently, the mobility of these molecular chains relative to one another is easily achievable through mechanical loading or thermal activation. For this reason, thermoplastic polymers creep and collapse easily under mechanical loads. They soften, melt, and viscously flow when heated, then solidify when cooled, and then melt again when heated again. In contrast to thermoplastics, the molecular chains of thermosets are cross-linked, forming a three-dimensional network. As a consequence, thermosets can carry mechanical loads higher than those that thermoplastics can withstand. Also, thermosets do not melt and flow but rather char and burn. Still, in both thermoplastics and thermosets, the backbone of the molecular chains is the element carbon, although it is replaced by silicon in a few cases where high-temperature applications are required.

A.3 PROPERTIES OF MATERIALS/STANDARD TESTS FOR OBTAINING MECHANICAL PROPERTIES

Hundreds of properties can be measured accurately in the laboratories. The properties of a certain material are a determining factor when considering the applications and processing methods of that material. During the process of selecting a material for certain applications, the properties play a major role. In order to handle a wide variety of properties, they are usually classified into three main categories: physical, chemical, and mechanical properties. Physical properties are those that pertain to the science of physics and include, for example, color, density, and magnetic characteristics. Chemical properties involve how a material reacts with the various acids and alkaline, as well as with the aqueous solutions of salts. Mechanical properties are the characteristics of a material that are revealed when that material is subjected to mechanical loading. As you may have expected, standard tests and procedures are performed in a materials laboratory in order to obtain the mechanical properties. Following are some of these standard tests.

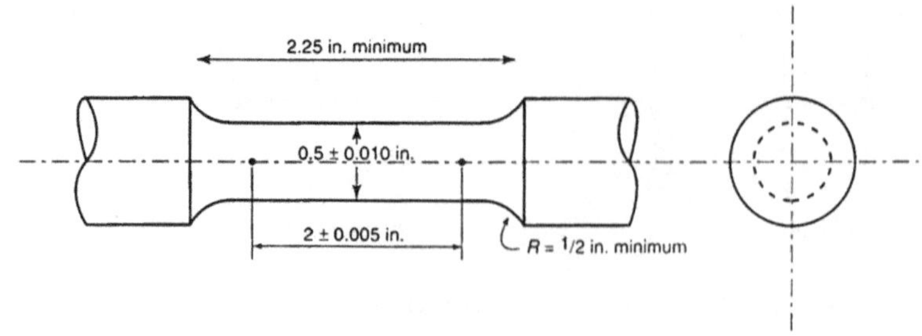

FIGURE A.3 A standard tensile test specimen.

A.3.1 Tension Test

The simplest way to obtain the important mechanical properties of a material is by a tension test. It is carried out by employing a tensile testing machine that incorporates a means of applying a tensile force and a means for measuring that force. A standard test specimen like that shown in Figure A.3 is used. It has enlarged ends to facilitate gripping in the machine and a uniform cross section in the middle. Gauge marks are scribed on the middle section so that the extension can be measured during loading. The load–extension curve can, therefore, be obtained for any desired material. A typical graph for mild steel is shown in Figure A.4. As can be easily realized, the cross-sectional area of the specimen markedly affects its load-carrying capacity. For this reason, we rely on using the concept of stress or load intensity when making comparisons between the mechanical properties of materials. The engineering stress is given by

$$S = \frac{P}{A_0} \tag{A.1}$$

where A_0 is the original cross-sectional area of the specimen.

On the other hand, it is not difficult to see that, for the same level of stress, the extensions of rods are dependent upon the initial length. It is, therefore, appropriate to consider the extension per unit

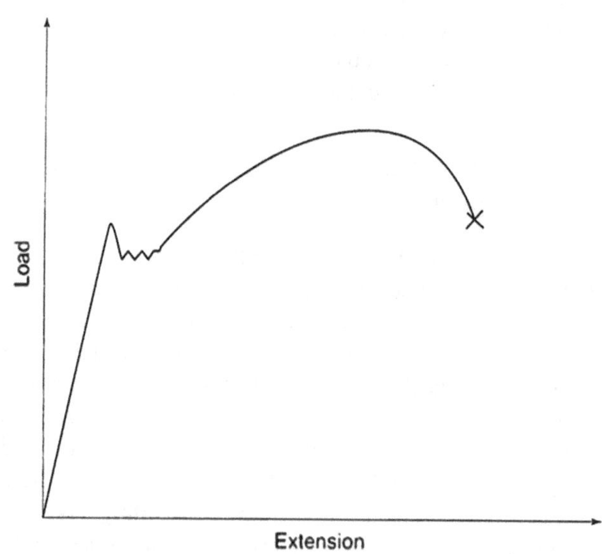

FIGURE A.4 A typical load–extension curve for mild steel.

length rather than the total extension. The extension per unit length is referred to as the engineering strain and is given by

$$e = \frac{\Delta l}{l_0} = \frac{l - l_0}{l_0} \tag{A.2}$$

where

l is the current length
l_0 is the initial length

Now, it is time to plot the relationship between the engineering stress and the engineering strain. It will look like the load–extension curve; the difference is the scale of the X- and Y-axes, as shown in Figure A.5. Looking at the graph in Figure A.5, we can easily distinguish three distinct regions as follows:

Region I, in which the relationship between the stress and the strain is linear, is referred to as the elastic region. As is evident, the strain is directly proportional to the stress. Elastic strain is also reversible (i.e., it disappears as soon as the stress is removed). The relationship between the stress and the strain can, therefore, be expressed as follows:

$$\frac{S}{e} = E \tag{A.3}$$

where E is Young's modulus of elasticity and has the same units as the stress. Region I ends at the elastic limit (i.e., the point beyond which the relationship between the stress and the strain is no longer linear). Finally, the bar (specimen) yields, as indicated by constant or dropping stress that is accompanied by appreciable strain. The point at which this process starts is called the yield stress (yield strength).

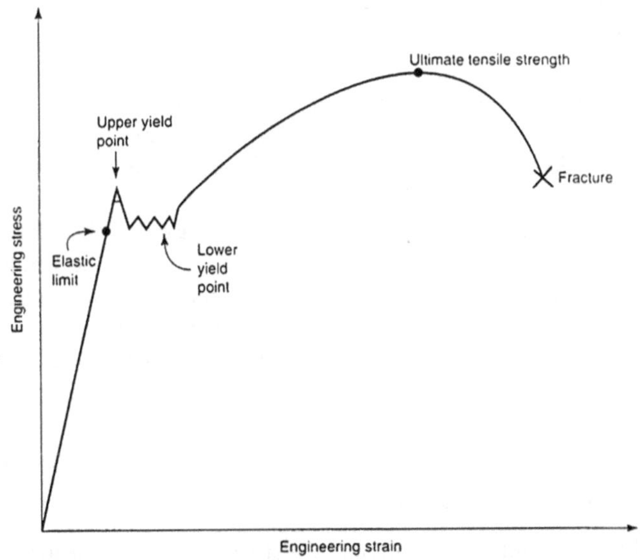

FIGURE A.5 The engineering stress–strain curve.

- Region II, in which major plastic deformation occurs and is accompanied by strain-hardening or work hardening (i.e., the stress beyond which the material deformation increases with increasing strain), is a result of the piling up of lattice faults or dislocation.
- Region III is where the specimen being tested collapses at the end. At the beginning of this stage, the middle section of the specimen contracts, resulting in necking. Consequently, the specimen cannot take the high level of stress. Stress decreases while the specimen continues to elongate, and, finally, fracture takes place.

The following are some specific tensile properties:

- Young's modulus of elasticity is the slope of the linear elastic curve.
- Yield stress is the stress at which the material first yields and undergoes plastic deformation.
- UTS is the maximum engineering stress that is observed during the test.
- Ductility is the ability to undergo large plastic deformation. Ductility has two common measures, as follows:

$$\text{Elongation percentage} = \frac{l_{\text{fracture}} - l_{\text{original}}}{l_{\text{original}}} \times 100 \tag{A.4}$$

where

l_{fracture} is the final length of the specimen after fracture
l_{original} is the original length

$$\text{Reduction in area percentage} = \frac{A_0 - A_f}{A_0} \times 100 \tag{A.5}$$

where

A_0 is the original cross-sectional area
A_f is the final area after fracture

- Resilience is the ability of the material to store elastic energy. It is more correctly called the modulus of resilience and is equal to the area under the linear elastic part of the engineering stress–strain curve. As you can see, this property is very important when selecting materials for springs.
- Toughness is the ability of the material to absorb energy until it breaks. A quantitative indication of toughness is the modulus of toughness, which is the whole area under the engineering stress–strain curve. The modulus of toughness is a measure of the ability of the material to withstand shock loading. Machine components that are subjected to sudden loading during their service life must be made from materials that possess high toughness.

A.3.2 Hardness Test

Hardness is defined as the ability of a material to resist scratching, abrasion, or indentation. It is probably the easiest criterion or measurement in acceptance tests and quality control of raw stock as well as manufactured products. Although there are numerous ways of measuring hardness, the most commonly used hardness-testing methods are the Brinell and the Rockwell hardness tests.

A.3.2.1 Brinell Hardness Test

The Brinell hardness test involves forcing a hardened-steel ball into a metal specimen under a definite static load and then measuring the size (diameter) of the impression produced by the ball. In this case, the hardness index, which is called the Brinell hardness number, is the static load acting on the ball divided by the spherical area of the impression (the unit is kilograms force per square millimeter). The hardened-steel ball has a diameter of 10 mm, the applied load is 3000 kg, and its duration is at least 10 seconds for ferrous metals. For nonferrous metals, the load is 500 kg, and its duration is at least 30 seconds. The diameter of the impression is usually obtained by optical magnification projected on a screen, and the Brinell hardness number corresponding to this value can be obtained from tables, thus eliminating the need for calculations. It is not difficult to realize that the Brinell hardness number and the UTS for a metal are correlated.

A.3.2.2 Rockwell Hardness Test

Similar to the Brinell hardness test, the Rockwell hardness test involves forcing an indentor into a test specimen under static load. However, in the Rockwell testing method, the hardness index is determined by measuring the increment of depth (of the impression) as a result of applying a primary and a secondary load instead of by measuring the diameter. Consequently, there is no need for optical measurements or calculations, and the Rockwell hardness number is readily shown on a dial indicator. The Rockwell hardness test is, therefore, commonly used in industry because of its simplicity and the ease with which it can be performed.

Two standard indentors are used with two commonly used hardness scales to determine the Rockwell hardness numbers for nearly all the common metals and alloys. These indentors are a hardened-steel ball having a diameter of 1/16 inch (1.59 mm) and a diamond cone having an apex angle of 120° and a rounded tip 0.2 mm in radius, called the brale, and they are used with scales designated as B and C, respectively. The working range for scale B, which is used for nonferrous metals and annealed low-carbon steels, is from R_b 0 to R_b 100. For the sake of measurement accuracy, when the hardness of the material being tested exceeds 100 R_b, you must switch to scale C; if the hardness is less than 0 R_b, another appropriate Rockwell hardness scale should be used. The useful range of the C scale, which is used for hardened and tempered steels, is from R_c 20 (equivalent to R_b 97) to slightly above R_c 70. Owing to inherent inaccuracies associated with shaping the brale, the C scale should not be used for measuring hardness below R_c 20; instead, the hardened-steel ball and scale B are usually employed.

Figure A.6 shows the procedure for performing a Rockwell hardness test on the C scale. First, the test specimen is placed on the anvil at the upper end of the elevating screw. The capstan wheel is then rotated so as to bring the surface of the test specimen in contact with the indentor. By further rotation of the wheel, the test specimen is forced against the indentor, and a minor load of 10 kg is slowly applied in order to seat the specimen firmly. At this moment, the dial indicator of the apparatus is set to zero. Next, an additional load of 140 kg (90 kg in a test on the B scale) is applied by means of a release handle mounted on the side of the apparatus. The total major load will now be 150 kg, and the duration of its application should be at least 10 seconds. Obviously, the application of this load forces the indentor into the specimen to an additional depth. Still, this depth must not be considered as an indication of hardness because it includes an elastic as well as plastic deformation. Therefore, the additional load is released without removing the minor load, and the hardness index is then shown on the dial indicator. The reading reflects the permanent or plastic increment of penetration depth resulting from the increment of load between the minor and major loads. It does not indicate the total depth of penetration of the indentor. As is the case in the Brinell hardness test, care must be taken to ensure that the surface conditions of the test specimen (its flatness and its thickness) are within the limits specified by the standards.

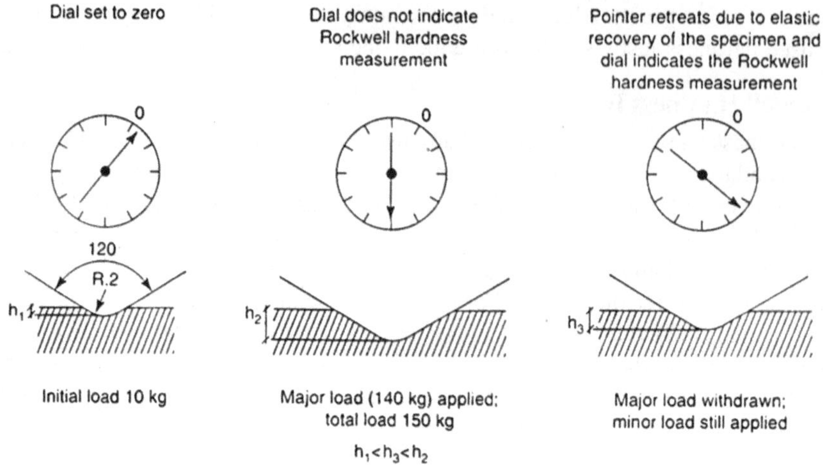

FIGURE A.6 Procedure for performing a Rockwell hardness test on the C scale.

A.3.3 Impact Test

Impact strength, a measure of a material's ability to withstand shock loading, is defined as the energy required to fracture a given volume of material (the common unit is joules per cubic centimeter). A pendulum-type impacting machine is commonly employed when measuring the impact strength of metals and polymers. There are, however, two types of impact tests that find widespread application: the Charpy and the Izod impact tests, which are shown in Figure A.7.

Usually, the impact specimen has a notch at the spot where it is desired to promote fracture. In such cases, the impact data are reported as Charpy V or notched Izod. This is not the case when testing brittle materials; there is no need to have a notch because the impact strength of a brittle material is naturally low. As you may have guessed, the notched impact data cannot be compared with the unnotched data.

Note that the impact strength of a metal is affected by temperature. Some metals experience a sharp drop in impact strength at low temperatures. In fact, it has been recently revealed that this was the main reason behind the sinking of the *Titanic*, as well as many other ships during World War II, in the chilly North Atlantic waters.

A.3.4 Fatigue Test

We can see that the results of a tensile test cannot be used when the part or machine component is to be subjected to dynamic alternating loads. The test used should emulate the service life conditions so as to be a good measure of how well a material will withstand the dynamic loading, such as that shown in Figure A.8. The number of cycles for which a component or a specimen can withstand an alternating load primarily depends upon the magnitude (or amplitude) of that load. The higher the magnitude of the alternating load, the smaller the number of cycles after which it fails. It has also been observed experimentally that the magnitude of the load that causes failure is much less than the yield stress of the material, indicating that the mechanism of failure, in this case, is different from one in which the specimen is subjected to uniaxial tension. This phenomenon is called metal fatigue, and the failure is due to the initiation and then propagation of a crack within the cross-sectional area of the part.

In order for a machine component to be designed properly to carry an alternating load, it must be capable of withstanding an infinite number of cycles at a particular level of loading. Consequently, it is important to determine the maximum alternating load that the desired material can withstand for

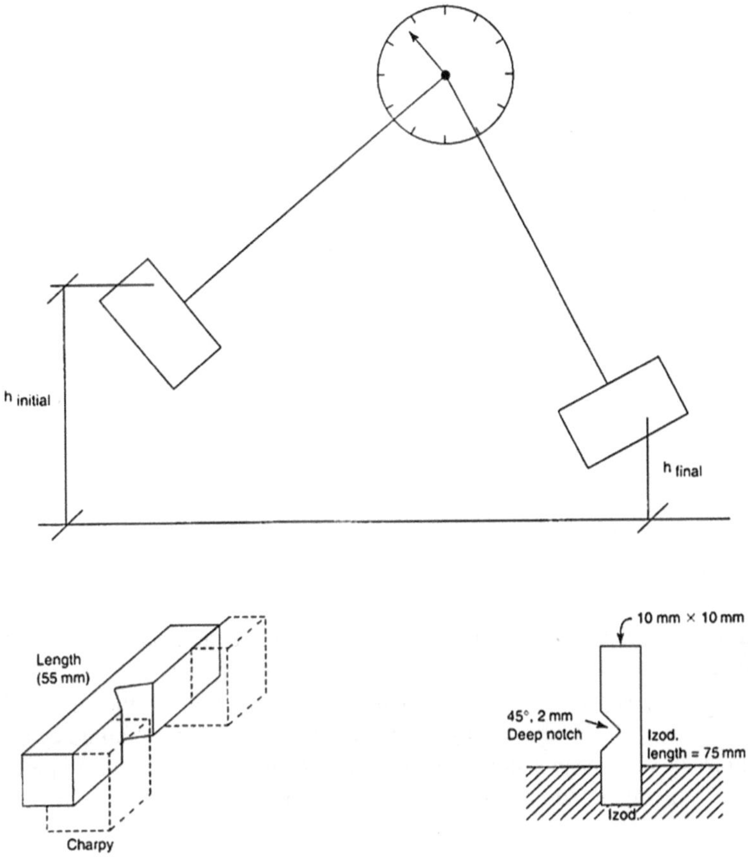

FIGURE A.7 Commonly used impact tests.

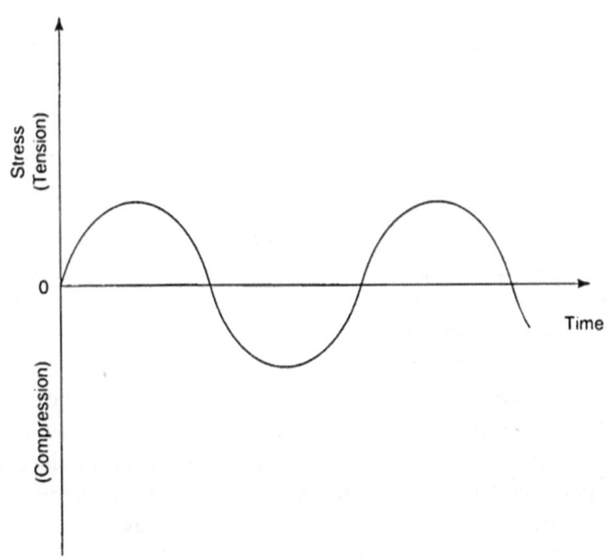

FIGURE A.8 A typical dynamic alternating load.

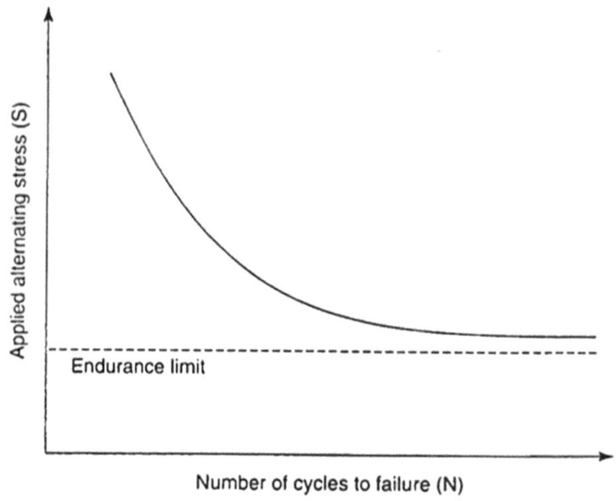

FIGURE A.9 The S-N curve.

an infinite number of cycles. Such a load is referred to as the endurance limit of the material. This property can be obtained experimentally by constructing an S-N curve for the desired material. A typical procedure involves preparing a set of identical test specimens, subjecting each to a different magnitude of alternating stress, and recording the number of cycles until failure occurs in each case. Higher magnitudes of load mean a smaller number of cycles and vice versa. By plotting the magnitude of the stress (S) versus the number of cycles to failure (N), we can obtain the S-N curve, as shown in Figure A.9.

The S-N curve is asymptotic to a certain value of stress, which is clearly the endurance limit of the material. The endurance limit is affected by many factors, such as the surface roughness of the test specimen or the presence of stress raisers that promote the initiation and propagation of fatigue cracks. Sometimes, the S-N curve is plotted on a semilogarithmic scale in order to facilitate the determination of the endurance limit. In addition, an increase in the material's temperature will result in a decrease in the endurance limit of the specimen. The frequency with which the load is applied also influences the endurance limit, especially in the case of polymers, where the dissipated energy reappears in the form of heat.

A.3.5 Creep Test

Creep can be defined as plastic deformation at elevated temperatures under constant load. It occurs even though the applied stress is less than the yield stress at that temperature. It is also interesting to note that some polymers undergo creep at room temperature. The creep characteristics of a material are determined by means of a creep test, which involves subjecting the specimen to different constant stresses at elevated temperatures and observing the corresponding strain. For each condition of a constant stress and a constant temperature, the elongation (or strain) is measured as a function of time and plotted to give a creep curve, as shown in Figure A.10. As can be seen, the curve involves three stages: an initial stage, a steady-state stage, and a final stage where deformation takes place at an accelerated rate after necking begins. The final stage continues until failure. For design purposes, the creep data are presented in the form of a family of straight lines on a log-log scale that indicates the relationship between stress and rupture life, with temperature as a parameter that determines the specific straight line to use, as shown in Figure A.11.

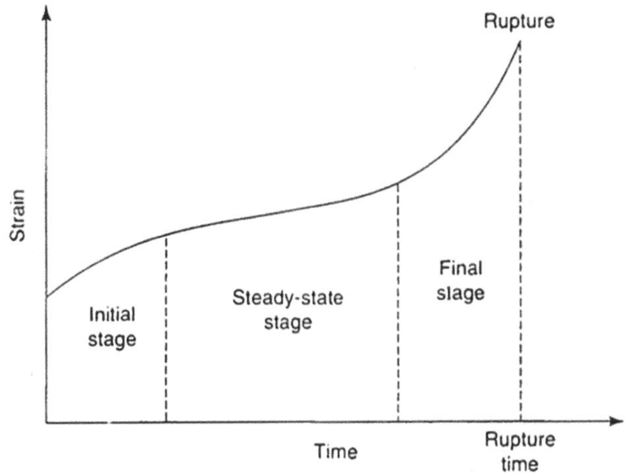

FIGURE A.10 A typical creep test.

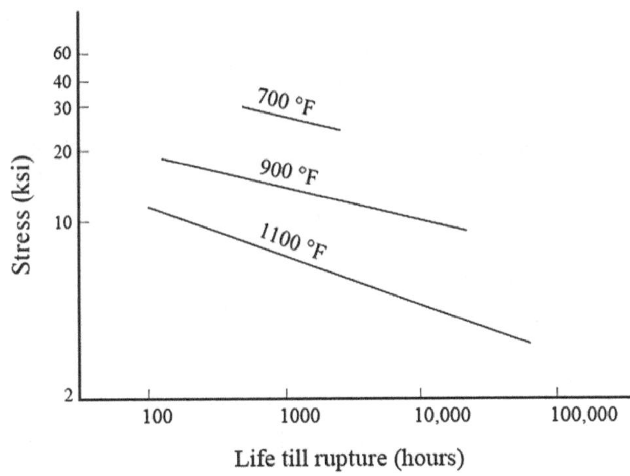

FIGURE A.11 Stress–rupture curves.

A.4 PHASE DIAGRAMS

Before we examine phase diagrams, we should fully understand what a phase is. A *phase* can be defined as a portion of material that is chemically and physically homogeneous and is separated from other portions by a well-defined surface (interface). Now, if we have ice and water in an isolated chamber, they must be considered as two phases and not one phase. Although the system is chemically homogeneous (both are H2O), it is not physically homogeneous (one is a solid, and one is a liquid). In fact, what we have is a system under *equilibrium* (i.e., the ratio between ice and water will continue to be the same until an external factor such as pressure or temperature acts to disturb that state of equilibrium). There are different types of systems based on the *components* (elements or compounds) that form the system. A *unary* system has a single component (like the preceding example), a *binary* system is generated or initiated by two components, and a *ternary* system involves three components.

A.4.1 Binary Phase Diagrams

In order to simplify the discussion of phase diagrams and alloying, only binary phase diagrams will be covered here. Interested readers are advised to consult a specialized text on materials science (see the reference list at the end of this book).

A phase diagram is like a road map in that it is a graphical illustration of the various phases present in equilibrium for different compositions of two alloying elements at different temperatures. By convention, the composition is always plotted horizontally and the temperature vertically. An alloy having a composition of x percent of component A (it has (100 – x) percent of component B) at a temperature T is represented by a point. Zones or regions in which specific phases exist are bounded by lines or curves. Phase diagrams are useful tools in studying alloying and alloys and in optimizing parameters in casting processes and heat treatment operations. The shape of a binary phase diagram depends upon whether the components are mutually soluble, partly soluble, or totally insoluble in their solid state. Let us discuss each of these cases in the following sections.

A.4.1.1 Isomorphous Phase Diagrams

An isomorphous phase diagram corresponds to the case of two elements that are completely mutually soluble in their liquid state as well as in their solid state. A typical example of this case is the copper–nickel (Cu-Ni) equilibrium phase diagram shown in Figure A.12. The first vertical line to the left indicates an alloy containing 100 percent Cu and 0 percent Ni (i.e., pure copper). The melting point of copper indicates the boundary between the two phases for pure copper: above it is the liquid copper phase, and below it is the solid copper phase. As can be seen in Figure A.12, the same applies to pure nickel. For all other alloys, the region above the upper-boundary curve represents a liquid solution of nickel in copper. That curve is, therefore, referred to as the liquidus. The lower curve is called the solidus because the area below it represents a solid solution of nickel in copper. The area between the two curves is a transition zone in which both a liquid and a solid phase exist. If we draw a vertical line representing a specific alloy (say, 60 percent Cu and 40 percent Ni), it will intersect with the liquidus at a point that indicates the temperature at which solidification starts and will intersect with the solidus at the temperature at which that alloy becomes completely solid. The difference between these two temperatures is the freezing range. Cast alloys usually must have a relatively wide freezing range in order to flow and fill the mold completely before solidification occurs.

Now, let us take a closer look at how various alloys containing different percentages of nickel solidify. This is achieved by plotting the temperature-time relationship when the molten alloy is left

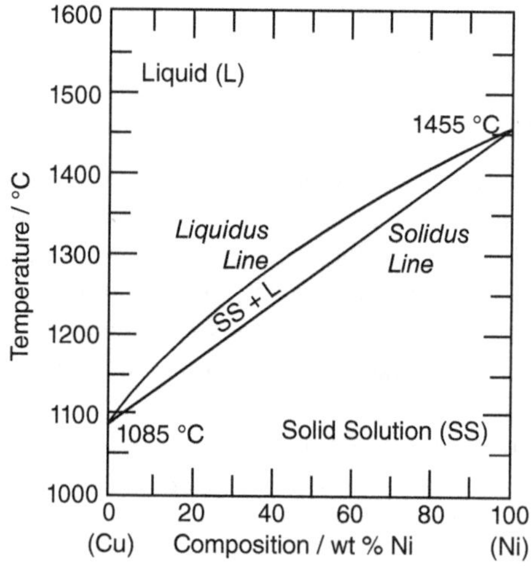

FIGURE A.12 Copper–nickel equilibrium phase diagram.

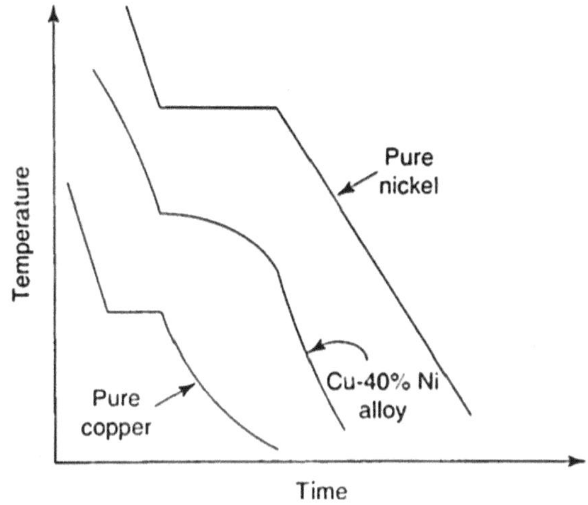

FIGURE A.13 Cooling curves for different alloys.

to cool down naturally. The graph is referred to as a cooling curve. As can be seen in Figure A.13, the cooling curve for a pure metal has a plateau, indicating that solidification takes place at a constant temperature over a period of time during which the molten metal loses the latent heat of fusion. This is not the case with the Cu-40 percent Ni alloy, where solidification is gradual and there is no plateau. In this case, crystals of a solid solution rich in copper precipitate first, and the remaining liquid phase is, therefore, richer in nickel. With further cooling, more crystals rich in copper precipitate, and the process continues, as explained earlier, until the alloy completely solidifies.

Next, we have to answer the following question: how do the mechanical properties of a copper–nickel alloy differ from those of either copper or nickel? In order to answer this question, let us consider the atomic arrangement. Copper has an FCC lattice, and so does nickel. When they form a solid solution, some nickel atoms replace copper atoms in the lattice. However, because a nickel atom is larger than a copper atom, the lattice becomes distorted. As a consequence, it is difficult for atoms to slide or slip over one another when the alloy is subjected to mechanical loading. In other words, the strength of the alloy will be higher than that of pure copper (or nickel), while the ductility will decrease. This type of solid solution is referred to as a substitutional solid solution, and it is common when both alloying elements have the same crystal lattice and more or less the same size of atoms. On the other hand, the phenomenon known as solid solution strengthening is shown in Figure A.14.

Note that there is another type of solid solution that does not necessarily yield the same type of phase diagrams, and it is known as the interstitial solid solution. As the name suggests, atoms of the alloying additive occupy the interstitial cavities in the crystal lattice of the base metal. In fact, this is always the case when carbon or nitrogen, which both have smaller atoms, dissolve into metals having FCC lattices, such as iron at elevated temperatures like 1673°F (slightly above 912°C).

A.4.1.2 Eutectic Phase Diagram

A eutectic phase diagram corresponds to the case of two components that are completely mutually soluble in their liquid state but exhibit negligible or no solubility in their solid state. A typical example of this case is the bismuth-cadmium (Bi-Cd) binary phase diagram shown in Figure A.15. As can be seen in the figure, alloy I at a temperature above the liquidus takes the form of a homogeneous liquid solution and is rich in bismuth. When that alloy is supercooled to a temperature just below the liquidus, solid crystals of bismuth start to precipitate (because the liquid alloy is supersaturated in bismuth). As a consequence, the remaining liquid alloy will have less bismuth. When it is further cooled, more solid bismuth crystals precipitate, and the concentration of cadmium

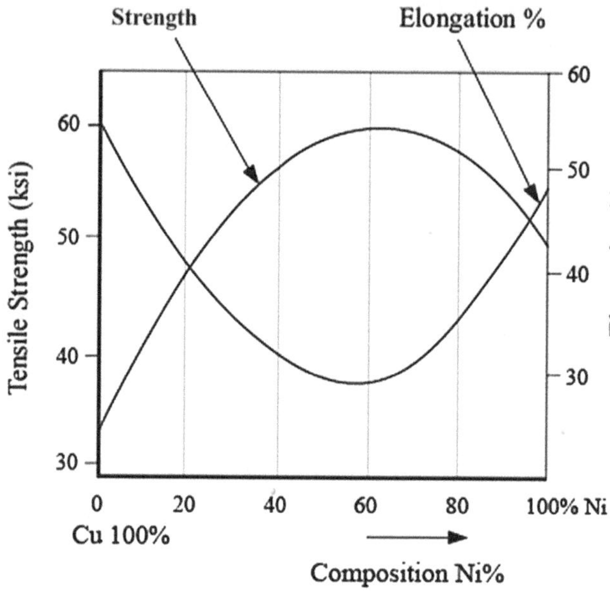

FIGURE A.14 Mechanical properties of copper–nickel alloys.

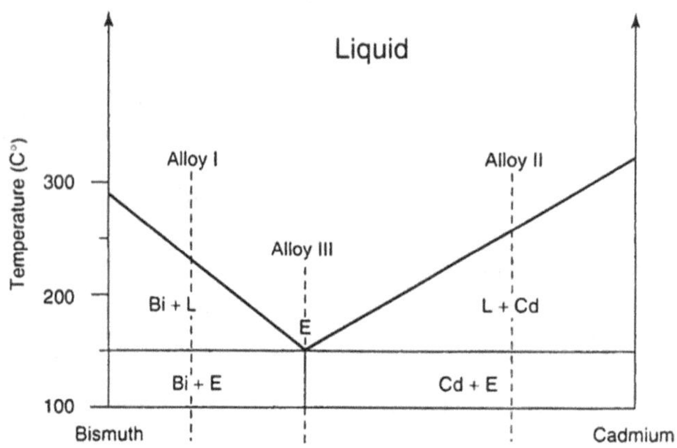

FIGURE A.15 Bismuth–cadmium binary phase diagram.

continually increases in the molten fraction that remains after the precipitation of the solid bismuth crystals. On the other hand, if a molten alloy rich in cadmium, such as alloy II, is supercooled, solid crystals of cadmium precipitate first, and the remaining liquid fraction will become richer in bismuth. As you may have guessed, there is a certain alloy with a specific concentration in which solid crystals of both bismuth and cadmium precipitate simultaneously when that molten alloy is supercooled below the liquidus. When solidification is completed, the solid constituents will form a mechanical mixture that can be revealed by microscopic examination.

A better understanding of the solidification of this type of alloy can be gained through examination of the cooling curves. It can be seen in Figure A.16 that alloy III solidifies at a constant temperature that is below the melting points of both bismuth and cadmium. For this reason, both the reaction and the alloy are referred to as eutectic (after a Greek word meaning “easily fusible”). After complete solidification, alloy I will consist of bismuth crystals and the eutectic, while alloy II will consist of cadmium crystals and the eutectic. Before leaving our discussion of eutectic systems, we need

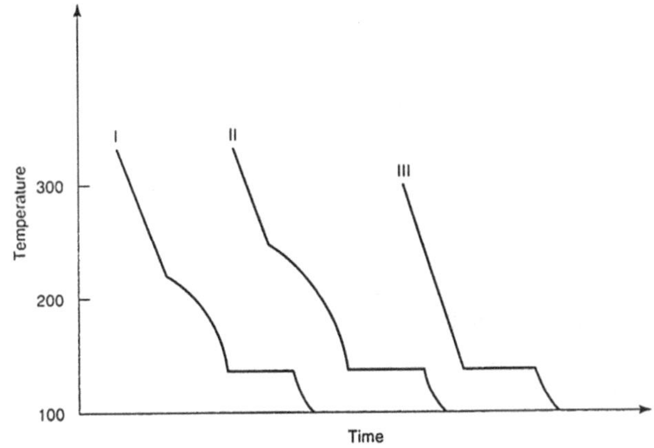

FIGURE A.16 Cooling curves for alloys of a eutectic system.

to see how the composition of an alloy affects the strength of such an alloy. The easiest alloy to consider first is the eutectic alloy. The very nature of the fine mechanical mixture impedes the slip of atoms (each phase forms an obstacle to the next one), thus resulting in an increase in the strength. As a consequence, alloys having a higher percentage of the eutectic possess strength higher than that of alloys containing less of the eutectic.

A.4.1.3 Alloys Showing Limited Solid Solubility in the Solid State

A typical example of this case is the lead–tin (Pb-Sn) equilibrium phase diagram shown in Figure A.17. As anticipated, this phase diagram is a hybrid of the preceding two types. Instead of two elemental components forming a mechanical mixture, the eutectic in this case is a mechanical mixture of two solid solutions. As can be seen in Figure A.17, there is a line below the solidus in the limited solid solubility region. This line indicates decreasing solid solubility with decreasing temperatures, and it is known as the solvus. At 362°F (183°C), the solubility of tin in lead is 19 percent and

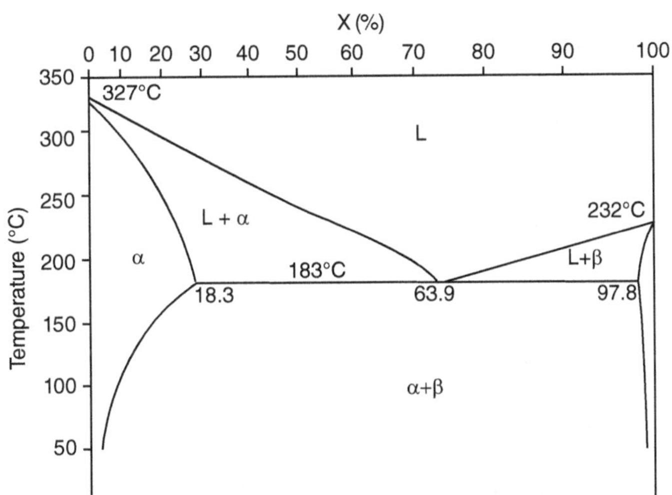

FIGURE A.17 Lead–tin equilibrium phase diagram. (Reprinted from *Physical Metallurgy: Principles and Design*, by G. N. Haidemenopoulos, CRC Press, by permission.)

decreases to only 2 percent at room temperature. As a consequence, if any alloy containing between 2 percent and 19 percent tin cools down past the solvus, the solubility limit is exceeded, and the surplus tin precipitates in the form of a solid solution of lead in tin, which is referred to as α. The properties of this type of alloy can be controlled by controlling the amount and characteristics of the phase. If we can make it take the form of tiny particles that are distributed all over, then slip and deformation will be inhibited, and the result will be an increase in the strength of the alloy. This mechanism for controlling the properties of the alloy is known as dispersion strengthening. This phenomenon is very important for the heat treatment of nonferrous alloys. Now, it is not difficult to see that any alloy containing from 19 percent to 61.9 percent will have a microstructure consisting of a solid solution surrounded by a solidified eutectic mixture called a microconstituent. This type of alloy is known as a hypoeutectic alloy, as opposed to one that contains from 61.9 percent to 99 percent tin, which is called a hypereutectic alloy and has a microstructure consisting of β solid solution surrounded by a eutectic.

A.4.2 Iron–Carbon Phase Diagram

In order to thoroughly understand ferrous alloys, we first must become familiar with the phase diagram. The iron–carbon (Fe-C) phase diagram is slightly more complicated than the ones we have covered so far. Iron and carbon form an intermetallic compound (iron carbide, Fe_3C) that contains 6.7 percent carbon. The phase diagram has, therefore, two components: iron and iron carbide. It is actually an iron-iron carbide equilibrium phase diagram. Nevertheless, for all practical purposes, it is the percentage of carbon that is indicated and not the percentage of iron carbide. We just trim the phase diagram at 6.7 percent carbon (i.e., the percentage at which the intermetallic compound is formed).

Figure A.18 shows the iron–carbon equilibrium phase diagram. Alloys containing more than 2.1 percent carbon are known as cast irons. They have a low melting point of 2100°F (1148°C) and a

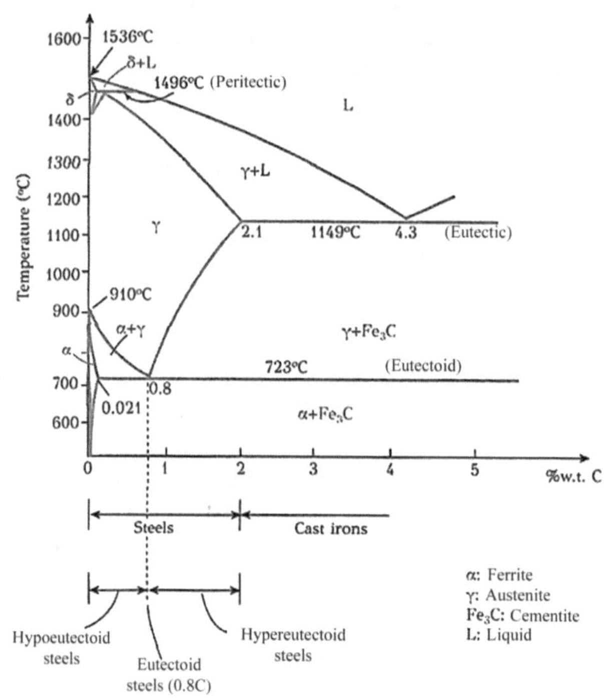

FIGURE A.18 Iron–carbon phase diagram. (Reprinted from *Physical Metallurgy: Principles and Design*, by G. N. Haidemenopoulos, CRC Press, by permission.)

wide freezing range because of the eutectic reaction that takes place at 2100°F (1148°C) and 4.3 percent carbon. Alloys containing less than 2.1 percent carbon are referred to as steels. They are shown in the lower left-hand section of the phase diagram. (A good engineering student will memorize this section and know it by heart.)

Now, let us discuss each of the phases shown in the iron–carbon phase diagram. The first phase is α iron, or ferrite. It has a BCC lattice and is stable up to 1673°F (912°C), where an allotropic reaction takes place. Ferrite is soft and ductile, is ferromagnetic below 1418°F (770°C), and can dissolve a maximum of 0.02 percent carbon (because the center of the BCC unit cell is occupied by an iron atom). The other phase is γ iron, or austenite. It has an FCC lattice and can, therefore, dissolve up to 2 percent carbon (because the center of the unit cell has no atoms). Austenite appears as a result of an allotropic reaction at 1673°F (912°C). It is stable up to 2543°F (1394°C), where another allotropic reaction takes place. Austenite is soft like ferrite, but it is not ferromagnetic. The intermetallic compound Fe_3C (iron carbide) is called cementite by metallurgists. It is hard and brittle because of its complex crystal structure. As can be seen from Figure A.18, the microstructure of carbon steels at room temperature is usually a combination of these phases. There is, however, a eutectic-like reaction at a temperature of 1340°F (727°C) and a carbon content of about 0.8 percent. This reaction involves the decomposition of austenite into a mechanical mixture consisting of lamellar alternate layers of cementite and ferrite. This microconstituent is known as pearlite because, under the microscope, it looks like mother-of-pearl. Because this reaction takes place in the solid state, it is referred to as a eutectoid reaction (to distinguish it from the eutectic reaction that involves liquid phases). Pearlite has fairly good strength and toughness because the cementite layers impede deformation while the ferrite layers are soft and ductile. In fact, all carbon steels contain pearlite to varying degrees. Steel that contains less than 0.8 percent carbon is known as hypoeutectoid steel and has a microstructure that consists of pearlite surrounded by ferrite. Steel that contains more than 0.8 percent carbon is called hypereutectoid steel and has a microstructure that consists of pearlite surrounded by cementite.

These microstructures can be obtained only when the cooling rate is very slow and diffusion of carbon atoms can take place, thus emulating equilibrium conditions. However, with the sudden cooling of austenite, the crystal structure changes from FCC to body-centered tetragonal (distorted cube). Carbon atoms do not have enough time to diffuse and therefore get trapped in the lattice, producing a supersaturated solid solution of carbon in α iron. Because the lattice is not BCC but rather tetragonal, the resulting phase is metastable (i.e., does not appear on the equilibrium phase diagram) and is known as martensite. In practice, sudden cooling is achieved by quenching (i.e., dropping the heated steel part into water or oil at room temperature). Only pearlite undergoes the martensite transformation. If hypoeutectoid steel is quenched, its microstructure will consist of ferrite and martensite; if hypereutectoid steel is quenched, its microstructure will include cementite and martensite.

A.5 HEAT TREATMENT OF STEEL

The main reason for heat-treating steels is to control their mechanical properties (such as hardness, strength, and ductility). These properties are adjusted so as to enable the structural component under consideration to withstand the conditions to which it is going to be subjected during its service life. Let us now briefly review the various heat treatment operations.

A.5.1 Hardening

Hardening involves heating steel, keeping it at an appropriate temperature until all pearlite is transformed into austenite, and then quenching it rapidly in water (or oil). The temperature at which austenitizing rapidly takes place depends upon the carbon content in the steel used (and can be obtained from the iron–carbon equilibrium phase diagram shown earlier in Figure A.18). The

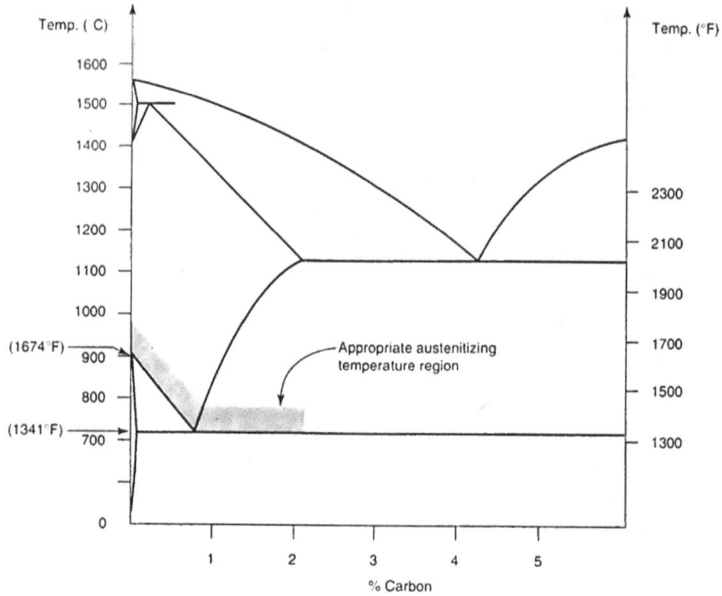

FIGURE A.19 Temperatures appropriate for austenitizing.

heating time should be increased with increasing component size to ensure that the core will also be fully transformed into austenite. (See Figure A.19.)

As previously mentioned, the microstructure of a hardened-steel part is ferrite and martensite for hypoeutectic steel, martensite for eutectoid steel, and cementite and martensite for hypereutectoid steel. As a consequence, there is no real gain in hardness when quenching hypoeutectic steels having a low carbon content (e.g., 0.1 percent or 0.2 percent carbon). Because martensite is a metastable phase, the hardening operation must be followed by another heat treatment in order to control the strength and hardness of the component, as will be explained later.

A.5.2 Tempering

Tempering involves heating steel that has been quenched and hardened to a temperature lower than 1340°F (727°C) for an adequate period of time so that the metal can be equilibrated. During the operation, martensite is decomposed, resulting in spheroidal particles of carbide embedded and distributed in a matrix of ferrite. The hardness and strength obtained depend upon the temperature at which tempering is carried out. Higher temperatures, those just below 1340°F (727°C), will result in high ductility but low strength and hardness; low tempering temperatures will produce low ductility but high strength and hardness. In practice, metallurgical engineers make use of an experimentally obtained curve showing hardness versus tempering temperature in order to select the appropriate tempering temperature that will produce the desired level of hardness and strength. This operation is performed on all carbon steels that have been hardened in order to reduce their brittleness so that they can be used.

A.5.3 Annealing

Annealing involves treating steel up to a temperature high enough to transform all the pearlite into austenite and then cooling it very slowly to room temperature. The resulting microstructure will be lamellar pearlite possessing high ductility and toughness but low hardness. In industry, annealing is performed by heating a component to the appropriate temperature, soaking it at that temperature, and then shutting off the furnace while the piece is in it. It is always recommended that stock steel be subjected to annealing before being processed by cold forming in order to reduce the load and energy requirements and to enable the metal to undergo large strains without failure.

A.5.4 Normalizing

Normalizing involves heating steel to the austenite temperature range, keeping it at that temperature for a period of time, and then cooling it in air. The resulting microstructure is a fine and feathery mechanical mixture of ferrite and cementite and is known as bianite. It has higher strength and hardness than pearlite but lower ductility. Normalizing is performed on structures and structural components that will be subjected to machining because it improves the machinability of carbon steels.

A.6 SURFACE HARDENING

In many engineering applications, it is necessary to have the surface of the component hard enough to resist wear and erosion, while having the core ductile and tough in order to withstand impact and shock loading. This can be achieved through one of two mechanisms: local austenitizing and quenching for medium- and high-carbon steels and diffusion of hardening elements like carbon or nitrogen into the surface of low-carbon steels. Let us now discuss the various processes that are used in industry.

A.6.1 Flame Hardening

Flame hardening uses a combustible gas flame as the source of heat for austenitizing the outer layer of the part. This is followed by rapid water quenching. Flame-hardened parts must then be tempered after hardening. The tempering temperatures used depend upon the desired hardness and the carbon content. This process is usually recommended for plain-carbon steels with carbon contents ranging from 0.4 percent to 0.95 percent carbon.

A.6.2 Induction Hardening

Induction hardening is similar to flame hardening; the difference is the source of heat for austenitizing. In this case, a coil in which an alternating current flows induces current in the part to be hardened. This induced current alternates at frequencies of thousands of cycles per second. As a result of the "skin effect" as well as the resistance to current flow, the outer layer of the component heats up very rapidly. Water quenching and tempering produce the desired level of hardness.

A.6.3 Carburizing

Carburizing involves the diffusion of carbon into the surface layer of a component of steel that has a low carbon content (less than 0.4 percent). This is achieved in industry by placing the component in a carbon-rich medium at elevated temperatures. There are three versions of this process: pack, gas, and salt.

Pack carburizing involves packing the component in a steel container and completely surrounding it by a mixture of granular charcoal and barium carbonate. At elevated temperatures, a series of chemical reactions take place and finally produce atomic carbon that penetrates and diffuses into the surface of the low-carbon steel component. As a consequence, the surface layer becomes rich in carbon to a depth that depends upon the carburizing time as well as the temperature at which the part is soaked. Once the carbon content reaches the predetermined level, subsequent quench-hardening is performed.

- Gas carburizing is similar to pack carburizing, except that the source of atomic carbon is a hydrocarbon gas such as natural gas or propane. The carburizing time is more or less the same as in pack carburizing, but this process provides the advantage of allowing quenching to be done directly at the carburizing temperature. It is for this reason that conveyor furnaces are often used for gas carburizing.
- Salt carburizing is performed in a heated molten-salt bath. The typical salt used is sodium cyanide (NaCN), which is an extremely poisonous chemical compound and represents the major drawback of this process. On the other hand, its main advantage is the short heating cycle achieved through liquid convection. With today's strict environmental regulations, environmentally safe disposal of cyanide compounds creates a further problem in using salt-bath carburizing. The process is also sometimes referred to as cyaniding or cyanide casehardening.

A.6.4 Nitriding

Nitriding involves the diffusion of monatomic nitrogen into the surface of the steel being treated. A chemical reaction takes place at elevated temperatures, resulting in iron-nitrogen or iron-alloy-nitrogen chemical compounds. The latter compounds form an extremely hard case, one that is by far harder than tool and carburized steels. A major advantage of this operation is that it is performed at subcritical temperatures and no quenching or heat treatment is required. Because the nitrogen atmosphere surrounding the workpiece is relatively inert, scaling or discoloration does not take place.

The most commonly used source of the nitrogen atmosphere is dissociated ammonia. At the nitriding temperature, which is between 925°F and 1050°F (500°C and 570°C), ammonia dissociates, resulting in atomic nitrogen that diffuses into steel and hydrogen that is exhausted and pumped out of the system.

The presence of some alloying elements with steel enhances the formation of a continuous, tenaciously hard case of nitrides. Such elements include aluminum, chromium, molybdenum, vanadium, and tungsten. Plain-carbon steels (not having these alloying elements) are, therefore, not well suited to this surface-hardening operation. It can, however, be applied to stainless steels, although it may reduce their corrosion resistance.

A.6.5 Carbonitriding

Carbonitriding is a surface-hardening operation that involves the diffusion of both nitrogen and carbon into the steel surface. The atmosphere surrounding the workpiece is a mixture of a carbon-rich gas such as propane or methane and ammonia. Because of the presence of nitrogen, the temperature at which this operation takes place is lower than that used in carburizing operations. In addition, unlike the nitriding operation, the process is suitable for low-carbon steels as well as low-carbon alloy steels.

A.7 STEELS

Let us now review the different types of steels that are used in industry and discuss the classification system that is used in the United States. International readers are advised to contact their national standards organizations for information regarding the systems used in their countries.

A.7.1 Carbon and Alloy Steels

Carbon steels are alloys of iron and carbon, with up to 2 percent carbon and only residual amounts of other elements except those added for deoxidization (such as aluminum). There are also limits for various additives, such as silicon 0.6 percent, copper 0.6 percent, and manganese 1.65 percent. Carbon steels are also referred to as plain-carbon steels and low-carbon steels. These steels comprise the largest fraction of steel production and are available in all forms, such as sheet, bar, pipe, slab, and wire.

Alloy steels can be defined as those having carbon content up to about 1 percent and having total alloy content below 5 percent. These steels are commonly used for structural components that are required to possess superior wear, strength, and toughness properties. Note that the ability of alloy steels to be hardened (i.e., their hardenability) is far superior to that of plain-carbon steels. Some of these steels can be quenched in air instead of water or oil.

In the United States, the commonly used designation system for carbon and alloy steels is the one adopted by the American Iron and Steel Institute (AISI) and the Society of Automotive Engineers (SAE). Four digits are usually employed for identifying any steel. The first digit indicates the major alloying element, as shown in Table A.1. If the first digit is 1, the major alloying element is carbon, and the alloy is accordingly plain-carbon steel. The second digit indicates the relative percentage of a primary alloying element. If steel is identified by 23XX, the primary alloying element is nickel, and its percentage in the alloy is about 3. The last two digits indicate the carbon content in hundredths of a percent. In addition to these numbers, letters are used as prefixes and suffixes to provide information about production methods and additives.

A.7.2 Tool Steels

Tool steels are actually alloy steels having a high content of alloying elements. The amount of impurities in tool steels is much lower than that in ordinary alloy steels because tool steels are always melted in electric furnaces. Tool steels are also subjected to a rigorous quality control and inspection process that is not applied to other kinds of steel, resulting in superior alloy-content control as well as excellent cleanliness.

The AISI classification system for tool steels is based on use or application. There are basically four categories, plus one for special purposes. A prefix (letter) is used to indicate the use category. The prefix is followed by one or two digits that identify the specific alloy within the category. The first category is shock-resisting tool steel, and it is identified by the prefix S. The second category is hot-work tool steel for use in forging dies and plastic molds; the prefixes H and P identify this group. The third category is cold-work tool steel, which is divided into four subgroups depending on the quenching media. The prefixes W, O, and A are used for water, oil, and air, respectively; the prefix D identifies the fourth subgroup as high-carbon, high-chromium cold-work tool steel. This subgroup contains a very large chromium-carbide content when hardened. As a consequence, it possesses excellent abrasion resistance and is, therefore, the most commonly used steel for cold-work tooling. The last category is high-speed tool steel, which derives its name from its intended application (i.e., machining other metals at high cutting speeds). HSSs can have either molybdenum or tungsten as the major alloying element, and they are identified as the M series and the T series, respectively. In both cases, they contain the highest percentage of alloying elements of any of the tool steels. Their alloying elements provide resistance to softening at elevated temperatures. HSSs can be hardened up to 67 R_c and maintain their hardness at temperatures up to 1000°F (540°C).

A.7.3 Stainless Steels

Stainless steels are steels with at least 10.5 percent chromium that resist corrosion from oxidizing environments by exhibiting remarkable passivity. There are different types of stainless steels, and each has its own characteristics and applications. The first type is ferritic stainless steel, which has a carbon content of less than 0.2 percent and a chromium content of up to 27 percent. As the name suggests, the structure of this steel is ferrite at room temperature. Ferritic stainless steels are nonhardenable, are notch-sensitive, and exhibit poor weldability. Still, they are used for cutlery and cookware.

TABLE A.1
Major Groups in the AISI-SAE Steel Designation System

Class	AISI Series	Major Constituents
	10XX	Carbon steel, resulfurized carbon steel
Carbon steels	11XX	
Alloy Steels		
Manganese	13XX	Manganese 1.75%
15XX	Manganese 1.00%	
Nickel	23XX	Nickel 3.50%
25XX	Nickel 5.00%	
Nickel–chromium	31XX	Nickel 1.25%–chromium 0.65 or 0.80%
33XX	Nickel 3.50%–chromium 1.55%	
Molybdenum	40XX	Molybdenum 0.25%
41XX	Chromium 0.95%–molybdenum 0.20%	
43XX	Nickel 1.80%–chromium 0.50 or 0.80%–molybdenum 0.25%	
46XX	Nickel 1.80%–molybdenum 0.25%	
48XX	Nickel 3.50%–molybdenum 0.25%	
Chromium	50XX	Chromium 0.30% or 0.60%
51XX	Chromium 0.80%, 0.95%, or 1.05%	
5XXXX	Carbon 1.00%–chromium 0.50%, 1.00%, or 1.45%	
Chromium–vanadium	61XX	Chromium 0.80% or 0.95%–vanadium· 0.10% or 0.15% min.
Multiple alloy	86XX	Nickel 0.55%–chromium 0.50%–molybdenum 0.20% Nickel 0.55%–chromium 0.50%–molybdenum 0.25% Manganese 0.85%–silicon 2.00%
87XX	Nickel 3.25%–chromium 1.20%–molybdenum 0.12% Manganese 1.00%–nickel 0.45%–chromium 0.40%–molybdenum 0.12%	
92XX	Nickel 0.55%–chromium 0.17%–molybdenum 0.20% Nickel 1.00%–chromium 0.80%–molybdenum 0.25%	
93XX	Nickel 3.25%–chromium 1.2%–molybdenum 0.12%	
94XX	Manganese 1.00%–nickel 0.45%–chromium 0.40%–molybdenum 0.12%	
97XX	Nickel 0.55%–chromium 0.17%–molybdenum 0.2%	
98XX	Nickel 1.00%–chromium 0.80%–molybdenum 0.25%	

The second type is martensitic stainless steel, which has a carbon content as high as 1.2 percent and a chromium content of 12 percent to 18 percent. This steel can be hardened by heat treatment. Sharp tools and knives can be made of martensitic stainless steel.

The third type is austenitic stainless steel, which contains nickel as a major alloying element, in addition to chromium, and carbon. Nickel is added as an austenite-stabilizing element that promotes the formation of an austenitic structure (γ-iron) at room temperature. Austenitic stainless steel is hardenable only by cold working and is used for parts that require good chemical resistance, such as piping and tanks.

A fourth type of stainless steel includes the PH alloys. PH stainless steels can be martensitic, semiaustenitic, and austenitic. These steels possess very high strength and hardness and are, therefore, used in structural components and springs.

A.7.4 Other Types of Steel

Other types of steel include high-strength low-alloy steels, ultra-high-strength steels, and austenitic manganese steels.

A.8 ALUMINUM ALLOYS

Aluminum possesses high electrical and thermal conductivity, as well as good ductility and formability. It has poor tensile properties and poor rigidity (low value of E), which limits its use in building bridges or high-rise metallic structures. It has a low density of 2.7 g/cm^3, and its specific strength (i.e., strength-to-weight ratio) is, therefore, excellent, which is why it finds widespread application in the aerospace and automotive industries.

Aluminum alloys can be divided into wrought and cast alloys, corresponding to the method of fabrication. Further, we can divide each of these groups into heat-treatable and non-heat-treatable alloys. Each group (i.e., wrought and cast alloys) has a four-digit alloy-designation system that indicates the major alloying element (e.g., 2024 is a wrought aluminum alloy with copper as the major alloying element). For wrought alloys, the alloy designation is followed by a suffix that can have a number after it. The suffix indicates the kind of thermal treatment or degree of work hardening. This system was developed by the Aluminum Association and is used mainly in the United States. International readers should consult with the national standards organizations in their countries.

The most important alloying elements in aluminum alloys are copper, manganese, silicon, magnesium, and zinc. An aluminum–copper alloy with 4 percent copper is a typical example of a heat-treatable alloy. The procedure involves heating the alloy to a temperature of 930°F (500°C) in order to form a homogeneous α solid solution. The piece is then quenched, resulting in a metastable phase because copper does not have the time to diffuse out of the solid solution. Next, the piece is subjected to artificial aging by holding it at some low temperature, such as 400°F (200°C), for some time. Copper will be separated and form an intermetallic compound ($CuAl_2$). The presence of tiny particles impedes the deformation of the matrix and increases the strength. This phenomenon is, therefore, referred to as dispersion strengthening, and the process is known as precipitation hardening.

A.9 COPPER ALLOYS

Copper was one of the first metals to be used by humans because it could be found in its metallic form. Now, it is seldom to find copper in its pure form, and it is usually extracted from ores that contain only 5 percent copper by weight. Copper and its alloys have excellent electrical and thermal conductivity as well as corrosion resistance. In addition, copper (and many of its alloys) has excellent ductility. The preceding properties make copper the ideal metal for electrical conductors. In fact, more than 80 percent of all copper produced is used in the form of pure copper; the remainder is used in many alloy forms.

The major alloying elements in copper alloys include zinc, tin, and nickel. When zinc is the principal alloying element, the alloy is known as brass. Depending upon the percentage of zinc, this alloy can be α brass (70 percent Cu, 30 percent Zn) or β brass (60 percent Cu, 40 percent Zn). Whereas α brass is ductile and can be formed by various forming processes, β brass is usually produced by casting because of its lack of ductility.

When the principal alloying element is tin, the copper alloy is known as bronze. In fact, bronze represents a family of alloys with varying percentages of tin that may also have elements other than tin as the major alloying additive. Bronze products are usually manufactured by casting. They have excellent corrosion resistance and are, therefore, used in marine applications.

Nickel is completely soluble in copper. If it is the only alloying element, a single-phase solid solution will be formed, no matter what the percentage of nickel is. This family of solid solutions is referred to as cupronickels. They possess excellent corrosion resistance and good ductility and can be hardened only by cold working (because the copper–nickel phase diagram is isomorphous). A common cupronickel alloy is one containing 70 percent copper and 30 percent nickel.

Nickel and zinc together are added to copper as alloying elements. With the right combination of nickel and zinc, alloys can be obtained with the appearance of silver. They are, therefore, called nickel silvers and are used for silverware, cutlery, and fake jewelry. Some of these alloys have good strength in cold-worked conditions and are, therefore, used in mechanical components.

Bibliography

Ashton, K. 2009. That "internet of things" thing. *Radio Frequency Identification (RFID) Journal, 22* June.

Askeland, D. R., and W. J. Wright. 2016. *The science and engineering of materials.* 7th ed. Boston: Cengage Learning.

ASM International Handbook. Committee and J. R. Davis. 1998. *Metals handbook: Desk edition.* 2nd ed. Metals Park, OH: American Society of Metals.

Avitzur, B. 1968. *Metal forming: Processes and analysis.* New York: McGraw-Hill.

Avitzur, B. 1980. *Metal forming: The application of limit analysis.* New York: Marcel Dekker.

Avitzur, B. 1983. *Handbook of metal-forming processes.* New York: John Wiley.

Beck, R. D. 1970. *Plastics products design.* New York: Van Nostrand Reinhold.

Bender, S., S. D. El Wakil, and V. B. Chaliventra. 2011. Fabrication and characterization of powder metallurgy parts having porosity gradient. *Powder Metallurgy 54(5)*: 599–603.

Blazynski, T. Z. 1976. *Metal forming: Tool profiles and flow.* New York: Halsted Press.

Boothroyd, G. 1975. *Fundamentals of metal machining and machine tools.* New York: McGraw-Hill.

Boothroyd, G. 2005. *Assembly automation and product design.* 2nd ed. Boca Raton, FL: CRC Press/Taylor & Francis.

Boothroyd, G., and P. Dewhurst. 1991. *Product design for assembly.* Wakefield, RI: Boothroyd Dewhurst, Inc.

Boothroyd, G., C. Poli, and L. E. Murch. 1982. *Automatic assembly.* New York: Marcel Dekker.

Bowman, B. 1980. *Teledyne Rodney metals handbook: Handbook of precision sheet, strip, and foil.* Metals Park, OH: American Society of Metals.

Brady, G. S., H. R. Clauser, and J. A. Vaccari. 2014. *Materials handbook.* 15th ed. New York: McGraw-Hill.

Brown, R. L. E. 1980. *Design and manufacture of plastic parts.* New York: John Wiley.

Budinski, K. G., and M. K. Budinski. 2009. *Engineering materials: Properties and selection.* 9th ed. Reston, VA: Reston-Prentice-Hall.

Calladine, C. R. 1969. *Engineering plasticity.* New York: Pergamon Press.

Callister, W. D. Jr, and D. G. Rethwisch. 2010. *Materials science and engineering: An introduction.* 8th ed. New York: John Wiley.

Campbell, S. A. 2013. *Fabrication engineering at the micro- and nanoscale.* 4th ed. New York: Oxford University Press.

Cary, H. B., and S. Helzer. 2004. *Modern welding technology.* 6th ed. Englewood Cliffs, NJ: Prentice-Hall.

CASA/SME Technical Council. 1987. *A program guide for CIM implementation.* 2nd ed. Dearborn, MI: Society of Manufacturing Engineers.

Chawla, K. K. 2003. *Ceramic matrix composites.* 2nd ed. Norwell, MA: Kluwer Academic.

Chawla, K. K. 2012. *Composite materials: Science and engineering.* 3rd ed. New York: Springer.

Chawla, N., and K. K. Chawla. 2013. *Metal matrix composites.* 2nd ed. New York: Springer.

Cheremisinoff, N. P. 1990. *Product design and testing of polymeric materials.* New York: Marcel Dekker.

Computers in manufacturing *by the Editors of American Machinist.* 1983. New York: McGraw-Hill.

Critchlow, A. J. 1985. *Introduction to robotics.* New York: Macmillan.

Datsko, J. 1997. *Materials selection for design and manufacturing.* New York: Marcel Dekker.

Davies, R., and E. R. Austin. 1970. *Developments in high-speed metal forming.* New York: Industrial Press.

Dieter, G. E. 1983. *Engineering design: A materials and processing approach.* New York: McGraw-Hill.

Dieter, G. E. 1986. *Mechanical metallurgy.* 3rd ed. New York: McGraw-Hill.

Donaldson, C., G. H. LeCain, and V. C. Goold. 1973. *Tool design.* 3rd ed. New York: McGraw-Hill.

Eary, D. F., and E. A. Reed. 1974. *Techniques of press-working sheet metal.* 2nd ed. Englewood Cliffs, NJ: Prentice-Hall.

El Wakil, S. D. 1993. *Laboratory manual for materials science and engineering.* Boston, MA: PWS Publishing.

El Wakil, S. D. 2014. Friction welding of sintered-iron compacts to wrought iron. *International Journal of Powder Metallurgy 50(4)*: 55–59.

El Wakil, S. D. 2016. Metallurgy of friction welding of porous stainless steel-solid iron billets. *International Journal of Materials and Metallurgical Engineering 10(12)*: 1421–1424.

El Wakil, S. D., and N. Azzab. 2008. Effect of process parameters on the grinding of polymeric composites. *Paper presented at the 13th European Conference on Composite Materials*, June 2008, Stockholm, Sweden. www.swereasicomp.se

El Wakil, S. D., and G. Fares. 2006. The grinding of epoxy-graphite composites. *Paper presented at the 9th International AVK Conference*, September 2006, Essen, Germany.

El Wakil, S. D., and M. Pladsen. 2017. Minimizing the drilling damage in graphite-epoxy composites. *International Journal of Materials and Metallurgical Engineering 11(12)*: 804–807.

El Wakil, S. D., and K. Srinagesh. 2008. Effect of the physical and mechanical properties of composites on their grinding characteristics. *High Performance Structures and Materials* IV:149–155.

Farag, M. M. 2014. *Materials and process selection in engineering design*. 3rd ed. Boca Raton, FL: CRC Press/Taylor & Francis.

Flinn, R. A., and P. K. Trojan. 1990. *Engineering materials and their applications*. 4th ed. New York: John Wiley.

Frazelle, E., ed. 1992. *Materials handling systems and terminology*. Atlanta, GA: Lionheart Publishers.

Gershwin, S. B. 1993. *Manufacturing systems engineering*. Englewood Cliffs, NJ: Prentice-Hall.

Grayson, M. 1984. *Recycling, fuel, and resources recovery: Economic and environmental factors*. New York: John Wiley.

Greenwood, D. C., ed. 1982. *Product engineering design manual*. Malabar, FL: R. E. Krieger.

Groover, M. P. 1987. *Automation, production systems, and computer-aided manufacturing*. 2nd ed. Englewood Cliffs, NJ: Prentice-Hall.

Groover, M. P., et al. 1986. *Industrial robotics: Technology, programming, and applications*. New York: McGraw-Hill.

Halevi, G. 1980. *The role of computers in manufacturing processes*. New York: John Wiley.

Harman, R. C., ed. 1967. *Handbook for welding design*. London: Pitman.

Hein, R. W., C. R. Loper Jr., and C. Rosenthal. 1967. *Principles of metal casting*. 2nd ed. New York: McGraw-Hill.

Hill, R. 1950. *The mathematical theory of plasticity*. New York: Oxford University Press.

Hocheng, H., ed. 2011. *Machining technology for composite materials*. Cambridge, UK: Woodhead.

Holland, J. R. 1984. *Flexible manufacturing systems*. Dearborn, MI: Society of Manufacturing Engineers.

Hosford, W. F., and R. M. Caddell. 2011. *Metal forming: Mechanics and metallurgy*. 4th ed. New York: Cambridge University Press.

Housner, H. H., and M. K. Mai. 1982. *Handbook of powder metallurgy*. New York: Chemical Publishing.

Ian, G., D. Rosen, and B. Stucker. 2015. *Additive manufacturing technologies: 3D printing, rapid prototyping, and direct digital manufacturing*. 2nd ed. New York: Springer.

Iredale, R. 1964. Automatic assembly—components and products. *Metalworking Production 108*. (8 April).

Jambro, D. 1976. *Manufacturing processes: Plastics*. Englewood Cliffs, NJ: Pearson College Division.

Jamnia, A. 2018. *An introduction to product design and development for engineers*. Boca Raton, FL: CRC Press/Taylor & Francis.

Johnson, W., and P. B. Mellor. 1983. *Engineering plasticity*. London: Ellis Harwood Limited, Van Nostrand Reinhold.

Jones, W. D. 1960. *Fundamental principles of powder metallurgy*. London: Edward Arnold.

Kant Vajpayee, S. 1998. *Principles of computer-integrated manufacturing*. Englewood Cliffs, NJ: Prentice-Hall.

Kim, G., et al. 1992. A shape metric for design-for-assembly. In *Proceedings of the 1992 IEEE, International Conference on Robotics and Automation*. Nice, France.

Kissel, T. E. 1986. *Understanding and using programmable controllers*. Englewood Cliffs, NJ: Prentice-Hall.

Kobayashi, A. 1967. *Machining of plastics*. New York: McGraw-Hill.

Koeningsberger, F. 1964. *Design principles of metal-cutting machine tools*. New York: Pergamon Press.

Koren, Y. 1985. *Robotics for engineers*. New York: McGraw-Hill.

Koren, Y. 1993. *Computer control of manufacturing systems*. New York: McGraw-Hill.

Krick, E. V. 1969. *An introduction to engineering and engineering design*. 2nd ed. New York: John Wiley.

Kronenberg, M. 1966. *Machining science and application*. New York: Pergamon Press.

Lancaster, J. F. 1999. *Metallurgy of welding*. 6th ed. London: Woodhead Publishing/Elsevier.

Landau, I. D. 1979. *Adaptive control*. New York: Academic Press.

Lane, J. D., ed. 1986. *Automated assembly*. 2nd ed. Dearborn, MI: Society of Manufacturing Engineers.

Lenel, F. V. 1980. *Powder metallurgy: Principles and applications*. New York: American Powder Metallurgy Institute.

Levy, S., and J. H. Davis. 1976. *Plastics product design engineering handbook*. New York: Van Nostrand Reinhold.

Lindbeck, J. R., and R. M. Wygant. 1997. *Product design and manufacture*. Englewood Cliffs, NJ: Prentice-Hall.

Machining data handbook. 3rd ed. 1980. Cincinnati, OH: Machinability Data Center.

McDonald, A. C. 1986. *Robot technology: Theory, design, and applications*. Englewood Cliffs, NJ: Reston-Prentice-Hall.
McMahon, C., and J. Browne. 1993. *CAD/CAM from principles to practice*. Reading, MA: Addison-Wesley.
Metal caster's reference and guide. 2nd ed. 1989. Des Plaines, IL: American Foundrymen Society.
Metal casting and molding processes. 1981. Des Plaines, IL: American Foundrymen Society.
Metals handbook (multivolume) 9th ed. 1982. Metals Park, OH: American Society of Metals.
Miles, B. L. 1989. Design for assembly—A key element within design for manufacture. In *Proceedings from the Institution of Mechanical Engineering*, London, England.
Milewski, J. O. 2017. *Additive manufacturing of metals: From fundamental technology to rocket nozzles, medical implants, and custom jewelry*. New York: Springer.
Miyakawa, S., and T. Ohashi. 1986. The Hitachi assemblability evaluation method (AEM). In *Proceedings of the International Conference on Product Design for Assembly*, Newport, RI.
Miyazawa, A. 1993. Productivity evaluation system. *Fujitsu Science Technology Journal 29(4)*: 425–431.
Modern plastics encyclopedia (n.d.). New York: McGraw-Hill, annual.
Nadai, A. 1963. *Theory of flow and fracture of solids*. 2nd ed. New York: McGraw-Hill.
Nagey, F., and A. Siegler, eds. 1987. *Engineering foundations of robotics*. Englewood Cliffs, NJ: Prentice-Hall.
Nanfara, F., T. Uccello, and D. Murphy. 2001. *The CNC workbook: An introduction to computer numerical control*. 2nd ed. Englewood Cliffs, NJ: Prentice-Hall.
Niebel, B. W., and A. B. Draper. 1974. *Product design and process engineering*. New York: McGraw-Hill.
Olson, D. L., et al., eds. 1993. *ASM Handbook: Vol. 6A. Welding: Fundamentals and processes*. Metals Park, OH: American Society of Metals International.
Ostwald, P. F. 1992. *Engineering cost estimating*. 3rd ed. Englewood Cliffs, NJ: Prentice-Hall College Division.
Ostwald, P. F., and J. Munoz. 1997. *Manufacturing processes and systems*. 9th ed. New York: John Wiley.
Pohlandt, K. 1989. *Materials testing for the metal forming industry*. Berlin: Springer-Verlag.
Ranky, P. G. 1986. *Computer integrated manufacturing*. Englewood Cliffs, NJ: Prentice-Hall.
Raji, R. S. 1994. Smart networks for control. *IEEE Spectrum, 31(6)*.
Redwood, B. 2017. *The 3D printing handbook: Technologies, design and applications*. New York: Springer.
Reed-Hill, R. E., and R. Abbaschian. 2010. *Physical metallurgy principles*. 4th ed. Boston, MA: Cengage Learning.
Rembold, V., B. O. Nanaji, and A. Storr. 1993. *Computer-integrated manufacturing and engineering*. Reading, MA: Addison-Wesley.
Rowe, G. W. 1979. *Elements of metalworking theory*. London: Edward Arnold.
Rowe, G. W. 2005. *Principles of industrial metalworking processes*. New Delhi: CBS Publishers.
Rowe, G. W., et al. 2005. *Finite-element plasticity and metal forming analysis*. London: Cambridge University Press.
Sava, M., and J. Pusztai. 1990. *Computer numerical control programming*. Englewood Cliffs, NJ: Prentice-Hall.
Schaffer, J. P. 1994. *Materials engineering*. Homewood, IL: Irwin.
Seames, W. S. 2001. *Computer numerical control: Concepts and programming*. 4th ed. Boston: Cengage Learning.
Shina, S. G. 1991. *Concurrent engineering and design for manufacture of electronics products*. New York: Van Nostrand Reinhold.
Sindo, Kou. 1996. *Transport phenomena and materials processing*. New York: John Wiley.
Singh, N. 1996. *Systems approach to computer-integrated design and manufacturing*. New York: John Wiley.
Slocum, A. H. 1992. *Precision machine design*. Dearborn, MI: Society of Manufacturing Engineers.
Starrgaard, A. C., Jr. 1987. *Robotics and AI: An introduction to applied machine intelligence*. Englewood Cliffs, NJ: Prentice-Hall.
Teicholz, E. 1985. *CAD/CAM handbook*. New York: McGraw-Hill.
Thomsen, E. G., C. T. Yang, and S. Kobayaski. 1964. *Mechanics of plastic deformation in metal processing*. New York: Macmillan.
Tipping, W. V. 1965. Component and product design for mechanized assembly. In *Conference on Assembly, Fastening, and Joining Techniques and Equipment*, P.E.R.A.
Tobias, S. A. 1965. *Machine tool vibrations*. Glasgow: Blackie & Son.
Tool and manufacturing engineers handbook. 4th ed. 1984. Dearborn, MI: Society of Manufacturing Engineers.
Trucks, H. E., and G. Lewis, eds. 1987. *Designing for economical production*. 2nd ed. Dearborn, MI: Society of Manufacturing Engineers.
Turning handbook of high-efficiency metal cutting. 1980. Detroit, MI: Carboloy Systems, General Electric.
Ulrich, K. T., and D. Steven. 1994. *Product design and development*. 5th ed. New York: McGraw-Hill.

Ulsoy, A. G., and W. DeVries. 1989. *Microcomputer applications in manufacturing*. New York: John Wiley.
Valliere, D. 1990. *Computer-aided design in manufacturing*. Englewood Cliffs, NJ: Prentice-Hall.
Van Vlack, L. H. 1997. *Materials for engineering: Concepts and applications*. Reading, MA: Addison-Wesley.
Wagoner, R. H., and J. Chenot. 1997. *Fundamentals of metal forming*. New York: John Wiley.
Welding and fabricating data book. 1979. Cleveland, OH: Penton/IPC.
Wieser, P. F. 1980. *Steel castings handbook*. 5th ed. Rocky River, OH: Steel Founders Society of America.
Yang, L., et al. 2017. *Additive manufacturing of metals: The technology, materials, design and production*. New York: Springer.

Index

A

B

C

D

N

O

P

Q

R

S

V

W

Y

Z

MIX
Papier | Fördert gute Waldnutzung
FSC® C083411

Kolibri 360 GmbH
Ferdinand-Jühlke-Straße 7
99095 Erfurt, Deutschland
produktsicherheit@kolibri360.de